Handbook of Fermented Meat and Poultry

Handbook of Fermented Meat and Poultry

Editor
Fidel Toldrá

Associate Editors
Y. H. Hui
Iciar Astiasarán
Wai-Kit Nip
Joseph G. Sebranek
Expedito-Tadeu F. Silveira
Louise H. Stahnke
Régine Talon

Blackwell Publishing

Fidel Toldrá, Ph.D., is a Research Professor and Head of the Laboratory of Meat Science, Instituto de Agroquímica y Tecnologia de Alimentos (CSIC), Valencia, Spain. Dr. Toldrá has co-edited and/or authored more than 14 books in food chemistry and food biochemistry, food technology, and meat and poultry processing.

Blackwell Publishing Professional
2121 State Avenue, Ames, Iowa 50014, USA

Orders: 1-800-862-6657
Office: 1-515-292-0140
Fax: 1-515-292-3348
Web site: www.blackwellprofessional.com

Blackwell Publishing Ltd
9600 Garsington Road, Oxford OX4 2DQ, UK
Tel.: +44 (0)1865 776868

Blackwell Publishing Asia
550 Swanston Street, Carlton, Victoria 3053, Australia
Tel.: +61 (0)3 8359 1011

First edition, 2007

Library of Congress Cataloging-in-Publication Data

Handbook of fermented meat and poultry/editor, Fidel Toldrá; associate editors, Y. H. Hui ... [et al.].—1st ed.
p. cm.
Includes bibliographical references and index.
ISBN 13: 978-0-8138-1477-3 (alk. paper)
ISBN 10: 0-8138-1477-4 (alk. paper)
1. Fermented foods—Handbooks, manuals, etc. 2. Meat—Preservation—Handbooks, manuals, etc.
3. Fermentation—Handbooks, manuals, etc. I. Toldrá, Fidel. II. Hui, Y. H. (Yiu H.)

TP371.44.H357 2008
664'.9028—dc22

2007013113

The last digit is the print number: 9 8 7 6 5 4 3 2 1

Contents

Contributors List

EDITOR

Fidel Toldrá, Ph.D.
Instituto de Agroquímica y Tecnología de Alimentos (CSIC)
P.O. Box 73, 46100 Burjassot
Valencia, Spain
Phone: +34 96 3900022, ext. 2112
Fax: +34 96 3636301
E-mail: ftoldra@iata.csic.es

ASSOCIATE EDITORS

Iciar Astiasarán, Ph.D.
University of Navarra
Faculty of Pharmacy
Department of Food Science, Technology and Toxicology
C/Irunlarrea s/n, 31080 Pamplona, Spain
Phone: +34 948 425600, ext. 6405
Fax: +34 948 425649
E-mail: iastiasa@unav.es

Y. H. Hui, Ph.D.
Science Technology System
P.O. Box 1374, West Sacramento, CA 95691, USA
Phone: 1-916-372-2655
Fax: 1-916-372-2690
E-mail: yhhui@aol.com

Wai-Kit Nip, Ph.D.
University of Hawaii at Manoa
Department of Molecular Biosciences & Bioengineering
1955 East-West Road
Honolulu, HI 96822, USA
Phone: 1-808-955-6942
Email: wknip@hawaii.edu

Joseph G. Sebranek, Ph.D.
Iowa State University
Department of Food Science and Human Nutrition
2312 Food Sciences Building
Ames, IA 50011, USA
Phone: 1-515-294-1091
Fax: 1-515-294-5066
E-mail: sebranek@iastate.edu

Expedito-Tadeu F. Silveira, Ph. D.
Instituto de Tecnología de Alimentos
Centro de Pesquisa e Desenvolvimento de Carnes—CTC/ITAL Av. Brasil
2880—Caixa Postal 139—CEP
13073-001 Campinas—SP, Brazil
Phone: +55 19 3743 1886
Fax: +55 19 3743 1882
E-mail: tfacco@ital.sp.gov.br

Louise H. Stahnke, Ph.D.
Chr. Hansen A/S
Boege Allé 10-12
DK-2970 Hoersholm, Denmark
Phone: +45 45 74 74 74
Fax: +45 45 74 88 16
E-mail: Louise.Stahnke@dk.chr-hansen.com

Régine Talon, Ph. D.
Unité Microbiologie
Institut National de la Recherche Agronomique (INRA)
Centre de Clermont-Ferrand-Theix
63122 Saint-Genès Champanelle, France
Phone: +33 4 73 62 41 70
Fax: +33 4 73 62 42 68
E-mail: talon@clermont.inra.fr

CONTRIBUTORS

Dong U. Ahn
Iowa State University
2276 Kildee Hall
Ames, IA 50011-3150, USA
Phone: 1-515-294-6595
Fax: 1-515-294-9143
E-mail: Duahn@iastate.edu

Ana Andrés
Polytechnical University of Valencia
Institute of Engineering for Food Development
P.O. Box 22012, Camino de Vera s/n
46020 Valencia, Spain
Phone: +34 96 3877364
Fax: +34 96 3877956
E-mail: aandres@tal.upv.es

Diana Ansorena
University of Navarra
Faculty of Pharmacy
Department of Food Science, Technology and Toxicology
C/Irunlarrea s/n, 31080 Pamplona, Spain
Phone: +34 948 425600, ext. 6405
Fax: +34 948 425649
E-mail: dansorena@unav.es

M-Concepción Aristoy
Instituto de Agroquímica y Tecnología de Alimentos (CSIC)
Department of Food Science
P.O. Box 73
46100 Burjassot
Valencia, Spain
Phone: +34 96 3900022, ext. 2103
Fax: +34 96 3636301
E-mail: mcaristoy@iata.csic.es

Teresa Aymerich
Centre de Tecnologia de la Carn, IRTA
Finca Camps i Armet s/n
E-17121 Monells (Girona), Spain
Phone: +34 972 630052
Fax: +34 972 630373
E-mail: teresa.aymerich@irta.es

José M. Barat
Polytechnical University of Valencia
Institute of Engineering for Food Development
P.O. Box 22012
Camino de Vera s/n, 46020 Valencia, Spain
Phone: +34 96 3877365
Fax: +34 96 3877956
E-mail: jmbarat@tal.upv.es

Shai Barbut
University of Guelph
Department of Food Science
Guelph, Ontario, CAN 1G2W1
Phone: +1519 824 4120, ext. 53669
Fax: +1519 824 6631
E-mail: sbarbut@uoguelph.ca

Silvana Barbuti
Department of Microbiology, SSICA
Viale Tanara 31/A, 43100 Parma, Italy
Phone: +39 0521 795 267
Fax: +39 0521 771829
E-mail: silvana.barbuti@ssica.it

Antonio S. Barreto
FMV, Av. da Universidade, Universidade Tecnica, Polo Universitário Alto da Ajuda
1300-477 Lisboa, Portugal
Phone: +35 121 365 2880
Fax: +35 121 365 2880
E-mail: asbaretto@fmv.utl.pt

Lopa Basu
The Ohio State University
015 Animal Science Building
2029 Fyffe Road,
Columbus, OH 43210, USA
Phone: 1-614-292-4317
Fax: 1-614-292-2929
E-mail: basul@uwstout.edu

Elettra Berni
Stazione Sperimentale per L'Industria delle Conserv
Viale F. Tanara 31/A
43100 Parma, Italy
Phone: +39 0521 1795262
Fax: +39 0521 1771829
E-mail: elettra.berni@ssica.it

Nicholas L. Berry
Iowa State University
Department of Animal Science
109 Kildee Hall, Ames, IA 50011-3151, USA
Phone: 1-515-295-4683
Fax: 1-515-294-5698

Sara Bover-Cid
Universitat de Barcelona
Departament de Nutrició i Bromatologia
Facultat de FarmaciaAvinguda
Avinguda Joan XXIII s/n
08028 Barcelona, Spain
Phone: +34 93 402 45 13
Fax: +34 93 402 18 96
E-mail: sara.bovercid@ub.edu

Elizabeth Boyle
Animal Sciences & Industry
224 Weber Hall, Manhattan, KS 66506, USA
Phone: 1-785-532-1232
Fax: 1-785-532-7059
E-mail: lboyle@k-state.edu

Marie Champomier-Vergès
INRA Jouy-en-Josas
Domaine de Vilvert Unite Flore Lactique et Environment Carné
78350 Jouy-en-Josas, France
Phone: +33 01 34 65 22 89
E-mail: Marie-Christine.Champomier-Verges@jouy.inra.fr

Suey-Ping Chi
Chung Hwa University of Medical Technology
College of Human Science and Technology
Department of Restaurant and Hospitality Management
89, Wunhwa 1st Street
Jen-Te County, Tainen, Taiwan
E-mail: spch@mail.hwai.edu.tw

Pier Sandro Cocconcelli
Universita Cattolica del Sacro Cuore
Istituto di Microbiologia
via Emilia Parmense 84, 29100 Piacenza, Italy
Phone: +39 523 599251
Fax: +39 523 599246
E-mail: pier.cocconcelli@unicatt.it

Antoine Collignan
Pôle agroalimentaire, CIRAD MRST
Station de la Bretagne BP 20
97408 Saint Denis cedex 9, France
Phone: +33 02 62 92 24 47
Fax: +33 02 62 92 24 31
E-mail: Antoine.collignan@cirad.fr

Anne-Marie Crutz-Le Coq
INRA Jouy-en-Josas
Domaine de Vilvert Unite Flore Lactique et Environment Carné
78350 Jouy-en-Josas, France
E-mail: LeCoq@jouy.inra.fr

Lorenzo de la Hoz
University Complutense
Department of Nutrition, Bromatology and Food Technology
Faculty of Veterinary
Avda Puerta del Hierro s/n
28040 Madrid, Spain
Phone: +34 913943744
Fax: +34 913943743
E-mail: delahoz@vet.ucm.es

Daniel Demeyer
University of Ghent
Department Animal Production
Proefhoevestraat 10, B-9090 Melle, Belgium
Phone: +32 9 2649000
Fax: +32 9 2649099
E-mail: Daniel.Demeyer@ugent.be

Emilie Dordet-Frisoni
INRA Clermont-Ferrand-Theix Unité Microbiologie
63122 Saint-Genès Champanelle, France
Phone: +33 04 73 62 41 70
Fax: +33 004 73 62 42 68
E-mail: edordet@clermont.inra.fr

Mario Estévez
University of Extremadura
Faculty of Veterinary
Ctra. Trujillo s/n, 10071 Cáceres, Spain
Phone: +34 927 257122
Fax: +927 257110
E-mail: mariovet@unex.es

Silvina Fadda
Centro de Referencia para Lactobacillos
CERELA, CONICET
Chacabuco 145
40000 Tucuman, Argentina
Phone: +54 381 4311720
Fax: +54 381 4005600
E-mail: sfadda@cerela.org.ar

Pedro Fito
Polytechnical University of Valencia
Institute of Engineering for Food Development
P.O. Box 22012
Camino de Vera s/n, 46020 Valencia, Spain
Phone: +34 96 3877364
Fax: +34 96 3877956
E-mail: pfito@tal.upv.es

Mónica Flores
Instituto de Agroquímica y Tecnología de Alimentos (CSIC)
Department of Food Science
P.O. Box 73
46100 Burjassot, Valencia, Spain
Phone: +34 96 3900022, ext. 2116
Fax: +34 96 3636301
E-mail: mflores@iata.csic.es

Maria Joao Fraqueza
FMV, Faculdade de Medicina Veterinária de Lisboa
CIISA, DTIA, Rua Prof
Cid dos Santos, Polo Universitário
Alto da Ajuda
1300-477 Lisboa, Portugal
Phone: +35 121 3652880
Fax: +35 121 3652880
E-mail: mjoaofraqueza@fmv.utl.pt

Reinhard Fries
Institut fur Fleischhygiene und Technologie
Meat Hygiene and Technology
Free University of Berlin
Brummerstr, 10
14195 Berlin, Germany
Phone: +49 83852791
Fax: +49 83852792
E-mail: fries.reinhard@vetmed.fu-berlin.de

M-Luisa García
University Complutense
Dep. Nutrición
Bromatología y Tecnología de Alimentos
28040 Madrid, Spain
Phone: +34 91 3943745
Fax: +34 91 3943743

Margarita Garriga
Centre de Tecnologia de la Carn, IRTA
Finca Camps i Armet s/n
E-17121 Monells (Girona), Spain
Phone: +34 972 630052
Fax: +34 972 630373
E-mail: Margarita.Garriga@irta.es

Raúl Grau
Polytechnical University of Valencia
Institute of Engineering for Food Development
P.O. Box 22012, Camino de Vera s/n
46020 Valencia, Spain
Phone: +34 96 3877365
Fax: +34 96 3877956
E-mail: rgrau@tal.upv.es

Dana J. Hanson
North Carolina State University
Food Science Department
Raleigh, NC 27695-7624, USA
E-mail: dana_hanson@ncsu.edu

Torunn T. Haseth
Norwegian Meat Research Centre
P.O. Box 396 Økern
N-0513 Oslo, Norway
Phone: +47 22092399
Fax: +47 22220016
E-mail: torunn.haseth@matforsk.no

Karl O. Honikel
Institute for Chemistry and Physics
Bundesforschungsanstalt für Ernährung
EC Baumann-Strasse 20
D-95326 Kulmbach, Germany
Phone: +49 9221 803200
Fax: +49 9221 803303
E-mail: karl-otto.honikel@bfel.de

Melvin C. Hunt
Animal Sciences & Industry
224 Weber Hall
Manhattan, KS 66506, USA
Phone: 1-785-532-1232
Fax: 1-785-532-7059
E-mail: hhunt@ksu.edu

Kálmán Incze
Hungarian Meat Research Institute
Gubacsi ut 6/b
1097 Budapest, Hungary
Phone: +36 1 2150946
Fax: +36 1 2150626
E-mail: ohki@interware.hu

Jean Labadie
INRA Clermont-Ferrand-Theix Unité Microbiologie
63122 Saint-Genès Champanelle, France
Phone: +33 04.73.62.41.57
Fax: +33 04.73.62.42.68
E-mail: labadie@sancy.clermont.inra.fr

M. Luz Latorre-Moratalla
Universitat de Barcelona
Departament de Nutrició
i Bromatologia
Facultat de Farmacia
Avinguda Joan XXIII s/n
08028 Barcelona, Spain
Phone: +34 93 402 45 13
Fax: +34 93 402 18 96

Isabelle Lebert
Unité Microbiologie
INRA Centre Clermont-Ferrand-Theix
63122 Saint-Genès Champanelle, France
Phone: +33 04 73 62 41 70
Fax: +33 04 73 62 42 68
E-mail: ilebert@clermont.inra.fr

Sabine Leroy
Unité Microbiologie
INRA Centre Clermont-Ferrand-Theix
63122 Saint-Genès Champanelle, France
Phone: +33 04 73 62 41 70
Fax: +33 04 73 62 42 68
E-mail: sleroy@clermont.inra.fr

Friedrich-Karl Lücke
University of Applied Sciences
FB Oecotrophologie
Department of Household Management
Nutrition and Food Quality
Fachhochschule Fulda Marquardstr. 35
D-36039 Fulda, Germany
Phone: +49 661 9640376
Fax: +49 661 9640399
E-mail: friedrich-karl.luecke@he.fh-fulda.de

Robert Maddock
North Dakota State University
Department of Animal & Range Sciences
Hultz 174
P.O. Box 5727, Fargo, North Dakota 58105, USA
Phone: 1-701-231-8975
Fax: 1-701-231-7590
E-mail: Robert.Maddock@ndsu.edu

William Mikel
Mississippi State University
Department Food Science, Nutrition and
Health Promotion
P.O. Box 9805
Mississippi State, MS 37962-9805, USA
Phone: 1-662-325-5508
Fax: 1-662-325-8728
E-mail: wmikel@fsnhp.msstate.edu

Byungrok Min
Iowa State University
2276 Kildee Hall
Ames, IA 50011-3150, USA
Phone: 1-515-294-6595
Fax: 1-515-294-9143
E-mail: mbrok@iastate.edu

Jens K.S. Møller
University of Copenhagen
Faculty of Life Science
Department of Food Science, Food Chemistry
Rolighedsvej 30
Frederiksberg C, 1958, Denmark
E-mail: jemo@life.ku.dk

David Morcuende
University of Extremadura,
Faculty of Veterinary
Ctra. Trujillo s/n,
10071 Cáceres, Spain
Phone: +34 927 257122
Fax: +34 927 257110

Ralf Neidhardt
Chr. Hansen GmbH
Giessener Strasse 94
Pohlheim, D-35415 Germany
Phone: +49 6403 95010
Fax: +49 6403 9501 30
E-mail: ralf.neidhardt@de.chr-hansen.com

Asgeir Nilsen
Matforsk AS, Osloveien 1
Aas, 1430, Norway
Phone: +47 64970100
Fax: +47 64970333
E-mail: asgeir.nilsen@matforsk.no

George-John E. Nychas
Agricultural University of Athens
Department of Food Science and Technology
Iera Odos 75
Athens 11855, Greece
Phone/fax: +30 210 5294693
E-mail: gjn@aua.gr

Herbert W. Ockerman
The Ohio State University
015 Animal Science Building
2029 Fyffe Road, Columbus, OH 43210, USA
Phone: 1-614-292-4317
Fax: 1-614-292-2929
E-mail: ockerman.2@osu.edu

Juan A. Ordoñez
University Complutense
Department of Nutrition, Bromatology and Food Technology
Faculty of Veterinary
Avda Puerta del Hierro s/n
28040 Madrid, Spain
Phone: +34 913943744
Fax: +34 913943743
E-mail: pereda@vet.ucm.es

Emin Burçin Özvural
Hacettepe University
Department of Food Engineering
Beytepe Ankara, 06532, Turkey
Phone: +90 312 2977116
Fax: +90 312 2992123

Esko Petäjä-Kanninen
Viikki Food Science
Department of Food Technology, Viikki EE
P.O. Box 66, FIN-00014 Helsinki, Finland
Phone: +358 5 046 27402
Fax: +358 9 191 58460

Colin Pierre
AFSSA (site de Brest)
Technopôle Brest-Iroise,
29280 Plouzane, France
Phone: +33 (0)2 98 22 45 34
Fax: +33 (0)2 98 05 51 65
E-mail: p.colin@brest.afssa.fr

Stella Planchon
INRA Clermont-Ferrand-Theix Unité Microbiologie
63122 Saint-Genès Champanelle, France
Phone: +33 04 73 62 41 70
Fax: +33 04 73 62 42 68

Eero Poulanne
Viikki Food Science
Department of Food Technology
Viikki EE
P.O. Box 66
FIN-00014 Helsinki, Finland
Phone: +358 9 191 58458
Fax: +358 9 191 58460
E-mail: eero.puolanne@helsinki.fi

Dorota Puszczewicz
Department of Biotechnology and Food Science
Plaza Misael Bañuelos s/n
09001 Burgos, Spain
Phone: +34 947 258814
Fax: +34 947 258831

Stefania Quintavalla
Department of Microbiology, SSICA
Viale Tanara 31/A, 43100 Parma, Italy
Phone: +39 0521 795 267
Fax: +39 0521 771829
E-mail: stefania.quintavalla@ssica.it

Milagro Reig
Instituto de Agroquímica y Tecnología de Alimentos (CSIC)
Department of Food Science
P.O. Box 73
46100 Burjassot
Valencia, Spain
Phone: +34 96 3900022, ext. 2111
Fax: +34 96 3636301
E-mail: mreig@iata.csic.es

Antonio M. Ribeiro
FMV, AV. da Universidade, Universidade Tecnica, Polo Universitário Alta da Ajuda
1300-477 Lisboa, Portugal
Phone: +35 121 365 2880
Fax: +35 121 365 2880
E-mail: mribeiro@fmv.utl.pt

Pedro Roncalés
Department Productión Animal y Ciencia de los Alimentas
Facultad de Veterinaria
University of Zaragoza
Miguel Servet 177
50013 Zaragoza, Spain
Phone: +34 976 761582
Fax: +34 976 761612
E-mail: roncales@unizar.es

Jordi Rovira
Department of Biotechnology and Food Science
Plaza Misael Bañuelos s/n
09001 Burgos Spain
Phone: +34 947 258814
Fax: +34 947 258831
E-mail: jrovira@ubu.es

Marit Rødbotten
Matforsk AS
Osloveien 1
Aas, 1430, Norway
Phone: +47 64970100
Fax: +47 64970333
E-mail: marit.rodbotten@matforsk.no

Jorge Ruiz
Tecnología de los Alimentos
Facultad de Veterinaria UEx
University of Extramadura
Campus Universitario
10071 Cáceres, Spain
Phone: +927 257 122
Fax: +927 257 110
E-mail: jruiz@unex.es

Robert E. Rust
118 E. 16th St.
Ames, IA 50010-5561, USA
Phone/fax: 1-515-232-1575 (H)
1-515-233-9207 (O)
1-515-450-0533 (M)
E-mail: rrust@iastate.edu
rustre@qwest.net

Sunita J. Santchurn
Pôle agroalimentaire, CIRAD MRST
Station de la Bretagne BP 20
97408 Saint Denis cedex 9, France
Phone: +33 02 62 92 24 47
Fax: +33 02 62 92 24 31
E-mail: sunita.santchurn@cirad.fr

Jürgen Schwing
Chr. Hansen GmbH
Giessener Strasse 94
Pohlheim, D-35415 Germany
Phone: +49 6403 95010
Fax: +49 6403 9501 30
E-mail: juergen.schwing@de.chr-hansen.com

M-Dolores Selgas
University Complutense
Facultad de Veterinaria
Dep. Nutrición
Bromatología y Tecnología de Alimentos
28040 Madrid, Spain
Phone: +34 91 3943745
Fax: +34 91 3943743
E-mail: selgar@vet.ucm.es

Miguel A. Sentandreu
Instituto de Agroquímica y Tecnología de Alimentos (CSIC)
Department of Food Science
P.O. Box 73
46100 Burjassot, Valencia, Spain
Phone: +34 96 3900022, ext. 2113
Fax: +34 96 3636301
E-mail: ciesen@iata.csic.es

Maan S. Sidhu
Norwegian Meat Research Centre
P.O. Box 396 Økern
N-0513 Oslo, Norway
Phone: +47 22092399
Fax: +47 22220016
E-mail: maan.singh.sidhu@fagkjott.no

Panagiotis Skandamis
Agricultural University of Athens
Department of Food Science & Technology
Iera Odos 75
Athens 11855, Greece
Phone/fax: +30 210 5294693
E-mail: pskan@aua.gr

Leif H. Skibsted
University of Copenhagen
Faculty of Life Science
Department of Food Science, Food Chemistry
Rolighedsvej 30
Frederiksberg C, 1958, Denmark
Phone: +45 35 33 3212
Fax: +45 35 33 3190
E-mail: ls@kvl.dk

Elisabetta Spotti
Stazione Sperimentale per L'Industria delle Conserv
Viale F. Tanara 31/A
43100 Parma, Italy
Phone: +39 0521 1795262
Fax: +39 0521 1771829
E-mail: elisabetta.spotti@ssica.it

Kenneth J. Stalder
Iowa State University
Department of Animal Science
109 Kildee Hall
Ames, IA 50011-3151, USA
Phone: 1-515-295-4683
Fax: 1-515-294-5698
E-mail: stalder@iastate.edu

Gudjon Thorkelsson
Iceland Fisheries Laboratories
Skulagata 4, 101 Reykjavik, Iceland
Phone: +354 414 8087
E-mail: gudjon@naust.is

Karsten Tjener
Chr. Hansen A/S
Boege Alle 10-12
Hoersholm, 2970, Denmark
Phone: +45 45747474
Fax: +45 45748816
E-mail: karsten.tjener@dk.chr-hansen.com

M. Teresa Veciana-Nogués
Universitat de Barcelona
Departament de Nutrició i Bromatologia
Facultat de Farmacia
Avinguda Joan XXIII s/n
08028 Barcelona, Spain
Phone: +34 93 402 45 13
Fax: +34 93 402 18 96
E-mail: veciana@ub.edu

Jesús Ventanas
University of Extremadura
Faculty of Veterinary
Ctra. Trujillo s/n, 10071 Cáceres, Spain
Phone: +34 927 257122
Fax: +34 927 257110
E-mail: ventanas@unex.es

Sonia Ventanas
University of Extremadura
Faculty of Veterinary
Ctra. Trujillo s/n
10071 Cáceres, Spain
Phone: +34 927 257122
Fax: +34 927 257110
E-mail: sanvenca@unex.es

M. Carmen Vidal-Carou
Universitat de Barcelona
Departament de Nutrició i Bromatologia
Facultat de Farmacia
Avinguda Joan XXIII s/n
08028 Barcelona, Spain
Phone: +34 93 402 45 13
Fax: +34 93 402 18 96
E-mail: mcvidal@ub.edu

Graciela Vignolo
Centro de Referencia para Lactobacillos
CERELA, CONICET
Chacabuco 145
40000 Tucuman, Argentina
Phone: +54 381 4311720
Fax: +54 381 4005600
E-mail: vignolo@cerela.org.ar

Halil Vural
Hacettepe University
Department of Food Engineering
Beytepe Ankara, 06532, Turkey
Phone: +90 312 2977116
Fax: +90 312 2992123
E-mail: ghalil@hacettepe.edu.tr

Yun-Chu Wu
Tunghai University
College of Agriculture
Taichung, Taiwan, ROC
Phone: +886 0 3573 1866
E-mail: ycwu@thu.edu.tw

Monique Zagorec
INRA Jouy-en-Josas, Domaine de Vilvert Unité Flore Lactique et Environment Carné
78350 Jouy-en-Josas, France
Phone: +33 01 34 65 22 89
Fax: +33 01 34 65 21 05
E-mail: Monique.Zagorec@jouy.inra.fr

Peter Zeuthen
Hersegade 7G
Roskilde, DK 4000, Denmark
Phone: +45 46355665
E-mail: Peter.Zeuthen@image.dk

Gai-Ming Zhao
Henan Agricultural University
College of Food Science and Technology
Zhengzhou 450002, China
Phone: +86 25 84395376
Fax: +86 25 84395939
E-mail: gmzhao126@yahoo.com.cn

Gian H. Zhou
Nanjing Agricultural University
College of Food Science and Technology
Nanjing 210095, PR China
Phone: +86 25 84395376
Fax: +86 25 84395939
E-mail: ghzhou@njau.edu.cn

Preface

Fermented meat products have been consumed for centuries in many different parts of the world and constitute one of the most important groups of food. Based on the natural meat flora, a wide range of products have been prepared since ancient times by varying the mixture of meats and salt as well as the addition of spices and seasonings. Thus, fermented meat products represent a great variety of flavors and textures and are receiving increased interest from consumers all over the world who are seeking new gustatory experiences. Most of these products still rely primarily on local, traditional manufacturing processes since little scientific information is available; but scientific knowledge has become an important tool for consistent production of high quality and safe products.

This book contains 50 chapters that are grouped into 11 parts covering most aspects of fermented meat and poultry. Part I deals with the history, production, and consumption of fermented meats as well as general aspects and principles of processing (curing, fermentation, drying, and smoking). Part II describes the main characteristics and uses of raw materials and ingredients. Part III is focused on the microbiology involved in meat fermentation and describes the most commonly applied starter cultures. Part IV looks into the sensory properties of fermented meat products. The composition and nutritional quality, packaging, and international standards are covered in Part V. The description of manufacturing processes and characteristics of semidry-fermented and dry-fermented sausages are covered in Parts VI and VII, respectively. Fermented poultry sausages and other fermented meats are described in Part VIII. Part IX covers the manufacture and characteristics of ripened meat products, especially dry-cured hams. Part X covers biological and chemical safety aspects, and, finally, Part XI is focused on sanitation and quality assurance.

This handbook provides an updated and comprehensive overview of meat fermentation. It includes important developments that have occurred during the last few decades including the role of microorganisms naturally present or added as starter cultures; important safety aspects in today's world; a description of the primary chemical, biochemical, physical, and microbiological changes that occur during processing; and a summary on how they influence the final product quality. The book also provides a detailed description of the major typical fermented meat products around the world and the processing technologies currently applied in meat processing plants.

This book is the result of the expertise of 93 international contributors from 17 different countries. These experts from industry, government, and academia have been led by an editorial team of 8 members from 5 different countries. The editorial team wishes to thank all the contributors for making this book possible. We also thank the production team at Wiley-Blackwell with special recognition to Susan Engelken, senior editorial assistant and coordinator of the project, Judi Brown, the production project manager, and Nancy Albright, the copy editor.

We sincerely hope that you will find this book enlightening and that the information provides a better understanding of fermented meat and poultry products.

Fidel Toldrá
Y. H. Hui
Iciar Astiasarán
Wai-Kit Nip
Joseph G. Sebranek
Expedito-Tadeu F. Silveira
Louise H. Stahnke
Régine Talon

Part I

Meat Fermentation Worldwide: History and Principles

1
A Historical Perspective of Meat Fermentation

Peter Zeuthen

INTRODUCTION

One of the most important prerequisites for the development of civilization was to devise methods on how to preserve foods for storage and transport in order to meet the need for food for increasing concentration in communities. Drying of foodstuffs was probably the first development in this direction, followed by smoking, which in many cases was a natural consequence; drying often was accelerated by hanging up the raw material near the open fire. Nobody had the idea that other processes such as enzymatic breakdown and product changes caused by microorganisms or endogenous enzymes could also be the reason for extension of product life.

However, the transformation of raw materials to more-or-less stable foods by drying and fermentation is well known in many ancient cultures and used for many different foods. Actually, drying as an initial step of preservation, or the "wet way," such as steeping often followed by heating, is well known in beer production, for example, where the breakdown of raw materials takes place because the enzymes are brought into a suitable environment regarding pH and water activity.

The very word *fermentation*, derived from the Latin (*to boil*), means among other things to simmer or bubble, or leaven as a process, was probably not well understood, except that the effect was certainly used when it came to baking, wine making, beer brewing, and production of dairy products or certain meat products. However, because it was poorly understood, this frequently resulted in faulty production. This happens today. When the author of this chapter visited a factory for producing Parma Hams about 25 years ago, the production manager admitted that up to 25% of the raw material never reached a stage qualifying for the stamp of *Parma Ham* because it was putrid before it was ready for sale.

Today, some uncertainties still exist regarding how to define different fermented meats. Adams (1986) thus describes *fermented food* in three categories: 1) those in which microorganisms play no or little part, such as tea fermentation; 2) those in which microbial growth, though an essential feature, does not involve fermentative metobolism, such as the production of tempeh; and 3) the true fermentations, which produce, e.g., lactic acid in products such as salami. If Adams had included raw ham, he would probably have placed hams under category 2, although it now is under discussion whether it rightly should be placed under category 1. Lücke (1986) writes that fermented sausages, and to some extent raw ham, are produced through the participation of microorganisms. Both products should therefore be considered fermented meat. The history of fermented meat products is very old. In the opinion of the author, it includes both raw, dried ham and sausages, because both categories of products alter qualities during production and storage due to fermentation.

EARLY RECORDS OF FERMENTED MEAT PRODUCTS

Raw Cured Ham

The Chinese (Houghton 1982) are claimed to be the first ones who mentioned the production of raw cured ham, but other authors of the ancients also discussed the process. Cato (234–149 B.C.) (1979) explains in details in his book *On Agriculture* how raw ham was manufactured and stored: After 5 days of dry-cure, the hams are rotated and restacked in the vat for 12 days. The remaining salt is brushed off. They are then hung for 2 days in a draught. Then, after a further cleaning they are finally rubbed with oil and smoked for 2 days. They are then rubbed with a mixture of oil and vinegar to avoid "flees and worms" and hung in a meat house.

Similarly, Leistner (1986b) gives an account about how the Roman Emperor Diocletian in 301 A.D. in a public notification distinguishes between highly cured and smoked hams and the lightly salted and dried (unsmoked) hams. Also, Leistner, citing Lissner (1939), mentions that Varro, who lived in the 1st century B.C., wrote about a substantial import to Rome from Gaul of hams, sausages, and other meat products. Leistner is also of the opinion that the technology of salted ham production was derived from the Romans.

Dry Sausage

From ancient times sausages were invented as a means of making the most of leftovers of meat and entrails. One of the most common names today for fermented sausages is *salami*. The name appears to have originated in the ancient Greek town of Salamis, on the Cyprian east coast (Pederson 1979). Although Salamis was destroyed about 450 B.C., the salami sausage style was apparantly widely known and appreciated by then, and clearly seems to have been the forerunner of the many popular European varieties (Smith 1987).

Homer (about 800–700 B.C.) talks about "sausages" in the Odyssey. Sausages were also well known in the Roman Empire (Leistner 1986a). Those sausages were mainly made of blood, fat, and meat scrap and were cooked, so they had nothing to do with dry, fermented sausages. Pederson (1979) citing Breasted (1938) stated that the success of Caesar's legions in the conquest of Gaul can be attributed to their use of dry sausage for their meat supply and that this aided the retention of their vigor and health. The Roman butchers cut their beef and pork into small pieces, added salt and spices, packed the blend into skins, and placed these in special rooms to dry. It was their experience that their sausages would keep better if stored and dried that way. Presumably, besides the favorable extrinsic factors, the sausages were inoculated by lactobacilli and micrococci (now we know that those are a mixture of Kocuria and staphylococci) from equipment and shelves, etc., and a fermentation took place. Sausage making thus spread throughout the Roman Empire.

Leistner (1986a) describes a Chinese type of sausage, *Lup Cheong* from the North and Southern Dynasty (589–420 B.C.) made from goat and lamb meat with salt and flavored with green onion, bean sauce, ginger, and pepper. The early type of Chinese sausage can hardly be regarded as a fermented sausage, but the contemporary type of Lup Cheong has a comparatively long keepability, mainly because of a high content of lactobacilli—so high that it is considered sour by many (Ho and Koh, cited by Leistner 1986a).

The art of producing fermented sausages with a long keepability spread to the rest of Europe. In Germany, the manufacture of fermented sausages commenced only some 150 years ago. Most of the fermented sausages are smoked; in the Mediterranean countries, France, Hungary, and the Balkan countries they are air dried. Spicy sausages predominate. Other types of fermented sausages emerged later as a consequence of advanced meat processing techniques and the availability of refrigeration (Lücke 1985).

One of the most famous types of fermented sausages was developed in Hungary from 1830 to 1840 by Italians (Incze 1986). Although the climate in Hungary is less ideal for fermenting and drying, the Italians succeeded in producing their type of dry sausage during the winter months. Characteristic for the original Hungarian sausage is that it is matured for a long time at a relatively low temperature—around 10°C. It is heavily smoked and is covered by a layer of mold. The latter has caused some problems because of the danger of toxin-producing molds, a problem which today by and large has been overcome.

TECHNICAL AND TECHNOLOGICAL DEVELOPMENTS

Raw Cured Ham

As earlier mentioned, drying, as a means of extending shelf life of food, was probably the earliest technique employed by man. This method, however, could be used only if the food was very easy to dry or semidry to begin with, e.g. seeds, or if the climatic conditions were suitable, such as charque and carne seca, which are South American air-dried meat specialities. In most other parts of the world, some kind of salting had to be used in addition to prevent rot of the raw material. The combination was well known in several ancient civilizations throughout the world—Rome and even Gaul and China (Leistner 1986a, 1986b). It is a method used today for the production of traditional dried hams.

Several attemps have been made to speed up the process. Thus, Puolanne (1982) reports on the pickle injection of raw ham. The method is also used in several other countries today. It reduces the production time considerately, from up to a year to a couple of weeks. Puolanne further reports that these kinds of ham have much less aroma than hams manufactured the traditional way. This observation is reported from several countries. The attempts to hasten the production and improve the safety precautions in ham production have led to the development of inoculation of microorganisms in dry ham.

To store dry ham at a higher temperature or to expose it at a higher temperature for a short time during the later part of manufacture is also used in some cases. Puolanne (1982) thus reports on a Finnish-made dry-cured ham, called a *sauna ham*, which was manufactured earlier. These hams were made as follows: After slaughter, which took place in the autumn, the hams were dry-salted and placed in wooden barrels for 1 to 4 months. Then they were hot-smoked at approximately 40–70°C or applied a cold smoke at approximately 15–20°C for several days.

Smoking, a cold smoke at 20–25°C, has traditionally been used in many countries. Thus, Cato (1979) mentions smoking. Various types of German dry hams are smoked, whereas the Parma type ham is never smoked; neither is the Iberian ham (Garcia et al. 1991).

Dry Sausage

Cutting up meat in smaller pieces to enable a more uniform distribution of the salt and other ingredients has been a well-known technique since the Greek and Roman days. However, to comminute meat also opened it up for a thorough contamination, not unly of a desired flora, but also of bacteria, which sometimes could be pathogens. It was especially important to exclude air. This often caused faulty productions, especially blowing and discolorations. Before artificial refrigeration and control of humidity was available, it was almost impossible to produce dry sausages of satisfactory quality. Because of these difficulties, and because of the demand for a dry sausage, which had at least some keepability, the summer sausage was developed (Leistner 1986b). The name *summer sausage* is still used in some countries; in others, it is not. Thus, Kinsman (1980) writes that in the U.S., dry and/or semidry sausages are often called *summer sausages*, but the term certainly covers a lot of names, which in other parts of the world would be called *typical fermented sausages*, meant to have shelf lives far beyond a few weeks. Rather, the classical dry sausage will have a keepability of a year or more.

The modern way of production of dry sausage with the use of starter cultures and full control of temperatures and humidity has completely changed the manufacture of today's fermented sausage.

Classification Systems for Keepable Sausages

Several authors have attempted to classify the abundance of sausage varieties. However, when trying to do so, the confusion is increased because sausages of same or similar names are very different according to the geographical area in which they are produced. According to Kinsman (1980), a cervelat sausage in the U.S. is listed as five different types, with some of them fermented, some uncooked, others cooked. In Ireland, a cervelat sausage is always uncooked and smoked. In Germany, a mettwurst from Westfalia is fermented and dry, whereas a mettwurst from Braunschwchweig is unfermented and classified moist. Kinsman presents several other examples, showing that the name of a certain sausage in different countries is not unambiguous.

The tradition for consuming fermented sausages in Europe varies, and it relates traditionally to the area. Thus, probably because of the climate, fermented sausages were practically not produced or consumed on the British Isles, and in Ireland in earlier times, only whole, dried meats such as ham were known. On the European continent, fermented sausages were produced abundantly, a tradition starting in Southern Europe.

STARTER CULTURES

The use of starter cultures in various types of foods has been used for a long time, probably for as long as fermentation has been used by man; but it was used simply as a kind of backslopping of remains of earlier production charges. No doubt the dairy industry has been using backslopping purposely, e.g., for making junket and other kinds of fermented milk long before anybody knew anything about bacteria. The process was derived from simple experience in the home. As Pederson (1979) describes it, "the mothers observed that when milk soured with a smooth curd and pleasing acid odor, the milk could be consumed without causing distress or illness. When instead, the milk had an unpleasing odour and was spoiled or gassy or oherwise exhibited unsatisfactory character, the infant sometimes became ill after consuming such milk." Pasteurization made even backslopping much more safe, but the use of defined starter cultures made the production a much more foolproof processes.

With meat, it was different. Microorganisms were added through natural contamination, and to this very day backslopping is still commonly used in many places and regions of the world (Bacus and Brown 1981).

Lücke (1985) states that many productions are still made successfully without addition of starter cultures or reinoculation—backslopping of finished sausages. This is due to sausage makers being able to design formulations and ripening conditions that favor the desired microorganisms so strongly that the products are safe and palatable, even if only few lactobacilli and micrococci (i.e., *Kocuria* and staphyloccoci) are present in the fresh mixture. No doubt the interest in the use of starter cultures arose parallel with the trend toward industrial production, short ripening times, and standardization of the mode in which the sausages were made. It was also very helpful that contemporary refrigeration and air-conditioned facilities became available. However, due to the increasing use of starter cultures, which had taken place with such great effect in the dairy industries, attempts were made to develop starter cultures for other foods, such as meats.

Lactobacilli

Jensen and Paddock (1940) were the first to have published and patented the idea. The inventors used lactobacilli in their patent, which was a mixture of *Leuconostoc mesenterioides*, *Leuconostoc dextranicum*, and *Leuconostoc pleofructi*, but they also set forth a whole range of organisms, all lactobacilli, which would be satisfactory for the purposes of the invention. They included *Lactobacillus casei*, *Lactobacillus plantarum*, *Lactobacillus pentosus*, *Lactobacillus arabinosus*, *Lactobacillus leichmannii*, and several other species. In their claims, they directly warned against gas-forming lactobacilli. The use of a defined microbial culture thus originated in the U.S. with a series of patents issued during the period 1920–1940.

Pediococci

Bacus and Brown (1985) report that before the introduction and subsequent use of the strains of pediococci as the predominant meat starter cultures, most natural isolates from fermented meats consisted of various species and strains of lactobacilli. They add that lactobacilli are still the predominant microflora in products that are made from chance inoculation (or backslopping) and that chance inoculation was depended on for nitrate reduction. However, the Americans (and other researchers) working with pure cultures failed to achieve successful lyophilized preparations of *lactobacillus* isolates with the technology used at that time, so the lactobacilli inoculation was replaced with *Pediococcus* strains that were resistant to lyophilization. When frozen culture concentrates for sausage were introduced, the interest in *Lactobacillus* strains was renewed. Lücke (1985) noted that in Europe, where low fermentation temperatures are common and nitrate is used much more frequently than nitrite, sausage makers are faced with a problem: A fast lactic acid fermentation suppresses the organisms containing catalase and nitrate reductase to such an extent that defects in color and flavor are common. For a long time, European researchers and sausage manufacturers therefore tried to avoid lactobacilli, and lactobacilli were considered detrimental in spite of their ability to form lactic acid. Not until later did they begin experimenting with suitable micrococci (i.e., *Kocuria* and staphylococci). A Finnish researcher, Nurmi (1966), isolated a strain of *Lactobacillus plantarum* from fermented sausage that turned out to be superior to *Pediococcus cerevisiae* at 20–22°C, a temperature normal for fermentation of dry sausage in Northern Europe.

Pediococci became very popular in the U.S. because of the difficulties there with the shelf life of freeze-dried lactobacilli some years ago. Another reason was that the employed fermentation temperature was usually higher than temperatures used in Europe, which meant that it was imperative that the onset of the pH decrease took place very fast during fermentation in order to avoid growth and toxin formation from *Staphylococcus aureus*. In fact, the problem of staphylococcal food poisoning caused by defective, fermented dry or semidry sausage was very well known in the U.S. (USDA 1977, cited by Bacus and Brown 1985). According to these authors, *pediococcus* starter cultures have proven very effective in inhibiting not only *S. aureus*, but also other undesirable microorganisms, such as *Salmonellae*, *Clostridium botulinum*, and other species, including bacilli, gram-negative enterics, and yeasts. Control of the natural fermentation in sausages also appears to prevent histamine accumulation. Higher histamine levels have been found in dry sausage where a natural fermentation process is employed for an extended aging period (Rice et al. 1975; Rice and Koehler 1976).

Staphylococci/Micrococci (Kocuria)

In his thesis, Niinivaara (1955) proposed a *micrococcus* strain, which he called M53. It was selected because it rapidly reduced nitrate, improved color and flavor, and inhibited undesired microorganisms. The strain was later replaced by a fermentative *micrococcus* isolated by Pohja (1960). Gram-negative bacteria have also been attempted to be used as starter cultures in fermented meats. Keller and Meyer (1954) reported that gram-negative bacteria have a favorable effect on dry sausage aroma. Buttiaux (1957) isolated *Vibrio costicolus*, which was used in meat curing in France, primarily as a nitrate reducer, but also because the strain was considered important for aroma and flavor formation. These were properties that also were emphasised by Hawthorn and Leich (1962).

Yeasts and Molds

Yeasts are also used as starter cultures. Following some investigations by Hammes et al. (1985), in which *Debaryomyces hansenii* was examined, other investigators also identified *D. kloeckeri* as useful in dry sausage ripening.

One might say that to use molds as starter cultures is rather to make a virtue out of a necessity. However, with the frequent occurrence of molds on sausages, especially at a time where climate rooms were not available during cold and humid weather, undesirable mold growth was a frequent menace. In southeast Europe, it was a well known problem. Incze (1986) mentions that Hungarian salami was originally made without molds when it was manufactured in Italy. Originally, the so-called *house flora* just contaminated the product, but when the significance of mycotoxins was realized, development proceeded to avoid the use of toxiogenic molds. Much development work took place, especially in Germany at the Federal Centre for Meat Research. Mintzlaff and Leistner (1972) reported on a strain of *Penicillium nalgiovense* as a starter culture, later known as *Edelschimmel Kulmbach*, which does not form mycotoxins.

STARTER CULTURES FOR CURED HAM

Regarding dry-cured hams, the use of bacterial starter cultures is very different. This is partly because the production takes place for a long time at a much lower temperature than is used for fermented sausages and partly because the dry-cured ham is subjected to a much more salty environment on the surface. The untreated ham is not opened, so essentially the internal ham is sterile from the beginning. Contamination can therefore take place only during the liquid exchange that takes place.

Giolitti et al. (1971) conducted a large investigation on dry ham. They concluded that the ripening of hams seems to follow a completely different pattern from that of dry sausages. In their opinion, the microflora, at least in Italian hams, exerts only a minor role in ripening. Souring seems to depend mainly on enzymatic hydrolysis, and this does not lead to end products qualitatively different from those present in normal ham. True putrefaction, however, is the result of microbial growth, starting from internal and external surfaces. Buscailhon et al. (1993) were of the opinion that the aroma of dry-cured ham muscles could be traced back to lipid oxidation or amino acid degradation. However, Hinrichsen and Pedersen (1995) concluded in their study on Parma hams, representing six different stages in the manufacturing process, that microorganisms are important for the development of flavor in Parma ham because all of the correlating volatile compounds they found can be generated by secondary metabolism of microorganisms, especially amino acid catabolism.

CONCLUDING REMARKS

It is impossible to cover the whole history of fermented meats in one short chapter, but going through some of the major epochs, the technical discoveries have resulted in far more rational procedures and thus changed many fermented meat products. These items sometimes were successful, but far from always, because many of the secrets of fermentation were unveiled. On the other hand, many discoveries still wait to be made, especially as far as ways in which more refined aroma and taste compounds are formed during processing. A good example is shown in an article by Bolzoni et al. (1996), which shows the development of volatiles in Parma ham. No doubt there is still much to discover in the future about the mechanisms and the control of volatile development.

REFERENCES

MR Adams. 1986. Fermented flesh foods. Progr Indust Microbiol 23:159–198.

JN Bacus, WL Brown. 1981. Use of microbial cultures: Meat products. Food Technol 47(1):74–78.

———. 1985. The pediococci: Meat products. In: SE Gilliland, ed. Bacterial Starter Cultures for Foods. Boca Raton, Florida: CRC Press, pp. 86–96.

L Bolzoni, G Barbieri, R Virgili. 1996. Changes in volatile compounds of Parma ham during maturation. Meat Sci 43:301–310.

JH Breasted. 1938. The Conquest of Civilization. New York: Harper & Row, pp. 542–556.

S Buscailhon, JL Berdagué, G Monin. 1993. Time-related changes in volatile compounds of lean tissue during processing of French dry-cured ham. J Sci Food Agric 63:69–75.

R Buttiaux. 1957. Technique simple d'examens bacteriologique des saumures de jambon et ses résultats. Proceedings II International Symposium on Food Microbiology, HMSO, London, 1957, pp. 137–146.

MP Cato. 1979. De agricultura (On Agriculture). Translated by WD Hooper. Harvard University Press, pp. 155–157.

C Garcia, JJ Berdagué, T Antequera, C López-Bote, JJ Córdoba, J Ventanas. 1991. Volatile components of dry cured Iberian ham. Food Chem 41:23–32.

G Giolitti, CA Cantoni, MA Bianchi, P Renon. 1971. Microbiology and chemical changes in raw hams of Italian type. J Appl Bacteriol 34:51–61.

WP Hammes, I Röl, A Bantleon. 1985. Mikrobiologische Untersuchungen der auf dem deutschen Markt vorhandenen Starterkulturpräparate für die Rohwurstbereitung. Fleischwirtsch 65:629–636, 729–734.

J Hawthorn, JM Leitch. 1962. Recent advances in Food Science Vol 2. London: Butterwords, pp. 281–282.

LL Hinrichsen, SB Pedersen. 1995. Relationship among flavour, volatile compounds, chemical changes, and microflora in Italian type dry cured ham during processing. J Agric Food Chem 43:2932–2940.

Homer. The Odyssey. XX:24–27.

CG Houghton. 1982. The Encyclopedia Americana International edition Vol 24. Danbury, Connecticut: Grolier Inc, p. 309.

K Incze. 1986. Technnologie und Mikrobiologie der ungarischen Salami. Fleischwirtsch 66:1305–1311.

LB Jensen, LS Paddock. 1940. Sausage treatment. US Patent 2,225,783.

H Keller, E Meyer. 1954. Die bakterielle Aromatiseierung von Rohwurst. Fleischwirtsch 6:125–126, 453–454.

DM Kinsman. 1980. Principal Characteristics of Sausages of the World. Connecticut: Storrs College of Agriculture and Natural Resources, University of Connecticut.

L Leistner. 1986a. Allgemeines über Rohschinken. Fleischwirtsch 66:496–510.

———. 1986b. Allgemeines über Rohwurst. Fleischwirtsch 66:290–300.

———. 1995. Fermented meats. London: Chapman & Hall, pp. 160–175.

E Lissner. 1939. Wurstologia oder Es geht um die Wurst. Frankfurt am Main: Hauserpresse Hans Schaefer.

FK Lücke. 1985. Fermented sausages. In: BJB Wood, ed. Microbiology of Fermented Foods Vol II. London: Elsevier Applied Science, pp. 41–83.

———. 1986. Mikrobiologische Vorgänge bei der Herstellung von Rohwurst und Rohschinken. Fleischwirtsch 66:302–309.

HJ Mintzlaff, L Leistner. 1972. Untersuchungen zur Selektion eines technologisch geeigneten und toxikologisch unbedenklichen Schimmelpilz-Stammes für die Rohwurst-Herstellung. Zeitblatt Veterinär Medicin B 19:291–300.

FP Niinivaara. 1955. über den Einfluss von Bakterien-Reinkulturen auf die Reifung und Umrötung der Rohwurst. Acta Agrialia Fennica 85:1–128.

E Nurmi. 1966. Effect of bacterial inoculation on characteristics and microbial flora of dry sausage. Acta Agralia Fennica 108:1–77.

CS Pederson. 1979. Microbiology of Food Fermentation. 2nd ed. Westport, Connecticut: The AVI Publishing Company Inc, p. 212.

MS Pohja. 1960. Micrococci in fermented meat products. Acta Agralia Fennica 96:1–80.

E Puolanne. 1982. Dry cured hams—European style. Proceedings of the Reciprocal Meat Conference 35:49–52.

SL Rice, RR Eitenmiller, PE Koehler. 1975. Histamine and tyramine content of meat products. J Milk Food Technol 38:256–258.

SL Rice, PE Koehler. 1976. Tyrosine and histidine decarboxylase activities of *Pediococcus cerevisiae* and *Lactobacillus* species and the production of tyramine in fermented sausages. J Milk Food Technol 39:166–169.

DR Smith. 1987. Sausage—A food of myth, mystery and marvel. CSIRO Food Research Quarterly 47:1–8.

USDA. 1977. The staphylococcal enterotoxin problem in fermented sausage, Task Force report, Food Safety and Wuality Service, USDA, Washington, D.C.

2
Production and Consumption of Fermented Meat Products

Herbert W. Ockerman and Lopa Basu

INTRODUCTION

Meat fermentation is a low-energy, biological acidulation, preservation method, which results in unique and distinctive meat properties, such as flavor and palatability, color, microbiological safety, tenderness, and a host of other desirable attributes of this specialized meat item. Changes from raw meat to a fermented product are caused by "cultured" or "wild" microorganisms, which lower the pH. Because this is a biological system, it is influenced by many environmental pressures that need to be controlled to produce a consistent product. Some of these factors include a fresh, low-contaminated, consistent raw material; a consistent inoculum; strict sanitation; control of time, temperature, and humidity during production; smoke; and appropriate additives.

Lactic acid, which accounts for the antimicrobial properties of fermented meats, originates from the natural conversion of glycogen reserves in the carcass tissues and from the added sugar during product fermentation. A desirable fermentation product is the outcome of acidulation caused by lactic acid production and lowering the water activity (a_w) caused by the addition of salt (curing) and drying. Both natural and controlled fermentations involve lactic-acid bacteria (LAB). Their growth must be understood to produce a safe and marketable product. Most starter cultures, today, consist of lactic acid bacteria and/or micrococci, selected for their metabolic activity, which often improves flavor development. The reduction of pH and the lowering of water activity are both microbial hurdles that aid in producing a safe product. Fermented sausages often have a long storage life due to added salt, nitrite, and/or nitrate; low pH due to lactic acid production by LAB organisms in the early stages of storage; and later drying, which reduces the water activity.

Production and composition figures for fermented products are difficult to obtain, particularly because many of these products are produced and consumed locally and quantities are not recorded. The limited number of references available would suggest that the production and consumption is sizeable.

CURRENT PRODUCTS

Definitions

The characteristics and types of fermented products can be found in Tables 2.1 and 2.2. Guidelines proposed in the U.S. (American Meat Institute 1982; Hui et al. 2004) for making fermented dry or semidry sausages include a definition of dry sausage as chopped or ground meat products, which, due to bacterial action, reach a pH of 5.3 or less. The drying removes 20 to 50% of the moisture resulting in a moisture to protein ratio (MP) of no greater than 2.3 to 1.0. Dry salami (U.S.) has an MP ratio of 1.9 to 1, pepperoni 1.6 to 1, and jerky 0.75 to 1.

Semidry sausages are similar except that they have a 15 to 20% loss of moisture during processing. Semidry sausages also have a softer texture and a different flavor profile than dry sausages. Because of the higher moisture content, semidry sausages are more

Table 2.1. Characteristics of different types of fermented sausages.

Type of Sausage	Characteristics
Dry; long ripening, e.g., dry or hard salami, saucission, pepperoni; shelf-stable	Chopped and ground meat Commercial starter culture or back inoculum U.S. fermentation temperature 15–35°C for 1–5 days Not smoked or lightly smoked U.S. bacterial action reduces pH to 4.7–5.3 (0.5–1.0% lactic acid; total acidity 1.3%, which facilitates drying by denaturing protein resulting in a firm texture; moisture protein ratio <2.3:1, moisture loss 25–50%, moisture level ,35% European bacterial action reduces pH to 5.3–5.6 for a more mild taste than U.S.; processing time 12–14 weeks Dried to remove 20–50% of moisture; contains 20–45% moisture, fat 39%, protein 21%, salt 4.2%, a_w 0.85–0.86, yield 64% Moisture protein ratio no greater than 2.3:1.0 Less tangy taste than semidry
Semidry; sliceable, e.g., summer sausage, Holsteiner, Cervelat (Zervelat), Tuhringer, Chorizos; refrigerate	Chopped or ground meat Bacterial action reduces pH to 4.7–5.3 (lactic acid 0.5–1.3%, total acidity 1%), processing time 1–4 weeks Dried to remove 8–30% of moisture by heat; contains 30–50% moisture, 24% fat, protein 21%, salt 3.5%, a_w 0.92–0.94, yield 90% Usually packaged after fermentation/heating Generally smoked during fermentation No mold Moisture protein ratio no greater than 2.3–3.7 to 1.0
Moist; undried; spreadable, e.g., Teewurst, Mettwurst, Frishe Braunschweiger	Contains 34–60% moisture, production time 3–5 days Weight loss ~10%, a_w 0.95–096 Usually smoked No mold Highly perishable, refrigerate, consume in 1–2 days

Source: Modified from American Meat Institute (1982); Gilliland (1985); Campbell-Platt and Cook (1995); Doyle et al. (1997); Farnworth (2003).

susceptible to spoilage and are usually fermented to a lower pH to produce a very tangy flavor. Semidry products are generally sold after fermentation (pH of 5.3 or less). These are heated, and they do not go through a drying process (water activity is usually 0.92 or higher). They are also usually smoked during the fermentation cycle and have a maximum pH of 5.3 in less than 24 hours. If the semidry sausage has a pH of 5.0 or less and a moisture protein ratio of 3.1 to 1 or less, it is considered to be shelf-stable, but most semidry products require refrigeration (2°C). In Europe, fermented meat with a pH of 5.2 and a water activity of 0.95 or less is considered shelf-stable (USDA FSIS Food Labeling Policy Manual).

To decrease the pH (below 5.0) with limited drying, the U.S semidry products are often fermented rapidly (12 hours or less) at a relatively high temperature (32–46°C). In Europe, fermentation is slower (24 hours or more) at a lower temperature and results in a higher pH. These differences in speed of fermentation and final pH, result in products having different flavors.

Product Ingredients

Raw Meat

Beef, mechanically separated beef (up to ~5%), pork, lamb, chicken, mechanically separated chicken (up to ~10%), duck, water buffalo, horse, donkey, reindeer, gazelle, porcupine, whale, fish, rabbit, by-products, and other tissue from a variety of species are used to make fermented meat products. Fermented meat is

Table 2.2. Types of fermented sausages and areas of production.

Type	Area	Production	Sausages	Sensory
Dry fermented, mixed cultures	Southern and Eastern Europe, psychotrophic lactic acid, optimum growth 10–15°C	40–60 mm, pork, 2–6 mm particle, nitrate, starters, fungi; ferment 10–24°C, 3–7 days; ripen 10–18°C, 3–6 weeks; weight loss higher than 30%, low water activity, higher pH, 5.2 to 5.8, lower lactate, 17 mmol/100 g of dry matter	Italian Salami Spanish Salchichon, Chorizo (usually no starter) French Saucisson, 9 mg/g of lactate/g	Fruity, sweet odor, medium buttery, sour and pungent, more mature
	Northern Europe, LAB (*Lactobacillus plantarum* or *Pediococcus*) and Micrococcaceae (*Staphylococcus carnosus* or *Micrococcus*)	90 mm pork/beef, 1–2 mm particle, nitrite, starters; smoked, ferment 20–32°C, 2–5 days; rapid acidification to below 5, smoking ripen 2–3 weeks; weight loss higher than 20%, lower pH, 4.8 to 4.9, higher lactate 20–21 mmol/100 g of dry matter	German Salami, 15 mg/g of lactate/g Hungarian Salami Nordic Salami	Buttery, sour odor, low levels of spice and fruity notes, more acid More acid More acid
Mold ripened	Europe, U.S.	Usually with starters Usually with starters, bowl chopper Without starters, bowl chopper Usually with starters, ground Usually with starters Usually with starters	French Salami Germany Salami Hungarian Salami Italian Salami California Salami Yugoslavian Salami	
Semidry	U.S.	Usually with starters	Summer sausage, M/P ratio 2.0–3.7:1.0 Thuringer Beef sticks	
Dry		Usually with *Pediococcus acidilactici* Usually with *Pediococcus acidilactici*	Beef sticks Pepperoni, M/P ratio 1.6:1.0	

Source: Hui et al. (2004); Schmidt and Berger (1998); Demeyer et al. (2000); Stahnke et al. (1999).

often divided into two groups: products made from whole pieces of meat, such as hams; and products made from meat chopped into small pieces, such as various sausage types. Details on producing these products will be treated in succeeding chapters.

The predominant bacteria that appear in fresh meat are typically gram-negative, oxidase-positive, aerobic rods of psychrotrophic pseudomonads along with psychrotrophic *Enterobacteriaceae,* small numbers of lactic acid bacteria, and other gram-positive bacteria. The lactic and other gram-positive bacteria become the dominant flora if oxygen is excluded and is encouraged during the fermentation stages. Because the production of fermented meats depends on microorganism growth, it is essential that these products are hygienically processed and chilled prior to use and maintained under refrigeration prior and during the curing operation.

Starter Cultures

Traditionally, fermented products depend on wild inoculum, which usually do not conform to any specific species but are usually related to *Lactobacilli plantarum*. However, other species such as *Lb. casei* and *Lb. leichmanii*, as well as many others, have been isolated from traditionally fermented meat products (Anon 1978). In the U.S., *Lb. plantarum*, *Pediococcus pentosaceus*, or *P. acidilactici* are the most commonly used starter cultures. In Europe, the starter cultures used most include *Lb. sakei*, *Lb. plantarum*, *Pediococcus pentosaceus*, *Staphylococcus xylosus*, *S. carnosus*, and to a lesser extent *Micrococcus* spp. Reliance on natural flora results in products with inconsistent quality. The advantage of a starter culture is that the same microorganisms can be used repeatedly, which cuts down on variation of the finished product, and a larger number of organisms can be added. Now, combined starter cultures are available in which one organism produces lactic acid (e.g., *Lactobacilli*) and another improves desirable flavors (*Micrococcaceae*, *Lb. brevis*, *Lb. buchneri*). This translates into a lot of very good and acceptable product and almost no undesirable fermented product. However, very little extremely excellent product is produced, because most starter cultures are a combination of just a few species of microorganisms and they cannot produce as balanced a flavor as sometimes can be obtained when many species are included.

Particularly in the south European countries, dry sausages are applied with atoxinogenic yeast and fungi to produce products with specific flavor notes. This is done by dipping or spraying. Mold cultures tend to suppress natural molds and, consequently, reduce the risk of mycotoxins. Due to the extended ripening and drying for these products, the final pH is usually higher (pH >5.5), even if the pH was lower after fermentation, because molds can utilize lactic acid and produce ammonia. This requires the final water activity to be low enough for preservation.

Other Ingredients

Salt is the major additive in fermented meat products. It is added in levels of 2–4% (2% minimum for desired bind, up to 3% will not retard fermentation), which will allow lactic acid bacteria to grow and will inhibit several unwanted microorganisms.

Nitrite is used at 80 to 240 mg/kg for antibacterial, color, and antioxidant purposes. Nitrate and nitrite are often used in combination, but nitrate is usually not necessary except as a reservoir for nitrite, which could be useful in long-term processing. Nitrite is also a hurdle, which inhibits bacterial growth and retards *Salmonellae* multination. Fermentation can be produced with only salt, but there is a greater microbial risk if no nitrite is used.

Simple sugars such as glucose (dextrose, 0.5 total %, a minimum of 0.75% is often recommended), is the fermentation substrate that can be readily utilized by all lactic acid bacteria. The quantity of sugar influences the rate and extent of acidulation, and also contributes favorably to flavor, texture, and product yield properties. The amount of dextrose added will directly influence the final product pH, and additional sugar will not decrease pH further since bacterial cultures cannot grow in excess acid.

Spices (e.g., black, red, and white pepper; cardamom; mustard; allspice; paprika; nutmeg; ginger; mace; cinnamon; garlic; and various combinations) are often included in the fermented meat formula. Spices are used for flavor, antioxidant properties, and to stimulate growth of lactic bacteria.

Sodium ascorbate (also in the U.S., sodium erythorbate) or ascorbic acid (also in the U.S. erythorbic acid) is used for improvement and stability of color and retardation of oxidation.

Processing

Formulations are numerous, even for products with the same name, and some are held in strict security. Examples of formulations and processing procedures can be found in Komarik et al. (1974), Rust (1976), Ockerman (1989), Campbell-Platt and Cook (1995), and Klettner and Baumgartner (1980). Time, temperature, humidity, and smoke are also variables that control the quality of the final product.

Types of casings available include natural casings, artificial casings (made from cotton linters), collagen casings, fibrous casings (synthetic, inedible) for use in the smokehouse or cooker and which are sometimes netted or prestuck (pin-pricked) to allow for better smoke penetration and elimination of air pockets, and cloth bags.

Temperature, time, and relative humidity combinations are quite variable in industrial productions. In general, the higher the fermentation temperature and water activity, the faster the lactic acid production. In Europe, the fermentation temperatures range from 5 to 26°C, with lower temperatures used in the Mediterranean area and higher temperatures in northern Europe. In the U.S., semidried products are

Table 2.3. Nutritive composition of examples of fermented meat products.

	Cervelat		Salami, Pork, Beef	Pepperoni, Pork, Beef
	Soft	Dry	Dry	Dry
Moisture	48.4%	29.4%		30.5%
Calorie (kcal)	307	451	424	466
Protein	18.6%	24.6%	23.1%	20.3%
Fat	25.5%	37.6%	34.8%	40.3%
Monounsaturated	13.0 g/100 g	–	17 g/100 g	19 g/kg
Polyunsaturated	1.2 g/100 g	–	3.2 g/100 g	2.6 g/100 g
Saturated	12.0 g/100 g		12.4 g/100 g	16.1 g/100 g
Ash	6.8%	6.7%	5.5%	4.8%
Fiber	0	0	–	1.5%
Carbohydrate	1.6%	1.7%	–	5 g/100 g
Sugar	0.85%		–	0.7%
Calcium	11 mg/100 g	14 mg/100 g	–	21 mg/110 g
Iron	2.8 mg/100 g	2.7 mg/100 g	1, 5 mg/100 g	1.4 mg/100 g
Magnesium	14.0 mg/100 g		17.8 mg/100 g	18 mg/100 g
Phosphorus	214 mg/100 g	294 mg/100 g	143 mg/100 g	176 mg/100 g
Potassium	260 mg/100 g	–	379 mg/100 g	315 mg/100 g
Sodium	1242 mg/100 g	–	1881 mg/100 g	1788 mg/100 g
Zinc	2.6 mg/100 g	–	3.3 mg/100 g	2.7 mg/100 g
Copper	0.15 mg/100 g	–	0.07 mg/100 g	0.07 mg/100 g
Manganese	–	–	0.4 mg/100 g	0.3 mg/100 g
Selenium	20.3 mg/100 g	–	–	21.8 μ/100 g
Vitamin C (ascorbic acid)	16.6 mg/100 g	–	–	0.7 mg/100 g
Thiamine (B_1)	0.3 mg/100 g	–	0.6 mg/100 g	0.5 mg/100 g
Riboflavin (B_2)	0.2 mg/100 g	–	0.3 mg/100 g	0.2 mg/100 g
Niacin (B_3)	5.5 mg/100 g	–	4.9 mg/100 g	5.4 mg/100 g
Pantothenic Acid (B_5)	–	–	1.1 mg/100 g	0.6 mg/100 g
Vitamin B_6	0.26 mg/100 g	–	0.5 mg/100 g	0.4 mg/100 g
Folate	2 DFE/100 g	–	–	6 DFE/100 g
Vitamin B_{12}	–	–	1.9 mg/100 g	1.6 mg/100 g
Vitamin E (alpha–tocopherol)	5.5 mcg/100 g	–		0.3 μ/100g
Cholesterol	75 g/100 g	–	78 mg/100 g	118 mg/100 g

– = data not reported.

Source: Calorie-Counter.net (2003); Nutrition Info (2005); United States Department of Agriculture (1963); National Livestock and Meat Board (1984).

usually fermented with slowly rising temperatures to over 35°C to shorten the fermentation time, which is frequently 12 hours or less. In the U.S., half-dried summer sausage is fermented 3 days at 7°C, 3 days at 27 to 41°C, and 2 days at 10°C, and then heated to 58°C for 4 to 8 hours. Smoking depends on tradition and product type in the area the product is produced, and it can vary to from no smoke to heavy smoke. If smoked, it is used to contribute flavor and to retard surface bacteria, molds, and yeast. Many semidry sausages are heated after fermentation and/or smoking and this often increases the pH. Often, an internal temperature of 58.3°C is used. European drying temperatures of 14°C, and 78 to 88% relative humidities are often used. Air velocity of 1 m/s is also used. This results in a final water activity of <0.90 in the finished fermented product (Klettner and Baumgartner 1980).

Composition

Composition varies widely due to the number of fermented product types. Products with the same name vary considerably when made in different countries or even in the same country or made by the same company in different locations. The major factors influencing composition of the finished product are the composition and ratio of raw materials used, processing procedures utilized, and the quantity of drying. Because most meat composition data are usually expressed on a percentage basis, the dryer (less moisture) the sausage, the higher will be the other individual ingredients in the finished product. A few selected examples of average composition can be found in Table 2.3.

FUTURE

Health and spoilage will be the center focus of research in the future and a better understanding of the geological mechanisms will help tremendously. Hazard Analysis Critical Control Point (HACCP) and hurdle technology control are becoming commonplace in the food industry, and both are particularly important in fermented sausages (see succeeding chapters of this volume). Easily controlled fermentation chambers are helpful and gaining in popularity to accurately control the environmental conditions during processing.

Genetic determinants and transfer mechanisms will be utilized to develop superior lactic acid bacteria strains. These genetically modified microorganisms have the potential to improve food safety and quality, but they must be evaluated on a case-by-case basis. Genetic manipulation to "tailor-make" cultures to be used as starters and fungal starter cultures are being evaluated for elimination of toxin production, regulation of fermentation including proteolytic and lipolytic activity, and biopreservative properties such as production of natural antimicrobial compounds for elimination of pathogens.

Muscle enzyme activity of a carcass and its influence on meat quality has been demonstrated (Toldrá and Flores 2000; Russo et al. 2002). Therefore, enzymes and genes can be used to select breeding animals and raw material for production of fermented sausages. For more value-added fermented products, muscle protease, muscle and fat lipase (activities), and spice influence on fermented sausage flavor (Fournaud 1978; Claeys et al. 2000) will be evaluated. If these can be altered, a flavor modification is probably possible.

New product developments with new ingredients will evolve. For example, making fermented product from ostrich (Bohme et al. 1996), carp (Arslan et al. 2001), and utilization of olive oil (Muguerza et al. 2001) have been suggested. Acceptance of fermented meat products in additional international food items seems to be gaining in popularity, such as increased consumption of pizza containing fermented meat in Asia. With the past progress and the possible future developments, the future looks bright for fermented meat products.

REFERENCES

American Meat Institute. 1982. Good Manufacturing Practices, Fermented Dry and Semi-dry Sausages. Washington D.C.: American Meat Institute AMI.

Anon. 1978. Some aspects of dry sausage manufacture. Fleischwirtsch 58:748–749.

A Arslan, AH Dincoglu, Z Gonulalan. 2001. Fermented Cyrinus carpio L. sausage. Turkish J Vet Anim Sci 25:667–673.

HM Bohme, FD Mellett, LMT Dicks, DS Basson. 1996. Production of salami from ostrich meat with strains of *Lactobacillus sakei*, *Lactobacillus curvatus* and Micrococcus. Meat Sci 44:173–180.

Calorie-Counter.net. 2003. http://calorie-net/meat-calories/pepperoni.htrm.

G Campbell-Platt, PE Cook. 1995. Fermented Meats. London, UK: Blackie Academic and Professional.

E Claeys, S De Smet, D Demeyer, R Geers, N Buys. 2000. Effect of rate of pH decline on muscle enzyme activities in two pig lines. Meat Sci 57:257–263.

PE Cook. 1995. Fungal ripened meats and meat products. In: G Campbell-Platt, PE Cook, eds. Fermented

Meats. Glasgow: Blackie Academic and Professional, pp. 110–129.

D Demeyer, M Raemaekers, A Rizzo, A Holck, Ad Smedt, Bt Brink, B Hagen, C Montel, E Zanardi, E Murbrekk, F Leroy, F Vandendriessche, K Lorentsen, K Venema, L Sunesen, L Stahnke, L Vuyst, R Talon, R Chizzolini, S Eerola. 2000. Control of bioflavour and safety in fermented sausages: First results of a European project. Food Res Int 33:171–180.

MP Doyle, LR Deuchat, TJ Montville. 1997. Food Microbiology. Washington D.C.: American Society for Microbiology, pp. 610–629.

ER Farnworth. 2003. Handbook of Fermented Functional Foods. Boca Raton, Florida: CRC Press, pp. 251–275.

J Fournaud. 1978. La microbiologie du saucisson sec. L'Alimentation et la Vie 64:382–392.

SE Gilliland. 1985. Bacterial Starter Cultures for Foods. Boca Raton, Florida: CRC Press, pp. 57–183.

YH Hui, LM Goddik, AS Hansen, WK Nip, PS Stanfield, F Toldrá. 2004. Handbook of Food and Beverage Fermentation Technology. New York: Marcel Dekker, Inc., pp. 353–368, 385–458.

PG Klettner, PA Baumgartner. 1980. The technology of raw dry sausage manufacture. Food Technol Austral 32:380–384.

SL Komarik, DK Tressler, L Long. 1974. Food Products Formulary. Westport, Connecticut: The AVI Publishing Company Inc, pp. 33–51.

E Muguerza, O Gomeno, D Ansorena, JG Bloukas, I Astiasarán. 2001. Effect of replacing pork back fat with pre-emulsified olive oil on lipid fraction and sensory quality of Chorizo de Pamplona. Meat Sci 59:251–258.

National Livestock & Meat Board. 1984. Nutritive Value of Muscle Foods. Chicago.

Nutrition Info. 2005. http://www.nutri.info/nutrition_facts/Pepperoni,%20pork,%20beef.htm.

HW Ockerman. 1989. Sausage and Processed Meat formulations. New York: Van Nostrand Reinhold, pp. 1–603.

V Russo, L Fontanesi, R Davoli, LN Costa, M Cagnazzo, 1 Buttazoni, R Virgili, M Yerle. 2002. Investigation of genes for meat quality in dry-cured ham production: The porcine cathepsin B (CTSB) and cystatin B (CSTB) genes. Anim Genet 33:123–131.

RE Rust. 1976. Sausage and Processed Meat Manufacturing. Washington, D.C.: American Meat Institute, pp. 97–101.

S Schmidt, RG Berger. 1998. Microbial formed aroma compounds during the maturation of dry fermented sausages. Adv Food Sci 20:144–152.

LH Stahnke, LO Sunesen, A de Smedt. 1999. Sensory characteristics of European dried fermented sausages and the correlation to volatile profile. Thirteenth Forum for Applied Biotechnology. Med Fac Landbouw Univ 64/5b, pp. 559–566.

F Toldrá, M Flores. 2000. The use of muscle enzymes as predictors of pork meat quality. Food Chem 69:387–395.

United States Department of Agriculture. 1963. Composition of Foods. Agricultural Handbook 8. Washington D.C.

3
Principles of Curing

Karl O. Honikel

INTRODUCTION

Curing meat means the addition of nitrite and/or nitrate together with salt to meat. The effective substance is nitrite or its derivatives. Nitrate becomes an active curing agent only after reduction to nitrite. Nitrite at the usual pH value range of meat and during meat processing at about pH 5–6.5 forms NO which binds to myoglobin. The NO-myoglobin or NO-hem is heat-stable and gives the meat the red curing color. Nitrite oxidizes to nitrate in meat products and acts as an oxygen scavenger or as an antioxidative substance.

By this and other reactions, the nitrite concentration is reduced in a ready-to-eat meat product to less than half of the ingoing amount. During storage the concentration is further reduced. Due to the lack of the reaction partner of secondary amines in fresh meat, the low nitrite concentration and the high pH-value, cooked meat products do not contain nitrosamines. Raw, long-aged meat products may contain traces of various nitrosamines.

DEFINITION OF CURING

Curing meat means the addition of nitrite and/or nitrate together with salt (NaCl) to meat in various degrees of comminution and at different processing steps.

The most basic method of curing is the covering of meat cuts (e.g., part of hind legs of pigs) in a mixture of salt and curing agent. The salt and curing agents penetrate slowly into the muscular tissue. This dry-curing process lasts up to several weeks. The process can be accelerated by wet curing, in which the meat cuts are either inserted into a brine of salt and curing agents and/or brine is injected into the meat by needles or via blood vessels.

In small pieces and in minced or comminuted meat like sausages, salt and the curing agent are mixed in or comminuted with the meat in a bowl chopper. This way of curing is fast, and within a short time (1–2 days) depending on the processing technique, leads to the characteristic appearance of a cured product like red color and a "cured" flavor.

HISTORY OF CURING

In ancient times, salt was used as one way of preservation of many foods, meat cuts, and fish. This process was usually combined with drying and/or smoking. Salt addition and drying result in a reduction of the free available water for microorganisms, lowering the water activity (a_w). Smoking preserves by covering the surface of meat with bacteriostatic and mycostatic compounds. Thus, smoking also adds to the preservation process.

Salt in its pure state is sodium chloride (NaCl). Many sources of salt, such as rock salt or marine salt, contain small amounts of other salts. In the 19th century, it was discovered that saltpeter (potassium and sodium nitrate) was a constituent of many preserving salts. The formation of the red curing color was identified as depending on the presence of nitrate.

Polenske, from the German Imperial Health Office, published a paper in 1891 where he tried to explain the red compound in meat products. He had observed that cured meat and pickling solutions contained nitrite even though only saltpeter (potassium nitrate) had been used in the fresh solution of the salt. A heated sterilized solution did not produce nitrite, but an unheated solution did so after a few days of storage. The solution was simple but nevertheless revolutionary: Nitrate (NO_3^-) was reduced to nitrite (NO_2^-) by microorganisms.

This observation was followed in 1899 by Lehmann. He observed that an acidified nitrite solution turned meat to a bright red color. Nitrate and nitric acid did not produce the bright red color known from cured meat. Kisskalt (1899) observed in the same year that cooking meat in nitrite solution turned it red, but not if nitrate was applied. Nitrate turned meat bright red only after the meat was left in nitrate solution at ambient temperatures for a few days.

So three scientists confirmed within a decade that nitrite was responsible for the curing color and not nitrate.

Haldane (1901) brought light into the chemistry of the process. He found out that the change in color by cutting the surfaces of fresh meat with its formation of oximyoglobin turned back to the less-red state on covering it afterward with a glass plate. This behavior was different when meat was treated with nitrite. So, salt (nitrate/nitrite containing)-treated meat in the raw state was bright red after a fresh cut. It turned brown upon exposure to air. Covering this meat surface with a glass plate turned it bright red again. Extracting the colored substances, he found that the bright red salted/cured color was nitrogen oxide–myoglobin (NO-myoglobin). NO-myoglobin was also existing in cured and cooked meat.

Hoagland (1910, 1914) found out that only an acidified solution led to the red cured color. He concluded that the nitrous acid (HNO_2) or a metabolite of it like NO formed the bright red cured meat color. The pH of meat is usually around 5.5; the sequence of events known in the second decade of the 20th century is described in Figure 3.1.

So the principle of the mechansim for the red color of cured meat was understood. The antimicrobial action was still thought to be due to the salt. Nitrite/nitrate were not considered to be responsible for the preservation itself.

LEGISLATION

After it was discovered that nitrite was the genuine curing agent, it took only a few years until nitrite was introduced into meat products manufacturing. But nitrite itself is rather toxic in comparison to nitrate. As a rule of thumb, nitrite is 10 times more toxic than nitrate. The lethal oral doses for human beings are established in 80–800 mg nitrate/kg body weight and 33–250 mg nitrite/kg body weight (Schuddeboom 1993).

Nitrite was added to meat products sometimes in too high amounts and—e.g., in Germany—some people died in the 3rd decade of the 20th century due to intoxication by nitrite in meat products. Germany solved the problem in 1934 with the Nitrit-Pökelsalz-Gesetz (nitrite-curing salt law). It says that nitrite is allowed only in premixes with table salt; its content should be 0.5% and must not exceed 0.6%. Only nitrate could be added directly to meat batters. 0.5% means that with 20 g nitrite-curing salt/kg of batter (2%) 100 mg nitrite/kg batter (100 ppm) would be added.

In the 1950s, the Fleisch-Verordnung (meat regulation) limited the residual amount to 100 mg sodium nitrite/kg in ready-to-eat meat products. In raw hams, 150 mg $NaNO_2$/kg were permitted. Nitrate restrictions also were applied. In the Fleisch-Verordnung of 1982, nitrate was limited to some nonheated products with ingoing amounts of 300–600 mg/kg and residual amounts of 100–600 mg/kg product.

$$(KNO_3)\text{ nitrate} \rightarrow \text{reduction by micro organisms} \rightarrow \text{nitrite } (KNO_2)$$

$$KNO_2 + H^+ \leftrightarrows HNO_2 + K^+$$

$$2HNO_2 \leftrightarrows N_2O_3 + H_2O$$

$$N_2O_3 \leftrightarrows NO + NO_2$$

$$NO + \text{myoglobin} \rightarrow \text{NO-myoglobin}$$

Figure 3.1. Scheme of the proposal of Hoagland (1910, 1914) for the action of nitrate in cured meat products.

This regulation was followed by the European Parliament and Council Directive 95/2/EC on food additives other than colors and sweeteners (Official Journal EU 1995) where indicative ingoing amount 150 mg nitrite/kg and 300 mg nitrate/kg were permitted in more-or-less all meat products (Table 3.1). The residual amounts could go up to 50 mg nitrite/kg in nonheat–treated meat products and 100 mg nitrite/kg in all other meat products, except Wiltshire bacon and some others, to 175 mg nitrite/kg. With nitrate the residual amounts were 250 mg nitrate/kg in all meat products.

Denmark opposed this and excluded the use of nitrate from all meat products except Wiltshire bacon and some other raw hams with ingoing 300 mg nitrate/kg. Nitrite up to 150 mg ingoing amounts of nitrite/kg were written down in the Danish regulation 1055/95 of December 18, 1995 (DRL 1995A).

The European Court in Luxembourg decided on March 20, 2003, that the European union must reconsider their regulation 95/2/EC.

A new directive 2006/52/EC (official J. of the EU 2006) amending directive 95/2/EC (official J. of the EU 1995) on food additives other than colors and sweeteners restricts the use of nitrates to unheated products with 150 mg/kg. Nitrate may be added with 100 or 150 mg/kg to all meat products. For both compounds a number of exceptions for named products with residual amounts are permitted.

In summary: In the European Union the use of nitrite and nitrate will be limited to about 150 mg/kg for each salt. Nitrate is permitted in nonheated products only.

Many countries have similar regulations. As an example, the regulations of the U.S. should be reported here. In their Code of Federal Regulations (2005) they state:

> The food additive sodium nitrite may be safely used in or on specified foods in accordance with the following prescribed conditions:
>
> 3) As a preservative and color fixative, with sodium nitrate, in meat curing preparations for the home curing of meat and meat products (including poultry and wild game), with directions for use which limit the amount of sodium nitrite to not more than 200 parts per million in the finished meat product, and the amount of sodium nitrate to not more than 500 parts per million in the finished meat product.

All regulations, directives, and laws take into account that nitrite is a toxic substance and that, contrary to other additives, nitrite does not remain unchanged in the product during processing. Some of the nitrite is converted to nitrate.

Also the discoveries of the early 20th century are taken into consideration. Nitrate is effective only after being reduced to nitrite. This happens only in products that are not heat treated early after manufacturing; i.e., in raw hams and raw sausages. The intake of nitrite plus nitrate is thus limited to the necessary minimal requirements.

CHEMISTRY OF NITRITE AND NITRATE

Chemical and Physical Characteristics of Nitrites and Nitrates and Related Compounds

Nitrite and nitrate are used in cured meat products in the form of their sodium and potassium salts:

Sodium nitrite, $NaNO_2$, sodium salt of nitrous acid, CAS No. 7632-00-0

Table 3.1. Nitrate and nitrite in meat products; extract from EU Directive 95/2/EC (1995).

E No	Name	Foodstuff	Indicative Amounts that May Be Added During Manufacturing
E 249	Potassium nitrite	Meat products	150 mg/kg
E 250	Sodium nitrite	Sterilized meat products (e >3.00)	100 mg/kg
		Wiltshire cured bacon and ham	175 mg/kg as a residue
		Dry cured bacon and ham	175 mg/kg as a residue
		Cured tongue, jellied veal, brisket	10 mg/kg as a residue expressed as $NaNO_2$
E 251	Sodium nitrate	Nonheat–treated meat products	150 mg/kg
E 252	Potassium nitrate	*Wiltshire cured bacon and ham*	250 mg/kg as a residue
		Dry cured bacon and ham	250 mg/kg as a residue
		Cured tongue, jellied veal, brisket	10 mg/kg as a residue expressed as $NaNO_2$

molecular weight 69 Dalton
white or yellowish white crystals
odorless
density 2.17 g/cm^3
solubility 85.2 g/100 g water at 20°C, 163 g/100 g water at 100°C
pH in aqueous solution around 9
melting point 271°C
hygroscopic on air
oxidizes in air slowly to nitrate
strong oxidizing agent with organic matter and inorganic material, especially ammonia compounds
toxicity oral rat LD50: 85 mg/kg; for humans 50 mg/kg body weight; irritating to skin and other body surfaces

Potassium nitrite, KNO_2, potassium salt of nitrous acid, CAS No. 7758-09-0
molecular weight 85.1 Dalton
white or yellowish white crystals
odorless
density 1.92 g/cm^3
solubility ca. 300 g/100 g water at 20°C, 413 g/100 g water at 100°C
pH in aqueous solution around 9
melting point 387°C; unstable above 350°C
strong oxidizing agent with organic matter
toxicity as $NaNO_2$

Sodium nitrate (saltpeter), $NaNO_3$, the sodium salt of Nitric acid, CAS No. 7631-99-4, used as gun powder
molecular weight 84.99 Dalton
white crystals
odorless
hygroscopic on moist air
density 2.16 g/cm^3
solubility 73 g/100 ml water at 0°C, 180 g/100 ml water at 100°C
pH in aqueous solution around 7
melting point 307°C
stable under normal conditions
may oxidize reduced organic matter (gun powder)

Potassium nitrate (saltpeter), KNO_3, the potassium salt of nitric acid, CAS No. 7757-79-1
molecular weight 101.1 Dalton
white crystals
odorless, salty taste
density 2.11 g/cm^3
solubility 31.6 g/100 ml water at 20°C, 247 g/100 ml water at 100°C
pH in aqueous solution around 7
melting point 334°C
stable under normal conditions
may oxidize organic matter and react vigorously with ammonia salts

Nitrous acid, HNO_2, known only in solution
molecular weight 47.02 Dalton
anhydride is N_2O_3, which exists only in the solid state at temperatures below –102°C
above the liquid is a mixture of $N_2O_3 \rightarrow NO + NO_2$
middle strong acid, pK 3.37

Nitric acid, HNO_3, CAS No. 7697-37-2
molecular weight 63.01 Dalton
colorless liquid in closed containers
melting point –42°C
boiling point 86°C
density: 1.52 g/ ml at 20°C
mixable in any quantity with water
strong acid; totally dissociated in water

Nitric oxide, Nitrogenoxid NO, CAS No. 10102-43-9
molecular weight 30 Dalton
colorless gas
melting point –163.6°C
boiling point –151.8°C
oxidizes in air $2NO + O_2 \rightarrow N_2O_4 \rightarrow 2NO_2$

NO can bind to myoglobin forming NO-myoglobin. Also this compound (NO) can easily loose an electron and form a NO^+ cation which may react with amines to nitrosamines. NO is formed in our body to function as a physiological messenger playing a role in cardiovascular, neurologic, and immune systems by suppressing pathogens, vasodilation, and neurotransmission (Lowenstein and Snyder 1992; Moncada et al. 1991; Nathan 1992; Luscher 1990, 1992; Marletta 1989; Ignarro 1990; Stamler et al. 1992).

Nitrogendioxide NO_2 (N_2O_4)
molecular weight NO_2 46.01 Dalton
N_2O_4 is a colorless solid, turns brown by dissociation into NO_2 on warming up, liquid at ambient temperature
melting point –11.2°C
boiling point 21.15°C
N_2O_4 is formed in the solid state only, at melting point and higher, more and more NO_2 is formed
N_2O_4 reacts with KI:

$$\overset{4+}{2N_2O_4} + \overset{-1}{2KI} \rightarrow \overset{5+}{2KNO_3} + \overset{2+}{2NO} + \overset{0\pm}{I_2}$$

Iodide is oxidized to iodine, NO_2 is oxidized to nitrate and reduced to NO. For this reason the addition of iodide as a iodine fortifying substance to salt is not possible. Iodate (IO_3^-) must be used.

Nitrite and Nitrate in Meat

If nitrites are added to meat, the salt is dissolved due to its good solubility in the aqueous solution of pH around 5.5 of the meat. Due to the pK of nitrous acid of 3.37, which means that a pH 3.37 around 50% of the acid is dissociated, it can be expected that about 99% of the nitrite exists of pH 5.5 as an anion, NO_2^- (Figure 3.2). The small amount of undissociated nitrous acid is in equilibrium with its anhydride N_2O_3, which again is in equilibrium with the 2 oxides nitric oxide and nitric dioxide (Figure 3.3).

The small amount of NO can react with oxygen forming NO_2. In the sum it means that from two HNO_2 molecules one HNO_2 and one HNO_3 molecule are formed (Figure 3.3). This reaction is rather important in meat batter because nitrite acts in this way as an antioxidant. Metal ions seem to accelerate the oxidation process. Nitrate added to meat will fully dissociate into $Na^+(K^+) + NO_3^-$. No detectable amount of undissociated HNO_3 will be found.

$$2HNO_2 \rightleftharpoons N_2O_3 + H_2O$$
$$N_2O_3 \rightleftharpoons NO + NO_2$$
$$NO + \tfrac{1}{2} O_2 \rightarrow NO_2$$
$$2NO_2 + H_2O \rightarrow HNO_2 + HNO_3$$

overall: $$2HNO_2 + \tfrac{1}{2} O_2 \rightarrow HNO_2 + HNO_3$$

Figure 3.3. Reaction of nitrous acid and its derivatives.

NITRITE AND NITRATE IN MEAT PRODUCTS

Concentration in Meat Products

The oxidation of nitrite to nitrate in meat also explains why in meat products to which only nitrite has been added nitrate will be found in considerable concentrations. In Figures 3.4 and 3.5, the nitrite and nitrate concentrations of German meat products are shown. The emulsion-type and cooked sausages and cooked hams are manufactured with nitrite only, but they contain a mean of 20–30 mg nitrate/kg, as also shown in a very recent survey (Table 3.2). Nitrite is in most cases lower than nitrate in the finished product, with concentration below 20 mg nitrite/kg in the mean value. Only a few samples of cooked and raw sausages and raw ham contain above 60 mg nitrite/kg (Figure 3.4), which also have higher nitrate concentrations. In the raw products nitrate may have been added, in cooked (liver and blood sausages) the ingoing nitrite is apparently strongly oxidized to nitrate by meat enzyme systems or metal ions due to their high pH values between 6.0 to 6.8.

In the last three decades the use of ascorbic acid or ascorbate or isoascorbate (erythrothorbate) has become common practice. There are added to meat batters with 500 mg/kg. Ascorbates and isoascorbates are antioxidative substances, which may also sequester oxygen (Figure 3.6). In this way, they retard the oxidation of NO to NO_2 and the formation of nitrate.

But ascorbate seems also to react with nitrite or one of its metabolites. Dahn et al. (1960), Fox et al. (1968), and Izumi et al. (1989) show that ascorbate reacts with nitrite and binds the resulting NO. The

Meat with 100 ppm (mg/kg) of nitrite added = 1.45×10^{-3} M

$$Na^+ + NO_2^- + H^+ \rightleftharpoons HNO_2 + Na^+ \quad (a)$$

pH of meat 5.5 = 3×10^{-6} M H^+

pK of nitrous acid = 3.37; $K = 4.27 \times 10^{-4}$

reaction (a) is with more than 99 % at the left side at pH 5.5

Figure 3.2. State of nitrite in meat homogenates.

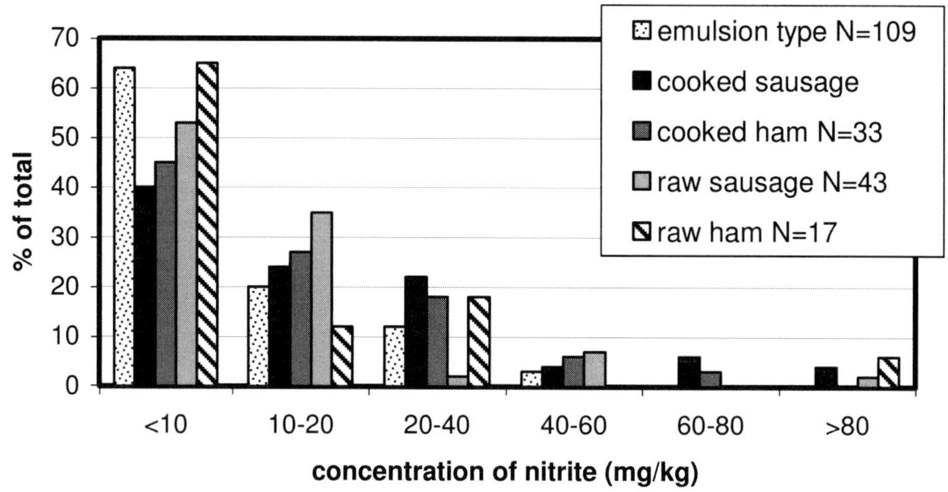

Figure 3.4. Nitrite in German meat products (1996–2001) (Dederer 2006).

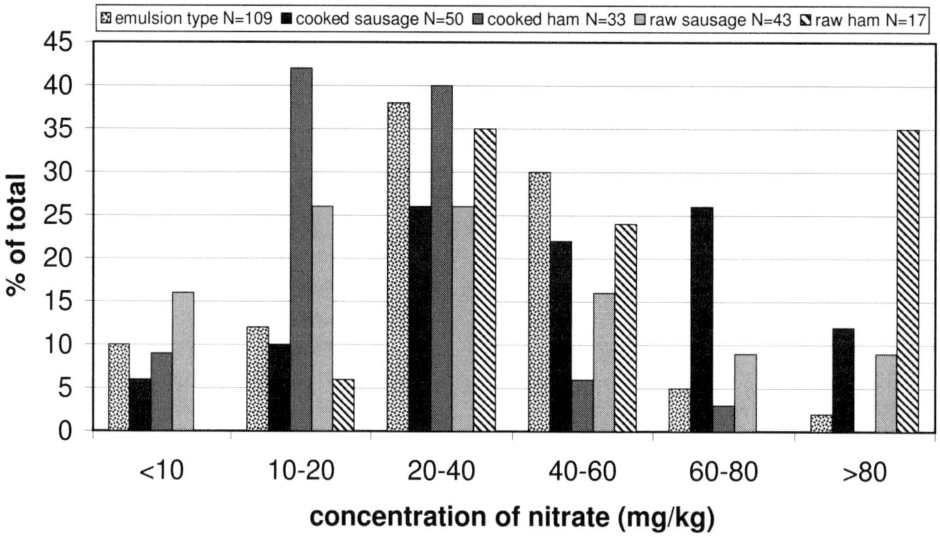

Figure 3.5. Nitrate in German meat products (1996–2001) (Dederer 2006).

nitrite is reduced to NO, the ascorbate oxidized to dehydroascorbate. The bound NO seems to be able to react as NO with other meat ingredients. Ascorbate is also added to reduce the formation of nitrosamines, which is discussed later.

But not only ascorbate and myoglobin react with nitrite derivatives; amino acids (Figure 3.7) and unsaturated fatty acids also react. These products are rather unstable and either release NO or are oxidized to NO_2 and/or nitrocompounds. These reactions of NO with other substances are another reason for the low residual amounts of nitrite in ready-to-eat meat products.

After the reported reactions it could be assumed that the concentration of nitrate in a sausage where only nitrite is added is related to the nitrite content.

Figure 3.8 shows that with emulsion-type sausages (only nitrite curing salt is used) the residual amounts of nitrite and nitrate exhibit no relationship above 20 mg residual nitrite/kg. There is no generally recognizable increase of nitrate with

Table 3.2. Nitrite and nitrate concentrations in sausages of Germany (2003–2005) (Dederer 2006).

			Median mg/kg	
	Year	n	Nitrite	Nitrate
Emulsion-type sausage	2003	30	13.2	23.4
	2004	32	12.65	20.5
	2005	29	19.9	30.0
Raw sausages	2003–2005	15	17.9	59.2
Raw ham	2003–2005	14	19.2	16.9
Liver/blood sausages (cooked sausages)	2003–2005	16	12.1	43.3

increasing residual amounts of nitrite. Without nitrite addition, a residual amount of nitrate is probably due to the added drinking water and spices.

Changes with Time of Storage

When does the nitrite disappear in the product? Table 3.3 shows results from Russian colleagues (Kudryashow 2003). The largest decrease is observed during the manufacturing up to the end of the heating process. This early loss exceeds usually 65% independent of the ingoing concentrations. Within 20 days of cold storage the concentrations drop further to a third of the concentration after heating.`

The disappearance continues until 60 days of cold storage. Table 3.4 confirms the results. It furthermore shows that a higher pH value retards the disappearance of nitrite. It also confirms the results of Table 3.3 that nitrate is already high at day 0 after heating. Nitrate also falls in concentrations with time of storage. The reduction during storage is slower with increasing pH.

Table 3.5 shows the influence of different heat treatment. The higher the heating, the greater the loss of nitrite. The formation of nitrate is also reduced. A higher pH value and/or a different muscle show less nitrite loss and a higher nitrate concentration. The addition of ascorbate and polyphosphate show that the disappearance of nitrite is accelerated by ascorbate in the raw batter (Table 3.6). Heating for 7 minutes to 80°C leads to a slower loss of nitrite. Heating for additionally 1 hour at 70°C retards the loss even longer. This is probably due to the inactivation of microorganisms and inactivation of enzymes by heating. With ascorbate and even more with polyphosphates, the retarding by heating is also observed.

Influence of High Pressure on Nitrite and Nitrate Concentration

The main loss of nitrite occurs in the heating process and without heating by microorganisms. The application of high pressure up to 800 MPa, which inactivates microorganisms but does not exceed 35°C shows that nitrite is oxidized to about 20% to nitrate, but nitrite + nitrate add up nearly to the added amount of nitrite (Table 3.7). During storage, the nitrite decreases to about half of the ingoing amount, and the nitrate increases to about 40%. The sum of nitrite plus nitrate is even at 21 days of storage at 90% of the ingoing amount of nitrite. These results show, in comparison to the results of Tables 3.3 through 3.5, that only small amounts of nitrite react with other substances than oxygen to nitrate, on applying high pressure. The reason for this observation is unknown.

Nitrite is in all cases described here partially oxidized to nitrate. In many experiments (see Tables 3.4, 3.5 and 3.7), about 10–40% of the nitrite is oxidized

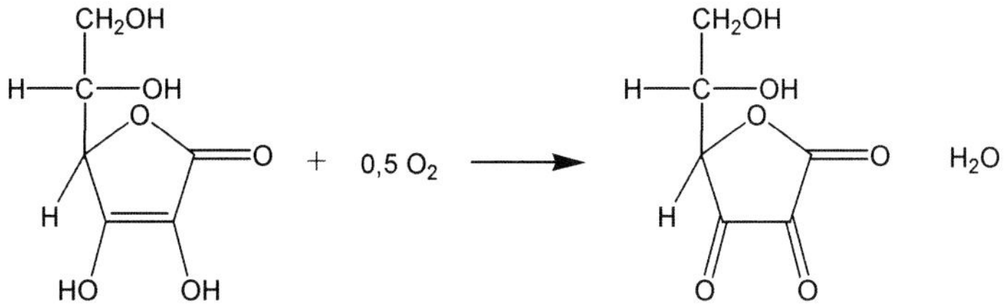

Figure 3.6. Oxygen sequestering by ascorbate (ascorbic acid).

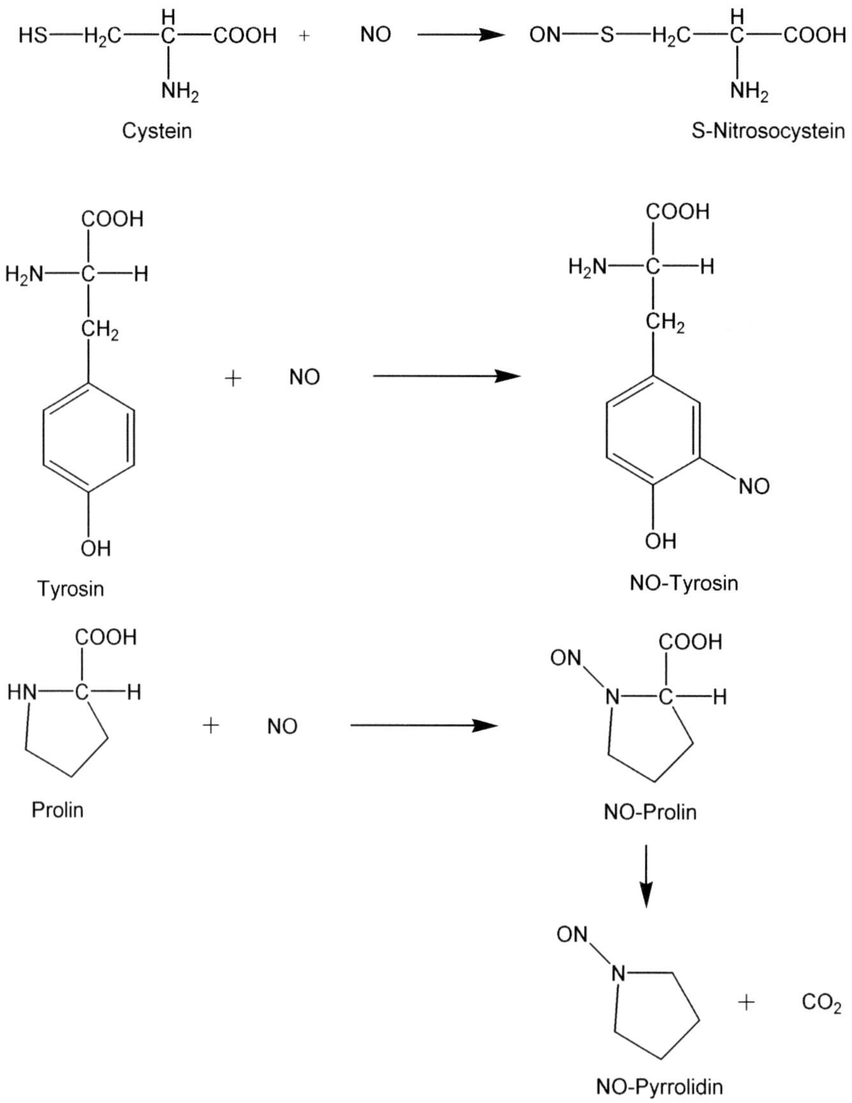

Figure 3.7. Reactions of NO with amino acid.

to nitrate. But only in the results of Table 3.7 nitrite plus nitrate add up to about 90% of more of the nitrite addition. This has been known for decades; in 1978, Cassens et al. postulated that nitrite is bound to various meat constituents (Table 3.8).

NITROSOMYOGLOBIN

The red color of cured meat products is one of the important effects of nitrite in meat products. The red color is developing in a number of complicated reaction steps until NO-myoglobin (Fe^{2+}) is formed.

Myoglobin exists in a muscle in three states, in which the cofactor hem, a porphyrin ring with an iron ion in its center binds different ligands or in which the iron exists in the Fe^{2+} or Fe^{3+} state. In the native myoglobin, the porphyrin moiety (Figure 3.9) is supported in the ligand binding by amino acids of the protein in the neighborhood.

In the "original" state, myoglobin with Fe^{2+} in the porphyrin cofactor does not bind any ligand except maybe a water molecule. In the presence of oxygen the porphyrin can bind an O_2 molecule and it becomes bright red. The iron ion is in the Fe^{2+} state. But oxygen and other oxidizing agents like nitrite

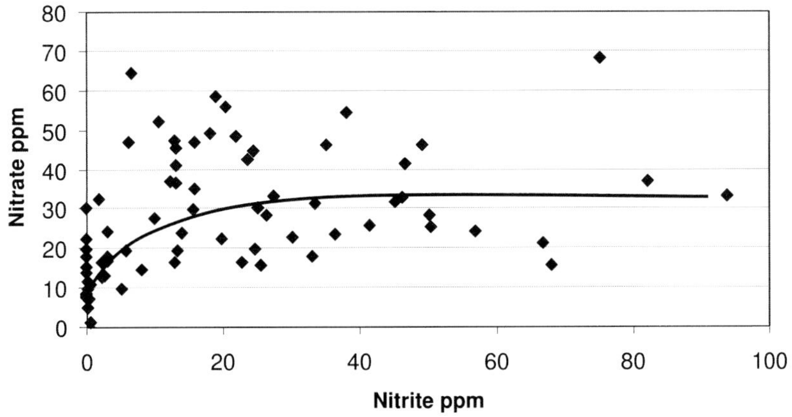

Figure 3.8. Concentration of nitrite and nitrate in emulsion-type sausages (N = 48) (Dederer 2006).

Table 3.3. Remaining nitrite (mg/kg) during storage at 2°C of an emulsion-type sausage (Kudryashow 2003).

	Concentration of Nitrite Added (mg/kg)			
Storage	75	100	150	200
After heating	21.9	30.5	59.5	53.7
20 days	7.5	9.3	10.2	15.4
40 days	3.6	6.4	7.6	7.7
60 days	0.5	0.9	4.0	5.8

can oxidize the Fe^{2+} to Fe^{3+} (Figure 3.10). This metmyoglobin is brown.

The "original" myoglobin (Mb), the oximyoglobin (MbO2) and the metmyoglobin are occurring together in meat. In a muscle in a live animal, there is very little metmyoglobin, which increases postmortem with the disappearance of oxygen except when meat is MAP-packed with high oxygen.

The three states of myoglobin have three characteristic absorbance spectra between 400 and 700 nm. As the three are in a kind of equilibrium to each other, the spectra have an isospestic point at 525 nm where all three absorption curves cross each other.

Table 3.4. Nitrite breakdown and nitrate appearance after nitrite addition to meat of various pH values after heating and storage (Dordevic et al. 1980).

		Nitrite		Nitrate	
pH	Days of Storage	100 mg/kg Added	200 mg/kg Added	100 mg/kg Nitrite Added	200 mg/kg Nitrite Added
5.3	After heating				
	0	28	70	20	50
	6	20	41	16	27
	12	5	18	9	20
5.8	After heating				
	0	45	120	30	64
	6	24	110	22	40
	12	13	21	8	17
6.3	After heating				
	0	58	135	18	40
	6	41	112	17	30
	12	31	90	10	22

Table 3.5. Mean values of nitrite and nitrate content in pasteurized and sterilized groups of homogenates (Dordevic et al. 1980).

Homogenate 100 mg/kg of $NaNO_2$ Added	Way of Heat Treatment*	After Heat Treatment		After 12 Days at 2–4°C	
		$NaNO_2$ mg/kg	$NaNO_3$ mg/kg	$NaNO_2$ mg/kg	$NaNO_3$ mg/kg
M. longissimus dorsi	P	38.6	27.4	10.9	7.9
pH 5,8	S	12.9	13.1	3.5	6.0
M. quadriceps femoris	P	49.2	35.8	26.7	12.8
pH 6,15	S	15.6	19.4	7.4	9.8

*P = pasteurization (75°C). S = sterilization (>110°C).

Table 3.6. Approximate number of days for residual nitrite to fall below 10 mg/kg in a pork slurry of pH 5.5–6.2 at 2.5–4.5% NaCl at a storage temperature of 15°C; (N = 5) (Gibson et al. 1984).

	Heat Treatment		
Nitrite Added (mg/kg)	Unheated	80°C/7 min	80°C/7 min + 70°C/1 hour
	Days	Days	Days
A) No further addition			
100	5	12	63
200	10	12	68
300	21	21	>168
B) As in A + ascorbate (1000 mg/kg)			
100	5	9	10
200	5	9	9
300	5	21	48
C) As in A + ascorbate (1000 mg/kg) + polyphosphate 0.3% (w/v)			
100	5	10	21
200	10	21	21
300	5	5	12

The absorbance of this wavelength can be used for detecting the percentage of each form in meat. Nitrosomyoglobin has a spectrum that has similar maxima as oximyoglobin (Figure 3.11).

Oxygen and NO are biatomic molecules. A similar biatomic molecule CO also binds to myoglobin, and is also very tight. In some countries, MAP packaging of meat with 1–2% CO is permitted.

In the last 2 years the riddle about the red color of non-nitrite/nitrate–cured raw hams like Parma ham or Jamon Iberico has been solved. Various authors could show and prove that the Fe^{2+} in the porphyrin ring is exchanged with Zn^{2+}, which gives a pleasant red color. Nitrite addition prevents the exchange (Morita et al., 1996; Møller et al., 2003; Parolari et al., 2003; Wakamatsu et al., 2004; Christina et al., 2006).

By reducing enzymes or chemical reactions with a reducing agent like ascorbate, the Fe^{3+} is reduced to Fe^{2+} (Figure 3.12). The NO formed from N_2O_3 can bind to the myoglobin (Fe^{2+}) and forms a heat-stable NO-myoglobin. Oximyoglobin is not heat-stable and dissociates. The meat turns grey or brown.

The NO-myoglobin is heat-stable. On heating the protein moiety is denatured, but the red NO-porphyring

Table 3.7. Nitrite and nitrate concentration in emulsion-type sausage (Lyoner) (72 ppm nitrite added) after ultra-high-pressure application and storage of the unheated batter.

Days	0	7	14	21	0	7	14	21	0	7	14	21
Treatment	Nitrite (mg/kg)				Nitrate (mg/kg)				Nitrite + Nitrate (mg/kg)			
Control	54,3	47,1	42,8	39,3	15,15	19,5	23,8	26,2	69,45	66,55	66,6	65,5
400 MPa	53,3	46,4	41,5	37,7	15,8	22	25,9	26,95	69,05	68,4	67,4	64,6
600 MPa	53,0	44,8	40,7	37,2	16,15	23,85	27,35	29,1	69,1	68,65	68	66,25
800 MPa	52,3	44,7	41,6	37,7	17,5	23,95	26,55	26,6	69,75	68,65	68,15	64,25

Source: Ziegenhals (2006).

Table 3.8. Nitrite and metabolites in meat products.

Bound to or From	% of Total[a]	% of Total[b]
Nitrite	5–20	
Nitrate	1–10	10–40[a]
Myoglobin	5–15	
Bound to −SH	1–15	
Bound to lipids	1–15	
Bound to proteins	20–30	
Gas	1–5	
Sum	~70	90

[a] Cassens et al. (1978).

[b] Assumption realized by the author (K. O. Honikel) according to results presented in Figures 3.4, 3.5, and 3.6.

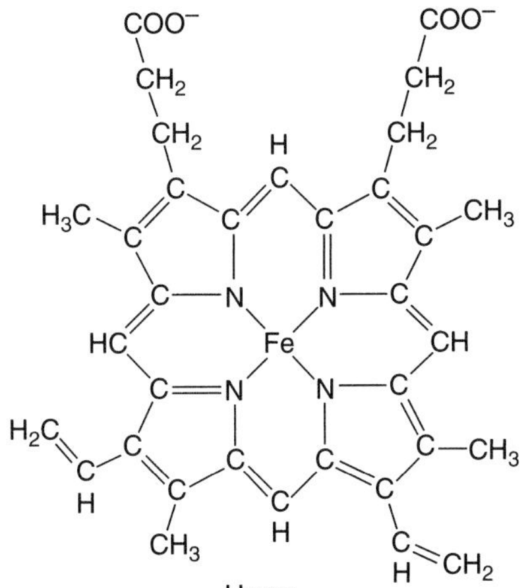

Haem
(Fe-protoporphyrin IX)

Figure 3.9. Hem in the myoglobin molecule.

3^{+}
$2H^{+} + NO_2^{-} + Fe^{2+}$ - myoglobin →
2^{+}
$H_2O + NO + Fe^{3+}$ - myoglobin(metmyoglobin)

Figure 3.10. Nitrite reaction with myoglobin in meat at pH 5.5.

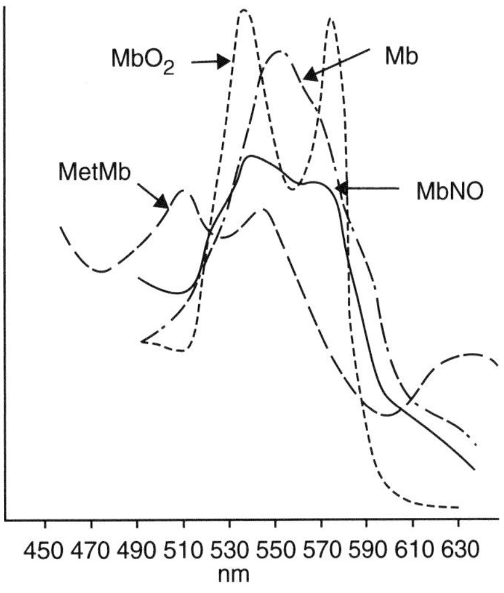

Figure 3.11. Spectra of myoglobin.

metmyoglobin (Fe^{3+}) + NO + ascorbate →
NO-myoglobin (Fe^{2+}) + dehydroascorbate

Figure 3.12. Formation of nitrosomyoglobin.

system (often called *nitroso-haemochromogen*) still exists and is found in meat products heated to 120°C.

This heat-stable red color will change on bacterial spoilage, and it fades on UV light. The first one is advantageous because the consumer recognizes spoilage, such as in fresh meat, which also changes color on spoilage.

NITROSAMINE FORMATION

In the 1970s in the U.S., a discussion came up about the formation of nitrosamines in cured meat products, especially fried bacon. Fiddler et al. (1978) showed that bacon and its cookout on frying contained considerable amounts of nitrosopyrrolidine.

Nitrosamines are formed by amines with nitrite at higher temperatures according to the reactions in Figure 3.13. But there are some prerequisites:

- Amines must be present. In fresh meat, there are very minute amounts of amines, which are the decarboxylation products of amino acids. During aging and fermentation amines are formed.
- Only secondary amines form stable nitrosamines (Figure 3.13). Primary amines are immediately degraded to alcohol and nitrogen. Tertiary amines cannot react. Most amines in meat are primary amines derived from α-amino acids.
- The pH must be low enough to produce NO^+ or metal ions must be engaged to form NO (Figure 3.13).

As heated meat products are produced from fresh meat (chilled or frozen) no amines are available. In raw meat products, the nitrite concentration is rather low (see Figure 3.4) and thus the formation of NO^+ is rather unlikely. In products heated above 130°C, nitrosamines can be formed. Bacon frying, cured sausage, grilling, or frying cured meat products, such as pizza toppings, may result in such conditions that nitrosamines are formed. Table 3.9 shows the results of an investigation by Deierling et al. (1997). In German foods, only beer and pizza exhibited dimethylnitrosamine in detectable amounts. Thus nitrosamines occur only in small amounts and are easily avoidable by proper frying, grilling, and pizza baking.

Table 3.9. NO-dimethylamine in foods (μg/kg) (Deierling et al. 1997).

		Content		
Food	N	>0.5	Min	Max
Beer	195	3	0.5	1.2
Pizza	57	6	0.5	8.7
Meat products	17	0	0	0
Milk products	6	0	0	0

CONCLUSIONS

Curing meat is a process known since ancient times. The curing agents nitrite and nitrate react due to easily varying oxidation status of nitrogen into many derivatives with meat ingredients.

Both are those additives in foods that change to a considerable extent. The curing agents give the products an esteemed and stable red color; they act as antioxidants, nitrite prevents or retards microbial growth, and the curing agents give a pleasant flavor. The positive effects are overwhelming against the small possibility of the formation of nitrosamines. The intake of curing agents by meat products is small (few percent) in comparison with other foods.

$$NaNO_2 + H^+ \rightarrow HNO_2 + Na^+$$
$$HNO_2 + H^+ \rightarrow NO^+ + H_2O$$
$$2HNO_2 \rightarrow N_2O_3 + H_2O$$
$$N_2O_3 \rightarrow NO + NO_2$$
$$NO + M^+ \rightarrow NO^+ + M$$

primary amine $RNH_2 + NO^+ \rightarrow RNH\text{-}N=O + H^+ \rightarrow ROH + N_2$

secondary amine $R_2NH + NO^+ \rightarrow R_2N\text{-}N=O + H^+$

tertiary amine $R_3N + NO+ \rightarrow$ no nitrosamine formation

Figure 3.13. Formation of nitrosamines, chemical reactions (M/M^+ are transition metal ions such as $Fe2^+/Fe3^+$).

REFERENCES

RG Cassens, I Ito, M Lee, D Buege. 1978. The Use of Nitrite in Meat. Biosci 28, 10:633–637.

E Christina, JKS Adamsen, K Møller, K Laursen, LH Olsen, Skibsted. 2006. Zn-porphyrin formation in cured meat products: Effect of added salt and nitrite. Meat Sci 72:672–679.

Code of Federal Regulations (USA). 2005. Title 21, Volume 3, Food and Drugs. Food Additives, 21CFR170.60, 172.175, 172.170.

H Dahn, Lotte Loewe, CA Bunton. 1960. 42. Über die Oxydation von Ascorbinsäure durch salpetrige Säure. Teil VI: Übersicht und Diskussion der Ergebnisse. 18. Mitteilung über Reduktone und 1,2,3-Tricarbonylverbindungen. Helvetica Chimica Acta 53:320–333.

Danish Regulation Lovtidende. 1995A. p. 5571 of 18.12.1995.

I Dederer. 2006. Personal communication.

H Deierling, U Hemmrich, N Groth, H Taschan. 1997. Nitrosamine in Lebensmitteln. Lebensmittelchemie 51:53–61.

V Dordević, B Vukksan, P Radetić, H Durdica, M Mitković. 1980. Prilog ispitivanju uticaja pojedinih faktora na promene sadrzaja nitrita u mesu. Tehnologija mesa 21, 10:287–290

W Fiddler, JW Pensabene, EG Piotrowski, JG Philips, J Keating, WJ Mergens, HL Newmark. 1978. Inhibition of formation of volatile nitrosamines in fried bacon by the use of cure-solubilized α-tocopherol. J Agric Food Chem 26, 3:653–656.

JB Fox, JR, SA Ackerman. 1968. Formation of nitric oxide myoglobin: Mechanisms of the reaction with various reductants. J Food Sci 33:364–370.

AM Gibson, TA Roberts, A Robinson. 1984. Factors controlling the growth of Clostridium botulinum types A and B in pasteurized cured meats. VI. Nitrite monitoring during storage of pasteurized pork slurries. J Food Technol 19:29–44.

J Haldane. 1901. The red color of salted meat. J Hygiene I:115–122.

R Hoagland. 1910. The action of saltpeter upon the color of meat. In: 25th Annual Report of the Bureau of Animal Industry. U.S. Department of Agriculture. Washington, D.C., Government Printing Office, 1910, pp 301–316.

———. 1914. Coloring matter of raw and cooked salted meats. J Agric Res III, Secretary of Agriculture. Washington, D.C., 3 (1914/15) 3. 211–225, 2 Taf., Heft 3, 15.12.1914.

LJ Ignarro. 1990. Biosynthesis and metabolism of endothelium-derived nitric oxide. Ann Rev Pharmacol Roxicol 30:535–560.

K Izumi, RG Cassens, ML Greaser. 1989. Reaction of nitrite with ascorbic acid and its significant role in nitrite-cured food. Meat Sci 26:141–153.

K Kisskalt. 1899. Beiträge zur Kenntnis der Ursachen des Rotwerdens des Fleisches beim Kochen nebst einigen Versuchen über die Wirkung der schwefeligen Säure auf der Fleischfarbe. Arch Hyg Bakt 35:11–18. Kudryasho. 2003. personal communication.

KB Lehmann. 1899. Über das Haemorrhodin. Ein neues weit verbreitetes Blutfarbstoffderivat. Über Phys Med Gesellschaft Würzburg 4:57–61.

CJ Lowenstein, SH Snyder. 1992. Nitric oxide, a novel biologic messenger. Cell 70:705–707.

TF Luscher. 1990. Endothelial control of vascular tone and growth. Clin Exp Hypertens (A) 12:897–902.

———. 1992. Endogenous and exogenous nitrates and their role in myocardial ischaemia. Br J Clin Pharmacol 34(Supp. 1):29–35.

MA Marletta. 1989. Nitric oxide: Biosynthesis and biological signfcance. Trends Biochem Sci 14:488–492.

JKS Møller, CE Adamsen, LH Skibsted. 2003. Spectral characterisation of red pigment in Italian-type dry-cured ham. Increasing lipophilicity during processing an maturation. European Food Res Technol 216:290–296.

S Moncada, RM Palmer, EA Higgs. 1991. Nitric oxide: Physiology, pathophysiology and pharmacology. Pharmacol Rev 43:109–142.

H Morita, J Niu, R Sakata, Y Nagata. 1996. Red pigment of Parma ham and bacterial influence on its formation. J Food Sci 61(5):1021–1023.

C Nathan. 1992. Nitric oxide as a secretory product of mammalian cells. FASEB J 6:3051–3064.

Official Journal of the EU. 1995. Directive 95/2/EC. L61, 18.3. Food Additives Other Than Colours and Sweeteners, pp. 1–40.

———. 2006. Directive 2006/52/EC. L204, 26.7. Food Additives Other than Colours and Sweeteners, pp. 10–22.

G Parolari, L Gabba, G Saccani. 2003. Extraction properties and absorption spectra of dry cured hams made with and without nitrite. Meat Sci 64:483–490.

E Polenske. 1891. Über den Verlust, welchen Rindfleisch und Nährwert durch das Pökeln erleidet sowie über die Veränderungen salpeterhaltiger Pökellaken. Arbeiten aus dem kaiserlichen Gesundheitsamt 7:471–474.

LJ Schuddeboom. 1993. Nitrates and nitrites in foodstuffs. Council of Europe Press, Publishing and Documentation Service. ISBN 92-871-2424-6.

JS Stamler, DJ Singel, J Loscalzo. 1992. Biochemistry of nitric oxide and its redox-activated forms. Sci

258:1898–1902.Verordnung über Fleisch und Fleischerzeugnis (Fleisch-Verordnung) in der Fassung der Bekanntmachung vom 21. Januar 1982. Bundesgesetzblatt I. 1982. Nr. 3, 89–101.

J Wakamatsu, T Nishimura, A Hattori. 2004a. A Zn-porphyrin complex contributes to bright red colour in Parma ham. Meat Sci 67:95–100.

J Wakamatsu, J Okui, Y Ikeda, T Nishimura, A Hattori. 2004b. Establishment of a model experiment system to elucidate the mechanism by which Zn-protoporphyrin IX is formed in nitrite-free dry-cured ham. Meat Sci 68 (2):313–317.

K Ziegenhals. 2006. Personal communication.

4
Principles of Meat Fermentation

Esko Petäjä-Kanninen and Eero Puolanne

INTRODUCTION

The process of fermenting and drying is believed to be one of the oldest techniques for preserving meat. First notions about fermented sausages are from 3000 B.C., and more information is from China and the Mediterranean area from 2000 years ago (Varnam and Sutherland 1995). Although the inner parts of animal tissues are sterile, in practical terms it is impossible to keep meat sterile during slaughtering and cutting. Once contaminated, the deterioration of quality and health hazards advance quickly unless the meat is consumed or preserved. Consequently, meat has been preserved since ancient times by cooking, salting, drying, and—mostly in colder areas—by smoking, alone or in combination. Early, it was also discovered that preservation resulted in meat products with sensory properties that are well liked.

A fermented dried sausage consists of muscular tissue, from which fat and thick connective tissue membranes have been trimmed off, and fatty tissue, which is most often pork back fat. The other ingredients are salt (NaCl) 2.2% or more, sugars (glucose, sucrose, oligosaccharides derived from starch, sometimes lactose), and occasionally foodstuffs of vegetable origin (like tomato or onion) or cheese. Spices, nitrite, and/or nitrate and ascorbate are used as well. Traditionally, the fermentation relied on the natural microbial flora, but in modern production pure microbial cultures are added. The ingredients are chopped, meat and fat in frozen stage, in a chopper more-or-less coarsely. As a result of salt, muscular tissue partly melts and at the end of chopping, a homogeneous, slightly sticky mince is formed which is then stuffed into casings.

During the first hours of processing, the temperature of the sausage mince will increase from approximately −2°C to the level of 20°C or more depending on the sausage type. In the initial fermentation step, sugars in the mince are converted into lactic acid by lactic acid bacteria, and water starts migrating to the surface where it evaporates. Later on during drying and ripening, staphylococci, yeasts, and molds form compounds from other sources than sugars that influence taste and aroma. These processes result in a characteristic flavored sausage with a lowered pH and a reduced water activity that makes the final sausage stable with long shelf life, even if not being subjected to heat treatment.

Fermentation of meat means, in the context of this chapter, the action of microorganisms and meat enzymes resulting in the desired quality and keepability. Along with fermentation of dry-fermented products (fermentation at ambient temperatures) drying and/or smoking are usually applied. In semidry products (quick fermentation at high temperatures), cooking is used to stop the fermentation, especially in the U.S. (Varnam and Sutherland 1995).

The utilization of fermentation, drying, and smoking may have developed originally by accident. Spontaneous fermentation made the meat stable at ambient temperatures and improved the sensory quality of the product. At first, salt, which included natural nitrate, was added to meat; later, more sophisticated comminuting-smoking-drying procedures were introduced.

A well-known practice has been to add a piece of good sausage from a previous batch to the ingredients of a new batch (backslopping). In the early 1900s it was discovered that it was beneficial bacteria that caused the positive reactions in the sausage batter, and the utilization of inoculated bacteria, starter cultures, was introduced in the 1930s and 1940s; lactobacilli (Jensen and Paddock 1940; Nurmi 1966), and later in the 1950s pediococci (Niven et al. 1955) and staphylococci (Niinivaara 1955) opened the way for industrial applications. The fundamental role of microorganisms was then clearly shown.

The general principles of the fermentation of dry sausages have remained the same over the millennia, and the aspects of preparation been reviewed many times (e.g., recently by Toldrá [2004] and Demeyer and Toldrá [2004]). The purpose of this chapter is only to present the basic principles of meat fermentation. Fermentation results in acid production influencing, e.g., firmness, keepability/safety, color, lipolysis, proteolysis, and formation of aroma and taste compounds. These aspects will be reviewed also in other chapters of the book.

FERMENTATION

Sausage Fermentation

The fermentation step in the sausage production includes the period in the sausage process where pH decreases from approximately 5.7 to its lowest value, which could vary from 5.5 (hard salamis) to 4.6 (or even 4.2 in high-temperature fermented sausages) depending on the sausage type. The fermentation lasts from less than 12 hours to several days depending on the sausage style. Sausages fermented at high temperatures (37°C or higher) reach quickly the lowest pH value (Varnam and Sutherland 1995), but temperatures of around 24°C result in a pH of 4.6 to 5.0, and lower temperatures usually in a higher pH at a slower speed. The amount of added sugar is the parameter of major influence on the lowest attainable pH but not on the speed (Puolanne 1977). Generally, pH increases during the subsequent maturation period, which lasts from a couple of weeks to several months. Fermentation is achieved by natural contaminating bacteria or by addition of lactic acid bacteria. In former times, backslopping was used to start the fermentation, but now well-defined starter cultures of lactic acid bacteria in combination with other microbial pure cultures are applied.

Biochemistry of Fermentation

Adenosine triphosphate (ATP) is a form of chemical energy that is necessary for life. The primary reason for fermentation is to produce energy anaerobically, e.g., to produce ATP using the substrate level phosphorylation. In the reaction sequence, there is for each reduction a corresponding oxidation, and no oxygen or any other molecule is needed for terminal electron recipient. Practically any organic substance that contains oxygen and hydrogen can be used as starting material, but usually carbohydrates or amino acids are fermented. Substances such as lactate, acetate, ethanol, acetoin, ammonia, etc., will be produced as end products, but they are not used any further inside the cells and will be secreted to the outside (Tortora et al. 2001).

Biochemically, fermentation is defined as a reaction where ATP is produced in substrate level phosphorylation, i.e., oxygen is not needed, and consequently no carbon is lost as CO_2. Microorganisms utilize ATP for maintaining cell functions and for growth (Tortora et al. 2001). The main part of the use is related to the pH gradient (proton motive force, which is "a biological accumulator") across the cell membrane, i.e., the proton flow into the cell fuels the ATP synthesis. Protons are also excreted from the cells to establish the pH difference between the external and internal surface. The energy required to maintain this increases, when pH is below 5.0, because it becomes difficult to transport protons out of the cell against a proton concentration gradient of higher than 100 (Konings 2002). In meat systems, it is especially difficult to obtain much lower pHs, because the buffering capacity of meat increases quickly at low pHs (buffering capacity is about 30 mmmol H^+/(pH*kg) at pH 5.5 and about 70 mmmol H^+/(pH*kg) at pH 4.8 (Puolanne and Kivikari 2000). The amount of ATP [expressed as moles ATP/(g*h)] needed for growth is, however, much higher than the consumption required for maintenance, the relationship depending on growth rate.

Each microbial strain has its own way of utilizing glucose and other substrates for energy production. The most frequent is glycolysis, i.e., the Embden-Meyerhof mechanism (see textbooks of biochemistry or microbiology for the detailed reaction cascades). Glucose is converted to pyruvate (2 CH_3COCOO^-), which is in turn converted into lactate (2 $CH_3CHOHCOO^-$), and simultaneously 2 ($NADH+H^+$) are regenerated to 2 NAD^+. It should be mentioned here that pyruvate and lactate remain dissociated (not in the form of an undissociated acid) during these conversions and that the above reactions bind two protons. The protons that reduce pH during fermentation are derived from the reaction ATP → ADP + Pi + H^+ (Robergs et al. 2004). Other carbohydrates can also be utilized, but the relative availability of carbohydrates other than added glucose is low. Sometimes other sugars, such as lactose, are added. Meat itself contains glycogen and its derivatives, mainly glucose phosphates. Glycogen is hardly decomposed by meat bacteria, and very little is known about the degradation of glycogen by meat enzymes in salted meat.

Microbial Fermentation

The starter cultures used in meat products are homofermentative lactic acid bacteria (LAB), i.e., the main end product of carbohydrate fermentation is lactate. The amounts of other end products produced by LAB are usually very small, but on the other hand, the flora may provide a wide diversity of substances, thus influencing the flavor and aroma of the product (Tortora et al. 2001; Sunesen et al. 2001). Even greater diversity will result from the fermentation of other substrates (amino acid, fatty acid), as well as the secondary reactions of these metabolic products in the food matrix outside the microbial cells. In sausage, the fermentation usually stops after lactic acid fermentation, and subsequent fermentations such as propanoic (propionic) acid production in cheese are not that relevant in sausages. In sausages prepared with molds or yeasts, however, these organisms further metabolize lactate. One factor that stops fermentation is low pH, but the decrease of a_w will also inhibit further fermentation.

Microbial and meat enzymes also affect other constituents of meat than carbohydrates. The metabolic activity of microorganisms aims at fulfilling the energy and nutrient requirements of the microorganisms, as well as creating a favorable environment for further growth, such as production of biogenic amines (may be health hazards) that increase pH thus resulting in a more favorable pH, or the production of bacteriocins that inhibit growth of microorganisms of similar nutritional requirements, or lactic acid that prevents the growth of acid-sensitive strains (see the section "Antagonistic Effects"). Microbial exo-enzymes, aimed at providing small molecular weight substances to be absorbed by the microorganisms, cause proteolytic and lipolytic changes during fermentation, and other enzymatic and nonenzymatic reactions increase the complexity of the ripening process (see sections "Proteolysis" and "Lipolysis"). As a consequence, the fermentation is much more influential than simply acid production.

The LAB most commonly used as starter cultures in sausage fermentation belong to the genera *Lactobacillus* and *Pediococcus* (Demeyer and Toldrá 2004). The main product of glucose fermentation is lactate, and to a lesser extent acetic acid, propanoic acid, or ethanol. The genera are facultative or microaerophilic. *Lactobacillus* and *Pediococcus* are catalase-negative and do not reduce nitrate, and they may thus cause discoloration during fermentation if they produce hydrogen peroxide (Nurmi 1966), which is not always the case. In recent years, there has been interest also on the probiotic properties of LAB starter cultures for meat products (Erkkilä and Petäjä 2000).

Staphylococci used as starter cultures are weak acidulants, produce catalase, and are able to reduce nitrate, which makes them important for color formation and gives them some degree of antioxidative capacity (see Chapter 20). In addition, *Staphylococci* is deeply involved in the flavor-forming reactions of fermented meat products (see Chapter 22).

FACTORS INFLUENCING FERMENTATION

The fermentation process is more complex than most other processes in meat technology. The fermentation, the energy-producing metabolism in microorganisms, although taking place inside the microbial cell, is not an isolated procedure, but is closely related to intrinsic and extrinsic factors and their interactions in the ripening sausage. There is no decisive factor ruling the fermentation/ripening (unlike in cooked sausage where cooking stabilizes the gel and destroys most microorganisms in a very consistent way). The natural laws provide the limits of the activities, and each individual aspect will follow these. The complexity is, however, a result of so many independent and dependent factors functioning at the same time. The main influencing factors are intrinsic, and only little can be done after the preparation of the sausages, except extrinsically varying the temperature-humidity-air velocity programming and smoking. Individual, and more or less interrelated, factors such as meat enzymes, the dynamic microbial flora, salt, nitrate/nitrite, temperature, drying, carbohydrates, smoke, etc., have an influence on the fermentation/ripening. The process is characterized by lactate production resulting in a decrease in pH, an increase in salt content, nitrate/nitrite reduction, and changes in color, firmness, taste, aroma, weight loss, etc.

The process can be controlled by careful selection of various ingredients, including meat raw materials, with special attention paid to their natural microbial flora, sugar, food additives, and spices. Reliable control of the atmosphere and air circulation, as well as temperature in the ripening chamber, suitable casings, and smoking conditions are necessary for successful and safe production. These aspects are exceptionally difficult to control theoretically, but long experience makes possible the fine-tuning of the process for good results. Finally, the utilization of bacterial pure cultures and surface yeasts and molds provides the final technological security and also safety concerning health hazards. There will, always remain large uncontrolled areas. A large part of the process remains more or less uncontrolled, but the processors try to keep the uncontrolled part as small as possible.

Starter cultures are not used only for safety reasons, but they also provide significantly added value. The process is accelerated compared to the case where only natural (accidental) flora is responsible for fermentation. Starter cultures create faster fermentation and nitrate/nitrite reduction, and usually better aroma and color will be the result.

For successful fermentation, suitable raw materials should be selected, especially microbial flora and pH (no meat with high pH). Sugar content (natural carbohydrates and added glucose with derivates) and other nutrients, like amino acids, fatty acids, minerals, vitamins, etc., create the driving forces that cause the fermentation. Starter cultures (and natural flora) and meat enzymes start the fermentation, the rate of which is controlled by external factors such as temperature and, to a lesser extent, by humidity, air circulation, and smoking. The above-mentioned external factors and intrinsic factors, such as salt content and nitrate/nitrite contents (Puolanne 1977), can be used as regulators of the process. Temperature is a decisive factor, because it has a strong and also selective effect on the microorganisms present (Puolanne 1977). As a consequence, the relative numbers of different strains in the flora may vary, as well as the end products of the meat and microbial metabolism.

PROTEOLYSIS

During ripening of fermented sausage a small portion of the proteins is hydrolyzed (Niinivaara et al. 1964). According to Diedrick et al. (1974), nonprotein nitrogen increases 20% in smoked sausages. The results of Verplaetse et al. (1989) and Demeyer (1992) indicate that tissue proteases (especially catepsin D) are the main agents of proteolysis early in fermentation. The pH 4.8–5.2 of fermented sausage and the ripening temperature of 15–20°C were optimal for the activity of such enzymes; this fact confirms the role of tissue proteases in meat fermentation.

LIPOLYSIS

Lipolysis is also taking place during fermentation and ripening of dry sausages. Lipolysis means the hydrolyzation of fat, the hydrolysis products being free fatty acids, glycerol, monoglycerides, and diclycerides. Release of long-chain fatty acids from neutral lipids and phospholipids and from cholesterol has long been a central theme in investigations on chemical changes during sausage fermentation (Nurmi and Niinivaara 1964; Cantoni et al. 1966; Demeyer et al. 1974; Roncales et al. 1989). Levels of up to about 5% of total fatty acid content have typically been encountered in fermented meat products.

ANTAGONISTIC EFFECTS

The microorganisms influence the growth of other species, and there is a hard competition in the microbial ecological environment during fermentation. The antagonism is caused by the competition for nutrients as well as the effects of antimicrobial substances produced by the microorganisms. The starter cultures or natural dominating flora need antagonism to maintain their dominance.

The microbiological flora of meat raw material primarily consists of gram-negative bacteria, most being psychrotrophs. The predominating genera are *Pseudomonas*, *Acinetobacter*, and *Moraxella*. *Brochothrix thermosphacta*, enterobacteria, staphylococci, micrococci, spore-forming bacteria, lactic acid bacteria (LAB), and yeasts are also encountered, but the counts are several orders of magnitude lower than the counts of the predominating genera (Lawrie 1998). Because the added salt contents may vary from 2.2% to over 3% in fermented sausages (in fermented whole meat products, from 1.7% to very high concentrations), this is enough to suppress growth of the most bacteria, especially gram-negative ones (Petäjä 1977). On the other hand, many LAB and staphylococci and micrococci are not sensitive to the salt concentrations prevailing during the fermentations in meat products. These bacteria also can grow at fermenting temperatures of 20–25°C. The counts of these bacteria increase to the level of over 8 log CFU/g of LAB and to over 5 log CFU/g of staphylococci and micrococci. Often, they hardly grow unless the production is very traditional, but they are added in levels of 6–7 logs. The pH decreases to 5.0 or below, which finally suppresses the growth of most other bacteria, but not staphylococci and micrococci, meaning that bacterial flora is formed of LAB and staphylococci and micrococci after the fermentation of about 1 week. Hydrogen peroxide production of LAB may also suppress the growth of other bacteria (Lücke 1986). In conclusion, LAB and staphylococci/micrococci compete well with and thus suppress the growth of other bacteria, but also thrive better in the circumstances of meat fermentation than the other bacteria.

The growth or survival of pathogens, such as *Salmonella*, *Yersinia enterocolitica*, and *Campylobacter*, are mainly suppressed in the fermented meat products because of low pH value in combination with the produced lactate and the low a_w value. Listeria survives better, but in most cases they also disappear during drying. In a dry sausage, listeria disappeared at latest during the third week of ripening (Työppönen et al. 2003). Neurotoxin production and growth of *Clostridium botulinum* are controlled by nitrite, low pH, and low a_w value in these products (Peck and Stringer 2005). The only pathogens, which could be harmful, are *Staphylococcus aureus* and *Escherichia coli*, such as O157:H7. The enterotoxin production and growth of *S. aureus* are suppressed by using starter cultures (LAB) (Niskanen and Nurmi 1976). If inoculated into dry sausage, the counts of *E. coli* O157:H7 (EHEC) decrease 3 log CFU/g to the level of 2 log CFU/g, but because of low level of contamination, it can be regarded that risk in fermented meat products is not particularly large (Erkkilä et al. 2000).

REFERENCES

C Cantoni, MR Molnar, P Renon, G Giolitti. 1966. Investigations on the Lipids of Dry Sausages. Proceedings of the 12th European Meeting of Meat Research Workers, Sandefjord, E-4.

DI Demeyer. 1992. Meat fermentation as an integrated process. In: FJM Smulders, F Toldrá, J Flores, M Prieto, eds. New Technologies for Meat and Meat Products. Nijmegen, NL: Audet Tijdschriften, pp. 21–36.

D Demeyer, J Hoozee, H Mesdom. 1974. Specificity of lipolysis during dry sausage ripening. J Food Sci 39:293–296.

D Demeyer, F Toldrá. 2004. Fermentation. In: WK Jensen, C Devine, M Dikeman, eds. Encyclopedia of Meat Sciences. Amsterdam: Elsevier Academic Press, pp. 467–474.

N Diedrick, P Vandekerchove, D Demeyer. 1974. Changes of non-protein nitrogen-compounds during dry sausage ripening. J Food Sci 39:301–304.

S Erkkilä, E Petäjä. 2000. Screening of commercial starter cultures at low pH and in the presence of bile salts for potential probiotic use. Meat Sci 55:297–300.

S Erkkilä, M Venäläinen, S Hielm, E Petäjä, E Puolanne, T Mattila-Sandholm. 2000. Survival of Escherichia coli O157:H7 in dry sausage fermented by probiotic lactic acid bacteria. J Sci Food Agric 80:2001–2004.

L Jensen, L Paddock. 1940. US Patent 2,225,783. (Ref. Niinivaara 1955.)

N Konings. 2002. The cell membrane and struggle for life of lactic acid bacteria. Antonie van Leeuwenhoek 82:3–27.

RA Lawrie. 1998. Meat Science. Cambridge, UK: Woodhead Publ. Ltd. Sixth ed., pp. 119–142.

FC Lücke. 1986. Microbiological processes in the manufacture of dry sausage and raw ham. Fleishwirtsch 66:1505–1509.

FP Niinivaara. 1955. Über den Einfluss von Bakterienreinkulturen auf die Reifung and Umrötung der Rohwurst. Acta Agralia Fennica 84:1–126.

FP Niinivaara, MS Pohja, SE Komulainen. 1964. Some aspects about using bacterial pure cultures in the manufacture of fermented sausages. Food Technol 18:25–31.

A Niskanen, E Nurmi. 1976. Effects of starter culture on staphylococcal enterotoxin and thermonuclease production in dry sausage. Appl Environ Microbiol 31:10–12.

C Niven, R Deibel, Wilson. 1955. The use of pure bacterial cultures in the manufacture of summer sausage. Annual Meeting of American Meat Institute. 5p. (Ref. Nurmi 1966.)

E Nurmi. 1966. Effect of bacterial inoculation on characteristics and microbial flora of dry sausage. Thesis, University of Helsinki. Acta Agralia Fennica 108:1–77.

E Nurmi, FP Niinivaara. 1964. Lipid changes of fats in dry sausages. Proceedings of the 10th European Meeting Meat Research Workers, Roskilde, G-8.

MW Peck, SC Stringer. 2005. The safety of pasteurised in-pack chilled meat products with respect to the foodborne botulism hazard. Meat Sci 70: 461–475.

E Petäjä. 1977. The effect of some gram-negative bacteria on the ripening and quality of dry sausage. J Sci Agric Soc of Finland 49:107–166.

E Puolanne. 1977. Der Einfluss von verringerten Nitrit und Nitratzusätzen auf den Eigenschaften der Rohwurst. J Sci Agric Soc of Finland 49:1–106.

E Puolanne, R Kivikari. 2000. Determination of the buffering capacity of postrigor meat. Meat Sci 56:7–13.

RA Robergs, F Ghiasvand, D Parker. 2004. Biochemistry of exercise-induced metabolic acidosis. Am J of Physiology. Regulatory, Integrative and Comparative Physiology 287:R502–516.

P Roncales, M Aguilera, JA Beltran, I Jaime, JM Piero. 1989. Effect of the use of natural or artificial casings on the ripening and sensory quality of dry sausage. Proceedings 35th International Congress Meat Science and Technology, Copenhagen, p. 825–832.

LO Sunesen, V Dorigoni, E Zanardi, L Stahnke. 2001. Volatile compounds released during ripening in Italian dried sausage. Meat Sci 58:93–97.

F Toldrá. 2004. Meat: Fermented meats. In: JS Smith, YH Hui, eds. Food Processing. Principles and Applications. Ames, Iowa: Blackwell Publishing, pp. 399–415.

G Tortora, R Funke, C Case. 2001. Microbiology. An Introduction. Chapter 5. Microbial metabolism. San Francisco: Addison Wesley Longman Inc., pp. 113–150.

S Työppönen, A Markkula, E Petäjä, M-L Suihko, T Mattila-Sandholm. 2003. Survival of *Listeria monocytogenes* in North European type dry sausage fermented by bioprotective meat starter cultures. Food Contr 14:181–185.

AH Varnam, JP Sutherland. 1995. Meat and Meat Products: Technology, Chemistry and Microbiology. London, UK: Chapman & Hall, pp. 314–354.

A Verplaetse, M De Bosschere, D Demeyer. 1989. Proteolysis during dry sausage ripening. Proceedings of the 35th International Congress of Meat Science and Technology, Copenhagen, pp. 815–818.

5
Principles of Drying and Smoking

Ana Andrés, José M. Barat, Raúl Grau, and Pedro Fito

INTRODUCTION

Food drying is one of the oldest methods of preserving food for later use that is simple, safe and easy to learn. It removes enough moisture from the food so bacteria, yeasts and molds can't grow. Drying also slows down the action of enzymes because it removes moisture, the food shrinks, and it becomes lighter in weight (Gould 1920).

Sun drying is the most ancient method for food preservation and even now is still used for fruit drying. In the 18th century, drying included the use of smoke to accelerate the drying processes, but this system implies some important flavor changes (Gould 1920).

Drying is used in the food industry as a method to preserve a large amount of products. The great weight and volume reduction of the dried products imply an important reduction of storage and shipping costs. However, drying is a process that largely affects the sensorial properties of the food products (Baker 1997; Heldman and Hartel 1997).

Because the presence of water in foods directly affects its spoilage, removing water from a product will reduce the possibility of its biological alteration, as well as the kinetics of other spoilage mechanisms.

Removing water from a food implies two important problems: it carries the risk of altering the nutritional and sensorial quality, and it is sometimes a high energy-consuming process. The lack of selectivity in water-removing processes can produce an important loss of aromas, even more when working at vacuum pressures (Bimbenet and Lebert 1990).

BASIC PRINCIPLES OF DRYING AND SMOKING

Drying and/or dehydration is a method of food preservation wherein the moisture is evaporated by exposure of the product to heat (solar, microwave, etc.) and air. It is a process governed by the fundamental principles of physics and physical chemistry. Drying is an energy-intense process using heat applied externally or generated internally by microwave or radio frequency energy to evaporate water on the surface and drive trapped moisture to the surface. Efficient drying is maintained only by heat that penetrates to the center of the food, forcing moisture to the surface for evaporation or the point of lower vapor pressure area of the food particle (Mujumdar 1995).

There are different types of driers depending on the way of heating (convection, radiation, dielectric, or conduction). The most common driers used in the food industry are convective or air driers, working continuously or by batches depending on the type of product. Designing a drier requires understanding the mechanism and the kinetics of the food product to be dried—that is, it is necessary to establish the quantitative relationships between the time of the process and the drying conditions (Bimbenet and Lebert 1990). The variables that determine the time of drying of a material are variables related to air-drying (temperature, velocity, humidity, flow characteristics, etc.) and variables related to the product (moisture, size, shape, structure, etc.). For this reason, the experimental determination of the drying curves (moisture versus time) for fixed experimental conditions is necessary. Those experimental conditions must be as close as possible to the industrial device in terms of heat transfer mechanism, airflow, product density, etc. In other words, the industrial process must be imitated to make the scaling-up easier (Baker 1997).

Different types of experimental data are necessary to design a drying process; some of them are related to quality aspects (evolution of color, shape and size, texture, flavor, etc.) and others are related to the kinetics of the process (evolution of the mass and moisture throughout the drying process). The determination of the evolution of moisture in a food product under a drying process gives the well-known drying curve. Figure 5.1 shows a typical drying curve for wet products, where three zones can be distinguished corresponding to three typical steps occurring in a drying process: induction period, constant rate period, and falling rate period of drying.

The induction period corresponds to the beginning of the drying process during which a heat transfer occurs from the air to the product, increasing the surface temperature until the wet-bulb temperature. The duration of this period depends on many factors, but it is negligible as compared to the following periods, and for this reason it is not taken into account for design purposes.

When the surface reaches the wet-bulb temperature the total amount of heat from the air is used to evaporate water, and because the rate of evaporation is lower than the rate of water transport to the surface, the drying rate is constant (constant rate period). In this period, it is assumed that the surface remains wet and behaves like a liquid mass. As the drying process proceeds, the product dries and the water transport rate toward the surface decreases. This is the end of the constant rate period, and the product moisture at this point is known as critical moisture content (X_c) and the time needed to reach this point is called *critical time* (t_c). The constant rate period is the drying step where the most important changes in terms of volume reduction take place. Internal moisture is more difficult to remove because it is more tightly bound and is protected by the insulating effect of the already dried material close to the surface. The water in the center is very difficult to remove and the removal of this water is done by diffusion. Thus, it is called the falling rate period or diffusion period of drying—that is, this water must be diffused from the center outward for evaporation to accomplish a dried product. The drying rate in this period is very low as a typical diffusion process, and product moisture decreases to reach equilibrium moisture (X_e) or in other words, the equilibrium of water activity with the surrounding airflow.

These typical drying periods can also be explained taking into account the shape of the sorption isotherm of a high-moisture food product (Figure 5.2). It is well known that the drying rate depends on the driving force acting during the process, that is, the difference between the water activity (a_w) of the product and the water activity of the drying air. Another consideration to take into account is that the equilibrium condition is accomplished at the interface; the water activity at the product surface equals the water activity of the drying air. During the constant rate period, the water

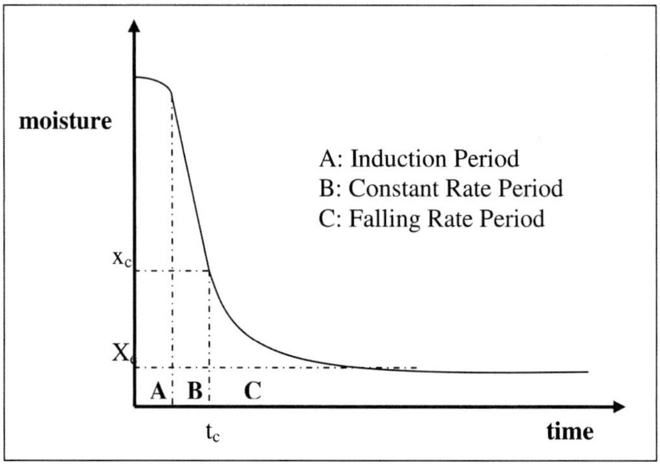

Figure 5.1. Drying periods in a typical drying curve.

activity of the product is near 1, and as the product dries, big changes on the moisture product lead to small changes on the water activity, as can be appreciated in the isotherm (Figure 5.2). Because a_w remains nearly constant, the driving force can be considered constant as well as the drying rate. However, during the falling rate period, small changes in moisture content lead to big changes in water activity.

Food properties (composition, structure, etc.) influence the a_w parameter, which is much related to water distribution in the different food phases, and will influence on the drying kinetics. For this reason, it is interesting to know the relationship between a_w and equilibrium moisture content (sorption isotherm).

The driving force for water evaporation is the water activity gradient between the food and the forced hot air, but other properties of drying air will also affect the drying kinetics, as was mentioned before. One of the variables affecting the drying rate is the air velocity; the higher the air velocity, the higher the drying rate. However, above a certain value of air velocity, the drying rate will not depend on the air velocity because the factor controlling the drying kinetic is the water diffusion rate from the center to the surface.

On the other hand, it can be said that the higher the air temperature, the higher the drying rate, although sometimes, if water is removed too fast, the food material may *case harden*, that is, seal up the outer surface area so that the water diffusion from the center outward is hindered. Some of the food components like pigments, vitamins, etc., are temperature-sensitive and these quality aspects will limit the maximum air temperature usable in the drying process. In some cases, the drying process has other purposes than removing water, allowing some biochemical reactions to develop taste and flavor, and it is known as a curing process. Because these reactions are time-dependent, low air temperatures are used. This is the case of fish and meat products, and sometimes a mixture of air and smoke is used to get a characteristic smoked flavor.

Smoke curing is a typical combined treatment, based on the concerted action of enzymes and heat, which promote protein and lipid changes in the previously salted raw material. The treatment has nutritional implications and affects also the sensory quality, safety, and shelf life of the product, but the extension of these changes will depend on many factors, such as the type of smoking; the relative humidity, velocity, temperature, density, and composition of the smoke; and the time of smoking (Yean et al. 1998).

The smoking technology has experienced an important evolution in the last years. In the past, products to be smoked were directly exposed to the smoke in the same chamber where it was produced. In the modern systems, the chamber for smoke production and the chamber for food smoking are different devices. The main advantage of separating these two steps is based on the smoke cooling during the pathway from the smoke production chamber to

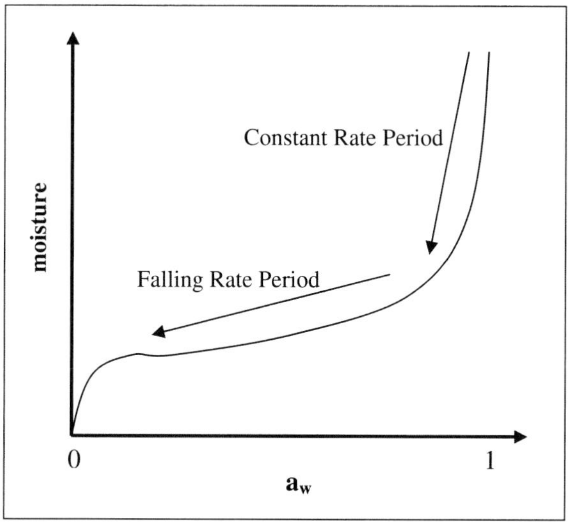

Figure 5.2. Relationship between the drying periods and the sorption isotherm.

the smoking chamber; thus, the aromatic polycyclic hydrocarbons (nondesirable substances from the point of view of consumers' health, with a demonstrated carcinogen action), together with the tars, settle in the conduction walls, thereby avoiding in some extent the meat products.

Today, the smoking chambers are air-conditioned rooms useful for meat processes other than smoking, like curing, drying, cooking, etc. There are various smoking procedures that are classified in the following sections (Cava-López and Andrés-Nieto 2001).

Traditional Smoking

Traditional smoking is divided into two categories:

- *Cold smoking*, with smoke temperatures between 20–25°C and relative humidity of 70–80%. The duration of this process can be days or weeks because the smoke used is poor in aromatic and preserving components.
- *Hot smoking*, usually applied to meat products previously cooked or blanched. Temperatures can be about 75–80°C, and the relative humidity must be high to avoid dehydration. With this procedure, the smoking time is reduced to hours.

Electrostatic Smoking

In this procedure, an electric field is applied between the smoke particles and the meat product. Electrically charged particles quickly precipitate over the product, reducing the smoking time, although the quality of the products obtained with this method differs from the traditional products.

Treatment with Smoke Aromas or Condensates

This process is considered a revolutionary smoking procedure because it allows giving the typical properties of a smoked product to new meat products or some meat products to which the traditional method is difficult to apply, without existing contact between the smoke and the meat product.

HURDLE TECHNOLOGY APPLIED TO DRIED AND SMOKED MEAT AND POULTRY PRODUCTS

As stated in the previous section, meat drying and smoking are techniques used originally for meat preservation due to the water activity reduction and additionally the bacteriostatic effect of substances present in smoke, in the case of smoking (Doe et al. 1998).

If drying and smoking should have to be used as unique meat preservation techniques, the final moisture of the dried products should have to be very low. That moisture would be below a "comfortable" value from a sensory point of view. Figure 5.3 shows the water activity value limiting the microbial growth and the typical water activity of the most common

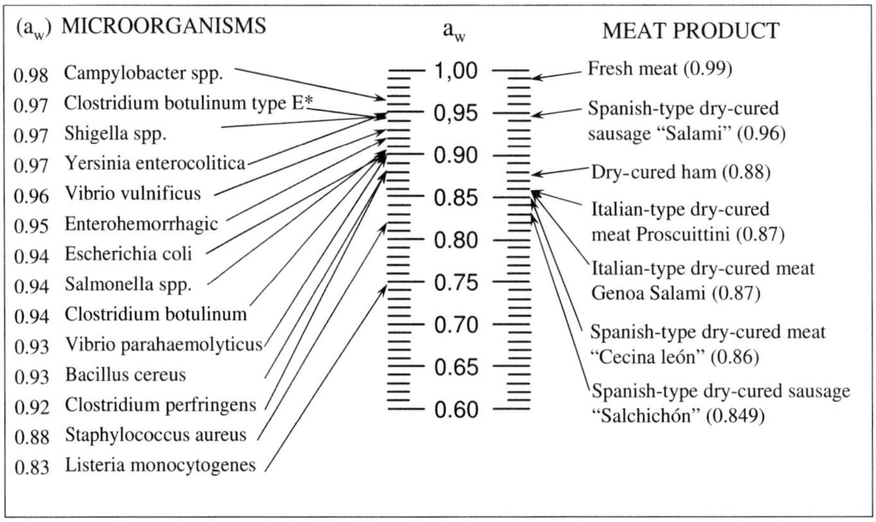

Figure 5.3. Water activity values limiting the growth of the most common spoilage microorganisms and of some meat products.

meat products (Smith et al. 1989; IFT/FDA 2003; Hoz et al. 2004; Garriga et al. 2004; Serra et al. 2005; Soriano et al. 2006; Muthukumarasamy and Holley 2006; Rubio et al. 2006a). Figure 5.3 shows that a water activity (a_w) value of 0.8 seems to be a reasonable value to get a shelf-stable product.

For instance, to get that value at room temperature (25°C) (Figure 5.4) for smoked chicken sausages, the moisture value should have to be around 24%, which is obviously too low for direct consumption.

That's the reason why drying and smoking are usually employed in combination with other preserving techniques, and thus the hurdle technology principles (Leistner 1995) are present in most of the meat products developed by those techniques.

The main preserving techniques employed in combination with drying and/or smoking are described in the following sections.

Refrigeration

All the raw materials are perishable at room temperatures in their original form, because of the use of low temperatures during storage, transportation, and processing. In some cases, the use of low temperatures at the beginning of the drying process is necessary, and is more important the higher the meat sample size. The temperature range used for meat products drying varies depending on the product, but is within a narrow range from 3–20°C. When using cold, smoking the maximum temperature increases to 30°C.

In the case of Spanish cured ham, the piece begins to dry during the postsalting stage and the process temperature is kept close to 3°C at the beginning of the postsalting. Further increases of temperature throughout the processing occur associated with the water activity decrease, until a shelf-stable product at room temperatures is achieved at the end of the process (Barat et al. 2005; Toldrá 2002).

Curing

Curing is basic in order to obtain an adequate shelf life. The curing process is well known and explained in Chapter 3. The most active components of the curing salts are the nitrites, which prevent the microbial growth, mainly the *Chlostridium botulinum* and *Listeria monocytogenes*.

Salting

Salting is always present in case of meat products. NaCl plays many technologic roles: salty taste and flavor perception, control of microbial growth and enzymatic activity, and modification of the adherence and the water-holding capacity of the protein matrix. In the case of sausages, NaCl is added during the mixing phase; in the case of whole meat pieces

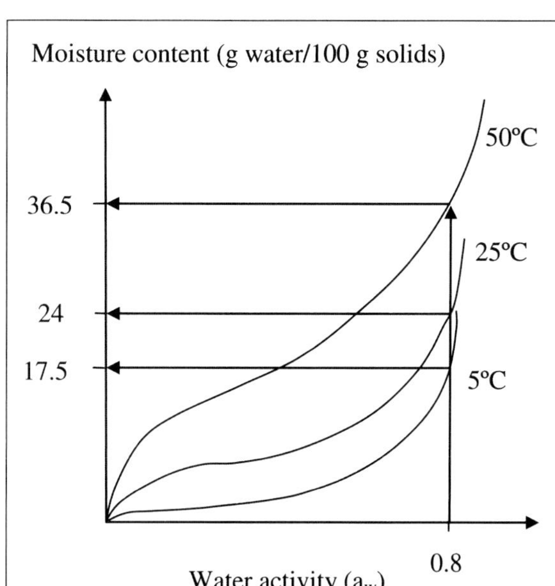

Figure 5.4. Moisture desorption isotherms of smoked chicken sausages (adapted from Singh et al. 2001).

processing, salt can be added by means of dry or wet salting. The use of dry salting without drainage with a small food/salt ratio could imply that at the end of the salting period the product could be surrounded by a brine. In this former case it would be considered as pickling (Barat et al. 2006).

Spicing

The use of spices is very frequent in the case of sausages. The addition of spices not only contribute to color and flavor development, but also become a source of external contamination and antioxidant effect, and in some cases they have an inhibitory activity on microbial growth (Shelef 1983; Zaika 1988; Martínez et al. 2006; Careaga et al. 2003; García et al. 2001).

Fermentation and Biopreservation

Fermentation is one of the operations involved in many manufactured food products (Adams and Robert-Nout 2001; Lücke 1999). The key point of fermentation consists in the decrease of the pH of the sausage as a consequence of the conversion of sugar into lactic acid due to the action of bacteria. The decrease of pH in fermented meat products not only improves its preservation but also the drying process due to the decrease in the water-holding capacity of the meat proteins when pH approaches the isoelectric value.

Specific bacteriostatic activity of some of the microorganisms added through the starter culture has been described (e.g., lactic acid bacteria [LAB]), contributing to the microbial stabilization of the final product (Hammes and Hertel 1998; Aymerich et al. 2000; Coffey et al. 1998).

Packaging and Storage

Vacuum and modified atmosphere packaging (MAP) are the most frequent ways for packaging. These techniques employ gas mixtures and packaging materials technology to extend the shelf life of food. For meat products, the atmosphere in the package is modified by pulling a vacuum and then replacing the package atmosphere with a gas mixture of oxygen (O_2) and carbon dioxide (CO_2) or nitrogen (N_2) and carbon dioxide mixtures. In dry-cured meat products, the gas mixture more utilized is formed by 0% O_2, 20–35% CO_2, and 65–80% N_2 (García-Esteban et al. 2004; Rubio et al. 2006b; Valencia et al. 2006). In these conditions, the time of storage at 10–15°C is 3 months.

FUNDAMENTALS OF DRYING AND SMOKING MEAT AND POULTRY PRODUCTS

The general principles related with food drying and smoking were introduced in the Introduction of the chapter. Nevertheless, some particular aspects must be taken into account when working with meat and poultry products.

The protein matrix suffers dramatic changes at a certain temperature, which implies the protein denaturation. These changes are usually undesirable except in certain smoked products where proteins coagulate as a consequence of the high temperatures. The before mentioned temperatures are in the range of 40–80°C (Bertram et al. 2006; Stabursvik and Martens 1980).

Drying and/or smoking not only contributes to decrease the a_w of the product but also affects the hardness and stability of the protein matrix as a consequence of the moisture reduction.

Another of the phenomenon associated to the drying process is the case hardening (Gou et al. 2005), which not only implies an increase in the hardness of the surface, but also a dramatic decrease in the drying rate and thus leads to a heterogeneous dried product with a highly dried surface and a poorly dried core. Recent studies have been done to correlate the moisture content and texture to characterize the crust formation and prevent case hardening by controlling the drying conditions (Beserra et al. 1998b; Gou et al. 2005; Ruiz-Ramírez et al. 2005; Serra et al. 2005).

The presence of fat in meat products contributes to a decrease in the drying rate because of its barrier effect, but on the other hand prevents the food surface from an excessive drying. In fact, in the case of ham drying, a layer of fat is usually applied on its surface in the final processing period—among other reasons, for preventing drying. The fat content of the sausage products range is very wide, ranging from 5–70%.

The higher the fat content, the lower the drying rate. Another thing that must be taken into account is that the excessive mincing of fat and lean gives place to a phenomenon called *smearing*, which leads to case hardening—a product with a highly moist core region.

One of the strategies followed by the industry to prevent case hardening is to fluctuate from lower to higher water activity of the air and vice versa. When the water activity of the surrounding air decreases (a_w = RH/100) the drying process accelerates, while the subsequent a_w increase enables the water to migrate from the interior to the surface fast enough to

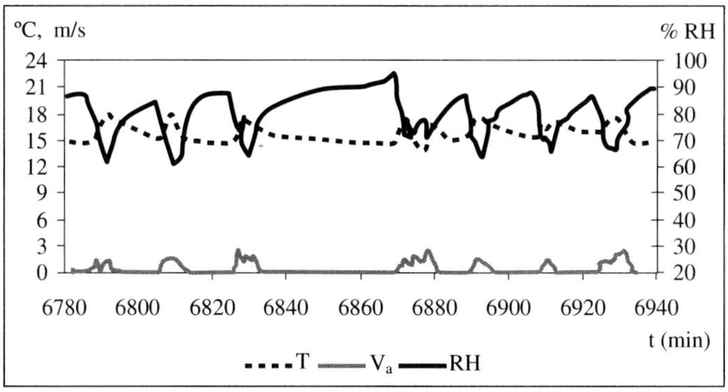

Figure 5.5. Relative humidity changes throughout a typical sausage drying process (adapted from Comaposada et al. 2004).

increase again the moisture of the surface. Figure 5.5 shows the relative humidity changes throughout a typical sausage-drying process.

In general, the risk of case hardening increases when the demand of water increases from the product by air. This demand mainly depends on the air relative humidity and the air speed inside the cabinet. The lower the relative humidity of the air, the higher the water activity gradient between the meat product surface and the bulk air, which implies an increase in the drying rate. The other parameter affecting the drying rate is the air velocity; when the air rate increases, a higher turbulent flow is created in the neighborhood of the product, which implies a decrease of the boundary layer thickness, which also implies an increase in the effective a_w gradient in the surroundings of the product.

On the other hand, the temperature increase can contribute to a decrease in the tendency for case hardening. The effective diffusion of water inside the meat product usually increases exponentially with temperature, as predicted by Arrhenius law (Palmia et al. 1993; Mittal 1999). The higher the transport rate of water inside the product, the later the case hardening will occur or the less intensity it will have, because surface water is constantly replaced by water migration from inside the product, keeping the product surface moist enough to avoid that problem (Beserra et al. 1998a). Nevertheless, the higher the temperature, the higher the microbial risks and the possibility of protein denaturing due to temperature. Another possible effect of the increase in temperatures is the fat melting, which can contribute to creating a hydrophobic barrier in the product, favoring the smearing effect. This is the reason why low melting point temperature fats are avoided in the manufacturing of this kind of product.

DRYING KINETICS MODELING

Modeling drying behavior of meat and poultry products is a very complex task (Kottke et al. 1996). First, the phase controlling the drying process has to be defined (surrounding air, meat product, or both of them). Under real processing conditions, both the internal and external phase control the drying process (Simal et al. 2003). The Biot number for mass transfer should have to be used to define under which conditions the process is accomplished. Nevertheless, it is usually assumed that the internal phase is controlling the drying process, and the diffusion equations obtained by integrating the Fick's law for Diffusion (Crank 1975) are used to model the experimental results.

Water diffusion inside the food depends on many factors: temperature (Gou et al. 2004), structure (Gou et al. 2002), fat and moisture content (Kottke et al. 1996), pH (Gou et al. 2002), and shrinkage (Trujillo et al. 2007). It even depends on the equation used to obtain such value (Merts et al. 1998; Trujillo et al. 2007) and the size and shape of the sample. That implies that a wide range of effective diffusion (D_e) values can be found in literature, ranging from 0.9×10^{-11} m^2/s to 5.3×10^{-10} m^2/s. (Motarjemi 1988; Mittal 1999), even for the same product working under the same drying conditions (Trujillo et al. 2007).

AIR-CONDITIONING AND CIRCULATING IN MEAT DRYING AND SMOKING

As stated in the previous sections, in addition to the temperature and the speed of the air, the air relative humidity control is fundamental to accomplish an

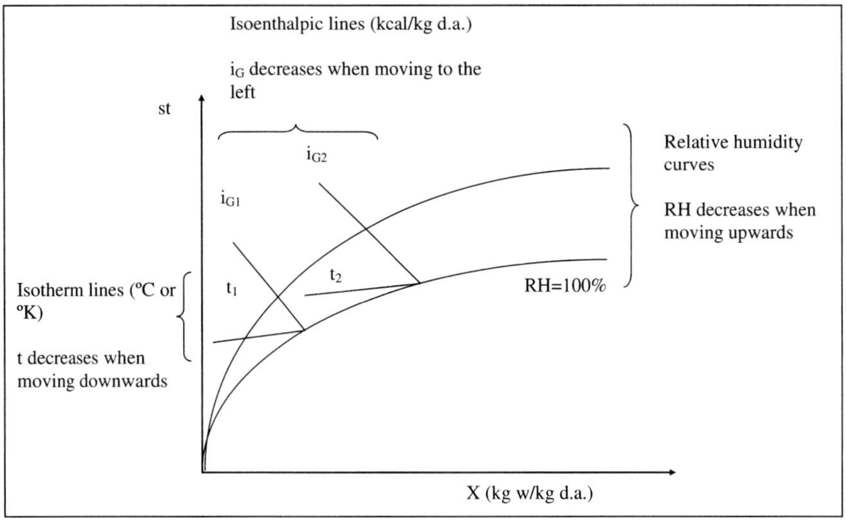

Figure 5.6. Scheme of the Mollier diagram.

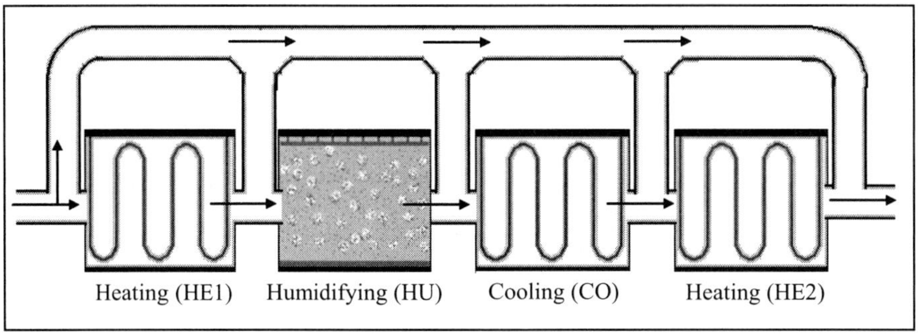

Figure 5.7. Basic scheme for air-conditioning.

adequate drying process—one that is fast, safe, and minimizes the case hardening.

The aim of this section is to give the fundamentals in air-conditioning for a better understanding of the drying chambers used in the processes.

All the given explanations are supported on a psychometric chart, in this case the Mollier diagram shown schematically in Figure 5.6. The *x* axis indicates the absolute humidity of the air (kg water/kg dry air), the curves in the diagram indicate the relative humidity of the air (RH), being the highest possible value corresponding to saturation (100%). The lines crossing the diagram diagonally from above the left to below the right correspond to isoenthalpic values (kcal/kg dry air), and the lines close to the horizontal ones correspond to isotherms (°C or °K).

In air-conditioning, there are four main operations: humidifying, dehumidifying, cooling, and heating. By means of the combination of all or any of them, any combination of relative humidity and temperature in the regular range of work can be achieved. Figure 5.7 shows a basic scheme of the combination of those operations that enables the desired T and RH combination inside the drying chamber.

When working the first heater (HE1), temperature of the air increases, through a vertical line in the Mollier diagram (X constant, i_G increasing, and RH decreasing) (Figure 5.8). This operation is usually accomplished to increase the water retaining capacity of air.

The following step consists of increasing air moisture (HU). The usual way of introducing water vapor in the air consists of spraying water (mechanically or using ultrasound humidifiers). This way of humidification is isoenthalpic, which means that no net energy

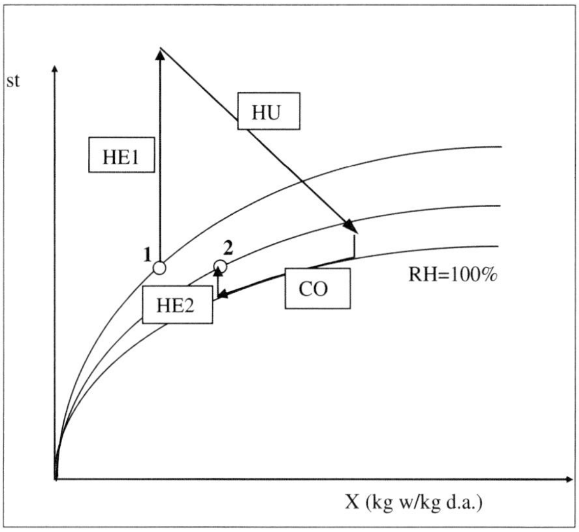

Figure 5.8. Scheme of the changes in the air conditions throughout a conditioning process.

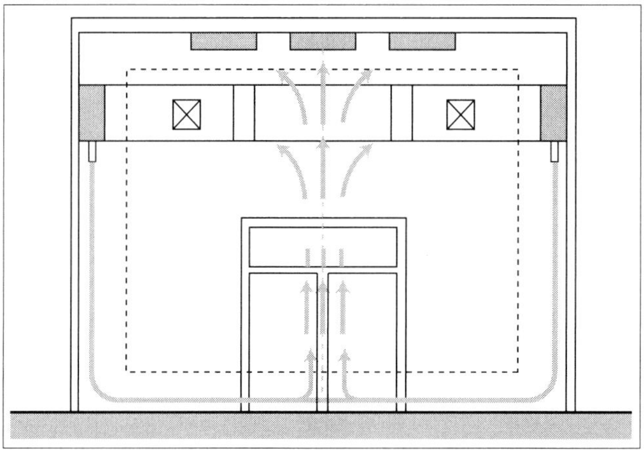

Figure 5.9. Normal way of air circulation inside a drying chamber.

changes in the air exist, and thus the changes in the air properties occur following an isoenthalpic line.

The following operation is air cooling (CO). This is the way to reach a relative humidity equal to 100%, reaching the X value of the final air conditions by controlling the final air temperature (dehumidifying process).

Finally, the air is heated (HE2) to reach the desired relative humidity.

The combination of these operations (HE1+HU+CO++HE2) is needed when the initial air conditions are placed on the left of the final air conditions in the Mollier diagram (Figure 5.8). The overhumidifying is needed because the accurate control of the X value after the humidifying is not possible; the temperature control of the air doesn't imply any problem at industry.

It must be kept in mind that these are the basic principles of air-conditioning, and at an industrial level more complex situations exist because the air recycling and the air-conditioning devices vary depending on the equipment supplier.

The homogeneity of air conditions inside the drying chambers is very important. The regular way of circulation is to introduce the air downstream through orifices joined to the walls and to suction the air through holes placed at the center of the ceiling (Figure 5.9).

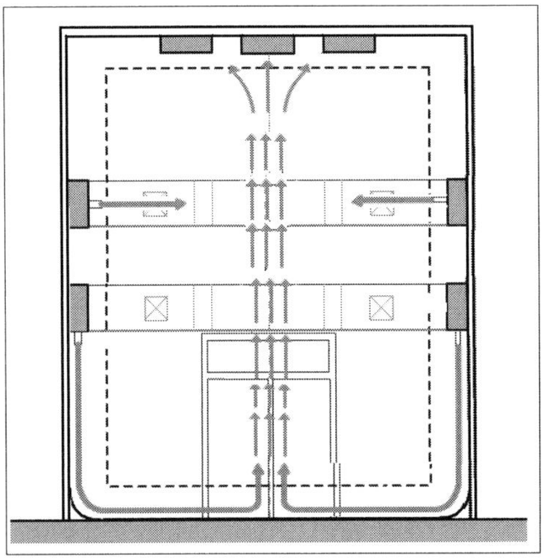

Figure 5.10. Air circulation arrangement in very high drying chambers.

Higher homogeneous drying conditions can be achieved by changing the air direction in the chamber, by introducing the air in the chamber at two different points (in case of very high chambers) (Figure 5.10), or by alternating the holes used to suction the air through the ceiling.

Finally, most of the air in the chamber is reused, and some is permanently renewed.

REFERENCES

MR Adams, MJ Robert-Nout. 2001. Fermentation and food safety. Dordretch: Kluwer Academic Publishers, pp. 300.

MT Aymerich, M Garriga, JM Monfort, I Nes, M Hugas. 2000. Bacteriocin-producing lactobacilli in Spanish-style fermented sausages: Characterization of bacteriocins. Food Microbiol 17(1):33–45.

CGJ Baker. 1997. Industrial drying of foods. London: Blackie Academic & Professional.

JM Barat, R Grau, P Fito, A Chiralt. 2006. Vacuum salting treatment for accelerating processing of dry-cured ham. In: Advanced Technologies for Meat Processing. L Nollet, F Toldrá, eds. Marcel Dekker, Inc., pp. 353–369.

JM Barat, R Grau, JB Ibáñez, P Fito. 2005. Post-salting studies in Spanish cured ham manufacturing. Time reduction by using brine thawing-salting. Meat Sci 69:201–208.

HC Bertram, A Kohler, U Böcker, R Ofstad, HJ Andersen. 2006. Heat-induced changes in myofibrillar protein structures and myowater of two pork qualities. A combined FT-IR spectroscopy and low-field NMR relaxometry study. J Agric Food Chem 54(5):1740–1746,

FJ Beserra, P Fito, JM Barat, A Chiralt, J Martínez-Monzó. 1998a. Drying kinetics of Spanish Salchichón. In: Proceedings of the 44th International Congress of Meat Science and Technology, Vol. II, 872–873, Barcelona, Spain.

FJ Beserra, P Fito, A Chiralt, JM Barat, J Martínez-Monzó. 1998b. Influence of ripening conditions on the pH, colour and texture development of Spanish Salchichón. In: Proceedings of the 44th International Congress of Meat Science and Technology, Vol. II, 870–871, Barcelona, Spain.

JJ Bimbenet, A Lebert. 1990. Some practical questions and remarks about drying. In: Engineering and Food Preservation Processes and Related Techniques. Vol.2. WEL Spiess, H Schubert, eds. London: Elsevier Applied Science.

MO Careaga, E Fernández, L Dorantes, L Mota, ME Jaramillo, H Hernandez-Sanchez. 2003. Antibacterial activity of Capsicum extract against Salmonella typhimurium and Pseudomonas aeruginosa inoculated in raw beef meat. Intl J Food Microbiol 83(3): 331–335.

R Cava-López, A Andrés-Nieto. 2001. EL ahumado de los productos cárnicos. In: Enciclopedia de la carne y de los productos cárnicos, Vol. II. M Bejarano, Plasencia, eds. Cáceres: Martín & Macías, pp. 983–993.

A Coffey, M Ryan, RP Ross, C Hill, E Arendt, G Schwarz. 1998. Use of a broad-host-range bacteriocin-producing Lactococcus lactis transconjugant as an alternative starter for salami manufacture. Intl J Food Microbiol 43(3):231–235.

J Comaposada, P Gou, I Muñoz, J Arnau. 2004. Caracterización y análisis de distribución de temperaturas, humedades relativas y velocidades de aire en un secadero industrial de embutidos. Eurocarne 1–13.

J Crank. 1975. The Mathematics of Diffusion. London: Oxford University Press, Ely House.

P Doe, Z Sikorski, N Haard, J Olley, BS Pan. 1998. Basic principles. In: Fish Drying and Smoking. Production and Quality. PE Doe, ed. Lancaster, Pennsylvania: TECHNOMIC, Pub. Co., Inc., pp. 13–45.

SJ Eilert. 2005. New packaging technologies for the 21st century. Meat Sci 71:122–127.

S García, F Iracheta, F Galván, N Heredia. 2001. Microbiological survey of retail herbs and spices from Mexican markets. J Food Protect 64(1):99–103.

M García-Esteban, D Ansorena, I Astiasarán. 2004. Comparison of modified atmosphere packaging and vacuum packaging for long period storage of dry-cured ham: effects on colour, texture and microbiological quality. Meat Sci 67:57–63.

M Garriga, N Grèbol, MT Aymerich, JM Monfort, M Hugas. 2004. Microbial inactivation after high-pressure processing at 600 MPa in commercial meat products over its shelf life. Innov Food Sci Emerg Technol 5:451–457.

PJ Gou, J Comaposada, J Arnau. 2002. Meat pH and meat fibre direction effects on moisture diffusivity in salted ham muscles dried at 5°C. Meat Sci 61(1):25–31.

———. 2004. Moisture diffusivity in the lean tissue of dry-cured ham at different process times. Meat Sci 67(2):203–209.

P Gou, J Comaposada, J Arnau, Z Pakowski. 2005. On-line determination of water activity at the lean surface of meat products during drying and its relationship with the crusting development. Drying Technol 23(8):1641–1652.

WA Gould. 1920. Fundamentals of Food Processing and Technology. Maryland: CTI Publications, Inc.

WP Hammes, C Hertel. 1998. New developments in meat starter cultures. Meat Sci 49(SUPPL. 1): S125–S138.

DR Heldman, RW Hartel. 1997. Principles of Food Processing. New York: Chapman and Hall.

L Hoz, MD Arrigo, I Cambero, JA Ordóñez. 2004. Development of an n-3 fatty acid and α-tocopherol enriched dry fermented sausage. Meat Sci 67:485–495.

IFT/FDA. 2003. Factors that Influence Microbial Growth. Comp Rev Food Sci Food Saf (2):21–32.

V Kottke, H Damm, A Fischer, U Leutz. 1996. Engineering aspects in fermentation of meat products. Meat Sci 43(1):243–255.

L Leistner. 1995. Principles and Applications of Hurdle Technology. In: GW Gould , ed. New methods of food preservation. London: Blackie Academic & Professional, pp. 1–21.

FK Lücke. 1999. Utilization of microbes to process and to preserve meat. Proceedings 45th ICoMST, pp. 538–547.

L Martinez, I Cilla, JA Beltrán, P Roncalés. 2006. Effect of Capsicum annuum (red sweet and cayenne) and piper nigrum (black and white) pepper powders on the shelf life of fresh pork sausages packaged in modified atmosphere. J Food Sci 71(1):S48–S53.

GS Mittal. 1999. Mass diffusivity of meat food products. Food Rev Intl 15(1):19–66.

I Merts, SJ Lovatt, CR Lawson. 1998. Diffusivity of moisture in whole muscle meat measured by a drying curve method. In IIR Proceedings Series, Refrigeration Science and Technology, Sofia, Bulgaria, pp. 473–479.

Y Motarjemi. 1988. A study of some physical properties of water in foodstuffs. Water activity, water binding and water diffusivity in minced meat products. PhD Thesis. Lund University, Lund, Sweden.

AS Mujumdar. 1995. Handbook of industrial Drying. Vol.1. New York: Marcel Dekker Inc.

P Muthukumarasamy, RA Holley. 2006. Microbiological and sensory quality of dry fermented sausages containing alginate-microencapsulated Lactobacillus reuterim. Intl J Food Microbiol 111:164–169

F Palmia, M Pecoraro, S Ferri. 1993. Essiccazione di prodotti carnei: calcolo del coefficiente di diffusione effettivo (D_e) dell'acqua in fette di lombo suino. Industria Conserve 68:238–242.

B Rubio, B Martínez, C González-Fernández, MD García-Cachán, J Rovira, I Jaime. 2006a. Influence of storage period and packaging method on sliced dry-cured beef "Cecina de Leon": Effects on microbiological, physicochemical and sensory quality. Meat Sci 74:710–717

———. 2006b. Development of an n-3 fatty acid and α-tocopherol enriched dry fermented sausage. Meat Sci 67:485–495.

J Ruiz-Ramírez, X Serra, J Arnau, P Gou. 2005. Profiles of water content, water activity and texture in crusted dry-cured loin and in non-crusted dry-cured loin. Meat Sci 69(3):519–525.

X Serra, J Ruiz-Ramírez, J Arnau, P Gou. 2005. Texture parameters of dry-cured ham m. *biceps femoris* samples dried at different levels as a function of water activity and water content. Meat Sci 69: 249–254.

LA Shelef. 1983. Antimicrobial effects of spices. J Food Saf 6:29–44.

S Simal, A Femenia, P Garcia-Pascual, C Rosselló. 2003. Simulation of the drying curves of a meat-based product: Effect of the external resistance to mass transfer. J Food Eng 58(2):193–199.

RRB Singh, KH Rao, AR Anjaneyulu, GR Patil. 2001. Moisture sorption properties of smoked chicken sausages from spent hen meat. Food Res Inter 34:143–148.

HJ Smith, S Messier, F Tittiger. 1989. Destruction of Trichinella spiralis spiralis during the preparation of the "dry cured" pork products proscuitto, proscuittini and Genoa salami. Can J Vet Res 53:80–83.

A Soriano, B Cruz, L Gómez, C Mariscal, A García-Ruiz. 2006. Proteolysis, physicochemical characteristics and free fatty acid composition of dry sausages made with deer (*Cervus elaphus*) or wild boar (*Sus scrofa*) meat: A preliminary study. Food Chem 96:173–184.

E Stabursvik, H Martens. 1980. Thermal denaturation of proteins in post rigor muscle tissue as studied by differential scanning calorimetry. J Sci Food Agric 31:1034–1042.

F Toldrá. 2002. Dry-Cured Meat Products. Trumbull, Connecticut: Food & Nutrition Press, pp. 27–62.

FJ Trujillo, C Wiangkaew, QT Pham. 2007. Drying modeling and water diffusivity in beef meat. J Food Eng 78(1):74–85.

I Valencia, D Ansorena, I Astiasarán. 2006. Stability of linseed oil and antioxidants containing dry fermented sausages: A study of the lipid fraction during different storage conditions. Meat Sci 73:269–277.

BHP Wilkinson, JAM Janz, PCH Morel, RW Purchas, WH Hendriks. 2006. The effect of modified atmosphere packaging with carbon monoxide on the storage quality of master-packaged fresh pork. Meat Sci 73:605–610.

YS Yean. R Pruthiarenun, P Doe, T Motohiro, K Gopakumar. 1998. Dried and smoked fish products. In: Fish Drying and Smoking. Production and Quality. PE Doe, et al., eds. Lancaster, Pennsylvania: TECHNOMIC, Pub. Co., Inc., pp. 47–87.

LL Zaika. 1988. Spices and herbs: Their antimicrobial activity and its determination. J Food Saf 9: 97–118.

Part II

Raw Materials

6
Biochemistry of Meat and Fat

Fidel Toldrá

INTRODUCTION: MUSCLE STRUCTURE

Skeletal muscle constitutes the most important part of meat. This is a voluntary and striated muscle with several nuclei located peripherally in the cell. The muscle is divided into sections by thin connective tissue layers, named *perimysium*, and each section is divided into fibers by thin collagen fibers, named *endomysium*. Each fiber contains about 1,000 myofibrils, arranged in a parallel way, which are responsible for the contraction of the muscle. The length of a muscle fiber may reach several centimeters and its diameter is about 50 μm. Myofibrils are composed of thick and thin filaments, which overlap in different alternative regions known as *dark (A)* and *light (I)* bands. These filaments are composed of proteins that are responsible for muscle contraction and relaxation. The Z and M lines bisect each I and A band, respectively. The *sarcomere* is the distance between two consecutive Z lines; its length is usually around 2–3 μm. Muscle fibers can be red and white, and each type of fiber has different biochemical properties. Depending on the proportion of both types of fibers, the appearance and properties of the muscle will vary significantly. Red muscle, which is rich in red fibers, has a larger content of myoglobin and lipid and higher oxidative enzymatic activity; white muscle, which is rich in white fibers, has larger glycogen content and glycolytic enzyme activity (Toldrá 2006a). Red muscle is more aerobic in metabolism but slower in contraction than white muscle.

MEAT COMPOSITION

Meat is basically composed of water, protein, lipid, minerals, and trace amounts of carbohydrate. An example of typical composition of pork muscle is given in Table 6.1. Water is the major constituent, followed by proteins, but proportions can vary depending on the amount of fattening. So, the content of protein and water decrease when the amount of fat increases. Proteins, lipids, and enzymes as major components of meat and fat are briefly described in the next sections.

Muscle Proteins

Proteins constitute the major fraction of meat, approximately from 15–22%. There are three main groups of proteins in the muscle: myofibrillar, sarcoplasmic, and stromal, which are briefly described in the next sections. Main proteins for each group and its localization are summarized in Table 6.2.

Myofibrillar Proteins

These proteins are the main constituents of the structure of the myofibrils. Major proteins are myosin and actin, which provide the structural backbone of the myofibril. Regulatory proteins like tropomyosin and troponins are involved in muscle contraction and relaxation. The Z line proteins serve as bridges between the thin filaments of adjacent sarcomeres. Desmin connects adjacent myofibrils at the level of the Z line. Longitudinal continuity and integrity of muscle cells are achieved with two very large proteins, titin and nebulin, that run in parallel to the long axis of the myofibril (Bandman 1987).

Sarcoplasmic Proteins

These proteins are water-soluble and comprise about 30–35% of the total protein in muscle. Sarcoplasmic proteins are quite diverse, including myoglobin (the natural pigment of meat) and many metabolic enzymes present in mithocondrias, lysosomes, microsomes, and nucleus and free in the cytosol. The amount of myoglobin depends on the fiber type, the age of the animal, and the animal species (Kauffmann 2001).

Table 6.1. Example of the approximate composition of pork lean muscle (expressed as g/100 g).

	Mean Range of Variation	
Moisture	72–76	
Proteins	15–22	Myofibrillar 9–10 Sarcoplasmic 9–10 Stromal 2–3
Lipids	1.5–4.0	Triacylglycerols 1.5–3.5 Phospholipids 0.5–0.6 Free fatty acids <0.05 Cholesterol <0.05
Carbohydrates	<1	
Ash	<1	

Stromal Proteins

Fibers and muscles are surrounded by connective tissue (epimysium, perimysium, and endomysium). Collagen, which is the basic protein of this tissue, provides strength and support to the muscle structure. Elastin is found in lower amounts, usually in arterial walls.

Peptides and Free Amino Acids

Three natural dipeptides, carnosine (β-alanyl-L-histidine), anserine (β-alanyl-L-1-methylhistidine), and balenine (β-alanyl-L-3-methylhistidine) are present in the muscle. Some physiological properties of these dipeptides in the muscle are buffering, antioxidant, and activity as neurotransmitter and as modulators of enzyme action (Chan and Decker 1994; Gianelli et al. 2000). Its content is specially higher in muscles with glycolytic metabolism but it varies with the animal species, age, and diet (Aristoy

Table 6.2. Description and localization of major muscle proteins (from Bandman 1987; Pearson and Young 1989; Toldrá 2002).

Proteins Group	Main Proteins	Main Localization in Muscle
Myofibrillar	Myosin	Thick filaments
	Actin	Thin filaments
	Tropomyosin	Between thick and thin filaments
	Troponins T, C, I	Between thick and thin filaments
	Titin and nebulin	Between Z-line and thin filaments
	α, β, γ, and eu-actinin	Z-line
	Filamin; synemin; vinculin; zeugmatin; Z nin; C, H, X, F, and I proteins	Z-line
	Desmin	Between adjacent myofibrils at Z-lines
	Myomesin, creatin kinase, and M protein	M-line
Sarcoplasmic	Mitochondrial enzymes	Mitochondria
	Lysosomal enzymes	Lysosomes
	Calpains	Z-line
	Lipases, glucohidrolases	Sarcoplasm
	ATP-ase	Thick filament
	Myoglobin	Sarcoplasm
	Hemoglobin	Remaining blood in sarcoplasm
Stromal	Collagen	Connective tissue
	Elastin	Capillaries, nerves, tendons, etc.

and Toldrá 1998). Beef and pork contain higher amounts of carnosine and lower amounts of anserine, lamb contains similar amounts of carnosine and anserine, and poultry is very rich in anserine. Balenine is present in minor amounts in pork muscle and at very low concentration in the rest of animals (Aristoy and Toldrá 2004).

Free amino acids are continuously generated in living muscle. The content of amino acids is significantly higher in oxidative muscles than in glycolytic ones (Aristoy and Toldrá 1998). The content of free amino acids is rather poor just after death but is substantially increased during postmortem storage as a consequence of the proteolytical chain during meat aging.

Muscle and Adipose Tissue Lipids

The amount of lipids in skeletal muscle is highly dependent on the degree of fattening and amount of adipose tissue and may range from 1–13% of total muscle. Lipids are mainly found within a muscle (intramuscular), between adjacent muscles (intermuscular), or around the external muscle (adipose tissue). The composition of lipids, mainly triacylglycerols and phospholipids, changes depending on its location. So, intramuscular lipids are mainly composed of triacylglycerols, which are stored in fat cells, and phospholipids, which form part of cell membranes. Intermuscular and adipose tissue lipids are mainly composed of triacylglycerols, with small amounts of cholesterol (40–50 mg/100 g).

Triacylglycerols

Triacylglycerols are the major constituents of fat. The fatty acid content mainly depends on the age of the animal, type of feed, and environment. In fact, the fatty acids composition of triacylglycerols is very important, because it defines its properties for processing. So, when the fat is rich in polyunsaturated fatty acids, such as linoleic acid (typical of feeds rich in corn or barley) or eicosapentanoic and docosahexanoic acids (feeds rich in fish oil), the fat appears to be soft (oily appearance) and prone to oxidation (rancid aromas and yellowish colors) (Irie and Sakimoto 1992; Morgan et al. 1992). The composition in fatty acids somehow reflects the diet of the animal, especially in pork and poultry (see some examples in Table 6.3). In the case of ruminants, the nutrients are somehow standardized due to the action of the microbial population of the rumen (Jiménez-Colmenero et al. 2006).

Table 6.3. Example of total fatty acids composition of pork muscle lipids expressed as % of total fatty acids as affected by type of feed (from Irie and Sakimoto 1992; Morgan et al. 1992; Miller et al. 1990).

Fatty Acids	6% Fish	Barley + Soya Bean Meal	Canola Oil
Myristic acid (C 14:0)	1.93	–	1.6
Palmitic acid (C 16:0)	26.89	23.86	20.6
Estearic acid (18:0)	16.30	10.16	9.8
Palmitoleic acid (C 16:1)	2.56	3	3.6
Oleic acid (C 18:1)	37.31	39.06	45.9
C 20:1	–	–	
Linoleic acid (C 18:2)	7.78	17.15	12.3
C 20:2	0.59	–	0.4
Linolenic acid (C 18:3)	2.08	0.91	3.0
C 20:3	0.08	0.21	0.1
Arachidonic acid (C 20:4)	0.46	4.26	0.74
Eicosapentanoic acid (C 22:5)	0.91	0.64	–
Hexadecanoic acid (C 22:6)	1.13	0.75	–
Total SFA	45.12	34.02	33.6
Total MUFA	39.87	42.06	49.5
Total PUFA	13.03	23.92	16.6

Phospholipids

Even though phospholipids are present in minor amounts, they have a strong influence on flavor development and oxidation. The reason is that phospholipids are relatively rich in polyunsaturated fatty acids in comparison to triacylglycerols. Phosphatidylcholine (lecithine) and phosphatidylethanolamine are the major constituents. The amount of phospholipids depends on the genetic type of the animal and anatomical location of the muscle and is higher in red oxidative muscles than in white glycolytic muscles (Hernández et al. 1998).

MUSCLE PROTEASES AND LIPASES

Muscle contains a large and varied number of enzymes, responsible for most of the biochemical changes observed during the processing of meat and meat products (Toldrá 2006b). Some of the most important muscle enzymes are related with protein breakdown (endopeptidases, also known as *proteinases*), generation of small peptides (peptidases), or free amino acids (aminopeptidases and carbxypeptidases) and lipids breakdown (lipases) (Toldrá 1992). The location of the enzymes may differ; some are located in lysosomes and others remain free in the cytosol or attached to membranes. A brief description of the main substrates and products for these enzymes is summarized in Table 6.4.

Endopeptidases

Lysosomal Proteinases

The main lysosomal proteinases are cathepsins B and L (cysteine proteinases that require a reducing environment for optimal activity and are very active at pH 6.0) and cathepsin D (aspartate proteinase with optimal acid pH in the range 3.0–5.0). These enzymes are small (20–40 KDa) and can penetrate into the myofibrillar structure (Etherington 1987). Cathepsin H, which is also a cystein proteinase with optimal pH at 6.8, may degrade myosin, although its main function seems to be as an aminopeptidase (Okitani et al. 1981). Cathepsins B, D, and L have shown good ability to degrade different myofibrillar proteins (Matsakura et al. 1981; Zeece and Katoh 1989) and are partially inhibited by sodium chloride (Toldrá et al. 1992). Cathepsin D was found to be major contributor to proteolysis in fermented sausages due to its stability and optimal activity at acid pH (Molly et al. 1997).

Neutral Proteinases: Calpains

Calpains I and II are cystein endopeptidases located in the cytosol but, very especially, in the Z-disc region. Calpains I requires 50–70 μM of Ca^{2+} for activity;

Table 6.4. Brief description of main substrates and products for most important muscle proteolytic and lipolytic enzymes.

Enzymes	Main Substrate	Main Product
Cathepsins	Myofibrillar proteins	Protein fragments
Calpains	Myofibrillar proteins	Protein fragments
20S proteasome	Myofibrillar proteins	Protein fragments
Tripeptidylpeptidases	Polypeptides	Tripeptides
Dipeptidylpeptidases	Polypeptides	Dipeptides
Dipeptidases	Dipeptides	Free amino acids
Aminopeptidases	Peptides (amino termini)	Free amino acids
Carboxypeptidases	Peptides (carboxy termini)	Free amino acids
Lysosomal acid lipase	Triacylglycerols	Free fatty acids
Acid phospholipase	Phospholipids	Free fatty acids
Esterases	Triacylglycerols	Short chain free fatty acids
Hormone sensitive lipase	Triacylglycerols	Free fatty acids
Monoacylglycerol lipase	Monoacylglycerols	Free fatty acids

calpain II requires 1–5 mM of Ca^{2+}. Both enzymes are active at neutral pH, around 7.5, but with poor activity below pH 6.0 (Etherington 1985). Calpains are heterodimers of 110 KDa composed of an 80 KDa catalytic subunit and a 30 KDa subunit of unknown function (Bond and Butler 1987). Calpains are able to degrade a wide variety of myofibrillar proteins except myosin and actin (Goll et al. 1983; Koohmaraie 1994). Calpain I has poor stability in postmortem muscle; calpain II seems to be stable for a few weeks (Koohmaraie et al. 1987). In addition, the acid pH values of fermented sausages makes rather unlikely any calpain activity after a few days of processing. The activity of calpains is regulated in postmortem muscle by the endogenous inhibitor calpastatin (Ouali 1991), which is destroyed by autolysis after a few days postslaughter (Koohmaraie et al. 1987).

Proteasome

The proteasome complex constitutes a large proteolytic enzyme exerting multiple activities such as a chymotrypsin-like activity, a trypsin-like activity and a peptidyl-glutamyl hydrolyzing activity (Coux et al. 1995). The 20S proteasome has shown ability to degrade myofibrils and affect M and Z lines, especially in high pH and slow-twitch oxidative muscles. So, this enzyme could have a role on tenderness in those specific muscles (Dutaud et al. 1996; Ouali and Sentandreu 2002).

Exopeptidases

Tripeptidylpeptidases

Tripeptidylpeptidases (TPP) are enzymes that hydrolyze the amino termini of polypeptides and release different tripeptides. TPP I is located in the lysosomes and has an optimal acid pH (4.0); TPP II, which is located in the cytosol, has an optimal neutral pH (7.0) and a wide substrate specificity (Toldrá 2006c).

Dipeptidylpeptidases

Dipeptidylpeptidases (DPP) are enzymes that hydrolyze the amino termini of polypeptides and release different dipeptides. DPP I and II are located in the lysosomes and have optimal acid pH (5.5). DPP III is located in the cytosol and DPP IV is linked to membranes. Both have basic optimal pH (7.5–8.0). DPP II and DPP IV have special preference for Gly-Pro. DPP III has preference for Arg-Arg and Ala-Arg and DPP I for Ala-Arg and Gly-Arg, These enzymes are able to generate a wide range of dipeptides during the processing of meat products (Sentandreu and Toldrá 2002).

Dipeptidases

Dipeptidases catalyze the hydrolysis of dipeptides into two single free amino acids. They receive different names depending on the preference for certain amino acids. This is the case of glycylglycine dipeptidase, which is very specific for dipeptides containing glycine, and cysteinylglycine dipeptidase, which is specific for the dipeptide Cys-Gly (McDonald and Barrett 1986).

Aminopeptidases

There are five different aminopeptidases reported in muscle: Leucyl, arginyl, alanyl, pyroglutamyl, and methionyl aminopeptidases. They are active at neutral or basic pH and receive different names depending on the preference or requirement for a specific N-terminal amino acid. However, these enzymes are able to hydrolyze other amino acids, although at a lower rate (Toldrá and Flores 1998). Alanyl aminopeptidase is the aminopeptidase with higher activity in postmortem muscle and accounts for most of the release of free amino acids (Toldrá 2002). The role of aminopeptidases in protein and peptide degradation during the processing of dry meat products has received attention in recent years due to the accumulation of free amino acids in the final products and its influence on flavor (Toldrá 2006b,c).

Carboxypeptidases

Carboxypeptidases are located in lysosomes, having optimal activity at acid pH. These enzymes generate free amino acids from the carboxy termini of peptides and proteins. Carboxypeptidase A has preference for hydrophobic amino acids, and carboxypeptidase B has a wide spectrum of activity (McDonald and Barrett 1986).

Muscle Lipases

There are several lipases in muscle, especially located in the lysosomes where lysosomal acid lipase and phospholipase A are located. Both enzymes are active at acid pH (4.5–5.5) and are able to release long chain free fatty acids from triacylglycerols and phospholipids, respectively. Lysosomal acid lipase hydrolyzes preferentially primary ester bonds of try-acylglycerols (Fowler and Brown

1984; Imanaka et al. 1985). Phospholipase A is responsible for the hydrolysis of phospholipids, at positions 1 or 2, at the water-lipid interface (Alasnier and Gandemer 2000). Acid and neutral esterases are located in the lysosomes and cytosol, respectively (Motilva et al. 1992). These enzymes hydrolyze short chain fatty acids from tri-, di- and monoacylglycerols, but their activity is restricted due to scarce available substrates.

Muscle Oxidative and Antioxidative Enzymes

Oxidative Enzymes

Lipoxygenase catalyzes the incorporation of molecular oxygen in polyunsaturated fatty acids giving a conjugated hydroperoxide as final product (Marczy et al. 1995). The different names of lipoxigenases depend on the position where oxygen is introduced. These enzymes are activated by mM concentrations of Ca^{+2} and by ATP (Yamamoto 1992). Lipoxygenase is stable during frozen storage and can accelerate rancidity development in meats like poultry, rich in unsaturated fats (Grossman et al. 1988).

Antioxidative Enzymes

Antioxidative enzymes constitute a defense system against oxidative susceptibility and physical stress (Young et al. 2003). Superoxide dismutase and catalase protect against the prooxidative effects by catalyzing the dismutation of hydrogen peroxide to less harmful hydroxides. Glutathione peroxidase, which has higher activity in glycolytic muscles than in oxidative muscles, catalyzes the dismutation of alkyl hydroperoxides by reducing agents like phenols (Daun et al. 2001).

ADIPOSE TISSUE LIPASES

Adipose tissue contains three important lipolytic enzymes, hormone-sensitive lipase, monoacylglycerol lipase, and lipoprotein lipase, which are active in the neutral/basic pH range (Toldrá 1992). The hormone-sensitive lipase hydrolyzes stored adipocyte lipids, with special preference for primary ester bonds of tryacylglycerols, and releases free fatty acids (Belfrage et al. 1984). The monoacylglycerol lipase, mainly located in the adipocytes, is able to hydrolyze medium and long chain monoacylglycerols with no positional specificity (Tornquist et al. 1978). Lipoprotein lipase, which is located in the capillary endothelium, hydrolyzes the acylglycerol components at the luminal surface of the endothelium (Smith and Pownall 1984). This enzyme has preference for the hydrolysis of fatty acids at position 1 (Fielding and Fielding 1980). Acid and neutral esterases are also present in adipose tissue (Motilva et al. 1992) and can participate in the degradation of lipoprotein cholesteryl esters (Belfrage et al. 1984).

POSTMORTEM MUSCLE METABOLISM AND QUALITY

Several enzymatic reactions that can affect the meat quality take place during early postmortem. Just after death, glucose is hydrolyzed to lactic acid by glycolytic enzymes. Due to the absence of blood circulation, lactic acid is accumulated into the muscle and produces a pH drop to values about 5.6–5.8 in a few hours. The pH drop rate depends on the glucose concentration, temperature of the muscle, and metabolic status of the animal previously to slaughter (Toldrá 2006a).

The rate and extent of pH drop has a great incidence on the quality of pork and poultry meats. So, pigs with accelerated metabolism as a consequence of some type of premortem stress result in rapid pH drops, below 5.8 after 2 hours postmortem, and almost full disappearance of ATP (Batlle et al. 2001), resulting in exudative meats. This is the case of the pale, soft, exudative pork meats (PSE) and red, soft, exudative pork meats (RSE) (Warner et al. 1997). Normal meat, known as red, firm, and normal meat (RFN), experiences a progressive pH drop down to values around 5.8–6.0 at 2 hours postmortem. Finally, the dark, firm, dry pork meat (DFD) and dark cutting beef meat are produced from exhausted animals (poor content of carbohydrates before slaughter), where no lactic acid can be generated during early postmortem. Glycolysis is almost negligible in this type of meats and the pH remains high—higher than 6.0, constituting a risk from the microbiological point of view.

Pork meat classification is usually based on pH measurement at early postmortem, color, and drip loss. Exudative meats, specifically PSE, which have L values higher than 50 and drip losses higher than 6%, must be controlled and detected when received at the factory because they can affect the production of fermented and ripened meat products. Some of the most important defects consist of excessive saltiness (in the case of ripened pieces where salt is added on the surface), poor red color (paleness), protein denaturation, and inadequate drying (Toldrá 2002). On the other hand, DFD meats that have a drip loss lower than 2%, may cause microbiological problems (prone to contamination by foodborne pathogens) due to high pH and may present drying defects due to high water retention (Cassens 2000).

REFERENCES

C Alasnier, G Gandemer. 2000. Activities of phospholipase A and lysophospholipases in glycolytic and oxidative skeletal muscles in the rabbit. J Sci Food Agric 80:698–704.

MC Aristoy, F Toldrá. 1998. Concentration of free amino acids and dipeptides in porcine skeletal muscles with different oxidative patterns. Meat Sci 50:327–332.

———. 2004. Histidine dipeptides HPLC-based test for the detection of mammalian origin proteins in feeds for ruminants. Meat Sci 67:211–217.

E Bandman. 1987. Chemistry of animal tissues. Part 1—Proteins. In: The Science of Meat and Meat Products. JF Price, BS Schweigert, eds. Trumbull, Connecticut: Food & Nutrition Press Inc., pp. 61–102.

N Batlle, MC Aristoy, F Toldrá. 2001. ATP metabolites during aging of exudative and nonexudative pork meats. J Food Sci 66:68–71.

P Belfrage, G Fredrikson, P Stralfors, H Tornquist. 1984. Adipose tissue lipases. In: Lipases. B Borgström, HL Brockman, eds.London (UK): Elsevier, pp. 365–416.

JS Bond, PE Butler. 1987. Intracellular proteases. Ann Rev Biochem 56:333–364.

RG Cassens. 2000. Historical perspectives and current aspects of pork meat quality in the USA. Food Chem 69:357–363.

KM Chan, EA Decker. 1994. Endogenous skeletal muscle antioxidants. Crit Rev Food Sci Nutri 34: 403–426.

O Coux, K Tanaka, A Goldberg. 1995. Structure and function of the 20S and 26S proteasomes. Ann Rev Biochem 65:801–847.

C Daun, M Johansson, G Onning, B Akesson. 2001. Glutathione peroxidase activity, tissue and soluble selenium content in beef and pork in relation to meat aging and pig RN phenotype. Food Chem 73:313–319.

D Dutaud, RG Taylor, B Picard, A Ouali. 1996. Le protéasome: une nouvelle protéase impliquée dans la maturation de la viande. Viandes Prod Carnés 17: 333–335.

DJ Etherington. 1984. The contribution of proteolytic enzymes to postmortem changes in muscle. J An Sci 59:1644–1650.

———. 1987. Conditioning of meat factors influencing protease activity. In: Accelerated Processing of Meat. A Romita, C Valin, AA Taylor, eds. London (UK): Elsevier Applied Sci, pp. 21–28.

CJ Fielding, PE Fielding. 1980. Characteristics of triacylglycerol and partial acylglycerol hydrolysis by human plasma lipoprotein lipase. Biochimica Biophysica Acta 620:440–446.

SD Fowler, WJ Brown. 1984. Lysosomal acid lipase. In: Lipases. B Borgström, HL Brockman, eds. London (UK): Elsevier Science Pub., pp. 329–364

MP Gianelli, M Flores, VJ Moya, MC Aristoy, F Toldrá. 2000. Effect of carnosine, anserine and other endogenous skeletal peptides on the activity of porcine muscle alanyl and arginyl aminopeptidases. J Food Biochem 24:69–78.

DE Goll, Y Otsuka, PA Nagainis, JD Shannon, SK Sathe, M Mugurama. 1983. Role of muscle proteinases in maintenance of muscle integrity and mass. J Food Biochem 7:137–177.

S Grossman, M Bergman, D Sklan. 1988. Lipoxygenase in chicken muscle. J Agric Food Chem 36:1268–1270.

P Hernández, JL Navarro, F Toldrá. 1998. Lipid composition and lipolytic enzyme activities in porcine skeletal muscles with different oxidative pattern. Meat Sci 49:1–10.

T Imanaka, M Yamaguchi, S Ahkuma, S Takano. 1985. Positional specificity of lysosomal acide lipase purified from rabbit liver. J Biochem 98:927–931.

M Irie, M Sakimoto. 1992. Fat characteristics of pigs fed fish oil containing eicosapentaenoic and docosahexanoic acids. J An Sci 70:470–477.

F Jiménez-Colmenero, M Reig, F Toldrá. 2006. New approaches for the development of functional meat products. In: Advanced Technologies for Meat Processing. LML Nollet, F Toldrá, eds. Boca Raton, Florida: CRC Press, pp. 275–308.

RG Kauffman. 2001. Meat composition. In: Meat Science and Applications. YH Hui, WK Nip, RW Rogers, OA Young, eds. New York: Marcel Dekker Inc., pp. 1–19.

M Koohmaraie. 1994. Muscle proteinases and meat aging. Meat Sci 36:93–104.

M Koohmaraie, SC Seideman, JE Schollmeyer, TR Dutson, JD Grouse. 1987. Effect of postmortem storage on Ca^{2+} dependent proteases, their inhibitor and myofibril fragmentation. Meat Sci 19:187–196.

JS Marczy, ML Simon, L Mozsik, B Szajani. 1995. Comparative study on the lipoxygenase activities of some soybean cultivars. J Agric Food Chem 43: 313–315.

U Matsukura, A Okitani, T Nishimura, H Katoh. 1981. Mode of degradation of myofibrillar proteins by an endogenous protease, cathepsin L. Biochim. Biophys Acta 662:41–47.

JK McDonald, AJ Barrett, eds. 1986. Mammalian Proteases. A Glossary and Bibliography. Vol 2. Exopeptidases. London (UK): Academic Press.

MF Miller, SD Shackelford, KD Hayden, JO Reagan. 1990. Determination of the alteration in fatty acid profiles, sensory characteristics and carcass traits of swine fed elevated levels of monounsaturated fats in the diet. J An Sci 68:1624–1631.

K Molly, DI Demeyer, G Johansson, M Raemaekers, M Ghistelinck, I Geenen. 1997. The importance of meat enzymes in ripening and flavor generation in dry fermented sausages. First results of a European project. Food Chem 54:539–545.

CA Morgan, RC Noble, M Cocchi, R McCartney. 1992. Manipulation of the fatty acid composition of pig meat lipids by dietary means. J Sci Food and Agric 58:357–368.

MJ Motilva, F Toldrá, J Flores. 1992. Assay of lipase and esterase activities in fresh pork meat and dry-cured ham. Z Lebensmittel Untersuchung und Forschung 195:446–450.

A Okitani, T Nishimura, H Kato. 1981. Characterization of hydrolase H, a new muscle protease possessing aminoendopeptidase activity. Eur J Biochem 115:269–274.

A Ouali. 1991. Sensory quality of meat as affected by muscle biochemistry and modern technologies. In: Animal Biotechnology and the Quality of Meat Production. LO Fiems, BG Cottyn, DI Demeyer, eds. Amsterdam, The Netherlands: Elsevier Science Pub, BV, pp. 85–105.

A Ouali, MA Sentandreu. 2002. Overview of muscle peptidases and their potential role in meat texture development. In: Research Advances in the Quality of Meat and Meat Products. F Toldrá, ed. Trivandrum, India: Research Signpost, pp. 33–63.

AM Pearson, RB Young. 1989. Muscle and Meat Biochemistry. San Diego, California: Academic Press, pp. 1–261.

MA Sentandreu, G Coulis, A Ouali. 2002. Role of muscle endopeptidases and their inhibitors in meat tenderness. Trends Food Sci Technol 13:398–419.

MA Sentandreu, F Toldrá. 2002. Dipeptidylpeptidase activities along the processing of Serrano dry-cured ham. European Food Res Technol 213:83–87.

LC Smith, HJ Pownall. 1984. Lipoprotein lipase. In: Lipases. B. Borgström, HL Brockman, eds. Amsterdam, The Netherlands: Elsevier, pp. 263–305.

F Toldrá. 1992. The enzymology of dry-curing of meat products. In: New Technologies for Meat and Meat Products. FJM Smulders, F Toldrá J Flores, M Prieto. Nijmegen, The Netherlands: Audet, pp. 209–231.

———. 1998. Proteolysis and lipolysis in flavour development of dry-cured meat products. Meat Sci 49:s101–s110.

———. 2002. Dry-cured Meat Products. Trumbull, Connecticut: Food & Nutrition Press, pp. 1–238.

———. 2006a. Meat: Chemistry and biochemistry. In: Handbook of Food Science, Technology and Engineering. YH Hui, JD Culbertson, S Duncan, I Guerrero-Legarreta, ECY Li-Chan, CY Ma, CH Manley, TA McMeekin, WK Nip, LML Nollet, MS Rahman, F Toldrá, YL Xiong, eds. Volume 1. Boca Raton, Florida: CRC Press, pp. 28-1–28-18.

———. 2006b. Biochemistry of processing meat and poultry. In: Food Biochemistry and Food Processing. YH Hui, WK Nip, ML Nollet, G Paliyath & BK Simpson, eds. Ames, Iowa: Blackwell Publishing, pp. 315–335.

———. 2006c. Biochemical proteolysis basis for improved processing of dry-cured meats. In: Advanced Technologies for Meat Processing. LML Nollet, F Toldrá, eds. Boca Raton, Florida: CRC Press, pp. 329–351.

F Toldrá, M Flores. 1998. The role of muscle proteases and lipases in flavor development during the processing of dry-cured ham. Crit Rev Food Sci Nutrit 38: 331–352.

F Toldrá, E Rico, J Flores. 1992. Activities of pork muscle proteases in cured meats. Biochimie 74:291–296.

H Tornquist, P Nilsson-Ehle, P Belfrage. 1978. Enzymes catalyzing the hydrolysis of long-chain monoacylglycerols in rat adipose tissue. Biochimica Biophysica Acta 530:474–486.

RD Warner, RG Kauffman, RL Russell. 1997. Muscle protein changes post mortem in relation to pork quality traits. Meat Sci 33:359–372.

S Yamamoto. 1992. Mammalian lipoxygenases: molecular structures and functions. Biochimica Biophysica Acta 1128:117–131.

JF Young, K Rosenvold, J Stagsted, CL Steffensen, JH Nielsen, HJ Andersen. 2003. Significance of preslaughter stress and different tissue PUFA levels on the oxidative status and stability of porcine muscle and meat. J Agric Food Chem 51:6877–6881.

MG Zeece, K Katoh. 1989. Cathepsin D and its effects on myofibrillar proteins: a review. J Food Biochem 13:157–161.

7
Ingredients

Jorge Ruiz

INTRODUCTION

Meat and lard are the two major ingredients in most fermented meat products and, consequently, their characteristics strongly influence the features of the final product, including sensory, nutritional, safety, and health aspects. For the purpose of this text, *meat* can be defined as the skeletal muscle of livestock, poultry, and game, mainly including muscle, fat, and connective tissues, but also others such as vascular, lymphatic, or nervous tissues. Nevertheless, most fermented meat products are produced using pork. As far as lard is concerned, it is constituted by the subcutaneous fat from swine. Only in a few regional types of fermented sausages are ovine and bovine subcutaneous fat added. Besides these two major components, salt is a main constituent in almost all types of fermented sausages, and only in recent years a few salt-free (or salt-reduced) products have been developed directed to potentially high blood pressure consumers. Other minor ingredients include offal, plant materials or water.

LEAN

Lean refers to commercial pieces of meat, including muscle, lard, and other tissues. Most fermented products are produced using pork obtained from barrows and gilts. Animals are usually slaughtered at 90–100 kg live weight when around 6 months old. Nevertheless, for some kind of products, heavier pigs are preferred and, thus, pigs with live weights as high as 150–160 kg are also used for fermented meat production (Mayoral et al. 1999). This leads to a much higher fat content both in the carcass and in the lean (Figure 7.1).

Bovine meat was frequently used as a supplement for pig meat in many fermented meat products. However, due to the high prices of this type of meat and to the better technological and sensory properties of pork for dry-fermented processing, the use of bovine meat for elaboration of fermented sausages has decreased. Ovine and goat meat are used to a lesser extent for the production of fermented products, although there is still a significant production of sausages from old ovine livestock whose meat cannot be aimed toward fresh meat consumption. The problem with these two types of meats (especially when animals are not young) is their strong and identifiable flavor, which is usually rejected by consumers. Nevertheless, in Muslim countries, fermented meat products are mainly based on bovine and ovine meats mixed with ovine fat, because pork is not allowed for religious reasons (Leistner 1992). The production of sausages with horse, mule, and donkey meats are strange and limited to some regional specialities. Finally, there is a growing production of fermented sausages made from game meat, especially those from wild boar and deer.

After decades in which frozen-thawed meat was the major form in which lean was used for fermented sausage production, there is currently a growing tendency to use fresh raw material (Solignat 1999). An inappropriate freezing process, frozen storage, or

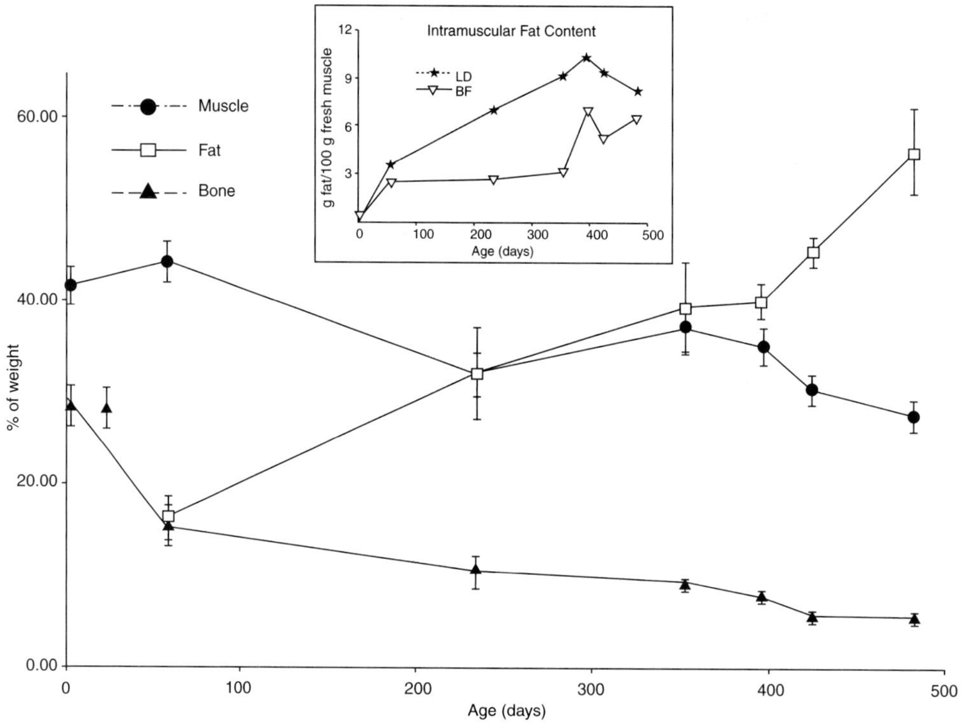

Figure 7.1. Mean (±SD) of the proportion of weight of muscle, dissectible fat, and bone in left hemicarcass of heavy pigs at slaughter throughout lifetime. The inset shows intramuscular fat content (g fat/100 g fresh muscle). LD: *Longissimus dors*; BF: *Biceps femoris* (taken from Mayoral et al. 1999).

thawing, might lead to protein damages which, in turn, produce drip and defects in meat texture (Miller et al. 1980; Lawrie 1995), affecting the final quality of fermented products. Moreover, frozen storage slows down oxidative processes, but does not totally avoid them (Novelli et al. 1998: Hansen et al. 2004). Thus, frozen meats stored for a long time might speed up lipid oxidative processes during the processing of fermented products. For these reasons, when using frozen meats for fermented meat products, it is generally recommended that storage at −18°C is not longer than 6 months for beef and 3 months for pork (Prändl et al. 1988).

FAT

Adipose tissue is a specialized type of connective tissue, constituted by fat cells called *adipocytes*, which show a drop of triglycerides comprising almost all the volume of the cell. As any connective tissue, adipose fat also shows connective fibers of collagen, although the number and strength of these fibers is much lower than in other types of connective tissues (such as those in tendons, fasciae, or muscle).

Fat is an essential component for fermented sausages, providing several features positively related to their sensory and technological quality, namely easing chewing, enhancing juiciness, contributing to aroma development and release, and slowing down the dehydration process (Wirth 1988).

The fat used for production of fermented sausages is mainly that from pig, especially lard, which is the subcutaneous adipose tissue. Other anatomical locations of fat, such as the fat associated with the kidney (perinephric) or the omental fat, are rarely used for elaboration of fermented meat products. It is currently getting more common to use fresh lard for fermented sausage production, although it is usually ground at −1.0°C to avoid melting and to achieve a well-defined cutting surface. When using frozen lard, storage should not be longer than 3 months, for the same reasons explained for frozen lean (Prändl et al. 1988).

In the last decades, there has been a growing interest in producing healthier meat products. Some of the most common strategies have been directed to a reduction of fat content and a decrease of saturated fatty acids with a concomitant increase in polyunsaturated fatty acids or monounsaturated fatty acids.

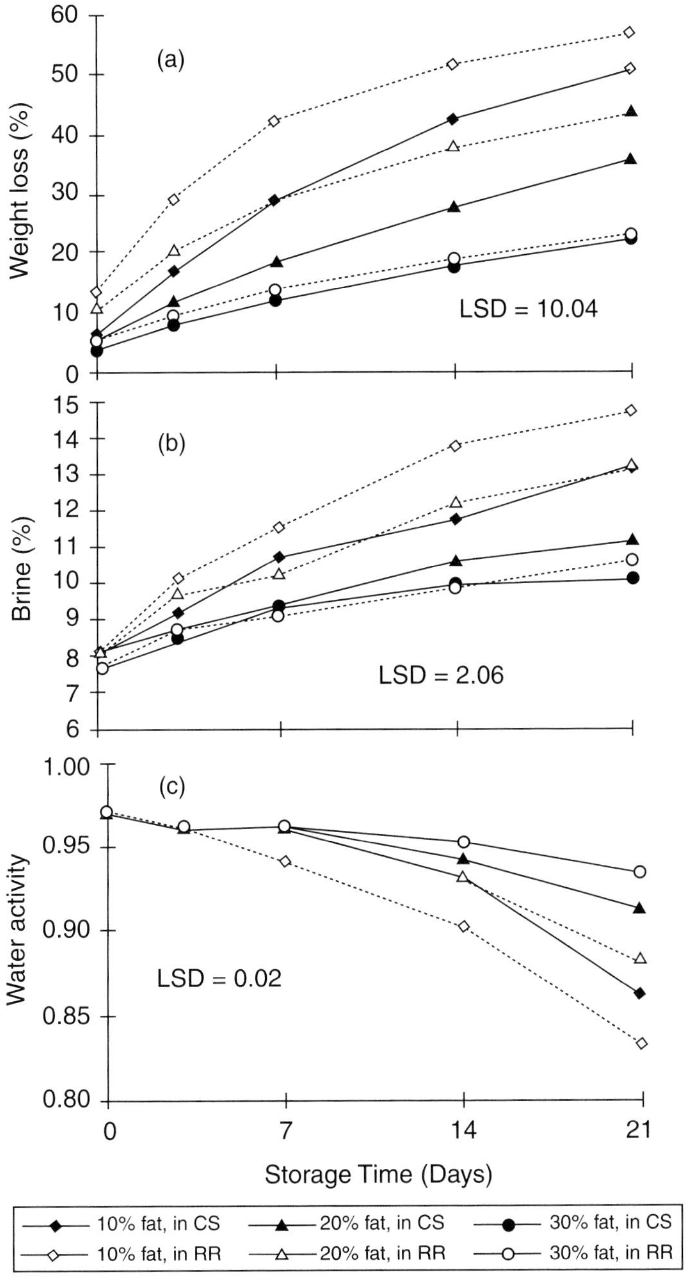

Figure 7.2. Effect of fat level and storage conditions on (a) weight loss, (b) brine content, and (c) water activity during storage of traditional sausages. CS = cold storage, RR = ripening room (taken from Papadima and Bloukas 1999).

The latter has been achieved through pig dietary means and by replacing lard with other oils during sausage production (discussed later in this chapter). As far as fat reduction is concerned, excessive reduction of fat content may lead to a very rapid dehydration and poor sensory features of the final product. Thus, Papadimas and Bloukas (1999) have shown much faster dehydration in 10% added fat sausages than in 30% ones, leading to huge differences in the water activity of the final product (Figure 7.2). Fat

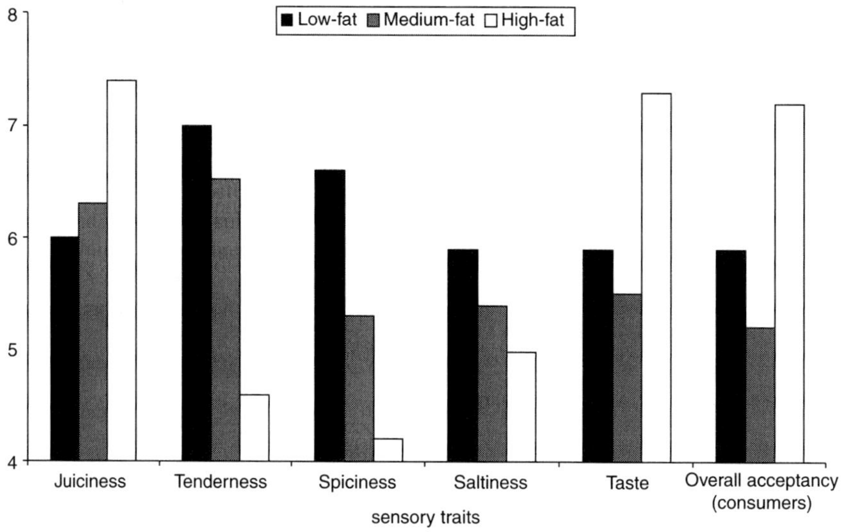

Figure 7.3. Sensory characteristics and overall acceptability of dry-cured fermented sausages with low (6.5%), medium (12.5%), and high (25%) added fat (elaborated with data from Mendoza et al. 2001).

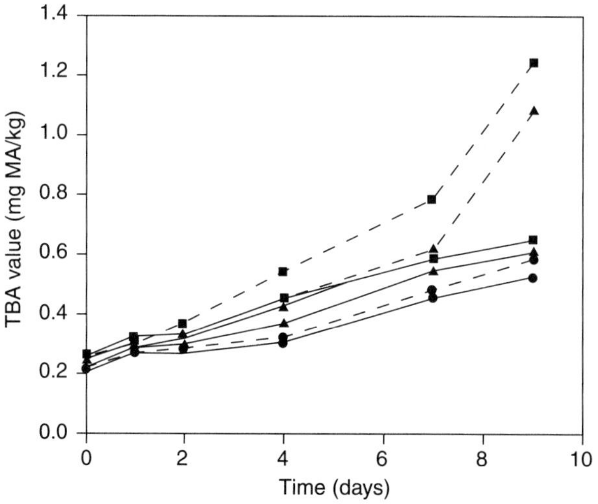

Figure 7.4. Effect of fat level and temperature on the TBA value during the ripening period of fermented sausages. Solid lines, 20–22°C, dashed lines, 24–26°C; ● 10% fat added, ▲ 20% fat added, ■ 30% fat added (taken from Soyer et al. 2005).

content is usually positively related to sensory features of the final product, as evidenced by Mendoza et al. (2001) comparing low, medium, and high fat content sausages (Figure 7.3) However, Soyer (2005) observed higher levels of free fatty acids and oxidation in fermented sausages elaborated with greater amounts of added fat (Figure 7.4). This author found the highest overall acceptability for sausages with intermediate (20%) levels of added fat.

FACTORS AFFECTING THE FEASIBILITY OF LEAN AND FAT FOR PROCESSING

The technological quality of the meat includes those features that make the meat optimum for production of good-quality fermented sausages, with as little spoilage as possible, as high yield as possible, and leading to a final product pleasant for consumers.

The economical profitability of the whole production depends on all these factors: directly on the yield and the proportion of spoiled product, and indirectly on the consumer acceptability, which in its turn determines how often the consumer buys the product and, thus, the total sales (Aaslying 2002).

pH and Water-Holding Capacity

Muscle contains approximately 75% water. The majority of water in muscle is held within the structure of the muscle itself, either within the myofibrils, between the myofibrils themselves, and between the myofibrils and the cell membrane. Most of this water is physically retained by capillarity within myofibrilar proteins; only a small amount is bound by the sarcoplasmic proteins and the connective tissue (Wismer-Pedersen 1987).

Water-holding capacity of meat is defined as the ability of the postmortem muscle (meat) to retain water even though external pressures (e.g., gravity, heating) are applied to it. Once the animal is slaughtered, the amount of retained water in muscle can change, depending on numerous factors related to the tissue itself and how the meat is handled. Some of these factors are the net charge of myofibrillar proteins and the structure of the muscle cell and its components (myofibrils, cytoskeletal linkages, and membrane permeability), as well as the amount of extracellular space within the muscle itself (Huff-Lonergan 2002). Changes undergone during rigor mortis (pH decline and contraction) lead to a decrease in the amount of water that can be retained by the myofibrilar structure, due to a reduction of the space within such structure, which in turn is due to both a pH close to the isoelectrical point of proteins (reducing the net charge of the proteins) and to a smaller size of the sarcomere as a consequence of contraction.

Due to different rates of pH decline during rigor mortis, and of the final pH reached during this process, marked differences in the ability for retaining water in muscle are observed (Huff-Lonergan 2002). Briefly, fast rates of pH decline lead to very exudative meats called PSE (pale, soft, and exudative, which summarize their features). Likewise, those meats with a low final pH (5.2–5.4) also show similar characteristics and exude a great amount of water. On the other hand, if the final pH value is higher than 6.2 (usually due to a low initial glycogen concentration when rigor mortis initiates), the amount of retained water is higher than in normal meats, leading to dark cuts with scarce exudates (DFD meats: dark, firm, and dry).

The use of exudative meats for fermented meat products processing may lead to several problems. First, as a consequence of the low water-holding capacity of these meats, dehydration occurs very fast (Prändl et al. 1988). It could be thought that this is not a problem for dry and semidry-cured fermented sausages, because dehydration is one of the main physical processes involved in the stabilization of the products. In fact, some authors have proposed the use of PSE meat for accelerating the drying during production of fermented sausages (Townsend et al. 1980), and Honkavaara (1988) has found slightly better sensory properties in dry-fermented sausages produced with PSE meat. However, if dehydration in the surface is too fast, water migration from the depth is not quick enough to equalize the moisture content between the surface and the depth of the product. If such a situation occurs, proteins in the surface may get too desiccated, forming a crust (personal observation). As a compromise for using PSE meat for the production of dry-fermented sausages, Wirth (1985) recommended a maximum of 30% PSE meat in the meat batter.

Another potential problem when using exudative meats is the formation of weak meat gels, due to the damage suffered by myofibrilar proteins. This leads to low consistency products with poor consumer acceptance (Kuo and Chu 2003) (Figure 7.5). Moreover, due to the low consistency of exudative meats, the surface of the ground meat obtained during the grinding step is not neat, giving rise to a fuzzy surface. As far as whole muscle products are concerned, the use of PSE shows many problems, including an excessive salting of hams, defective texture, and two-toning appearance (Guerrero et al. 1996; Garcia-Rey et al. 2004).

There is little scientific information about the potential problem of using DFD meats in fermented sausage production. Although water retention is quite high in DFD meats, and protein functionality is very good for gel formation and emulsification, their pH value above 6.2 makes them more prone to microbial spoilage. This might not be a problem in fermented sausage production, because microbial growth in the first steps of processing will lead to a pH decline that will probably compensate the intrinsic high pH of DFD meat. However, when starter cultures are not used for the production of fermented sausages, such as in small plants for traditional sausage production, the use of DFD meat could lead to microbial spoilage in the first steps of sausage production. With respect to whole meat fermented and dry-cured products, the use of high ultimate pH meats should be avoided because it leads to either texture problems or spoilage (Guerrero et al. 1999).

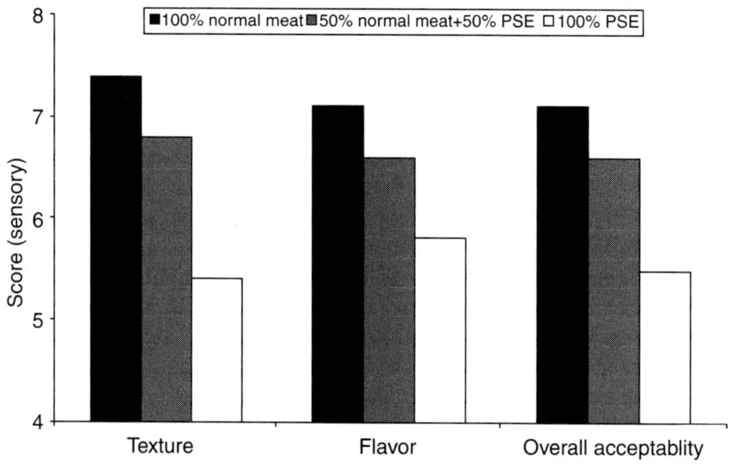

Figure 7.5. Sensory features of Chinese fermented sausages made with either normal or PSE pork. Hedonic scale: 1 = extremely dislike in texture, flavor, or overall acceptability; 5 = neither like nor dislike; 9 = extremely like in texture, flavor, or overall acceptability (elaborated with data from Kuo & Chu 2003).

Lean Color

The intensity of meat color is dependent upon the concentration of the sarcoplasmic protein myoglobin (Giddings 1977). The chemical state of iron in myoglobin determines the color of the pigment, and, thus, the color of the meat (Govindarajan 1973). The bright red fresh color of meat is due to the presence of the reduced form of myoglobin binding an oxygen molecule in the heme group. This form of the pigment is called *oxymyoglobin*. In the absence of oxygen, the pigment can also be in a reduced state, giving rise to the so-called *deoxymyoglobin form*, which is purple-red. Oxidation of heme iron to the ferric (Fe^{3+}) form results in the formation of metmyoglobin, which has a brownish color negatively valued by consumers. The oxidation of the reduced forms of myoglobin to metamyoglobin occurs as a consequence of several prooxidant factors, such as compounds from lipid oxidation, light, temperature, or metals.

During the curing process myoglobin is converted into nitrosomyoglobin in the presence of nitrite, forming a red stable color (Skibstead 1992). For most processed fermented sausages, intense colors are preferred to pale ones. This leads to the utilization of heavier animals in some cases, because animal age is directly related to muscle myoglobin content (Mayoral et al. 1999).

During storage the color might further change depending on both the package material and on the storage conditions (Andersen et al. 1988). This can cause discoloration due to the oxidation of the remaining pigment, which has not reacted with nitrite, leading to dark greyish colors in the surface. This process is very common during slicing and packaging, because the surface is exposed to the air and light, promoting such oxidation processes. In this sense, the use of meats with high levels of vitamin E through supplementation of pig diets with α-tocopherol is an effective strategy to reduce discoloration during the slicing of dry-fermented products (Ruiz and López-Bote 2002).

Intramuscular Fat Content

Intramuscular fat refers to the adipose tissue between muscle fiber bundles that constitutes marbling. The positive effect of marbling on the sensory quality of the meat is well known (Bejerholm and Barton-Gade 1986). Moreover, in dry-cured ham intramuscular fat has been pointed out as the main factor influencing juiciness (Ruiz et al. 2000), which in turn was highlighted as the main sensory trait related to acceptability (Ruiz et al. 2002). However, given that in most dry-cured fermented sausages supplementary fat is added, intramuscular fat is not so important as in fresh meat or dry-cured hams. Nevertheless, in those dry-cured fermented meat products in which the muscle is not ground, or in those in which the pieces of meat are of an appreciable size, intramuscular fat plays an important role in controlling the rate of desiccation, in the development of the flavor during the processing and in the sensory characteristics of the elaborated product (Muriel et al. 2004).

A decrease in water activity in fermented meat products is one of the main factors contributing to their microbial stability. Such a decrease is achieved essentially by adding sodium chloride to the meat batter and by dehydration during the processing. Water losses imply two different processes: water migration from the depth to the surface of the product and water evaporation in the surface. Diffusion coefficients of salt and water in meat products are inversely correlated to the fat content (Palumbo et al. 1977), due to lower diffusion in the fat than in the lean tissue (Wood 1966). Therefore, the higher the intramuscular fat content in those products in which muscle is intact, the longer the time required for salt diffusion and for the appropriate moisture loss.

We have recently found a close relationship between intramuscular fat content and most appearance and texture sensory traits of dry-cured loin, a dry-fermented meat product typical from Spain, pointing to a direct influence of this parameter on the eating quality of the product, and probably to others in which the whole muscle is processed (Ventanas et al. 2007).

Fatty Acid Composition

Not only the amount of fat, but its consistency is essential for an appropriate rate of moisture loss during the processing of fermented sausages (Ruiz and López-Bote 2002). In fact, there are some indications that dietary fat unsaturation affects water migration in dry-cured meat products (Girard et al. 1989; Ruiz and López-Bote 2002). In this sense, the levels of polyunsaturated fatty acids (PUFA) in fatty tissues of pigs used for production of fermented sausages are critical, because an excessive proportion will lead to impaired dehydration and poorer texture and microbial stability. Accordingly, Girard et al. (1989) fed diets to pigs in which the main energy source was provided by barley and copra oil (saturated fat), or maize and maize oil (unsaturated fat). These diets produced a range of linoleic acid (C18:2 n-6) concentration in pig adipose tissue from 7.58% to 30.95% (expressed as the percentage of total fatty acids). Subsequent processing of dry-fermented sausages from these pigs revealed a major effect of fat unsaturation on the drying process. Dry pork sausages manufactured from pigs fed highly unsaturated fatty acids did not permit adequate drying. According to these authors, impaired moisture loss became evident from the 10th day of manufacture and was most pronounced during the later stages of drying (Figure 7.6). Low weight losses during processing, higher water activity values (0.88 versus 0.75), and higher moisture content in the fat-free product (55% versus 45%) revealed impaired drying in the group of pigs fed highly unsaturated fats (maize and maize oil).

This effect is most likely due to the physical state of the fat during the processing. The physical state of the fat mostly depends on the fatty acid melting point, which is dependent on its turn on the number

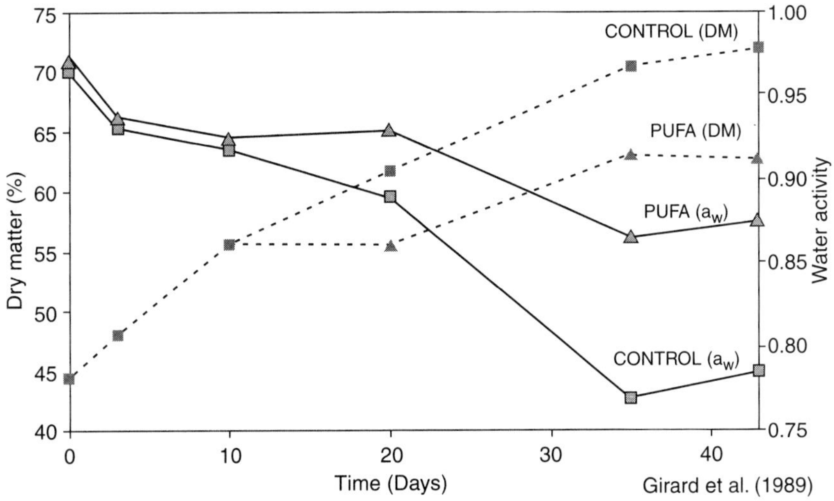

Figure 7.6. Variation in dry matter and water activity throughout the processing of dry-fermented sausages produced with meat from pigs fed a diet either high in SFA (CONTROL) or high in PUFA (PUFA) (elaborated with data from Girard et al. 1989).

of double bonds per fatty acid residue, the energy needed to melt the fat being lower as the number of unsaturations rises. Among fatty acids present in pig tissues, stearic acid (C18:0) and linoleic acid (C18:2 n-6) show the highest correlations (positive and negative, respectively) with fat consistency (Wood 1984; Davenel et al. 1999). However, from a practical point of view, linoleic acid is the main fatty acid determining fat consistency in pig fat.

Fat consistency may also affect firmness of dry-cured ham. It has been shown that an increased polyenoic fatty acid content in pig diets leads to significantly reduced product firmness in dry-cured fermented sausages (Stiebing et al. 1993; Warnants et al. 1998; Houben and Krol 1980). Due to this effect, some researchers (Stiebing et al. 1993; Boulard et al. 1995) have recommended a maximum proportion of linoleic acid of 12% when meat is used for sausage processing.

Fatty acid melting point also affects lard color: solid fat shows a white color, and melted fat shows a grey-dark appearance, probably because it becomes transparent and therefore allows the observation of capillary tubes and connective tissue (Zhou et al. 1993). This is generally considered a drawback because consumers relate darker colors in the pieces of fat to rancidity (Ruiz et al. 2002).

Finally, fat consistency shows important consequences on the sensory features of the final product. Products with fats rich in PUFA and MUFA show a bright and oily surface when they are sliced and may have an excessively soft texture (Townsend et al. 1980). This can be either a drawback or a quality characteristic depending on the type of product and the country: for most consumers worldwide, oiliness is not a desirable characteristic, but it is in dry-fermented sausages from Iberian pigs in Spain (Ruiz et al. 1998).

Oxidative Status

Lipid oxidation is one of the main causes of deterioration in the quality of meat products during storage and processing, due to generation of compounds showing rancid flavor (Grosch 1987). On the other hand, some of the volatile compounds with higher impact on the aroma found in meat and meat products arise from oxidation of unsaturated fatty acids (Gasser and Grosch 1988). Furthermore, compounds from autoxidation of unsaturated fatty acids may further react with amino compounds, leading to formation of very low threshold volatiles, which have been highlighted as being responsible for pleasant flavor notes in cooked meat and meat products (Mottram 1998). Lipid oxidative phenomena in muscle-based foods occur immediately after slaughter (even preslaughter), when the cellular mechanisms for controlling lipid oxidation no longer work. The rate and extent of lipid oxidation depends on the presence of PUFA and the ratio between prooxidant and antioxidant factors. Among the formers, the main one in muscle is the presence of heme proteins (myoglobin). Major antioxidants in muscle include several enzymes, some proteins, carnosine, tocopherols, ascorbic acid, and polyphenols.

Phospholipids are the primary substrates of lipid oxidation in muscle foods, and triglycerydes seem to play a minor role (Igene and Pearson 1979). The high sensitivity of phospholipids to oxidation has primarily two causes: the high proportion of long chain PUFA, which are very susceptible to oxidation, and the close contact of phospholipids with catalysts of lipid oxidation located in the aqueous phase of the muscle cell, because they are the major cell membrane components.

The overall mechanism of fatty acid oxidation includes three steps: initiation, propagation, and termination (Frankel 1984). Primary products of lipid autoxidation are hydroperoxides, which can further decompose to form a large variety of volatiles including alkanes, aldehydes, ketones, alcohols, esters, carboxylic acids, and so on. Both the nature and the relative proportion of the volatile compounds in the muscle food depend on several factors, among which the fatty acid profile of the raw material is the main one (Grosch 1987). Among the numerous volatiles formed from unsaturated fatty acids, the most important aroma compounds are aldehydes and several unsaturated ketones and furan derivatives (Grosch 1987). They include C3–C10 aldehydes, C5–C8 unsaturated ketones, and pentyl or pentenyl furans. In fact, saturated aldehydes have been extensively used as markers for lipid oxidation in food (Shahidi et al. 1987).

Among PUFA, the n-3 fatty acids have been reported to more severely affect the susceptibility of fresh meat to lipid oxidation than n-6 fatty acids, probably because some n-3 fatty acids are highly unsaturated (e.g., eicosapentaenoic acid or docosahexaenoic acid). In fact, the susceptibility of fatty acids to oxidation increases with the number of double bonds: the relative oxidation rates of fatty acids containing 1, 2, 3, 4, 5, or 6 double bonds are 0.025, 1, 2, 4, 6, and 8, respectively (Horwitt 1986). This explains why even small differences in the concentration of these long chain PUFA (particularly in the polar fraction of the fat) may be critically important in the development of oxidation. Accordingly, Eichenberger et al. (1982) revealed that lipid oxidation in isolated membranes occurred mainly in n-3 fatty acids containing 5 or 6 double bonds. Moreover, using a triangle test, de la Hoz et al. (2004)

evidenced noticeable sensory differences between dry-fermented sausages made with meat from pigs fed either on a diet rich in linolenic acid (C18:3 n-3) or on a control diet.

Warnants et al. (1998), studying the effect of feeding diets rich in PUFA on the sensory characteristics of salami, found that enriched products were easily detectable due to higher oxidative rancidity. These authors suggested a limit of 21% PUFA for fat aimed toward production of fermented sausages (equivalent to the inclusion of 26 g PUFA/kg concentrate used for pig feeding). Nevertheless, other authors have suggested considerably lower levels (14% PUFA) for fat dedicated to production of these kinds of products (Stiebing et al. 1993). A similar behavior has been evidenced in other dry products, and, thus, linoleic acid content in fresh meat is limited to 15% in Parma ham and to 9% in Iberian ham.

The use of diets rich in n-3 fatty acids shows similar or even more accentuated consequences. The main problem in this case is that products from oxidation of these fatty acids show low olfaction threshold and particular objectionable flavor notes. Thus, Pastorelli et al. (2003) found easily detectable those samples from Parma hams from pigs fed linseed enriched diets.

Some other compounds in the muscle may experience oxidative reactions, such as proteins or heme pigments. Very little is known about protein oxidation in fermented meat products. In a recent experiment (Ventanas et al. 2006), evidence of protein oxidation throughout processing of dry-cured loins was detected. This process has been linked to impaired water retention and decreased texture in other meat products (Xiong 2000), and so its importance in the quality of dry-fermented products should not be dismissed.

Compounds with antioxidative properties may reduce overall lipid oxidative phenomena, avoiding or decreasing the presence of rancid notes in fermented products. There are two possible strategies: feeding pigs with mixed diets rich in antioxidants or directly adding antioxidants to the meat batter. As far as pig dietary strategies are concerned, phenolic antioxidants (BHT, BHA, ethoxyquine) are extensively used in animal feeding due to their reducing lipid oxidation properties. However, only a small amount of these antioxidants are absorbed by animals and, therefore, they only show the effect of stabilizing fat in the feeding but not in animal tissues lipids. Some other antioxidants added to the feeding may be absorbed and accumulated in animal tissues, showing their effects in vivo and after slaughter. In fact, the use of α-tocopherol for controlling oxidative reactions in meat and meat products has gained great attention in the last decades (Ruiz and López-Bote 2002). The main advantage of this practice is that the effect is much higher than the direct use of the antioxidant in the meat product, since the antioxidant is distributed to every tissue and subcellular structures. For example, de la Hoz et al. (2004) found that dry-fermented sausages produced using meat enriched in n-3 fatty acids were easily detected by panelists. However, when animal feeding was supplemented with 250 mg/Kg of α-tocopherol, panelists were unable to detect such differences, and lipid oxidation was not significantly different to the control sausages. Such a procedure is especially interesting when producing whole dry-cured fermented products, like hams or loins, in which the direct addition of antioxidants is not possible or it is scarcely effective (Ruiz and López-Bote 2002).

Direct addition of antioxidants to the meat batter also contributes to a better control of oxidative deterioration of fermented meat products. In fact, Valencia et al. (2006) achieved similar levels of lipid oxidation in sausages enriched with n-3 PUFA through addition of a mixture of BHT and BHA. Nevertheless, the presence of such additives is usually not pleasing for some consumers and, thus, the use of other type of plant extracts with antioxidative properties is getting more common for elaboration of fermented meat products. The reduction in the number of "E" additives for elaborating meat products is currently a goal for many producers; this practice is known as "cleaning the label." Thus, in a recent paper, Sebranek et al. (2005) showed that the rosemary extract at 2500 ppm was equally effective as BHA/BHT for controlling lipid oxidation in sausages during refrigerated storage (Figure 7.7).

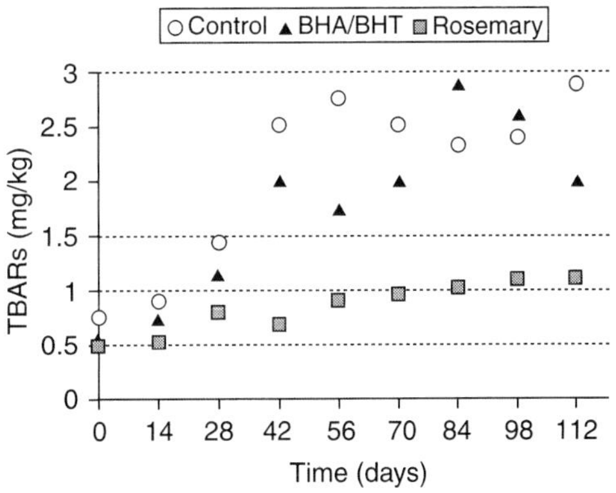

Figure 7.7. TBARS values during frozen storage of raw pork sausage with different antioxidants (taken from Sebranek et al. 2005).

Connective Tissue

Connective tissue in muscle is organized at different levels: epimysium surrounding the whole muscle, perymisium enclosing muscle fiber bundles, and endomysium encircling individual muscle fibers (Lawrie 1995). This tissue is essentially formed by collagen and elastin. It seems that from a meat quality point of view the former is more directly related with texture. Collagen is formed by tropocollagen subunits crosslinked by stable bonds forming highly resistant fibers. As a consequence of a higher number of stable bonds, collagen becomes more thermally stable and less soluble (Sims and Bailey 1981). These features have been linked to meat toughness. The number and stability of collagen crosslinking bridges is dependent upon a number of factors, such as animal age, sex, type of muscle, or exercise (Sims and Bailey 1981; Mayoral et al. 1999) (Figure 7.8).

From a practical point of view, real variations in the collagen stability between meats from commercial pigs of different ages are not relevant, nor that occasioned by sex, because pigs are slaughtered when they are not old enough for these factors to show any influence. However, there are large variations in collagen content between muscles, leading to great differences in texture. This in turn influences the texture and the quality of the final dry-fermented product. Thus, those sausages produced using the meat from the foreleg are considered of higher quality, and cuts including more fibrous muscles are more frequently used for lower-quality products. At any rate, legislation of some countries clearly establishes a limit for the hidroxyproline content of some type of sausages to control the addition of tendons, fasciae, or highly fibrous meat to high-quality sausages. In this way, consumers are protected against those who could potentially use low-quality cuts to produce expensive sausages through carrying out a very intense grinding of fresh lean.

The processing of fermented products does not produce significant changes in the structure of collagen, and, thus, does not reduce the fibrousness and toughness occasioned by this structural component (Cordoba 1990).

OTHER INGREDIENTS

Although meat and fat constitute most of the total volume of the fermented products, other ingredients are also crucial for their elaboration. In fact, from ancient times these products were produced through adding sodium chloride to lean meat, and, thus, salt is a key component for achieving the conservation and sensory characteristics in fermented meat products. Only in the last years have some attempts been

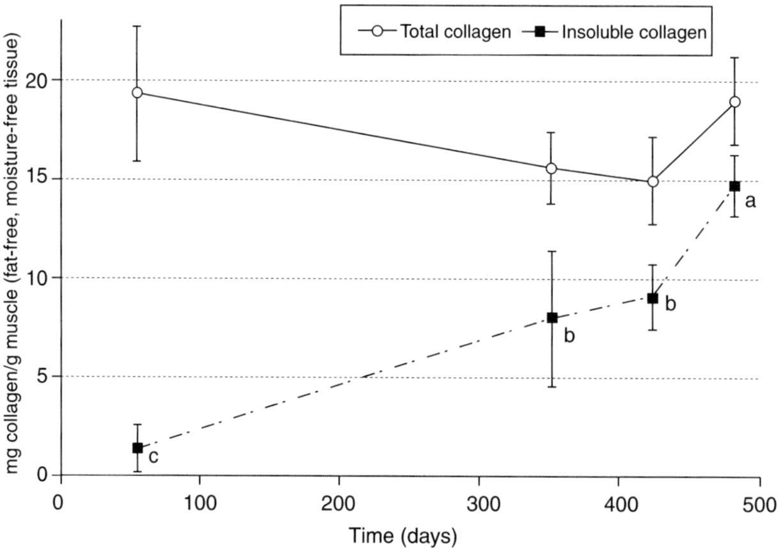

Figure 7.8. Means (±SD) of total and insoluble collagen content in longissimus dorsi throughout lifetime of pigs slaughtered at heavy weights. Means of the insoluble collagen with different letter differ significantly ($P < 0.05$) (taken from Mayoral et al. 1999).

made to produce sausages with reduced salt content or even the total absence of this ingredient, but their presence in the market is still scarce. Other common ingredients are discussed latter in this book. Finally, some other ingredients, such as offal, blood, vegetable starch, or exogenous proteins are less usual in common commercial sausages, but there are some regional varieties that may include them.

Water

Water is a common ingredient for most cooked meat products because it allows obtaining an adequate meat gel and increases yield. However, given that in fermented meat products there is a process of desiccation, water addition has none of these goals, but enables or makes possible a correct mixing of ingredients or dissolving of some of them, such as exogenous proteins or starches. Thus, in some whole fermented meat products, salting, seasoning, and additive addition is achieved through dissolving all these ingredients in water forming a brine and immersing the product in it for several days. At any rate, water does not exert any other technological function in fermented products, and, thus, any edible water can be used for the cited purposes.

Salt

For ages, meat and salt have been the only two strictly necessary ingredients for production of fermented meat products. Sodium chloride shows a number of effects on meat products, most of them desirable, which include imparting salty taste, the reduction of water activity and hence the control of microbial spoilage, the increase of water retention ability, the facilitation of proteins to get dissolved, the reduction of the activity of some enzymes, and the enhancement in the occurrence of oxidative processes.

The increasing occurrence of high blood pressure and cardiovascular diseases in developed countries together with a trend to foods with a milder taste during the last 50 years, have led to a considerable reduction in salt content in fermented meat products, and even to the production of salt-free products in the last decade, because sodium intake has been related to high blood pressure (Frost et al. 1991). Such a reduction might produce technological and sensory problems because the relevance of NaCl in solubilization and binding capacity of proteins, preservation, and taste makes it especially difficult to reduce its concentration in dry-fermented sausages.

Salty taste is an essential taste note in all fermented meat products. Saltiness is sensory defined as the taste occasioned by sodium chloride. In fact, the saltiness of other salts is not as pure as that of sodium chloride, and this is a major drawback when trying to reduce salt content in fermented meat products by using substitutes. For salt to produce a salty taste in the tongue it should be dissolved in the mouth. In dry-fermented products, chloride and sodium ions are tightly bound to proteins, and products with 5–6% salt do not show a deep salty taste; with lower salt concentration, other fresh products show a stronger saltiness. The high concentration of a compound necessary for partial replacement of NaCl gives rise in general to sensory problems.

A saturated sodium chloride solution in water (more than 26.5 g/100 g solution) produces a water activity (a_w) of 0.753. Therefore, the addition of salt to fermented meat products decreases a_w in the first steps of processing. Variation in the quantity of water and in the salt/water ratio during fermentation and storage of fermented sausages has an important role on microbial multiplication. Decreased a_w due to salt addition controls microbial spoilage of fermented products in the first steps of processing, when other hurdles to microbial growth still are not enough for limiting it. However, salt added to raw fermented meat products is usually below 4% (w/v), so it just slightly contributes to the global decrease in a_w. Nevertheless, such a slight decrease is crucial in fermented sausages stabilization because it takes place in the beginning of the process, contributing to the selection of a microbial population which will thereafter help stabilization through decreasing the pH. A derived effect of different microbial growth due to the different salt content is that concerning flavor formation by microorganisms, especially *Staphylococcus* (Tjener et al. 2004; Olesen et al. 2004). Similarly, Gardini et al. (2001) observed lower biogenic amine production due to lower microbial counts in sausages with higher salt content. In fact, with levels of 5% salt, presence of biogenic amines in the products was almost avoided.

One of the problems when using a salt substitute in fermented meat products is that the initial decrease in a_w is not as intense: the a_w of a saturated KCl solution is 0.843 (Greenspan 1977), and that of a saturated glycine solution is 0.921 (Na et al. 1994). Thus, selective growth of halotolerant microorganisms, such as Lactobacillaceae or Micrococcaceae bacteria, could be impaired, and other bacteria can reach higher counts (Gelabert et al. 2003). However, several authors have shown that with an appropriate mixture of salts such a problem can be solved

(Gimeno et al. 2001a). Vanburik and Dekoos (1990) noted the preservative qualities of the lactate ion, and suggested that it may be possible to use potassium lactate (K-lactate) as a substitute for salt, although the abnormal taste encountered could limit its use. Another approach has been the use of calcium ascorbate for substituting sodium chloride, with no detrimental effect on microbial counts and limited variations in instrumental measurements (Gimeno et al. 2001b), but these authors did not measure sensory traits.

Muscle water retention ability is directly related to pH and the extent to which myofibrilar proteins are denatured. As commented before, at the isoelectrical point of muscle proteins (around 5.0–5.2), water retention ability reaches the minimum. For a pH value that can be naturally found in meat (above 5.4), WHC is increased by the presence of sodium chloride, probably due to strong binding between the chloride ion and positive charged groups of meat proteins, leading to electrostatic repulsion of proteins, which in turn leads to a higher swelling of the muscle structure, conducting to a higher water retention (Durand 1999).

Myofibrilar protein solubility is directly related to the ionic strength of the solution. The presence of sodium chloride increases ionic strength and hence increases protein solubilization, leading to better gelling and emulsifying properties of meat proteins. This is especially important for the formation of a firm protein gel during subsequent acidification and dehydration processes undergone by meat fermented products, leading to their final desirable texture. Myosin solubilization was good enough for meat products processed at levels above 2% of NaCl and extremely good when initial salt content was around 4%, although this would lead to very salty products (Durand 1999).

In order to substitute salt in meat products without affecting protein functionality, some authors have proposed modifications in the processing, such as tumbling (Frye et al. 1986), emulsion coating (Thiel et al. 1986), or adding potassium sorbate (Sofos 1986) or certain protein hydrolysates on a collagen basis (Hofmann and Maggrander 1989), which compensate for the loss of binding produced by a reduction of NaCl.

The activity of several muscle and microbial enzymes involved in the development of the flavor in fermented meat products can be affected by the concentration of sodium chloride. Thus, as far as lipases are concerned, Toldrá et al. (1997) showed that NaCl concentrations from 25 to 60 g/l activate the acid esterase but inhibit lysosomal acid lipase. Other authors have reported that a high concentration of NaCl could inhibit lipolytic activity, showing that when the NaCl level decreases, the release of fatty acids increases (Stahnke 1995). On the other hand, proteolytic activity seems to be inhibited by high levels of salt. Accordingly, Toldrá et al. (1992) pointed out that the activity of cathepsins could be affected by different salt levels, cathepsin H and D activities being substantially inhibited by salt. Thus, Ibañez et al. (1997) showed a higher degradation of sarcoplasmic and myofibrilar protein and higher nonprotein nitrogen content in sausages with partial replacement of sodium chloride with potassium chloride, most likely due to a lower inhibition of the proteolytic activity (Figure 7.9). Accordingly, Waade and Stahnke (1997) observed a negative influence of salt content on the amount of several free amino acids in fermented sausages.

Several authors consider that sodium chloride acts as a prooxidant in meat and meat products (Coutron-Gambotti et al. 1999; Kanner et al. 1991), although other authors have not evidenced such an effect in a long-processing dry-cured product (Andrés et al. 2004). Salt accelerates lipid oxidation, although the mechanism is not fully understood. Sodium chloride may enhance the activity of oxidases (Lea 1937). Salt could modify the heme proteins catalysing lipid oxidation (Love and Pearson 1974), or this lipid oxidation may be due to the action of the reactive chloride ion on lipids. Osinchak et al. (1992) reported that a chloride ion may solubilize iron ions and stimulate the lipid peroxidation in liposomes. By contrast, sodium chloride has been also considered as an antioxidant (Chang and Watts 1950; Mabrouk and Dugan 1960).

Meat By-products

The use of animal by-products for production of fermented meat products is very limited. Nevertheless, there are a number of regional varieties in which different offal, blood, or porcine skin are used as ingredients. For example, in pepperoni, finely ground cooked porcine skin is added to the mixture to provide desirable textural characteristics. The procedure usually used for this inclusion is to cook the pork skins to a point in which they are somewhat translucent; subsequently skins are drained and chilled, and then ground to the desirable texture to be included in the finished product and mixed in a proportion usually lower than 10% (Rust 1988). Osburn et al. (1997) have studied the use of connective tissue from the rind of pigs for reducing costs in bologna sausages. Visessanguan

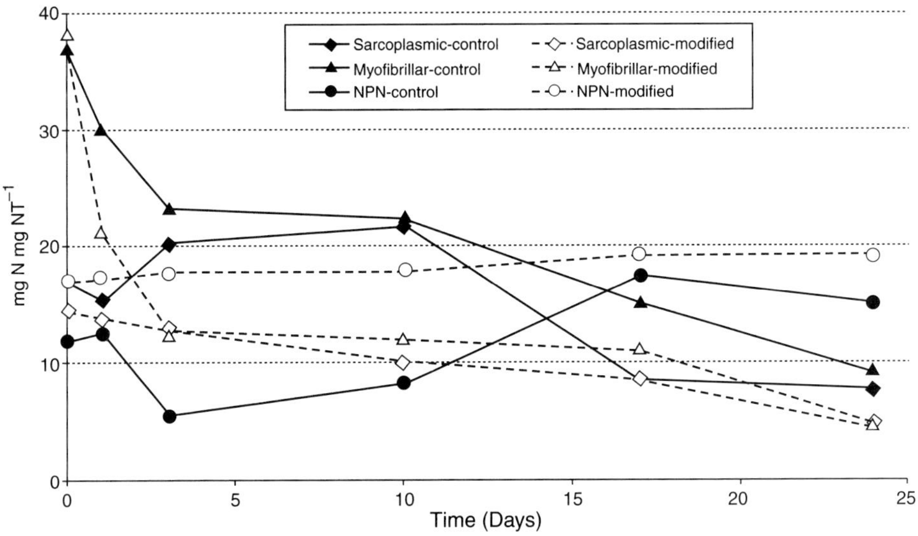

Figure 7.9. Evolution of sarcoplasmic, myofibrillar, and nonprotein nitrogen (NPN) throughout the ripening of fermented sausages made with 3% NaCl (control) and with 1.5% NaCl + 1.5% KCl (modified) (elaborated with data from Ibáñez et al. 1997).

et al. (2005) found that a minced pork to cooked rind ratio of 5:5 was the best for production of Nham, a Thai fermented pork sausage.

Blood is used in some Spanish fermented sausages, providing binding properties, due to its high protein content, and some color, due to hemoglobin. These types of sausages usually include some vegetables, such as cooked potatoes or pumpkin. Much less common is the use of offal, such as liver, heart, or lungs, for production of fermented sausages, although some minor varieties include them in their formula in different proportions. As meat, these organs usually show a high protein content. However, their protein structure, protein functionality, final pH, or fat distribution, are considerably different from those of meat (except, perhaps, in the heart).

Vegetables, Fruits, Beans, and Other Plant Ingredients

Only some fermented meat product specialities include vegetables, fruits, or beans. In most cases, these plant materials are boiled before added to the mixture of ingredients. These plant ingredients include onion, garlic (which could be also considered spices or condiments), potato, pumpkin, leeks, different types of nuts, and mushrooms. Their roles in meat products are to impart a characteristic flavor and achieve a distinguishing texture, attained by the formation of a polysaccharide gel through heating in the presence of water. However, the scientific information regarding the effects of these ingredients on the characteristics of the final product is scarce. Magra et al. (2006) claimed leek (*Allium porrum*) in Greek fermented sausages is a source of nitrates as well as a rich source of sulphur volatiles such as thiopropanal S-oxide, thiosulphinates, and related compounds (zwiebelanes, capaenes) in minor quantities, which participate in the flavor (Ferary and Auger 1996; Mondy et al. 2002). Moreover, flavonol glycosides present in leek show antioxidant activity (Fattorusso et al. 2001). In addition, allium plants also possess antifungal activity due to the chitinaces (Vergawen et al. 1998; Yin and Tsao 1999) and show high contents of inulin-type fructans with potential effects on texture (Mendoza et al. 2001).

In the last decade there has been an increasing interest in plant ingredients—or at least part of their components, such as protein, starch, or fiber—as fat replacers and functional ingredients for meat products. Thus, Garcia et al. (2002) reported dry-cured fermented sausages with similar physical and sensory properties to commercial ones with reduced added fat (only 10%) and added fruit fiber (1.5%). These authors claimed that using this procedure is possible to obtain healthier fermented sausages. Papavergou et al. (1999) used lupin flour and lupin protein isolate as ingredients in fermented sausages.

Although flour-added sausages showed very different properties to those of control sausages, the use of lupin protein isolate allowed the production of sausages with similar properties as those of control ones, and even lower lipid oxidation development. Other plant (and animal) proteins used for fermented meat products will be discussed later in this book.

The addition of different types of vegetable oils to the mixture of ingredients in fermented sausages production constitutes another approach to achieve a healthier fatty acid profile in this type of product (Muguerza et al. 2004). This procedure has been used for ages in the production of some dry-cured sausage varieties, but the amount of oil was rather low and the objective was not to modify the fatty acid profile but to facilitate mixing and slow down the subsequent dehydration process. Compared to changing the fatty acid profile of pig lard by dietary means, this procedure is more effective, is cheaper, and does not imply the coordination between farmers and meat sausages producers. However, the physical location of the fat in the meat batter is not the same, because when the fatty acid profile of animal fat is modified, the fat shows a normal histological distribution in fat drops constituted by triglycerides within adipocytes and in a double layer of phospholipids in cell and organelles membranes. On the other hand, when directly added to the meat batter, oils distribute around ground meat and fat. Such a distribution may alter dehydration, does not perfectly allow oil retention, and could make the fatty acids more easily oxidized. The use of different types of oils for modifying the fatty acid profile of fermented meat products will be discussed latter in this book (see Chapter 24, "Functional Meat Products").

Sugars are also commonly added to fermented meat products. Among most used sugars are sucrose, lactose, dextrose, glucose, corn syrups, different starches, and sorbitol (Rust 1987). The main role of sugars in fermented meat products is to act as substrates for lactic bacteria to produce lactic acid, which is essential in the elaboration of fermented meat products (as is thoroughly explained and discussed later in this book). The type of sugar influences the rate of pH decline during fermentation, because disaccharides should be first hydrolyzed to monosaccharides. Thus, fermented sausages produced using glucose or fructose show a more rapid acidification than those in which lactose or sucrose are used (Diaferia et al. 1995). Short-processing fermented products are usually added with 0.5–0.7% glucose or sucrose or 1% lactose; to long-processing products, common levels are around 0.3% glucose or sucrose or 0.5% lactose (Leistner, 1992). Sugars are also used to contribute to the overall taste of the product, counteracting salty taste (Prändl et al. 1988).

REFERENCES

MD Aaslying. 2002. Quality indicators for raw meat. In: J Kerry, J Kerry, D Ledward, eds. Meat Processing: Improving Quality. Boca Raton, Florida: CRC Press LLC and Woodhead Publishing Ltd. Chapter 8.

H Andersen, G Bertelsen, L Boegh-Soerensen, CK Shek, L Skibsted. 1988. Effect of light and package conditions on the colour stability of sliced ham. Meat Sci 22:283–292.

AI Andrés, R Cava, J Ventanas, E Muriel, J Ruiz. 2004. Lipid oxidative changes throughout the ripening of dry-cured Iberian hams with different salt contents and processing conditions. Food Chem 84:375–381.

C Bejerholm, P Barton-Gade. 1986. Effect of intramuscular fat level on eating quality of pig meat. Proceedings of the 32nd European Meeting of Meat Research Workers, Ghent, 1986, pp. 389–391.

J Boulard, M Bouyssiere, P Chevillon, R Kerisit, P Jossec. 1995. Le tri des jambons selon la qualité des gras en salle de découpe. Techni-Porc 21:31–32.

I Chang, BM Watts. 1950. Some effects of salt and moisture on rancidity in fats. Food Res 15:313–321.

JJ Cordoba. 1990. Transformaciones de los Componentes Nitrogenados Durante la Maduración del Jamón de Cerdo Ibérico. PhD dissertation, University of Extremadura, Spain.

C Coutron-Gambotti, G Gandemer, S Rousset, O Maestrini, F Casabianca. 1999. Reducing salt content of dry-cured ham: Effect on lipid composition and sensory attributes. Food Chem 64:13–19.

A Davenel, A Riaublanc, P Marchal, G Gandemer. 1999. Quality of pig adipose tissue: Relationship between solid fat content and lipid composition. Meat Sci 51:73–79.

L de la Hoz, M D'Arrigo, I Cambero, JA Ordóñez. 2004. Development of an n-3 fatty acid and α-tocopherol enriched dry fermented sausage. Meat Sci 67:485–495.

C Diaferia, G Pirone, L La Pietra, V Magliano. 1995. Impego di microorganismi autoctoni nella preparazione del salame stagionato. I. Caratteristiche chimico-fisiche e microbiologiche in funzione del tipo e del livello di zucchero. Industria Conserve 70:138–143.

P Durand. 1999. Ingrédients et additives. In: P. Durand, ed. Technologies des Produits de Charcuterie et des Salaisons. Paris: Tec&Doc Editions, pp. 81–124.

K Eichenberger, P Bohni, KH Winterhalter, S Kawato, C Richter. 1982. Microsomal lipid-peroxidation causes an increase in the order of the membrane lipid domain. FEBS Letters 142:59–62.

E Fattorusso, Y Lanzotti, O Taglialatela-Scafati, C Cicala. 2001. The flavonoids of leek, *Allium porrum*. Phytochem 57:565–569.

S Ferary, J Auger. 1996. What is the true odour of cut Allium? Complementarity of various hyphenated methods: Gas chromatography-mass spectrometry and high-performance liquid chromatography mass spectrometry with particle beam and atmospheric pressure ionization interfaces in sulphenic acids rearrangement components discrimination. J Chromatogr A 750:63–74.

EN Frankel. 1984. Lipid oxidation—Mechanisms, products and biological significance. J Am Oil Chemists Soc 61:1908–1917.

CD Frost, MR Law, NJ Wald. 1991. By how much does dietary salt reduction lower blood-pressure 2. Analysis of observational data within populations. Br Med J 302:815–818.

CB Frye, LW Hand, CR Calkins, RW Mandigo. 1986. Reduction or replacement of sodium-chloride in a tumbled ham product. J Food Sci 51:836–837.

ML García, R Domínguez, MD Galvez, C Casas, MD Selgas. 2002. Utilization of cereal and fruit fibres in low fat dry fermented sausages. Meat Sci 60: 227–236.

RM García-Rey, JA García-Garrido, R Quiles-Zafra, J Tapiador, MD Luque de Castro. 2004. Relationship between pH before salting and dry-cured ham quality. Meat Sci 67:625–632.

F Gardini, M Martuscelli, MC Caruso, F Galgano, MA Crudele, F Favati, ME Guerzoni, G Suzzi. 2001. Effects of pH, temperature and NaCl concentration on the growth kinetics, proteolytic activity and biogenic amine production of *Enterococcus faecalis*. Int J Food Microbiol 64:105–117.

U Gasser, W Grosch. 1988. Identification of volatile flavor compounds with high aroma values from cooked beef. Z Lebensm Unster Frosch 186:489–494.

J Gelabert, P Gou, L Guerrero, J Arnau. 2003. Effect of sodium chloride replacement on some characteristics of fermented sausages. Meat Sci 65:833–839.

GG Giddings. 1977. The basis of color in muscle foods. CRC Crit Rev Food Sci Nutr 9:81–114.

O Gimeno, I Astiasarán, J Bello. 2001a. Influence of partial replacement of NaCl with KCl and $CaCl_2$ on microbiological evolution of dry fermented sausages. Food Microbiol 18:329–334.

———. 2001b. Calcium ascorbate as a potential partial substitute for NaCl in dry fermented sausages: Effect on colour, texture and hygienic quality at different concentrations. Meat Sci 57:23–29.

JP Girard, C Bucharles, JL Berdague, M Ramihone. 1989. The influence of unsaturated fats on drying and fermentation processes in dry sausages. Fleischwirtsch 69:255–260.

S Govindarajan. 1973. Fresh meat colour. CRC Crit Rev Food Technol 1:117–140.

L Greenspan. 1977. Humidity fixed points of binary saturated aqueous solutions. J Res Nat Bur Stan-A. Physics and Chem 81A:89–96.

W Grosch. 1987. Reactions of hydroperoxides—Products of low molecular weight. In: HWS Chan, ed. Autoxidation of unsaturated lipids. London: Academic Press, pp. 95–140.

L Guerrero, P Gou, P Alonso, J Arnau. 1996. Study of the physicochemical and sensorial characteristics of dry-cured hams in three pig genetic types. J Sci Food Agric 70:526–530.

L Guerrero, P Gou, J Arnau. 1999. The influence of meat pH on mechanical and sensory textural properties of dry-cured ham. Meat Sci 52:267–273.

E Hansen, D Juncher, P Henckel, A Karlsson, G Bertelsen, LH Skibsted. 2004. Oxidative stability of chilled pork chops following long term freeze storage. Meat Sci 68:479–484.

K Hofmann, K Maggrander. 1989. Reducing the common salt content of meat-products by using collagen hydrolysates. Fleischwirtsch 69:1135–1138.

M Honkavaara. 1988. Influence of PSE pork on the quality and economics of cooked, cured ham and fermented dry sausage manufacture. Meat Sci 24: 201–207.

MK Horwitt. 1986. Interpretations of requirements for thiamin, riboflavin, niacin-tryptophan, and vitamin-e plus comments on balance studies and vitamin-B6. Am J Clin Nutr 44:973–985.

JH Houben, B Krol. 1980. Acceptability and storage stability of port products with increased levels of poly-unsaturated fatty-acids. Meat Sci 5:57–70.

E Huff-Lonergan. 2002. Water-Holding Capacity of Fresh Meat. Pork Checoff. National Pork Board, American Meat Science Association. http://www.porkboard.org/.

C Ibáñez, L Quintanilla, C Cid, I Astiasarán, J Bello. 1997. Dry fermented sausages elaborated with *Lactobacillus plantarum—Staphylococcus carnosus*. Part II: Effect of partial replacement of NaCl with KCl on the proteolytic and insolubilization of proteins. Meat Sci 46:277–284.

JO Igene, AM Pearson. 1979. Role of phospholipids and triglycerides in warmed-over flavor development in meat model systems. J Food Sci 44:1285–1290.

J Kanner, S Harel, R Joffe. 1991. Lipid peroxidation of muscle food as affected by NaCl. J Agric Food Chem 39:1017–1024.

CC Kuo, CY Chu. 2003. Quality characteristics of Chinese sausages made from PSE pork. Meat Sci 64:441–449.

R Lawrie. 1995. The structure, composition and preservation of meat. In: G Campbell-Platt, PE Cook, eds. Fermented Meats, London: Blackie Academic & Professional, pp. 39–52.

CH Lea. 1937. TITULO. J Soc Chem Ind 56:376.

F Leistner. 1992. The essential of production stable and safe raw fermented sausages. In: FJM Smulders, F Toldrá, J Flores, M Prieto, eds. New Technologies for meat and meat products. Utrecht: ECCEAMST, pp. 1–18.

JD Love, AM Pearson. 1974. Metmyoglobin and nonheme iron as pro-oxidants in cooked meat. J Agric Food Chem 22:1032–1034.

AF Mabrouk, LR Dugan. 1960. A kinetic study of the autoxidation of methyl linoleate and linoleic acid emulsions in the presence of sodium chloride. J Am Oil Chem Soc 37:486–490.

TI Magra, G Bloukas, GA Fista. 2006. Effect of frozen and dried leek on processing and quality characteristics of Greek traditional sausages. Meat Sci 72:280–287.

AI Mayoral, M Dorado, MT Guillén, A Robina, JM Vivo, C Vázquez, J Ruiz. 1999. Development of meat and carcass quality characteristics in Iberian pigs reared outdoors. Meat Sci 52:315–324.

E Mendoza, ML García, C Casas, MD Selgas. 2001. Inulin as fat substitute in low fat, dry fermented sausages. Meat Sci 57:387–393.

AJ Miller, SA Ackerman, SA Palumbo. 1980. Effect of frozen storage on functionality of meat for processing. J Food Sci 45:1466–1471.

N Mondy, D Duplat, JP Christides, I Arnault, J Auger. 2002. Aroma analysis of fresh and preserved onions and leek by dual solid phase microextraction-liquid extraction and gas chromatography-mass spectrometry. J Chromatogr A 963:89–93.

DS Mottram. 1998. The chemistry of meat flavour. In: F Shahidi, ed. Flavor of Meat, Meat Products and Seafoods, London: Blackie Academic & Professional, pp. 5–26.

E Muguerza, O Gimeno, D Ansorena, I Astiasarán. 2004. New formulations for healthier dry fermented sausages: A review. Trends Food Sci Technol 15:452–457.

E Muriel, J Ruiz, L Martin, MJ Petron, T Antequera. 2004. Physico-chemical and sensory characteristics of dry-cured loin from different Iberian pig lines. Food Sci Technol Int 10:117–123.

HS Na, A Arnold, AS Myerson. 1994. Cluster formation in highly supersaturated solution droplets. J Cryst Growth 139:104–112.

E Novelli, E Zanardi, GP Ghiretti, G Campanini, G Dazzi, G Madarena, R Chizzolini. 1998. Lipid and Cholesterol Oxidation in Frozen Stored Pork, Salame Milano and Mortadella. Meat Sci 48:29–40.

PT Olesen, AS Meyer, LH Stahnke. 2004. Generation of flavour compounds in fermented sausages—The influence of curing ingredients, *Staphylococcus* starter culture and ripening time. Meat Sci 66:675–687.

WN Osburn, RW Mandigo, KM Eskridge. 1997. Pork skin connective tissue gel utilization in reduced-fat bologna. J Food Sci 62:1176–1182.

JE Osinchak, HO Hultin, TZ Olver, S Kelleher, CH Huang. 1992. Effect of NaCl on catalysis of lipid oxidation by the soluble fraction of fish muscle. Free Rad Biol Med 12:35–41.

SA Palumbo, M Komanowsky, V Metzger, JL Smith. 1977. Kinetics of pepperoni drying. J Food Sci 42:1029–1033.

SN Papadimas, JG Bloukas. 1999. Effect of fat level and storage conditions on quality characteristics of traditional Greek sausages. Meat Sci 51: 103–113.

EJ Papavergou, JG Bloukas, G Doxastakis. 1999. Effect of lupin seed proteins on quality characteristics of fermented sausages. Meat Sci 52:421–427.

G Pastorelli, S Magni, R Rossi, E Pagliarini, P Baldini, P Dirinck, F Van Opstaele, C Corino. 2003. Influence of dietary fat on fatty acid composition and sensory properties of dry-cured Parma ham. Meat Sci 65:571–580.

O Prändl, A Fischer, T Schmidhofer, HJ Sinell. 1988. Fleisch. Technologie und Hygiene der Gewinnung und Verarbeitung. Stuttgart: Ulmer.

J Ruiz, C García, E Muriel, AI Andres, J Ventanas. 2002. Influence on sensory characteristics on the acceptability of dry-cured ham. Meat Sci 61: 247–354.

J Ruiz, CJ López-Bote. 2002. Improvement of dry-cured ham quality by lipid modification through dietary means. In: F Toldrá, ed. Research Advances in the Quality of Meat and Meat Products. Kerala, India: Research Signpost, pp. 255–271.

J Ruiz, J Ventanas, R Cava, AI Andrés, C García. 2000. Texture and appearance of dry-cured ham as affected by fat content and fatty acid composition. Food Res Int 33:91–95.

J Ruiz, J Ventanas, R Cava, ML Timon, C García. 1998. Sensory characteristics of Iberian ham: Influence of processing time and slice location. Food Res Int 31:53–58.

RE Rust. 1987. Sausage products. In: JF Price, BS Schweigert, eds. The Science of Meat and Meat

Products (3rd ed.) Westport, Connecticut: Food and Nutrition Press, Inc., pp. 457–485.

RE Rust. 1988. Formulated meat products using edible meat by-products. In: AM Pearson, TR Dutson, eds. Edible Meat By-products. Advances in Meat Research 5. London: Elsevier Applied Science, pp. 341–357.

JG Sebranek, VJH Sewalt, KL Robbins, TA Houser. 2005. Comparison of a natural rosemary extract and BHA/BHT for relative antioxidant effectiveness in pork sausage. Meat Sci 69:289–296.

F Shahidi, J Yun, LJ Rubin, DF Wood. 1987. The hexanal content as an indicator of oxidative stability and flavor acceptability in cooked ground pork. Can Inst Food Sci Technol 20:104–106.

TJ Sims, AJ Bailey. 1981. Connective Tissue. In: R Lawrie, ed. Developments in Meat Science 2. London: Applied Science Publishers, pp. 29–60.

LH Skibstead. 1992. Cured meat products and their oxidative stability. In: DA Ledward, DE Johnston, MK Knight, eds. The Chemistry of Muscle-Based Foods. Cambridge: The Royal Society of Chemistry, pp. 266–286.

JF Sofos. 1986. Antimicrobial activity and functionality of reduced sodium-chloride and potassium sorbate in uncured poultry products. J Food Sci 51:16–20.

G Solignat. 1999. Séchage—Maturation. In: P Durand, ed. Technologies des Produits de Charcuterie et des Salaisons. Paris: Tec&Doc Editions, pp. 279–401.

A Soyer. 2005. Effect of fat level and ripening temperature on biochemical and sensory characteristics of naturally fermented Turkish sausages (sucuk). Eur Food Res Technol 221:412–415.

A Soyer, AH Ertas, U Uzumcuoglu. 2005. Effect of processing conditions on the quality of naturally fermented Turkish sausages (sucuks). Meat Sci 69(1):135–141.

LH Stahnke. 1995. Dried sausages fermented with *Staphylococcus-xylosus* at different temperatures and with different ingredient levels 2. Volatile components. Meat Sci 41:193–209.

A Stiebing, D Kuhne, W Rodel. 1993. Fat qualityI—Its influence on the storage stability of firm dry sausage. Fleischwirtsch 73:1169–1172.

LF Thiel, PJ Bechtel, FK McKeith, J Novakofski, TR Carr. 1986. Effect of salt reduction on the yield, breaking force, and sensory characteristics of emulsion-coated chunked and formed ham. J Food Sci 51:1439–1443.

K Tjener, LH Stahnke, L Andersen, J Martinussen. 2004. The pH-unrelated influence of salt, temperature and manganese on aroma formation by *Staphylococcus xylosus* and *Staphylococcus carnosus* in a fermented meat model system. Int J Food Microbiol 97:31–42.

F Toldrá, M Flores, Y Sanz. 1997. Dry-cured ham flavour: Enzymatic generation and process influence. Food Chem 59:523–530.

F Toldrá, E Rico, J Flores. 1992. Activities of pork muscle proteases in model cured meat systems. Biochimie 74:291–296.

WE Townsend, CE Davis, CE Lyon, SE Mescher. 1980. Effect of pork quality on some chemical, physical, and processing properties of fermented dry sausage. J Food Sci 45:622–626.

I Valencia, D Ansorena, I Astiasarán. 2006. Nutritional and sensory properties of dry fermented sausages enriched with n-3 PUFAs. Meat Sci 72:727–733.

AMC Vanburik, JT Dekoos. 1990. Sodium lactate in meat products. Fleischwirtsch 70:1266–1268.

S Ventanas, M Estevez, JF Tejeda, J Ruiz. 2006. Protein and lipid oxidation in *Longissimus dorsi* and dry cured loin from Iberian pigs as affected by crossbreeding and diet. Meat Sci 72:647–655.

S Ventanas, J Ventanas, J Ruiz. 2007. Sensory characteristics of Iberian dry-cured loins: Influence of crossbreeding and rearing systems. Meat Sci 75:211–219.

R Vergawen, F Van Leuven, A Van Laere. 1998. Purification and characterization of strongly chitin-binding chitinases from salicylic acid treated leek (*Allium porrum*). Physiol Plantarum 104:175–182.

W Visessanguan, S Benjakul, A Panya, C Kittikun, A Assavanig. 2005. Influence of minced pork and rind ratios on physico-chemical and sensory quality of Nham—A Thai fermented pork sausage. Meat Sci 69:355–362.

C Waade, LH Stahnke. 1997. Dried sausages fermented with *Staphylococcus xylosus* at different temperatures and with different ingredient levels. Part IV. Amino acid profile. Meat Sci 46:101–114.

N Warnants, MJ Van Oeckel, CC Boucque. 1998. Effect of incorporation of dietary polyunsaturated fatty acids in pork backfat on the quality of salami. Meat Sci 49:435–445.

F Wirth. 1985. The technology of processing meat not of standard quality. Fleischwirtsch 66:998.

———. 1988. Technologies for making fat-reduced meat products. Fleischwirtsch 68:1153–1156.

J Wismer-Pedersen. 1987. Water. In: JF Price, BS Schweigert, eds. The Science of Meat and Meat Products. Westport, Connecticut: Food & Nutrition Press Inc., pp. 141–154.

FW Wood. 1966. Diffusion of salt in pork muscle and fat tissue. J Sci Food Agric 17:138–140.

JD Wood. 1984. Fat deposition and the quality of fat tissue in meat animals. In: J Wiseman, ed. Fats in Animal Nutrition. London: Butterworths, pp. 407–435.

YL Xiong. 2000. Protein oxidation and implications for muscle foods quality. In: EA Decker, C Faustman, CJ López-Bote, eds. Antioxidants in Muscle Foods. New York: Wiley, pp. 85–111.

MC Yin, SM Tsao. 1999. Inhibitory effect of seven Allium plants upon three *Aspergillus* species. Int J Microbiol 49:49–56.

GH Zhou, A Yang, RK Tume. 1993. A relationship between bovine fat color and fatty-acid composition. Meat Sci 35:205–212.

8
Additives

Pedro Roncalés

INTRODUCTION

Dry-cured and/or fermented sausages are traditional meat products that may be stored even at ambient temperatures without risk of spoilage. This means that the use of additives either for manufacturing purposes or for preservation is not necessary. In fact, a high percentage of sausages made following traditional procedures do not contain any additives. However, a larger part of commercial dry-cured or fermented sausages do include a variable number of additives in their formulation. Why is this? Manufacturers pursue one or more of the following advantages with the use of additives: (1) to facilitate, enhance, accelerate or standardize the manufacturing processes; (2) to ensure product safety regarding microbial toxins; (3) to avoid the presence of undesirable surface molds; (4) to meet consumer requirements regarding sensory attributes; and (5) to increase the product yield. Only the latter purpose may be associated with low-quality products, but this is not usually a preferential aim in this type of sausages.

In accordance with the statements of the previous paragraph, most legal regulations concerning the use of additives in dry-cured and fermented sausages are rather restrictive; the number of permitted items is normally low. An example of this is provided by European Union regulations (EPC 1995), which allow the use of a few acids, antioxidants, colorants, flavor enhancers, and preservatives, as well as the multipurpose phosphates. It appears that such a short additive list fulfills the requirements of both industrials and consumers. In addition, a few nontraditional supplementary ingredients not considered properly as additives, such as nonmeat proteins, sugars, hydrolysates, and smoke extracts, are also used depending upon particular regulations. This chapter deals with the brief description of usage, advantages, and shortcomings of the most commonly used additives and supplementary ingredients in the manufacturing of dry-cured and/or fermented sausages. Sugars are addressed in Chapter 7, which deals with ingredients. For further information on the use and properties of additives, see Ash and Ash (2002), Martín-Bejarano (2001), Ranken (2000), Saltmarsh (2000), and Toldrá (2002).

ACIDS AND RELATED ADDITIVES

Acidification is common in most dry/fermented sausages; indeed, they usually have pH values within the range 4.5–5.5. Reasons underlying its convenience are many and varied: It selects the microbiota, inhibiting pathogens and undesirable nonpathogenic microorganisms; it favors the emulsifying and gelling ability of meat proteins, enhancing particle binding and soft texture perception; it facilitates the formation of curing pigments; it enhances proteolysis and lipolysis involved in maturation; and it provides a light acid taste. In summary, acidification greatly helps to conduct and accelerate the whole process, as well as to achieve product quality and safety. A convenient acidification may be obtained following sugar fermentation by lactic acid bacteria (LAB), either naturally present in the meat or added as starter cultures to the ingredients mixture (see Part III). This is the way chiefly used for acidifying sausages by

Table 8.1. Acids and related additives: CAS (Chemical Abstract Service) and EU numbers, and common concentration range added to sausage formulations.

Additive	Molecule/Form	CAS Number	EU Number	Common Concentration Range
Lactic acid		50-21-5	E270	0.5–3 g/100 g
	L-	79-33-4		
	D-	10326-41-7		
	DL-	598-82-3		
	sodium salt	72-17-3	E325	
	potassium salt	996-31-6	E326	
Citric acid	Anhydrous	77-92-9	E330	0.5–3 g/100 g
	monohydrate	5949-29-1		
Glucono-delta-lactone		90-80-2	E575	0.5–2 g/100 g

both traditional and large industries. Nevertheless, in many cases, particularly when no starter cultures are used, direct addition of acids is preferred in order to assure pH lowering within a very short time, i.e., as initial value.

Common organic acids are used for this purpose, mainly lactic and citric acids, as well as their sodium and potassium salts, which show much less ability to lower pH values. Besides these, an acid-related molecule may also be used: glucono-delta-lactone (GDL). GDL is the cyclic 1,5-intramolecular ester of D-gluconic acid. In aqueous media, it is hydrolyzed to an equilibrium mixture of D-gluconic acid (55–66%) and the delta- and gamma-lactones. Therefore, GDL was envisaged as a slower acid-releasing agent, with a view to mimicking gradual acidification by LAB fermentation.

Table 8.1 gives the Chemical Abstract Service (CAS) (JECFA 2006) and EU (EPC 1995) numbers of molecules most commonly used as acidification additives, as well as the common concentration range added to sausage formulations. Lowering of pH by added acids depends chiefly on the concentration used, but the presence of other molecules that may act as buffering agents, such as phosphates, must also be taken into account because they decrease their acidification ability. Lactic acid concentrations of 1–3% are used for a pH lowering of 0.5–1.5 units. Sodium and potassium salts of lactate cause only a minimum pH decrease; nevertheless, addition of those salts has been described to be very effective in microbial inhibition (Shelef 1994). Most legal regulations allow the addition of all those additives on the basis of the quantum satis principle, i.e., there is no other limitation to the concentration added than good manufacturing practices.

Problems regarding the addition of acids, or even GDL, have been related to the high speed of acidification, which may result in early dehydration, particularly within the outer part of the sausage mass. This early drying will cause a rapid mass shrinkage and, as a consequence, separation and wrinkling of the casing, together with undesirable crust formation. Therefore, strict control of temperature and relative humidity is needed in order to avoid these shortcomings.

ANTIOXIDANTS

Oxidative reactions may be deleterious for dry/fermented sausages; in fact, intense oxidation leads to rancidity development. However, most synthetic antioxidant additives are not permitted by legal regulations, so rancidity becomes usually the shelf-life limiter of many meat products. Nevertheless, a limited number of "soft" antioxidants are allowed by most regulations. The reason for this is that sausage curing, particularly concerning nitrate/nitrite action, cannot be achieved under prevalent oxidant conditions; well on the contrary, reducing conditions are needed for color formation and antimicrobial activity. Accordingly, the permitted "soft" antioxidants find their use rather as reductant agents than as proper antioxidants.

In agreement with this view, the antioxidants most commonly used include only ascorbic acid (vitamin C) and tocopherols (vitamin E) and some related molecules, such as erythorbic acid, as well as their salts.

Table 8.2. Antioxidant additives: CAS (Chemical Abstract Service) and EU numbers, and common concentration range added to sausage formulations.

Additive	Molecule/Form	CAS Number	EU Number	Common Concentration Range
Ascorbic acid		50-81-7	E300	Quantum satis
	Sodium salt	134-03-2	E301	~500 mg/kg
	Calcium salt	5743-27-1	E302	
	Fatty acid esters	25395-66-8	E304	
Erythorbic acid		89-65-9	E315	<500 mg/kg
	Sodium salt	6381-77-7	E316	
Tocopherols	Extract mix	59-02-9	E306	Quantum satis
	Alpha-	1406-18-4	E307	~500 mg/kg
	Gamma-		E308	
	Delta-		E309	
Extracts of a variety of herbs and spices	Phenols			Quantum satis Highly variable

All these molecules have been proven to exert an effective steady reducing effect while they suffer oxidation themselves. Some fatty acid esters of ascorbic acid (palmitate and stearate) may also be considered as reductant additives, particularly for midterm reducing ability. Furthermore, the fact that many spices and condiments possess antioxidant properties must not be overlooked, because they contribute to the oxidative stability of meat products, too (Sánchez-Escalante et al. 2003; Martínez et al. 2006a). Phenolic constituents present in those seasonings protect them with varying intensity against oxidative deterioration. There is the opportunity, too, of using a variety of industrial extracts of those herbs and spices (rosemary, oregano, tea, citrus fruits, etc.) as antioxidant additives; furthermore, they are not considered as proper additives by legal regulations.

Table 8.2 gives the CAS (JECFA 2006) and EU (EPC 1995) numbers of molecules most commonly used as antioxidant additives, as well as the common concentration range added to sausage formulations. As in the case of acids, those antioxidants are allowed on the basis of "the quantum satis" principle, with the exceptions of erythorbic acid and sodium erythorbate, which are limited to 500 mg/kg. In any case, this amount appears to be the average of those used ordinarily. Some of the antioxidants are more indicated for steady midterm reducing ability, particularly the fatty acid esters of ascorbic acid and the tocopherols; formulations with ascorbic acid and its salts are designed to provide mainly a reliable and rapid reducing capacity. Many formulations include both types of reductants in order to assure both initial and midterm antioxidant effects. Besides this, seasoning with herbs and spices, either raw or in the form of extracts, also contributes to a midterm antioxidant effect.

No deleterious or negative effects have been described following the addition of those reductants within the common concentration range. Conversely, almost all formulations for industrial manufacturing of dry/fermented sausages include one or more of those additives (Balev et al. 2005).

COLORANTS

Color is by far the most important sensory attribute in relation to the visual appearance and acceptability of most food items. Dry/fermented sausages are not an exception to this; appealing red colors of cured meat strongly determine product acceptance and purchase decision. Appropriate nitrate and/or nitrite curing provides a rather stable nitroso-compound of myoglobin, but the stability of this pigment is not guaranteed for long periods of time if prooxidant conditions are prevailing. Addition of colorants may afford a convenient solution to this problem. Seasoning with red pepper powder has been traditional in many countries (Spain, Hungary, Mexico, etc.), not only because of its desired flavor, but also because it strongly helps to avoid sausage discoloration. A huge number of natural and artificial colorants have been proposed for use in food coloring, although only a few of them were approved in legal regulations.

Table 8.3. Colorant additives: CAS (Chemical Abstract Service) and EU numbers, and common concentration range added to sausage formulations.

Additive	Molecule/Form	CAS Number	EU Number	Common Concentration Range
Cochineal extract	Extract powder Carminic acid	1343-78-8 1260-17-9	E120	<200 mg/kg
Paprika extract	Oleoresin	68917-78-2	E160c	<10 mg/kg
Ponceau 4R (cochineal red)	Trisodium-2-hydroxy-1-(4-sulfonato-1-naphthylazo)-6,8-naphthalenedisulfonate	2611-82-7	E124	<250 mg/kg
Beet red	Betanine	7659-95-2	E162	Quantum satis <200mg/kg

In the case of dry/fermented sausages, the variety of useful colorants is rather short; in fact, only those mimicking the red color of cured-meat pigment should be considered, despite the fact that some yellow hues or caramels are also useful, but only in special products.

Only a few red colorants are commonly used in dry/fermented sausage manufacture. These include both natural and artificial. Cochineal and paprika extracts are from natural origin. Cochineal extract is the powder of dried cochineals (*Dactylopius coccus*), an insect growing on nopal cactuses (*Opuntia ficus indica*). Mexico and Spain are large producers of this dye, well known and appreciated for its extremely high content of intense-red carmine (or carminic acid) in America since the pre-Hispanic times. Paprika extract is the oleoresin of any of the varieties of red pepper (*Capsicum annuum*), also of American origin, which are currently grown in large amounts all over the world. This extract is a rich source of red colorants capsorubin and capsanthin, together with capsaicin, and some varieties may also contain hot-pungent capsaicinoids. Cochineal red (Ponceau 4R) and beet red (betanine) are from artificial origin. Ponceau 4R is an industry-synthesized carmine, a closely related molecule to the red colorant of cochineal. Its color is quite similar to that of cochineal extract, but its price is significantly lower. Betanine is also an industry-synthesized colorant, which is known as a major constituent of red beet root (*Beta vulgaris*) (Martínez et al. 2006b).

Table 8.3 gives the CAS (JECFA 2006) and EU (EPC 1994) numbers of molecules or extracts most commonly used as colorant additives, as well as the common concentration range added to sausage formulations. All the colorants normally have limitations for their use; as an example, EU regulations establish, with a few exceptions, maximum concentrations (see Table 8.3). Among the colorants considered for dry/fermented sausages, only beet red is allowed without any limit other than good practices. The use of natural colorants is usually preferred for traditional sausage manufacture, although there appear to be no perceptible differences in the final products; all of them provide a stable red color of convenient appeal for consumers quite similar to that of cured meat. Taking into account that the price of artificial colorants is much lower, manufacturers prefer its use in formulations designed for mid-price sausages; natural colorants are used mainly in traditional and high-price sausages. Paprika extract seems to be more appropriate for use only in paprika-sausages because its addition results in an orange-red color typical of this meat products.

No shortcomings derived from the use of colorants have been described. Nevertheless, it is recommended that they be used at rather low concentrations in order to provide a natural appearance, as similar as possible to colorant-free, nitrite-cured sausages.

EMULSIFIERS

Dry/fermented sausages, unlike most cooked sausages, are not essentially made of an emulsion but of meat and fat fragments of varying sizes within the range of, approximately, 2 to 20 mm. However, a certain amount of emulsion is needed for binding and tightening those pieces into forming a proper mass. The emulsifying ability is provided by meat proteins, especially myosin, which are first extracted from the tissue and solubilized by added salt; following mechanical work, they form thereafter an aqueous phase able to emulsify droplets of solid fat. The emulsion formed is stabilized by slow gellation of

Table 8.4. Emulsifier nonmeat proteins: CAS (Chemical Abstract Service) and EU numbers, and common concentration range added to sausage formulations.

Additive	Molecule/Form	CAS Number	EU Number	Common Concentration range
Milk proteins	Defatted milk powder			5–30 g/kg
	Whey proteins	9005-46-3		
	Caseinates (sodium)			
Blood proteins	Red blood cells powder			5–30 g/kg
	Serum powder			
Soja proteins	Soja meal			5–30 g/kg
	Soja proteins			
Cereal proteins	Wheat proteins			5–30 g/kg
	Corn proteins			

proteins throughout the sausage-drying process. This imperfect gel fills all the spaces among meat and fat chunks, giving rise to a sliceable mechanically homogeneous mass. Naturally, the higher the amount of soluble protein, the larger the volume of interchunk matrix.

There exists a variety of emulsifiers of very different origins available for their use in foods and particularly in meat products. Among them, only proteins are of interest for manufacturing dry/fermented sausages; the rest find their use chiefly in cooked meat emulsions. Emulsifying nonmeat proteins may proceed from a large number of animal and vegetal sources. Major animal sources are milk and blood, either raw preparations or isolated proteins or protein groups. They possess an emulsifying ability close to that of meat proteins, so they are well liked by industrials. Major vegetal sources are soja and cereals, either raw preparations or isolated protein groups. They also possess a good emulsifying ability, although it is lower than that of animal proteins (Stiebing 1998).

Table 8.4 points out that most emulsifier nonmeat proteins are natural products containing a variable number of proteins and even other nonrelated molecules, as well as that they are not considered as additives by EU and many other regulations. Their use is normally regulated in local dispositions for the diverse meat products, often varying according to product quality categories. In a strict sense, added proteins are not at all necessary for dry/fermented sausage manufacture; meat proteins should be enough for forming the small amount of emulsion required for sausage sliceability. However, small amounts of exogenous proteins may be sometimes convenient for assuring rapid emulsion formation and early stabilization, as well as for achieving a homogeneous and soft texture of the sausage even after drying. But they are also added frequently with the aim of increasing the product yield by embedding meat and fat chunks in a large amount of a protein-water matrix. In any case, addition of nonmeat proteins is be related to the manufacturing of medium- to low-quality products.

The preferred nonmeat proteins are those having a higher emulsifying ability, which are mainly of animal origin, particularly caseinates. Nevertheless, the price of those proteins is rather high, and the higher the degree of protein isolation, the higher the price. This is why many industrials are using defatted milk powder; actually, it is used not only because of its low price but also because milk powder contains lactose, a sugar that may serve both as a selective substrate for lactic acid bacteria and as a fairly good humectant. All proteins from vegetal sources have a medium emulsifying ability, but their low prices are highly convenient for industrials. Protein concentrates are a better choice than raw meals because the latter may confer extraneous flavors to sausages. Furthermore, new protein preparations are regularly appearing in the additive market. Addition of nonmeat proteins must be based on sausage quality criteria; therefore, the concentrations used must remain low in any case, ranging between 0.5% and 3%. No shortcomings other than quality lowering have been described.

FLAVOR ENHANCERS

Flavor is the most appreciated sensory attribute of foods, and this is particularly true regarding meat products. Consumers expect dry/fermented sausages to be very savory. Normally, this is achieved by traditional seasonings and appropriate processing; flavor of dry/fermented sausages is the result of the interaction of spice and condiment flavors and those generated by the fermentation-curing process of meat and fat ingredients. On many occasions, however, one or more of the following problems occur: seasoning is poor, ingredients are not of the best quality, or the fermenting-curing process is not sufficiently intense. Consumers also may demand extremely savory products, hardly achieved by traditional manufacturing. In all those cases, industrials may solve the problem, at least in part, by using flavor enhancers. These additives, very well known in oriental cuisines, have the ability of making more intense basic flavors of meat, meat products, and other foods of animal origin. Most researchers acknowledge those molecules as actors of the perception of a particular taste known by its Japanese name *umami*.

Glutamic acid is the best known flavor enhancer; this common amino acid, as well as some of its salts, has been used for this purpose for many years and it is a principal ingredient in oriental cooking practices and the oriental food industry. Now it is a highly used food additive all over the world. More recent developments in food additive research demonstrated that a number of compounds very active in biological systems, such as nucleosides and nucleotides, also had the ability to be perceived as umami by taste receptors, i.e., to be flavor enhancers. These include many derivatives of guanylic acid, inosinic acid, and a variety of ribonucleotides. So, guanylates, inosinates, and ribonucleotides greatly broadened the spectrum of available flavor enhancers.

Table 8.5 gives the CAS (JECFA 2006) and EU (EPC 1995) numbers of molecules most commonly used as flavor enhancers, as well as the common concentration range added to sausage formulations. Glutamic acid and its salts, particularly monosodium glutamate (MSG), have been used for many years as exclusive flavor enhancers in meat products. Results have been good, but criticism arose regarding the flavor homogeneity brought about in quite different meat products. Currently, the large number of available flavor enhancers helps overcome this problem; the diverse molecules contribute with fairly different flavors to overall sausage flavor. So, flavor spectra have increased, especially with the use of mixtures of variable but small amounts of some of the available flavor enhancers. The added concentrations of all the additives included in this group are normally limited by most legal regulations, but they are high enough to achieve intense flavors. In fact, they are very commonly used at low concentrations in a variety of medium-quality dry/fermented sausages. No shortcomings other than flavor homogeneity have been so far described.

Table 8.5. Flavor enhancers: CAS (Chemical Abstract Service) and EU numbers, and common concentration range added to sausage formulations.

Additive	Molecule/Form	CAS Number	EU Number	Common Concentration Range
Glutamic acid		56-86-0	E620	
	Monosodium salt	142-47-2	E621	<10 g/kg
	Monopotassium salt	19473-49-5	E622	
	Calcium diglutamate	19238-49-4	E623	
	Magnesium diglutamate	18543-68-5	E625	
	Monoammonium glutamate	7558-63-6	E624	
Guanylic acid		85-32-5	E626	<500 mg/kg
	Calcium salt	38966-30-2	E629	
	Dipotassium salt	3254-39-5	E628	
	Disodium salt	5550-12-9	E627	
Inosinic acid		131-99-7	E630	<500 mg/kg
	Calcium salt	38966-29-9	E633	
	Dipotassium salt	20262-26-4	E632	
	Disodium salt	4691-65-0	E631	
Ribonucleotides	Mixtures of calcium or disodium salts of guanylate and inosinate		E634 E635	<500 mg/kg

FLAVORING AGENTS

The convenience of using flavor enhancers has been discussed in the previous section. The use of flavoring agents is partly related to the same arguments, the need of fulfilling consumer demands on intense flavor. This is the case for protein or yeast hydrolysates, which might be considered as either flavor enhancers or flavoring agents. These hydrolysates are very complex mixtures of amino acids and different-sized peptides in the first case, and of amino acids, peptides, nucleotides, nucleosides and many other biologically active compounds in the second one. But, besides this, there are specific flavor deficiencies that can be overcome by using particular flavoring agents; those agents are mainly the following: replacing raw preparations of spices and condiments with their extracts or replacing natural smoking with smoke extracts. So, both herb and spice extracts, chiefly oleoresins, and liquid smoke preparations are to be included in this additive group.

Table 8.6 points out that most flavoring agents are industrial preparations of natural products containing a highly variable number of many different molecules, as well as that they are not considered as additives by EU and many other regulations, with the exception of paprika oleoresin. Their use is sometimes regulated in local dispositions for the diverse meat products, often varying according to product quality categories, but they are more frequently considered as mere ingredients within the group of seasonings. In summary, regulations on flavorings are rather confusing. In any case, all of them confer high-intensity flavors, together with color in the case of paprika, smoke, and other extracts, so that the concentrations added are subjected to achieving the desired level of flavor and, therefore, to good industrial practices. No shortcomings following their use have been described, with the exception of liquid smoke. This extract contains a large number of molecules formed through wood combustion. Because uncontrolled combustion may lead to formation of carcinogenic benzopyrenes, smoke extracts must be prepared under conditions avoiding the presence of those molecules, especially by using low-temperature combustion.

PRESERVATIVES

Dry/fermented sausages are the result of processing meat by using technologies that ensure a convenient preservation and safety from a microbial perspective. In agreement with this view, legal regulations are very restrictive in preservative additives allowance. Only two groups of preservatives are usually considered for use: 1) those inhibiting specifically toxin-producing, spore-former microbes; and 2) those inhibiting surface molds and yeasts. The former have been used traditionally for meat curing; they are highly recognized for their ability to effectively avoid risks associated to microbial toxins produced by *Clostridium* and *Bacillus*. In addition, they are essential substrates for the development of cured-meat color and flavors (see Chapter 3). Regarding mold and yeast inhibitors, their use must be considered if growth of those organisms on sausage surface is not desired, particularly when sausages are not subject to smoking. It is well known that smoke components have an intense fungistatic effect, but many dry/fermented sausages, especially in Southern Europe, are not smoked.

Nitrates and nitrites are the only integrants of the first group of antimicrobials. These inorganic salts, especially nitrates, are present as contaminants in nonpurified crystals of common salt, sodium chloride. They have been present in sausage formulations in this form since ancient times, giving rise to the

Table 8.6. Flavoring agents: CAS (Chemical Abstract Service) and EU numbers, and common concentration range added to sausage formulations.

Additive	Molecule/Form	CAS Number	EU Number	Common Concentration Range
Herbs and spices extracts	Paprika oleoresin Many other oleoresins Many other extracts	68917-78-2	E160c	<10 mg/kg
Smoke extracts	Liquid smoke			Highly variable
Hydrolysates	Protein Yeast			<1000 mg/kg

desired, though involuntary, curing. Modern scientific evidence led to current knowledge on the involvement of nitrates/nitrites in curing chemistry and resulted in the use of those salts as technological additives (see Chapter 3).

Fungistatic additives for surface treatment of sausages include a variety of molecules. Most of them are organic acids such as sorbic, benzoic, and para-hydroxybenzoic acids, as well as their salts. These had been used in meat products as broad-spectrum antimicrobials, and they still have this use in many other foods. But they have been demonstrated to be highly effective fungistatic agents, too; so this is their only use in dry/fermented sausages according to many legal regulations, in which they are allowed to be used only on the sausage surface. Two other fungistatic agents come usually into consideration: nisin and natamicin. They are both antibiotics of microbial origin and they are very active against molds and yeasts.

Table 8.7 gives the CAS (JECFA 2006) and EU (EPC 1995) numbers of molecules most commonly used as preservatives, as well as the common concentration range added to sausage formulations. The use of nitrates and nitrites is highly controversial. They have been demonstrated to be related to some forms of cancer, particularly to stomach tumors, due to the formation of nitrosamines (see Chapter 43). However, all legal regulations allow use of nitrates and nitrites in meat products on the basis of their highly beneficial effects; but maximum concentrations are continuously subject to revision, with a steady trend to being diminished. Usually, unlike most other additives, both an indicative added concentration and a residual maximum concentration are established. This is due to the well-known fact that both nitrates and nitrites are very reactive. In fact, nitrate is transformed into nitrite under reducing conditions and nitrite, in turn, may form a variety of molecules. Some of them will react to produce nitroso-myoglobin if reducing conditions prevail, together with many other nitroso-compounds, including those exerting the antimicrobial effect (Cassens 1990). Therefore, both nitrates and nitrites may be used as curing/preservative agents. Nitrates must be added at higher concentrations and their action is delayed throughout curing; nitrites must be added at lower concentrations and their effect is achieved earlier. So, industrials prefer nitrates for long-curing sausages and nitrites for short-curing ones, but the use of mixtures of both salts is common, too.

Regarding fungistatic preservatives, there is a large number of available molecules, but not all of them have the same interest in dry/fermented sausages. They are all highly effective against mold and yeast growth, but they do not have the same pH optimum for activity. P-hydroxybenzoates (parabens) are more efficient at higher pH values, i.e., at 6.0 or above; sorbates are more active at relatively acidic pH values,

Table 8.7. Preservatives: CAS (Chemical Abstract Service) and EU numbers, and common concentration range added to sausage formulations.

Additive	Molecule/Form	CAS Number	EU Number	Common Concentration Range
Nitrates/nitrites	Potassium nitrate	7757-79-1	E252	250 (300) mg/kg*
	Sodium nitrate	7631-99-4	E251	250 (300) mg/kg*
	Potassium nitrite	7758-09-0	E249	50 (150) mg/kg*
	Sodium nitrite	7632-00-0	E250	50 (150) mg/kg*
Sorbic acid		110-44-1	E200	Quantum
	Potassium salt	24634-61-5	E202	satis
	Calcium salt	7492-55-9	E203	(surface)
Benzoic acid		65-85-10	E210	Quantum
	Potassium salt	582-25-2	E212	satis
	Sodium salt	532-32-1	E211	(surface)
	Calcium salt		E213	
P-hydroxybenzoates	Ethyl-PHB	120-47-8	E214	Quantum
	Methyl-PHB	99-76-3	E218	satis
	Propyl-PHB	94-13-3	E216	(surface)
Natamicin	Pimaricin	7681-93-8	E235	<1 mg/dm2

*Figures in parentheses are recommended maximum added concentrations.

i.e., at about 5.0, and benzoates are more active at lower pH values (Calvo 1991). Therefore, sorbates are more appropriate for their use in most dry/fermented sausages. The antibiotic natamicin is also active within the usual pH range of these sausages. Because all these preservatives are mainly used on the sausage surface (this is compulsory in EU regulation), they are not added to the mass mix; they are applied by immersion in—or nebulization with—a concentrated preservative solution.

MULTIPURPOSE ADDITIVES: PHOSPHATES

Some additives are not easy to classify according to their possible uses and benefits as additives in meat products; this is the case with phosphates. They may have activity as thickeners, stabilizers, humectants, gelling agents, or even bulking agents; in fact, their function includes all of these features. Accordingly, they may be qualified as multipurpose additives. Phosphates are considered convenient additives by many industrials, not only for the manufacturing of cooked meat products, in which they perform a valuable enhancing effect on water-holding capacity, but also in dry/fermented sausage manufacturing. Why is this? The answer is easily understandable: Addition of phosphates results in a better binding of meat and fat particles, a more equilibrated drying, a longer shelf life, a smoother texture, and a higher juiciness. So, processing and product quality benefits appear to be doubtless. Then, this is the question: Are they necessary? The answer is not so easily understandable, but it is simple: No, they are not necessary. All those apparent benefits may be achieved without adding phosphates by carefully optimizing every phase of the manufacturing process. In conclusion, working without phosphates is hard work; but the ingredients label will be free of an extra additive.

The number of available phosphate molecules is high. All are inorganic salts of single, double, triple, or multiple phosphate anion groups. They can be classified in mono (orto-), di- (pyro-), tri- and polyphosphates. All of them have the ability to bind strongly to the outer parts of proteins, causing a sort of "swelling" of the protein matrix, which results in an enhancement of the beneficial actions of meat

Table 8.8. Multipurpose phosphates: CAS (Chemical Abstract Service) and EU numbers, and common concentration range added to sausage formulations.

Additive	Molecule/Form	CAS Number	EU Number	Common Concentration Range
Monophosphates	Sodium salts (mono, di and tri)	7558-80-7 7601-54-9	E339 (i, ii, iii)	<5 g/kg usually ca.1 g/kg (total phosphate)
	Potassium salts (mono, di and tri)	7778-77-0 7778-53-2	E340 (i, ii, iii)	
	Calcium salts (mono, di and tri)	7757-93-9	E341 (i, ii, iii) E343 (i, ii)	
	Magnesium salts (mono and di)	7757-86-0		
Diphosphates	Sodium salts (di, tri and tetra)	7758-16-9 7722-88-5	E450 (i-vii)	<5 g/kg usually ca.1 g/kg (total phosphate)
	Potassium salt (tetra)	7320-34-5		
	Calcium salts (mono and di)	14866-19-4 35405-51-7		
Triphosphates	Sodium salt (penta)	7758-29-4	E451 (i, ii)	<5 g/kg usually ca.1 g/kg (total phosphate)
	Potassium salt (penta)	13845-36-8		
Polyphosphates	Sodium salt	68915-31-1	E452 (i-iv)	<5 g/kg usually ca.1 g/kg (total phosphate)
	Potassium salt	7790-53-6		
	Calcium salt			
	Sodium and calcium salt			

proteins on sausage evolution and properties. Naturally, the different phosphate types confer different properties to the protein matrix, depending upon the degree of polymerization and the type of salt. Besides this, all of them act as buffering agents, showing a trend to increasing pH values to different degrees.

Table 8.8 gives the CAS (JECFA 2006) and EU (EPC 1995) numbers of phosphates most commonly used, as well as the common concentration range added to sausage formulations. It has been said previously that the diverse phosphates interact differently with meat proteins; this means that final effects on sausage properties also will be different. In summary, though highly simplifying, the higher the degree of polymerization, the higher the additive "multipurpose" effect and the better the results. Of course, the effect of phosphates is concentration-dependent. As a consequence, polyphosphates are much more valued than monophosphates, with tri- and diphosphates in intermediate ranking. Nevertheless, it has been demonstrated that the mixture of two or more kind of phosphates has synergistic effects (Klettner 2001). Therefore, industrials tend to use mixtures containing poly- and any of tri-, di- or monophosphates.

The concentration of added phosphates is normally limited by most regulations. As an example, the EU regulations establish a maximum residual concentration of 5 mg/kg in any meat product. This seems to be a sufficient amount to obtain the benefits of phosphates addition, but it must not be overlooked that meat itself contains a large amount of phosphates. In fact, the concentration of inorganic phosphate may pile up to circa 4 mg/kg, which proceeds from the postmortem degradation of organic phosphates ATP, ADP, and IMP, among others. Consequently, industrials have a narrow concentration range at free disposal. In any case, relatively low concentrations of about 1 mg/kg, particularly in the form of mixtures of poly- and other phosphates, are enough to bring about the desired beneficial effects on dry/fermented sausages. In conclusion, industrials may decide the use of phosphates or not as a function of their technological needs and the display of its presence in the product label. No shortcomings have been reported related to the use of phosphates as a food additive, despite some consumer groups resistance to its use.

REFERENCES

M Ash, I Ash. 2002. Handbook of Food Additives. Endicott, New York: Synapse Information Resources Inc.

D Balev, T Vulkova, S Dragoev, M Zlatanov, S Bahtchevanska. 2005. A comparative study on the effect of some antioxidants on the lipid and pigment oxidation in dry-fermented sausages. Intern J Food Sci Technol 40:977–983.

M Calvo. 1991. Aditivos Alimentarios; Propiedades, Aplicaciones y Efectos Sobre la Salud. Zaragoza: Mira Editores.

R Cassens. 1990. Nitrite-cured Meat; A Food Safety Issue in Perspective. Trumbull, Connecticut: Food and Nutrition Press.

EPC (European Parliament and Council). 1994. Directive No 94/36/EC, of 30/06/94, on colours for use in foodstuffs. Official Journal L237, of 10/09/94, pp. 13–29.

———. 1995. Directive No 95/2/EC, of 20/02/95, on food additives other than colours and sweeteners. Official Journal L061, of 18/03/95, pp. 1–40.

JECFA (Joint FAO/WHO Expert Committee on Food Additives). 2006. Compendium of Food Additives Specifications. FAO. http://www.fao.org/ag/agn/jecfa/.

PG Klettner. 2001. Effect of phosphates on technological parameters by dry fermented sausage. Fleischwirtshaft 81 (10):95–98.

S Martín-Bejarano. 2001. Enciclopedia de la Carne y de los Productos Cárnicos. Plasencia: Martín & Macías.

L Martínez, I Cilla, JA Beltrán, P Roncalés. 2006a. Effect of *Capsicum annuum* (red sweet and cayenne) and *Piper nigrum* (black and white) pepper powders on the shelf-life of fresh pork sausages packaged in modified atmosphere. J Food Sci 71:48–53.

———. 2006b. Comparative effect of red yeast rice (*Monascus purpureus*), red beet root (*Beta vulgaris*) and betanin (E-162) on colour and consumer acceptability of fresh pork sausages packaged in modified atmosphere. J Sci Food Agric 86:500–508.

MD Ranken. 2000. Handbook of Meat Product Technology. Oxford: Blackwell.

M Saltmarsh. 2000. Essential Guide to Food Additives. Leatherhead: Leatherhead Publishing.

A Sánchez-Escalante, D Djenane, G Torrescano, JA Beltrán, P Roncalés. 2003. Antioxidant action of borage, rosemary, oregano and ascorbic acid in beef patties packaged in modified atmosphere. J Food Sci 68:339–344.

LA Shelef. 1994. Antimicrobial effect of lactates; A review. J Food Protect 57:445–450.

A Stiebing. 1998. Influence of proteins on the ripening of fermented sausages. Fleischwirtshaft 78: 1140–1143.

F Toldrá. 2002. Dry-cured meat products. Ames, Iowa: Blackwell Publishing.

9
Spices and Seasonings

Suey-Ping Chi and Yun-Chu Wu

INTRODUCTION

The terms *spices* and *herbs* are usually used together to describe a group of aromatic plant parts, e.g., bark (cinnamon), buds (cloves), flowers (saffron), leaves (bay leaf, sage), fruit (allspice, chilies), bulbs (garlic, onion), roots (ginger), or seeds (caraway, mustard). In general, spices are highly aromatic due to their high contents of essential oils, whereas herbs are low in essential oils and are usually used to produce delicate or subtle flavors in food preparation in contrast to the aromatic flavors imparted by spices.

Spices and herbs are not clearly differentiated because various countries have differences of opinion on their definitions. Even within the same country, different agencies and trade organizations have different definitions for them:

- The American Spice Trade Association define spices in very broad terms—dried plant products used primarily to season food.
- The USDA defines the term *spice* as any aromatic vegetable substance in the whole, broken, or ground forms, with the exception of onions, garlic, and celery, whose primary function in food is seasoning rather than nutritional and from which no portion of any volatile oil or other flavoring principle has been removed.
- The USFDA also has a broad definition for spices with one important exception: it excludes dehydrated vegetables such as onions, garlic powder, and celery powder from the spice list. According to the FDA's definition, a *spice* is *any aromatic vegetable substance in the whole, broken, or ground form that is used to season food rather than to contribute nutrients*. This definition also requires spices to be true to the name and unmodified so that no volatile oil or other flavoring principle has been removed. FDA regulations state that spices may be labeled as *spices*; however, color-contributing spices—paprika, turmeric and saffron—must be declared as *spice and coloring* or by their common names. Essential oils, oleoresins, and their natural plant extractives containing flavor constituents may be declared as *natural flavor*. Dehydrated vegetables must be declared by their usual or common names.

This chapter discusses (a) ethnic preferences of spices and herbs (collectively called *spices* below), (b) botanical properties of spices, (c) product forms and appearance of spices, (d) sensory properties of spices, (e) chemical properties of spices, and (f) application of spices in fermented meats and poultry with emphasis on meat and poultry products.

ETHNIC PREFERENCES OF SPICES

Spices and herbs have been used by different cultures for centuries to season their foods. Originally, they used the spices and herbs available in their own regions. Later, spices and herbs from other regions penetrated into other cultures. Only until the last two to three centuries, with the frequency of international trade and preference, have spices and herbs available in one culture become easily accessible to another culture. Historically, Europeans sailed out to the "Indies" searching for tropical spices, with Columbus "discovering" America as the best known example. However, various cultures still maintain their own cultural heritage with typical preferences for spices and herbs, even though some of these spices are not native to their region. The following are some of the spices preferred by selected cultures (Giese 1994b; Lewis 1984; Moore 1979):

- *Chinese*: Anise, cinnamon, clove, garlic, onion, red pepper, star anise
- *East Indian*: Black pepper, coriander,
- *French*: Black pepper, marjoram, thyme
- *German*: Ginger, nutmeg, onion, paprika, white pepper
- *Greek*: Black pepper, cinnamon, garlic, onion, thyme
- *Hungarian*: Cinnamon, dill seeds, garlic, onion, paprika, white pepper
- *Indonesia*: Cinnamon, clove, curry, garlic, ginger, red pepper, cardamom, celery, cinnamon, clove, nutmeg
- *Italian:* Anise, black pepper, garlic, marjoram, onion, oregano, red pepper, sage, thyme
- *Jamaican*: Allspice, garlic, red pepper, thyme, onion
- *Mexican*: Allspice, anise, bay leave, chili pepper, cinnamon, cloves, coriander, garlic, onion, cumin seeds, red pepper
- *Moroccan*: Anise, cinnamon, coriander, cumin seeds, red pepper
- Southwestern U.S.: Chili pepper, cinnamon, clove, cumin, garlic, onion
- *Spanish*: Cinnamon, cumin seeds, garlic, hot pepper, nutmeg, onion, paprika, sweet pepper, thyme
- *Spanish Caribbean*: Black pepper, cinnamon, cloves, cumin seed, garlic, ginger, nutmeg, onion, paprika, thyme
- *Thai*: Black, white, and red peppers; cardamom; cinnamon; cloves; coriander root; curry paste (red and green); garlic; ginger
- *Vietnamese*: Chilies; black, white and red peppers; clove; curry powder; five spice powder; star anise.

This cultural heritage offers typical flavors for food products of individual ethic backgrounds.

COMMONLY USED SPICES IN PROCESSED MEATS

In general, about 50 different spices are commonly used in the food industry (Charalambous 1994; Farrell 1990; Hirasa and Takemasa 1998; Lewis 1984; Tainter and Grenis 2001; Uhl 2000; Underrinder and Hume 1994). About 25 to 35 of these spices are used in the processing of meats and poultry. Table 9.1 lists the names of these spices and their applications in meat processing.

BOTANICAL PROPERTIES OF SPICES

Origins of Various Spices

The spices used in processing meats originate from many countries in Asia, Europe, Africa, and the Americas (Table 9.2). They have spread all over the world and are now cultivated in their original countries as well as in other countries that can produce them in large quantities for international trade.

Botanical Classification of Spices

Spices we use now come from various plant families with growth characteristics that include evergreens, annuals, biannuals, or perennials. Plant parts may be berries, seeds, leaves, inner bark, fruits, fruit pericarp, fruit pods, unopened flower buds, bulbs, rhizomes, aril of seed, or flowing tops. Table 9.3 is a summary of the botanical characteristics of common spices used in processing meat.

PRODUCT FORMS AND APPEARANCES OF SPICES

Spices used in processing meat and poultry may exist in various forms depending on the product to be processed. They may be whole seeds, whole ground, whole pods, leaf flakes, slices, chopped, granules, or powder. They can also be applied in the form of spice oleoresin, essential oil, encapsulated oil or oleoresin, or essential oil plus oleoresin (Anonymous 1981; Charalambous 1994; Deline 1985; Farrell 1990; Giese 1994a; Hirasa and Takemasa 1998; Lewis 1984; Pszzola 1999; Tainter and Grenis 2001; Uhr 2000, Underrinder and Hume 1994). Table 9.4 is a summary of product forms and appearance of some common spices used in processing meats and poultry.

Table 9.1. Spices commonly used in processed meats.

Spice	Use
Allspice	Bologna, pickled pigs feet, head cheese
Anise seed	Dry sausage, mortadella, pepperoni
Bay (laurel) leaves	Pickle for pigs feet, lamb, pork tongue
Cardamom	Frankfurters, liver sausage, head cheese
Cassia	Bologna, blood sausage
Celery seed	Pork sausage
Cinnamon	Bologna, head sausage
Clove	Bologna, head cheese, liver sausage
Coriander seed	Frankfurters, bologna, Polish sausage, luncheon specialties
Garlic	Pork sausage, many types of smoked sausage
Ginger	Pork sausage, frankfurters, corned beef
Mace	Veal sausage, liver sausage, frankfurters, bologna
Marjoram	Liver sausage, Polish sausage, head cheese
Mustard	Good in all sausage
Nutmeg	Veal sausage, bologna, frankfurters, liver sausage, head cheese
Onion	Liver sausage, head cheese, baked loaf
Paprika	Frankfurters, Mexican sausage, dry sausage
Pepper (black)	Bologna, Polish sausage, head cheese
Pepper (Cayenne or red)	Frankfurters, bologna, veal sausage, smoked sausage, smoked country sausage
Pepper (white)	Good in all sausage
Pimento (pimiento)	Loaves
Sage	Pork sausage, baked loaf
Savory	Good in all sausages
Thyme	Good in all sausages

Source: Coggins (2001); Forrest et al. (1975); Ockerman (1989).

CHEMICAL PROPERTIES OF SPICES

MAJOR CONSTITUENTS

Each spice has its own major flavor constituent(s). Usually it exists in its essential oil and in some cases also in its oleoresin. Together these constituents provide the spice's typical flavor. Table 9.5 summaries the principal constituents in essential oils and significant constituents in oleoresins of various spices used in processed meats and poultry.

ANTIOXIDATIVE EFFECTS

It is now known that some spices have antioxidative effects. Some spices are more effective than the others. The effectiveness also depends on the medium in which they are tested. For example, in lard medium, sage is remarkably effective and mace, nutmeg, and thyme are effective; in oil-in-water medium, clove is strongest, followed by allspice, cassia, cinnamon, ginger, sage, and savory being strong, and anise seed, cardamom, marjoram, and pepper (black and white) showing some effectiveness

Table 9.2. Origin and current major sources of some common spices.

Spice	Places of Origin	Current Major Sources
Allspice	Western Hemisphere	Caribbean and Central America
Anise seed	Middle East	Turkey, Egypt, Syria, Spain, China, Mexico
Bay leaf	Mediterranean countries	Greece, Turkey
Caraway seed	Europe and Western Asia	Canada, The Netherlands, Egypt, Poland, Denmark
Cardamom	Southern India and Sri Lanka (Ceylon) Guatemala	Guatemala, India, Sri Lanka
Cinnamon	China	China, Indonesia, Vietnam
Clove	Madagascar, Zanzibar, Pemba	Madagascar, Indonesia, Brazil, Zanzibar, Sri Lanka
Coriander	Southern Europe, Asia Minor, Southwest Russia	Canada, Mexico, Morocco, Rumania, Argentina
Cumin seed	Egypt and Mediterranean region	Middle Eastern countries, India, Pakistan, Switzerland
Dill seed	Mediterranean region	Canada, India, Indonesia
Dill weed	Mediterranean region	Canada, India, Indonesia
Garlic	Central Asia	U.S., China
Ginger, dehydrated	China	India, Jamaica, Nigeria, China
Mace	Moluccas Islands and East Indian Archipelago	East Indies, Indonesia, West Indies
Marjoram, sweet	Mediterranean region and Western Asia	Egypt, France,
Mustard seed	Middle East	Canada, U.S., U.K., Denmark
Nutmeg	Moluccas Islands and East Indian Archipelago	East Indies, Indonesia, West Indies
Onion	West or Central Asia	U.S.
Paprika	Western hemisphere	U.S., Spain, Hungary, Morocco
Parsley	Mediterranean region	U.S., Israel, Hungary
Pepper, black	Malabar Coast of Southern India	Indonesia, India, Brazil, Malaysia
Pepper, red	Western hemisphere	Africa, China, India, Pakistan, Mexico
Pepper, white	Malabar Coast of Southern India	Indonesia, India, Brazil, Malaysia
Sage	Mediterranean region	Albania, Croatia, Germany, Italy Macedonia, Turkey
Star anise	China	China
Savory, sweet summer	Southern Europe and Mediterranean region	Yugoslavia, France
Thyme	Mediterranean region	Spain, France

Source: Charalambous (1994); Curry and Nip (2005); Farrell (1990); Lewis (1984); Tainter and Grenis (2001); Uhl (2000); Underrinder and Hume (1994).

Table 9.3. Botanical characteristics of some common spices used in processed meats.

Spice	Scientific Name	Plant Family	Growth Characteristics	Plant Part
Allspice	*Pimenta doica*	Myrtle	Evergreen	Berries
Anise seed	*Pimpineela anisum*	Parsley	Annual	Seeds
Basil	*Ochimum basilicum*	Mint	Annual	Leaves
Bay leaf	*Laurus nobilis*	Laurel	Evergreen	Leaves
Cassia	*Cinnamonum cassia*	Laurel	Evergreen	Inner bark
Cardamom	*Elettraia cardamomum*	Ginger	Perennial	Seeds
Celery seed	*Apium graveolens*	Parsley	Biannual or annual	Fruits
Cinnamon	*Cinnamonum zelancium*	Laurel	Evergreen	Inner bark
Clove	*Syzgium aromaticum*	Myrtle	Evergreen	Unopened flower buds
Coriander	*Coriandrum sativum*	Parsley	Annual	Fruits
Cumin seed	*Cumium cyminum*	Parsley	Annual	Ripe fruits
Garlic, dehydrated	*Allium sativum*	Onion	Perennial	Bulbs
Ginger, dehydrated	*Zingiber officinale,* Roscoe	Ginger	Perennial	Rhizomes
Mace	*Myristica fragans*	Nutmeg	Evergreen	Aril of seed
Marjoram, sweet	*Origanum majorana*	Mint	Perennial	Leaves
Mustard seed, black	*Brassica hira B. alba*	Cabbage	Annual	Seeds
Mustard seed, brown	*Brassica juncea*	Cabbage	Annual	Seeds
Mustard, white	*Brassica nigra*	Cabbage	Annual	Seeds
Nutmeg	*Myristica fragans*	Nutmeg	Evergreen	Seeds
Onion, dehydrated	*Allium cepa*	Onion	Biannual	Bulbs
Oregano	*Origanum* spp.	Mint	Perennial	Leaves
Paprika (pepper, red)	*Capsicum annum*	Nightshade	Annual or perennial	Fruit Pericarp
Parsley	*Petrosalinum crispum*	Parsley	Biannual	Leaves
Pepper, black	*Piper nigum*	Pepper	Perennial	Unripened berries or peppercorns
Pepper, chili	*Capsicum minimum*	Pepper	Annual or Perennial	Fruit
Pepper, red (Capsicum)	*Capsicum fructescens*	Nightshade	Perennial	Fruit pods
Pepper, white	*Piper nigum*	Pepper	Perennial	Skinless, ripe berries
Sage	*Salvia officinalis*	Mint	Perennial	Leaves
Savory, sweet summer	*Satureja indicum*	Mint	Annual	Leaves
Star anise, Chinese	*Illicum vercum*	Illiciaceae	Perennial	Fruit
Thyme	*Thymus vulgaricus*	Mint	Perennial	Leaves and flowering tops

Underrinder and Hume (1994).

Table 9.4. Product forms and appearance of some common spices used in processed meats.

Spice	Product Forms (Dried)	Product Appearance (Dried)
Allspice	Whole and ground	Dark reddish-brown pea-sized fruit
Anise seed	Whole and ground	Greyish brown, oval-shaped, and about 3/16 in long
Bay leaf	Whole and ground	Deep green leaves up to 3 in long
Caraway seed	Whole only	Brown, hard seeds of 3/16 in long
Cardamom	Whole pods (bleached and green), decorticated (seed alone), and ground	Irregularly round pod with seeds about 3/32 in long
Celery seed	Light brown seeds up to 1/16 in long	Whole, ground, and ground mixed with table salt (celery salt)
Celery flakes	Flakes, granulated, powdered	Light green dehydrated celery leaves
Cinnamon	Whole and ground	Brown bark
Clove	Whole and ground	Brown nail-shaped, 1/2 to 3/4 in long
Coriander	Whole and ground	Tan to light brown, globular with vertical ridges up to 3/16 in diameter
Cumin seed	Whole and ground	Yellow-brown seeds of 1/8 to 1/4 in long
Dill, seed	Whole and ground	Light brown, oval-shaped, up to 1/16 in long
Garlic, dehydrated	Large sliced, chopped, minced, ground, granulated, powder; also powder or granulated with table salt (garlic salt)	Off-white
Ginger, dehydrated	Whole, cracked or ground	Even, light buff hands or fingers
Mace	Whole or ground	Yellow-orange aril or skin
Marjoram, sweet	Whole (bits of leaves) and ground	Grey-green leaves
Mustard seed, black	Whole, mustard flour, and ground	Black seeds
Mustard seed, brown	Whole, mustard flour, and ground	Brown seeds
Mustard, white	Whole, mustard flour, and ground	White seeds
Nutmeg	Whole, ground	Greyish-brown, oval fruit, up to 1-1/4 in long
Onion, dehydrated	Sliced, chopped, diced, minced, ground, granulated, powdered, toasted (all sizes), powder, or granulated onion with table salt (onion salt)	Off-white
Paprika	Powder	Bright rich red stemless pods
Pepper, black	Whole, ground, cracked, decorticated	Black seeds
Pepper, red	Whole and ground	Various-sized red-colored pods
Pepper, white	Whole and ground	White seeds
Sage	Whole, cut, rubbed, and ground	Silver-grey leaves
Savory, sweet summer	Whole (bits of leaves) and ground	Brown-green leaves less than 3/8 in long
Thyme	Whole (bits of leaves) and ground	Brownish-green leaves, seldom exceed 1/4 in long

Source: Charalambous (1994); Curry and Nip (2005); Farrell (1990); Lewis (1984); Tainter and Grenis (2001); Uhl (2000); Underrinder and Hume (1994).

Table 9.5. Principal constituents in some common spices used in processed meats.

Spice	% Essential Oil (Range)	Principal Constituents in Essential Oil	Significant Constituents in Oleoresin
Allspice	3.3 to 4.5	Eugenol (60–75%), eugenol methyl ether, cineole, phellandrene, caryophylene	None
Anise seed	1.5 to 3.5	Anethole (major), anisaldehyde, anisketone, methyl chavicol	None
Bay leaf	1.5 to 2.5	Cineole	None
Cardamom	2 to 10	Cineole, *alpha*-terpinyl acetate, liminone, linalyl linalool, borneol, *alpha*-terpineol, *alpha*-pinene, limonene, myracene	None
Celery seed	1.5 to 3	*d*-Limonene, *beta*-selinene	None
Cinnamon	1.5 to 3.0	Eugenol (clovelike flavor), cinnamic aldehyde (cinnamonlike flavor)	None
Clove	Up to 20%	Eugenol, eugenol acetate, *beta*-caryophyllene	None
Coriander	0.1 to 1.5	*d*-Linalool (coriandrol), Pinenes, terpinenes, geranol, vorneal, decylaldehyde	None
Cumin seed	2 to 5	Cumaldehyde (cumic aldehyde), dihydrocumaldehyde, ciminyl alcohol, *dl*-pinene, *p*-cumene, diterpene	None
Ginger	??	Gingerols (pungency)	Zingerone, shogoals
Ginger, dehydrated	1.5 to 3.0	(−)-*alpha*-zingiberene, (−)-*beta*-bisabolene, (+)-ar-curcumene (ginger flavor), farnesene, *beta*-sequiphellandrene (ginger flavor), *alpha*-terpineol (lemony flavor), citral (lemony flavor)	None
Mace	15 to 25	*Alpha*-pienene, *Beta*-pinene, sabinene, myristicin (main flavor)	None
Marjoram, sweet	0.7 to 3	Terpen-4-ol, *alpha*-terpineol	None
Mustard seed, black	0.6 (av.)	Allyisothiocyanate	None
Mustard seed, yellow	None	None	Sinabin mustard oil (pungent)
Nutmeg	6 to 15	*Alpha*-pienene, *Beta*-pinene, sabinene, myristicin (main flavor)	None
Onion, dehydrated	Trace	Ethyl and propyl disulfides, vinyl sulfide	None
Paprika	None	None	Capsanthin, carotene, capsorubin (pigments)
Pepper, black	0.6 to 2.6	*alpha*-pinene, *beta*-pinene, 1-*alpha*-pellandrene, *beta*-caryophyllene, limonene, sabine-delta-3-carene pungency:piperine, piperettine, peperyline, piperolein A and B, piperanine	None

(*Continues*)

Table 9.5. (*Continued*)

Spice	% Essential Oil (Range)	Principal Constituents in Essential Oil	Significant Constituents in Oleoresin
Pepper, red		Capsaicin, Dihydrocapsaicin	None
Pepper, white	1.0 to 3.0	*alpha*-pinene, *beta*-pinene, 1-*alpha*-pellandrene, *beta*-caryophyllene, limonene, sabine-delta-3-carene pungency: piperine, piperettine, peperyline, piperolein A and B, piperanine	None
Sage	1.5 to 2.5	Thujone, borneol, cineole	None
Savory, sweet summer	<1	Carvacrol, thymol, *p*-cymene	None
Sesame seed	N/A	Pyrazines (in toasted seeds)	Sesamin, sesamolin
Star anise	5 to 8	Anethol	None
Thyme	0.8 to 2.0	Thymol, *p*-cymene, *d*-linalool	None

Source: Charalambous (1994); Deline (1985); Farrell (1990); Lewis (1984); Pszzola (1999); Tainter and Grenis (2001); Uhl (2000); Underrinder and Hume (1994).

(Balententine et al. 1989; Charalambous 1994; Duke 1994; Farrell 1990; Mandsen et al. 1997; Mansour and Khalil 2000; Nakatani 1994; Tainter and Grenis 2001; Uhl 2000; Underrinder and Hume 1994). These findings may explain why some of the processed meats and poultry are fairly stable during storage without development of rancidity.

Antibacterial Effects

It is also now know that some spices provide antibacterial properties. Table 9.6 summarizes some examples in which bacteria are being inhibited by various spices. This may explain why processed meats and poultry containing various spices are fairly safe, provided they are properly packaged and stored.

Naturally Occurring Toxicants

Some spices have been known for a long time to show adverse effects if consumed in large doses or when they are exposed to consumers for an extended period. It is now known that some constituents in some spices may be estrogenic, carcinogenic, goitrogenic, mutagenic, and hallucinogenic (Ames 1983; Conner 1993; Dalby 2000; Farrell 1990; Hirasa and Takemasa 1998; Miloslav 1983; Tainter and Grenis 2001; Uhl 2000). Consumers should not be scared from consuming the very minute amounts of spices present in processed meats and poultry, but processors should be careful in handling large amount of the following spices:

- *Anise volatile oil*: Anthethole (estrogenic activity)
- *Black pepper*: Safrole (carcinogenic activity), Piperine (carcinogenic activity)
- *Black pepper + nitrate*: *n*-Nitropiperine (carcinogenic activity)
- *Mustard (brown)*: Sinalbin, then allylisocyanate (lacinmater, mutagenic in Ames test, carcinogenic in rats)
- *Mustard (white)*: Glucosinolates (Goitrogenic and antithroid activity), Sinalbin, then allylisocyanate (lacinmater, mutagenic in Ames test, carcinogenic in rats)
- *Paprika + nitrate*: *n*-Nitrosopyrrolidine (carcinogenic activity)
- *Nutmeg*: Myristicin (Hallucinogen)
- *Sage*: Evidence of estrogenic activity.

SPICE QUALITY STANDARDS

Chemical Tests Used in Quality Control

Spices used in the food industry come from various parts of the world with diversified quality standards.

Table 9.6. Common foodborne bacterial pathogens inhibited by selected spices used in meat processing.

Spice	*Staphylococcus aureus*	*Clostridium botulinum*	*Clostridium perfrigens*	*Salmonella enterica*	*E. coli*	*Camphobacter jejuni*
Allspice		X			X	
Bay leaves	X	X			X	
Cinnamon	X				X	
Cloves	X	X	X	X	X	
Cumin	X	X		X		
Dill	X					
Fennel	X			X		
Garlic	X		X	X	X	X
Lemongrass	X			X	X	
Mint					X	
Onion	X				X	
Oregano	X	X			X	
Rosemary	X	X				
Sage	X	X				
Tarragon	X				X	
Thyme	X				X	

Source: Beuchat (1994); Beuchat and Golden (1989); Billing and Sherman (1998); CAST (1998); Curry and Nip (2005); Davidson et al. (1983); Deans and Richie (1987); Dorman and Deans (2000); Giese (1994a); Jay and Rivers (1984); Kim et al. (1995); Nakatani (1994); Shelef (1984); Smith-Palmer et al. (1998); Synder (1997); Zaika (1988).

This is expected because the spices are produced in various locations in the world. Their active constituents are not the same due to different growing conditions and handling practices. However, some measures can be set up to provide some standards for producers and users to follow. For example, standard chemical tests can be used to determine the potency of some spices. The following are some of the standard tests that can be conducted for spices commonly used in processing meats and poultry (Charalambous 1994; Farrell 1990; Hirasa and Takemasa 1998; Lewis 1984; Tainter and Grenis 2001; Uhl 2000; Underrinder and Hume 1994):

- *Allspice*: Total acid, acid-soluble ash, water-soluble ash
- *Capsicum*: Scoville heat units
- *Capsicum oleoresin*: Capsaicin
- *Nutmeg or mace*: Phenol
- *Paprika*: Carotenoids
- *Paprika, defatted*: Sample structure
- *Pepper (Black)*: Piperine
- *Residual solvent tolerance*: Various solvents.

Sanitation Criteria

Most spices are produced in developing countries with a poor sanitation environment. It is not uncommon to have insect infestation and bacterial contamination in spices arriving at the port of entry. USFDA has Defective Action Levels set up for various spices, their probable courses, and their significance. Readers interested in this topic should visit the USFDA's website at www.fda.gov for detailed information.

Spices are also easily contaminated with bacteria in the production phase. Several measures are available to reduce bacterial counts in spices. Each procedure has its own advantages and disadvantages. They are summarized as follows (Charalambous 1994; Farrell 1990; Hirasa and Takemasa 1998; Leistritz

Table 9.7. Flavor characteristics and commercial applications of some common spices used in processed meats.

Spice	Flavor Characteristics	Commercial Applications
Allspice (pimento)	Warm, sweet, and slightly peppery, reminiscent of clove	In almost all meats and sausages
Clove	Very aromatic, warm and astringent	In hams, tongue, and pork products
Coriander	Mild and pleasantly aromatic	In meat stuffings
Cumin seed	Salty sweet, similar but coarser than that of caraway, bitter aftertaste	Basic ingredient of processed meats
Garlic, dehydrated	Strong and characteristic	With care in most types of products
Ginger, dehydrated	Aromatic, hot, and biting	In most meat products
Mace	Finer than, but similar to, that of nutmeg	In processed meat loaf and sausages
Marjoram, sweet	Distinctive, delicately sweet and spicy with a slightly bitter back note	In most meat dishes, soups, and stuffings
Nutmeg	Sweet and aromatic adding a certain richness to spice mixes	In many meat products
Onion, dehydrated	Characteristic and penetrating	In many meat dishes requiring an onion flavor
Paprika (red pepper)	Piquant and moderately pungent	In all foods to give added zest
Paprika, rosen	Characteristic and stronger than normal paprika with a definite pungency	In many specialty meat products
Pepper	Penetrating, strong, and characteristic	In almost all dishes, a small quantity of pepper improves the overall flavor
Pepper, chile	Intense pungent	Used with discretion in any product requiring heat
Pepper, red	Hot and pungent	Used with discretion in any product requiring heat
Sage, Dalmatian	Strongly aromatic, warm with slightly bitter aftertaste	With care in all pork dishes and poultry stuffings
Sage, English	Milder and much more pleasant flavor than that of Dalmatian sage, lacks the objectionable "thujone" note	With care in all pork dishes and poultry stuffings
Savory, sweet summer	Strongly aromatic	In mixtures for sausages and patés
Star anise	Warm, sweet, aromatic	In meat dishes, also as a component for Chinese Five Spices
Thyme	Penetratingly warm and pungent, the aroma is long lasting	In meat stuffings and many meat dishes

Source: Charalambous (1994); Curry and Nip (2005); Farrell (1990); Hirasa and Takemasa (1998); Geise (1994b); Lewis (1984); Moore (1979); Tainter and Grenis (2001); Underrinder and Hume (1994); Uhl (2000).

1997; Lewis 1984; Tainter and Grenis 2001; Uhl 2000; Underrinder and Hume 1994.):

- *Ethylene oxide*: Proven method for whole spices and seeds; safety of using ethylene oxide is a concern; long treatment time; banned in some countries
- *Irradiation*: Cost effective; overall effectiveness; can process packed products; consumer concerns on irradiated products; not universally accepted
- *Dry steam*: Highly effective; more expensive than irradiation process, but less expensive than ethylene oxide and wet steam process; absence of residual fumigant
- *Wet steam*: Absence of residual fumigant; loss of volatile oils; lost of color in some spices; development of unaccepted flavor.

SENSORY PROPERTIES OF SPICES

The major purpose of adding various spices to meats in their preparation or processing is to provide distinct flavor(s) to the final products. Each spice has its own sensory characteristic(s), and the same flavor characteristic may also be provided by several spices, allowing choices of spices in the formulation. Adding spices to processed meats also provides other minor purposes, such as texture and masking/deodorizing undesirable flavors. Basic flavoring attributes provided by various spices are summarized as follows (Charalambous 1994; Deline 1985; Farrell 1990; Giese 1994b; Hirasa and Takemasa 1998; Lewis 1984; Moore 1979; Tainter and Grenis 2001; Uhl 2000; Underrinder and Hume 1994):

- *Bitter*: Bay leaf, celery, clove, mace, thyme
- *Cooling*: Anise
- *Earthy*: Black cumin
- *Floral aromatic*: Black pepper, ginger
- *Fruity*: Coriander root, star anise
- *Herbaceous*: Sage
- *Hot*: Chili pepper, mustard, black pepper, white pepper, wasabi
- *Nutty*: Mustard seeds
- *Piney*: Thyme
- *Pungent*: Garlic, ginger, onion, wasabi
- *Spicy*: Clove, cumin, coriander, ginger
- *Sulfury*: garlic, onion
- *Sweet*: Allspice, anise, cardamom, cinnamon, star anise
- *Woody*: Cassia, cardamom, clove.

Garlic, mustard, and onion also provide texture to the final product. Bay leaf, clove, coriander, garlic, onion, sage, and thyme can make or deodorize some undesirable flavor. Paprika is an oil-soluble colorant.

Table 9.7 is a summary of flavor characteristics and commercial applications of various spices in processing meats and poultry.

It should be noted that sensory attributes recognized in one culture may be different in another culture. For example, cinnamon is considered as woody and sweet in the western culture, the same spice is considered bitter in Indian culture and pungent and sweet in Chinese culture. Table 9.8 presents several examples showing discrepancies in sensory attributes in various cultures.

APPLICATIONS OF SPICES IN FERMENTED MEAT PROCESSING

Components in Spice Blends

Spice blends are available in the market for the convenience of consumers or processors. The composition of each blend generally contains some "standard" components that have been well recognized. It should be noted that producers usually have their proprietary formulations for their products, even though the spices may or may not be the same. Table 9.9 summarizes some published information on spice blend components for groups of processed meats.

Amount of Spices in Various Fermented Sausages and Meats

Individual fermented sausage and meat processors usually have their own formulations to attract their customers. They either blend their own spices or have the spice companies provide proprietary formulations for them. There are also published formulations available. Table 9.10 shows some published formulations for some common fermented sausages and meats.

CONCLUSION

The blending of various spices with different sensory attributes provides the characteristic flavor for each product. It is this characteristic flavor that consumers enjoy and that provides momentum for meat processors to come up with new formulations to keep their business moving. Formulation of spice blends is still an art that challenges the processors and spice providers.

Table 9.8. Sensory differences of some common spices between cultures.

Spice	Basic Sensory Characteristics	Indian	Chinese
Cinnamon	Woody, sweet	Bitter	Pungent, sweet
Clove	Bitter	Bitter	Pungent
Coriander	Spicy	Sweet	Pungent
Cumin seed	Spicy	Sweet	Pungent
Garlic, dehydrated	Pungent, spicy	Spicy/pungent	Pungent
Ginger, dehydrated	Spicy	Spicy/pungent	Pungent
Pepper, black	Hot	Spicy/pungent	Pungent
Pepper, chili	Hot	Spicy/pungent	Pungent

Source: Charalambous (1994); Caurry and Nip (2005); Deline (1985); Farrell (1990); Giese (1994b); Hirasa and Takemasa (1998); Lewis (1984); Lu (1986); Moore (1979); Tainter and Grenis (2001); Uhl (2000); Underrinder and Hume (1994).

Table 9.9. Components in some common spice mixtures or blends for processed meats.

Name of Mixture (Blend)	Components
Bologna or wiener seasoning	Pimento, nutmeg, black pepper, clove leaf or stem, capsicum, paprika, coriander, cassia, mustard, onion, garlic
Five spices, Chinese	Black pepper, star anise or anise, fennel, clove, cinnamon
Ham seasoning	Cassia, clove stem or leaf, pimento, capsicum, celery, bay leaf,garlic
Italian sausage seasoning	Anise, fennel, paprika, black pepper, red pepper
Italian seasoning	Oregano, basil, red pepper, rosemary, garlic (optional)
Japanese seven spices	Perilla, basil, sesame, poppy seed, hemp, red pepper, Japanese pepper
Mixed pickling spice	Mustard seeds, bay leaves, black and white peppercorns, red peppers, ginger, cinnamon, mace, allspice, coriander seeds, etc.
Pork sausage seasoning	Sage, black pepper, red pepper, nutmeg
Poultry seasoning	Sage, thyme, marjoram, savory, rosemary (optional)
Roast beef or corn beef rub	Coriander, onion, garlic, celery, oregano, basil, black pepper, paprika
Smoked sausage seasoning	Garlic powder, black pepper, coriander, pimento

Source: Charalambous (1994); Curry and Nip (2005); Farrell (1990); Giese (1994b); Hirasa and Takemasa (1998); Lewis (1984); Lu (1986); Moore (1979); Pszzola (1999); Tainter and Grenis (2001); Uhl (2000); Underrinder and Hume (1994).

Table 9.10. Amount of spices used in selected fermented sausages and smoked meats.

Sausage or Meat	Amount of Spice Used per 100 lbs of Raw Meat	Sausage or Meat	Amount of Spice Used per 100 lbs of Raw Meat
Semidry or Summer Sausage		**Dry Sausage**	
Thuringer #1	Ground black pepper, 4 oz	Genoa Salami	White ground pepper, 3 oz
	Whole mustard seed, 2 oz		Whole white pepper, 1 oz
	Coriander, 1 oz	B. C. Salami (Hard)	White pepper, 3 oz
Thuringer #2	Caraway, 0.75 oz		Garlic pepper, 0.25 oz
	Celery seed, 0.75 oz	Pepperoni #1	Sweet paprika, 12 oz
	Coriander, 1 oz		Decorticated ground pepper, 6 oz
	Ginger, 1oz		Capsaicin, 4 oz
	Mace, 1 oz		Whole fennel seed, 4 oz
	Pepper, black, 6 oz	Pepperoni #2	Allspice, 2 oz
Cerevelat #1	Ground black pepper, 4 oz		Anise seed, 5 oz
	Whole black pepper, 2 oz		Pepper, red, 2 oz
Cerevelat #2	Cardamom, 1 oz	**Ham**	
	Coriander, 2.8 oz	Deviled ham	Mustard seed, ground, 7.5 oz
	Nutmeg, 1 oz		Paprika, 0.3 oz
	Pepper, white, 7.5 oz		Pepper, red, 0.4 oz
Lebanon bologna #1	Mustard, 8 oz		Pepper, white, 0.1 oz
	White pepper, 2 oz		Tumeric, 0.4 oz
	Ginger, 1 oz	Prosciutto	All spice, 8 oz
	Mace, 1 oz		White pepper 5 oz
Lebanon bologna #2	Allspice, 2 oz		Black pepper, 2 oz
	Cinnamon, 1 oz		Nutmeg, 2 oz
	Cloves, 1 oz		Mustard seed, 0.5 oz
	Nutmeg, 2 oz		Coriander, 0.5 oz
	Pepper, black, 5 oz	**Beef**	
	Pepper, white, 1 oz	Corned beef	Celery seed, 0.5
Chorizos #1	Sweet red pepper, 6 oz		Onion powder, 2 oz
	Chili powder, 6 oz		Pepper, black, 2 oz
	Red pepper (hot), 2 oz	Pastrami	
Chorizos #2	Cardamom, 0.4 oz	Curing solution	Garlic juice, 1.5
	Coriander, 0.4 oz	Rub #1	Black pepper and coriander
	Garlic powder, 0.1 oz	Rub #2	Paprika
	Ginger, 0.6 oz	Rub #3	65% black pepper and 35% all spice
	Oregano, 0.1 oz	Rub #4	60% coriander, 25% all spice, 15% white pepper, and garlic flour
	Paprika, 2.5 oz		
	Pepper, black, 1.6 oz		

Source: Adapted from Kramlich et al. (1973); Ockerman (1989).

REFERENCES

BN Ames. 1983. Dietary carcinogens and anticarcinogens. Science 21:1256–1264.

Anonymous. 1981. The Oleoresin Handbook. 3rd ed. New York: Fritzsche Dodge & Olcott, Inc.

DA Balententine, MC Albano, MG Nair. 1999. Role of medicinal plants, herbs, and spices in protecting human health. Nutr Rev 57(9):S41–S45.

LR Beuchat. 1994. Antimicrobial properties in spices and their essential oils. In: VM Dillon, RG Board, eds. Natural Antimicrobial Systems and Food

Preservation. Wallingford, U.K.: CAB International, pp. 167–179.
LR Beuchat, DA Golden. 1989. Antimicrobials occurring naturally in food. Food Technol 43(1):134–142.
J Billing, PW Sherman. 1998. Antimicrobial functions of spices: Why some like it hot. Quart Rev Biol 73(1):3–49.
CAST. 1998. Naturally Occurring Antimicrobials in Food. Ames (Iowa): Council for Agricultural Science and Technology.
G Charalambous. 1994. Spices, Herbs, and Edible Fungi. New York: Elsevier.
PC Coggins. 2001. Spices and flavorings for meat and meat products. In: Yh Hui, WK Nip, RW Rogers, OA Young, eds. Meat Science and Applications. New York: Marel Dekker, Inc., pp. 371–401.
DE Conner. 1993. Naturally occurring compounds. In: PM Davidson, AL Bransen, eds. Antimicrobials in Foods. New York: Marcel Dekker, Inc., pp. 441–468.
JC Curry, WK Nip. 2005. Spices and herbs. In: YH Hui, JD Culbertson, S Duncan, I Guerrero-Legarreta, ECY Li-Chan, CY Ma, CH Manley, TA McMeekin, WK Nip, LML Nollar, MS Rahman, F Toldrá, YL Xiong, eds. Handbook of Food Science, Technology and Engineering, Vol. 2. Boca Raton, Florida: Taylor & Francis Group (CRC), pp. 89–1 to 89–28.
A Dalby. 2000. Dangerous Tastes—The Story of Spices. Berkeley, California: University of California Press.
PM Davidson, LS Post, AL Branen, AR McCurd. 1983. Naturally occurring and miscellaneous food antimicrobials. In: AL Branen, PM Davidson, eds. Antimicrobials in Food. New York: Marcel Dekker, Inc., pp. 371–419.
SG Deans, G Richie. 1987. Antibacterial properties of plant essential oils. Intl J Food Microbiol. 5:165–180.
GD Deline. 1985. Modern spice alternatives. Cereal Foods World 30(10):697–698, 700.
HJD Dorman, SG Deans. 2000. Antimicrobial agents from plants: Antibacterial activity of plant essential oils. J Appl Microbiol 88:308–316.
JA Duke. 1994. Biologically active compounds in important spices. In: Spices, Herbs, and Edible Fungi. G Charalambous, ed. New York: Elsevier, pp. 225–250.
KT Farrell. 1990. Spices, Condiments, and Seasonings. New York: Van Nostrand Reinhold.
JC Forrest, ED Aberle, HB Hederick, MD Judge, RA Merkel. 1975. Principles of Meat Science. San Francisco: Freeman.
J Giese. 1994a. Antimicrobials: Assuring food safety. Food Technol. 48(6):101–110, 1994.
———. 1994b. Spices and seasoning blends: A taste for all seasons. Food Technol 48(4):87–90, 92, 94–95, 98.
K Hirasa, KM Takemasa. 1998. Spice Science and Technology. New York: Marcel Dekker, Inc.
JM Jay, GM Rivers. 1984. Antimicrobial activity of some food flavoring compounds. J Food Saf 6:129–139.
J Kim, MR Marshall, C Wei. 1995. Antibacterial activities of some essential oil components against five foodborne pathogens. J Agric Food Chem 43: 2839–2845.
WE Kramlich, AM Pearson, FW Tauber. 1973. Processed Meats. Westport: AVI.
W Leistritz. 1997. Methods of bacterial reduction in spices. In: SJ Risch, and CH Ho, eds. Spices: Flavor Chemistry and Antioxidant Properties. Washington, D.C.: American Chemical Society, pp. 7–10.
YS Lewis. 1984. Spices and Herbs for the Food Industry. Orpington, England: Food Trade Press.
HC Lu. 1986. Chinese System of Food Cures—Prevention and Remedies. New York: Sterling Publishing Co., Inc.
HL Mandsen, G Beretelsen, LH Skibsted. 1997. Antioxidative activity of spices and spice extracts. In: SJ Risch, and CH Ho, eds. Spices: Flavor Chemistry and Antioxidant Properties. Washington, D.C.: American Chemical Society, pp. 176–187.
EH Mansour, AH Khalil. 2000. Evaluation of antioxidant activity of some plant extracts and their application to ground beef patties. Food Chem 69:15–141.
R Miloslav, Jr., ed. 1983. CRC Handbook of Naturally Occurring Food Toxicants. Boca Raton, Florida: CRC Press.
K Moore. 1979. The correct spices—Keys to ethnic products. Food Prod Dev May:18–23.
N Nakatani. 1994. Antioxidant and antimicrobial constituents of herbs and spices. In: Spices, Herbs, and Edible Fungi. G. Charalombous, ed. New York: Elsevier, pp. 251–271.
HW Ockerman. 1989. Sausage and Processed Meat Formulations. New York: Van Nostrand Reinhold.
D Pszzola. 1999. Engineering ingredients: Believe it or not. Food Technol 53(7):98–105.
LA Shelef. 1984. Antimicrobial effect of spices. J Food Saf 6:29–44.
A Smith-Palmer, J Stewart, L Fyfe. 1998. Antimicrobial properties of plant essential oils and essences against five important foodborne pathogens. Letters in Appl Microbiol 26:118–122.
OP Synder. 1997. Antimicrobial Effects of Spices and Herbs. St. Paul (Minnesota): Hospitality Institute of Technology and Management.
DR Tainter, AT Grenis. 2001. Spices and Seasonings—A Food Technology Handbook. 2nd ed. New York: Wiley-VCH.
R Uhl. 2000. Handbook of Spices, Seasonings, and Flavorings. Lancaster, Pennsylvania: Technomic Pub.
EW Underrinder, IR Hume, eds. 1994. Handbook of Industrial Seasonings. London: Blackie Academic & Professional.
LL Zaika. 1988. Spices and herbs: Their antimicrobial activity and its determination. J Food Saf 9:97–117.

10
Casings

Yun-Chu Wu and Suey-Ping Chi

INTRODUCTION

The word *sausage* originally came from the Latin word *salsus*, which means salted or preserved. In those days, people did not have ways to preserve their meat, so making sausage was one of the ways to overcome this problem.

Dry sausage was born as a result of the discovery of new spices, which helped enhance flavor and preserve the meat. Different countries and different cities within those countries started producing their own distinctive types of sausage, both fresh and dry. These different types of sausage were mostly influenced by the availability of ingredients as well as the climate.

Basically, people living in particular areas developed their own types of sausage and these sausages became associated with this particular area. For example, bologna originated in the town of Bologna in Northern Italy, Lyons sausage from Lyons in France, and Berliner sausage from Berlin in Germany. Casings were first used to contain and form chopped meat products. Ground and comminuted meat often require some type of casing to hold the product intact while it is being processed, precooked, distributed, and finally cooked. Today, casings are used not only to contain and form, but also to flavor, season, and protect products. Consumers around the world are so used to sausages being packed into some type of casings that a variety of casings can now be found in marketed products. Today there are numerous types of sausage casings, including natural and artificial such as collagen, cellulose, and plastic. Collagen, cellulose, and plastic casings are relative newcomers to the artificial field, mainly born out of market demand during the early twentieth century (Anonymous 2004; Forrest et al. 1975; INSCA 1987; Lawrie 1974; Ockerman 1996; Pearson and Tauber 1984; Pearson and Dutson 1988).

TYPES OF CASINGS

Natural Casings

Natural casings are made from the submucosa, largely a collagen layer of the intestine. The fat and the inner mucosa lining are removed. Because small intestines are collagen in nature, they have many of the same characteristics common to all types of collagen, particularly the unique characteristic of variable permeability.

Natural casings are hardened and rendered less permeable through drying and smoking processes. Moisture and heat make casings more porous and tend to soften them, which explains why smoking, cooking, and relative humidity must be carefully controlled.

Animal casings have the advantage of contributing to the old-world appearance. They often do not have to be peeled prior to consumption, and have many of the physical and chemical properties that make them very desirable containers for sausage items; but they are usually the most expensive type of casings. Therefore, animal casings are usually used for the more expensive sausage items, such as fermented sausages. Areas of the animal used to manufacture casings are the small and large intestines, weasand, urinary bladder, stomach, and rectum from most meat-producing animals. The utilization of these products varies tremendously from country to country and for different sausage types in which they are going to be utilized. This also indicates the areas that are removed from the intestinal tract in order to transfer the product into usable casings (Table 10.1).

As received from the slaughter floor, the intestinal tract always is highly microbiologically contaminated. The cleaning and removal of various internal, and sometimes external layers are necessary to convert this product into a useful casing. Many factors influence the quality of the casing, such as health of the animal, species, age of the animal, breed, fodder consumed, conditions under which the animal was raised, portion of the intestinal tract utilized, and how the product is handled and processed after the animal is slaughtered. Some determinants for evaluating casings include

- *Cleanliness*: Casings should be clean, free from stains, odor, fat particles, parasites, nodules, and ulcers, and should be sound and free from pin holes.
- *Strength*: Casings should be strong enough to withstand the pressures that will be put on them during filling, stuffing, cooking, processing, storage, and consumer cooking. Only the submucosal part of the intestine has the required strength to make this possible.
- *Length*: The number of pieces per hank or bundle often varies according to the country in which the casings were collected. The country in which the product is going to be utilized also influences the desirable length. On the average, sheep and hog casings are 299.8 ft (91.4 m), beef rounds are packaged in bundles of 54 ft (18 m), and tennis rackets and surgical strings in lengths of 9.24 ft (6 m).
- *Caliper*: Diameter of the casing desired is determined by the country of use and also according to type of sausage ingredient that is going to be placed in the casing. Modern sausage processing equipment requires, however, that the caliper of the casing be uniform in order to have adequate machinability. The greatest demand for sheep casing is in the 0.78 to 0.95 in (20 to 24 mm) diameter range. In some cases, small caliber hog casings can be substituted for the larger sheep casings. The greatest demand for hog casings is in the 1.38 in (35 mm) and over range. Beef rounds are normally 3.09 in (33 mm) and over, and middles are 2.17 in (55 mm) and over. Oversizes are in greatest demand.
- *Curing*: Casings are normally cleaned, and then salted, and only in a few cases are they dried. Also, in some cases, some of them may be frozen, but this primarily is utilized for surgical cat gut. Some countries, even in this area, use salted products. The curing should be accomplished with high-quality, fresh, and small-particle–sized salt grades.
- *Packaging*: The type of packaging is either wooden or plastic containers or tins. Often these containers are lined with plastic bags. Sometimes plastic bags are utilized and protected by sacks. However, the trend seems to be in favor of plastic containers because of cost and ease of sanitation.

Removal of the Viscera

The first step in casing preparation (Table 10.2) is removing the viscera and separating them from the internal organs. This is primarily a hand and/or knife operation. The viscera are placed on a table and

Table 10.1. Composition of natural casings.

Natural casings are acquired from the alimentary tract of meat animals. The construction of the intestine is composed of five layers, which from inside to outside, are shown, along with the layers that are remaining after it is processed into a casing.

Hog and sheep casing	1. Mucosa glands aid in secretion, digestion, and absorption
Beef casing	2. Submucosa—collagen and often fat
	3. Circular muscle
	4. Longitudinal muscle
	5. Serosa—collagen, elastin

Source: Adapted from Ockerman (1996).

Table 10.2. Steps for the conversion of the digestive tract into sausage casing.

1. Remove the intestine from the animal.
2. Running—remove loose mesentery fat.
3. First stripping—squeeze to force out intestinal contents.
4. Wash and cool (50°F water).
5. Mechanically strip or hand-strip.
6. Use brushes to remove fat.
7. Slimming—remove tissue layers—revolving drums and warm (115°F) water or hand-slimming.
8. Use strippers or hand-remove appropriate tissue layers.
9. Store overnight in ice and 15–20% saturated salt solution.
10. Grade—species, size, quality.
11. Cure—rub with salt; allow to set for 1 week.
12. Remove from cure, shake free of excess salt, rub with fine salt, and pack (40% salt).
13. Flush prior to use.

Source: Adapted from Ockerman (1996).

separated from the mesentery fat. Again, this is a hand operation that can be facilitated by using an air-operated knife. The puller usually will start at the stomach and pull the casing away from the ruffle fat.

The next step in casing manufacturing is to run the casing through a manure stripper to squeeze out the liquid and manure using large rollers similar to a laundry ringer. These rollers are usually rubber and sometimes are wrapped with burlap. This operation can also be done by hand and by pulling the casing through the fingers. In either case, a great quantity of potable water is needed to wash the casings and to keep the operation clean. The casing should be soaked for approximately 30 minutes in 100–108°F (38–42°C) water.

In some areas of the world, casings now go through a fermentation cycle, but in many areas, processing casings by fermentation is no longer legal (e.g., U.S.). If fermentation is to be conducted, the casings usually are soaked overnight in 72°F (22°C) water or until the mucosa and muscular coatings are loosened. Excessive fermentation will soften the casings. If fermentation is used, the casings usually are stripped after fermentation, soaked, restripped, and often restripped again. Next, they are run through a cleaning machine (which is often a drum with a revolving scraper blade).

If the casing is not fermented, the next step is to run it through a crushing machine and soaking tank. The purpose of this is to break the intermucosal membrane and separate it from the casing. This machine also has two adjustable rollers with eccentric bearings. Again, a great quantity of 108°F (42°C) water is used in this operation. Crushing also can be accomplished by a hand operation wherein the casing is scrapped with a dull-bladed knife or an oyster shell. The casing then goes through a mucosa stripper, which looks essentially like the manure stripper. Again, 108°F (42°C) water is used to keep the operation sanitary. If labor is very economical, this operation also can be accomplished by hand scraping.

Next, the casing goes through a finishing machine to remove any stringlike material and remaining mucosa. Rollers again are used in this operation. Finishing also can be accomplished by hand if a great deal of attention is paid to detail. In either system, large quantities of 108°F (42°C) potable water are essential to keep the operation clean.

After finishing, the casings are soaked again in 50–60°F (10–16°C) water and/or a salt brine tank to remove excess blood. This tank normally has continuously flowing water to remove the stained liquid. If continuously running water is used, salt normally is not used at this stage. The soaking time usually ranges from 30 minutes to overnight.

After the soaking operation, casings usually are salted either by hand or again by machine. The salting and shaking of the casing usually are continued until the casings absorb 40% salt, at which point they are packed into a container. If the casings are packed in a slush container, the container will hold 10–15% salt brine. The advantage of using a dry pack is that the casings become less tangled. Also, the casings will be darker in color.

Items such as hog bung, hog stomach, blind end or cecum, bladder, beef bung, and weasand usually are handled by hand, trimmed of excess material, and salted again.

If attention is paid to detail, a large quantity of cool water is used to keep the casings clean. When labor is inexpensive and strict sanitation is maintained, suitable casings can be produced by hand operation and will make very desirable casings for the sausage industry. However, a breakdown in any one of these processes will result in an unsanitary product that has little use in modern sausage processing.

Table 10.2 summarizes the various steps in the manufacture of natural casings.

Casing names and locations, as well as sausage products for which they are utilized and yield per animal can be found in Table 10.3. Table 10.4 is a summary of how to use sausage casings in the meat packing industry.

If all steps of the process are performed satisfactorily, a very gourmet sausage product can be produced by utilizing value-added and upgraded animal casings which, in their native state, have almost no value but can be transformed to a very desirable container for high-quality sausage products.

Generally speaking, the natural-casing markets include hog stomachs; bungs (hog, beef, and lamb cecum); beef bladders; hog, sheep, and lamb casings (small intestine); beef rounds (small intestine of cattle); beef middles (large intestine of cattle); and hog middles (large intestine in hogs). On the average, the small casings used in breakfast sausages are edible, but the larger casings such as bungs, bladders, and stomachs, are generally not considered as edible.

The benefits of natural casings are the remarkably beautiful, smoked, and cooked appearance. They take on a beautiful color when they're smoked. Natural casings add to the nutritional value of the product because they are a protein-based material. They tend to be flavor-neutral, but they certainly do not create off-flavors in the sausage product. Natural casings allow products to be placed in upscale markets primarily due to the eye appeal of the sausage.

Natural casings are expensive. Unit price of casing per tonnage output of sausage would be at a manufacturing cost disadvantage when compared to similar products manufactured in regenerated collagen and cellulose casing. Handling of natural casings in preparation for sausage manufacture is more

Table 10.3. Classifications of natural casings used for meat purposes.

Natural Casing	Location	Appearance and Comments	Sausage	Average Yield/Animal
Beef				
Round	Small intestine	Ringlike, tougher, easily handled, less breakage	Ring bologna, Polish sausage	90 to 135 ft long
Bung	Cecum		Capocolla, salami	4 to 5 ft long
Bladder	Bladder	Oval or molded	Minced specialty, Mortadella	7 to 14 in wide
Middles	Large intestine	Sewed, most expensive, adds uniformity	Bologna, salami	20 to 25 feet long
Weasand	Windpipe			18 to 26 in long
Hog				
Round	Small intestine	May be eaten or peeled	Large frankfurter, hog sausage	42 to 52 ft long
Bung	Cecum	May be sewed	Braunschweiger	30 to 72 in long
Middles	Large intestine	Curly	Chitterlings	12 to 16 ft long
Stomach	Stomach		Head cheese, souse	
Bladder	Bladder			5 to 9 in wide
Sheep				
	Small intestine	Most tender, most breakage	Small frankfurter, pork sausage	90 ft long

Source: Adapted from Ockerman (1996).

Table 10.4. How to use casings.

In using natural casing, there should be a drying period in the smoking cycle to obtain uniform smoke color, and great care should be exercised in cooking and smoking these products. Because these products are primarily collagen they behave in the following manner.
Dry: They become brittle and shrink; also, smoke will not penetrate. Some drying is necessary to reduce smoke penetration to keep from forming a skin under the casing.
Wet: Smoke goes through the casing to the sausage.
Acid (smoke): Firm, harder, and tougher.
Heat in presence of moisture: They become soft and hydrolyze slowly (sheep is the worst).
If these casings are handled properly, the meat product can stand more abuse (less fatting out) than with cellulose casings.

Source: Adapted from Ockerman (1996).

labor intensive and is a bit more delicate than the handling of collagen or cellulose casings. Natural casings must be rinsed because they have been packed and stored with salt—the salt must be washed out of them. It takes a little more finesse and artisan skill to handle natural casings for filling sausage product (Anonymous 2004; Dewied 2001; Naghaski and Rust 1987; Naghaski 1971; Ockerman 1988, 1996; Pearson and Dutson 1988).

Hog Casings

Hog casing is the most popular casing, which can actually be used for almost any sausage. It is used for the manufacture of cooked sausage, country-style sausage, fresh pork sausage, Pepperoni, Italian sausage, large frankfurters, Kishka, Kielbasa, and Bratwurst—to name just some of the best-selling items (Dewied 2001; INSCA 1987; Ockerman 1996; RAMP 2004; Sausagemaking 2004).

Hog casings are used commonly for manufacturing of Bratwurst, Saucisson (chitterlings), liver sausage (hog bungs), pork sausage, and Polish sausage. Typically, hog casings are used for fresh and smoked sausages, as well as breakfast links. Hog bungs are used for liver sausages and certain varieties of dry sausages. Hog middles are used for liver sausage and certain types of Italian salamis.

Common origins of hog casings include China, Europe, and the U.S.

Hog Bungs: Regular and Sewn, Hog Bungs, and Hog Bung Ends

Regular hog bungs are sold as individual pieces and are used primarily for liver sausage and Branschweiger. Sewn (or *Sewed*) hog bungs are produced in double-walled and single-walled varieties; they are made by sewing two or more pieces of smaller sizes of regular hog bungs together. To obtain a larger, more uniform finished product, these casings are custom-made and can be purchased in almost any shape or size suitable to the needs of the processors. Most of the products are used exclusively for liver sausage, Genoa or Thuringer, Braunschweiger, summer sausage, and Cervelats.

Hog Middles/Chitterlings

Hog middles/chitterlings are put up in three calibers: wide, medium, or narrow. The size is determined by the location of the item within the animal. There are normally nine to ten 1 m pieces to a bundle. Hog middles are easily recognizable by their curly appearance. Chitterlings are also available and selected into 5 mm calibers.

Beef and Ox Casings

Beef casings, such as rounds, middles and bungs are ideal for boiled, cooked, and cured sausages of highest quality. Beef bladders are used for specialty sausages and mortadellas. Beef rounds are used for bolognas, liver sausage, and certain types of fermented sausages. Beef middles are used for liver sausages and dry sausages. Beef bungs are used for specialty pork products, bolognas, and salamis. They are also used for sausages that require a very thick casing, such as bologna and salami.

The three most used beef casings are beef bung caps, beef rounds, and beef middles. Beef bung caps—are used for Capocolla, veal sausage, large bologna, and Lebanon and cooked salami.

Origins of beef bungs, rounds, and casings are South America, especially Brazil, and Europe.

Beef Rounds and Middles

These casings derive their name from their characteristic “ring” or “round” shape. Beef rounds are used for ring bologna, ring liver sausage, Mettwurst, Polish sausage, blood sausage, Kishka, and Holsteiner. Beef rounds and middles are mainly being used for Servelats, round Bologna, Fleischwurst, and Holsteiner Italian salami. Beef middles can be used for Leona style sausage, all other types of Bologna, dry and semidry Cervelats, dry and cooked salami and veal sausage.

Beef middles are measured in sets or bundles of 9 and 18 m (29–30 ft and 57–60 ft) each. Beef middles can be sewn so that they have a uniform diameter and a uniform length, with or without a hanger (stitching loop).

Beef Bladders

Beef bladders, the largest diameter casings from cattle, are oval in shape, and will hold from 2.5–6.5 kg (5–14 lb) of sausage. They are used chiefly for minced specialty sausage and mortadella, either in their natural oval form, in square molds for sandwich slices, or in the flat, pear-shaped styles.

Natural Ox Bung

Natural ox bungs are ideal for making haggis, mortadella, or any sausage or salami that requires a large casing.

Ox Runners

Ox runners are excellent for traditional salami.

Lamb and Sheep Casings

Sheep casings are very tender, and used in the processing of fresh and smoked sausages. Sheep casings are the highest-quality, small-diameter casings used for the finest in sausages, such as Bockwurst, frankfurters, and Port Sausage.

Origins of lamb and sheep casings are Turkey, Iran, China, Mongolia, Australia, New Zealand, Afghanistan, Pakistan, Chile, Peru, Egypt, and Syria.

Horse Casings

The high standards of Italian salami are best fulfilled by using horse casings.

Origins of horse casings are South America (mainly Brazil and Uruguay) and Europe (mainly France and Eastern Europe).

Glossary of Terms for Natural Casings

See Table 10.5 for a summary of terms for natural casings.

Table 10.5. Glossary of terms for casings.

Bundles	A measured unit of casings ready for sale in salted, preflushed, or tubed form; bundles will be either hog casings or sheep casings consisting of 91 m (100 yd). Bundles can also refer to a customer-defined specification.
Green Weights	Represents approximate stuffing capacity of casings before cooking or smoking, per 91 m lengths.
Hanks	Another essentially interchangeable term with the same meaning as *bundles*, applying to hog and sheep casings.
Nodules	Pimples that appear on some beef rounds or beef bung caps.
Sets	A unit of beef casings ready for sale in salted form, consisting of 18–30 m for beef rounds and 9–18 m for beef middles.
Shirred	Refers to the result of applying a casing to a dummy transfer horn or to a flexible plastic sheath to expedite the stuffing process.
Tierce	A shipping container made of plastic with a packing volume to 208 l (approximately 55 gal).
Windows	Damage to casings caused by overcrushing. They are approximately half the thickness of the casing.
Whiskers	The capillary that holds the intestine in the fat and provides a flow of blood to the intestine. When removing the intestine with a knife, the capillary is not completely removed, creating a hairlike appearance on the surface of the casing.

Source: Adapted from INSCA (1987); Rust (1988).

COLLAGEN CASING

Collagen casings are made from the gelatinous substance (collagen) found in the connective tissue, bones, and cartilage of all mammals. The substance is harvested from the animals and reconstructed in the form of a casing. Most sausage in the U.S. is stuffed into this kind of casing (Devro 2006; Hood 1987; Mid-Western Research & Supply 2006; RAMP 2004).

Sausages produced by this method are easier to pack than those in variable animal casings due to their uniform size and weight. This is advantageous in both retail and professional markets.

The collagen is derived from the middle corium layer of cattle hide, obtained from hides by the mechanical removal of the epidermal and flesh layers. After a pretreatment, the collagen-rich corium layer is referred to normally as the *limed collagen split*. The process involves an extensive washing, decalcification, and buffering process of the limed collagen split to give a purified food-grade collagen material. This is further reduced into a fibrous slurry, which is blended with a food-grade acid and vegetable cellulose fiber to produce a homogeneous collagen protein gel. The collagen fibers are arranged to give a crisscross structure that contributes significantly to the strength and caliber stability of the finished product.

The following are advantages of collagen sausage casings (Devro 2006):

- Elastic (shrinks with the product)
- Digestible (edible)
- Permeable to smoke and moisture strong
- Uniform in size
- Nonrefrigerated storage
- Consistent casing size equals reduced giveaway
- Fresh casings allow the "real bloom" of product to show through

Collagen casings can be used for any comminuted, coarse, or fine-chopped sausage products. There are very few sectioned-and-formed meats placed in collagen casings. Sausage products that are manufactured in regenerated collagen casings have the snap, bite, and attractiveness that natural casings bring to the product. However, there is often a production cost advantage of choosing natural versus collagen casings in ease of handling and cost of product.

There is another development in regenerated collagen casing products brought to the market by one casing manufacturer. This method uses the approach of a coextrusion of collagen dough with a sausage batter.

CELLULOSE CASING

This is an artificial casing made from solubilized cotton or wood pulp. It is very uniform, strong, and not quite as susceptible to bacteria as other types of casings. Skinless hotdogs are made with cellulose casings (Anonymous 2004; NOSB 2001; RAMP 2004; Viskase 2006).

Small-diameter cellulose casings are used to manufacture most skinless frankfurters in the United States. They probably have captured more than 90% of the small-diameter frankfurter market in this country.

Large cellulose casings are larger in diameter and thicker-walled cellulose casings. They are used to contain the deli-style bolognas. These products are typically about 2 in diameter, shaped nicely, glossy, and have an attractive appearance.

The third type of regenerated cellulose casing is the very large casing used to shape and hold boneless and semiboneless hams. These very large cellulose casings are thick-walled and will be sufficiently robust for the ham-filling process.

The fourth type of regenerated cellulose casing is the fibrous casing used to make dry and semidry sausage. They are cellulose impregnated paper. The fibrous casing is extremely strong and is used to stuff sausage that is very tightly packed because it will not break. The inside of this casing is coated with protein that allows it to shrink with the meat as it dries out. Fibrous casing is manufactured by bonding regenerated soluble cellulose (viscose) to a highly refined paper. Together they become very robust; they are resistant to stretching and breakage during the sausage filling process. Fibrous casings are manufactured with uniform diameters.

The last type of regenerated cellulose casing is called *moisture-proof* (*MP*) casing. This casing is a fibrous casing with a barrier latex coating. It is used for sausage products that must be cooked or need an extensive shelf life. It is an expensive but effective product for water cooking and extending product shelf life. MP casings, however, are expensive. Nylon casing is also a barrier and less expensive. The difference is that MP casings are coated-fibrous casings, and they offer exact-diameter control. The nylon casing, although less expensive and a barrier, doesn't offer the same exact-diameter control.

The small-diameter cellulose casing is used to manufacture a variety of smoked sausages and some types of fresh sausages, such as chorizo or Italian sausage. The large cellulose casing would be used for bolognas and salamis. The very large cellulose casings are typically for boneless and semiboneless hams. The fibrous casing would be used for all the fermented sausages, large-diameter cooked sausages, and deli-style meats.

The small-diameter cellulose casing is manufactured and formed in shirred sticks, just as the small-diameter collagen casing. The difference is that the small-diameter cellulose is much stronger and elastic than collagen and is easier to use to manufacture products.

Productivity is a strength for cellulose casings, and the cost is lower than collagen casings. Cellulose is a moisture-sensitive membrane, and it will allow water-soluble material such as smoke to move across it to the surface of the sausage. Products manufactured in small-diameter cellulose casings are usually smoked and cooked, and the casings are removed before the product goes to market. The large-cellulose casings are used for deli bolognas, as well as boneless and bone-in hams.

Fibrous casings are strong, smoke permeable, and can be peeled easily after products have been cooked and cooled. All fibrous and cellulose casings must be removed before product is consumed. Spent casing is the one key downside of cellulose casings. The following are advantages of fibrous sausage casings (Viskase 2006):

- Excellent size uniformity, giving uniform product and increased yields
- Consistent casing color due to a controlled dying process
- Wide range of colors available
- Can be printed, clipped, string-tied, cut, and shirred
- Nonrefrigerated storage
- Some permeable to smoke and moisture
- Strength

SYNTHETIC CASING

This casing is made from alginates and requires no refrigeration. It is used by mass producers and can be made in different colors: red for bologna, clear for some salamis, and white for liverwurst. Much like the cellulose casing, it is uniform and strong.

PLASTIC/NYLON CASING

Nylon casing is a relatively inexpensive and attractive barrier casing. It allows a way to produce a high volume of items that need shelf life or products that must be water-cooked and then enter the market—and the product does not have to be repackaged.

There are both cost and high-productivity benefits. They are also delivered in shirred casing units with a first clip attached to begin the stuffing process. They are barrier casings, and they offer tremendous graphics for attractiveness to the package. They can be offered in various sizes. They typically range in sizes from 1–5 in diameter. The nylon casing's challenge has been the inability to be smoked because nylon/plastic casing is a barrier.

Table 10.6 elucidates the advantages and disadvantages of different casing types.

REGULATORY COMPLIANCE

In the U.S., all casing suppliers are under the jurisdiction of the Food and Drug Administration (FDA) because they are manufacturing a food container that is in direct contact with food. All components used in their casing materials must comply with Title 21 Code of Federal Regulation—whether it is a natural casing and the way a natural casing is manufactured; the collagen casings and the procedures approved for regenerated collagen, or the regenerated cellulose casing and procedures approved for regenerated cellulose; and all materials used to plasticize those casings to keep them flexible, supple, and damp.

The second step is complying with United States Department of Agriculture (USDA) regulations, because casings are used in USDA establishments. The casing manufacturers must supply to the establishment owner a Continuing Product Guarantee that meets the requirements of Title 7. It is different from an individual guarantee in that an individual guarantee is for a particular item with a particular specification. A Continuing Product Guarantee meets the requirement of a nonchanging item.

The USDA occasionally does a compositional review check. USDA offices will put a call out to casing suppliers for them to update their files to make sure that everything is in compliance. This happens every so often.

Once casings come into a USDA establishment, the establishment must comply with USDA regulations. Cellulose casings can not be eaten by the consumer and must be removed before consumption. The wording on the cellulose casing box must identify that the casings remaining on the product must be removed. Any collagen casings that cannot be eaten must be identified as inedible and removed. Those natural casings that should not be consumed must also be identified *To be removed*.

Any product that goes to the retail market cannot mislead or cause the end-user to assume that it is something different. In other words, if a processor chooses to send a smoked sausage to retail in a fibrous casing and the color of the casing is brown,

Table 10.6. Advantages and disadvantages of casing types.

	Natural	Collagen	Cellulose
Most expensive, cost/lb product	Most expensive	Less expensive	Least expensive
Refrigeration storage	Yes	Yes	No
Degree of tenderness	Most tender	Less tender	Peeled
Break during processing	Most likely	Less likely	Least likely
Casing preparation cost	Most expensive	None	None
Soaking and flushing before use	Yes	No	Sometimes soaking
Ease of smoke penetration	Most penetration	Less penetration	Least penetration
Best machinability	Least	Less	Best
Best product yield, per ft of casing	Least	Less	Best
Finished product uniformity	Least	Less	Best
Cost of casing removal	None	None	Most
Printability	None	Limited	Best
Old World appearance	Best	Less	None
Ease of plant storage	Least storage	Less storage	Best storage

Source: Adapted from Ockerman (1996).

the end-user might not necessarily associate that brown color with the casing—and may think that the brown color is the color of the product. That is considered misleading, which the USDA does not accept (Anonymous 2004; INSCA 1987).

HANDLING CASINGS

Today, casing recovery—most often done in large slaughtering facilities—is both a precise science and an elaborate process. It requires high-level expertise, state-of-the-art machinery, and maximum sanitation and quality control procedures. Because the intrinsic value of the raw material represents a large part of the finished casing product, every inch of tract needs to be utilized. In the slaughterhouse, the viscera of the animal are removed, and the various parts of the intestinal tract are separated. This separation of parts is instrumental in creating a variety of products ranging all the way from pig chitterlings to sheep appendixes for pharmaceutical products (Anonymous 2004; INSCA 1987; Ockerman 1996; Pearson and Dutson 1988).

The casings are prepared by removing the manure; mucosa (raw material for the anticoagulant heparin); and any undesirable elements such as fat, threads, and animal fluids. This removal, facilitated in a series of both hot- and cold-water soaks, is accomplished by machine crushing under close "hands-on/eyes-on" scrutiny. The fully cleaned casings are placed in a saturated salt environment to prepare for further processing. The casings are then sorted into various grades and diameters. The selection process is dictated by such factors as type of animal and criteria set by the casing processor, and ultimately the sausage producers.

DETERMINING QUALITY

Qualities of casing are determined in several precise and labor-intensive ways. In sheep, for example:

- *A Quality* casing is determined during selection, and is defined as a casing with no holes or weakness. This casing can be used for the finest frankfurter emulsion.
- *B Quality* casings are of acceptable strength and quality for coarse ground emulsions, such as those used in pork sausage. *Export Quality* is sometimes used. This term describes casings as free of nodules (pimples) or scores (windows).

With hog casings, there is a single quality standard with several specifications for length. Where the various hog casings originate from—taking into consideration factors of species, climate, and diet—will generally determine the different characteristics of the casings. Some will be white or virtually transparent/clear; others will be darker and more opaque and will have more visible veining. These characteristics also have an effect on the tenderness or bite of the casing.

Clear hog casings are generally used for fresh products. Thicker and stronger casings such, as Chinese hog casings, are generally best suited for smoked products, because these casings better withstand the smoking process and because casing appearance is not as critical a selling feature, due to the smoking process itself (INSCA 1987).

TEST PROCEDURES

The traditional methods for grading and testing natural casings are

- Water testing for sheep and hog casings.
- Air testing for beef casings.

The casings are appropriately filled with water or air and periodically expanded under pressure to check for size and quality. The casings are then cut to final sizes, and quality specifications are confirmed during quality control.

It is of utmost importance to adhere to the normal practice of filling product, and follow proper handling procedures of Good Manufacturing Practices (GMPs). As casings are removed from meat products, there is the potential for contamination or recontamination. Significant attention has been focused on minimizing recontamination at the point of casing removal of hotdogs and frankfurters in meat processing plants. It has been a real issue, but processors have made a great effort to deal with it.

There is also a risk in food preparation of sausages. For example, almost all hotdogs are fully cooked when they enter commercial trade. Many times the end-user will not reheat the hotdogs or will do a quick microwave. Any low-level contamination that could have occurred during packaging could create risk for the consumer eating this product (INSCA 1987; Mid-Western Research & Supply 2006; Naghaski and Rust 1987).

CONCLUSION

Natural casings give the best flavor and appearance to the final product. A natural casing enhances and complements the natural juices and quality of the meat and spices. Natural casings permit deep smoke penetration. However, artificial casings offer uniformity and cost savings. They also offer opportunities that are not available to natural casings. Both types of casings complement each other in the manufacturing of very acceptable products for the consumers.

REFERENCES

Anonymous. 2004. Casings 101. The National Provisioner. NP/2004/08, p. 26.

Devro. 2006. Collagen casings. http://www.devro.plc.uk/products/index.htm.

Dewied. 2001. Casing Products. http://.www.dewied.com/casing/htm.

JC Forrest, ED Aberle, HB Hederick, RA Merkel. 1975. Principles of Meat Science. San Francisco: Freeman.

LL Hood. 1987. Collagen in sausage casing. In: Advances in Meat Research. Vol. 4. New York: Elsevier.

INSCA. 1987. Facts about Natural Casings. International Natural Sausage Casings Association.

RA Lawrie. 1974. Meat Science. New York: Pergamon.

Mid-Western Research & Supply, 2006. Collagen sausage casings. http://www.midwesternresearch.com/CSAING_COLLAGEN.htm.

J Naghaski, RE Rust. 1987. Natural casings. In: JF Price, BS Schweigert, eds. The Science of Meat and Meat Products. Westport, Connecticut: Food and Nutrition Press.

J Naghaski. 1971. By-products—Hide, skins, and natural casings. In: JF Price, BS Schweigert, eds. The Science of Meat and Meat Products. Westport, Connecticut: Food and Nutrition Press.

NOSB. 2001. Cellulose processing. National Organic Standards Board Technical Panel Review.

HW Ockerman. 1996. Meat Chemistry. Columbus, OH: The Ohio State University.

HW Ockerman, CL Hansen. 1988. Animal By-product–processing. Ltd., Cambridge, U.K.: VCH Publisher.

AM Pearson, TR Dutson. 1988. Edible meat by-products. In: Advances in Meat Research. Vol. 5. New York: Elsevier.

AM Pearson, FW Tauber. 1984. Processed Meats. Westport: AVI.

RAMP. 2004. Natural and artificial casings. http://ramp.ch/English/Produkte_E.htm.

Sausagemaking. 2004. Casings. http://www.sausage-making.org/acatalog/Casings.html.

Viskase. 2006. Products/Applications. http://www.viskase.com/products.php.

Part III

Microbiology and Starter Cultures for Meat Fermentation

11
Microorganisms in Traditional Fermented Meats

Isabelle Lebert, Sabine Leroy, and Régine Talon

INTRODUCTION

In Europe, natural fermented sausages have a long tradition originating from Mediterranean countries during Roman times (Comi et al. 2005). Then the production spread to Germany, Hungary, and other countries including the United States, Argentina, and Australia (Demeyer 2004). Today's Europe is still the major producer and consumer of dry-fermented sausages (Talon et al. 2004).

There are a wide variety of dry-fermented products on the European market as a consequence of variations in the raw materials, formulations, and manufacturing processes, which come from the habits and customs of the different countries and regions. But from a global viewpoint, two categories can be distinguished: Northern and Mediterranean products with some specific characteristics. In Northern products, rapid acidification by lactic acid bacteria to a final pH below 5 and smoking ensure safety, improve shelf life, and contribute largely to the sensory quality. In Mediterranean products, acidification reaches a final pH above 5. Safety and shelf life are mainly ensured by drying and lowered water activity (Demeyer 2004). In both categories, the industrial development in the second half of the nineteenth century has led to using starter cultures to control sausage manufacturing. However, considering the recent abundant literature, it appears that in Mediterranean countries there are still a lot of traditional naturally fermented meats as outlined in this chapter. These traditional products are often known at local or regional levels. They are characterized by a great diversity in the methods of production leading to diversity in the size of the sausages and in their organoleptic characteristics. This diversity is observed between the countries but also between different regions in a country.

As a consequence, the type of microflora that develops in traditional sausages is closely related to the diversity in formulation, and to the fermentation and ripening practices. The knowledge and the control of this typical microflora during the processing are essential in terms of the microbiological quality, organoleptic characteristics, and food safety.

The objective of this paper is to review the diversity of manufacturing practices and microflora in traditional fermented sausages. The variety of lactic acid bacteria (LAB) and gram-positive catalase-positive cocci (GCC+), which are the two categories of bacteria that play a significant role in the final organoleptic and hygienic properties of the fermented sausages is also reported.

TRADITIONAL SAUSAGE MANUFACTURE

Traditional fermented sausage can be defined as a meat product made of a mixture of meat (often pork); pork fat and salt, including eventually sugar; nitrate; and/or nitrite. The meat to fat ratio is 2 to 1 in most industrial sausages mixes, whereas in traditional ones, the ratio is variable. The fat percentage could be very low (about 10%), or high (40%), or difficult to evaluate in mixes prepared with raw materials in which lean and fat are not separated (Fontana et al. 2005a; Lebert et al. 2007). The mixture is seasoned with various spices depending on the recipe. Contrary to industrial products inoculated by starter culture, traditional ones are frequently not inoculated, and fermentation relies on the indigenous microbiota. Backslopping, mentioned by only few authors (Leroy et al. 2006), seems to be a marginal practice for traditional producers.

The mixture is stuffed usually into natural casings. The resulting sausages are often processed in two consecutive and separate stages, referred to as *fermentation* and *ripening/drying*, respectively (Demeyer 2004). Greek sausages are often smoked (Drosinos et al. 2005; Papamanoli et al. 2003; Samelis et al. 1998).

Industrial manufacturers, to standardize their production, carry out the fermentation and the drying in rooms with well-controlled conditions of temperature, relative humidity (RH), and air velocity. Traditional producers carried out the process in less-controlled rooms for temperature and RH (Parente et al. 2001). Some of these producers dry their sausage in a natural dryer depending on local climatic conditions (Corbiere Morot-Bizot et al. 2006; Lebert et al. 2007).

Conditions of fermentation vary in terms of temperature and duration. When the fermentation is carried out at high temperature (between 18–24°C), usually it lasts between 1 and 2 days (Cocolin et al. 2001a; Comi et al. 2005; Mauriello et al. 2004). However, even at high temperature, the fermentation can last 7 days for Greek (Drosinos et al. 2005; Papamanoli et al. 2003; Samelis et al. 1998), Argentinean (Fontana et al. 2005a; Fontana et al. 2005b), and some Italian sausages (Cocolin et al. 2001b). Fermentation can be carried out at low temperature (10–12°C), and lasts approximately 1 week (Chevallier et al. 2006; Lebert et al. 2007; Mauriello et al. 2004).

Similarly, conditions of drying vary in terms of temperature and duration. The variation of temperature is less important than for fermentation; it ranges mainly between 10–14°C for French (Chevallier et al. 2006; Lebert et al. 2007), Italian (Cocolin et al. 2001a; Cocolin et al. 2001b; Comi et al. 2005; Coppola et al. 2000; Rantsiou et al. 2005b), and Greek sausages (Papamanoli et al. 2003). The duration of drying for these products varies from 4 to 12 weeks. Higher drying temperature between 15–19°C can be observed for Argentinean (Fontana et al. 2005a; Fontana et al. 2005b) or Greek sausages (Drosinos et al. 2005; Samelis et al. 1998); it is accompanied by short ripening from 1 to 3 weeks.

The relative humidity during the fermentation varies from a minimum of 63–75% to maximum values of 86–95% and during drying from 64–69% to 70–85% (from references in Tables 11.1a and 11.1b). This diversity leads to variable water content of traditional sausages at the end of drying. Thus, water activity varies from 0.83 to 0.93 in French, Spanish, Portuguese, and Italian sausages, and it was higher (approximately 0.95) for most of the Greek traditional sausages (Talon 2006).

DESCRIPTION OF ECOSYSTEMS

Technological Microflora

In industrial production, lactic acid bacteria (LAB) and *Staphylococcus* or *Kocuria* (GCC+) are usually used as starter cultures. They have been developed to improve the hygienic and sensory qualities of the final product (Talon and Leroy-Sétrin 2006; Talon et al. 2002).

In traditional sausages, LAB constitute the major microflora at the end of the ripening stage and their final levels are similar in the different sausages presented in Tables 11.1a and 11.1b. Even if their initial levels vary and are lower than that of industrial sausages inoculated, their final levels are close to the one of industrial manufacturers. LAB usually increase the very first days of fermentation (Comi et al. 2005), and they remain constant during ripening at about 7 to 9 log cfu/g (Cocolin et al. 2001a; Comi et al. 2005). Rantsiou et al. (2005b) observed that in one fermentation of naturally fermented northeast Italian sausages, LAB population increased more slowly. The initial population can be low, as observed in Salame Milano in Italy (Rebecchi et al. 1998) and in a French sausage manufactured at low temperature (Chevallier et al. 2006), but it is generally comprised of between 3.2 and 5.3 log cfu/g (Tables 11.1a and 11.1b). LAB growth is often correlated with the decrease in pH in the first stage of maturation (Cocolin et al. 2001a; Cocolin et al. 2001b). Because of the good adaptation of the LAB to the meat environment and their fastest growth rate during fermentation and ripening of sausages, they become the dominant microflora.

In traditional sausages, GCC+ constitute the second microflora at the end of ripening. In the majority of the sausages, GCC+ have a population of 6–8 log cfu/g, which is generally inferior to that of the LAB and close to the population reached in industrial sausages. GCC+ are poor competitors in the presence of active aciduric bacteria. Their initial level varies from 3.1 to 4.4 log cfu/g (Tables 11.1a and

Table 11.1a. Microbial ecosystems of traditional fermented sausages.

Country	Greece		Greece		Spain		France			
Reference	(1)		(2)		(3)		(4)			
Sausages			Dry Salami		Fuet	Chorizo	Product F01		Product F07	
Time of Process	Day 0	28 Days	Day 0	28 Days	Final	Final	Day 0	88 Days	Day 0	55 Days
TVC	5.3 ± 0.8	7.6 ± 1.1	6.1 ± 0.5	8.1 ± 0.1						
LAB	4.5 ± 1.0	7.8 ± 0.5	5.3 ± 0.7	8.0 ± 0.2	7.9 ± 0.7	8.5 ± 0.7	3.5	8.6	4.2	7.9
GCC+	4.4 ± 0.6	3.0 ± 1.4	4.4 ± 0.5	5.2 ± 0.7	7.0 ± 0.9	6.3 ± 0.9	3.5	6.6	4.2	6.3
Yeasts/molds	4.1 ± 0.6	2.8 ± 1.4	3.9 ± 0.3	4.2 ± 0.6			4.0	5.8	4.5	5.9
Enterococci	4.1 ± 1.2	5.1 ± 0.6	2.8 ± 0.5	5.0 ± 0.2	2.7 ± 1.0	2.6 ± 0.5	3.3	3.3	2.6	5.7
Pseudomonas	5.1 ± 0.8	<2	4.5 ± 1.0	<2			4.4	4.0	5.2	4.5
Enterobacteria	2.5 ± 0.3	<1	3.2 ± 1.0	<2	0.3 ± 0.6	0.8 ± 1.0	3.1	1.0	4.4	2.6
S. aureus	<2	<2	1.3 ± 1.6	0.7 ± 1.3	<2	<2	<2	<2	<2	<2
Clostridia	<1	<1	<1	<1	<1	0.3 ± 0.5				
Salmonella	0/9	0/9	abs	abs	abs	abs	abs	abs	abs	abs
L. monocytogenes	9/9	0/9	4/4	abs	1/10	2/7	abs	abs	abs	abs
pH	6.3 ± 0.1	4.9 ± 0.2	6.2 ± 0.1	5.1 ± 0.1	6.0 ± 0.2	5.5 ± 0.1	5.6	5.4	5.7	5.8

TVC: Total Viable Count; LAB: Lactic acid bacteria; GCC+: Gram-positive catalase-positive cocci; *Clostridia*: Sulfite reducing *Clostridia.* Data are expressed in log cfu/g except for *Salmonella* and *L. monocytogenes*: number of positive samples out of total samples or abs: absence in 25 g; ±: standard deviation; (1) Drosinos et al. (2005); (2) Samelis et al. (1998); (3) Aymerich et al. (2003); (4) Lebert et al. (2007).

Table 11.1b. Microbial ecosystems of traditional fermented sausages.

Country	Argentina		Italy							
Reference	(5)		(6)		(7)					
Sausages					Product C		Product L		Product U	
Day	Day 0	14 Days	Day 0	28 Days	Day 0	120 Days	Day 0	45 Days	Day 0	28 Days
TVC	5.3	7.2	5.4 ± 0.7	7.8 ± 1.2	4.5 ± 0.3	7.0 ± 0.0	4.7 ± 0.4	4.2 ± 0.2	5.1 ± 0.2	9.1 ± 0.5
LAB	4.1	7.1	4.6 ± 0.8	8.4 ± 0.0	3.2 ± 0.9	8.8 ± 0.2	5.3 ± 0.2	8.3 ± 0.1	4.2 ± 0.2	8.5 ± 0.1
GCC+	4.0	5.5	4.0 ± 0.7	5.2 ± 0.8	3.2 ± 0.1	5.2 ± 0.2	3.1 ± 0.2	4.6 ± 0.2	3.7 ± 0.1	5.0 ± 0.1
Yeasts/molds	3.9	4.7	2.7 ± 1.6	1.6 ± 1.4	2.0 ± 0.0	6.7 ± 0.1	3.6 ± 0.8	<2	3.0 ± 0.4	2.3 ± 0.3
Enterococci	2.4	5.5	3.9 ± 0.2	6.1 ± 0.2	2.1 ± 0.3	4.1 ± 0.2	3.2 ± 0.1	5.5 ± 0.5	3.7 ± 0.4	6.3 ± 0.3
Pseudomonas			1.5 ± 2.7	NP						
Enterobacteria	2.5	2.7	3.2 ± 0.9	3.8 ± 1.5	1.7 ± 0.2	<1	1.9 ± 0.2	2.6 ± 0.1	2.4 ± 0.1	4.7 ± 0.5
S. aureus	abs	abs	<2	<2	<2	<2	<2	<2	<2	<2
Clostridia										
Salmonella			abs	abs	abs	NP	abs	NP	abs	NP
L. monocytogenes			abs	abs	abs	NP	abs	NP	abs	NP
pH	5.9	5.2		5.7 ± 0.1	5.5	5.6	5.6	5.7	5.8	5.6

TVC: Total Viable Count; LAB: Lactic acid bacteria; GCC+: Gram-positive catalase-positive cocci; *Clostridia*: Sulfite reducing *Clostridia.* Data are expressed in log cfu/g except for *Salmonella* and *L. monocytogenes*: number of positive samples out of total samples or abs: absence or NP: not performed; ± : standard deviation; d.: days
(5) Fontana et al. (2005a); (6) Comi et al. (2005); (7) Rantsiou et al. (2005b).

11.1b). In a French sausage, their level was lower in winter (low temperature) than in spring (higher temperature) (Corbiere Morot-Bizot et al. 2006). GCC+ sometimes grew during the fermentation period to about 10^6–10^8 cfu/g (Mauriello et al. 2004) but they usually grew during ripening (Comi et al. 2005). An increase in the pH to 5.5–5.7 and sometimes 6.2 at the end of the ripening period (Chevallier et al. 2006) could be explained by the proteolytic activity of various microorganisms including the GCC+.

Other Microflora

Yeasts and Molds

Yeast and molds are usually detected in all sausages in the batter at levels varying from 2.0 to 4.5 log cfu/g (Tables 11.1a and 11.1b). Some authors observed slow growth during the ripening period, with levels not increasing above 5 log cfu/g (Drosinos et al. 2005), and then a decrease in the population (Comi et al. 2005). Yeasts and molds were not detected in some products at the end of ripening (Rebecchi et al. 1998).

Enterococci

Enteroccocci have an initial level between 2 and 4 log cfu/g (Tables 11.1a and 11.1b). Enterococci usually grow during early fermentation and remain constant at a level of 4–6 log until the end of the whole process (Comi et al. 2005; Drosinos et al. 2005; Rebecchi et al. 1998). In few cases their counts declined. Chevallier et al. (2006) observed that the counts were dependent of the season. In some French sausage, their levels reached 6 to 7.5 log cfu/g (Lebert et al. 2007). The persistence of enterococci during ripening can be attributed to their wide range of growth temperature and to their high tolerance to salt. Moreover, enterococci are poor acidifiers, and in traditional sausages of high pH they find good conditions for survival and growth (Hugas et al. 2003). There is still controversy over considering them as GRAS (Generally Recognized as Safe) microorganisms (Giraffa 2002). However, studies point out that meat enterococci, especially *Enterococcus faecium*, have a much lower pathogenicity potential than clinical strains, and some strains of *E. faecium* are already used as starter culture or probiotic (Hugas et al. 2003; Martin et al. 2005).

Spoilage Microflora

Spoilage bacteria such as *Pseudomonas* and enterobacteria have far different initial levels according to the type of sausages (Tables 11.1a and 11.1b). They ranged from 1.7 to 4.4 log cfu/g for enterobacteria and from 1.5 to 5.2 log cfu/g for *Pseudomonas*. In Greek sausages, they were progressively eliminated regardless of their initial population (Drosinos et al. 2005; Samelis et al. 1998). Other authors found that enterobacteria and *Pseudomonas* increased during the fermentation (Comi et al. 2005). Then enterobacteria remained constant until the end while *Pseudomonas* remained constant or disappeared (Chevallier et al. 2006; Comi et al. 2005; Lebert et al. 2007).

Pathogenic Microflora

Salmonella and sulphite-reducing clostridia were not detected in any traditional sausages.

Pathogenic staphylococci were detected only in some Italian and French sausages. In Salame Milano, *Staphylococcus aureus* was observed at the beginning of the production process and decreased during ripening until it was undetectable at the end of the process (Rebecchi et al. 1998). It was still enumerated in sausages produced in Italian artisanal plants in levels ranging from 2 to 4 log cfu/g (Blaiotta et al. 2004). In French sausages, *S. aureus* was present in sporadic cases, as, for example, presence in winter samples and not detected in spring (Chevallier et al. 2006).

Listeria monocytogenes was sometimes present in initial samples, but it was diminished or was not detected by the end of fermentation in Greek sausages (Drosinos et al. 2005; Samelis et al. 1998). In the European project *Tradisausage* (Talon 2006), *L.monocytogenes* was enumerated in 3 French traditional producers out of the 10 studied. In the first one, the level of *L. monocytogenes* decreased from the batter until the end of the ripening. In the second producer, *L. monocytogenes* was detected in the final product at a level under the hygienic tolerable level (2.0 log CFU/g), and in the third producer *L. monocytogenes* level was above this limit (Lebert et al. 2007). In Spanish sausages, sporadic cases of contamination by *L. monocytogenes* were mentioned (Tables 11.1a and 11.1b [Aymerich et al. 2003]).

IDENTIFICATION OF THE FLORA

In the last decade, it was shown that classical microbiological techniques did not give a real view of microbial diversity. To improve identification of LAB and GCC+, molecular methods have been developed (Aymerich et al. 2003; Comi et al. 2005;

Table 11.2. LAB species identified in traditional fermented dry sausages at the end of ripening.

Country (Reference)/Product	Species	Percentage
Spain (1)	*L. sakei*	74.0
	L. curvatus	21.2
	Leuc. mesenteroïdes	4.8
Greece (2)	*L. plantarum*	37.2
	L. plantarum/pentosus	25.0
	LAB n.i.	9.0
	L. curvatus	7.3
	L. pentosus	5.9
	Other species	<5.0
	L. rhamnosus, L. sakei, L. paracasei, L. salivarius, L. brevis	
Greece (3)/Product I	*L. sakei*	42.8
	L. curvatus	18.7
	L. buchneri	13.3
	L. plantarum	8.0
	Other species	<5.0
	L. paracasei, L. casei, L. coryniformis	
	L. rhamnosus	
Greece (3)/Product II	*L. paracasei*	29.2
	L. sakei	23.6
	L. buchneri	16.7
	L. curvatus	13.8
	Other species	<5.0
	L. plantarum, L. casei, L. paraplantarum	
Greece (4)	*L. curvatus*	48.2
	L. sakei	19.2
	L. plantarum	14.9
	L. paraplantarum	5.3
	L. casei/paracasei	5.3
	L. alimentarius	<5.0
Hungary (4)	*L. sakei*	70.8
	W. viridescens	7.3
	L. curvatus	7.3
	W. paramesenteroides/hellenica	5.8
	Other species	<5.0
	Leuc. mesenteroïdes, L. paraplantarum/plantarum, L. plantarum	
Italy (4)	*L. sakei*	48.8
	L. curvatus	29.8
	L. plantarum	6.6
	L. paraplantarum	5.8
	Other species	<5.0
	L. paraplantarum/pentosus, W. paramesenteroides/hellenica	

Table 11.2. *(Continued)*

Country (Reference)/Product	Species	Percentage
Italy (5)/Product 1	*L. sakei*	42.0
	L. curvatus	22.0
	L. plantarum	18.0
	L. paraplantarum/pentosus	8.0
	W. paramesenteroides/hellenica	6.0
	Other species	<5.0
	L. paraplantarum, Leuc. citreum	
Italy (5)/Product 2	*L. curvatus*	52.0
	L. sakei	34.0
	L. paraplantarum	12.0
	W. paramesenteroides/hellenica	<5.0
Italy (5)/Product 3	*L. sakei*	52.0
	L. curvatus	34.0
	Leuc. mesenteroïdes	8.0
	L. brevis	<5.0
Italy (6)	*L. sake**	34.0
	L. sakei	28.8
	Leuc. mesenteroïdes	9.4
	L. curvatus	8.4
	L. plantarum	5.2
	Other species	<5.0
	L. casei, LAB spp.*, L. alimentarius, L. sanfranciscensis, L. maltaromicus*	
Italy (7)	*L. sakei*	43.3
	L. plantarum	16.6
	L. carnis	15.0
	L. curvatus	13.3
	Other species	<5.0
	L. casei, L. farciminis, L. sharpeae, L. delbrueckii, L. amilophylus	
Italy (8)/Product C	*L. sakei*	88.8
	L. casei	5.9
	L. curvatus	<5.0
Italy (8)/Product L	*L. sakei*	85.5
	L. curvatus	8.3
	L. plantarum	<5.0
Italy (8)/Product U	*L. sakei*	52.0
	L. curvatus	31.0
	L. plantarum	6.7
	L. brevis	<5.0
	L. paraplantarum	<5.0

L. *Lactobacillus*; Leuc. *Leuconostoc*; LAB: Lactic Acid bacteria;* *L. sake* former *L. bavaricus*; W. Weissella. n.i. not identified. (1) Aymerich et al. (2006); (2) Drosinos et al. (2005); (3) Papamanoli et al. (2003); (4) Rantsiou et al. (2005a); (5) Comi et al. (2005); (6) Coppola et al. (2000); (7) Greco et al. (2005); (8) Urso et al. (2006).

Table 11.3. GCC+ species identified in traditional fermented dry sausages at the end of ripening.

Country (Reference)/Product	Species	Percentage
Greece (1)	*S. saprophyticus*	31.1
	S. xylosus	19.2
	S. simulans	11.4
	S. capitis	8.2
	S. haemolyticus	7.8
	S. sciuri	7.8
	S. caprae	5.9
	Other species	<5.0
	S. aureus/intermedius, S. hominis, S. auricularis, S. warneri, S. cohnii, S. epidermidis, S. carnosus	
Greece (2)/Product I	*S. saprophyticus*	27.6
	S. carnosus	25.9
	S. cohnii	13.8
	S. xylosus	5.2
	S. epidermidis	5.2
	S. hominis	5.2
	K. varians	5.2
	Other species	<5.0
	S. capitis, S. warneri, S. auricularis, S. hyicus, S. sciuri	
Greece (2)/Product II	*S. xylosus*	16.7
	S. saprophyticus	14.3
	S. carnosus	11.9
	S. haemolyticus	11.9
	S. simulans	9.5
	S. epidermidis	7.1
	Other species	<5.0
	S. cohnii, S. auricularis, S. hyicus, K. varians, S. lentus	
Spain (3)/Fuet	*S. xylosus*	83.3
	S. warneri	10.8
	S. epidermidis	5.0
	S. carnosus	<5.0
Spain (3)/Salchichon	*S. xylosus*	72.9
	S. warneri	12.5
	S. epidermidis	12.5
	K. varians	<5.0
Spain (3)/Chorizo	*S. xylosus*	80.8
	S. warneri	8.3
	S. epidermidis	5.8
	S. carnosus	<5.0
Spain (4)/Chorizo	*S. xylosus*	94.6
	Other species	<5.0
	S. intermedius, S. saprophyticus, S. hominis S. epidermidis, S. aureus	

Table 11.3. (*Continued*)

Country (Reference)/Product	Species	Percentage
Italy (5)/Salami Naples	*S. xylosus*	44.8
	S. saprophyticus	17.2
	S. lentus	17.2
	S. warneri	13.8
	S. succinus	<5.0
Italy (5)/Soppressata Ricigliano	*S. saprophyticus*	32.0
	S. xylosus	29.5
	S. succinus	14.1
	S. equorum	11.5
	S. vitulus	5.1
	Other species	<5.0
	S. pasteuri, S. warneri, S. haemolyticus	
Italy (5)/Sopressata Gioi	*S. xylosus*	25.4
	S. saprophyticus	24.8
	S. equorum	20.9
	S. succinus	14.7
	Other species	<5.0
	S. warneri, S. lentus, S. vitulus, S. pasteuri, S. epidermidis, S. haemolyticus	
Italy (6)/Salame Milano	*S. xylosus*	65.0
	S. sciuri	35.0
Italy (7)	*S. xylosus*	100.0
Italy (8)/Plant C	*S. warneri*	27.0
	S. xylosus	25.0
	S. pasteuri	25.0
	M. caseolyticus	12.0
	S. epidermidis	5.0
	S. saprophyticus, S. equorum	<5.0
Italy (8)/Plant L	*S. xylosus*	62.0
	S. equorum	12.0
	S. warneri	8.0
	M. caseolyticus	8.0
	S. pasteuri	7.0
	S. epidermidis, S. saprophyticus	<5.0
Italy (8)/Plant U	*S. xylosus*	40.0
	S. equorum	18.0
	S. carnosus	11.0
	S. warneri	8.0
	S. epidermidis	8.0
	M. caseolyticus	8.0
	S. pasteuri	5.0
	S. saprophyticus, S. cohnii	<5.0
Italy (9)/sausages from Basilica	*S. xylosus*	51.2
	S. pulvereri/vitulus	13.4
	S. equorum	10.2
	S. saprophyticus	10.0
	Other species	<5.0
	S. pasteuri, S. succinus, S. epidermidis	

(*Continues*)

Table 11.3. (*Continued*)

Country (Reference)/Product	Species	Percentage
France (10)	*S. equorum*	82.0
	S. succinus	12.0
	Other species	<5.0
	S. warneri, S. saprophyticus	

(1) Drosinos et al. (2005); (2) Papamanoli et al. (2002); (3) Martin et al. (2006); (4) Garcia-Varona et al. (2000); (5) Mauriello et al. (2004); (6) Rebecchi et al. (1998); (7) Cocolin et al. (2001b); (8) Iacumin et al. (2006); (9) Blaiotta et al. (2004); (10) Corbière Morot-Bizot et al. (2006).

Corbiere Morot-Bizot et al. 2004; Giammarinaro et al. 2005; Rantsiou and Cocolin 2006; Urso et al. 2006).

Lactic Acid Bacteria

Table 11.2 shows the diversity of LAB identified in naturally fermented sausages at the end of ripening originated from different countries, regardless of the identification method. The LAB species most commonly identified in traditional fermented sausages are *Lactobacillus sakei*, *Lactobacillus curvatus,* and *Lactobacillus plantarum*, with *L. sakei* as the dominant one. When *L. sakei* is dominant, it represents 42–88.8% of the isolates (Table 11.2). Aymerich et al. (2006) showed that *L. sakei* was identified in all of the Spanish sausages and represented 89.3% in chorizo and 76.3% in fuet. In a French sausage, *L. sakei* represented 100% of the isolates on the final products, although it was minor in the raw materials (Ammor et al. 2005). *L. curvatus* is the second species identified; it is dominant in some Greek or Italian sausages. *L. plantarum* is the third one; it dominates the LAB flora in a Greek sausage. Many other species are identified, as listed in Table 11.2, but represent a minor population.

Gram-positive Catalase-positive Cocci

Staphylococcus carnosus and *Staphylococcus xylosus* are used as starter cultures for manufacturing industrial dry sausages in Europe. In traditional products at the end of ripening, *S. xylosus* is the most common species and is identified in all the sausages (Table 11.3). It represents from 17–100% of the isolates according to the type of sausages. Even if it was not the dominant species in the batter, *S. xylosus* became rapidly dominant in Salame Milano (Rebecchi et al. 1998). *S. saprophyticus* is the second dominant species identified. It is dominant in some Greek and Italian sausages (Drosinos et al. 2005; Mauriello et al. 2004; Papamanoli et al. 2002). In the Italian products, *S. equorum* and *S. succinus* were also isolated (Table 11.3). In a French small producer, Corbiere Morot-Bizot et al. (2006) showed that the staphylococal microflora of the products and the environment was dominated by *S. equorum* (49% of the isolates) and *S. succinus* (33%) of the processing unit. Many other minor species were identified in the different traditional sausages (Table 11.3).

CONCLUSION

This review outlines the diversity in the manufacturing and the microflora of naturally fermented sausages, but also the similarity in the microorganism development. Indeed, LAB and GCC+ always became the dominant flora in the final products. Traditional fermented sausages generally did not present sanitary risk. This study also outlined the diversity in the LAB and GCC+ species, which was certainly due to the formulation, fermentation, and ripening practices.

REFERENCES

S Ammor, C Rachman, S Chaillou, H Prévost, X Dousset, M Zagorec, E Dufour, I Chevallier. 2005. Phenotypic and genotypic identification of lactic acid bacteria isolated from a small-scale facility producing traditional dry sausages. Food Microbiol 22: 373–382.

T Aymerich, B Martín, M Garriga, M Hugas. 2003. Microbial quality and direct PCR identification of lactic acid bacteria and nonpathogenic Staphylococci from artisanal low-acid sausages. Appl Environ Microbiol 69:4583–4594.

T Aymerich, B Martín, M Garriga, MC Vidal-Carou, S Bover-Cid, M Hugas. 2006. Safety properties and molecular strain typing of lactic acid bacteria from slightly fermented sausages. J Appl Microbiol 100:40–49.

G Blaiotta, C Pennacchia, F Villani, A Ricciardi, R Tofalo, E Parente. 2004. Diversity and dynamics of communities of coagulase-negative staphylococci in traditional fermented sausages. J Appl Microbiol 97:271–284.

I Chevallier, S Ammor, A Laguet, S Labayle, V Castanet, E Dufour, R Talon. 2006. Microbial ecology of a small-scale facility producing traditional dry sausage. Food Contr 17:446–453.

L Cocolin, M Manzano, D Aggio, C Cantoni, G Comi. 2001b. A novel polymerase chain reaction (PCR)—Denaturing gradient gel electrophoresis (DGGE) for the identification of *Micrococcaceae* strains involved in meat fermentations. Its application to naturally fermented Italian sausages. Meat Sci 58: 59–64.

L Cocolin, M Manzano, C Cantoni, G Comi. 2001a. Denaturing gradient gel electrophoresis analysis of the 16S rRNA gene V1 region to monitor dynamic changes in the bacterial population during fermentation of Italian sausages. Appl Environ Microbiol 67:5113–5121.

G Comi, R Urso, L Iacumin, K Rantsiou, P Cattaneo, C Cantoni, L Cocolin. 2005. Characterisation of naturally fermented sausages produced in the North East of Italy. Meat Sci 69:381–392.

S Coppola, G Mauriello, M Aponte, G Moschetti, F Villani. 2000. Microbial succession during ripening of Naples-type salami, a southern Italian fermented sausage. Meat Sci 56:321–329.

S Corbiere Morot-Bizot, S Leroy, R Talon. 2006. Staphylococcal community of a small unit manufacturing traditional dry fermented sausages. Int J Food Microbiol 108:210–217.

S Corbiere Morot-Bizot, R Talon, S Leroy. 2004. Development of a multiplex PCR for the identification of *Staphylococcus* genus and four staphylococcal species isolated from food. J Appl Microbiol 97:1087–1094.

D Demeyer. 2004. Chapter 20: Meat fermentation: Principles and applications. In: YH Hui, L Meunier-Goddik, AS Hansen, J Josephsen, W-K Nip, PS Stanfield, F Toldrá, eds. Handbook of Food & Beverage Fermentation Technology. New York: Marcel Decker Inc., pp. 353–368.

EH Drosinos, M Mataragas, N Xiraphi, G Moschonas, F Gaitis, J Metaxopoulos. 2005. Characterization of the microbial flora from a traditional Greek fermented sausage. Meat Sci 69:307–317.

C Fontana, PS Cocconcelli, G Vignolo. 2005a. Monitoring the bacterial population dynamics during fermentation of artisanal Argentinean sausages. Int J Food Microbiol 103:131–142.

C Fontana, G Vignolo, PS Cocconcelli. 2005b. PCR-DGGE analysis for the identification of microbial populations from Argentinean dry fermented sausages. J Microbiol Methods 63:254–263.

M García-Varona, EM Santos, I Jaime, J Rovira. 2000. Characterisation of Micrococcaceae isolated from different varieties of chorizo. Int J Food Microbiol 54:189–195.

P Giammarinaro, S Leroy, JP Chacornac, J Delmas, R Talon. 2005. Development of a new oligonucleotide array to identify staphylococcal strains at species level. J Clin Microbiol 43:3673–3680.

G Giraffa. 2002. Enterococci from foods. FEMS Microbiol Rev 26:163–171.

M Greco, R Mazette, EPL De Santis, A Corona, AM Cosseddu. 2005. Evolution and identification of lactic acid bacteria isolated during the ripening of Sardinian sausages. Meat Sci 69:733–739.

M Hugas, M Garriga, MT Aymerich. 2003. Functionality of enterococci in meat products. Int J Food Microbiol 88:223–233.

L Iacumin, G Comi, C Cantoni, L Cocolin. 2006. Ecology and dynamics of coagulase-negative cocci isolated from naturally fermented Italian sausages. Syst Appl Microbiol 29:480–486.

I Lebert, S Leroy, P Giammarinaro, A Lebert, JP Chacornac, S Bover-Cid, MC Vidal-Carou, R Talon. 2007. Diversity of micro-organisms in environments and dry fermented sausages of French traditional small units. Meat Sci 76:112–122.

F Leroy, J Verluyten, L De Vuyst. 2006. Functional meat starter cultures for improved sausage fermentation. Int J Food Microbiol 106:270–285.

B Martin, M Garriga, M Hugas, T Aymerich. 2005. Genetic diversity and safety aspects of enterococci from slightly fermented sausages. J Appl Microbiol 98:1177–1190.

B Martín, M Garriga, M Hugas, S Bover-Cid, MT Veciana-Nogues, T Aymerich. 2006. Molecular, technological and safety characterization of gram-positive catalase-positive cocci from slightly fermented sausages. Int J Food Microbiol 107:148–158.

G Mauriello, A Casaburi, G Blaiotta, F Villani. 2004. Isolation and technological properties of coagulase negative staphylococci from fermented sausages of Southern Italy. Meat Sci 67:149–158.

E Papamanoli, P Kotzekidou, N Tzanetakis, E Litopoulou-Tzanetaki. 2002. Characterization of *Micrococcaceae* isolated from dry fermented sausage. Food Microbiol 19:441–449.

E Papamanoli, N Tzanetakis, E Litopoulou-Tzanetaki, P Kotzekidou. 2003. Characterization of lactic acid bacteria isolated from a Greek dry-fermented sausage in respect of their technological and probiotic properties. Meat Sci 65:859–867.

E Parente, M Martuscelli, F Gardini, S Grieco, MA Crudele, G Suzzi. 2001. Evolution of microbial populations and biogenic amine production in dry sausages produced in Southern Italy. J Appl Microbiol 90:882–891.

K Rantsiou, L Cocolin. 2006. New developments in the study of the microbiota of naturally fermented sausages as determined by molecular methods: A review. Int J Food Microbiol 108:255–267.

K Rantsiou, EH Drosinos, M Gialitaki, R Urso, J Krommer, J Gasparik-Reichardt, S Toth, I Metaxopoulos, G Comi, L Cocolin. 2005a. Molecular characterization of *Lactobacillus* species isolated from naturally fermented sausages produced in Greece, Hungary and Italy. Food Microbiol 22:19–28.

K Rantsiou, R Urso, L Iacumin, C Cantoni, P Cattaneo, G Comi, L Cocolin. 2005b. Culture-dependent and -independent methods to investigate the microbial ecology of Italian fermented sausages. Appl Environ Microbiol 71:1977–1986.

A Rebecchi, S Crivori, PG Sarra, PS Cocconcelli. 1998. Physiological and molecular techniques for study of bacterial community development in sausage fermentation. J Appl Microbiol 84:1043–1049.

J Samelis, J Metaxopoulos, M Vlassi, A Pappa. 1998. Stability and safety of traditional Greek salami—A microbiological ecology study. Int J Food Microbiol 44:69–82.

R Talon. 2006. Assessment and improvement of safety of traditional dry sausages from producers to consumers "Tradisausage" European project, QLK1 CT-2002-02240. Summary available at: http://www.clermont.inra.fr/tradisausage/index.htm.

R Talon, S Leroy-Sétrin. 2006. Chapter 16: Latest developments in meat bacterial starters. In: LML Nollet, F Toldrá, eds. Advanced Technologies for Meat Processing. New York: CRC Press, Taylor and Francis group, pp. 401–418.

R Talon, S Leroy-Sétrin, S Fadda. 2002. Chapter 10: Bacterial starters involved in the quality of fermented meat products. In: F Toldrá, ed. Research Advances in the Quality of Meat and Meat Products. Research Signpost, pp. 175–191.

———. 2004. Chapter 23: Dry Fermented sausages. In: YH Hui, L Meunier-Goddik, AS Hansen, J Josephsen, W-K Nip, PS Stanfield, F Toldrá, eds. Handbook of Food and Beverage Fermentation Technology, 1st Ed. New York: Marcel Dekker Inc., pp. 397–416.

R Urso, G Comi, L Cocolin. 2006. Ecology of lactic acid bacteria in Italian fermented sausages: isolation, identification and molecular characterization. Syst Appl Microbiol 29(8):671–690.

12
The Microbiology of Fermentation and Ripening

Margarita Garriga and Teresa Aymerich

INTRODUCTION

Raw meat is a nutrient matrix rich in protein with a higher water activity (a_w = 0.96–0.97) and a favorable pH (5.6–6.0) for optimal microbial growth. It allows a rapid spoilage process when microorganism contamination occurs and preservation measures are not correctly applied. Gastrointestinal tract, feet, hides, or skins of slaughtered animals are primary sources of contamination. Instruments and surfaces in the slaughterhouse, infected personnel, or healthy carriers may contaminate the meat at all stages of meat processing. The contaminating microflora includes technologically important microorganisms but also spoilage and pathogenic bacteria, resulting in a varied and complex microflora.

Fermentation and drying are one of the oldest methods used to preserve food for long periods of time. The origin of fermented sausages seems to have come from Mediterranean countries in Roman times (Lücke 1974), although a fermented pork sausage from Thailand called *Nham* has also been reported as an example of other areas of local production (Adams 1986). Production was established in America, South Africa, and Australia by immigrants from European countries. In France in the fifth century, the *charcutiers* prepared different pork specialities; in fact the *chair cuite*, which originated from the French word *charcuterie*, is closely related to the Italian word *salsamenteria/salumeria*. The word *salami* comes from *salumen,* which in old Latin means a conjugation of salted things (Cantoni et al. 1987). Pederson (1971) suggested the name could come from Salamis, a city on the cost of Cyprus destroyed by an earthquake in 450 B.C. *Sausage*, *saucisse*, and *salchicha* (English, French, and Spanish, respectively) derive from the Italian *salsiccia*.

Although typical preparations based on indigenous microflora still exist, fermentations should now be considered more than just a preservation method. Across the centuries, people learned by empirical practice and experimentation how to control these processes, and fermented sausages became an independent class of foodstuff with different regional specialities, appreciated for their rich diversity in sensorial attributes. They comprise a great variety of products that vary between countries and different regions of the same country in the batter composition, fermentation, and ripening processes. Europe is the major producer of fermented sausages (Campbell-Platt 1995). Germany, which is known to have been manufacturing fermented sausages since the 1850s, is probably the country producing the greatest variety, with differences in ripening times, flavor, firmness, acidification, meat/fat grinding, and diameter of products. In general, Mediterranean countries produce naturally dried, nonsmoked, spiced sausages with a limited sour taste.

THE MANUFACTURE OF FERMENTED SAUSAGES

Pork and pork/beef, and rarely mutton or chicken meat, are used in the manufacture of fermented sausages. Fat comes from pork, backfat being the choice for producing high-quality products. The proportion of meat and fat varies depending on the product as well as the method and degree of mincing. Fermented sausage consists of comminuted meat and fat, mixed with salt, sugar, curing agents, and spices, and then stuffed into casings and subjected to fermentation by indigenous microflora or by selected starter cultures. After the fermentation period, which lasts between 1–3 days depending on the temperature and relative humidity applied, a ripening-drying process transforms a highly perishable product into a stable cured product that does not need refrigeration.

The main additives added to the meat batter are carbohydrates, curing salts, and spices:

- *Carbohydrates*: Glucose is usually added to the mixture because it is the primary growth substrate for all bacteria. Generally it is dosed at 1–8 g/kg to reach a pH value under 5.2 after the fermentation period, which ensures firmness and microbiological stability. Excess amounts of easily fermentable sugars may result in a product that is too sour and the development of undesired Lactic Acid Bacteria (Kröckel 1995).
- *Curing salts*: Sodium chloride is added to the mixture at 20–30 g/kg. Together with the antimicrobial effect against spoilage bacteria by the decrease of the a_w, NaCl participates in the solubilization of meat proteins, which determine the cohesiveness of the product. Current meat-curing practice involves the addition of nitrate and/or nitrite salts, which contribute to the formation of the characteristic cured-meat color and flavor. Nitrite participates in the inhibition of oxidative processes, thus preventing rancidity as well as the inhibition of pathogens like *Clostridium botulinum*. The added salt and sugars in the comminuted meat favors the growth of indigenous fermentative microorganisms, which will grow at relatively high osmotic pressure. Sodium ascorbate is usually added at 0.5 g/kg. It is an antioxidant substance that favors the transformation of nitrite into nitric oxide, improving the stability of the cured-meat color.
- *Spices*: Black and white pepper are usually added to practically all types of fermented sausages. Other spices such as paprika and garlic are used in several specialities, such as the Spanish chorizo. Wine is used in some specialities.

According to Buckenhüskes (1993), the main objectives of the fermentation and ripening process are to achieve the formation of the desired and typical cured color and fermentation flavor, the development of firmness to permit sliceability, the inhibition of pathogenic and spoilage bacteria, and an extended shelf life. The intrinsic factors of the sausage mix, together with the final a_w (generally under 0.9) and pH (between 6.2 and 4.5) exert a natural selectivity for promoting the development of the desired salt-tolerant microflora, which contributes to the complexity of the reactions that take place in order to fulfil these objectives.

TECHNOLOGICAL MICROFLORA

The first studies on the ecology of fermented sausages back in the 1970s (Lücke 1974) established two groups of microorganisms as being mainly responsible for the transformation involved during fermentation and ripening: lactic acid bacteria (LAB), in particular *Lactobacillus*, and the gram-positive coagulase-negative cocci (GCC+), especially *Staphylococcus*. Yeasts and molds were also often isolated from fermented meats. LAB are responsible for lactic acid production and the subsequent decrease of pH, which causes the coagulation of meat proteins and the tangy flavor of sausages for the necessary reactions for color formation and for microbiological stability. They also produce small amounts of acetic, ethanol, acetoin, carbon dioxide, and pyruvic acid that are produced during fermentation, depending on the endogenous microflora, the carbohydrate used, and the sources of meat proteins and additives used (Demeyer et al. 2000; Thornill and Cogan 1984; Bacus 1986). Some LAB strains may also act as bioprotective cultures by the production of antimicrobial compounds (bacteriocins), thus enhancing the safety of fermented sausages (Hugas 1998). Some authors have also pointed out their ability to hydrolyze the pork muscle sarcoplasmic proteins, releasing peptides that could play a role in the taste (Fadda et al. 1999). *Staphylococcus* and *Kocuria* are important for nitrite and nitrate reductase activity, promoting the desired red color development and its stabilization (Liepe 1983), decomposition of peroxides (Schleifer 1986; Montel et al. 1998), limiting lipid oxidation and preventing rancidity (Barrière et al. 1998; Talon et al. 1999), and contributing to the flavor by formation of esters and other aromatic compounds from amino acids through their proteolytic and lipolytic activities (Cantoni et al. 1967; Debevere et al. 1976; Berdagué et al. 1993; Johansson et al. 1994; Miralles et al. 1996; Montel et al. 1998; Cai et al. 1999). Yeast and molds may also contribute to the product quality by the development of typical flavors and tastes through the lactate oxidation, proteolysis, degradation of amino acids, and lipolysis (Sorensen and Samuelsen 1996; Sunesen and Stahnke 2003). The interconnection of their metabolic activities is considered essential to healthiness, texture, desired flavor, and color of fermented sausages.

Biodiversity and Behavior of the Technological Microflora

Sausage fermentation is characterized by an increase in the number of LAB from 10^3–10^5 CFU/g to 10^6–10^9 CFU/g within the first days of fermentation (1–3 days) or maturation/ripening (up to 14 days), which generally remains stable for the rest of the process and becomes the dominant microflora. The evolution of the LAB population on traditional Spanish fermented sausage is reported in Figure 12.1. The initial number of GCC+, between 10^3–10^5 CFU/g, increased to 10^5–10^8 CFU/g during the first 14–20 days of fermentation and ripening (Schillinger and Lücke 1987; Torriani et al. 1990; Hugas et al. 1993; Samelis et al. 1994; Cocolin et al. 2001; Metaxopoulos et al. 2001; Aymerich et al. 2003; Fontana et al. 2005). Evolution of the GCC+ population in traditional Spanish fermented sausages is reported in Figure 12.1. Short ripening time and higher fermentation temperature lead to higher numbers of lactobacilli from the early stages of fermentation and produce a final product with acid flavor and little aroma. In contrast, sausages with longer maturation times and low fermentation temperatures contain higher numbers of GCC+ in the early stages and produce a less tangy product with more flavor (Demeyer et al. 2000). Yeasts showed slow growth during the first half of the ripening and then decreased at different rates (Mauriello et al. 2004).

Lactic Acid Bacteria

Although meat batter composition, together with final pH and process technology, may influence the microflora, the species of LAB most commonly found in different kinds of fermented sausages processed with different technologies in Germany, Hungary, Greece, Spain, Italy, and Argentina were *L. sakei* and *L. curvatus*. In the majority of the studies, *L. sakei* was the dominating species in the final products, exceeding 55% of the total LAB population. Some other species, such as *L. plantarum*, *L. alimentarius*, *L. casei*, *L. paraplantarum*, *L. pentosus*, *L. paracasei*, *Pediococcus* spp., *Lactococcus lactis*, *Lactococcus garviae*, *Leuconostoc citreum*, *Leuconostoc mesenteroides*, *Weisella paramesenteroides/hellenica*, *Weissella viridescens*, *Enterococcus faecium*, *Enterococcus faecalis*, *Enterococcus durans*, and *Enterococcus pseudoavium* have also been described (Schillinger and Lücke 1987; Torriani et al. 1990; Montel et al. 1991; Hugas et al. 1993; Samelis et al. 1994; Santos et al. 1998; Cocolin et al. 2000; Parente et al. 2001; Aymerich et al. 2003; Rantsiou et al. 2005a, Rantsiou et al. 2005b, Aymerich et al. 2006). In some slightly fermented Italian sausages, *Enterococcus* and the lactobacilli population have been reported to be highly balanced (Dellapina et al. 1994).

Although only a few species seem to be able to implant and dominate at the end of the ripening

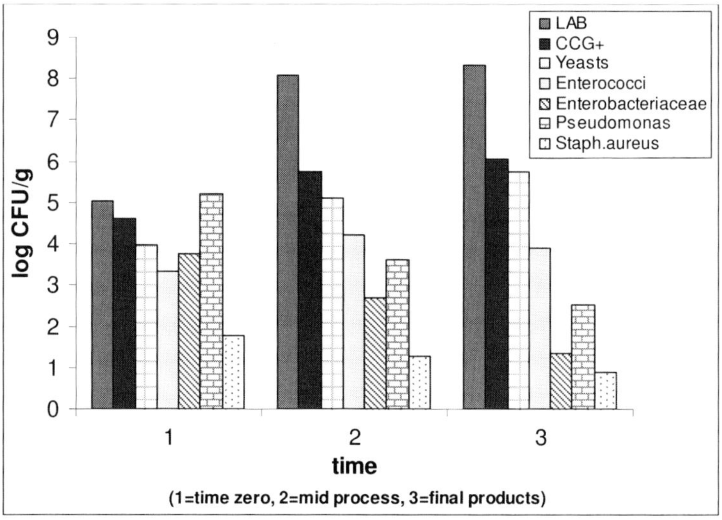

Figure 12.1. Microflora of traditional Spanish fermented sausages (pH 5.79) during the process. Values are the median of 10 different products from different producers (TRADISAUSAGE QLK1-CT-2002-02240).

process, biodiversity at strain level inside the predominant species has been reported. In Figure 12.2, biodiversity of 12 different strains isolated from a slightly fermented Spanish sausage is reported, and 6 different clusters at 70% homology could be differentiated through a composite RAPD-profile. Santos et al. (1998) detected 4 different biochemical groups of *L. sakei* and 4 different groups of *L. curvatus* among the isolates from different kinds of chorizo, the group S1 of *L. sakei* being the most competitive during the whole fermentation/ripening process. Rantsiou et al. (2005a) reported a high biodiversity inside the 295 isolates analyzed from the predominant LAB species found in different kinds of sausages produced in Hungary, Italy, and Greece. The clustering analysis at 70% of homology gave 5 different clusters for the 27 isolates of *L. plantarum*, 9 clusters for the 100 isolates of *L. curvatus* and 19 clusters for the 168 isolates of *L. sakei*. The majority of the clusters grouped country-specific. Aymerich et al. (2006) were able to distinguish 144 different RAPD/plasmid profiles among the 250 isolates from 21 different Spanish slightly fermented sausages. *Leuconostoc mesenteroides* displayed the higher intraspecificic variability, 9 profiles out of 12 isolates, and *L. sakei* showed 112 different strains out of 185 isolates. *L. curvatus* presented the lower intraspecific variability.

Enterococci are ubiquitous microorganisms that inhabit the gastrointestinal tract of humans and animals. They are frequently isolated from fermented meat products with counts of up to 10^6 CFU/g (Papa et al. 1995; Samelis et al. 1998; Metaxopoulos et al. 2001; Aymerich et al. 2003; Fontana et al. 2005) due to their tolerance to sodium chloride and nitrite, allowing them to survive and even to multiply during the first period of fermentation and ripening (Cocolin et al. 2001; Giraffa 2002). The enterococci evolution through the fermentation/ripening process of the traditional Spanish fermented sausages is reported in Figure 12.1. The species most frequently isolated are *Enterococcus faecium* and *Enterococcus faecalis*, followed by *Enterococcus hirae* and *Enterococcus durans*. (Devriese et al. 1995; Peters et al. 2003; Martín et al. 2005). A high intraspecific biodiversity has been reported in the enterococci population of fermented sausages by RAPD-PCR. Sixty strains out of 106 isolates were identified from 21 different

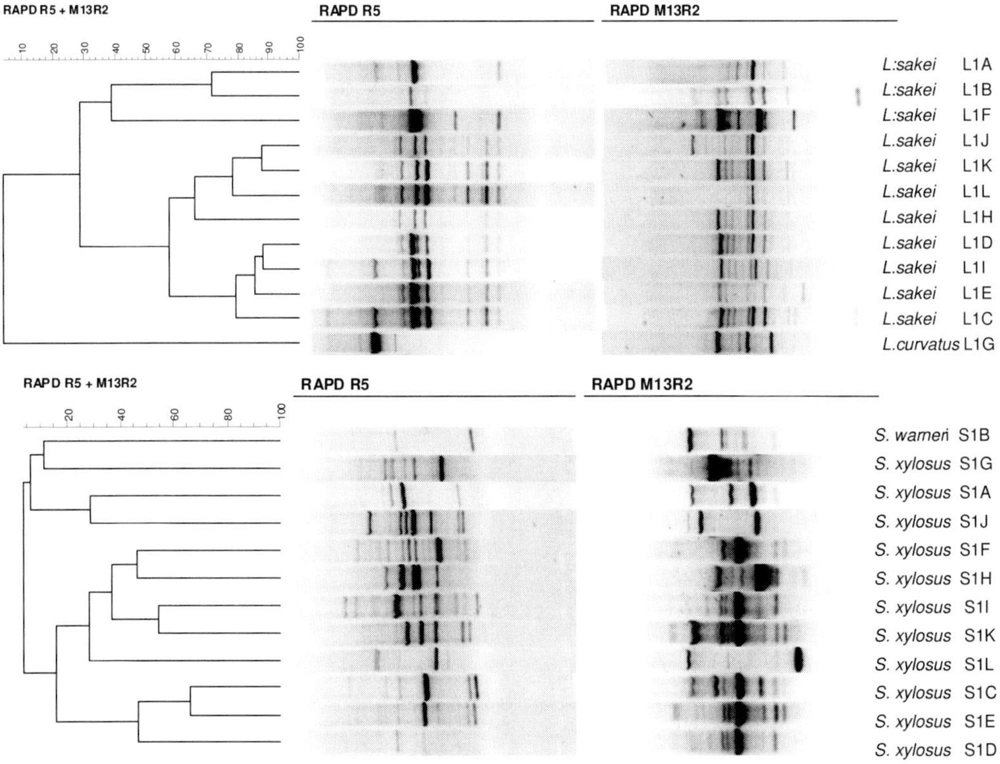

Figure 12.2. Strain biodiversity of LAB and GCC+ population of a traditional Spanish fermented sausage.

fermented Spanish sausages by RAPD profiles, thus indicating a high genetic variability (Martín et al. 2006).

The presence of enterococci in foods is highly controversial. Although some authors consider them undesirable, indicators of fecal contamination, and responsible for the spoilage of meat products (Franz et al. 1999), they are also producers of toxic substances such as biogenic amines (Tham 1988), and sausage alterations (Knudtson and Hartman 1992); others report their important role in flavor development and bioprotection (Coppola et al. 1988; Centeno et al. 1996; Aymerich et al. 2000; Callewaert et al. 2000). They also produce L(+) lactate, which contributes to pH drop. In recent decades, enterococci have unfortunately acquired clinical relevance, although foodborne enterococci had never been proved to be the direct cause of clinical infections (Adams 1999). The resistance of enterococci to a wide variety of antimicrobials (Murray 1990; Mundy et al. 2000) together with their ability to exchange genes by conjugation (Clewell 1990; Franz et al. 1999) are a cause for concern to the health authorities.

Gram-positive Coagulase-negative Cocci

Regarding GCC+, *Staphylococci*, mainly *Staphylococcus xylosus*, have been described as the predominant species in Spanish, German, and Italian fermented sausages (Fischer and Schleifer 1980; Simonetti and Cantoni 1983; Seager et al. 1986; Coppola et al. 1996; García-Varona et al. 2000; Aymerich et al. 2003; Martín et al. 2006). In some fermented Greek and Argentinean sausages, a dominance of *S. saprophyticus* has been reported (Samelis et al. 1998; Papamanoli et al. 2002; Fontana et al. 2005). A codominance of *S. saprophyticus*, *S. xylosus*, and *S. equorum* has been reported in three different types of Italian fermented sausages by Mauriello et al. (2004). Other species such as *S. carnosus*, *S. succinus*, *S. warneri*, *S. lentus*, *S. vitulus*, *S. pasteuri*, *S. epidermidis* and *S. haemolyticus*, *S. intermedius*, *S. pulverei*, *S. sciuri*, *S. cohnii*, *S. hyicus*, *Kocuria varians*, *Kocuria kristiniae*, *Micrococcus luteus*, and *Macrococcus caseolyticus* have been described (Fischer and Schleifer 1980; Seager et al. 1986; Cocolin et al. 2001; Aymerich et al. 2003; Rantsiou et al. 2005b, Martín et al. 2006). Underestimation of some staphylococci species such as *S. equorum* has been suggested by misidentification when only biochemical methods were used (Blaiotta et al. 2003).

A high biodiversity inside staphylococci from fermented sausages has been described by genotyping and/or biochemical methods. The strain biodiversity of the 12 different isolates from the GCC+ population is reported in Figure 12.2, where 12 different clusters could be differentiated at 70% of homology when a composite RAPD profile was considered. By genotypic clustering based on comparison of RAPD-PCR profiles, Rossi et al. (2001) were able to distinguish 22 different clusters at 70% of homology. The combination of RAPD-PCR and plasmid profiling allowed the discrimination of 208 different profiles among the 240 GCC+ characterized from the slightly fermented sausages (Martín et al. 2005). Fontana et al. (2005) reported a high variability at the first stage of Argentinean sausage processing, and a strong selection of the microflora occurred after day 5 with only two strains of *S. saprophyticus* that became dominant.

Yeasts and Molds

Yeast microbiota in sausages depends on the nature and the processing conditions of the product. The main role of yeasts in fermented sausages seems to be related to the increase of ammonia and reduction of lactic acid contents (Gehlen et al. 1991) together with proteolytic activity (Durá et al. 2004), thus contributing to the final sensory properties of the fermented sausages.

In traditional Spanish fermented sausages, an increase in the population from 4 log CFU/g to nearly 6 log CFU/g at the end of ripening has been observed (Figure 12.1). Rantsiou et al. (2005b) observed three different behaviors of the yeast population in three different local products. In the first product, the yeast population decreased during the central period of fermentation with an increase in the counts at the last sampling point, reaching 10^6 to 10^7 CFU/g. In the second product yeasts were detected only on the first 3 days. An undulating pattern was observed in the third, with initial counts of 10^3–10^4 CFU/g turning out to be undetectable for a 2-week period; they increased to 10^3–10^4 CFU/g at day 14 and decreased to 10^2–10^3 CFU/g at the end of ripening.

Debaryomyces, *Rhodotorula*, *Hansenula*, and *Torulopsis* have been described as being the most representative genera (Comi and Cantoni 1980). *D. hansenii* has been reported as the dominant species in Spanish products and has been isolated from all the stages of processing (Encinas et al. 2000). In other products, the *D. hansenii* population has been reported to decrease during processing (Gehlen et al. 1991; Olesen and Stahnke 2000; Rantsiou et al. 2005b). Other species such as *Candida tropicalis*, *Candida krisii*, *Candida sake*, *Candida fermentati*, *Willopsis saturnus*, and *Entyloma dahliae* have also been described (Rantsiou et al. 2005b).

The development of fungal mycelia is commonly observed on the casings of sausages, mainly those of Southern Europe, which have a long ripening period. Fungi have been described as protecting against lipid oxidation and extending the volatile profile of the final product (Hierro et al. 2005). The species of molds most commonly found are *Penicillium*, *Aspergillus*, *Mucor*, and *Cladosporium*.

SPOILAGE MICROFLORA

Gram-negative aerobic bacteria (e.g., *Pseudomonas*, *Acinetobacter*) may be involved in the spoilage of the products. Proteolytic activity and/or catabolism of sulphur-containing amino acids can cause defects in texture and flavor. High counts of *Pseudomonas* up to 10^4 CFU/g have been reported in freshly prepared sausages (Metaxopoulos et al. 2001) (Figure 12.1), but they are reported to disappear (Metaxopoulos et al. 2001) or decrease (Figure 12.1) during fermentation and ripening. *Brochothrix thermosphacta*, a gram-positive bacterium that causes cheesy odors, can grow to 10^4 CFU/g during the fermentation period, but it decreases during ripening.

The gram-negative facultative anaerobic *Enterobacteriaceae* may grow to values of up to 10^3–10^6 CFU/g during the first days of fermentation/ripening when pH, a_w, and salt/moisture coefficient are compatible with these microorganims. From days 5–7, they progressively drop down to 10^2 CFU/g (Figure 12.1) (Lücke 1986; Domínguez et al. 1989; Lizaso et al. 1999; Castaño et al. 2002). Moreover, *Enterobacteriaceae* are aminogenic bacteria that may produce biogenic amines as cadaverin, putrescine, and histamine, thus being a concern in relation to food quality (Vidal-Carou et al. 1990; Mariné-Font et al. 1995). *Escherichia coli*, *Hafnia alvei*, *Serratia liquefaciens*, *Serratia odorifera* and *Enterobacter amnigenus*, *Enterobacter intermedius*, *Providencia alcalifaciens*, *Proteus vulgaris*, *Providencia penneri*, and *Morganella morganii* (Silla-Santos 1998; Castaño et al. 2002; Mavromatis and Quantick 2002; Chevallier et al. 2006) have been isolated from fermented sausages.

FOODBORNE PATHOGENS

Veterinary inspection can detect and remove raw meat affected by *Bacillus anthracis*, *Mycobacterium tuberculosis*, or *Brucella abortis*, but infections produced by *Salmonella* spp., *Listeria monocytogenes*, *Campylobacter jejuni*, *Escherichia coli* O157:H7, and *Yersinia enterocolitica* are not readily detected in vivo and have been involved in several outbreaks. Proper cooling of carcasses at chill temperatures and food handling is essential to avoid hazards such as *Salmonella*, *Shigella*, *E. coli*, *Staphylococcus aureus*, and *Clostridium perfringens*. Nevertheless, some foodborne pathogens such as *Pseudomonas*, *Clostridium botulinum* type E, *Listeria monocytogenes* can even grow.

Salmonella spp. and *L. monocytogenes* have been reported as the major causes of deaths by foodborne pathogens, followed by *Campylobacter* and enterohemorragic *E.coli* (Mead et al. 1999; Anonymous 2001; Le Loir et al. 2003). *Salmonella* spp. has frequently been isolated on pork carcasses and pork cuts (Giovannacci et al. 2001; Pala and Sevilla 2004). *Salmonella* Arizonae have also been isolated from black pepper (Kneifel and Berger 1994). Absence of *Salmonella* in final products has been observed in the study of 10 traditional Spanish workshops, as has also been reported in other products from Greece, Spain, and Italy (Metaxopoulos et al. 2001; Aymerich et al. 2003). *S. aureus* and *L. monocytogenes* are ubiquitous in processing environments (Salvat et al. 1995; Borch et al. 1996) and their presence in raw pork and meat products has been reported, (Anonymous 1999; Atanassova et al. 2001; Pala and Sevilla 2004) although the reported levels in the final products are generally low (Aymerich et al. 2003; Chevallier et al. 2006). In a survey of 10 traditional Spanish workshops, *L. monocytogenes* was detected in some of the workshop equipment and products, but generally with counts of less than 1.0 log CFU/100 cm^2 and 0.7 log CFU/g, respectively. Counts of *S. aureus* through the ripening process of traditional Spanish fermented sausages are reported in Figure 12.1. Levels under 100 CFU/g are usually reported in final products (Metaxopoulos et al. 2001; Aymerich et al. 2003; Rantsiou et al. 2005b).

Few outbreaks associated with *Salmonella* and *S. aureus* have been reported in fermented sausages from Europe (Smith and Palumbo 1980; van Netten et al. 1986; Hartog et al. 1987). In 1994, an outbreak of *E.coli* O157:H7 related to sliced salami was reported in the U.S. (Anonymous 1995).

STARTER CULTURES

Although fermented sausages can be produced without the use of inoculated strains, the use of starter cultures for sausage production is increasing so as to guarantee safety and standardize product properties, including consistent flavor and color, and to shorten the ripening period. Nevertheless, proper hygienic and safety strains, properly identified at species and strain level, which preserve typical characteristics, are essential in order to assure the hygienic and safety quality together with the regional tradition for fermented sausages. A positive implantation of the starter culture used has to be controlled and guaranteed. In recent years, molecular methods such as Pulse Field Gel Electrophoresis (PFGE), plasmid profiling, amplified fragment length polymorphisms (AFLP), and randomly amplified polymorphic DNA (RAPD)-PCR have been used as efficient molecular techniques that enable strain typing in a rapid and sensitive way (Cocconcelli et al. 1995; Farber et al. 1996; Garriga et al. 1996; Descheemaeker et al. 1997; Rebecchi et al. 1998; Andrighetto et al. 2001; Rossi et al. 2001; Di Maria et al. 2002; Martín et al. 2005; Aymerich et al. 2006).

Relevant safety aspects such as bacteriocin production, absence of amino acid-decarboxylase activity, nontoxicogenic and nonantibiotic resistance profile, and mainly transferable antimicrobial resistance (European Commission 2003) are important properties to evaluate potential starter cultures and probiotic strains. Bacteriocins, natural antimicrobial peptides, together with the lactic acid produced from glucose could improve quality and safety of meat products by avoiding the presence of pathogens, such as *L. monocytogenes* and spoilage microorganisms, and improving the competitiveness of their producers for survival (Nes and Tagg 1996). Several *L. sakei* and *L. curvatus* have been reported as bacteriocin producers and have been used as protective cultures, and their activity against *L. monocytogenes* has been proven in meat products (Foegeding et al. 1992; Hugas et al. 1995). Bacteriocinogenic lactic acid bacteria and selected strains of *S. xylosus* and *S. carnosus* to improve safety, color, and flavor of the final products are commercially available. The growth of toxicogenic molds could be prevented by the surface inoculation of commercially available nonmycotoxigenic strains of *Penicillium chrysogenum* and *Penicillium nalgoviensis*.

REFERENCES

Adams MR. 1986. Fermented flesh foods. Progr Indust Microbiol 23:159–198.

———. 1999. Safety of industrial lactic acid bacteria. J Biotechnol 68:171–178.

Andrighetto C, Zampese L, Lombardi A. 2001. RAPD-PCR characterization of lactobacilli isolated from artisanal meat plants and traditional fermented sausages of Veneto region (Italy). Lett Appl Microbiol 33:26–30.

Anonymous. 1995. Centers for disease control and prevention. Escherichia coli 0157:H7 outbreak linked to commercially distributed dry-cured salimi. Morb Mortal Weekly Rcp 44:157–160.

Atanassova V, Meindl A, Ring C. 2001. Prevalence of Staphylococcus aureus and staphylococcal enterotoxins in raw pork and uncooked smoked ham—a comparison of classical culturing detection and RFLP-PCR. Int J Food Microbiol 68:105–113.

Aymerich MT, Garriga M, Jofré A, Martín B, Monfort JM. 2006. The use of bacteriocins against Meat-borne pathogens. In:Advanced technologies for meat processing. LML Nollet, F Toldrá. Marcel Dekker, Inc.

Aymerich MT, Garriga M, Ylla J, Vallier J, Monfort JM, Hugas M. 2000. Application of Enterocins as biopreservatives against Listeria innocua in meat products. J Food Protect 63:721–726.

Aymerich MT, Martín B, Garriga M, Hugas M. 2003. Microbial quality and direct PCR identification of lactic acid bacteria and nonpathogenic staphylococci from artisanal low-acid sausages. Appl Environ Microbiol 69:4583–4594.

Bacus JN. 1986. Fermented meat and poultry products. In: Advances in Meat and Poultry Microbiology. AMD Pearson, T.R. Macmillan. pp. 123–164.

Berdagué JL, Monteil P, Montel MC, Talon R. 1993. Effect of Starter Cultures on the Formation of flavour Compounds in Dry Sausage. Meat Sci 35:275–287.

Blaiotta G, Pennachia C, Parente E, Villani F. 2003. Design and evaluation of specific PCR primers for rapid reliable identification of Staphylococcus xylosus strains isolated from dry fermented sausages. Appl Microbiol 26:601–610.

Borch E, Kant-Muermans ML, Blixt Y. 1996. Bacterial spoilage of meat and cured meat products. Int J Food Microbiol 33:103–120.

Buckenhüskes HJ. 1993. Selection criteria for lactic acid bacteria to be used as starter cultures for various food commodities. FEMS Microbiol Rev 12:253–271.

Cai Y, Kumai S, Ogawa M, Benno Y, Nakase T. 1999. Characterization and identification of Pediococcus species isolated from forage crops and their application for silage preparation. Appl Environ Microbiol 65:2901–2906.

Callewaert R, Hugas M, De Vuyst L. 2000. Competitiveness and bacteriocin production of Enterococci in the production of Spanish-style dry fermented sausages. Int J Food Microbiol 57:33–42.

Campbell-Platt G. 1995. Fermented meats—A world perspective. In: Fermented meats. G Campbell-Platt, PE Cook. Black academic & professional. pp. 39–52.

Cantoni C, Comi G, Bresciani C. 1987. Ruolo di Streptococcus faecalis nel rammollimento di wurstel. Industrie Alimentari 766–770.

Cantoni C, Molnar MR, Renon P, Gioletti G. 1967. Lipolytic micrococci in pork fat. J Appl Bacteriol 30:190–196.

Castaño A, García Fontán MC, Fresno JM, Tornadijo ME, Carballo J. 2002. Survival of Enterobacteriaceae during processing of Chorizo de cebolla, a Spanish fermented sausage. Food Contr 13:107–115.

Centeno JA, Menéndez S, Rodríguez-Otero JL. 1996. Main microbial flora present as natural starters in Cebreiro raw cow's-milk cheese (Northwest Spain). Int J Food Microbiol 33:307–313.

Chevallier I, Ammor S, Laguet A, Labayle S, Castanet V, Dufour E, Talon R. 2006. Microbial ecology of a small-scale facility producing traditional dry sausage. Food Contr 17:446–453.

Clewell DB. 1990. Movable genetic elements and antibiotic resistance in enterococci. European Journal of Clinical Microbiology & Infectious Diseases : Official Publication of the European Society of Clinical Microbiology 9:90–102.

Cocconcelli PS, Porro D, Galandini S, Senini L. 1995. Development of RAPD protocol for typing of strains of lactic acid bacteria and enterococci. Lett Appl Microbiol 21:376–379.

Cocolin L, Manzano M, Cantoni C, Comi G. 2001. Denaturing Gradient Gel Electrophoresis Analysis of the 16S rRNA Gene V1 Region To Monitor Dynamic Changes in the Bacterial Population during Fermentation of Italian Sausages. Appl Environ Microbiol 67:5113–5121.

Cocolin L, Manzano M, Comi G. 2000. Development of a rapid method for the identification of Lactobacilus spp. isolated from naturally fermented Italian sausages using a polymerase chain reaction—Temperature gradient gel electrophoresis. Lett Appl Microbiol 30:126–129.

Comi G, Cantoni C. 1980. I lieviti in insacatti crudi stagionati. Industrie Alimentari 19:857–860.

Coppola R, Iorizzo M, Sorrentino A, Sorrentino E, Grazia L. 1996. Resistenza al congelamento di lattobacilli mesofili isolati da insaccati e paste acide. Industrie Alimentari 35:349–351, 356.

Coppola S, Parente E, Dumontet S, La Peccerella A. 1988. The microflora of natural whey cultures utilized as starters in the manufacture of Mozzarella cheese from water-buffalo milk. Lait 68:295–310.

Debevere JM, Voets JP, De Schryver F, Huyghebaert A. 1976. Lipolytic activity of Micrococcus sp. isolated from a starter culture in pork fat. Lebensmittel-Wissenchaft und Technology 9:160–162.

Dellapina G, Blanco D, Pancini E, Barbuti S, Campanini M. 1994. Microbiological evolution in Italian Felino, Milan and Hungarian-style salami. Industrie Conserve 69:85–90.

Demeyer D, Raemaekers M, Rizzo A, Holck A, De Smedt A, Ten Brink B, Hagen B, Montel C, Zanardi E, Murbrekk E, Leroy F, Vandendriessche F, Lorentsen K, Venema K, Sunesen L, Stahnke Lh, De Vuyst L, Talon R, Chizzolini R, Eerola S. 2000. Control of bioflavour and safety in fermented sausages: first results of a European project. Food Res Intern 33:171–180.

Demeyer D, Todorov N, Van Nevel C, Vets, J. 1982. alpha-epsilon Diaminopimelic acid (DAPA) in a sheep rumen infused with a synthetic diet of sugars and urea: evidence for degradation of bacteria. J Anim Physiol Anim Nutr 48(1–2):21–32.

Descheemaeker P, Lammens C, Pot B, Vandamme P, Goossens H. 1997. Evaluation of arbitrarily primed PCR analysis and pulsed-field gel electrophoresis of large genomic DNA fragments for identification of enterococci important in human medicine. Int J Syst Bacteriol 47:555–561.

Devriese LA, Pot B, Van Damme L, Kersters K, Haesebrouck F. 1995. Identification of Enterococcus species isolated from foods of animal origin. Int J Food Microbiol 26:187–197.

Di Maria S, Basso AL, Santoro E, Grazia L, Coppola R. 2002. Monitoring of Staphylococcus xylosus DSM 20266 added as starter during fermentation and ripening of soppressata molisana, a typical Italian sausage. J Appl Microbiol 92:158–164.

Domínguez MC, Gutiérrez LM, López A, Seco F, Zumalácarregui JM. 1989. Evolución de los principales grupos de microorganismos durante la maduración del chorizo de León. Alimentaria 26:11–15.

Durá MA, Flores M, Toldrá F. 2004. Effect of Debaryomyces spp. on the proteolysis of dry fermented sausages. Meat Sci 68:319–328.

Encinas JP, López Díaz TM, García López ML, Otero A, Moreno B. 2000. Yeast population on Spanish fermented sausages. Meat Sci 54:203–208.

Fadda S, Sanz Y, Vignolo G, Aristoy MC, Oliver G, Toldrá F. 1999. Hydrolysis of pork muscle sarcoplasmic proteins by Lactobacillus curvatus and Lactobacillus sake. Appl Environ Microbiol 65:578–584.

Farber JM, Cai Y, Ross WH. 1996. Predictive modeling of the growth of Listeria monocytogenes in CO2 environments. Int J Food Microbiol 32:133–144.

Fischer U, Schleifer KH. 1980. Vorkommen von staphylokokken un mikrokokken in Rohwurst. Fleischwirtschaft 60:1046–1049.

Foegeding PM, Thomas AB, Pilkington DH, Klaenhammer TR. 1992. Enhanced control of Listeria monocytogenes by in situ-produced pediocin during dry fermented sausage production. Appl Environ Microbiol 58:884–890.

Fontana C, Cocconcelli PS, Vignolo G. 2005. Monitoring the bacterial population dynamics during fermentation of artisanal Argentinean sausages. Int J Food Microbiol 103:131–142.

Franz CM, Holzapfel W, Stiles ME. 1999. Enterococci at the crossroads of food safety?. Int J Food Microbiol 47:1–24.
García-Varona M, Santos EM, Jaime I, Rovira J. 2000. Characterisation of Micrococcaceae isolated from different varieties of chorizo. Int J Food Microbiol 54:189–195.
Garriga M, Hugas M, Gou P, Aymerich MT, Arnau J, Monfort JM. 1996. Technological and sensorial evaluation of Lactobacillus strains as starter cultures in fermented sausages. Int J Food Microbiol 32:173–183.
Garriga M, Marcos B, Martín B, Veciana-Nogués MT, Bover-Cid S, Hugas M, Aymerich MT. 2005. Starter cultures and high pressure processing to improve the hygiene and safety of slightly fermented sausages. J Food Protect 68:2341–2348.
Gehlen KH, Meisel C, Fischer A, Hammes WP. 1991. Influence of the yeast Debaryomyces hansenii on dry sausages fermentation. Kulmbach (Germany).
Giovannacci I, Queguiner S, Ragimbeau C, Salvat G, Vendeuvre JL, Carlier V, Ermel G. 2001. Tracing of Salmonella spp. in two pork slaughter and cutting plants using serotyping and macrorestriction genotyping. J Appl Microbiol 90:131–147.
Giraffa G. 2002. Enterococci from foods. FEMS Microbiol Rev 26:163–171.
Hartog BJ, De Boer E, Lenssick JB, De Wilde GJ. 1987. The microbiological status of dry sausage in East Nederlands. Tijdschr Dieegeneeskd 112:322–333.
Hierro E, Ordoñez J, Bruna JM, Pin C, Fernandez M, De La Hoz L. 2005. Volatile Compound generation in dry fermented sausages by the surface inoculation of selected mould species. Eur J Appl Microbiol 220:494–501.
Hugas M. 1998. Bacteriocinogenic lactic acid bacteria for the biopreservation of meat and meat products. Meat Sci 49:S139–S150.
Hugas M, Garriga M, Aymerich MT, Monfort JM. 1995. Inhibition of Listeria in dry fermented sausages by the bacteriocinogenic Lactobacillus sake CTC494. J Appl Bacteriol 79:322–330.
Hugas M, Garriga M, Aymerich T, Monfort JM. 1993. Biochemical characterization of lactobacilli from dry fermented sausages. Int J Food Microbiol 18: 107–113.
Johansson G, Berdagué JL, Larsson M, Tran N, Borch E. 1994. Lipolysis, proteolysis and formation of volatile components during ripening of a fermented sausage with Pediococcus pentosaceus and Staphylococcus xylosus as starter cultures. Meat Sci 38:203–218.
Kneifel W, Berger E. 1994. Microbiological criteria of random samples of spices and herbs retailed on the Austrian market. J Food Protect 57:893–901.
Knudtson LM, Hartman PA. 1992. Routine procedures for isolation and identification of Enterococci and fecal Streptococci. Appl Environ Microbiol 58: 3027–3031.
Kröckel L. 1995. Bacterial fermentation of meats. In: Fermented meats. G Campbell-Platt, PE Cook. Black academic & professional. pp. 69–109.
Le Loir Y, Baron F, Gautier M. 2003. Staphylococcus aureus and food poisoning. Genetics and Molecular Research 2:63–76.
Liepe HU. 1983. Starter cultures in meat production. In: Biotechnology and food Processing. SD Kung, Bills, D.D., Quatrano, R. pp. 273–286.
Lizaso G, Chasco J, Beriain MJ. 1999. Microbial and biochemical changes during ripening of salsichón, a Spanish dry-cured sausage. Food Microbiol 16: 219–228.
Lücke FK. 1974. Fermented sausages. In: Microbiology of fermented foods. BJB Wood. Applied Science Publishers. pp. 41–49.
———. 1986. Microbiological processes in the manufacture of dry sausages and raw ham. Fleischwirtschaft 66:1505–1509.
Mariné-Font A, Vidal-Carou MC, Izquierdo-Pulido M, Venciana-Nogués MT, Hernández-Jover T. 1995. Les amines biogènes dans les aliments: leur signification, leur analyze. Ann Fals Exp Chim Toxicol 88:119–140.
Martín B, Garriga M, Hugas M, Aymerich T. 2005. Genetic diversity and safety aspects of enterococci from slightly fermented sausages. J Appl Microbiol 98:1177–1190.
Martín B, Garriga M, Hugas M, Bover-Cid S, Veciana-Nogués MT, Aymerich MT. 2006. Molecular, technological and safety characterization of Gram-positive catalase-positive cocci from slightly fermented sausages. Int J Food Microbiol 107:147–158.
Mauriello G, Ercolini D, La Storia A, Casaburi A, Villani F. 2004. Development of polythene films for food packaging activated with an antilisterial bacteriocin from Lactobacillus curvatus 32Y. J Appl Microbiol 97:314–322.
Mavromatis P, Quantick PC. 2002. Modification of Niven's medium for the enumeration of histamine-forming bacteria and discussion of the parameters associated with its use. J Food Protect 65:546-551.
Mead PS, Slutsker L, Dietz V, Mccaig LF, Bresee JS, Shapiro C, Griffin PM, Tauxe RV. 1999. Food-related illness and death in the United States. Emerg Infect Dis 5:607–625.
Metaxopoulos J, Samelis J, Papadelli M. 2001. Technological and microbiological evaluation of traditional processes as modified for the industrial manufacturing of dry fermented sausage in Greece. Ital J Food Sci 1:3–18.

Miralles MC, Flores J, Perez-Martinez G. 1996. Biochemical tests for the selection of Staphylococcus strains as potential meat starter cultures. Food Microbiol. 13:227–236.

Montel MC, Masson F, Talon R. 1998. Bacterial role in flavour development. Meat Sci 49:S111–S123.

Montel MC, Talon R, Fournaud J, Champomier MC. 1991. A simplified key for identifying homofermentative Lactobacillus and Carnobacterium spp. from meat. J. Appl. Bacteriol. 70:469–472.

Mundy LM, Sahm DF, Gilmore MS. 2000. Relationship between enterococcal virulence and antimicrobial resistance. Clin Microbiol Rev 13:513–522.

Murray BE. 1990. The life and times of the Enterococcus. Clin Microbiol Rev 3:46–65.

Nes IF, Tagg JR. 1996. Novel lantibiotics and their prepeptides. Antonie Van Leeuwenhoek 69:89–97.

Olesen PT, Stahnke LH. 2000. The influence of Debaryomyces hansenii and Candida utilis on the aroma formation in garlic spiced fermented sausages and model minces. Meat Sci 56:357–368.

Pala TR, Sevilla A. 2004. Microbial contamination of carcasses, meat, and equipment from an Iberian pork cutting plant. J Food Protect 67:1624–1629.

Papa F, Zambonelli C, Grazia L. 1995. Production of Milano style salami of good quality and safety. Food Microbiol 12:9–12.

Papamanoli E, Kotzekidou P, Tzanetakis N, Litopoulou-Tzanetaki E. 2002. Characterization of Micrococcaceae isolated from dry fermented sausage. Food Microbiol 19:441–449.

Parente E, Griego S, Crudele MA. 2001. Phenotypic diversity of lactic acid bacteria isolated from fermented sausages produced in Basilicata (Southern Italy). J Appl Microbiol 90:943–952.

Pederson CS. 1971. Microbiology of food fermentations. IW AVI Publishing Company, Conneecticut. pp. 153–172.

Peters J, Mac K, Wichmann-Schauer H, Klein G, Ellerbroek L. 2003. Species distribution and antibiotic resistance patterns of enterococci isolated from food of animal origin in Germany. Int J Food Microbiol 88:311–314.

Rantsiou K, Drosinos EH, Gialitaki M, Urso R, Krommer J, Gasparik-Reichardt J, Tóth S, Metaxopoulos J, Comi G, Cocolin L, Cantoni C. 2005a. Molecular characterization of Lactobacillus species isolated from naturally fermented sausages produced in Greece, Hungary and Italy. Food Microbiol 22:19–28.

Rantsiou K, Urso R, Iacumin L, Cantoni C, Cattaneo P, Comi G, Cocolin L. 2005b. Culture-dependent and -independent methods to investigate the microbial ecology of Italian fermented sausages. Appl Environ Microbiol 71:1977–1986.

Rebecchi A, Crivori S, Sarra PG, Cocconcelli PS. 1998. Physiological and molecular techniques for the study of bacterial community development in sausage fermentation. J Appl Microbiol 84: 1043–1049.

Rossi F, Tofalo R, Torriani S, Suzzi G. 2001. Identification by 16S–23S rDNA intergenic region amplification, genotypic and phenotypic clustering of Staphylococcus xylosus strains from dry sausages. J Appl Microbiol 90:365–371.

Salvat G, Toquin MT, Michel Y, Colin P. 1995. Control of Listeria monocytogenes in the delicatessen industries: The lessons of a listeriosis outbreak in France. Int J Food Microbiol 25:75–81.

Samelis J, Maurogenakis F, Metaxopoulos J. 1994. Characterisation of lactic acid bacteria isolated from naturally fermented Greek dry salami. Int J Food Microbiol 23:179–196.

Samelis J, Metaxopoulos J, Vlassi M, Pappa A. 1998. Stability and safety of traditional Greek salami- a microbiological ecology study. Int J Food Microbiol 44:69–82.

Santos EM, González-Fernández C, Jaime I, Rovira J. 1998. Comparative study of lactic acid bacteria house flora isolated in different varieties of "chorizo". Int J Food Microbiol 39:123–128.

Schillinger U, Lücke FK. 1987. Identification of lactobacilli from meat and meat products. Food Microbiol 4:199–208.

Schleifer KH. 1986. Gram positive cocci. In: Bergey's Manual of Systematic Bacteriology. PHA Sneath, NS Mair, JG Holt. Williams & Wilkins. pp. 999–1003.

Seager MS, Banks JG, Blackburn C, Board RG. 1986. A taxonomic study of Staphylococcus spp. isolated from fermented sausages. J Food Sci 51:295–297.

Silla-Santos MH. 1998. Amino acid decarboxylase capability of microorganisms isolated in Spanish fermented meat products. Int J Food Microbiol 39:227–230.

Simonetti P, Cantoni C. 1983. Coagulase negative staphylococci for dry sausage ripening. Industrie Alimentari 22:262–264.

Smith JL, Palumbo SA. 1980. Inhibition of aerobic and anaerobic growth of Staphylococcus aureus. J Food Saf 4:221–233.

Sorensen BB, Samuelsen H. 1996. The combined effects of environmental conditions on lipolysis of pork fat by lipase of the meat starter culture organisms Staphylococcus xylosus, S. carnosus and S. equorum—a comparitive study in model systems. Int J Food Microbiol 35:59–71.

Sunesen LO, Stahnke LH. 2003. Mould starter culture for dry sausages—Selection, application and effect. Meat Sci 65:935–948.

Talon R, Walter D, Chartier S, Barrière C, Montel MC. 1999. Effect of nitrate and incubation conditions on the production of catalase and nitrate reductase by staphylococci. Int J Food Microbiol 52:47–56.

Tham W. 1988. Histamine Formation by Enterococci Isolated from Home-Made Goat Cheeses. Int J Food Microbiol 7:103–108.

Thornill PJ, Cogan TM. 1984. Use of gas-liquid chromatografy to determine the end-products of growth of lactic acid bacteria. Appl Environ Microbiol 47:1250–1254.

Torriani S, Dellaglio F, Palummeri M. 1990. Characterization of Lactobacilli Isolated from Italian Salami. Ann Microbiol 40:225–233.

Van Netten P, Leenaerts J, Heikant GM, Mossel DA. 1986. A small outbreak of salmonellosis caused by Bologna sausage. Tijdschr Dieegeneeskd 24: 1271–1275.

Vidal-Carou MC, Izquierdo-Pulido M, Martin-Morro MC, Mariné-Font A. 1990. Histamine and tyramine in meat products: relationship with meat spoilage. Food Chem 37:239–249.

13
Starter Cultures: Bacteria

Pier Sandro Cocconcelli

INTRODUCTION

Bacteria have long been an unrecognized agent responsible for fermentation and aroma formation of meat products. Traditional meat fermentation occurs without the addition of selected microbial starter cultures, and fermentation was driven by strains from the indigenous microbiota deriving from raw materials, mainly meat, and from the environment. Still today several fermented meat products are produced without starter cultures, and indigenous homofermentative lactobacilli and coagulase-negative staphylococci (CNS), selected by the physicochemical conditions imposed by production technology, dominate the fermentation process. The identification and the selection of bacteria and their use as starter cultures began in the second half of the twentieth century. In 1940, Jensen and Paddock (1940, US Patent 2,225,783) developed the idea of inoculating the raw material for fermented sausages using lactobacilli, with the aim to govern and accelerate meat fermentation.

In the fifties, cultures of *Micrococcus* sp. and *Pediococcus cerevisiae* were developed in Europe by Niinivaara (1955) and in the U.S. by Deibel and Niven (1957). The combination of these first experiences on meat starter cultures led Nurmi (1966) to develop the first mixed culture composed of lactobacilli and micrococci. Now, in the industrial process for fermented sausage production, bacterial starter cultures are widely used to drive the fermentation process in the desired direction, reducing the variability in the quality of the product, limiting the growth of spoilage bacteria by accelerating fermentation, and improving the sensorial properties of fermented meat products. The first generation of starter bacteria, generally derived from cultures for vegetable fermentation, such as *L. plantarum* and pediococci, was selected mainly for their acidification properties. The second generation has been developed isolating strains of meat origin, such as *L. sakei* and coagulase-negative staphylococci, harboring phenotypic traits of technological relevance. Increased knowledge on the physiological properties of meat-fermenting bacteria, comprising information derived from genomic and proteomic analysis, and the use of molecular tools for strain characterization, now allows the development of a third generation of functional starter cultures, selected for the presence of technologically relevant traits. The aim of this chapter is to provide the most recent information on bacterial starters, and to delineate their functional properties to improve the quality of fermented meat products.

BACTERIAL STARTER CULTURES FOR FERMENTED MEAT: CLASSIFICATION AND PHYSIOLOGY

The bacterial starter cultures used for meat fermentations are preparations of viable bacteria that develop desired metabolic activities, primarily acidification and the development of aroma, during meat fermentation. They are composed of lactic acid bacteria strains, primarily belonging to *Lactobacillus* or *Pediococcus* genus, of coagulase-negative staphylococci (CNS) and members of *Micrococcaceae*. The most frequently used bacterial species and their

Table 13.1. Bacteria species most frequently used as meat starter culture.

Species	Functional Properties for Meat Fermentation	Safety Qualification—Lack of
Lactobacillus sakei	Acidification Catalase activity Flavor development Amino acid metabolism Antioxidant properties: catalase and SOD Bacteriocin production*	Acquired antimicrobial resistance** Biogenic amine production
Lactobacillus curvatus	Acidification Proteolytic activity Antioxidant properties: catalase Bacteriocin production*	Acquired antimicrobial resistance** Biogenic amine production
Lactobacillus plantarum	Acidification Antioxidant properties: catalase Bacteriocin production*	Acquired antimicrobial resistance** Biogenic amine production
Lactobacillus rhamnosus	Probiotic	Acquired antimicrobial resistance** Biogenic amine production
Pediococcus acidilactici	Acidification Bacteriocin production*	Acquired antimicrobial resistance** Biogenic amine production
Pediococcus pentosaceus	Acidification Bacteriocin production*	Acquired antimicrobial resistance** Biogenic amine production
Staphylococcus xylosus	Flavor development Antioxidant properties: catalase and SOD Amino acid catabolism Fatty acid metabolism Nitrate reduction	Acquired antimicrobial resistance Enterotoxin determinants
Staphylococcus carnosus	Flavor development Antioxidant properties: catalase and SOD Amino acid catabolism Fatty acid metabolism Nitrate reduction	Acquired antimicrobial resistance Enterotoxin determinants
Staphylococcus equorum	Flavor development Nitrate reduction	Acquired antimicrobial resistance Enterotoxin determinants
Kocuria varians	Nitrate reduction	

*Bacteriocin production is discussed in Chapter 14.

**Heterofermentative lactobacilli and pediococci are intrinsically resistant to vancomycin.

functions during sausage fermentation are reported in Table 13.1.

LACTOBACILLUS

The genus *Lactobacillus* is a wide and heterogeneous taxonomic unit, composed of rod-shaped lactic acid bacteria. This genus encompasses more than 100 different species with a large variety of phenotypic, biochemical, and physiological properties (Axelsson 2004), but only few of them have been isolated from meat fermentations and are used as starter cultures. *L. sakei*, *L. curvatus* and *L. plantarum*, belonging to the subgroup of facultative heterofermentative lactobacilli, are generally used for this purpose. The main energetic metabolism of these bacteria is the dissimilation of sugar to organic acid, by means of glycolysis and phosphoketolase pathways. When hexoses are the energy source, lactic acid is the major fermentation end product.

L. sakei is the predominant species in fermented meat products and its use as a starter culture for sausage production is widespread. The recent analysis of the 1.8 Mb genome sequence (Chaillou et al. 2005) has revealed that this organism has evolved to adapt itself to the meat environment, harboring the genetic function that confers the ability to grow and survive in the meat environment. The adaptation to the meat, an environment rich in amino acids due to the activity of endogenous proteases, has led to a lack of biosynthetic pathways for amino acid synthesis. Thus, *L. sakei* is auxotrophic for all amino acids but not aspartic and glutamic acid (Champomier-Vergès et al. 2002). Amino acid metabolism can provide an alternative energy source for *L. sakei* when glucose is exhausted, and this affects the sensorial properties of the sausage, as discussed later. Moreover, this species has psychrotrophic and osmotolerant properties, being able to grow at low temperature and in the presence of up to 10% of sodium chloride (NaCl). The physiological features are associated to the presence in the genome of a higher number of genes coding for stress response proteins, such as cold shock and osmotolerance proteins, when compared to other Lactobacilli.

Although *L. plantarum* has been identified as part of the meat microbiota and it is used as starter cultures for meat fermentation, this species lacks the meat specialization found in *L. sakei*. *L. plantarum* is a versatile bacterium, encountered in a variety of different environment and a genome size of 3.3 Mb (Kleerebezem et al. 2003), the wider genome of *Lactobacillus* genus, reflects its metabolic and environmental flexibility.

Much less information is available on the physiology and genetics of *L. curvatus*, other than the production of antimicrobial peptides, discussed in Chapter 14.

Pediococcus

Pediococci are gram-positive, coccus-shaped, lactic acid bacteria, showing the distinctive characteristic of tetrads formation, via cell division in two perpendicular directions in a single plane. Pediococci present a homofermentative metabolism where lactic acid is the major metabolic end product (Axelsson 2004). Phylogenetically *Pediococcus* species belong to the *L. casei –Pediococcus* subcluster of the *Lactobacillus* cluster. The genus consists currently of nine species, but only *P. pentosaceus* is generally used as a starter culture for meat fermentation. The species *P. cerevisiae,* frequently mentioned as a starter culture, has been now reclassified as *P. pentosaceus*. The genome sequencing project of this species is ongoing (http://genome.jgipsf.org/draft_microbes/pedpe/pedpe.info.html).

Staphylococcus and *Kocuria* Genera

Although *Micrococcaceae* are frequently mentioned as components of meat starter cultures, this term generally refers to members of *Staphylococcus* genus, which belong to the family *Staphylococcaceae*. Staphylococci were originally grouped with other gram-positive cocci, such as *Micrococcus* genus, because these two genera often cohabit the same habitats. However, molecular taxonomy has revealed that these genera are phylogenetically separate and distant: *Staphylococcus* belongs to the cluster of low GC gram-positive bacteria, and *Micrococcus* is part of the *Actinomycetales*.

The staphylococci present a spherical shape and cell often grouped to form clusters. *Staphylococcus* genus can be separated into 38 species and 21 subspecies, consisting of 6 coagulase-positive or variability coagulase and 32 coagulase-negative species (CNS).

Staphylococci are widespread in nature; their major habitats are skin, skin glands, and mucous membranes of mammals and birds. Some species, mainly coagulase-positive such as *S. aureus,* are pathogens responsible for infections and foodborne intoxications. Staphylococci are facultative anaerobes capable of metabolizing a number of different sugars. Under anaerobic conditions, the major end product is lactic acid, but acetate and pyruvate and acetoin are also formed. These organisms show the ability to survive environmental stress, such as high salt and low temperatures, encountered during meat fermentation. Moreover, they reduce nitrate to nitrite, a technologically relevant physiological feature.

Many CNS, such as *Staphylococcus xylosus*, *S. carnosus*, *S. equorum* and *S. saprophyticus*, have been isolated from dry-fermented sausages, but other species occur too. The genome sequencing of *S. xylosus* is ongoing and information can be found at www.cns.fr/externe/English/Projets/Projet_NN/NN.html.

Kocuria varians, formerly classified as *Micrococcus varians*, is a member of *Micrococcaceae* and is used as meat starter cultures for its nitrate reductase ability.

FUNCTIONAL PROPERTIES OF BACTERIAL STARTER CULTURES FOR FERMENTED MEAT

Colonization and Establishment in the Bacterial Community of Meat Fermentation

One of the fundamental properties of bacterial starter cultures is the ability to compete with the indigenous microbiota of meat, to colonize this environment, and to dominate the microbial community of the fermented products. Thus successful colonization of meat products by starter cultures is a prerequisite for their contribution to sensorial qualities of the final product.

In recent years, the development of a variety of molecular methods has allowed the study of the growth dynamics of bacterial strains during meat fermentation. Staphylococcal community development in traditional dry-fermented sausage has been studied by means of Multiplex PCR, Pulse Field Gel Electrophoresis (PFGE) and sequencing analysis of the *sodA*gene (Corbière Morot-Bizot et al. 2006). Fontana et al. (2005) showed that RAPD is a valid method to monitor the population dynamics in sausage fermentation, and this technique has more recently been applied to assess the performance of bacterial strains from commercial starter cultures (Cocolin et al. 2006).

Acidification

Acidification in sausage manufacture is mainly the result of carbohydrate fermentation by lactic acid bacteria. This process has a central role for flavor, color, and texture development and for control of undesired microbiota during fermentation. The production of lactic acid is primarily derived from the dissimilation of endogenous glucose or other sugars added to the batter. The addition of sugar to the batter and the manufacture technology (temperature, salting and ripening time) influence the acidification rate, the final pH, and the sensorial properties of fermented sausage. The level of acidification required depends on the sensorial properties of the product: lactic acid taste is desired in north European sausages but not in typical south European fermented meats.

Acidification could also be the result of alternative pathways. In *L. sakei*, the presence of genes involved in the energetic catabolism of nucleoside, such as adenosine and inosine, is an example of the adaptation of this organism to the meat environment. Glucose, the favourite carbon source of *L. sakei*, is rapidly consumed in meat; adenosine and inosine are abundant, reaching twice the concentration of glucose. *L. sakei* harbors genes coding for adenosine deaminase, for inosine hydrolase, and for nucleoside phosphorilase, which allow the release of ribose moiety from nucleoside and its subsequent metabolism (Chaillou et al. 2005). Moreover, the presence of methylglyoxal synthase, a novel genetic trait in lactic acid bacteria, has been proposed as a pathway to counteract frequent glucose starvation and modulate metabolism of alternative carbon sources (Chaillou et al. 2005).

Nitrate Reduction

The nitrate is added to the fermented sausages by its capacity to fix and to obtain the typical color of cured products other than by its antimicrobial properties. To be effective, the added nitrate must be reduced to nitrite. Besides contributing to flavor, staphylococci and *Kocuria* have a role for their nitrate reductase and antioxidant activities (Talón et al. 1999). These microorganisms reduce nitrate to nitrite, which is important for the formation of nitrosylmyoglobin, the compound responsible for the characteristic red color of fermented meats. The nitrate reductase activity is widespread in CNS: It has been detected in *S. xylosus*, *S. carnosus*, *S. epidermidis S. equorum*, *S. lentus*, and *S. simulans* (Talón et al. 1999; Mauriello et al. 2004). In *S. carnosus*, the molecular genetic determinants for nitrogen regulation, the *nreABC* genes, were identified and shown to link the nitrate reductase operon (*narGHJI*) and the putative nitrate transporter gene *narT.* The data provide evidence for a global regulatory system with oxygen as the effectors molecule (Fedtke et al. 2002).

Flavor Formation

The flavor and the aroma of fermented meats is formed in a combination of several elements, such as batter ingredients, manufacture technology, activity of tissular enzymes, proteases and lipases, and microbial metabolism. Bacterial community contributes to flavor development through carbohydrate dissimilation, peptide metabolism, and lipolysis. Lactic acid bacteria produce lactic acid and small amounts of acetic acid, ethanol, and acetoin, and together with staphylococci, are involved in peptide and amino acid metabolism; only CNS show an effective metabolism of lipids.

Metabolism of meat protein is of major importance for flavor characteristics of fermented meats. In these products, tissular proteases as well as microbial enzymes are responsible for the proteolytic changes that occur during meat processing, leading to the generation of small peptides and free amino acids, which are considered to be flavor compound precursors (Aristoy and Toldrá 1995; Sanz and Toldrá 1997; Sanz et al. 1998). It has been shown that several *Lactobacillus* spp. exhibit proteolytic activity on porcine muscle myofibrillar and sarcoplasmic proteins; even the processing conditions during meat fermentation did not favor meat protein breakdown by bacterial proteases. Fadda et al. (2001) reported the contribution of curing conditions, in the generation of hydrophilic peptides and free amino acids by the proteolytic activity of *L. curvatus* CRL 705. Moreover, it has been demonstrated that *L. sakei* play an important role in amino acid generation (Fadda et al. 1999a,b; Sanz et al. 1999; Champomier-Verges et al. 2002). Sanz and Toldrá (2002) reported an arginine-specific aminopeptidase activity in *L. sakei* important for the release of the free amino acid, because it could be further channeled into the arginine deiminase pathway. The genes encoding the proteins required for arginine catabolism in *L. sakei* are organized in a cluster (Züriga et al. 2002) and their transcription is repressed by glucose and induced by arginine. Arginine in particular is an essential amino acid for *L. sakei* and specifically promotes its growth in meat, being used as an energy source in absence of glucose (Champomier-Verges et al. 1999). The concentration of free arginine in raw meat is low, although it is relatively abundant in muscle myofibrillar proteins. Moreover, the genome analysis have shown that *L. sakei* harbors a second putative arginine deiminase pathway, containing two peptydil-arginine deaminases, enzymes that can contribute to the metabolism of arginine residues not released from peptides (Chaillou et al. 2005), increasing the competitiveness of *L. sakei* in meat environment.

To ensure sensory quality of fermented sausages, the contribution of staphylococci with their proteolytic and lipolytic activities is fundamental. The volatiles so far recognized as being produced by staphylococci are primarily amino acid catabolites, pyruvate metabolites, and methylketones from β-oxidation of fatty acids (Stahnke et al. 2002). In particular, *S. xylosus* and *S. carnosus* modulate the aroma through the conversion of amino acids (particularly the branched-chain amino acids leucine, isoleucine, and valine) and free fatty acids. They also were able to enhance the dry-salami odor and to increase the concentration of methylketones, ethyl esters (Larrouture et al. 2000; Beck et al. 2002).

Antioxidant Properties

The metabolism of most lactic acid bacteria, such as the indigenous lactobacilli which contaminate raw meat, could lead to the formation of hydrogen peroxide, a compound that interferes with the sensorial properties of meat products, being involved in discoloration of nitroso-heme-pigment and lipid oxidation. Bacterial strains used in meat cultures can produce catalase, antioxidant enzymes that disproportionate hydrogen peroxide to oxygen and water, preventing the risk of reduced quality of fermented meat. Thus, catalase production is considered a relevant technological property of starter cultures for meat fermented products (Leroy 2006). Production of this antioxidant enzyme is a common trait in aerobic bacteria, such as CNS. The characterization of catalase and superoxide dismutases in *S. carnosus* and *S. xylosus* have been reported. The catalase gene *katA* of *S. xylosus* has been studied in detail (Barrière 2001a,b; 2002). Transcriptional activity of this gene is induced upon entry into stationary phase by oxygen and hydrogen peroxide. Moreover, a second gene coding for heme-dependent catalase has been detected in *S. xylosus*.

Although lactic acid bacteria have long been considered as a catalase-negative microorganism, in the last decade two groups of catalase activity have been reported in genera *Lactobacillus*, *Pediococcus*, and *Leuconostoc*. The first group is defined as heme catalase and the second group nonheme Mn-containing catalase. The presence of a heme-dependent catalase has been demonstrated in *L. plantarum* (Igarashi et al. 1996) and *L. sakei* (Noonpakdeea et al. 1996). Moreover, the genome analysis of *L. sakei*, revealed that this meat organism harbors several system for the protection against reactive oxygen species, such as Mn-dependent SOD, heme-dependent catalase (Chaillou et al. 2005).

SAFETY OF BACTERIAL STARTER CULTURES FOR FERMENTED MEAT

The safety of bacterial strains intentionally added to food, such as starter cultures used for meat products, is becoming an issue for their application in food. Although meat starter cultures have a long history of apparent safe use, safety concerns can be associated with lactic acid bacteria and, more frequently, with CNS.

Antimicrobial Resistance

A risk factor potentially associated with all bacterial groups used as starter cultures for sausage is the presence of acquired genes for antimicrobial resistance.

The emergence and spread of resistance to antimicrobials in bacteria poses a threat to human and animal health, and commensal bacteria has been proposed to act as reservoirs of resistance genes. Antibiotic resistance bacteria has been isolated from raw meat and meat products and genetic determinants for antimicrobial resistance, such as *tet*(M) conferring resistance to tetracycline, has been identified in *L. plantarum* and *L. sakei* (Gevers et al. 2003). Antimicrobial resistance in CNS has been studied in detail, due to its clinical relevance. These bacteria display a high prevalence of antibiotic resistance (Archer et al. 1994; Agvald-Ohman et al. 2004) and can constitute a reservoir of antibiotics resistance genes that can be transferred to other staphylococci (Wieders et al. 2001). Moreover, antibiotic resistance strains were found in fermented meat products (Gardini et al. 2003; Martin et al. 2006) and genes for antimicrobial resistance to tetracycline, *tet*(M), *tet*(K), to erythromycin, *ermB* and *ermC*, and to β-lactams (*blaZ* and *mecA*), has been detected in CNS isolated from fermented meat (S. Gazzola, personal communication 2006).

An additional concern is that, even in the absence of selective pressure, mobile genetic elements carrying antibiotic resistance can be transferred at high frequency among the microbial community during sausage fermentation (Cocconcelli et al. 2003).

Enterotoxins Production

Risk concerns are related to CNS, because members of this group, primarily *S. epidermidis*, have been identified as the third most common cause of intrahospital infections and most frequent cause of bacteremia (CDC NNIS System 1998). Moreover, staphylococci could be the cause of foodborne intoxication by contaminated food consumption for the production of termostable enterotoxins (Balaban and Rasooly 2000). Although CNS of food origin have not been found to produce nosocomial infections, some strains have been described that produced enterotoxins, but to date, the coagulase-positive *S. intermedius* is the only species in addition to *S. aureus* that clearly has been related to staphylococcal intoxication. Vernozy-Rozand et al. (1996) reported enterotoxin E as the most frequent enterotoxin found in *S. equorum* and *S. xylosus*, however, Martin et al. (2006), reported that the occurrence of staphylococcal enterotoxins genes in CNS from slightly fermented sausages was rare, detecting only *ent*C in *S. epidermidis*. It is essential that the CNS strains selected to be used as starter culture do not produce enterotoxins and do not harbor gene coding for enterotoxin or enterotoxin-like superantigens.

Biogenic Amines Production

Biogenic amines (BA) such as cadaverine, putrescine, spermidine, histamine, phenethylamine, agmatine, and tyramine, are basic compounds present in living organisms at low levels where they are responsible for many essential functions. The presence in food of these bioactive compounds is concern for health, for their biological effect, which can lead to toxicological symptoms, such as pseudoallergenic reactions, histaminic intoxication, and interaction with drugs. (Mariné-Font et al. 1995). The formation of BA in fermented foods requires the presence of decarboxylase-producing microorganisms, which may be part of the associated microbiota of a particular food product or can be introduced by contamination before, during, or after food processing. The final BA contents in fermented sausages depend on the microbial composition of meat used as raw material, but also on the type and activity of the starter culture inoculated (Bover-Cid et al. 2001). High levels of BA, especially tyramine, but also histamine and the diamines putrescine and cadaverine, have been described in fermented sausages (Hernández-Jover et al. 1997a,b; Bover-Cid et al. 2000a,b).

The use of highly competitive decarboxylase-negative starter cultures was shown to prevent the growth of biogenic amine producers and leads to end products nearly free of BA, as long as the raw material is of sufficient quality (Bover-Cid et al. 2001). González-Fernández et al. (2003), reported that to avoid the presence of high concentrations of BA in chorizo, it is advisable to use a competitive starter culture of *L. sakei*, a negative-decarboxylate strain, which may decrease the pH quickly during the fermentation step and be predominant throughout the process; thus it would prevent the growth of bacteria that can produce BA. Similarly, the presence of a selected starter culture *L. sakei* CTC494 reduced BA accumulation during ripening, but only if raw meat was characterized by good quality. Addition of the same strain of *L. sakei*, along with proteolytic *S. carnosus* and *S. xylosus*, decreased BA accumulation in the production of fuet (Bover-Cid et al. 2000a) and reduced the total BA content 80–90% with respect to the sausages without starter cultures added. Also, the introduction of starter strains that possess amine oxidase activity might be a way of further decreasing the amount of BA produced during meat fermentation (Martuscelli et al. 2000; Fadda et al. 2001; Gardini et al. 2002; Suzzi and Gardini 2003).

MEAT STARTER CULTURES AND PROBIOTICS

Probiotics, live microorganisms, which when administered in adequate amounts confer a health benefit on the host (FAO 2006), are added to a variety of food. Recently, attention has been directed to the use of fermented sausages as a food carrier because these products could contain high numbers of viable lactic acid bacteria. To use probiotics as starter cultures for fermented sausages, in addition to the demonstrated probiotic features (FAO 2006), other properties are demanded. The probiotic culture should be well adapted to the conditions of fermented sausage to become dominant in the final product, competing with other bacterial populations from meat and from starter culture. In addition, the culture should not develop off-flavors in the product. Commercial probiotic cultures, such as strains *L. rhamnosus* GG, *L. rhamnosus* LC-705, *L. rhamnosus* E-97800, and *L. plantarum* E-98098 have been tested as functional starter culture strains in Northern European sausage fermentation without negatively affecting the technological or sensory properties, with an exception for *L. rhamnosus* LC-705 (Erkkila et al. 2001a, b). Klingberg et al. (2005), identified *L. plantarum* and *L. pentosus* strains, originated from the dominant NSLAB of fermented meat products, as promising candidates of probiotic meat starter cultures suitable for the manufacture of the Scandinavian-type fermented sausage.

PROTECTIVE MEAT STARTER CULTURES

The production of bacteriocins and other antimicrobial compounds, one of the most promising technological features of starter cultures, is discussed in Chapter 14.

CONCLUSION

A new generation of starter cultures, with distinctive properties for quality and safety of fermented meat products, could be expected from advancement in the study of lactic acid bacteria and CNS. New tools, such as genomic, proteomic, transcriptomic (the analysis of gene expression at genome level), and high throughput screening, combined with the increased knowledge of the interaction between technology and the bacterial community, may allow the identification of new strains from fermented meat with improved functional properties and lacking undesirable traits, such as virulence and antibiotic resistance determinants. Moreover, the microbial biodiversity of traditional high-quality fermented meat products, characterized by a complex microbial community highly adapted to meat environment, could be exploited to provide the new strains of lactic acid bacteria and CNS for development of new functional starter cultures.

REFERENCES

C Agvald-Ohman, B Lund, C Edlun. 2004. Multiresistant coagulase-negative staphylococci disseminati frequently between intubated patient into a multidisciplinary intensive care unit. Crit6 Care 8:42–47.

G Archer, D Niemeyer, J Thanassi, J Pucci. 1994. Dissemination among staphylococci of DNA sequences associated with methicillin resistance. Antmicrob Agents Chemother 38:447–454.

MC Aristoy, F Toldrá. 1995. Isolation of flavour peptides from raw pork meat and dry-cured ham. In: G Charalambous, ed. Food Flavours: Generation, Analysis and Process Influence. Amsterdam, The Netherlands: Elsevier Science B.V., pp. 1323–1344.

L Axelsson. 2004. Lactic acid bacteria: Classification and physiology. In: S Salminen, A Ouwehand, A Von Wright, eds. Lactic Acid Bacteria: Microbiology and Functional Aspects, 3rd ed. New York: Marcel Dekker, Inc.

N Balaban, A Rasooly. 2000. Staphylococcal enterotoxins. Int J Food Microbiol 61:1–10.

C Barriere, R Bruckner, D Centeno, R Talon. 2002. Characterisation of the katA gene encoding a catalase and evidence for at least a second catalase activity in *Staphylococcus xylosus*, bacteria used in food fermentation. FEMS Microbiol Lett 216(2): 277–283.

C Barriére, D Centeno, A Lebert, S Leroy-Setrin, J Berdagué, R Talon. 2001a. Roles of superoxide dismutase and catalase of *Staphylococcus xylosus* in the inhibition of linoleic acid oxidation. FEMS Microbiol Lett 201:181–185.

C Barriére, S Leroy-Sétrin, R Talon. 2001b. Characterization of catalase and superoxide dismutase in *Staphylococcus carnosus* 833 strain. J Appl Microbiol 91:514–519.

H Beck, A Hansen, F Lauritsen. 2002. Metabolite production and kinetics of branched-chain aldehyde oxidation in *Staphylococcus xylosus*. Enzyme and Microbial Technol 31:94–101.

S Bover-Cid, M Hugas, M Izquierdo-Pulido, M Vidal-Carou. 2000a. Reduction of biogenic amine formation using a negative amino acid-decarboxylase starter culture for fermentation of “Fuet” sausages. J Food Protect 63:237–243.

S Bover-Cid, M Izquierdo-Pulido, M Vidal-Carou. 2000b. Mixed starter cultures to control biogenic amine production in dry fermented sausages. J Food Protect 63:1556–1562.

S Bover-Cid, M Izquierdo-Pulido, M Vidal-Carou. 2001a. Effectiveness of a *Lactobacillus sakei* starter culture in the reduction of biogenic amine accumulation as a function of the raw material quality. J Food Protect 64:367–373.

S Chaillou, MC Champomier-Verges, M Cornet, AM Crutz-Le Coq, AM Dudez, V Martin, S Beaufils, Darbon-Rongere E Bossy, R Loux, M Zagorec. 2005. The complete genome sequence of the meat-borne lactic acid bacterium *Lactobacillus sakei* 23K. Nat Biotechnol 23:1527-1533.

MC Champomier-Verges, A Marceau, T Mera, M Zagorec. 2002. The pepR gene of Lactobacillus sakei is positively regulated by anaerobiosis at the transcriptional level. Appl Environ Microbiol. 68(8):3873–3877.

MC Champomier-Verges, M Zúñiga, F Morel-Deville, G Pérez-Martinez, M Zagorec, SD Ehrlich. 1999. Relationships between arginine degradation, pH and survival in *Lactobacillus sakei*. FEMS Microbiol Lett 180:297–304.

PS Cocconcelli, D Cattivelli, S Gazzola. 2003. Gene transfer of vancomycin and tetracycline resistances among *Enterococcus faecalis* during cheese and sausage fermentation. Int J Food Microbiol 88:3 15–323.

L Cocolin, R Urso, K Rantsiou, C Cantoni, G Comi. 2006. Multiphasic approach to study the bacterial ecology of fermented sausages inoculated with a commercial starter culture. Appl Environ Microbiol 72:942–945.

R Deibel, F Niven. 1957. *Pediococcus cerevisiae*: A starter cultures for summer sausage. Bacteriol Proc 14–15.

S Erkkilä, E Petajä, S Eerola, T Lilleberg, T Mattila-Sandholm, ML Suihko. 2001a. Flavour profiles of dry sausages fermented by selected novel meat starter cultures. Meat Sci 58:111–116.

S Erkkilä, ML Suihko, S Eerola, E Petajä, T Mattila-Sandholm. 2001b. Dry sausage fermented by Lactobacillus rhamnosus strains. Int J Food Microbiol 64:205–210.

S Fadda, Y Sanz, G Vignolo, MC Aristoy, G Oliver, F Toldrá. 1999a. Hydrolysis of pork muscle sarcoplasmic proteins by Lactobacillus curvatus and Lactobacillus sakei. Appl Environ Microbiol 65: 578–584.

———. 1999b. Characterization of pork muscle protein hydrolysis caused by Lactobacillus plantarum. Appl Environ Microbiol 65:3540–3546.

S Fadda, G Vignolo, G Oliver. 2001. Tyramine degradation and tyramine/histamine production by lactic acid bacteria and Kocuria strains. Biotech Lett 23:2015–2019.

FAO. 2006. Probiotics in Food. Health and nutritional properties and guidelines for evaluation. FAO food and nutrition paper. 85.

I Fedtke, A Kamps, B Krismer, F Gotz. 2002. The nitrate reductase and nitrite reductase operons and the *nar*T gene of *Staphylococcus carnosus* are positively controlled by the novel two-component system *Nre*BC. J Bacteriol 184:6624–6634.

C Fontana, PS Cocconcelli, G Vignolo. 2005. Monitoring the bacterial population dynamics during fermentation of artisanal Argentinean sausages. Int J Food Microbiol 103:131–142.

F Gardini, M Matuscelli, MA Crudele, A Paparella, G Suzzi. 2002. Use of Staphylococcus xylosus as a starter culture in dried sausages: Effect on biogenic amine content. Meat Sci 61:275–283.

F Gardini, R Tofalo, G Suzzi. 2003. A survey of antibiotic resistance in *Micrococcaceae* isolated from Italian dry fermented sausages. J Food Prot 66:937–945.

D Gevers, L Masco, L Baert, G Huys, J Debevere, J Swings. 2003. Prevalence and diversity of tetracycline resistant lactic acid bacteria and their tet genes along the process line of fermented dry sausages. Syst Appl Microbiol 26:277–283.

C González-Fernández, E Santos, I Jaime, J Rovira. 2003. Influence of starter cultures and sugar concentrations on biogenic amine contents in chorizo dry sausage. Food Microbiol 20:275–284.

T Hernández-Jover, M Izquierdo-Pulido, MT Veciana-Nogués, A Mariné-Font, MC Vidal-Carou. 1997a. Effect of starter cultures on biogenic amine formation during fermented sausage production. J Food Protect 60:825–830.

———. 1997b. Biogenic amine and polyamine contents in meat and meat products. J Agric Food Chem 45:2098–2102.

T Igarashi, Y Kono, K Tanaka. 1996. Molecular cloning of manganese catalase from *Lactobacillus plantarum*. J Biol Chem 271:29521–29524.

L Jensen, L Paddock 1940. US Patent 2,225,783. In: P Zeuthen. 1995. Historical aspects of meat fermentations. In: Fermented Meats. G Campbell-Platt, PE Cook, eds. pp. 53–68. London: Blackie Academic and Professional.

M Kleerebezem, J Boekhorst, R van Kranenburg, D Molenaar, OP Kuipers, R Leer, R Tarchini, SA Peters, HM Sandbrink, MW Fiers, W Stiekema, RM Lankhorst, PA Bron, SM Hoffer, MN Groot, R Kerkhoven, M de Vries, B Ursing, WM de Vos, RJ Siezen. 2003. Complete genome sequence of *Lactobacillus plantarum* WCFS1. Proc Natl Acad Sci USA 100:1990–1995.

TD Klingberg, L Axelsson, K Naterstad, D Elsser, B Budde. 2005. Identification of potential probiotic starter cultures for Scandinavian-type fermented sausages. Int J Food Microbiol 105:419–431.

C Larrouture, V Ardaillon, M Pepin, MC Montel. 2000. Ability of meat starter cultures to catabolize leucine and evaluation of the degradation products by using an HPLC method. Food Microbiol 17:563–570.

F Leroy, J Verluyten, L De Vuyst. 2006. Functional meat starter cultures for improved sausage fermentation Int J Food Microbiol 106:270–285.

A Mariné-Font, M Vidal-Carou, M Izquierdo-Pullido, M Veciana-Nogués, T Hermandez-Jover. 1995. Les amines biogénes dans les aliments: Leur signification, leur analyse. Ann Fals Exp Chim Toxicol 88:119–140.

B Martin, M Garriga, M Hugas, S Bover-Cid, MT Veciana-Nogues, T Aymerich. 2006. Molecular, technological and safety characterization of Gram-positive catalase-positive cocci from slightly fermented sausages. Int J Food Microbiol 107:148–158.

M Martuscelli, MA Crudele, F Gardini, G Suzzi. 2000. Biogenic amine formation and oxidation by Staphylococcus xylosus from artisanal fermented sausages. Lett Appl Microbiol 31:228–232.

G Mauriello, A Casaburi, G Blaiotta, F Villani. 2004. Isolation and technological properties of coagulase negative staphylococci from fermented sausages of Southern Italy. Meat Sci 67:49–158.

SC Morot-Bizot, S Leroy, R Talon. 2006. Staphylococcal community of a small unit manufacturing traditional dry fermented sausages. Inter J Food Microbiol 108:210–217.

F Niinivaara. 1955. Über den Einfluss von Bacterienreinkulturen auf die Reifung und Umrötung der Rohwurst. Avta Agr Fenn 84:1–128.

W Noonpakdeea, K Pucharoen, R Valyasevi, S Panyim. 1996. Molecular cloning, DNA sequencing and expression of catalase gene of *Lactobacillus sake* SR911. AsPac J Mol Biol Biotechnol 4:229–235.

E Nurmi. 1966. Effect of bacterial inoculations on characteristics and microbial flora of dry sausage. Acta Agr Fenn 108:1–77.

Y Sanz, S Fadda, G Vignolo, MC Aristoy, G Oliver, F Toldrá. 1999. Hydrolysis of muscle myofibrillar proteins by *Lactobacillus curvatus* and *Lactobacillus sakei*. Int J Food Microbiol 53:115–125.

Y Sanz, F Mulholland, F Toldrá. 1998. Purification and characterization of a tripeptidase from *Lactobacillus sake*. J Agric Food Chem 46:349–353.

Y Sanz, F Toldrá. 1997. Aminopeptidases activities from *Lactobacillus sake* in models of curing ingredients and processing conditions for dry sausage. J Food Sci 62:1211–1213.

Y Sanz, F Toldrá. 2002. Purification and characterization of an arginine aminopeptidase from *Lactobacillus sakei*. Appl Environ Microbiol 68:1980–1987.

LH Stahnke, A Holck, A Jensen, A Nilsen, E Zanardi. 2002. Maturity acceleration of Italian dried sausage by *Staphylococcus carnosus*—Relationship between maturity and flavor compounds. J Food Sci 67: 1914–1921.

G Suzzi, F Gardini. 2003. Biogenic amines in dry fermented sausages: A review. Int J Food Microbiol 88:41–54.

R Talón, D Walter, S Chartier, C Barriere, MC Montel. 1999. Effect of nitrate and incubation conditions on the production of catalase and nitrate reductase by staphylococci. Int J Food Microbiol 52:47–50.

C Vernozy-Rozand, C Mazuy, G Prevost, C Lapeyre, M Bes, Y Brun, J Fleurette. 1996. Enterotoxin production by coagulase-negative staphylococci isolated from goats' milk and cheese. Int J Food Microbiol 30:271–80.

C Wielders, M Vriens, S Brisse, L de Graaf-Miltenburg, A Troelstra, A Fleer, F Schmitz, J Verhoof, A Fluit. 2001. Evidence for in vivo transfer of mecA DNA between strains of *Staphylococcus aureus*. Lancet 357:1674–1675.

M Zúriga, M Miralles, G Pérez- Martínez. 2002. The product of arcR, the sixth gene of the arc peron of *Lactobacillus sakei*, is essential for expression of the arginine deiminase pathway. Appl Environ Microbiol 68:6051–6058.

14
Starter Cultures: Bioprotective Cultures

Graciela Vignolo and Silvina Fadda

INTRODUCTION

In the last few years, concerns over food safety have increased their importance due to its dramatic impact on public health. Over the past decade, a series of food scandals have erupted involving meat and meat products, which feature prominently in the food safety crisis. At present, bovine spongiform encephalitis and the rapid spread of avian influenza from eastern to western countries have triggered a sudden lack of consumer confidence in meat products and led to a dramatic fall in demand. The globalization of commerce, the gradual increase in world population and the change in lifestyles have resulted in consumer claims for safety oriented to foods of animal origin. Undoubtedly, the major threat to food safety is the emergence of "new" pathogens. The recent role of *Listeria monocytogenes*, *Escherichia coli O157:H7*, *Campylobacter jejeuni*, *Yersinia enterocolitica* and *Vibrio parahemolyticus* as food-borne microorganisms has been related to the increase in food poisoning outbreaks, compared to traditional food pathogens. Changes in the food chain will continue to create opportunities for the emergence of new diseases and the reemergence of old ones (Elmi 2004; Church 2004). In addition, the presence in meat products of chemical additives and residues of agrochemical and veterinary drugs are also perceived by consumers as a health risk. Even when the level of these residues seldom exceeds regulatory limits in meat products (Tarrant 1998), the use of antibiotics in intensive animal production poses the additional risk of bacterial resistance, which constitutes a microbiological hazard. On the basis of these data, the need emerges for solutions to the problem of food hygienic quality. Consumers are increasingly demanding pathogen-free foods with minimal processing, fewer preservatives and additives, high nutritional value, and intact sensory quality. In response to these conflicting demands, current trends in the meat industry include the investigation of alternatives for safer and healthier products. Biopreservation has gained increasing attention as a means of naturally controlling the shelf life and safety of meat products. The application of bioprotective cultures to ensure hygienic quality is a promising tool. However, it should be considered only as an additional hurdle to good manufacturing, processing, storage, and distribution practices. In the present contribution, the use of antagonist microorganisms to inhibit and/or inactivate pathogens and spoilage flora in meat fermented products is discussed, with particular reference to bacteriocin-forming bacterial strains. A new concept of starter cultures for fermented sausages is analyzed where biopreservative and probiotic features complete the recently established essential criteria for a meat starter culture.

DRY-CURED SAUSAGES

Microorganisms Involved

The fermentation of foodstuffs can be traced back thousands of years. It comes forth as a food preservation procedure although it was only in the more recent past that microorganisms were recognized as being responsible for the fermentation process. Traditional dry sausages rely on natural contamination by environmental flora, which occurs during slaughtering and increases during manufacturing. Each workshop has a specific "house flora" composed of useful microorganisms for fermentation and flavor development of sausages, but also of spoilage and pathogenic flora. Different species of lactic acid bacteria (LAB) and gram-positive, coagulase-negative cocci (GCC) are the microorganisms primarily responsible for sausage fermentation. LAB, in particular lactobacilli, contribute to the hygienic and sensory quality of meat products mainly through their carbohydrate and protein catabolism resulting in sugar depletion, pH reduction, production of antimicrobial agents, and generation of flavor compounds (Lücke 2000; Talon et al. 2002). GCC participate in color development and stabilization through a nitrate reductase activity that leads to the formation of nitrosomyoglobin. The antioxidant potential due to catalase and superoxide dismutase activity, the role in aroma formation via GCC proteolytic and/or lipolytic properties, and the removal of excess nitrate in the meat batter contribute to a stable and safe fermented product (Stahnke 2002). Molds and yeast traditionally play an important role by bringing about a characteristic surface appearance and flavors due to their well-known proteolytic and lipolytic activity. Inoculation of sausage batter with a starter culture composed of selected microorganisms improves the quality and safety of the final product and standardizes the production process (Lücke 2000; Demeyer 2003). The use of starter cultures in meat products has increased over the past 25 years, mainly in large-scale industrial processes. Commercial starter cultures in Europe are generally made up of a balanced mixture of LAB (*Lactobacillus, Pediococcus*) and GCC (*Staphylococcus, Kocuria*). In Mediterranean countries, yeasts and molds are also inoculated onto the surface, in contrast to northern technology in which smoke is applied. In spontaneously fermented European sausages, homofermentative lactobacilli constitute the predominant flora throughout ripening, *Lactobacillus sakei* and *Lactobacillus curvatus* dominating the fermentation process. Recently, studies carried out on small-scale processing plants confirmed on the bases of their genetic fingerprints that *Lb. sakei* was the predominant species isolated from artisanal fermented sausages (Fontana et al. 2005).

Target Pathogens and Spoilage Microbiota in Dry-Fermented Sausages: Shelf Stability and the Hurdle Effect

Fermented dry sausages are known as *shelf-stable products*, this term referring to those products that do not require refrigeration or freezing for safety and acceptable organoleptic characteristics because most often they are stored at room temperature. Shelf stability is due to a combination of factors known as the hurdle effect (Leistner 1996, 2000). Dry-fermented sausages (salami and salami-type sausage) become stable and safe through a sequence of hurdles, some of which are specifically included (NaCl, $NaNO_2$/$NaNO_3$, ascorbate) but others are indirectly created in the stuffed mix (low Eh, antagonistic substances, low water activity). By means of these hurdles, spoilage and food-poisoning bacteria are inhibited, whereas the desirable organisms, especially LAB, are hardly affected. Apart from LAB, GCC, and molds and yeasts involved in sausage fermentation, beef and pork meat as the major components of dry-cured sausages regularly contain pathogenic bacteria and are often implicated in the spread of foodborne diseases. Raw dry sausage materials (meat and casings) are the principal vehicles for pathogens and contaminating microorganisms. After dressing of carcasses, the microbiota comprises a mixture of mesophiles and psychotrophs. During meat chilling a selection is produced and mesophiles growth will no longer occur. Because most pathogens are mesophiles, meat obtained in good hygienic conditions would presumably not be implicated in sanitary risk. Still, the growth of pathogenic bacterial species overcoming the existent natural hurdles can occur. Food-poisoning *Staphylococcus aureus*, *Salmonellae*, and *Clostridium perfringes* have been traditionally implicated in fermented dry sausage contamination, and species within the genera *Campylobacter* and *Yersinia* as well as *L. monocytogenes* and shiga toxin-producing *E. coli* have recently emerged as pathogens of major public health concern. Dry-fermented sausage conditions, curing additives, and the presence of LAB starter cultures may act as significant hurdles for the control of these pathogens (Table 14.1). However, these hurdles are not sufficient to prevent the survival of *L. monocytogenes* or *E. coli* O157:H7 during the manufacturing process. An additional hurdle to reduce the risk of *L. monocytogenes* would be the use of competitive bacteriocin-producing starter cultures referred to as bioprotective cultures. Besides, strategies to overcome the barrier presented by the outer membrane in gram-negative bacteria, such as

Table 14.1. Effective hurdles inhibitory to pathogens present in dry fermented sausages.

Pathogen	Hurdles
Staphylococcus aureus	pH < 5.1; a_w < 0.86; bacteriocins
Salmonella	pH < 5.0; a_w < 0.95; NaCl/NaNO2
Clostridium perfringes	LAB (acid and bacteriocins)
Yersinia enterocolítica	LAB (acid)
Campylobacter jejuni	LAB (acid)
Listeria monocytogenes	a_w < 0.90; bacteriocins
Escherichia coli O157:H7	LAB (acid)

the use of chelating agents, will improve bacteriocin efficiency.

BIOPROTECTIVE CULTURES

Bioprotection: Definition

Preservation of foods using antagonistic microorganisms or their antimicrobial metabolites has been termed *bioprotection* or *biopreservation*. Antagonistic cultures added to foods to inhibit pathogens and/or extend the shelf life, while changing the sensory properties as little as possible, are referred to as *protective cultures* (Lücke 2000). The main objectives of bioprotection are to extend storage life, to enhance food safety, and to improve sensory qualities. LAB have a major potential for use in biopreservation because they are safe for consumption, and during storage they naturally dominate the microbiota of many foods. Due to their typical association with food fermentations and their long tradition as food-grade bacteria, LAB are Generally Recognized as Safe (GRAS). In addition, antimicrobial peptides produced by LAB can be easily broken down by digestive proteases so as not to disturb gut microbiota. LAB can exert a bioprotective or inhibitory effect against other microorganisms as a result of competition for nutrients and/or of the production of bacteriocins or other antagonistic compounds such as organic acids, hydrogen peroxide, and enzymes. A distinction can be made between starter cultures and protective cultures in which metabolic activity (acid production, protein hydrolysis) and antimicrobial action constitute the major objectives, respectively. Food processors face a major challenge, with consumers demanding safe foods with a long shelf-life, but also expressing a preference for minimally processed products, less severely damaged by heat and freezing and not containing chemical preservatives. Bacteriocins constitute an attractive option that could provide at least part of the solution.

Antimicrobial Peptides Produced by LAB Associated with Meat Products

Acid production as a result of carbohydrate catabolism is a common feature among LAB, although not all LAB can produce antimicrobial peptides during growth. The production of these antagonistic substances seems to be a common phenotype among LAB because numerous bacteriocins have been isolated over the last three decades, varying in size from small (<3 kDa), heavily post-translationally modified peptides, to large heat-labile proteins. Bacteriocins produced by LAB are a heterogeneous group of peptides and proteins. The latest revised classification scheme divides them into two main categories: the lanthionine-containing lantibiotics (class I) and the non-lanthionine–containing bacteriocins (class II); the large, heat-labile murein hydrolases (formerly class III bacteriocins) constitute a separate group called *bacteriolysins* (Cotter et al. 2005). Even though there has been a dramatic increase in the number of novel bacteriocins discovered in the past two decades, biochemical and genetic characterization has been carried out only in some of them. The majority of bacteriocins identified to date and produced by LAB belong to class I and II and the present description will principally focus on the one produced by LAB associated with meat and meat products (Table 14.2). Lactococcal bacteriocins are produced by several species of *Lactococcus lactis* isolated from dairy, vegetable, and meat products (Guinane et al. 2005). In favorable conditions, nisin has a wide spectrum of inhibition against gram-positive microorganisms. It has been extensively characterized and the precise structure of its molecule and the mechanism of action have been determined. Although nisin is the only commercially exploited lantibiotic to date, efforts have been made to develop applications for other lantibiotics. Among them, lacticin 3147, a two-peptide lantibiotic produced by *L. lactis* subsp. *lactis* DPC3147 isolated from Irish kefir grains, exhibits a bactericidal mode of action against food spoilage and pathogenic bacteria. The high heat stability and broad pH range of activity of lacticin 3147 make it attractive for use in the food industry. Even though most lactococcal bacteriocins were isolated from dairy and vegetable products, several nisin-producing *L. lactis* strains were isolated from fermented sausages, indicating the potential use of lactococci in meat fermentation. Nisin-producing *L. lactis* strains from Spanish fermented sausages and from traditional Thai fermented sausage were effective in inhibiting closely related LAB, *L. monocytogenes*, *Cl. perfringes*, *B. cereus*, and *St. aureus* (Leroy at al. 2006). Moreover, *Lb. sakei* L45 isolated from Norwegian dry sausages and *Lb. sakei* 148 from Spanish fermented sausages secrete lactocin S, a lantibiotic whose moderate spectrum of activity comprises LAB and *Clostridium* (Aymerich et al. 1998). The abundance of LAB strains producing lantibiotic bacteriocins shows the special relevance of these substances in fermented products; they are present in very different products and climatic environments. Class II bacteriocins include a very large group of small (<10 kDa) heat-stable peptides but, unlike lantibiotics, they are unmodified bacteriocins. Four different groups have been recently suggested (Cotter et al. 2005): class IIa include pediocinlike bacteriocins, class IIb two-peptide bacteriocins, class IIc cyclic bacteriocins, and class IId non-pediocinlike single linear peptides. Class IIa bacteriocins have a narrow spectrum of action but display a high specific activity against *L. monocytogenes*. Pediocinlike bacteriocins can be considered as the major subgroup among non-lantibiotic peptides, not only because of their large number, but also because of their significant biological activities and potential applications (Fimland et al. 2005). Pediocin PA-1 is produced by *P. acidilactici* isolated from America-style sausages, *P. pentosaceous* Z102 from Spanish-style sausages, *P. parvulus* from vegetable products, and *Lb. plantarum* WHE92 isolated from cheese. Sakacin A and K produced by *Lb. sakei* strains (Lb706 and CTC494), curvacin A produced by *Lb. curvatus* LTH1174, and enterocin A produced by *Enterococcus faecium* CTC492, all of them strains isolated from meat and fermented sausages, demonstrated to be active against other LAB, *L. monocytogenes* and *Clostridium*. In addition, *Leuconostoc gelidum* and *Leuconostoc mesenteroides* isolated from chill-stored vacuum-packaged meat secrete leucocin A, inhibitory against LAB, *L. monocytogenes*, and *Enterococcus faecalis*. *Carnobacterium piscicola* LV17B and V1 isolated from vacuum-packaged meat also produce bacteriocins such as carnobacteriocin B1 and B2, which are highly effective against *L. monocytogenes* and other LAB. *Lb. curvatus* L442, isolated from Greek traditional fermented sausages producing curvaticin L442, was also demonstrated to inhibit *L. monocytogenes*. Due to their high antilisterial potential, these bacteriocin producer LAB are of considerable interest as biopreservative cultures (Table 14.2). The two-peptide bacteriocins (class IIb) require the combined activity of both peptides to exert their antimicrobial activity. Lactococcin G and lactacin F produced by *L. lactis* and *Lactobacillus johnsonii* of dairy origin, respectively, isolated and characterized in the early 1990s, are the first reported two-peptide bacteriocins (Garneau et al. 2002). These bacteriocins have a narrow spectrum of inhibition against other lactobacilli, *Clostridium* and *E. faecalis*. Plantaricins EF and JK produced by *Lb. plantarum* C11 from vegetable origin and plantaricin S produced by *Lb. plantarum* LPCO10 from fermented green olives are among the best characterized two-peptide systems. On the other hand, lactocin 705, a two-peptide bacteriocin produced by *Lb. curvatus* CRL705 (formerly *Lb. casei*) isolated from Argentine fermented sausages, showed to be antagonistic toward other LAB and *Brochothrix thermosphacta* when tested in meat systems (Castellano et al. 2004). Class IIc bacteriocins comprise a few cycle peptides and class IId, also called non-pediocin single linear peptides, include the remaining isolated antimicrobial substances. Bacteriocins from class I and II are among the best characterized biochemically and genetically and the most likely to be used in food applications due to their high target specificity.

Use of Bacteriocins as Biopreservatives in Fermented Sausages

As previously described, fresh meat and fermented meat products provide an excellent environment for the growth of pathogenic and spoilage organisms. Properly processed meat should have a low pH and minimum numbers of contaminating bacteria, even though spoilage will occur if methods to restrict microbial growth are not applied. Microbial antagonism is used empirically in sausage fermentations where LAB accumulate lactic acid that inhibits meat foodborne pathogenic bacteria while coagulating soluble meat proteins, thereby reducing water-binding capacity and facilitating the drying of the product. However, the use of bacteriocinogenic strains as starter cultures in fermented sausages has been suggested as an extra hurdle to ensure product safety and quality. Bacteriocins can be introduced into food in at least three different ways: (1) as purified or semipurified bacteriocins added directly to food,

Table 14.2. Examples of bacteriocins produced by LAB isolated from meat and fermented sausages.

Producer strain	Bacteriocin	Class	Source	Active Against[a]
L. lactis BB24	Nisin	Ia	Spanish fermented sausages	4,5,7,9
L. lactis WNC20	Nisin Z	Ia	Thai fermented sausages	1,2,4,6,9
Lb. sakei 148, V18	Lactocin S	Ia	Spanish fermented sausages	1,4,5,6
Lb. sakei L45	Lactocin S		Norwegian fermented sausages	1,4,5,6
Lb. sakei LTH673, 674	Sakacin K, P	IIa	Beef	1,6,7
Lb. sakei I151	Sakacin P	IIa	Italian fermented sausages	7
Lb. sakei Lb706	Sakacin A	IIa	Beef, meat products	1,6,7
Lb. sakei CTC494	Sakacin K	IIa	Spanish fermented sausages	1,6,7
Lb. sakei MN	Bavaricin MN	IIa	Beef	7
Lb. brevis SB27	Brevicin 27	IId	Fermented sausages	1,2
Lb. curvatus LTH1174	Curvacin A	IIa	German meat products	1,2,7
Lb. curvatus CRL705	Lactocin 705	IIb	Argentine fermented sausages	1,2,3,8
Lb. curvatus FS47	Curvaticin FS47	IId	Minced beef products	1,2,6
Lb curvatus L442	Curvaticin L442	IIa	Greek fermented sausages	1,7
Lb. plantarum CTC305	Plantaricin A	IId	Spanish fermented sausages	1,7
Lc. gelidum UAL187	Leucocin A	IIa	Vacuum-packed meat	1,6,7
Lc. mesenteroides TA33a	Leucocin A	IIa	Vacuum-packed meat	1,6,7
Lc. mesenteroides L124		IIa	Greek fermented sausages	1,7
Lc. mesenteroides E131			Greek fermented sausages	7
Lc. carnosum TA11a	Leucocin A	IIa	Vacuum-packed meat	1,6,7
P. acidilactici PAC1.0	Pediocin PA-1/AcH	IIa	American-style sausages	1,4,5,7,8
P. acidilactici L50	Pediocin L50	IId	Spanish fermented sausages	1,4,5,6,7,8,9
P. pentosaceous Z102	Pediocin PA-1	IIa	Spanish fermented sausages	1,4,5,7,9
C. piscicola LV17B	Carnobacteriocin B2	IIa	Vacuum-packed meat	7
C. piscicola V1	Piscicocin V1a	IIa	Fish	1,6,7
C. piscicola LV17A	Carnobacteriocin A	IId	Processed meat	1,7
C. piscicola JG126	Ciscicolin 126	IIa	Spoiled ham	7
C. piscicola KLV17B	Carnobacteriocin B1/B2	IIa	Vacuum-packed meat	1,6,7
C. divergens 750	Divergicin 750	IId	Vacuum-packed meat	1,5,6,7
C. divergens LV13	Divergicin A	IId	Vacuum-packed meat	1
E. faecium CTC492	Enterocin B	IId	Spanish fermented sausages	1,5,7
E. faecium CTC492	Enterocin A	IIa	Spanish fermented sausages	7
E. casseliflavus IM416K1	Enterocin 416K1	IIa	Italian fermented sausages	7

[a]1: other LAB; 2: *B. cereus*; 3: *B. thermosphacta*; 4: *Cl. botulinum*; 5: *Cl. perfringes*; 6: *E. faecalis/faecium*; 7: *L. monocytogenes*; 8: Propionibacteria; 9: *Staph. aureus*.

(2) as bacteriocins produced in situ by bacterial cultures that substitute for all or part of the starter culture, and (3) as an ingredient based on a fermentate of a bacteriocin-producing strain. At present only nisin and pediocin PA-1/AcH are widely used in foods. Nisin is used mostly as Nisaplin (Danisco), a preparation containing 2.5% nisin. Pediocin PA1 for food biopreservation has also been commercially exploited as ALTA 2431 (Quest), based on a fermentate from a pediocin PA-1-producing strain of *P. acidilactici*, its use being covered by several U.S. and European patents. The major criteria to be taken into account for food application of bacteriocins are summarized in Table 14.3. Among class Ia bacteriocins, nisin has not been quite successful in meat products because of its low solubility, uneven distribution, and lack of stability. Lactococcal bacteriocins are not particularly adapted to sausage technology because they display sensitivity to some ingredients and manufacture conditions. However, the use of lacticin 3147-producing strains as starter cultures in the production of salami showed that the bacteriocin was produced throughout manufacture and that final overall characteristics were similar to the control. Its protective ability was also demonstrated in spiked sausages because there was a significant reduction of *L. innocua* and *Staph. aureus* (Scannell et al. 2001). Nevertheless, nisin producers isolated from fermented meat products seem to have some ecological advantages that may improve their effectiveness in these products. The narrow-spectrum *Listeria*-active bacteriocins seem to have a niche that provides a solution for foods in which the presence of *L. monocytogenes*, mainly as a postcontamination event, can be a problem. *Lb. sakei* sakacin producers, have been assayed in in situ experiments, in which they demonstrated a strong activity against *Listeria*. In fermented dry sausages in particular, *Lb. sakei* CTC494 sakacin K producer, was able to suppress the growth of *Listeria*, whose numbers significantly decreased compared to the initial values and to the standard starter non-bacteriocin producer (Hugas et al. 2002). *Lb. sakei* Lb706, which secretes sakacin A, was also able to reduce the number of *Listeria* in fresh Mettwurst. On the other hand, the inoculation in sausages as a starter culture of *Lb. sakei* LTH673, a sakacin K producer isolated from meat, failed to lead the fermentation, although it reduced the number of *Listeria* when compared to the non-bacteriocinogenic starter (Hugas et al. 1996). The meat isolate *Lb. curvatus* LTH1174, which secretes curvacin A, also proved to be highly competitive in sausage fermentations, dominating the fortuitous microbiota by reducing counts of artificially added *Listeria* compared to a nonbacteriocinogenic control (Hugas et al. 2002). On the basis of these results, *Lb. curvatus* LTH1174 shows a promising potential for use in European sausage fermentation. Pediocin-producing strains of *P. acidilactici* tested for their ability to reduce the counts of *L. monocytogenes* during chicken summer sausages fermentation showed significant reductions compared to non-pediocin–producing strains. Moreover, when the survival of *L. monocytogenes* was compared in German, American, and Italian-style fermented sausages using pediococci as a starter culture, higher reductions in *Listeria* numbers were observed in American-style sausages after fermentation and drying, with respect to German-style sausages prepared with a non-bacteriocinogenic culture. The higher fermentation temperature used in American-style sausages (25–35°C) may have a positive influence on the production of bacteriocins by *P. acidilactici*. Although the use of bacteriocin-producing enterococci in cheese manufacture is well

Table 14.3. Major criteria for the selection of bacteriocins and/or biopreservative cultures for dry sausage application.

GRAS status of producer strain
Adaptation to meat environment (preferably sausage isolates)
No gas or polysaccharide production
Noninhibitory to GCC involved in the mixed culture
Selection according to sausage formulation and technology. Recipe (NaCl, $NaNO_2/NaNO_3$, spices) and temperature used
Broad spectrum of inhibition or high specific activity against a specific pathogen (*L. monocytogenes*)
Heat stability of the bacteriocin produced
No associated health risks
Activity in food must lead only beneficial effects (improved safety, quality, flavor, improved health).

documented in the literature, information on their potential as biopreservatives in the meat industry is scarce. Bacteriocin-producing *E. faecium* strains have been included in the starter culture in Spanish-style dry-fermented sausage to effectively inhibit *Listeria*. The bacteriocin-producing *Enterococcus casseliflavus* IM416K1 isolated from Italian cacciatore sausage showed a high effectiveness in *L. monocytogenes* elimination; a listeriostatic effect was obtained when enterocin CCM4231 produced by a strain of *E. faecium* was added to the sausage batter during the manufacture of Hornád salami, a dry-fermented traditional Slovakian product (Lauková et al. 1999). In contrast, *E. faecium* CTC492, an enterocin A and B producer, was unable to exert any positive antilisterial effect on different meat products compared to the batches treated with enterocins A and B, this being attributed to an insufficient bacteriocin production (Aymerich et al. 2000). Because enterococci are hardly competitive in the early stages of the fermentation reducing listerial counts, bacteriocin-producing enterococcal strains may be used as adjunct cultures during sausage manufacture.

Even when many bacteriocinogenic strains inhibited *L. monocytogenes* in vitro and in meat systems, most studies on the in situ effect were carried out on single strains of the pathogen or have used high levels of contamination. Significant variations in the sensitivity of *Listeria* strains to the same bacteriocin as well as the emergence of spontaneous resistant mutants to individual bacteriocins were reported (Cotter et al. 2005). For improved control of target microorganisms and bacteriocin-resistant strains and species inhibition, the combined use of bacteriocins was proposed. The combined effect of nisin (class Ia), lactocin 705 (class IIb), and enterocin CRL35 (class IIa) against different *L. monocytogenes* strains in meat slurry showed no viable counts after incubation (Vignolo et al. 2000). The simultaneous addition of a nisin-curvaticin 13 combination also led to the absence of viable *Listeria* cells in broth, preventing the regrowth of bacteriocin-resistant mutants (Bouttefroy and Milliere 2000). It is generally assumed that a mixture containing more than one bacteriocin will be bactericidal to more cells in a sensitive population, because cells resistant to one bacteriocin would be killed by another.

Gram-negative Bacteria Inhibition

Although lantibiotics have wider spectra of inhibition than nonlantibiotics, none of the bacteriocins produced by LAB were effective in killing gram-negative bacteria. This inability is due to the protective outer membrane, which covers the cytoplasmic membrane and peptidoglycan layer of the cells. This asymmetrical membrane contains glycerophospholipids and lipopolysaccharide (LPS) molecules in which anchoring divalent cations are involved. Treatment with metal-chelating agents as EDTA generally results in removal by chelation of divalent cations with a consequent disruption of the outer membrane (Alakomi et al. 2003). Lactic acid and its salts are also potent outer membrane disintegrating agents as evidenced by their ability to cause LPS release. These permeabilizers increase the susceptibility to hydrophobic substances such as bacteriocins by enabling them to penetrate the cell wall. Accordingly, food-grade permeabilizers in combination with bacteriocins would be effective as part of the hurdle concept in inhibiting gram-negative bacteria. Various food-grade chelators (citrate, phosphate, EDTA, hexametaphosphate) have been examined for their ability to overcome the penetration barrier in *Salmonella enterica* and *E. coli* by rendering these species sensitive to hydrophobic antibiotics. When the deferred addition of bacteriocins produced by *Lb. curvatus* CRL705 (lactocin 705) and *L. lactis* CRL1109 (nisin) in combination with chelators was assayed, differences in *E. coli* strains sensitivity to bacteriocin/chelator combinations were reported, the sensitization by sodium lactate and EDTA combined with lactocin 705 being the most effective strategy (Belfiore et al. 2006). When the ability of chelators and nisin generated in situ to inhibit and inactivate Gram-negative bacteria was studied, *E. coli* was found to be more sensitive than salmonellas, but *Pseudomonas* showed the greatest susceptibility. The combination of sodium lactate and nisin also afforded increased protection against *Staph. aureus* and *Salmonella* species in fresh pork sausage, providing an alternative to sulphite addition. Moreover, the spectrum of nisin activity can be extended to a variety of pathogens including gram-negative bacteria when used together with food-grade enzymes such as lysozyme or combined with physical treatments. The effectiveness of high hydrostatic pressure to induce sublethal injuries to living foodborne bacteria cells was also assayed in combination with various bacteriocins (Hugas et al. 2002).

Influence of Curing Additives and Fermentation Conditions on Bacteriocin Effectiveness

The effect of bacteriocins as well as the bacteriocinogenic LAB in meat fermented products should consider the adequacy of the environment for bacteriocin stability and/or production. When purified or semipurified bacteriocins are added to fermented meat, a loss of activity is often detected as a direct consequence of meat components such as proteolytic enzymes, bacteriocin-binding protein, or fat particles. The endogenous enzymes present in raw meat would produce a degradation of the bacteriocin added or produced during ripening, thus impairing bacteriocin effectiveness. Inactivation by pasteurization of a meat slurry used to inoculate *L. monocytogenes* showed a cell count decrease by 2 logs in the presence of lactocin 705 compared to a nonpasteurized meat system, thus corroborating the meat protease effect (Vignolo et al. 1996). Lipids and phospholipids may also interfere with bacteriocin activity, as observed for nisin against *L. monocytogenes*, the activity of which decreases with increasing fat content, probably due to binding to fat globules. Fat was also shown to be responsible for a large loss of bacteriocin activity in a *Lb. sakei* CTC494 culture supernatant (Leroy and De Vuyst 2005). Even when nisin is apparently not the bacteriocin of choice in meat products, promising results were obtained when it was used as an alternative to reduce nitrite levels in cured sausages. Nisin application in meat products is affected by inadequate pH as well as low temperatures, both leading to poor solubility in the food matrix and less efficiency against pathogens, probably due to a decreased cell membrane fluidity. Moreover, because nisin is the only bacteriocin approved as a food ingredient by the Food and Drug Administration, the use of bacteriocin-producing LAB may be a better choice for introducing bacteriocins into meat products because they can provide a bacteriocin source over a longer period of time. In this regard, it seems important to apply strains from their own environment, naturally better adapted, more competitive, and more in agreement with the characteristics of the product and the technology used. Bacteriocin-producing LAB associated with meat and meat fermentations such as *Pediococcus*, *Leuconostoc*, *Carnobacterium,* and *Lactobacillus* spp. are likely to have greater potential as meat preservatives (Table 14.2). Bacteriocin-producing LAB have been used successfully in a number of cases to control adventitious populations in meat products of consistent quality and to delay spoilage. It has been shown that the previously described sausage isolates *Lb. curvatus* LTH1174, *Lb. curvatus* L442 and *Lb. sakei* CTC494 optimally produce bacteriocins under conditions of pH and temperature that prevail during European sausage fermentation. However, the performance of bacteriocinogenic LAB starter or adjuncts will be influenced by curing ingredients. When *E. faecium* CTC492 was used as a bioprotective culture in fermented sausages and hamburgers, it did not exert any positive antilisterial effects compared to enterocins A and B, this fact being attributed to a higher inhibition of the producer strain by refrigeration temperatures and sausage ingredients (Aymerich et al. 2000). Although $NaNO_2$ did not affect sakacin K produced by *Lb. sakei* CTC494, its presence decreased bacteriocin production because of its toxic effect on cell growth, this being enhanced by the lactic acid produced. Moreover, a protective effect of *L. monocytogenes* against the action of lactocin 705 by NaCl (3–5%) and $NaNO_2$ (200 µg/g) was demonstrated in a meat slurry at 20°C (Vignolo et al. 1998). Addition of spices to the sausage mixture also influences the effectiveness of bioprotective cultures. Pepper, nutmeg, rosemary, mace, and garlic decreased curvacin A production by *Lb. curvatus* LTH1174 in vitro; paprika was the only spice that increased it (Verluyten et al. 2004). Black pepper was shown to enhance the inhibitory activity of sakacin K against *L. monocytogenes* in vitro and in model sausages, a synergistic effect with nitrite, NaCl and Bac^+ starter cultures being observed (Hugas et al. 2002). However, the addition of some spices may have a positive influence on the metabolism of starter cultures due to the presence of micronutrients, especially manganese (Leroy and De Vuyst 2005).

NEW TRENDS FOR STARTER CULTURE UTILIZATION

Functional Meat Products: Functional Starter Cultures

As discussed above, a spectacular growth in the development of health products has occurred in the last decades as a result of consumer demands. The concept of health products includes what is known as *functional foods*. These are defined as foods that are used to prevent and treat certain disorders and diseases, in addition to their nutritional value per se (Jiménez-Colmenero et al. 2001). Meat and meat products are important sources of proteins, vitamins, and minerals, which could be considered functional compounds. However, they also contain fat, saturated fatty acids, cholesterol, and salt, which are responsible for the unfortunate image of meat and meat products because of their association with cardiovascular diseases, some types of cancer, and obesity. The use of functional ingredients to obtain healthy meat products constitutes a suitable strategy for meat processors by eliminating or reducing components considered harmful or by adding ingredients considered beneficial for health, such as modified fatty acid profiles, antioxidants, dietary fiber, and probiotics (Fernández-Ginés et al. 2005).

The performance of commercial starter cultures has been questioned because their behavior is not always the same when applied to other types of fermented meat products. It is crucial, therefore, to give traditional producers the means to produce safe and standardized products while preserving their typical sensory quality. As a response to these needs and to the demands for health products, the use of *new generation* starter cultures has already been suggested (Hansen 2002; Leroy et al. 2006). The so-called *functional starter cultures* contribute to food safety by producing antimicrobial compounds such as bacteriocins and also provide sensorial, technological, nutritional, and/or health advantages.

Starter Cultures with Probiotic Potentialities

Probiotics represent an expanding research area and the possibilities of the use of probiotics in dry sausage manufacturing processes have been addressed recently (Työppönen et al. 2003). Probiotics have been used mainly in dairy products such as yogurt and other fermented milks. The use of fermented sausages as probiotic food carriers has been postulated because these products are not heated and harbor high numbers of LAB. However, in order to use probiotics in fermented sausages, several characteristics of the culture, in addition to technological, sensory, and safety properties, should be taken into account. The evaluation of probiotic cultures suitable for use as starter cultures for dry-fermented sausages should involve the ability to become dominant in the final product because fermented meat products contain a natural background microbiota, this ability being mainly achieved by a fast acid and bacteriocin production. Specific probiotic properties such as bile tolerance, resistance to stomach acidity, and adhesion to the intestinal epithelium in addition to antimicrobial activity will ensure strain survival in the human gastrointestinal tract. Even though LAB are considered GRAS organisms, specification of origin, nonpathogenicity, and antibiotic resistance characteristics of the strains should also be assessed. Moreover, other LAB abilities that could improve safety and impart a healthy note to meat fermented products include moderate cholesterol-lowering action, bioactive peptide generation from food proteins and conjugated linoleic acid (CLA) production. In fact, studies feeding lactating ewes with selected strains of *Lactobacillus acidophilus* caused an important drop in the mean cholesterol level of meat cuts (Lubbadeh et al. 1999). The proteolytic system of LAB can contribute to the release of health-enhancing bioactive peptides from food proteins, thus improving absorption in the intestinal tract. They can also stimulate the immune system, exert antihypertensive or antithrombotic effects, display antimicrobial activity, or function as carriers for minerals, especially calcium. Bioactive peptide may be generated during meat aging as well as during meat fermentation by LAB (Korhonen and Pihlanto 2003). The ability to produce CLA, another health promoter, was described in vitro by propionibacteria, bifidobacteria, and lactobacilli. However, human clinical studies are needed to confirm the health-promoting effects of probiotics in sausages.

Traditional fermented sausages constitute a highly appreciated specialty with gastronomic value and are a rich source of biodiversity, the deliberate use of which in industrial processes could help to introduce quality and safety advantages. The use of bacteriocin-producing strains that are well adapted to the sausage environment will provide an extra hurdle in a multihurdle integrated system. Despite the above-mentioned advantages of biopreservation, it should be borne in mind that bacteriocins are not meant to be used as the sole means of food preservation.

REFERENCES

H Alakomi, M Saarela, J Helander. 2003. Effect of EDTA on *Salmonella enterica* serovar *Typhimurium* involves a component not assignable to lipopolysaccharide release. Microbiol 149:2015–2021.

T Aymerich, M Artigas, M Garriga, J Monfort, M Hugas.2000. Effect of sausage ingredients and additives on the production of enterocins A and B by *Enterococcus faecium* CTC492. Optimization of in vitro production and anti-listerial effect in dry fermented sausages. J Appl Microbiol 88:686–694.

T Aymerich, M Hugas, J Monfort. 1998. Review: Bacteriocinogenic lactic acid bacteria associated with meat products. Food Sci Technol Int 4:141–158.

C Belfiore, P Castellano, G Vignolo. 2007. Reduction of *Escherichia coli* population following treatment with bacteriocins from lactic acid bacteria and chelators. Food Microbiol 24:223–229.

A Bouttefroy, J Milliere. 2000. Nisin-curvaticin 13 combinations for avoiding the regrowth of bacteriocin resistant cells of *Listeria monocytogenes* ATCC15313. Int J Food Microbiol 62:65–75.

P Castellano, W Holzapfel, G Vignolo. 2004. The control of *Listeria innocua* and *Lactobacillus sakei* in broth and meat slurry with the bacteriocinogenic strain *Lactobacillus casei* CRL705. Food Microbiol 21:291–298.

D Church. 2004. Major factors affecting the emergence and re-emergence of infectious diseases. Clin Lab Med 24:559–586.

P Cotter, C Hill, P Ross. 2005. Bacteriocins: Developing innate immunity for food. Nature Rev Microbiol 3:777–798.

D Demeyer. 2003. Meat fermentation: Principles and applications. In: Y Hui, L Goddik, A Hansen, J Josephsen, W Nip, P Stanfield, F Toldrá, eds. Handbook of Food and Beverage Fermentation Technology. New York: Marcel Dekker, pp. 353–367.

M Elmi. 2004. Food safety: Current situation, unaddressed issues and the emerging priorities. East Mediterr Health J 10:794–800.

J Fernández-Ginés, J Fernández-López, E Sayas-Barberá, J Pérez-Alvarez. 2005. Meat products as functional foods: A review. J Food Sci 70: 37–43.

G Fimland, L Johnsen, B Dalhaus, J Nissin-Meyer. 2005. Pediocin-like antimicrobial peptides (class IIa bacteriocins) and their immunity proteins: Biosynthesis, structure, and mode of action. J Peptide Sci 11:688–696.

C Fontana, P Cocconcelli, G Vignolo. 2005. Monitoring the bacterial population dynamics during fermentation of artisanal Argentine sausages. Int J Food Microbiol 103:131–142.

S Garneau, N Martin, J Vederas. 2002. Two-peptide bacteriocins produced by lactic acid bacteria. Biochimie 84:577–592.

C Guinane, P Cotter, C Hill, R Ross. 2005. Microbial solutions to microbial problems; lactococcal bacteriocins for the control of undesirable biota in food. J Appl Microbiol 98:1316–1325.

E Hansen. 2002. Commercial bacterial starter cultures for fermented foods of the future. Int J Food Microbiol 78:119–131.

M Hugas, M Garriga, M Aymerich, J Monfort. 2002. Bacterial cultures and metabolites for the enhancement of safety and quality in meat products. In: F Toldrá, ed. Research Advances in the Quality of Meat and Meat Products. Kerala, India: Research Signpost, pp. 225–247.

M Hugas, B Neumeyer, F Pagés, M Garriga, W Hammes. 1996. Comparison of bacteriocin-producing lactobacilli on *Listeria* growth in fermented sausages. Fleischwirtschaft 76:649–652.

F Jiménez-Colmenero, J Carballo, S Cofrades. 2001. Healthier meat and meat products: Their role as functional foods. Meat Sci 59:5–13.

H Korhonen, A Pihlanto. 2003. Food-derived bioactive peptides, opportunities for designing future foods. Curr Pharm Des 9:1297–1308.

A Lauková, S Czikková, S Laczková, P Turek. 1999. Use of enterocin CCM 4231 to control *Listeria monocytogenes* in experimentally contaminated dry fermented Hornád salami. Int J Food Microbiol 52:115–119.

L Leistner. 1996. Food protection by hurdle technology. Bull Jpn Soc Res Food Prot 2:2–26.

———. 2000. Basic aspects of food preservation by hurdle technology. Int J Food Microbiol 55:181–186.

F Leroy, L, De Vuyst. 2005. Simulation of the effect of sausage ingredients and technology on the functionality of the bacteriocin-producing *Lactobacillus sakei* CTC494 strain. Int J Food Microbiol 100:141–152.

F Leroy, J Verluyten, L De Vuyst. 2006. Functional meat starter cultures for improved sausage fermentation. Int J Food Microbiol 106:270–285.

W Lubbadeh, M Haddadin, M Al-Tamimi, R Robinson. 1999. Effect on the cholesterol content of fresh lamb of supplementing the feed of Awassi ewes and lambs with *Lactobacillus acidophilus*. Meat Sci 52: 381–385.

F-K Lücke. 2000. Utilization of microbes to process and preserve meat. Meat Sci 56:105–115.

A Scannell, G Schwarz, C Hill, R Ross, E Arendt. 2001. Pre-inoculation enrichment procedure enhances the performance of bacteriocinogenic *Lactococcus lactis* meat starter culture. Int J Food Microbiol 64:151–159.

L Stahnke. 2002. Flavour formation in fermented sausage. In: F Toldrá, ed. Research Advances in the Quality of Meat and Meat Products. Kerala, India: Research Signpost, pp. 193–223.

R Talon, S Leroy-Sétrin, S Fadda. 2002. Bacterial starters involved in the quality of fermented meat products. In: F Toldrá, ed. Research Advances in the Quality of Meat and Meat Products. Kerala, India: Research Signpost, pp. 175–191.

P Tarrant. 1998. Some recent advances and future priorities in research for the meat industry. Meat Sci 49(Suppl 1):S1–S16.

S Työppönen, E Petäjä, T Mattila-Sandholm. 2003. Bioprotectives and probiotics for dry sausages. Int J Food Microbiol 83:233–244.

J Verluyten, F Leroy, L De Vuyst. 2004. Effects of different spices used in production of fermented sausages on growth of and curvacin A production by *Lactobacillus curvatus* LTH1174. Appl Environ Microbiol 70:4807–4813.

G Vignolo, S Fadda, M Kairuz, A R Holgado, G Oliver. 1996. Control of *Listeria monocytogenes* in ground beef by lactocin 705, a bacteriocin produced by *Lactobacuillus casei* CRL705. Int J Food Microbiol 29:397–402.

———. 1998. Effects of curing additives on the control of *Listeria monocytogenes* by lactocin 705 in meat slurry. Food Microbiol 15:259–264.

G Vignolo, J Palacios, M Farías, F Sesma, U Schillinger, W Holzapfel, G Oliver. 2000. Combined effect of bacteriocins on the survival of various *Listeria* species in broth and meat system. Curr Microbiol 41:410–416.

15
Starter Cultures: Yeasts

M-Dolores Selgas and M-Luisa Garcia

INTRODUCTION

Fermentation is one of the oldest and most important methods of preserving meat. The process involves the anaerobic breakdown of carbohydrates yielding several compounds (alcohols, organic acids, mainly lactic acid) whose nature depends on the type of food and the correct growth of responsible microorganisms. In meat, fermentation has been used especially in comminuted products such as sausages, which differ from fresh meat by the addition of salts or sugars. These compounds select the fermentative microbiota, which will grow at relatively high osmotic pressure.

Dry-fermented sausages are one of the most important meat products. They have a long storage life due to many factors, such as the presence of salts, the lactic acid produced by lactic microbiota, the nitrite produced from nitrate thanks to members of the genus *Kocuria*, and the drying that occurs during the ripening process. All these factors modify the original meat and contribute to the development of the texture and flavor and to the selection of the desired microbiota (Ordóñez et al. 1999).

However, although bacteria are the predominant factors in fermented sausages, molds and yeasts have also been tried for curing and fermenting sausages.

PRESENCE OF YEASTS ON MEAT SAUSAGES

Yeasts are found naturally on fresh meat, an excellent growth media, from where they are easily spread during slaughtering (Dillon and Board 1991). There, the yeast level is close to 10^2–10^4 cfu/g and coexists with gram-negative bacteria, which range between 10^3–10^7 cfu/g. During low-temperature storage of fresh meat, yeast counts may increase and eventually dominate the microbiota, reaching values close to 10^6 cfu/g. Competition between both groups of microorganisms favors the increase and predominance of yeasts. Although their counts are generally lower than those of spoilage bacteria, they can proliferate and reach amounts enough to form a visible surface slime (Cook 1995).

Yeasts can also be considered habitual components of microbiota growing on fermented sausages, and their origin is mainly related to the environment and the meat used as raw material. In fermented meats, the lactic acid produced by bacteria and the low a_w resulting from the presence of salts or a dehydration process constitute modified environmental factors that hinder the bacterial growth and favor the development of natural competitors. Thus, yeasts use all the nutrients and energy and grow quickly and easily (Dillon and Board 1991).

Yeasts can grow at pH, water activity, and temperature values habitual in fermented sausages (Monte et al. 1986; Hammes and Knauf 1994). Many species can even grow as easily at pH 4 as at pH 6 and are able to maintain a neutral intracellular pH when the external environmental is very acid. This ability has been attributed to a relative impermeability of the plasmatic membrane or the existence of a mechanism that drives out the protons that enter by passive diffusion. As a consequence, pH is not an active controller to yeast growth in these foods.

The decrease of a_w in meat products, as a consequence of either the increase in salts or the dehydration process, scarcely influences yeasts. In this sense, it has been described as *D. hansenii* and has been isolated from low a_w foods like sugar syrup (Marquina et al. 2001). However, growth slows when the external metabolite concentration is close to 50–60% and even when the solute concentration is higher than 70%. Tolerance at a high level of metabolites is due to the liberation in the cytoplasm of compatible metabolites, mainly polyalcohols (glycerol) that balance intra/extracellular a_w.

Although yeasts are able to grow in anaerobic media, the majority of species have a strictly aerobic metabolism. This favors yeast growth on the surface of meat products. When yeasts arrive to processed meats as sausages, they can grow on the outer surface, where they form a firm and white external coat. This coat controls water loss and permits uniform dehydratation. Moreover, this coat gives to the products a peculiar and characteristic appearance that is considered a criterion of quality due to the favorable whiteness aspect that it gives to sausages (Leistner 1995; Coppola et al. 2000; Selgas et al. 2003). The oxygen requirement restricts their development to near the surface. However, in several cases these microorganisms have been isolated in the innermost parts of the products (Grazia et al. 1986; Selgas et al. 2003). This is mainly related to the growth of facultative anaerobic metabolism of several yeast species, which require only oxygen for the production of cell wall constituents such as fatty acids (Deak and Beuchat 1996). It is possible to isolate them even into the original mince of ripened sausages. These yeasts are the most important because they can grow at low oxygen levels, and the final metabolite of the fermentation process (gas and volatiles products) can modify the food properties (Leyva et al. 1999).

Regarding temperature, it has been described that yeasts are sensitive to heat from 38–40°C; at 50–60°C, cellular death occurs in the majority of species. Thus, all are able to grow at the temperatures used in the fermentation of meat.

Hence, yeasts can grow on sausages. The most frequently described on meat and meat products are the following: *Candida*, *Cryptococcus*, *Debaryomyces*, *Hansenula*, *Hypopichia*, *Kluyveromyces*, *Leucosporidium*, *Pichia*, *Rhodosporidium*, *Rhodotorula*, *Trichosporon*, *Torulopsis*, and *Yarrowia*, with *Debaryomyces* being the most commonly isolated (Dillon and Board 1991; Cook 1995; Coppola et al. 2000; Osei Abunyewa et al. 2000).

The presence of yeasts in fermented raw sausages has been studied less than bacteria and molds due to the difficulties associated with isolation and identification and the continuous modifications in the taxonomy. Traditionally, the identification of yeast involves the use of numerous morphological and physiological tests (Encinas et al. 2000).

Samelis et al. (1994) founded a clear predominance of *Debaryomyces* in salami; lower levels of *Candida* and *Pichia*; and very low levels of *Cryptococcus*, *Torulopsis*, and *Trichosporon*. Metaxopoulos et al. (1996) isolated 100 yeast strains during the fermentation and ripening stages of Greek dry salami. The most common were ascomycetous yeasts, in both perfect and imperfect states. *Debaryomyces* was again the most abundant and represented 66% of the total. *Candida famata*, *C. zeylanoides*, *C. guilliermondii*, *C. parapsilosis*, and *C. kruisii* were the ascomycetous species identified. Basidiomicetous yeasts were represented by *Candida humicola*, *Cryptococcus albidus*, *Cr. skinneri*, and *Trichosporon pullulans*. Due to the prevalence of *D. hansenii* in all batches, these authors suggested its potential use as starter culture. Wolter et al. (2000) identified the same genera in salami and other dry sausages of South Africa; they also identified *Sporobolomyces*, *Torulaspora*, and *Saccharomyces*. Encinas et al. (2000) identified D. *hansenii* as the dominant species in Spanish fermented sausages; other species like *Trichosporon*, *Citeromyces*, and *Candida* were also present. Coppola et al. (2000) isolated 79 yeast strains from fermented sausages, which were grouped in four genera: *Debaryomyces*, *Candida*, *Trichosporon*, and *Cryptococcus*. Most of them (40%) were classified as *D. hansenii*, by far the most predominant yeast, especially during the first few days of ripening. Similar data were reported by Selgas et al. (2003), who isolated yeast strains from Spanish dry-fermented sausages manufactured without starter culture, with *C. famata*, *T. mucoides*, *Y. lipolytica*, and *D. hansenii* being the most abundant species.

In any case, the composition of the mycobiota changes during ripening of fermented sausages. A change from a yeast-dominated to a mold-dominated pattern has been observed. During ripening, yeasts undergo a selection process mainly related to the decrease of the a_w, but in general terms, yeast counts are lower than those of bacteria. During early ripening, yeasts are the most important component of mycobiota, but after 2 weeks the numbers of mold and yeast are in balance (Andersen 1995). Roncalés et al. (1991) studied the evolution of yeasts and molds for a period of 28 days. They observed initial values of 10^2–10^3 cfu/cm^2 and an increase during the 10 days of ripening that reached levels of

10^5–10^6 cfu/cm^2 after 25 days. Boissonnet et al. (1994) reported levels of 10^3–10^4 yeast/g after 3 weeks of ripening. Osei Abunyewa et al. (2000) described initial values of 10^3 cfu/g in commercial salami, but this number increased after 12 days of maturation, reaching a maximum of 10^5 cfu/g at day 20. Encinas et al. (2000) observed an initial count close to 10^4 cfu/g and a final count of 10^5 cfu/g in dry-fermented sausages. Coppola et al. (2000) obtained yeast from the core of the product that peaked at 10^4 cfu/g and remained steady until the end of ripening. By contrast, counts from external parts reached a maximum of 10^6 cfu/g after 3 weeks.

It is also important that the most common genera are present in low concentrations compared with the levels of general microbiota. For that, several authors considered that their participation on the ripening of sausages is very low (Cook 1995). Although these yeast levels are relatively low, 10^6 cfu/cm^2 is considered equivalent to a biomass of 10^8 bacterias/cm^2, a number considered sufficient to have a considerable effect in meat products. In this context, it should be remembered that the biomass of a yeast cell is approximately 100 times greater than the biomass of a bacterial cell (Gill and Newton 1978; Dillon and Board 1991).

The evolution of yeast population during the fermented sausages manufacture process is influenced by different variables in which it is possible to consider the type of products, the influence of the environment of the factory and other factors, like the degree of smoking. So, Encinas et al. (2000) observed that the type of manufacture (industrial and artisanal) and sausage diameter were the most influential variables, with differences in yeast counts statistically significant from the second stage of manufacture onward; they observed how the use of a lactic acid starter culture and sorbate control the yeast growth and contribute to the decrease of their counts in relation to those of artisanal manufacture. Sausage diameter is also an important variable; the availability of oxygen is higher in smaller sausage and explains the higher counts in these products.

Yeasts are sensitive to the smoking process (Leistner 1995), and hence lower counts have been observed in smoked sausages than in the homologous nonsmoked sausages (Encinas et al. 2000). Several studies have been carried out in order to know the effect of spices on yeast growth. These studies have found an inhibitory effect of garlic (Asehraou et al. 1997; Olesen and Stahnke 2000).

The presence of yeasts is also detected in dry-cured hams. Their evolution during the ripening has been extensively described. In Iberian ham, Núñez et al. (1996) detected the lowest yeasts counts before the salting period, probably due to the predominance of bacteria, mainly micrococci. The highest counts were founded at the end of the postsalting period and during the drying stage, due to increase of temperature. Initial counts close to 10^4 cfu/g increased to 10^6–10^7 cfu/g during these periods. However, the subsequent decrease of a_w, and probably that of some essential nutrients, led to a decrease until counts close to 10^4–10^6 cfu/g (Rodriguez et al. 2001). Similar results were obtained by Huerta et al. (1988) in other dry-cured Spanish hams.

The dominant species described belong to the genera *Debaryomyces*, *Pichia*, *Rhodotorula*, and *Cryptococcus*; *Candida* and *Saccharomycopsis* occurred in low numbers (Cook 1995). Molina et al. (1990) isolated species of *Hansenula* (67%), *Rhodotorula* (19%), *Cryptococcus* (9%), and *Debaryomyces* (5%). However, there were some differences with respect to manipulation methods, geographic area of manufacture, and contamination of the raw meat. In any case, due to the characteristics of this type of meat product, all yeast growing on ham must be able to grow at high NaCl levels (8%) (Cook 1995).

D. hansenii and *C. zeylanoides* are the predominant species in the Iberian ham, although *C. blankii*, *C. intermedia*, *Pichia corsonii*, and *Rhodotorula rubra* have also been detected (Núñez et al. 1996). Yeast levels generally decrease during ripening, but it is possible to find them at the end of long ripening time if a_w is higher than 0.85. During salting and postsalting, new yeasts appear on the ham surface. Most of them disappear during ripening, except *Debaryomyces* spp., which are the only yeasts able to grow at $a_w < 0.85$ (Rodriguez et al. 2001).

Saldanha-da-Gama et al. (1997) isolated several species from ham, with *D. hansenii* being the predominant strain and *D. polymorphus*, *Cryptococcus laurentii*, and *C. humicolus* presented in a low number.

Table 15.1 lists a selection of yeasts species that have frequently been isolated from meat products.

ROLE OF YEASTS ON MEAT PRODUCTS

Fermented sausages are a very complex system. During fermentation and ripening, many reactions occur in which the final products of some constitute the substrates of others. Most of the different compounds contribute to the flavor and give to final products sensory properties different from those of the meat used as raw material.

Table 15.1. Yeasts species frequently isolated from meat products.

Yeasts	Synonym	Cured Meats	Fermented Sausages	Ham
Candida spp.			1,14	16
C. blanckii				11,13
C. brumptii		14		
C. catenulata		14		15
C. curvata		14	14	14
C. famata			1,2,3	
C. gropengiesseri			9	
C. guilliermondii			3	
C. haemulonii			9	
C. humicola	*Cryptococcus humicola*		1,3	
C. iberica			14	
C. incommunis			7	
C. intermedia			5,8	11,13
C. kruisii			3	
C. lipolytica		14	1,14	14
C. parapsilosis		14	1,3,5,14	
C. rugosa		14	14	15
C. torulopsis			1	
C. tropicalis			14	14
C. zeylanoides			1,3,5,6,9,14	11,13
Citeromyces matritensis			5	
Cryptococcus spp.			4	16
C. skinneri			3	
C. laurentii			6,14	10
C. hungaricus			6	
C. albidus			3,7,9,14	
C. humicola		10		10
Debaryomyces hansenii		10,14	2,3,4,5,6,7,8,9,13	10,11,13,14,15,16
D. kloeckeri	*D. hansenii*	14	8,14	
D. marama			3	12
D. matruchoti	*D. hansenii*		8	
D. occidentalis			9	
D. nicotianae		14	14	
D. polymorphus			3,9	10
D. subglobosus		14	14	
D. vanriji			6,9	

Table 15.1. (*Continued*)

Yeasts	Synonym	Cured Meats	Fermented Sausages	Ham
Galactomyces geotrichum			9	
Geotrichum candidum			14	
Hansenula anomala var. anomala	*Pichia anomala*	14	8	
Hansenula citerii				16
H. holstii				16
Kluyveromyces spp.			1	
Pichia spp.			1,4	
P. carsonii	*Debaryomyces carsonii*		1	11,13
P. etchellsii	*Debaryomyces etchellsii*		1	
P. farinosa			9	
P. haplophila		14		
P. philogaea			9	
Rhodotorula spp.			1,14	16
R. glutinis			1,8,14	
R. rubra	*R. mucilaginosa*		6	11,13
R. minuta		14	9,14	
R. mucilaginosa	*R. rubra*	14	6,9	14
Saccharomyces spp.			1	
S. rosei	*Torulaspore delbrueckii*		8	
S. cerevisiae			6	
Saccharomycopsis spp.				16
Sporobolomyces roseus		14	6,9,14	14
Sterigmatomyces halophilus		9		
Torulaspora delbrueckii			6,9	
Torulopsis spp.			4,14	
T. apicola	*Candida apicola*		8	
T. candida		14	8,14	
T. etchellsii		14	14	
T. famata		14	14	14
T. pulcherrima	*Metschnikowia pulcherrima*		8	
T. sphaerica	*Kluyveromyces lactis*		8	
Trichosporon spp.			4,14	
T. mucoides			2	
T. pullulans		14	3,7,14	

(*Continues*)

Table 15.1. *(Continued)*

Yeasts	Synonym	Cured Meats	Fermented Sausages	Ham
T. beigelii	*T. cutaneum*		6,9	
T. terrestre	*Arxula terrestris*		7	
T. ovoides			5	
T. cutaneum			8	
Yarrowia lipolytica			2,5,6,9	15

Sources: 1: Boissonet et al. (1994); 2: Selgas et al. (2003); 3: Metaxopoulos et al. (1996); 4: Samelis et al. (1994); 5: Encinas et al. (2000); 6: Wolter et al. (2000); 7: Coppola et al. (2000); 8: Comi and Cantoni (1983); 9: Osei Abunyewa et al. (2000); 10: Saldanha-da-Gama et al. (1997); 11: Núñez et al. (1996); 12: Monte et al. (1986); 13: Rodríguez et al. (2001); 14: Dillon and Board (1991); 15: Huerta et al. (1988); 16: Molina et al. (1990).

The role played by bacteria is reasonably well known and their contribution to the flavor of fermented sausages has been established. In general terms, their implication in the biochemical changes depends on their enzymatic capacity against proteins and fat, the main substrate of these meat products (Hierro et al. 1997; Ordóñez et al. 1999; Tjener et al. 2003). Similar considerations can be made in relation to the role played by molds. A direct relationship between molds and flavor has been demonstrated in both sausages and hams (Toledo et al. 1997; Selgas et al. 1999; Bruna et al. 2001; García et al. 2001; Rodriguez et al. 2001).

Traditionally, it has been believed that the presence of yeasts enhances the appearance and the flavor of the meat products (Boissonnet et al. 1994; Cook 1995; Metaxopoulos et al. 1996), and a contribution to the color of fermented sausages has also been described, related to the consumption of oxygen that favors the curing reactions and the establishment of the appealing red pigment (Lücke and Hechelmann 1987).

Lipolytic and proteolytic activities have been described from several yeasts isolated from fermented sausages: *Debaryomyces*, *Pichia*, *Trichosporon*, and *Cryptococcus* (Comi and Cantoni 1983; Huerta et al. 1988; Cook 1995; Selgas et al. 2003). Miteva et al. (1986) described how the presence of *Candida utilis* enhances the flavor of both smoked and unsmoked fermented sausages, due probably to its lipolytic activity. Molina et al. (1991) studied the lipolytic activity of *Cryptococcus albidus* in model meat systems inoculated with several microorganisms. The volatile and nonvolatile acidity generation observed was much lower than *L. curvatus*, *P. pentosaceous*, or *S. xylosus*, which are commonly used in meat fermentation.

Changes in flavor are assumed to be a consequence of the formation of volatiles (Ordóñez et al. 1999). Many species of yeasts are able to produce volatiles esters (Stam et al. 1998). Peppard and Halsey (1981) associated yeasts with the reduction of carbonyls to alcohols, with 2-methylpropanol, 2- and 3-methylbutanol, and 2-phenylethanol being the main ones produced. It is also believed that yeasts delay the autoxidative reactions and the onset of rancidity that take place in the fat. Yeasts' contribution seems to be related to the consumption of oxygen and the catalase production with the consequent peroxides degradation (Lücke and Hechelmann 1987).

There is great interest in the metabolism of branched-chain amino acids by yeasts, due to the formation of branched-chain alcohols and their contribution to the flavor (Flores et al. 2004). Olesen and Stahnke (2000) studied the influence of *C. utilis* on the aroma formation in fermented sausages and model minces and observed high metabolic activity producing several volatile compounds, particularly esters. They also observed an ability to ferment valine, yielding important compounds for the aroma of sausages such as 2-methylpropanoic acid and the respective acetate and ethyl esters in addition to 2-methylpropanol. This metabolic fate also seems to be shared by the two other methyl branched amino acids, isoleucine and leucine. These authors indicated that these complicated results could be related to the microbial interaction between *C. utilis* and other bacteria, mainly staphylococci.

D. hansenii is the most commonly isolated yeast from sausages. This suggests that it could have an important effect on flavor (Gehlen et al. 1991), but contradictory data have been published in this regard. Jessen (1995) indicated that the presence of *D. hansenii* influences the final flavor because of the enzymatic activity against proteins and fat. This author also reported that it leads to higher oxygen consumption, which favors the color formation. Lücke and Hechelmann (1987) related the presence of *D. hansenii* with a protective effect against the influence of light and oxygen through the formation of a dense white superficial coat (Leistner 1995) that delays fat oxidation and enhances the flavor.

Encinas et al. (2000) found a relationship between the presence of this yeast and the development of the most desirable organoleptic characteristics of fermented sausages. Santos et al. (2001) studied the activity of *D. hansenii* on sarcoplasmic proteins isolated from pork muscle and observed that this yeast showed good potential for the hydrolysis of these proteins by generating several polar and nonpolar peptides and free amino acids. Increases in Tyr, Val, Met, Leu, Ile, Phe, and Lys in excess of 50% were observed.

Flores et al. (2004) manufactured fermented sausages inoculating *Debaryomyces hansenii* and a mixture of bacterial starter cultures. They observe significant increases in compounds like ethyl 2-methyl-propanoate and ethyl 2- and 3-methyl butanoate, related to the sausage aroma (Stahnke 1994), when sausages were inoculated with *D. hansenii* at the level of 10^6 cfu/g. They also observed that these sausages showed a marked decrease in volatile compounds that arise from lipid oxidation, such as aldehydes and hydrocarbons. This suggests that the yeast has an impact on aroma by inhibiting the development of rancidity and generating esters that contribute to the flavor. Durá et al. (2004) described how the generation of volatile compounds from branched-chain amino acids by *D. hansenii* was negatively affected by the presence of salt, and the decrease of pH raised the yield of alcohols and aldehydes. These studies seem to indicate that the incorporation of this yeast as starter culture could favorably modify the flavor pattern of dry-cured sausages. However, Olesen and Stahnke (2000) observed a little impact on the production of volatile compounds in model minces; these authors pointed out the possibility that the fungistatic effect of garlic present on the experimental sausages was responsible for an inhibition of growth with the corresponding decrease of volatile compounds.

Yeasts' influence on aroma has also been attributed to their ability to metabolize organic acids, mainly lactic and acetic acid. It has been observed that lactic acid levels were higher in sausages with the lowest yeast counts; in this sense, a definite synergistic relationship between LAB and yeast has been described (Campbell-Plat 1995; Wolter et al. 2000). Moreover, their deaminase activity produces a slight increase in pH that could have influence on the flavor (Gehlen et al. 1991; Larpent-Gourgaut et al. 1993; Durá et al. 2002).

Other authors disagree with the results above described. Lizaso et al. (1999) reported that the enzymatic action of yeasts can be considered of secondary importance in the manufacture of Spanish fermented sausages because yeast counts were lower than those of other microbial groups. Selgas et al. (2003) studied the technological ability of 172 yeast strains isolated from the surface of artisanal sausages. Four strains, belonging to *Debaryomyces*, *Trichosporon*, *Candida*, and *Yarrowia*, were selected because of their high hydrolytic activity against fat and proteins isolated from pork meat. These active strains were inoculated onto sausages or into the mince, and sensory properties were evaluated after ripening. The organoleptic characteristics, irrespective of inoculation site, did not differ significantly from those of the control noninoculated batches. These authors concluded that the influence of yeasts on the sensory characteristics of the products is, at least, doubtful. Hadorn et al. (2005) inoculated into salami a yeast starter culture originally used for cheese. In comparison with the uninoculated control, similar values were obtained in physicochemical parameters and microbial counts. Remarkable differences were determined in 2-heptanone and 2-nonanone, but no differences were observed in sensory profiles.

In the case of dry-cured hams, numerous studies have also been performed in order to find out the influence of yeasts on sensory characteristics. Comi and Cantoni (1983) found that the lipolytic activity of the yeasts isolated from Parma hams was low. Huerta et al. (1988) observed that most of the yeasts isolated from dry-cured Spanish hams were lipolytic but not proteolytic. Rodríguez et al. (1998) studied the proteolytic activity of *Debaryomyces* strains isolated from Spanish dry-cured ham using myosin and slices aseptically obtained from pork loins as substrate. After incubation at 25°C during 30 days, free amino acid fractions were analyzed and results were similar to those of a sterile control. Only small changes without statistical differences were observed. The authors considered that the contribution of this yeast to the proteolytic process occurring in dry-cured hams is very poor. Martín et al. (2002) studied a *D. hansenii* strain, isolated from dry-cured ham, which showed proteolytic activity against myosin in culture broth. They observed that the changes in the free amino acids detected on pork loins inoculated with this yeast seem to be negligible, and no statistical differences were obtained in relation to the uninoculated control. However, *D. hansenii* seems to improve the sensory properties of ham, contributing to the development of less yellow superficial fat as an index of a low rancidity (Rodriguez et al. 2001).

On the other hand, there are several studies that demonstrate the importance of the muscle proteinases (cathepsins) that remain active after almost 9 months of curing, favoring the breakdown of proteins. This important role minimizes the contribution of yeasts and even molds to the flavor of cured hams because they grow only on the surface (Toldrá and Etherington 1988; Molina and Toldrá 1992).

Finally, it is important to bear in mind that yeasts have an antagonistic effect on different spoilage bacteria. This has been observed in cheese with respect to *Clostridium tyrobutiricum* and *Cl. butyricum* (Fatichenti et al. 1983). It has also been demonstrated that *D. hansenii* is capable of inhibiting the growth of *Staphylococcus aureus* in fermented sausages through oxygen consumption (Meisel et al. 1989). Similar results have been described by Olesen and Stahnke (2000) in meat mince models. Nevertheless, this inhibitory action can also occur with other species of the genus *Staphylococcus* and could be detrimental to their role in the ripening process. A protective effect against the growth of unfavorable molds has also been described (Hadorn et al. 2005).

YEAST STARTER CULTURES

Starter cultures are considered one of the scientific and industry responses to the need for increase meat product quality and safety. They improve sensory quality, shelf life, and production safety, minimizing the risk of pathogenic microorganisms. They also help overcome problems with ripening, achieving the standardization of the production pattern.

Current starter culture contains mainly lactic acid bacteria and strains of *Kocuria* and *Staphylococcus*. The most commonly used bacteria are *L. plantarum*, *L. sake*, *L. curvatus*, *Pediococcus pentosaceus*, *P. acidilactici*, *K. varians*, *Staph. carnosus*, and *Staph. xylosus* (Toldrá 2002).

Pioneering works on the yeasts present in fermented sausages were conducted in the first decades of the twentieth century (Cesari 1919; Cesari and Guilliermond 1920) when the importance of the "fleur du saucisson" was recognized and the use of pure yeast cultures for flavoring in fermented sausages began to be recommended. Later works established that yeasts are commonly present on sausages and suggested their utilization as starter culture (Dowell and Board 1968; Rossmanith et al. 1972). They reported that curing, color, and flavor of sausages could be improved by adding selected *Debaryomyces* strains. Coretti (1977) observed that the use of a combination of *D. hansenii*, lactobacilli, and micrococci resulted in better flavor and taste of sausages.

The predominant strains used in commercial starter culture, alone or in combination with bacteria or molds, are classified as *D. hansenii* and its imperfect form *Candida famata* (Hammes and Hertel 1998; Toldrá 2002; Pozzi 2003). The species is characterized by a high salt tolerance, low pH and a_w resistance, and aerobic or weak fermentative metabolism (Wyder and Puhan 1999; Kurita and Yamazaki 2002). Also, it can grow at temperatures close to those used during the ripening of fermented sausages (Fleet 1990). Nitrate reduction ability is not an important characteristic of yeast starters; however, a slight inhibition of nitrate reduction in fermented sausages has been described (Meisel et al. 1989).

D. hansenii has been used habitually as a component of cultures offered for exterior and interior use (Jessen 1995). For surface treatment, yeasts can be used alone or in combination with mold strains to give an external appearance that the consumers judge favorably. The inoculation of the surface can be performed by immersion or spraying with a suspension containing the culture. Their antioxidant effect and the possibility of contributing the final flavor by enzymatic activity, and the degradation of lactic acid also justifies its use in starter cultures (Pozzi 2003). However, the amount of *Debaryomyces* spp. should be optimized in order to avoid a high generation of acids associated with off-flavors that can mask the positive effect. Flores et al. (2004) observed that a concentration of 10^6 cfu/g increases acid content (acetic, propanoic, and butanoic acids) to levels associated with off-flavors. However, no off-flavors have been detected by Selgas et al. (2003) using as inoculum a yeast suspension containing 10^7 cfu/g.

Independently of the considerations made above, the future of yeast starter cultures will depend on the knowledge of the metabolism of the yeast used as a means of enhancing and improving their technological capability. Exploratory studies on the natural diversity of wild yeast strains occurring in artisanal products will also be very important, especially studies on genetic aspects that could build a framework that will lead to the generation of new and enhanced starters with increased diversity, stability, and industrial performance. This work could lead to new genetically modified yeast strains with a tailored functionality.

REFERENCES

SJ Andersen. 1995. Compositional changes in surface mycoflora during ripening of naturally fermented sausages. J Food Prot 58:426–429.

A Asehraou, S Mobieddine, M Faid, M Serhrouchni. 1997. Use of antifungal principles from garlic for the inhibition of yeasts and moulds in fermenting green olives. Grasas y aceites 48:68–73.

JA Barnett, RW Payne, D Yarrow. 2000. Yeasts: Characteristics and Identification. 3rd ed. Cambridge: Cambridge University Press.

B Boissonnet, C Callon, M Larpent-Gourgaut, O Michaux, J Sirami, P Bonin. 1994. Isolement et sélection des levures lipolytiques a partir de saucissons secs fermentés. Viand Prod Carnés 15:64–67.

JM Bruna, EM Hierro, L de la Hoz, DS Mottram, M Fernández, JA Ordóñez. 2001. The contribution of *Penicillium aurantiogriseum* to the volatile composition and sensory quality of dry fermented sausages. Meat Sci 59:97–107.

G Campbell-Platt. 1995. Fermented meats: A world perspective. In: G Campbell-Platt, PE Cook, eds. Fermented Meats. London: Blackie Academic & Professional, pp. 39–52.

EP Cesari. 1919. La maturation du saucisson. Acad Sci Paris 168:802–803.

EP Cesari, A Guilliermond. 1920. Les levures du saucisson. Ann Inst Pasteur 34:229–248.

G Comi, C Cantoni. 1983. Presenza di lieviti nei prosciutti crudi stagionati. Ind Alim February:102–104.

PE Cook. 1995. Fungal ripened meats and meat products. In: G Campbell-Platt, PE Cook, eds. Fermented Meats. London: Blackie Academic & Professional, pp. 39–52.

S Coppola, G Mauriello, M Aponte, G Moschetti, F Villani. 2000. Microbial succession during ripening of Naples-type salami, a southern Italian fermented sausage. Meat Sci 56:321–329.

K Coretti. 1977. Starterculturen in der fleischwirtschaft. Fleischwirtschaft 3:545–549.

T Deak, LR Beuchat. 1996. Handbook of Food Spoilage Yeasts. Boca Raton: CRC Press, pp. 25–26.

VM Dillon, RG Board. 1991. Yeast associated with red meats. J Appl Microbiol 71:93–108.

MJ Dowell, RG Board. 1968. A microbiological survey of British fresh sausages. J Appl Bacteriol 31:378–396.

MA Durá, M Flores, F Toldrá. 2002. Purification and characterization of a glutaminase from *Debaryomyces* spp. Int J Food Microbiol 76:117–126.

———. 2004. Effect of growth phase and dry-cured sausage processing conditions on *Debaryomyces* spp. Generation of volatile compounds from branched-chain amino acids. Food Chem 86:391–399.

JP Encinas, TM López-Diaz, ML García-López, A Otero, B Moreno. 2000. Yeast populations on Spanish fermented sausages. Meat Sci 54(3):203–208.

E Fatichenti, JL Gergere, D Pietrino, AF Giovaunnii. 1983. Antagonistic activity of *Debaryomices hansenii* towards *Clostridium tyrobutyricum* and *Cl. butyricum*. J Dairy Res 80:449–457.

GH Fleet. 1990. Yeast in dairy products. J Appl Bacteriol 68:199–221.

M Flores, MA Durá, A Marco, F Toldrá. 2004. Effect of *Debaryomyces* spp. On aroma formation and sensory quality of dry-fermented sausages. Meat Sci 68:439–446.

ML García, C Casas, E Mendoza, VM Toledo, MD Selgas. 2001. Effect of selected moulds on the sensory properties of dry fermented sausages. Europ Food Res Technol 212(3):287–291.

KH Gehlen, C Meisel, A Fischer, WP Hammes. 1991. Influence of the yeast *Debaromyces hansenii* on dry sausage fermentation. Proceedings of 35th International Congress Meat Science and Technology, Kulmbach, 1991, pp. 871–876.

CO Gill, KG Newton. 1978. The ecology of bacterial spoilage of fresh meat at chill temperatures. Meat Sci 2:207–217.

L Grazia, P Romano, A Bagni, D Roggiani, G Guglielmi. 1986. The role of moulds in the ripening process of salami. Food Microbiol 3:19–25.

R Hadorn, P Eberhard, D Guggisberg, D Isolini, P Piccinali, H Schlichtherle-Cerny. 2005. An alternative surface culture for salami. Proceedings of 51st International Congress of Meat Science and Technology, Baltimore, 2005.

WP Hammes, C Hertel. 1998. New developments in meat starter cultures. Meat Sci 49(suppl 1): S125–S138.

WP Hammes, HJ Knauf. 1994. Starters in the processing of meat products. Meat Sci 36:155–168.

E Hierro, L de la Hoz, JA Ordóñez. 1997. Contribution of microbial and meat endogenous enzymes to the lipolysis of dry fermented sausages. J Agric Food Chem 45:2989–2995.

T Huerta, A Querol, J Hernández Haba. 1988. Yeasts of dry cured hams, qualitative and quantitative aspects. Microbiol Alim Nutr 6:289–294.

B Jessen. 1995. Starter culture for meat fermentations. In: G Campbell-Platt, PE Cook, eds. Fermented Meats. London, Blackie Academic & Professional, pp. 130–159.

O Kurita, E Yamazaki. 2002. Growth under alkaline conditions of the salt tolerant yeast *Debaryomyces hansenii* IFO10939. Current Microbiol 45:277–280.

M Larpent-Gourgaut, O Michaux, C Callon, O Brenot, J Sirami, C Bonnin, B Boissonnet. 1993. Etude comparée des flores d'aromatisation du saucisson de fabrication industrielle ou artisanale. Viand Prod Carnés 14:23–27.

I Leistner. 1995. Stable and safer fermented sausages world-wide. In: G Campbell-Platt, PE Cook, eds. Fermented meats. London, Blackie Academic & Professional, pp. 160–175.

JS Leyva, M Manrique, I Prats, MC Loureiro-Dias, J Peinado. 1999. Regulation of fermentative CO_2 production by the food spoiling yeast *Zygosaccharomyces bailii*. Enz Microb Technol 24:270– 275.

G Lizaso, J Chasco, MJ Beriain. 1999. Microbiological and biochemical changes during ripening of salchichón, a Spanish dry cured sausage. Food Microbiol 16(3):219–228.

FK Lücke, H Hechelman. 1987. Cultivos starter para embutidos secos y jamón crudo. Composición y efecto. Fleischwirtschaft español 1:38–48.

D Marquina, P Llorente, A Santos, JM Peinado. 2001. Characterization of the yeast population in low water activity foods. Adv Food Sci 23:249–253.

A Martín, MA Asensio, ME Bermúdez, MG Córdoba, E Aranda, JJ Córdoba. 2002. Proteolytic activity of *Penicillium chrysogenum* and *Debaryomyces hansenii* during controlled ripening of pork loins. Meat Sci 62:129–137.

C Meisel, KH Gehlen, A Fisher, WP Hammes. 1989. Inhibition of the growth of *Staphylococcus aureus* in dry sausages by *Lactobacillus curvatus*, *Micrococcus varians* and *Debaryomyces hansenii*. Food Biotechnol 3:145–168.

J Metaxopoulos, S Stravopoulos, A Kakouri, J Samelis. 1996. Yeast isolated from traditional Greek dry salami. Ital J Food Sci 1:25–32.

E Miteva, E Kirova, D Gadjeva, M Radeva. 1986. Sensory aroma and taste profiles of raw-dried sausages manufactures with lipolitically active yeasts culture. Nahrung 30(8):829–832.

I Molina, P Nieto, J Flores, H Silla, H. S Bermell. 1991. Study of the microbial flora in dry-cured hams. 5. Lipolytic activity. Fleischwirtschaft 71:906–908.

I Molina, H Silla, J Flores. 1990. Study of microbial flora in dry cured hams. 4: Yeasts. Fleischwirtschaft 70:74–76.

I Molina, F Toldrá. 1992. Detection of proteolytic activity in microorganisms isolated from dry cured ham. J Food Sci 57:1308–1310.

E Monte, JR Villanueva, A Domínguez. 1986. Fungal profiles of Spanish country-cured hams. Int J Food Microbiol 3:355–359.

F Núñez, MM Rodríguez, JJ Córdoba, ME Bermúdez, MA Asensio. 1996. Yeast population during ripening of dry-cured Iberian ham. Int J Food Microbiol 29:271–280.

PT Olesen, LH Stahnke. 2000. The influence of *Debaryomyces hansenii* and *Candida utilis* on the aroma formation in garlic spiced fermented sausages and model minces. Meat Sci 56:357–368.

JA Ordóñez, EM Hierro, JM Bruna, L de la Hoz. 1999. Changes in the components of dry-fermented sausages during ripening. Crit Rev Food Sci Nutr 39(4):329–367.

AA Osei Abunyewa, E Laing, A Hugo, BC Viljoen. 2000. The population change of yeasts in commercial salami. Food Microbiol 17:429–438.

TL Peppard, SA Halsey. 1981. Malt-flavour-Transformation of carbonyls compounds by yeast during fermentation. J Inst Brew 87:3860–3890.

W Pozzi. 2003. Advantages in dry sausages production. Fleischwirtschaft Int 2:14–16.

M Rodriguez, A Martín, F Núñez. 2001. Población microbiana del jamón y su contribución en la maduración. Cultivos iniciadores. In: J. Ventanas, ed. Tecnología del jamón ibérico. Madrid: Mundi Prensa, pp. 347–366.

M Rodríguez, F Núñez, JJ Córdoba, ME Bermúdez, MA Asensio. 1998. Evaluation of proteolytic activity of micro-organisms isolated from dry cured hams. J Appl Microbiol 85:905–912.

P Roncalés, M Aguilera, JA Beltrán, I Jaime, JM Peiro. 1991. The effect of natural or artificial casing on the ripening and sensory quality of a mould-covered dry sausage. Int J Food Sci Technol 26:83–89.

E Rossmanith, HJ Mintzlaff, B Streng, W Christ, L Leistner. 1972. Hefen als starterkulturen für rohwürste. Jahr der BAFF 147–148.

A Saldanha-da-Gama, M Malfeito-Ferreira, V Loureiro. 1997. Characterization of yeasts associated with Portuguese pork-based products. Int J Food Microbiol 37:201–207.

J Samelis, S Stavropoulos, A Kakouti, J Metaxopoulos. 1994. Quantification and characterization of microbial populations associated with naturally fermented Greek dry salami. Food Microbiol 11:447–460.

NN Santos, RC Santos-Mendoça, Y Sanz, T Bolumar, MC Aristoy, F Toldrá. 2001. Hydrolysis of pork muscle sarcoplasmic proteins by *Debaryomyces hansenii*. Int J Food Microbiol 68:199–206.

MD Selgas, C Casas, VM Toledo, ML García. 1999. Effect of selected mould strains on lipolysis in dry fermented sausages. Europ Food Res Technol 209(5):360–365.

MD Selgas, J Ros, ML García. 2003. Effect of selected yeast strains on the sensory properties of dry fermented sausages. Eur Food Res Technol 217:475–480.

LH Stahnke. 1994. Aroma components from dried sausages fermented with *Staphylococcus xylosus*. Meat Sci 38:39–53.

H Stam, M Hoogland, C Laane. 1998. Food flavours from yeast. In: BJ Wood, ed. Microbiology of Fermented Foods (2nd ed.). London: Blackie Academic & Professional, pp. 505–542.

K Tjener, LH Stahnke, L Anmdersden, J Martinussen. 2003. A fermented meat model system for studies of microbial aroma formation. Meat Sci 66:211–218.

F Toldrá. 2002. Fermentation and starter cultures. In: F Toldrá, ed. Dry-cured meat products. Trumbull: Food & Nutrition Press Inc., pp. 89–112.

F Toldrá, DJ Etherington. 1988. Examination of cathepsins B, D, H and L activities in dry-cured hams. Meat Sci 23:1–7.

VM Toledo, MD Selgas, C Casas, JA Ordóñez, ML García. 1997. Effect of selected mould strains on proteolysis in dry fermented sausages. Z Lebensm.Unters. Forsch204:385–390.

H Wolter, E Laing, BC Viljoen. 2000. Isolation and identification of yeasts associated with intermediate moisture meats. Food Technol Biotechnol 38(1):69–75.

MY Wyder, Z Puhan. 1999. Role of selected yeast in cheese ripening: An evaluation in aseptic cheese curd slurries. Int Dairy J 9:117–124.

16
Starter Cultures: Molds

Elisabetta Spotti and Elettra Berni

INTRODUCTION

Molds on Naturally Ripened Dry Sausages

The ripening techniques used to process meat products quickly lead to a distinctive surface colonization by a great number of filamentous fungi. Apart from molds, surface microflora can also include yeasts, which play a role similar to that of molds in sausage production (Samelis and Sofos 2003).

In general, at the beginning of the ripening process yeasts are the predominant microorganisms on the surface, totaling about 95% of the mycoflora. After the first 2 weeks, molds and yeasts are present in equal amounts. At the end of the sausage ripening process (from the 4th to 8th week) mycoflora is mainly represented by molds; yeasts may undergo one logarithmic decrement.

Molds usually prevail on yeasts; this is due to the progressive reduction of surface water activity in sausages, to the invasive way molds grow on the surfaces (by apical increase of cells called *hyphae*) and by substantial production of conidia, the molds reproductive units, and to the production of secondary toxic metabolites.

On the surface of many fermented products from several European countries (Italy, Rumania, Bulgaria, France, Hungary, Switzerland, Germany, Spain, Austria, and Belgium) mold growth is desirable. This is due to the fact that the mycelium:

- Prevents excessive drying, allowing water loss and therefore homogeneous dehydration of the product.
- Protects fat from oxidation because it metabolizes and consumes peroxides, thus preventing rancidity.
- Reduces the O_2 levels on the surface of the product, thus avoiding oxidative processes and improving meat color.
- Contributes to enhancing the flavor of the final product (especially when natural casing is used), because it breaks up fats, proteins, and lactic acid, thus favoring pH increase.
- Makes sausage peeling easier thanks to the differentiation of the fungal basal hyphae into a sort of root called *rhizoid*, which can penetrate the inner part of the mixture (Grazia et al. 1986).

Parameters Affecting Mold Growth

The physicochemical parameters recorded in ripening environments are directly related to the microbial growth. In particular, surface molding of fermented meat during ripening is influenced by: composition of the sausage mince, especially the fat particle size; the relative humidity (RH) of the air and the temperature conditions applied in industrial plants; drying profile (how surface water activity, a_w, decreases during the product ripening) and ventilation methods (alternating ventilation cycles at high air velocities with cycles in which the product is allowed to rest favors fungal growth). According to this, it has been demonstrated that matured sausages of a greater size and with longer ripening times usually show initial growth mainly characterized by hydrophilic molds and toward the end of the ripening period there is a prevalence of xerotolerant strains (Baldini et al. 2000).

ENVIRONMENT-CONTAMINATING MOLD SPECIES

For a long time, molding of sausages was left to contamination by environment-contaminating mold species, mainly belonging to the genera *Penicillium* and *Aspergillus*, which better adapt to the technological conditions to which these products are subjected.

Natural contamination results in a prevalence of xerotolerant strains belonging to the genus *Penicillium* subgenus *Penicillium* over those belonging to more xerophilic genera such as *Aspergillus* or *Eurotium*. This is due to the easier growth of most *Penicillium* species in environments with RH ranging from 85 to 92% and with temperatures from 10–20°C (see Table 16.1). A minor percentage of growing molds consists of hydrophilic fungi producing darkish mycelium and conidia, such as *Mucor* spp. and some *Scopulariopsis* spp. These species must be removed because the former do not allow homogeneous drying of the product in the first steps of the process and the latter can produce an unpleasant ammonia odor caused by their strong proteolytic activity in the final step of the ripening (Dragoni et al. 1997; Lücke 2000; Spotti and Baldini, personal communication 2005).

FUNGAL STARTER CULTURES

The ripening techniques considered positive and normally applied in industrial practice usually allow for the growth of characteristic whitish molds, which can also inhibit the multiplication of other molds, especially those that prove to be potentially toxinogenic and/or with a mycelium having undesirable color (Baldini et al. 2000). In particular, recent findings showing that many of the *Penicillium* and *Aspergillus* species frequently found on the surface of dry sausages have the ability to produce mycotoxins in the actual product, have stressed the importance of possibly controlling contaminating molds on raw sausages during ripening by using mold starters. Starters studied in the past were selected and are now available on the market (Leistner et al. 1989). They have been shown to impart a desirable appearance and good technological and sensory characteristics to the product, to be unable to synthesize antibiotics and/or toxic metabolites such as mycotoxins and to not cause well-known allergies or mycoses. For the reasons above, they can be used to inoculate fermented sausages by spraying or by immersion in a spore suspension.

The selection of fungal starter cultures among the molds that naturally occur on the surface of fermented and ripened sausages allows control of molding in the first steps of the process and makes the starter presence predominant.

Among the starters that can be considered both safe and effective, *P. nalgiovense* has been frequently isolated on the surface of ripened meat products as a "domesticated species" from *P. chrysogenum* (wild type) (Pitt and Hocking 1997). At present it is routinely used as starter culture in many traditional productions, and so it is often present in the air of environments. Apart from *P. nalgiovense*, various strains of *P. gladioli* are usually considered acceptable because they are nontoxinogenic and result in desirable flavor and appearance due to the grey color of mycelium (Grazia et al. 1986). Even *P. camemberti* should be considered a starter culture for the meat industry because of its ability to improve the sensory characteristics of dry-fermented sausages (Bruna et al. 2003).

P. chrysogenum has been mentioned too as a suitable starter for mold-fermented products (Hammes and Knauf 1994; Ministero della Sanità 1995), because some nontoxic strains have been shown to

Table 16.1. Fungal species isolated from typical salami from Northern Italy.

Most Significant Species Isolated			Uncommon Species
A. candidus (22)	*P. chrysogenum* (7)	*P. nalgiovense* (15)	*A. ochraceus* (3)
E. rubrum (11)	*P. commune* (3)	*P. olsonii* (3)	*Mucor* spp. (4)
P. aurantiogriseum (18)	*P. gladioli* (7)	*P. solitum* (7)	*Eupenicillium* spp. (10)
P. brevicompactum (24)	*P. griseofulvum* (15)	*P. verrucosum* (30)	*S. brevicaulis* (2)
P. camemberti (5)	*P. implicatum* (5)	*P. waksmanii* (9)	*S. candida* (5)

Source: E Spotti, Department of Microbiology, SSICA, Parma, Italy. Unpublished data.
Note: The number in parentheses indicates the occurrence in the past years.

produce proteases that contribute to the generation of characteristic flavor compounds in dry-fermented sausages (Rodríguez et al. 1998; Martín et al. 2002). However, *P. chrysogenum* is not always considered acceptable in industrial practice because its albino mutant may start producing green conidia, and most of the strains tested may produce antibiotics and toxic substances such as roquefortine C (Pitt and Hocking 1997). Other strains belonging to *Penicillium* species have also been excluded because they produce dark-green conidia that may impair sausage outward appearance.

LIPOLYTIC AND PROTEOLYTIC ACTIVITIES OF STARTERS

In the last few years a lot of studies have been carried out throughout the world concerning the lipolytic and proteolytic activity of molds involved in the ripening process, including *P. nalgiovense, P. chrysogenum* and *P. camemberti* (Geisen et al. 1992; Rodríguez et al. 1998; Ockerman et al. 2001; Bruna et al. 2003). A recent in vitro experiment focused on the biochemical activity of two starters, *P. nalgiovense* and *P. gladioli*, in comparison with that of the environment-contaminating molds grown on ripened products (*P. griseofulvum, P. brevicompactum, P. olsonii, P. implicatum, P. verrucosum, Talaromyces wortmannii, Mucor* spp.) (Spotti and Berni 2004). With regard to lipid breakdown, both *P. nalgiovense* and *P. gladioli* proved to have good lipolytic activity: The lipase production of the former was increased by lower temperature (more marked at 14°C than at 20°C), whereas that of the latter was greater at the higher temperatures. *P. nalgiovense* showed a strong proteolytic activity both at 14°C and at 20°C, whereas *P. gladioli* did not show this activity at the two temperatures tested. Despite their different metabolisms, *P. gladioli* used as starter culture in the SSICA pilot plants and in industrial practice was shown to supply ripened products with good technological and sensory features, as *P. nalgiovense* does. This fact highlights that lipolytic activity affects the molding process more than the proteolytic one, which seems to be exerted also by bacteria and proteases native to the meat.

GROWTH AND COMPETITION TESTS

Recently studies were carried out in the SSICA pilot plant in Parma, Italy, in order to evaluate a range of commercial and noncommercial fungal starter cultures for growth and competitiveness towards various undesirable species.

Growth

A research work was carried out in model systems reproducing a traditional Italian ripening process, and the studies were focused on the growth of five starter cultures (*P. camemberti M* and *P*, *P. gladioli P* and *B*, and *P. nalgiovense*) and four environment-contaminating strains (*P. solitum*, *P. verrucosum*, *P. chrysogenum*, and *A. ochraceus*) as a function of temperature (from 8–22°C) and water activity (from 0.78 to 0.92) (Spotti et al. 1999). For each strain, the average of growth radial rate (in mm/day, obtained from the linear correlation between colony diameters and growth time during linear phase) and the average duration of lag time (time during which no growth occurs) were determined. The most important results are reported in Tables 16.2 and 16.3.

Table 16.2. Optimal growth conditions with corresponding values of minimum lag time and growth rate for different fungal strains.

Strain	Optimum Temperature (°C)	Optimum RH (%)	Minimum Lag Time (days)	Growth Rate (mm/day)
P. camemberti M	17	91	2	1.7
P. camemberti P	12	91	5	1.2
P. gladioli P	19	88	4	1.0
P. gladioli B	17	90	4	1.2
P. nalgiovense	14	90	4	2.0
P. solitum	16	89	5	1.4
P. verrucosum	20	88	4	1.3
P. chrysogenum	16	89	3	1.1
A. ochraceus	19	86	1	2.2

Source: Spotti et al. (1999).

Table 16.3. Lag time and growth rate at 15°C for different fungal strains.

Strain	Lag Time (Days)	Growth Rate (mm/day)
P.camemberti M	4	1.6
P.camemberti P	5	1.2
P. gladioli P	5	0.7
P. gladioli B	5	1.0
P. nalgiovense	4	2.0
P. solitum	5	1.3
P. verrucosum	6	1.0
P. chrysogenum	3	1.0
A. ochraceus	14	1.1

Source: Spotti et al. (1999).

In order to make the results transferable to the industry, the results of the trial were described for the three primary stages of the salami production, taking into account the wide variability of temperature and RH applied in the industrial practice.

1st Stage—Heating

This phase is carried out at 20–24°C, at environmental RH values for a few hours (depending on salami size) until the temperature of the inner product reaches 18–20°C. This phase does not seem to have a significant effect on subsequent surface mold growth; instead, it is very important for the onset of correct product fermentation by lactic acid bacteria, staphylococci, and micrococci.

2nd Stage—Drying

This phase allows the fermentation to go on because the RH is maintained at high values. Processing temperatures usually range from 15–20°C, whereas surface a_w values are much more variable, ranging from 0.80 to 0.95, and most frequently between 0.88 and 0.91. Surface molding begins to appear in this phase, which lasts for a few days. The molds most likely to grow during this phase, based on temperature and relative humidity for optimum growth, are *P. gladioli*, *P. solitum*, *P. verrucosum*, and *P. chrysogenum* (see Table 16.2). However, it is also possible to find *P. camemberti* and *A. ochraceus* (Mutti et al. 1988). Under these conditions, the strains have a lag time of 1–5 days and a growth rate of 1–2.2 mm/day (Table 16.2); product surface can thus be indifferently invaded by *P. camemberti* and *A. ochraceus*, and the suitability of initial treatment of the products with starters cultures proves recommendable.

3rd Stage—Ripening

This last phase is carried out at temperatures ranging from 10–15°C, whereas product surface a_w remains within the same range as in the drying phase. Temperature must therefore be regarded as the parameter discriminating between the two processing stages.

On the basis of the temperature for optimum growth, as reported in Table 16.2, in particular *P. nalgiovense* and *P. camemberti* are able to proliferate during ripening. However, for a better comparison of the various strains, lag time and growth rate values were determined at 15°C, keeping the relative humidity at the optimal value for each individual strain (see Table 16.3).

Table 16.3 shows that the strain for which dominance over the "undesired species" is predictable is *P. nalgiovense*, because it has the highest growth rate under the temperature conditions indicated. Even though *P. chrysogenum* has shorter lag time (3 days) as compared to *P. nalgiovense* (4 days), its growth rate is half that of *P. nalgiovense*, and its mycelium is therefore less invasive. As to the other strains, evaluated by the same criterion, the following rating scale is obtained: *P. nalgiovense* > *P. camemberti M* > *P. chrysogenum* (wild type) > *P. solitum* > *P. camemberti P* > *P. gladioli B* > *P. verrucosum* > *P. gladioli P* > *A. ochraceus*.

The starters therefore show different dominance over "undesired molds." In particular, together with *P. camemberti* and the two *P. gladioli* strains, assuming equal contamination levels, also *P. chrysogenum*, *P. solitum*, and *P. verrucosum* may be present on the same product. Instead, *P. nalgiovense* and *P. camemberti M* prove to dominate during ripening.

Relative humidity proves to have greater influence than temperature for almost all strains except *A. ochraceus*; this accounts for the lag time at optimum temperature in Table 16.3 not being too different from those in Table 16.2. The only exception is *A. ochraceus*, which is much more sensitive to temperature changes than to humidity changes, as is shown by its long lag time at 15°C (14 days), whereas under optimum conditions it is only 1 day.

Recently, the course of growth rate of the fungal species that more frequently occur on the surface of molded meat products during ripening has also been investigated (Table 16.4) (Spotti and Berni 2005). Table 16.4 clearly shows how the growth of starter cultures such as *P. nalgiovense* and *P. gladioli* are favored by higher relative humidity in the first days of ripening.

Table 16.4. Trend in growth rate of most frequent fungal species during ripening at 14–15°C as a function of the surface water activity (a_w).

Fungal Species	$a_w = 0.85$	$a_w = 0.86$	$a_w = 0.88$	$a_w = 0.90$	$a_w = 0.92$
P. gladioli	→	→	→	→	MG
P. nalgiovense	→	→	→	MG	←
P.chrysogenum	MG	←	←	←	←
P. griseofulvum	MG	←	←	←	←
P. implicatum	MG	←	←	←	←
P. verrucosum	MG	←	←	←	←
P. solitum	MG	←		←	←

Source: E Spotti. Presentation at 2nd Annual Conference on the Research Project "Salumi piacentini," 25 October 2005, SSICA, Parma, Italy.
Note: MG = Maximum growth rate. → = Growth rate tends to increase if surface water activity increases. ← = growth rate tends to decrease if surface water activity increases.

Competition

According to techniques tested by Wheeler and Hocking (1993), several competition trials were carried out, where each starter was inoculated in combination with the undesired strain, each at a time (Spotti et al. 1994, 1999; Spotti and Berni 2004). The results of these experiments showed that *P. nalgiovense* may be overgrown by *A. ochraceus* during drying (T ≥ 15°C), whereas *P. nalgiovense* will dominate this undesired mold during ripening (T ≤ 15°C) as previously referred. On the contrary, *P. nalgiovense* is unable to dominate over *P. chrysogenum* (wild type), *P. solitum*, and *P. verrucosum* in the case of similar initial contamination levels. During the drying phase, *P. gladioli* is inhibited by *A. ochraceus* and *P. solitum*, whereas during ripening it prevails over *A. ochraceus*, but not over *P. chrysogenum* (wild type), P. *solitum*, and *P. verrucosum*. *A. ochraceus* may therefore dominate over the starter strains during drying and may grow invasively throughout the product surface even during ripening, if present in equal amounts. The results of the trials also suggested that the starter cultures could coexist with *P. chrysogenum* (wild type), *P. solitum*, and *P. verrucosum* on the product, if their initial levels are the same.

CONCLUSIONS

The poor competitiveness of surface starter cultures such as *P. nalgiovense* and *P. gladioli* compared with other contaminants requires a low rate of environmental contamination and higher average relative humidity; therefore, starter molds should be used at high inoculation levels on sausages to be ripened in order to avoid possible development of undesirable species. The above-mentioned rating scale, relating to the individual starters reported, appears to be valid and useful in the industrial practice only on the basis of these remarks.

Nevertheless, the use of selected cultures is not a widespread practice. In fact, it is not applied when the producers expect that the final product will obtain satisfactory technological and sensory results. In that case, the presence of occasional indigenous mycoflora on the surface of ripened products is favorably considered. Regarding this, Williams et al. (1985) observed that most spontaneous molds species subjected to consecutive subcultures, as it frequently occurs in the ripening rooms within subsequent productive cycles, proved to rapidly degenerate or to change their morphological appearance and adapt to environmental conditions. This is due to the fact that the genome of the species belonging to the genus *Penicillium* subgenus *Penicillium* isolated from sausages appears to be unstable and tends to cause a rapid adaptation of the above species to the nutritional niches available in order to take advantage of them.

To conclude, when the producers choose not to use starters, it is necessary to identify the fungal species occurring on the surface of the ripened product, regardless of the good appearance and the high sensory quality of the final product. Besides, the control on the growth of the selected species and their actual predominance over undesirable molds should be periodically planned by routine laboratory tests (i.e., isolation and identification of the species applied in the industrial process) to avoid the unexpected setting up of environment-contaminating mold species, which are morphologically similar to the selected ones.

REFERENCES

P Baldini, E Berni, C Diaferia, A Follini, S Palmisano, A Rossi, PG Sarra, GL Scolari, E Spotti, M Vescovo, C Zacconi. 2005. Materia prima flora microbica nella produzione dei salumi piacentini. Rivista di Suinicoltura 4:183–187.

P Baldini, E Cantoni, F Colla, C Diaferia, L Gabba, E Spotti, R Marchelli, A Dossena, E Virgili, S Sforza, P Tenca, A Mangia, R Jordano, MC Lopez, L Medina, S Coudurier, S Oddou, G Solignat. 2000. Dry sausages ripening: Influence of thermohygrometric conditions on microbiological, chemical and physico-chemical characteristics. Food Res Int 33:161–170.

JM Bruna, EM Hierro, L de la Hoz, DS Mottram, M Fernández, JA Ordoñez. 2003. Changes in selected biochemical and sensory parameters as affected by the superficial inoculation of *Penicillium camemberti* on dry fermented sausages. Int J Food Microbiol 85:111–125.

I Dragoni, C Cantoni, A Papa, L Vallone. 1997. Muffe alimenti e micotossicosi. 2nd ed. Milano: CittàStudiEdizioni di UTET, p. 109.

R Geisen, FK Lücke, L Kröckel. 1992. Starter and protective cultures for meat and meat products. Fleischwirtsch 72:894–898.

L Grazia, P Romano, A Bagni, D Roggiani, D Guglielmi. 1986. The role of molds in the ripening process of salami. Food Microbiol 3:19–25.

WP Hammes, HJ Knauf. 1994. Starter in the processing of meat products. Meat Sci 36:155–168.

L Leistner, R Geisen, J Fink-Gremmels. 1989. Mold-fermented foods of Europe: Hazards and Developments. In: S Natori, K Hashimoto, Y Ueno, eds. Mycotoxins and Phycotoxins 1988. Amsterdam: Elsevier Science Publishers B.V., pp. 145–160.

F-K Lücke. 2000. Fermented meats. In: BM Lund, TC Baird-Parker, GW Gould, eds. The Microbiological Safety and Quality of Food. Vol 1. Maryland: Aspen Publishers Inc, pp. 420–444.

A Martín, MA Asensio, ME Bermúdez, MG Cordoba, E Aranda, JJ Cordoba. 2002. Proteolytic activity of *Penicillium chrysogenum* and *Debaryomyces hansenii* during controlled ripening of pork loins. Meat Sci 62:129–137.

Ministero della Sanità. 1995. Decreto 28 Dicembre 1994. Gazzetta Ufficiale Repubblica Italiana. Serie generale n. 89, pp. 4–5, 15.04.1995.

P Mutti, MP Previdi, S Quintavalla, E Spotti. 1988. Toxigenity of mold strains isolated from salami as a function of culture medium. Industria Conserve 63:142–145.

HW Ockerman, FJ Céspedes Sánchez, MA Ortega Mariscal, F León-Crespo. 2001. The lipolytic activity of some indigenous Spanish molds isolated from meat products. J Muscle Foods 12:275–284.

JI Pitt, AD Hocking. 1997. Fungi and Food Spoilage. 2nd ed. Cambridge: University Press. Blackie Academic & Professional.

M Rodríguez, F Núñez, JJ Cordoba, ME Bermúdez, MA Asensio. 1998. Evaluation of proteolytic activity of micro-organisms isolated from dry cured ham. J Appl Microbiol 85:905–912.

J Samelis, JN Sofos. 2003. Yeasts in meat and meat products. In: Yeasts in Food. Cambridge: Woodhead Publishing Limited, pp. 243–247.

RA Samson, JC Frisvad. 2004. Polyphasic taxonomy of the Penicillium. A guide to identification of food and air-borne terverticillate Penicillia and their mycotoxins. In: Penicillium subgenus Penicillium: New taxonomic Schemes and Mycotoxins and Other Extrolites. Stud Mycol 49:1–173.

E Spotti, E Berni. 2004. Il controllo della flora fungina superficiale. Proceedings of 1st Annual Conference on the Research Project "Salumi piacentini," Università Cattolica del Sacro Cuore, Piacenza, Italy, 27 October 2004.

E Spotti, E Berni. 2005. Variazione della flora microbica superficiale nei salumi oggetto della ricerca. Proceedings of 2nd Annual Conference on the Research Project "Salumi piacentini", SSICA, Parma, Italy, 25 October 2005.

E Spotti, C Busolli, F Palmia. 1999. Sviluppo di colture fungine di importanza rilevante nell'industria dei prodotti carnei. Industria Conserve 74:23–33.

E Spotti, P Mutti, F Scalari. 1994. *P. nalgiovense*, *P. gladioli*, *P. candidum* e *A. candidus*: possibilità d'impiego quali colture starter. Industria Conserve 69:237–241.

KA Wheeler, AD Hocking. 1993. Interactions among xerophilic fungi associated with dried salted fish. J Appl Bacteriol 74:164–169.

AP Williams, JI Pitt, AD Hocking. 1985. The closely related species of subgenus Penicillium—A phylogenic exploration. In: RA Samson, JI Pitt, eds. Advances in Aspergillus and Penicillium Systematics. New York: Plenum Press, pp. 121–128.

17
Genetics of Microbial Starters

Marie Champomier-Vergès, Anne-Marie Crutz-Le Coq, Monique Zagorec, Sabine Leroy, Emilie Dordet-Frisoni, Stella Planchon, and Régine Talon

INTRODUCTION

The main bacterial species used in meat fermentation are *Lactobacillus sakei*, *Staphylococcus xylosus*, and *Staphylococcus carnosus*. These species have been applied for more than 50 years alone or in combination with each other or other microorganisms as starter cultures for the manufacturing of fermented sausages. Studies of these starter cultures were mainly focused on their role in the sensory quality or in the microbial safety of fermented meat products. Genetic studies have provided basic knowledge on main metabolic activities and stress-resistance functions. Appropriate genetic tools for molecular genetic analyses were developed, allowing the exchange of chromosomal genes or plasmids in these naturally nontransformable bacteria. Global approaches based on proteomics and transcriptomics were developed or are in progress in order to better understand their adaptation to the meat environment and their interactions with the ecosystem and the meat substrate.

CHROMOSOMAL ELEMENTS

Lactobacillus sakei

Although numerous *L. sakei* strains have been studied in different laboratories, most of the genetic information has been obtained from one strain, *L. sakei* 23K, initially isolated from a fermented dry sausage (Berthier et al. 1996; Champomier-Vergès et al. 2002). The size of its genome, determined by Pulse Field Gel Electrophoresis (PFGE), was estimated to be 1,845 kbp (Dudez et al. 2002), and a first chromosome map was obtained by the use of 47 genetic loci available in 2002. In 2005, Chaillou et al. published the complete genome sequence of *L. sakei* 23K. The exact size of this circular chromosome is 1,884,661 bp and the knowledge of the 1,883 putative genes highlighted the main functions of the species, especially those linked to the adaptation to technological environments encountered during sausage production (see also Eijsink and Axelsson 2005). The main features of this genome, when compared to three other lactobacilli genomes, are summarized in Table 17.1. From the known involvement of *L. sakei* as sausage starter, the genome analysis confirmed that the main role of *L. sakei* is to ferment sugars into lactic acid, and that it lacks main aroma production pathways. *L. sakei* 23K also lacks genes responsible for biogenic amine production. A battery of functions were identified, which might explain adaptation or resistance of this species to stressing environmental conditions used during sausage manufacturing (presence of curing agents, spices, smoke, low temperature).

Staphylococcus carnosus and *Staphylococcus xylosus*

The size of the circular chromosome of the *S. carnosus* TM300 strain was estimated by PFGE to be 2,590 kb (Wagner et al. 1998). The complete genome sequence of this strain allowed establishing

Table 17.1. Summary of the main characteristics of *L. sakei* genome with comparison to three other lactobacilli genomes.

	L. sakei 23K	*L. plantarum* WCFS1	*L. johnsonii* NCC533	*L. acidophilus* NCFM
Chromosome size	1,884,661	3,308,274	1,992,676	1,993,564
GC (%)	41.25	45.60	34.60	34.71
Coding sequences	1,883	3,009	1,821	1,864
Unique genes	254	293	150	186
rRNAs	7	5	4	4
tRNAs	63	62	79	61
IS elements	12	15	14	17
PTS[1] transporters	6	25	16	20
Proteinase/peptidases	0/22	0/19	PrtP/25	PrtP/20
Amino-acid transport/biosynthesis	24/2	22/17	25/3	24/10
Redox[2] functions	55	60	20	29
Compatible solute Transport[3]	4	3	1	1
CspA (cold shock Proteins A)	4	3	1	1

Source: from Chaillou et al. (2005).
1 = Phosphoenol pyruvate : carbohydrate phosphotransferase transport systems.
2 = Proteins involved in redox balance adaptation.
3 = Transporters of osmolytes compatible solutes are involved in osmolarity and cold temperature resistance.

the exact size of the chromosome to 2.56 Mb (Rosenstein et al. 2005). The size of the chromosome of *S. xylosus* strain C2a was estimated to be 2,868 kb (Dordet-Frisoni et al. 2007). So the size of the genome of *S. carnosus* and *S. xylosus* is comparable to that of other staphylococci. The size variability within the staphylococcal genomes is about 14%, ranging from 2,499 kb (*Staphylococcus epidermidis* ATCC 12228, Zhang et al. 2003) to 2,903 kb (*Staphylococcus aureus* MRSA252, Holden et al. 2004). Inter- and intraspecies variations of genome size were observed as summarized Table 17.2. *S. xylosus* exhibited also a relatively large intraspecies variation (11%) (Dordet-Frisoni et al. 2007). In contrast, *S. carnosus* species constituted a relatively homogeneous genetic group (Martin et al. 2006; Planchon et al. 2007).

A physical and genetic map of *S. carnosus* TM300 was constructed (Wagner et al. 1998). Thirty-one restriction fragments obtained from rare cutting restriction enzymes were positioned and 18 genetic markers were located, including four ribosomal operons (*rrn*) and genes related to sugar metabolism and to nitrate and nitrite reduction important for meat fermentation (Wagner et al. 1998). The chromosome of *S. carnosus* TM300 comprises 2,454 ORFs and the annotation of the completed sequence is under way (Rosenstein et al. 2005). The first comparative studies showed a high conservation in gene order between *S. carnosus* and other sequenced staphylococci. However, comparison of the gene products revealed that about 20% of them have no staphylococcal homolog (Rosenstein et al. 2005).

A physical and genetic map of *S. xylosus* C2a was established by locating 47 restriction fragments and 33 genetic markers. Twenty-three previously identified loci mainly concerned carbohydrate utilization and carbon catabolite repression (Brückner et al. 1993; Egeter and Brückner 1995; Bassias and Brückner 1998; Fiegler et al. 1999) and antioxidant capacities (Barrière et al. 2001, 2002), and 10 were newly identified (Dordet-Frisoni et al. 2007). The *S. xylosus* C2a chromosome contains six *rrn* operons. As also observed in other staphylococcal species, this number varies among strains from five to six (Dordet-Frisoni et al. 2007; Kuroda et al. 2001, 2005; Baba et al. 2002; Holden et al. 2004; Gill et al. 2005; Zhang et al. 2003; Takeuchi et al. 2005). A first comparison of *S. xylosus* C2a map with *S. aureus*, *S. epidermidis*, *S. haemolyticus*, and *S. saprophyticus* sequenced genome showed that the relative position of *rrn* operons and several genetic markers are well conserved between *S. xylosus* and the other staphylococci (Figure 17.1). The sequencing of the complete genome of *S. xylosus* is in progress (http://www.genoscope.cns.fr).

Table 17.2. Intra- and interspecies variation of chromosome size among staphylococcal sequenced genomes.

Species	Strains	Chromosome Size (kb)	References
S. aureus	Mu50	2,879	Kuroda et al. 2001
	N315	2,815	Kuroda et al. 2001
	MW2	2,820	Baba et al. 2002
	MRSA252	2,903	Holden et al. 2004
	MSSA476	2,800	Holden et al. 2004
	COL	2,809	Gill et al. 2005
	USA300	2,873	Diep et al. 2006
S. epidermidis	ATCC 12228	2,499	Zhang et al. 2003
	RP62A	2,617	Gill et al. 2005
S. saprophyticus	ATCC 15305	2,517	Kuroda et al. 2005
S. haemolyticus	JCSC1435	2,685	Takeuchi et al. 2005
S. carnosus	Tm300	2,560	Rosenstein et al. 2005

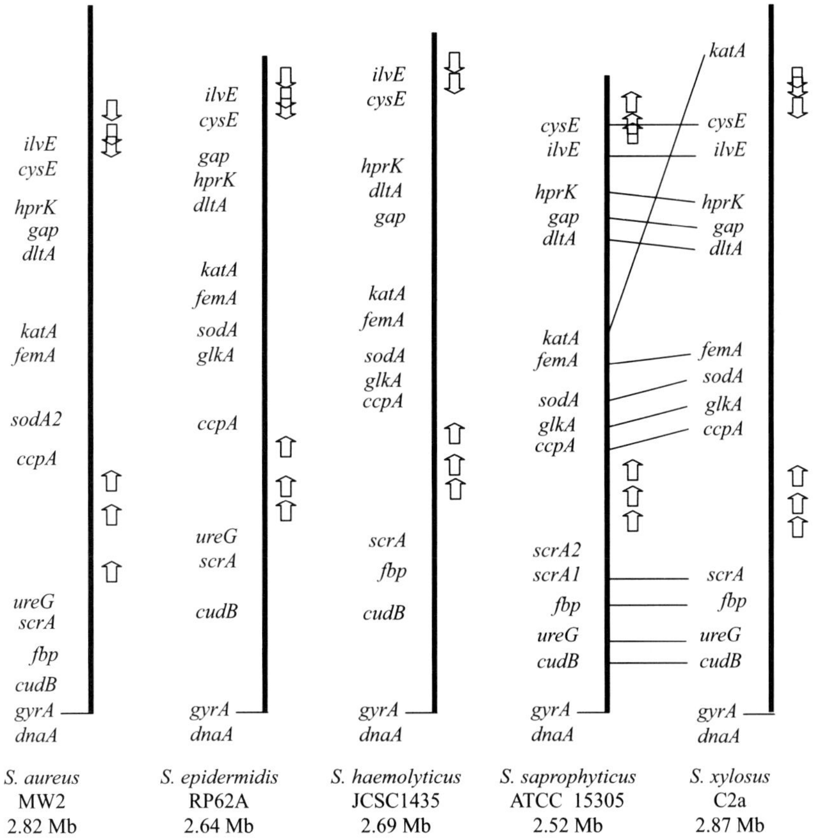

Figure 17.1. Comparison of the genetic map of *S. xylosus* C2a with the chromosomes of *S. aureus* MW2, *S. epidermidis* RP62a, *S. haemolyticus* JCSC1435, and *S. saprophyticus* ATCC 15305. The alignment was done from the putative origin of replication labeled by *dnaA* and *gyrA* genes. The arrows correspond to ribosomal operons.

CRYPTIC PLASMIDS

Lactobacillus sakei

L. sakei strains usually contain 1–5 plasmids, the size of which can range from about 1.5 to more than 50 kbp. The two main types of replication mode, rolling-circle (RC) and theta replication, exist. Plasmids homologous to the 2.1 kbp pLP1 from *Lactobacillus plantarum* or to the 2.6 kbp pLC2 from *Lactobacillus curvatus* (RC type) are known to be present in *L. sakei* (Bringel et al. 1989, Vogel et al. 1992). The replication region of different groups of plasmids has been sequenced: pSAK1 (theta-type, Genbank: Z50862), pLS141-1 (RC type, Genbank: AB109041), and pRV500, a 13 kbp theta-replicating plasmid (Alpert et al. 2003). The latter is the only plasmid isolated from *L. sakei* that has been totally sequenced so far.

Gene products other than those involved in replication have been identified by sequence homology, such as a restriction/modification system on pRV500 (Alpert et al. 2003) and a general stress protein on pLS141-1. The first correlation between the presence of plasmids and a phenotype was evidenced in 1988 by Shay et al. for cystein transport. Tetracycline resistance genes, detected in *L. sakei* isolated from fermented sausages, were mainly plasmid-borne (Gevers et al. 2003). Bacteriocin production, an important technological trait, has been given high attention. Although many gene clusters characterized so far are chromosomal, plasmid-encoded genes were reported for sakacin P (Vaughan et al. 2003), sakacin A (Axelsson and Holck 1995), sakacin G (Simon et al. 2002), and lactocin S (Skaugen et al. 1997). To our knowledge, no other potentially noteworthy technological traits have been linked to plasmids in *L. sakei* so far.

Staphylococcus carnosus and *Staphylococcus xylosus*

One cryptic plasmid, pSX267, issued from *S. xylosus* DSM20267, was described. It is a 29.5 kbp multiresistance plasmid, which contains an antimonite, arsenite, and arsenate operon (Rosenstein et al. 1992). The pSX267 replication region (1.8 kb) was sequenced and defined a new family of theta-replicating plasmid in staphylococci (Gering et al. 1996). Utilization of the pSX267 replicon was considered to develop a novel narrow host range vector for gram-positive bacteria (Gering et al. 1996). A large diversity of plasmid profiles was recently reported for *S. xylosus* strains isolated from slightly fermented sausages (Martin et al. 2006).

Although the presence of plasmids was recently reported for this species (Martin et al. 2006), none was well described for *S. carnosus*.

DNA TRANSFER, VECTORS, AND GENETIC TOOLS

Lactobacillus sakei

Plasmid or gene transfer exists naturally among bacteria and is also of great importance for genetic studies. Natural transformation essentially relies on competence and conjugation. Natural competence was never reported in *L. sakei*, although the complete genome sequence of *L. sakei* 23K revealed genes homologous to the well-known competence genes of *Bacillus subtilis*, but their exact function is unknown. Conjugative plasmids required for conjugation were not yet isolated from *L. sakei*. Nevertheless, conjugative transfers were observed under laboratory conditions with the broad host-range conjugative plasmid pAMβ1 (Langella et al. 1996), and during sausage fermentation in the closely related species *Lactobacillus curvatus* (Vogel et al. 1992).

For laboratory research, in the objective of understanding the molecular basis of *L. sakei* physiology, genetic tools and transformation protocols have been developed. Electroporation protocols give relative good efficiency, depending of the strains (Aukrust and Blom 1992, Berthier et al. 1996). A list of vectors developed for *L. sakei*'s molecular biology is given in Table 17.3. Construction of chromosomal mutants relies on homologous recombination and utilization of integrative or delivery vectors. This allowed the construction of point mutations for which no exogenous DNA persists (Stentz and Zagorec 1999), or deletion mutants (Malleret et al. 1998, Stentz et al. 2000). No genetic tools based on phage or transposition are yet available. This is the consequence of the lack of knowledge of these elements in *L. sakei*. Indeed, only one phage has been described, but its sequence was not determined (Leuschner et al. 1993) and the presence of IS elements is strain-dependent (Skaugen et al. 1997; Skaugen and Nes 2000; Chaillou et al. 2005). Inducible promoters to study gene functions (Stentz et al. 2000) or to overexpress proteins (Sorvig et al. 2005) have been described and reporter genes can also be used in *L. sakei* such as *lacZ* (Stentz et al. 2000), *gusA* (Hertel et al. 1998), and GFP (Gory et al. 2001), the activity of the latter being detected by fluorescence and allowing to follow strains in complex environments.

Staphylococcus carnosus and *Staphylococcus xylosus*

Natural competence has never been reported in *S. xylosus* and *S. carnosus*. An efficient transformation system was first reported for the *S. carnosus*

Table 17.3. Some of the plasmids developed for molecular genetics of *L. sakei.*

Plasmid Name	Replicon	Selection	Host-range	Function/Use
pRV300[1]	colE1[4]	erm	NA	Integrative vector Insertional mutagenesis
pRV566[2]	colE1, pRV500	erm	*E. coli* lactobacilli	Shuttle vector Replicative
pSIP vector[3] series	p256 including or not colE1	erm	*L. sakei*, *L. plantarum*	Expression vectors with sakacin inducible promoters

[1] Leloup et al. (1997).
[2] Alpert et al. (2003).
[3] Sorvig et al. (2005).
[4] colE1 is a replicon of *Escherichia coli* that does not sustain replication in gram-positive bacteria.

TM300 strain using competent protoplasts (Götz et al. 1983; Götz and Schumacher 1987). *S. carnosus* TM300 can be considered as a wild-type plasmid and cloning host strain (Götz, 1990) and has been used in many studies. The transformation procedure of this strain was simplified by the development of a procedure for electroporation (Augustin and Götz 1990). This method could be applied to other staphylococcal species but not to *S. xylosus*. A specific procedure for electroporation of *S. xylosus* C2a strain was developed (Brückner 1997). In the same time, a system for gene replacement in *S. carnosus* and *S. xylosus* was described using the temperature-sensitive *E. coli—Staphylococcus* shuttle plasmids, pBT1 and pBT2 (Brückner 1997). Transposon mutagenesis can be used to obtain *S. carnosus* and *S. xylosus* mutants (Wagner et al. 1995; Neubauer et al. 1999). Several expression vectors have been created for the display of heterologous proteins to the surface of *S. carnosus* (Samuelson et al. 1995; Wernérus and Ståhl 2002), and *S. xylosus* (Hansson et al. 1992). A reporter gene system based on the β-galactosidase gene of *S. xylosus* has been described (Jankovic et al. 2001). So, many genetic tools can be used to study the molecular basis of *S. carnosus* and *S. xylosus* physiology.

TOWARD POSTGENOMICS

Lactobacillus sakei

The determination of the whole chromosome sequence of *L. sakei* 23K allows the development of proteomics and transcriptomics, through large scale and global analyses of the response of this bacterium to its environment. The use of two-dimensional gel electrophoresis (2-DE) revealed to be a powerful tool to analyze protein expression under various environmental conditions (Figure 17.2).

As an example of analysis of process adaptation, proteins differentially expressed at cold temperature or in the presence of NaCl were identified. By the use of mutants constructed with the genetic tools described above, the involvement of these proteins in this adaptation was demonstrated. Two proteins belong to carbon metabolism and four could be clustered as stress proteins. Five of them appeared to be necessary for a correct survival during stationary phase (Marceau et al. 2004). Proteomics also revealed that high-pressure stress survival mechanisms of *Listeria monocytogenes* and *L. sakei* elicited different responses through the involvement of different set of proteins. Stress proteins were induced in *L. sakei*, whereas *L. monocytogenes* induced proteases to overcome high-pressure stress (Jofré et al. 2007).

Staphylococcus xylosus

Similarly, the knowledge of the whole chromosome sequence of *S. xylosus* C2a will allow the development of proteomic and transcriptomic studies for a better understanding of the physiology of *S. xylosus*.

A proteomic approach to study cell-envelope proteins of *S. xylosus* has been developed (Planchon 2006; Planchon et al. 2007). This method was based on lysostaphin treatment giving a first fraction of protein, and a second membrane protein fraction was efficiently recovered in a procedure involving a delipidation and solubilization steps. Proteins were separated by 2-DE and identified by mass spectrometry. A total of 90 distinct proteins were identified in the two fractions. Proteins with transmembrane domains, predicted as lipoproteins or homologous to

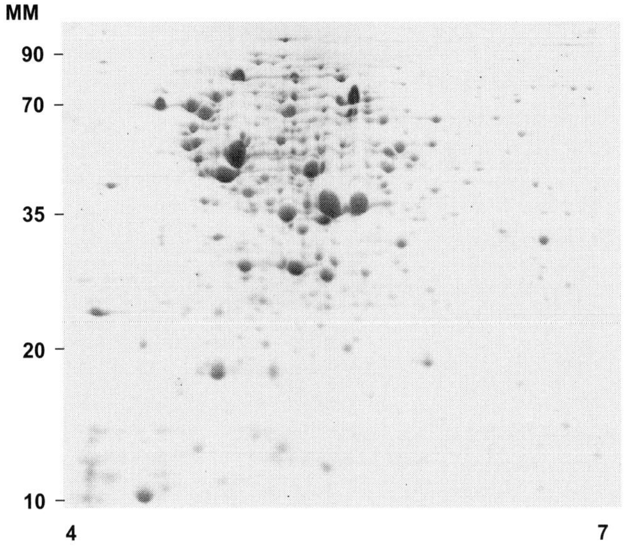

Figure 17.2. Two-dimensional gel electrophoresis of proteins of the cytoplasmic fraction after Commassie blue staining. Proteins expressed by *L. sakei* 23K during growth in MCD (Chemically Defined Medium, Lauret et al. 1996) at 30°C are shown. About 420 spots could be detected in the pI range 4–7 on the 1116 predicted proteins encoded by the genome in this range. Peptide Mass Finger Print analysis by MALDI Tof combined with *L. sakei* 23K genome database searching allowed protein identification.

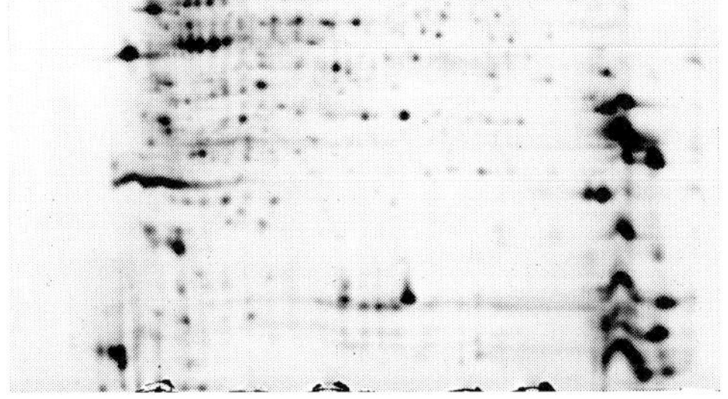

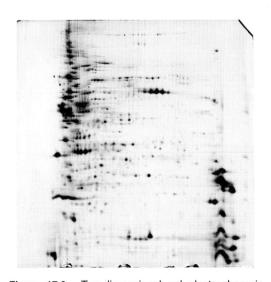

Figure 17.3. Two-dimensional gel electrophoresis of proteins of the membrane fraction from *S. xylosus* C2a grown in tryptic soya broth medium supplemented with 0.6 g/L yeast extract and 20 g/L sodium chloride under shaking (150 rpm) at 30°C during 48 h. Protein extracts were separated by IEF over a nonlinear pH gradient 3 to 10 followed by a 10% SDS-PAGE and revealed with silver staining.

components of membrane protein complex were identified in the membrane fraction (Figure 17.3). Proteins potentially secreted or primarily predicted as cytoplasmic, but that could interact temporarily with membrane components, were also identified. The method, allowing to recover a significant set of cell envelope proteins from *S. xylosus*, could be applied to study the physiology of this bacterium in different environmental meat conditions.

These proteomic approaches combined with the future analysis of the transcripts will allow having a more integrated and comprehensive knowledge of the mechanisms by which *L. sakei* and *S. xylosus* contribute to the fermentation of meat and how these bacteria interact with each other.

REFERENCES

CA Alpert, A-M Crutz-Le Coq, C Malleret, M Zagorec. 2003. Characterization of a theta-type plasmid from *Lactobacillus sakei*: A potential basis for low-copy vectors in lactobacilli. Appl Environ Microbiol 69:5574–5584.

J Augustin, F Götz. 1990. Transformation of *Staphylococcus epidermidis* and other staphylococcal species with plasmid DNA by electroporation. FEMS Microbiol Lett 54:203–207.

T Aukrust, H Blom. 1992. Transformation of *Lactobacillus* strains used in meat and vegetable fermentation. Food Res Int 25:253–261.

L Axelsson, A Holck. 1995. The genes involved in production of and immunity to sakacin A, a bacteriocin from *Lactobacillus sake* Lb706. J Bacteriol 177:2125–2137.

T Baba, F Takeuchi, M Kuroda et al. 2002. Genome and virulence determinants of high virulence community-acquired MRSA. Lancet 359:1819–1827.

C Barrière, R Brückner, D Centeno, R Talon. 2002. Characterization of the *katA* gene encoding a catalase and evidence for at least a second catalase activity in *Staphylococcus xylosus*, bacteria used in food fermentation. FEMS Microbiol Lett 216:277–283.

C Barrière, R Brückner, R Talon. 2001. Characterization of the single superoxide dismutase of *Staphylococcus xylosus*. Appl Environ Microbiol 67: 4096–4104.

J Bassias, R Brückner. 1998. Regulation of lactose utilization genes in *Staphylococcus xylosus*. J Bacteriol 180:2273–2279.

F Berthier, M Zagorec, M Champomier-Vergès, SD Ehrlich, F Morel-Deville. 1996. High-frequency transformation of *Lactobacillus sake* by electroporation. Microbiol 142:1273–1279.

F Bringel, L Frey, JC Hubert. 1989. Characterization, cloning, curing, and distribution in lactic acid bacteria of pLP1, a plasmid from *Lactobacillus plantarum* CCM 1904 and its use in shuttle vector construction. Plasmid 22:193–202.

R Brückner. 1997. Gene replacement in *Staphylococcus carnosus* and *Staphylococcus xylosus*. FEMS Microbiol Lett 151:1–8.

R Brückner, E Wagner, F Götz. 1993. Characterization of a sucrase gene from *Staphylococcus xylosus*. J Bacteriol 175:851–857.

S Chaillou, M Champomier-Vergès, M Cornet, A-M Crutz Le Coq, A-M Dudez, V Martin, S Beaufils, R Bossy, E Darbon-Rongère, V Loux, M Zagorec. 2005. Complete genome sequence of the meat-born lactic acid bacterium *Lactobacillus sakei* 23K. Nature Biotechnol 23:1527–1533.

M Champomier-Vergès, S Chaillou, M Cornet, M Zagorec. 2002. *Lactobacillus sakei*: Recent development and future prospects. Res Microbiol 153: 115–123.

BA Diep, SR Gill, RF Chang et al. 2006. Complete genome sequence of USA300, an epidemic clone of community-acquired meticillin-resistant *Staphylococcus aureus*. Lancet 367:731–739.

E Dordet-Frisoni, R Talon, S Leroy. 2007. Physical and genetic map of the *Staphylococcus xylosus* C2a chromosome. FEMS Microbiol Lett, 266:184–195.

AM Dudez, S Chaillou, L Hissler, R Stentz, M Champomier-Vergès, CA Alpert, M Zagorec. 2002. Physical and genetic map of the *Lactobacillus sakei* 23K chromosome. Microbiol 148:421–431.

O Egeter, R Brückner. 1995. Characterization of a genetic locus essential for maltose-maltotriose utilization in *Staphylococcus xylosus*. J Bacteriol 177:2408–2415.

VGH Eijsink, L Axelsson. 2005. Bacterial lessons in sausage making. Nature Biotechnol 23:1494–1495.

H Fiegler, J Bassias, I Jankovic, R Brückner. 1999. Identification of a gene in *Staphylococcus xylosus* encoding a novel glucose uptake protein. J Bacteriol 181:4929–4936.

M Gering, F Götz, R Brückner. 1996. Sequence and analysis of the replication region of the *Staphylococcus xylosus* plasmid pSX267. Gene 182:117–122.

D Gevers, M Danielsen, G Huys, J Swings. 2003. Molecular characterization of tet(M) genes in *Lactobacillus* isolates from different types of fermented dry sausage. Appl Environ Microbiol 69: 1270–1275.

SR Gill, DE Fouts, GL Archer et al. 2005. Insights on evolution of virulence and resistance from the complete genome analysis of an early methicillin-resistant *Staphylococcus aureus* strain and a biofilm-producing methicillin-resistant *Staphylococcus epidermidis* strain. J Bacteriol 187:2426–2438.

L Gory, MC Montel, M Zagorec. 2001. Use of green fluorescent protein to monitor *Lactobacillus sakei* in meat fermented products. FEMS Microbiol Lett 194:127–133.

F Götz. 1990. *Staphylococcus carnosus*. A new host for gene cloning and protein production. Soc Appl Bacteriol Symp Ser 19:49S–53S.

F Götz, B Kreutz, KH Schleifer 1983. Protoplast transformation of *Staphylococcus carnosus* by plasmid DNA. Mol Gen Genet 189:340–342.

F Götz, B Schumacher. 1987. Improvements of protoplast transformation in *Staphylococcus carnosus*. FEMS Microbiol Lett 40:285–288.

M Hansson, S Stahl, TN Nguyen, T Bachi, A Robert, H Binz, A Sjolander, M Uhlen. 1992. Expression of recombinant proteins on the surface of the coagulase-negative bacterium S*taphylococcus xylosus*. J Bacteriol 174:4239–4245.

C Hertel, G Schmidt, M Fischer, K Oellers, WP Hammes. 1998. Oxygen-Dependent Regulation of the Expression of the catalase gene katA of *Lactobacillus sakei* LTH677 Appl Environ Microbiol 64:1359–1365.

MT Holden, EJ Feil, JA Lindsay et al. 2004. Complete genomes of two clinical *Staphylococcus aureus* strains: evidence for the rapid evolution of virulence and drug resistance. Proc Natl Acad Sci USA 101:9786–9791.

I Jankovic , O Egeter, R Brückner. 2001. Analysis of catabolite control protein A-dependent repression in *Staphylococcus xylosus* by a genomic reporter gene system. J Bacteriol 183:580–586.

A Jofré, MC Champomier-Vergès, P Anglade, F Baraige, B Martin, M Garriga, M Zagorec, T Aymerich. 2007. Proteomic analysis of the response of lactic acid and pathogenic bacteria to high hydrostatic pressure treatment, submitted.

M Kuroda, T Ohta, I Uchiyama et al. 2001. Whole genome sequencing of meticillin-resistant *Staphylococcus aureus*. Lancet 357:1225–1240.

M Kuroda, A Yamashita, H Hirakawa et al. 2005. Whole genome sequence of *Staphylococcus saprophyticus* reveals the pathogenesis of uncomplicated urinary tract infection. Proc Natl Acad Sci USA 102:13272–13277.

P Langella, M Zagorec, SD Ehrlich, F Morel-Deville. 1996. Intergeneric and intrageneric conjugal transfer of plasmids pAMβ1, pIL205 and pIP501 in *Lactobacillus sake*. FEMS Microbiol Lett 139:51–56.

R Lauret, F Morel-Deville, F Berthier, M Champomier-Vergès, P Postma, SD Ehrlich, M Zagorec. 1996. Carbohydrate utilization in *Lactobacillus sake*. Appl Environ Microbiol 62:1922–1927.

L Leloup, SD Ehrlich, M Zagorec, F Morel-Deville. 1997. Single cross-over integration in the *Lactobacillus sake* chromosome and insertional inactivation of the *ptsI* and *lacL* genes. Appl Environ Microbiol 63:2127–2133.

RGK Leuschner, EK Arendt, WP Hammes. 1993. Characterization of a virulent *Lactobacillus sake* phage PWH2, Appl Microbiol Biotechnol 39: 617–621.

C Malleret, R Lauret, SD Ehrlich, F Morel-Deville, M Zagorec. 1998. Disruption of the sole *ldhL* gene in *Lactobacillus sakei* prevents the production of both L- and D-lactate. Microbiol 144:3327–3333.

A Marceau, M Zagorec, S Chaillou, T Méra, M Champomier-Vergès. 2004. Evidence for the involvement of at least six proteins in *Lactobacillus sakei* adaptation to cold temperature and addition of NaCl. Appl Environ Microbiol 70:7260–7268.

B Martin, M Garriga, M Hugas, S Bover-Cid, MT Veciana-Nogues, T Aymerich. 2006. Molecular, technological and safety characterization of gram-positive catalase-positive cocci from slightly fermented sausages. Int J Food Microbiol 107: 148–158.

S Morot-Bizot, R Talon, S Leroy-Sétrin. 2003. Development of specific PCR primers for a rapid and accurate identification of *Staphylococcus xylosus*, a species used in food fermentation. J Microbiol Methods 55:279–286.

H Neubauer, I Pantel, F Götz. 1999. Molecular characterization of the nitrite-reducing system of *Staphylococcus carnosus*. J Bacteriol 181:1481–1488.

S Planchon 2006. Aptitude de *Staphylococcus carnosus* et *Staphylococcus xylosus* à former des biofilms—Étude d'une souche *biofilm positif* par une approche protéomique. Thèse de Docteur de l'Université Blaise Pascal, Clermont-Ferrand II.

S Planchon, C Chambon, M Desvaux, I Chafsey, S Leroy, R Talon, M Hébraud. 2007. Proteomic analysis of cell envelope proteins from *Staphylococcus xylosus* C2a. J Proteomic Res, submitted.

R Rosenstein, C Nerz, A Resch, F Götz. 2005. Comparative Genome Analyses of Staphylococcal Species. 2nd European Conference on prokaryotic genomes. Prokagen.

R Rosenstein, A Peschel, B Wieland, F Götz. 1992. Expression and regulation of the antimonite, arsenite, and arsenate resistance operon of *Staphylococcus xylosus* plasmid pSX267. J Bacteriol 174: 3676–3683.

P Samuelson, M Hansson, N Ahlborg, C Andreoni, F Götz, T Bachi, TN Nguyen, H Binz, M Uhlen, S Stahl. 1995. Cell surface display of recombinant proteins on *Staphylococcus carnosus*. J Bacteriol 177:1470–1476.

BJ Shay, AF Egan, M Wright, PJ Rogers. 1988. Cystein metabolism in an isolate of *Lactobacillus sake*: plasmid composition and cystein transport. FEMS Microbiol Lett 56:183–188.

L Simon, C Frémaux, Y Cenatiempo, J-M Berjeaud. 2002. Sakacin G, a new type of antilisterial bacteriocin. Appl Environ Microbiol 68:6416–6420.

M Skaugen, CI Abildgaard, IF Nes. 1997. Organization and expression of a gene cluster involved in the biosynthesis of the lantibiotic lactocin S Mol Gen Genet 253:674–686.

M Skaugen, IF Nes, 2000. Transposition in *Lactobacillus sakei*: inactivation of a second lactocin S operon by the insertion of IS*1520*, a new member of the IS*3* family of insertion sequences. Microbiol 146: 1163–1169.

E Sorvig, G Mathiesen, K Naterstad, VG Eijsink, L Axelsson. 2005. High-level, inducible gene expression in *Lactobacillus sakei* and *Lactobacillus plantarum* using versatile expression vectors. Microbiol 151:2439–2449.

R Stentz, C Loizel, C Malleret, M Zagorec. 2000. Development of genetic tools for *Lactobacillus sakei*: disruption of the β-galactosidase gene and use of *lacZ* as a reporter gene to study regulation of the putative copper ATPase, AtkB. Appl Environ Microbiol 66:4272–4278.

R Stentz, M Zagorec. 1999. Ribose utilization in *Lactobacillus sakei*: analysis of the regulation of the *rbs* operon and putative involvement of a new transporter. J Mol Microbiol Biotechnol 1:165–173.

F Takeuchi, S Watanabe, T Baba et al. 2005. Whole-genome sequencing of *Staphylococcus haemolyticus* uncovers the extreme plasticity of its genome and the evolution of human-colonizing staphylococcal species. J Bacteriol 187:7292–7308.

A Vaughan, VG Eijsink, D van Sinderen. 2003. Functional characterization of a composite bacteriocin locus from malt isolate *Lactobacillus sakei* 5. Appl Environ Microbiol 69:7194–7203.

RF Vogel, M Becke-Schmid, P Entgens, W Gaier, WP Hammes. 1992. Plasmid transfer and segregation in *Lactobacillus curvatus* LTH1432 in vitro and during sausage fermentations. Syst Appl Microbiol 15: 129–136.

RF Vogel, M Lohmann, AN Weller, M Hugas, WP Hammes. 1991. Structural similarity and distribution of small cryptic plasmids of *Lactobacillus curvatus* and *Lactobacillus sake*. FEMS Microbiol Lett 84:183–190.

E Wagner, J Doskar, F Götz. 1998. Physical and genetic map of the genome of *Staphylococcus carnosus* TM300. Microbiol 144:509–517.

E Wagner, S Marcandier, O Egeter, J Deutscher, F Götz, R Brückner. 1995. Glucose kinase-dependent catabolite repression in *Staphylococcus xylosus*. J Bacteriol 177:6144–6152.

H Wernérus, S Ståhl. 2002. Vector engineering to improve a staphylococcal surface display system. FEMS Microbiol Lett 212:47–54.

YQ Zhang, SX Ren, HL Li et al. 2003. Genome-based analysis of virulence genes in a non-biofilm–forming *Staphylococcus epidermidis* strain (ATCC 12228). Mol Microbiol 49:1577–1593.

18
Influence of Processing Parameters on Cultures Performance

Louise H. Stahnke and Karsten Tjener

INTRODUCTION

During the processing of cured meat products the applied starter cultures—and the natural contaminating flora, not to forget—are affected by an immense number of changing parameters that invariably will influence their biochemical activity and subsequently be of importance for the quality of the final product. Thus, in order to produce cured meat products of high reproducible quality it is essential that intrinsic and extrinsic parameters that may influence the progress of the applied starter cultures during the different processing steps be well controlled.

As touched upon in previous chapters, the primary objective of using starter cultures is to ensure reproducible and correct microbial growth to avoid faulty productions caused by contaminating spoilage bacteria, pathogens, or other undesired microorganisms. In particular, it is essential to ensure that the fermentation step, and thereby the growth of the added lactic acid bacteria, progresses correctly, because the acidification process directly and indirectly influences the texture, color, and flavor formation as well as the drying-out (Bacus 1984; Lücke and Hechelmann 1987).

In the production of fermented sausages, particular mince ingredients and temperature could vary quite a lot depending on the actual recipe and processing technology, and this will influence the culture growth. Processing parameters do not vary quite as much within different recipes for cured whole muscles, but factors such as temperature and salt concentration are still of utmost importance to keep in control—although for controlling the spoilage flora rather than for enhancing the growth of a starter culture.

Parameters that affect the activity of starter cultures in the production of cured meat products are temperature; curing salts (NaCl, nitrate, nitrite, ascorbate); carbohydrates; and, to a lesser extent, microbial contamination and initial pH of the raw materials; spices; and other additives such as phosphate or binders. For dry sausages, the diameter is also of some importance, and for mold-inoculated products, the casing type, smoke, air speed, and air humidity (Roncales et al. 1991; Incze 1992; Dossmann et al. 1996).

TEMPERATURE

Fermentation and ripening temperature is a factor of most importance for all genera of starter cultures applied to cured meat products, because the metabolic pathways and regulation systems within the microbial cells are highly temperature-dependent due to temperature's strong influence on the involved enzymes. Bacteria grow within a temperature range whose average is approximately 30°C, with each species having a well-defined upper and lower limit (Nester et al. 1983). However, since the optimum temperature for growth of a species is also close to the upper limit of its growth, meat fermentation always takes place far below or just below the optimum temperature for the applied lactic acid bacteria.

Lactic Acid Bacteria

Table 18.1 shows the recommended fermentation temperatures for different species of lactic acid bacteria from various culture manufacturers. Apart from *Pediococcus acidilactici* being best suited for very high fermentation temperature, such as those applied in the U.S., there is no general trend showing that specific species are more suited than others at European fermentation temperatures. It is strain-dependent rather than species-dependent and relies on the actual technology applied. However, it has been shown that sausages made with *Lactobacillus* are able to reach a lower pH than sausages fermented with *Pediococcus* (Landvogt and Fischer 1991b; Chr. Hansen 2006).

For lactic acid bacteria, an increase in temperature below the optimum temperature may increase the fermentation rate greatly. It has been found that a 5°C increase in temperature approximately doubles the rate of acid formation (Baumgartner et al. 1980). Temperature affects the lag phase as well as the actual speed of acidification during the exponential growth phase (Landvogt and Fischer 1991 a,b; Chr. Hansen 2006).

Figure 18.1 shows typical curves of the temperature-related pH reduction caused by a commercial strain of *Lb. sakei*. Similar temperature dependencies have been shown as actual growth curves, acidification curves, or fitted mathematical models for other starter cultures in broth or in sausage fermentations (Baumgartner et al. 1980; Blickstad and Molin 1981; Landvogt and Fischer 1991a,b; Raccach 1992; Huang and Lin 1993; Wijtzes et al. 1995; Balduino et al. 1999; Leroy and De Vuyst 1999; Messens et al. 2003; Leroy et al. 2005).

Table 18.1. Recommended fermentation temperatures of typical commercial lactic acid bacteria for meat fermentation.

Genera	Species	Recommended Fermentation Temperature (°C)
Lactobacillus	*sakei*	21–32
	plantarum	25–35
	farciminis	22–32
	curvatus	22–37
Pediococcus	*pentosaceus*	20–37
	acidilactici	25–45

Source: Texel (1998); Quest (1999); Chr. Hansen (2003).

Staphylococcus and *Kocuria*

The optimum growth temperature for *Staphylococcus* and *Kocuria* strains for meat fermentation is around 21–32°C (Texel 1998; Chr. Hansen 2003), which is well suited for most fermentation steps during which the nitrate reductase activity must be optimal for color formation. However, in sausages applied with an acidifying culture, such high temperatures could lower pH too fast, depending on the acidifying strain and the sugar content, having a detrimental effect on the nitrate reductase activity (color formation) and the flavor-forming activity of the *Staphylococcus/Kocuria*. Both *Staphylococcus* and *Kocuria* are very sensitive to pH lowering (Sørensen and Jakobsen 1996; Guo et al. 2000; Søndergaard and Stahnke 2002), and even in sausages fermented at a low speed *Staphylococcus* often begin to die after a few days in the process (Johansson et al. 1994; Tjener et al. 2003, 2004a). Undoubtedly also the enzymatic activities of the cells are influenced negatively because the optimum pH of a cell represents the best average for the function of all its cellular enzymes (Nester et al. 1983). With respect to flavor formation by *Staphylococcus* it has been shown that temperature and, thus, pH-lowering both affect the level of important flavor compounds (Tjener et al. 2004b).

Yeast and Mold

The optimum temperature of *Debaryomyces* and *Penicillium* strains used in meat processing is around 18–25°C in a wide pH range (Texel 1998; Chr. Hansen 2003), but the species will grow well down to 8–15°C, which is most often applied during the ripening process for yeast- and mold-inoculated sausages (Texel 1998; Masoud and Jakobsen 2005). However, to ensure a fast onset of *Penicillium* growth on the sausage surface, both temperature and humidity must be high at the very beginning: above 20°C and around 90% relative humidity, respectively. If water activity in sausage surface drops below 0.80, neither germination nor growth will take place (Sunesen and Stahnke 2003).

CURING SALTS

NaCl

The primary curing agents include common salt, sodium nitrite, and potassium nitrate, as well as sodium ascorbate or sodium erythrobate (Lücke 1998; Adams 1986). Usually 2.0–3.5% common salt (NaCl) is added to the sausage mince, giving an initial water activity around 0.97–0.96 (salt-in-water of 5–7%) depending on the amount of fat or added binders. The lowered water activity inhibits growth of the spoilage flora normally associated with chilled meats, though undesired microorganisms such as *Staphylococcus aureus*, *Listeria monocytogenes*, *Salmonella*, and *E. coli* O157:H7 are still able to survive unless other hurdles are applied (Lücke 1998; Adams 1986; Roca and Incze 1990; Leistner 1991).

Just as for the contaminating flora, the activity of the applied starter cultures is affected by the concentration of NaCl. The inhibition brought about depends on the involved species and their sensitivity to the water activity (Landvogt and Fischer 1991a). In general, decreasing the water activity of a sausage mince, either by adding salt or decreasing the water content by drying, results in a continues decline in growth for most bacterial meat cultures (Landvogt and Fischer 1991a; Søndergaard and Stahnke 2002; Leroy and De Vuyst 2005), though some staphylococci seem to have maximum growth within 3.5–10% salt-in-water depending on strain (McMeekin et al. 1987; Hammes et al. 1995). In particular, the lag phase is prolonged by increased salt concentration, whereas the growth rate in the exponential phase is much less affected (Leroy et al. 2005). Also, the specific bacteriocin production of bacteriocin-positive lactic acid bacteria is lowered by increased salt concentration (Leroy and De Vuyst 1999). Among homofermentative lactobacilli, strains of *Lb. curvatus* seem more sensitive to NaCl than *Lb. sakei* (Korkeala et al. 1992).

The most commonly applied yeast in the meat industry, *Debaryomyces hansenii*, is quite salt-tolerant but species differences do exist. Optimum is normally around 0–4% NaCl, with a decline at higher salt levels (Prista et al. 1997; Sørensen and Jakobsen 1997; Texel 1998; Olesen and Stahnke 2000; Chr. Hansen 2003). Molds are very salt-tolerant and the most typical used strains of *Penicillium nalgiovense* will grow at water activities as low as 0.85, though optimum growth occur at approximately 0.90 or above (Lopez Diaz et al. 2002; see also Chapter 16, "Starter Cultures: Molds," in this volume). Table 18.2 shows the maximum recommended salt concentration for typical starter cultures; species differences are not outlined. Most cultures do have some activity at higher salt levels, but their function is seriously hampered.

Nitrite and Nitrate

Nitrite alone or together with nitrate is added to the sausage mince in order to produce the characteristic

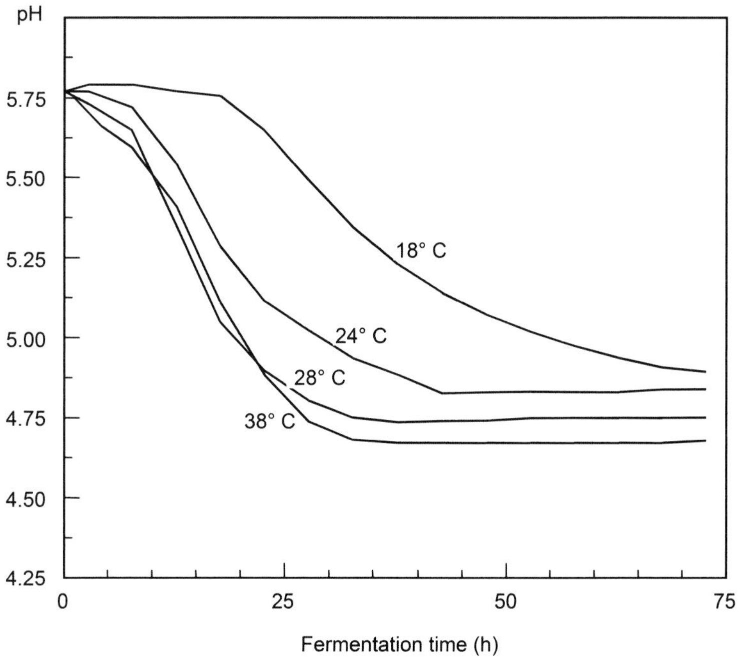

Figure 18.1. Influence of fermentation temperature on pH lowering by a commercial *Lb. sakei* strain growing in sausage mince, added 0.5% glucose (from Chr. Hansen 2003).

Table 18.2. Maximum recommended salt concentrations for typical meat starter cultures.

Genera	Maximum Recommended NaCl-concentration (%)
Lactobacillus	9–13
Pediococcus	7–10
Staphylococcus	15–16
Kocuria	No data available
Debaryomyces	16
Penicillium	18–20

Source: Texel (1998); Chr. Hansen (2003).

Table 18.3. Lactic acid production by the meat starter culture *Lb. pentosus* during growth in MRS-broth at 30°C for 12 hours.

Carbohydrate (1%)	Lactic Acid (%)
Glucose	0.98
Saccharose	0.86
Maltose	0.72
Maltodextrin	0.54
Galactose	0.31

Source: Chr. Hansen (2003).

cured color and to inhibit the growth of undesired bacteria. Nitrite together with salt promotes the establishment of *Lactobacillus* and *Micrococcaceae/Staphylococcaceae* by inhibiting species of *Enterobacteriaceae* and other bacteria (Lücke 1998; Vösgen 1992). If present in large amounts, nitrite could inhibit growth of lactic acid bacteria, in particular at low pH, but this is not the case with the amounts regularly used (Blickstad and Molin 1981; Korkeala et al. 1992; Dossmann et al. 1996; Leroy and De Vuyst 1999; Leroy et al. 2005). If inhibition takes place, nitrite seems to affect the lag phase rather than the exponential growth rate (Blickstad and Molin 1981). *Staphylococcus* and *Kocuria* are not affected by the nitrite levels used in meat fermentation (Guo et al. 2000; González and Díez 2002).

In the levels used for meat fermentation, nitrate does not influence the growth of lactic acid bacteria; however, *Staphylococcus* and *Kocuria* may exploit nitrate as an electron acceptor for respiratory growth at low redox potential, in this way retaining their metabolic activity during sausage processing (Metz and Hammes 1987; Neubauer and Götz 1996). Nitrate has a significant effect on the free amino acid profile and the aroma compound development of dry sausages through its influence on staphylococcal metabolism (Waade and Stahnke 1997; Olesen et al. 2004).

Neither nitrite nor nitrate seems to have a significant effect on growth of yeast and *Penicillium* species used as starter culture, but literature on the subject is very scarce (Lopez Diaz et al. 2002; Dura et al. 2004).

CARBOHYDRATES

Sugar Type

Various sugars, such as glucose, sucrose, lactose, corn syrups, etc., are commonly added to sausage mince as fermentation substrates for the lactic acid bacteria because the natural content of glucose is too low and variable to be reliable in modern sausage production (Lücke 1998). The sugar concentration and the type of sugar are directly related to the lowest pH achieved and the rate of the pH drop, respectively.

Glucose is fast utilized by all lactic acid bacteria, but other sugars, such as saccharose, lactose or maltose, are less easily fermented (Klettner and List 1980; Bacus 1984; Liepe et al. 1990; Huang and Lin 1994) or not fermentable at all by some strains (Chr. Hansen 2003). The enzymes involved in the glycolytic pathway that degrades glucose into lactic acid are constitutive, whereas proteins for transportation and degradation of sugars such as lactose or maltose are inducible, i.e., they are synthesized only when those sugars are present (Nester et al. 1983). However, these differences make it possible to control the profile of the pH drop. A mixture of fast and slowly fermentable carbohydrates can ensure a rapid, but relatively small, pH decrease at the beginning of the fermentation cycle—in order to inhibit unwanted bacteria without suppressing the staphylococci—followed by a slower fermentation step to achieve the final pH. Table 18.3 shows the amount of lactic acid produced from different sugars within a certain time span.

Sugar Content

Depending on the type, up to 2% carbohydrate is added to the sausage mince (Lücke 1998), but usually 0.3–0.8% glucose or saccharose proves sufficient (Pyrcz and Pezacki 1981; Leistner 1991). Some traditional sausages ripened for a very long time with very low water activities have no added sugar at all (Lois et al. 1987; Incze 1992). However, these days, industrial-produced sausages typically demand pH drops that make it essential to optimize the extent of

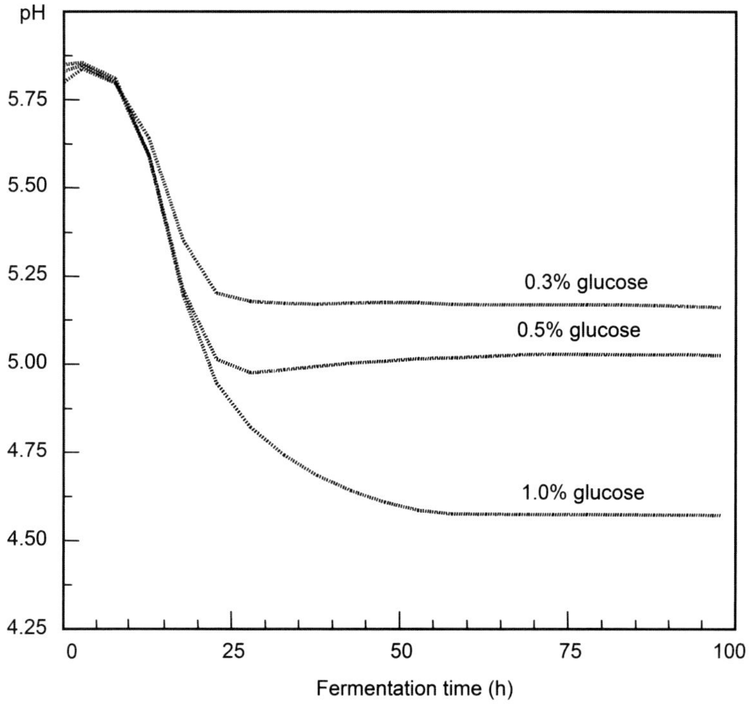

Figure 18.2. Influence of glucose concentration on pH lowering by a commercial *P. pentosaceus* strain growing in sausage mince at 24°C (from Chr. Hansen 2003).

acidification with the exact right amount of sugar. For Mediterranean sausages, pH is typically lowered to 5.0–5.1, whereas for North European- and U.S.-style sausages, pH is commonly reduced to around 4.8, though specific sausage types reach even lower values. Examples are some Swedish and Danish fermented sausages with final pH of 3.9–4.2.

When glucose concentration is above a certain level, increased amounts of fermentable sugar primarily affect the final pH but neither the lag phase nor the rate of lactic acid formation (see Figure 18.2). Landvogt and Fischer (1991b) found the critical level to be 0.15% glucose in their study. However, the extent of the pH drop also depends on the specific lactic acid bacteria and the water activity of the drying sausage. In general, for the same amount of fermentable sugar, pH will drop to a lower value in a fast fermentation than in a slow fermentation using the same strain (Chr. Hansen 2003). The less-pronounced pH drop in the slower fermentation is due to sugar being consumed by competing background flora producing other metabolic products than lactic acid or due to inhibition of the culture growth by the lowering in water activity over time.

Staphylococci are also able to degrade fermentable sugars into lactic acid, though part of the lactic acid is converted to acetic acid at high redox potential (Arkoudelos and Nychas 1995). However, the amount of acid produced by staphylococci is probably insignificant because meat fermentation conditions in general are less favorable for staphylococcal growth. No studies seem to have looked into this. It is possible, though, that sugar fermentation could matter for staphylococcal metabolism in traditionally fermented sausages with high pH and high amounts of residual sugars, such as certain Mediterranean-style sausages to which are added relatively high concentrations of lactose or skim milk powder. In basic investigations using broth and meat models, glucose seemed to have no, or even a slight negative effect on staphylococcal growth, but probably this was due to the lack of buffering capacity enabling the staphylococci to produce acid and reduce their own growth (Sørensen and Jakobsen 1996; Stahnke 1999; Søndergaard and Stahnke 2002).

To the authors' knowledge, the direct effect of sugars on the growth of yeasts and molds in sausages has not been studied. Yeasts and molds grow slowly as compared to bacteria in fermented meat products, and typically they develop late in the process where most sugar has been converted to acids unless very large amounts were added. However, yeasts and molds assimilate the produced lactic acid, in this way maintaining their metabolism and increasing the pH of the product. The lowering in pH that takes place during sausage fermentation is insignificant for their growth (Sørensen and Jakobsen 1997; Lopez Diaz et al. 2002; Masoud and Jakobsen 2005).

SPICES

A number of different spices are used for fermented meat products, the most typical being black and white pepper. Apart from the obvious function as flavorings, some spices have been found to stimulate the growth of lactic acid bacteria and thereby the rate of lactic acid formation (Nes and Skjelkvåle 1982; Zaika and Kissinger 1984). Black pepper, white pepper, mustard, garlic, allspice, nutmeg, ginger, coriander, clove, mace, cinnamon, cardamom, and red pepper all stimulate acid production to varying degrees depending on the concentration and the bacterial strain (Nes and Skjelkvåle 1982; Zaika and Kissinger 1984; Adams 1986). The content of manganese ions in the spices is the primary factor (Zaika and Kissinger 1984; Hagen et al. 2000). In addition, spices such as paprika contain amounts of fermentable sugars so high that the final pH drop may be significantly altered in sausages like the Spanish Chorizo (Lois et al. 1987; Incze 1992).

OTHER PARAMETERS

In order to obtain a safe dominance of the starter culture the size of inoculum must be 10^6–10^7 CFU/g so that the level of background flora is at least 2 log units lower than the added culture (Bacus 1984; Lücke 1998; Hammes et al. 1990). The performance of the starter lactic acid bacteria may be seriously affected if a strongly competitive population of another lactic acid bacteria is present in the raw materials. This is, for example, the case if vacuum-packed, cold-stored, nonfrozen meat is used. An increased level of inoculum brings about a shorter lag phase and thereby a shorter acidification time. However, size of inoculum does not affect the final pH unless water activity decreases very fast and inhibits the starter growth too soon (Landvogt and Fischer 1991b).

Initial pH of the meat could influence the fermentation time because high pH means higher buffering capacity and an extension of time to reach final pH (Incze 1992). In addition, the amount of sugar could be limiting if higher amounts of acid must be produced to reach the wanted pH. Meat pH is influenced by factors such as animal type and breed, slaughtering technology, and degree of spoilage (Lawrie 1998).

Sausage casing diameter has an effect on the growth of lactic acid bacteria because increasing diameter decreases the availability of oxygen for the microorganisms inside the sausage. With increasing diameter, lactic acid production increases due to less oxidative metabolism by the microorganisms involved (Incze 1992). Casing type is important for molded sausages because mycelium growth is significantly better on natural casings as compared to cellulose fiber and collagen casings (Roncales et al. 1991; Andersen 1994; Toledo et al. 1997).

REFERENCES

MR Adams. 1986. Fermented flesh foods. In: MR Adams, ed. Progress in Industrial Microbiology, vol 23. Amsterdam: Elsevier, pp. 159–198.

SJ Andersen. 1994. Anti-microbial characteristics of white mold cultures for fermentation of sausages. PhD thesis. Lyngby: Technical University of Denmark, pp. 1–165.

JS Arkoudelos, GJE Nychas. 1995. Comparative studies of the growth of *Staphylococcus carnosus* with or without glucose. Letters Appl Microbiol 20:19–24.

J Bacus. 1984. Utilization of Microorganisms in Meat Processing. A Handbook for Meat Plant Operators. Letchworth, England: Research Studies Press Ltd, pp. 1–170.

R Balduino, AS de Oliveira, MC de O Hauly. 1999. Influence of carbon source and temperature on the lactic acid fermentation developed by starter cultures. English Summary. Ciencia Tecnol Alim 19:363–366.

PA Baumgartner, PG Klettner, W Rödel. 1980. The influence of temperature on some parameters for dry sausage during ripening. Meat Sci 4:191–201.

E Blickstad, G Molin. 1981. Growth and lactic acid production of *Pediococcus pentosaceus* at different gas environments, temperatures, pH values and nitrite concentrations. Europ J Appl Microbiol Biotechnol 13:170–174.

Chr. Hansen. 2003. Bactoferm™ Meat Manual, vol. 1. Production of Fermented Sausages with Chr. Hansen Starter Cultures. 1st ed. Hørsholm, Denmark: Chr. Hansen A/S, pp. 1–31.

———. 2006. Brief Characterization of Bactoferm® Starter Cultures from Chr. Hansen. Bulletin no. 3. vs1. Hørsholm, Denmark: Chr. Hansen A/S, pp. 1–20.

MU Dossmann, RF Vogel, WP Hammes. 1996. Mathematical description of the growth of *Lactobacillus sake* and *Lactobacillus pentosus* under conditions prevailing in fermented sausages. Appl Microbiol Biotechnol 46:334–339.

M-A Dura, M Flores, F Toldrá. 2004. Effects of curing agents and the stability of a glutaminase from *Debaryomyces* spp. Food Chem 86:385–389.

B González, V Díez. 2002. The effect of nitrite and starter culture on micrological quality of "chorizo"—A Spanish dry cured sausage. Meat Sci 60:295–298.

HL Guo, MT Chen, DC Liu. 2000. Biochemical characteristics of *Micrococcus varians* and *Staphylococcus xylosus* and their growth on Chinese-style beaker sausages. Asian-Australasian J Anim Sci 13: 376–380.

BF Hagen, H Næs, AL Holck. 2000. Meat starters have individual requirements for Mn^{2+}. Meat Sci 55:161–168.

WP Hammes, A Bantleon, S Min. 1990. Lactic acid bacteria in meat fermentation. FEMS Microbiol Rev 87:165–174.

WP Hammes, I Bosch, G Wolf. 1995. Contribution of *Staphylococcus carnosus* and *Staphylococcus piscifermentans* to the fermentation of protein foods. J Appl Bacteriol Symp Suppl 79:76S–83S.

C-C Huang, C-W Lin. 1993. Drying temperature and time affect quality of Chinese-style sausage inoculated with lactic acid bacteria. J Food Sci 58:249–253.

———. 1994. Utilization of sugars in sausage meat by lactic acid bacteria. English Summary. J Chinese Soc Anim Sci 23:209–220.

K Incze. 1992. Raw fermented and dried meat products. Fleischwirtsch 72:58–62.

G Johansson, J-L Berdagué, M Larsson, N Tran, E Borch. 1994. Lipolysis, proteolysis and formation of volatile components during ripening of a fermented sausage with *Pediococcus pentosaceus* and *Staphylococcus xylosus* as starter cultures. Meat Sci 38:203–218.

P-G Klettner, D List. 1980. Beitrag zum einfluss der kohlenhydratart auf den verlauf der rohwurstreifung. Fleischwirtsch 60:1589–1593.

H Korkeala, T Alanko, T Tiusanen. 1992. Effect of sodium nitrite and sodium chloride on growth of lactic acid bacteria. Acta Vet Scand 33:27–32.

A Landvogt, A Fischer. 1991a. Dry sausage ripening, targeted control of the acidification achieved by starter cultures. Fleischwirtsch 71:902–905.

———. 1991b. Dry sausage ripening, targeted control of the acidification achieved by starter cultures. Part 2. Fleischwirtsch 71:1055–1056.

RA Lawrie. 1998. Lawrie's Meat Science, 6th ed. Cambridge, UK: Woodhead Publishing, pp. 1–352.

L Leistner. 1991. Stability and safety of raw sausage, part I. Fleischerei 42:III.

F Leroy, L De Vuyst. 1999. The presence of salt and curing agent reduces bacteriocin production by Lactobacillus sakei CTC 494, a potential starter culture for sausage fermentation. Appl Environ Microbiol 65:5350–5356.

———. 2005. Simulation of the effect of sausage ingredients and technology on the functionality of the bacteriocin-producing *Lactobacillus sakei* CTC 494 strain. Int J Food Microbiol 100:141–152.

F Leroy, K Lievens, L De Vuyst. 2005. Modeling bacteriocin resistance and inactivation of *Listeria innocua* LMG 13568 by *Lactobacilllus sakei* CTC 494 under sausage fermentation conditions. Appl Environ Microbiol 71:7567–7570.

H-U Liepe, E Pfeil, R Porobic. 1990. Influence of sugars and bacteria on dry sausage souring. Fleischwirtsch 70:189–192.

AL Lois, LM Gutierrez, JM Zumalacarregui, A Lopez. 1987. Changes in several constituents during the ripening of "Chorizo"—A Spanish dry sausage. Meat Sci 19:169–177.

TM Lopez Diaz, CJ Gonzalez, B Moreno, A Otero. 2002. Effect of temperature, water activity, pH and some antimicrobials on the growth of *Penicillium olsonii* isolated from the surface of Spanish fermented meat sausages. Food Microbiol 19:1–7.

F-K Lücke. 1998. Fermented sausages. In: BJB Wood, ed. Microbiology of Fermented Foods, vol 2, 2nd ed. London: Blackie Academic & Professional, pp. 441–483.

F-K Lücke, H Hechelmann. 1987. Starter cultures for dry sausages and raw ham, composition and effect. Fleischwirtsch 67:307–314.

W Masoud, M Jakobsen. 2005. The combined effects of pH, NaCl and temperature on growth of cheese ripening cultures of *Debaryomyces hansenii* and coryneform bacteria. Int Dairy J 15:69–77.

TA McMeekin, RE Chandler, PE Doe, CD Garland, J Olley, S Putros, DA Ratkowsky. 1987. Model for combined effect of temperature and salt concentration/water activity on the growth rate of *Staphylococcus xylosus*. J Appl Bacteriol 62:543–550.

W Messens, J Verluyten, F Leroy, L De Vuyst. 2003. Modelling growth and bacteriocin production by *Lactobacillus curvatus* LTH 1174 in response to temperature and pH values used for European sausage fermentation processes. Int J Food Microbiol 81:41–52.

M Metz, WP Hammes. 1987. Optimum conditions for growth and nitrate reductase activity of *Micrococcus varians*. In: H Chmiel, WP Hammes, JE Bailey, eds. Biochemical Engineering: A Challenge for Interdisciplinary Cooperation. New York: VCH Publishers Inc, pp. 355–360.

IF Nes, R Skjelkvåle. 1982. Effect of natural spices and oleoresins on *Lactobacillus plantarum* in the fermentation of dry sausage. J Food Sci 47:1618–1625.

EW Nester, CE Roberts, ME Lidstrom, NN Pearsall, MT Nester. 1983. Microbiology, 3rd ed. Philadelphia: Saunders College Publishing, pp. 1–875.

H Neubauer, F Götz. 1996. Physiology and interaction of nitrate and nitrite reduction in *Staphylococcus carnosus*. J Bacteriol 178:2005–2009.

PT Olesen, AS Meyer, LH Stahnke. 2004. Generation of flavour compounds in fermented sausages—The influence of curing ingredients, *Staphylococcus* starter culture and ripening time. Meat Sci 66:675–687.

PT Olesen, LH Stahnke. 2000. The influence of *Debaryomyces hansenii* and *Candida utilis* on the aroma formation in garlic spiced fermented sausages and model minces. Meat Sci 56:357–368.

C Prista, A Almagro, MC Loureiro-Dias, J Ramos. 1997. Physiological basis for the high salt tolerance of *Debaryomyces hansenii*. Appl Environ Microbiol 63:4005–4009.

J Pyrcz, W Pezacki. 1981. Einfluss sauerer gärungsprodukte der kohlenhydrate auf die sensorische qualität von rohwurst. Fleischwirtsch 61:446–454.

Quest. 1999. Product Information. Quest International, The Netherlands, July 1999.

M Raccach. 1992. Some aspects of meat fermentation. Food Microbiol 9:55–65.

M Roca, K Incze. 1990. Fermented sausages. Food Rev Int 6:91–118.

P Roncales, M Aguilera, JA Beltran, I Jaime, JM Peiro. 1991. The effect of natural or artificial casing on the ripening and sensory quality of mould-covered dry sausage. Int J Food Sci Technol 26:83–89.

AK Søndergaard, LH Stahnke. 2002. Growth and aroma production by *Staphylococcus xylosus*, *S. carnosus*, *S. equorum*—A comparative study in model systems. Int J Food Microbiol 75:99–109.

BB Sørensen, M Jakobsen. 1996. The combined effects of environmental conditions related to meat fermentation on growth and lipase production by the starter culture *Staphylococcus xylosus*. Food Microbiol 13:265–274.

———. 1997. The combined effects of temperature, pH and NaCl on growth of *Debaryomyces hansenii* analyzed by flow cytometry and predictive microbiology. Int J Food Microbiol 34:209–220.

LH Stahnke. 1999. Volatiles produced by *Staphylococcus xylosus* and *Staphylococcus carnosus* during growth in sausage minces. Part II. The influence of growth parameters. Lebensm-Wiss u-Technol 32:365–371.

LO Sunesen, LH Stahnke. 2003. Mould starter cultures for dry sausages—Selection, application and effects. Meat Sci 65:935–948.

Texel. 1998. Product information material. Groupe Rhône-Poulenc Denmark, February 1998.

K Tjener, LH Stahnke, L Andersen, J Martinussen. 2003. A fermented meat model system for studies of microbial aroma formation. Meat Sci 66:211–218.

———. 2004a. Growth and production of volatiles by *Staphylococcus carnosus* in dry sausages: Influence of inoculation level and ripening time. Meat Sci 67:447–452.

———. 2004b. The pH-unrelated influence of salt, temperature and manganese on aroma formation by *Staphylococcus xylosus* and *Staphylococcus carnosus* in a fermented meat model system. Int J Food Microbiol 97:31–42.

VM Toledo, MD Selgas, MC Casa, JA Ordonez, ML Garcia. 1997. Effect of selected mould strains on proteolysis in dry fermented sausages. Zeitschr Lebensm-Untersuch-Forsh A 204:385–390.

W Vösgen. 1992. Curing. Are nitrite and nitrate superfluous as curing substances? Fleischwirtsch 72: 1675–1678.

C Waade, LH Stahnke. 1997. Dried sausages fermented with *Staphylococcus xylosus* at different temperatures and with different ingredient levels. Part IV. Amino acid profile. Meat Sci 46:101–114.

T Wijtzes, JC de Wit, JHJ Huis in 't Veld, K van't Riet, MH Zwietering. 1995. Modelling bacterial growth of *Lactobacillus curvatus* as a function of acidity and temperature. Appl Environ Microbiol 61:2533–2539.

LL Zaika, JC Kissinger. 1984. Fermentation enhancement by spices: Identification of active component. J Food Sci 49:5–9.

Part IV

Sensory Attributes

19
General Considerations

Asgeir Nilsen and Marit Rødbotten

INTRODUCTION

Microbiological and nutritious quality are essential parameters when the value of food is considered, and so is the sensory quality. Several instrumental techniques are able to measure important quality attributes in meat products with impressive accuracy, and some of the measurements have a high correlation to sensory measurements. To this day, however, no instrument can fully replace the human subject when the purpose is to achieve information that is relevant to understand the consumers' preferences.

Sensory evaluation has been defined as a scientific method used to evoke, measure, analyze, and interpret those responses to products as perceived through the senses of sight, smell, touch, taste, and hearing (Stone and Sidel 1993). Amerine et al. (1965) have given an overview of the history of sensory science up to 1965, and Meilgaard et al. (1991), Lawless and Klein (1991), Stone and Sidel (1993), and Lawless and Heymann (1998) have published other recognized sensory textbooks more recently.

Sensory evaluation is a quantitative science in which numerical data are collected to establish specific relationships between product characteristics and human perception. Sensory practitioners are concerned both with analytical specification of perceived product differences as well as predicting consumer acceptance of the product. One major concern of anyone who is responsible for a scientific analysis is to ensure that the test method is appropriate to answer the question being asked about the product. This is also true when sensory parameters are to be measured. Thus, the sensory tests are often classified according to their primary purpose and most valid use. First, the sensory evaluation specialist has to decide upon the choice between analytical or affective analysis. Because the evaluation is conducted with human subjects as assessors or instrument, the panel (i.e., the sensory instrument) can be used to give analytical, objective information about the product, i.e., which sausage is the juiciest, or how juicy one sausage is compared to other sausages. This task will demand specific sensory methods and also special training of the panelists. A sensory panel may also give affective, subjective information about the meat product, i.e., which meat sample is preferred and whether the sample is acceptable. This task will also demand specific sensory methods, but in this situation the panel should be consumers untrained in sensory analysis.

In product development, the use of sensory analysis today is recognized worldwide, both the objective and the subjective methods. The quality of a product is often measured by the consistency of the perceived attributes of the product. The consumer expects the product he/she buys to be of the same (or close to) sensory quality tomorrow as it was when bought yesterday. Quality control by means of sensory analysis conducted by a well-trained panel that is supervised by a well-trained panel leader will inform the producer about the degree of perceived variation of the product. The quality factor is of the greatest importance to the consumer, and consequently of economic value for the producer.

SENSORY METHODS

It is important to discriminate between sensory evaluations made by trained assessors in laboratories or production facilities and the preference evaluation made by consumers. The Standards from the International Organization for Standards (ISO) describe all the difference test and descriptive tests commonly used. The brief overview of sensory and consumer testing in Figure 19.1 shows which sensory methods are proper for different tasks. Exact information on how they are carried out is among other aspects described in Lawless and Heymann (1998).

Analytical Methods

Analytical sensory evaluations are performed with trained assessors that give analytical, objective information about the product. Regular training regarding relevant products and maintenance of the assessors are necessary to trust the results from a sensory panel performing analytical methods. In the following sections, the major analyses are described further.

Difference Tests

Difference tests indicate whether a difference can be found between two samples. The samples can be presented in a paired comparison test (ISO 5495:2005), triangular test (ISO 4120:2004), or duo-trio test (ISO 10399:2004). These tests are used for revealing slight differences between samples and can also be used in selecting and training assessors for sensory panels. Difference tests can be interpreted by using statistical tables where significant levels are easily found. Other methods for difference testing, such as "2-out-of-five" and "A—not A" tests, exist, but they are not commonly used in the industry.

Ranking Tests

Ranking tests (ISO 8587:1988) are used to place samples in order of a defined sensory stimulus, e.g., increasing color intensity or increasingly ripened flavor in a fermented meat product. An advantage of this test is that more than two samples can be compared to each other in the same test. Friedman's statistical analysis determines the smallest difference between samples that are necessary to claim significant differences (Friedman 1937).

Profile Tests (Descriptive Tests)

Profile tests are powerful sensory tools used in product control and development to describe the sensory profile of products. Descriptive analysis is widely used to specify quantitative differences in appearance, odor, flavor, and texture characteristics of products. The "Sensory Spectrum" method (Lawless

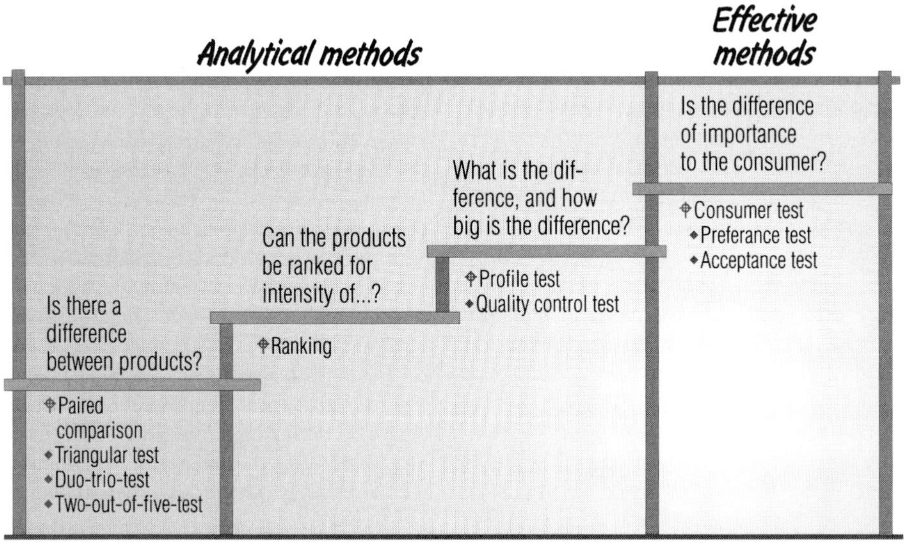

Figure 19.1. Overview of methods used in sensory analyses and consumer tests (from A. B. Tecator, Box 70, S-263 01, Höganäs, Sweden, 1985).

and Heymann 1998), "Quantitative Descriptive Analysis" (Lawless and Heymann 1998), and "Flavor Profile" method (ISO 6564:1985) are examples of other methods used for descriptive testing.

Common for all descriptive methods is the use of statistical methods such as analysis of variance (ANOVA) (Lea et al. 1997), Principal Component Analysis (Martens and Martens 2001), and others to describe significant differences between different product profiles.

Affective Methods (Preference Tests)

Affective tests are performed with untrained consumers who give information about preference, acceptance, liking, or appropriateness of products. ISO 6658:2005 gives general guidance on the use of sensory analysis and indicates which tests can be used for determining preferences.

Test Environment

It is suggested that evaluation of food products should be performed in specially designed lab booths. Assessors in a sensory panel are then free from outside distractions and in a room with controlled lighting, heating, and ventilation. Detailed guidelines for the layout of general sensory laboratories have been published in ISO 8589:1988. This standard includes suggested design for the testing room and lighting requirements. General guidance for the selection, training, and monitoring of assessors can be found in ISO 8586-1:1993.

SENSORY ANALYSIS OF FERMENTED MEAT PRODUCTS

During development of a fermented meat product or during research investigating the influence of different parameters on sensory quality of meat products, various sensory analytical methods come in handy. A preference test, preferably in addition to a sensory profile test, gives essential information about the product attributes of importance to the consumers, whereas a profile test—or in simpler cases, difference tests—are valuable to show objective differences during research and development. Thus, a number of publications show the use of sensory analysis on fermented meat products where appearance, flavor, and/or texture are described in various degrees. The attributes most often used scientifically are color, flavor, rancid flavor, salty taste, and texture, in addition to overall acceptance. Some research work does, however, describe the products in more detail—for example, with the following objective attributes:

- *Appearance*: whiteness, color hue, color intensity, shiny, uneven, greasy (Berdagué et al. 1993; Helgesen and Næs 1995; Andrés et al. 2004)
- *Odor*: meat (i.e., pork, goat, poultry), acidic, sour, gamy, off-odor, metallic, rancid, moldy, cheesy (Mielnik et al. 2002; Cosnenza et al. 2003a,b)
- *Taste*: salty, sweet, sour, bitter, umami (Rødbotten et al. 2004)
- *Flavor*: meat (i.e., pork, goat, poultry), spicy, gamy, metallic, acidic, moldy, rancid, off-flavor, matured, nutty, cheesy, smoky (Mielnik et al. 2002, Cosnenza et al. 2003a,b)
- *Texture*: juicy, dry, hard, fatty, sticky, grainy, mellow, firm, fibrous (Stahnke et al. 2002).

Appearance and Color

The first impression of a fermented meat product is usually perceived by sight. Depending on the actual product, the size, shape, color, and degree of gloss give the visual impression of the quality. When, for instance, slices of fermented, dry sausages are evaluated, it is important to consider the particle size, uniformity of particles, glistening of fat, color, stickiness, and more (Meilgaard et al. 1991). Sensory difference tests and descriptive tests have been used in both hedonic and analytical evaluation of appearance (Berdagué et al. 1993; Helgesen and Næs 1995; Andrés et al. 2004).

Color can be evaluated by describing degree of whiteness and blackness, hue and color intensity, and it is recommended, whenever possible, to use references to both attributes and samples (Mielnik et al. 2002). An example of an attribute reference is the Natural Color System (SS 19102:2004), a system based on resemblance of sample color to the six elementary colors: white, black, yellow, red, blue, and green. The intensity of the color may vary depending on the gloss of the product surface. For reliable results, the assessors should be instructed to evaluate every sample placed at the same distance and angle from the light source and the same distance from the eyes of the assessor.

One of the challenges when evaluating appearance of fermented meat products is the lack of uniformity in the product. A fermented sausage may vary, in all sensory aspects, from the center to the outside part, and a smoked, cured ham is naturally very uneven from the inside to the outside and from the thin to the thick part. When samples are uneven like the described examples, a decision has to be taken on whether the entire slice of the sausage or ham is going to be evaluated or whether the inner part and the outer part should be evaluated separately. Usually the whole sample is of interest, but the producer or the scientist may want to investigate effects of special processes or treatments on parts of the samples, and thus detailed evaluations have to be undertaken.

Odor and Flavor

Odor and flavor are often highly correlated, but in some situations it is observed that a specific flavor is present where no corresponding odor can be detected. The volatile compounds are dependent on moisture to be perceived as odor by the human nose (Engen 1982). A newly cut slice of ham or sausage will most presumably be more intense in odor than a slice being cut some time ago. It is therefore important when comparing various samples to ensure equal preparation time for all samples before serving to the panel. During the eating process, flavor is released from the product and the flavor is often perceived differently from the beginning to the end of the masticating process. This is partly explained by the fact that the various flavor compounds in the product are released over time due to the mechanical and chemical process in the mouth while chewing, in addition to the increased activation of the taste buds in the mouth. Thus it is often difficult to find a high correlation between sensory and instrumental measurements of the "same flavor" attribute. No instrument can fully simulate the process in the mouth.

Flavor wheels are developed for groups of foods and beverages, for example, for wine, cheese, and beer. The starting point (the center of the wheel) uses few descriptors, i.e., the following for meat description: meat flavor, acidic flavor, and rancid flavor. These words or attributes are not very detailed; each of them can be described in more detailed words. "Meat flavor" is a description to be used in a general meaning, and it gives no information about the special species being described. Assessors may be trained in recognizing the specific flavor of the species, i.e., pork, goat, or poultry (Mielnik et al. 2002; Cosnenza et al. 2003a,b), as well as discriminating between various species without using the species specific attribute, i.e., not using the attribute's pork flavor or goat flavor, but rather acidic, metal, bitter, gamy flavor, and more (Rødbotten et al. 2004).

A sensory panel may also be trained in discriminating between various origins of fat and the sensory variation between, for example, fresh contra-stored pork fat, lamb fat, and so on (Buscailhon et al. 1994; Coutron-Gambatti et al. 1999; Muguerza et al. 2001; Mielnik et al. 2002). In most situations, a rancid flavor is an undesirable flavor in the product, but in many of the smoked and cured ham products, a rancid flavor is a desired and an important attribute of the product. Most consumers would probably not describe a matured ham to be rancid even if we know that the fat in the product is partly oxidized. A certain degree of rancidity is a part of the flavor profile of cured ham and is consequently a positive attribute when present in the "correct" intensity.

Meat from some species is more naturally acidic in flavor than other meats (Rødbotten et al. 2004). In the fermentation and ripening process of meat products, additional flavors add to the sensory profile and one is a sour flavor. It is important to discriminate between sour and acidic. The intensity of the two attributes can give valuable information about the fermentation process, if it is successful or not.

Boar-taint is a sensory phenomenon well known in the pig industry. The flavor is very unpleasant and much stronger in intensity in warm products than cold products. One should be aware of the fact that the boar-taint is not recognizable to everyone (Amoore et al. 1977), so it is important to check whether the participants in the sensory panel conducting sensory quality control are able to detect this unpleasant flavor (Lundström and Bonneau 1996).

Texture

The mouth is a very special instrument not only for measuring flavor, but also for measuring texture. The jaw muscles, the molars, and the saliva give a unique setting for measuring texture. The jaw muscles are necessary to measure the force required to penetrate the sample with the molars. The geometric surface of the molars is important for the typical cutting and crushing of the product. The saliva adds to the product in the mouth and may change the sensory perception. Saliva together with melting fat in the mouth can change the textural sensation of a product from the start of eating a product to when it is ready for swallowing.

When evaluating texture properties of fermented meats it is important that each participant of the sensory panel has the same understanding of the various texture attributes. Definition of texture attributes are given in ISO 5492:1992. Several mechanical devices exist that can measure texture properties in fermented meats, based on principles such as shearing, biting, compression, and structural measures. The instrumental measurements are often necessary in practical evaluation of a product and will often show a high correlation to the sensory measurements.

However, a high correlation is not always the case and sensory evaluations must be used to develop a product that can obtain customer acceptance. Meat consumers rate tenderness as the most important factor determining the quality of meat. For fermented meat products the tenderness is important but not the most important issue because fermented products are known to be tender products in general.

REFERENCES

MA Amerine, RM Pangborn, EB Roessler. 1965. Principles of Sensory Evaluation of Food. New York: Academic Press, pp. 1–2.

JE Amoore, P Pelosi, LJ Forrester. 1977. Specific anosmia to 5-androst-16-ene-3one and pentadecalactone: The urinous and musky primary odors. Chem Senses Flav 3:401–425.

AI Andrés, R Cava, J Ventanas, V Thovar, J Ruiz. 2004. Sensory characteristics of Iberian ham: Influence of salt content and processing conditions. Meat Sci 68:45–51.

JL Berdagué, P Monteil, MC Montel, R Talon. 1993. Effect of starter cultures on the formation of flavor compounds in dry sausages. Meat Sci 35:275–287.

S Buscailhon, JL Berdagué, J Bousset, M Cornet, G Gandemer, C Touraille, G Monin. 1994. Relations between compositional traits and sensory quality of French dry-cured ham. Meat Sci 37:229–243.

GH Cosnenza, SK Williams, DD Johnson, C Sims, CH McGowan. 2003a. Development and evaluation of a fermented cabrito snack stick product. Meat Sci 64:51–57.

———. 2003b. Development and evaluation of a cabrito smoked sausage product. Meat Sci 64:110–124.

C Coutron-Gambatti, G Gandemer, S Rousset, O Maestrini, F Casabianca. 1999. Reducing salt content of dry-cured ham: effect on lipid composition and sensory attributes. Food Chem 64:13–19.

T Engen. 1982. The Perception of Odors. New York: Academic Press Inc, pp. 17–34.

M Friedman. 1937. The use of ranks to avoid the assumption of normality implicit in the analysis of variance. J Am Statist Assoc 32:675–701.

H Helgesen, T Næs. 1995. Selection of dry fermented lamb sausages for consumer testing. Food Qual Pref 6:109–120.

ISO 4120:2004 Sensory analysis—Methodology—Triangle Test. International Organization for Standardization, Geneve.

ISO 5492:1992. Sensory analysis—Vocabulary. International Organization for Standardization, Geneve.

ISO 5495:2005. Sensory analysis—Methodology—Paired comparison test. International Organization for Standardization, Geneve.

ISO 6658:2005. Sensory analysis—Methodology—General guidance. International Organization for Standardization, Geneve.

ISO 6564:1985. Sensory analysis—Methodology—Flavour profile methods. International Organization for Standardization, Geneve.

ISO 8586-1:1993. Sensory analysis—General guidance for the selection, training and monitoring of assessors—Part 1: Selected assessors. International Organization for Standardization, Geneve.

ISO 8587:1988. Sensory analysis—Methodology—Ranking. International Organization for Standardization, Geneve.

ISO 8589:1988. Sensory analysis—General guidance for the design of test rooms. International Organization for Standardization, Geneve.

ISO 10399:1991. Sensory analysis—Methodology—Duo—Trio test. International Organization for Standardization, Geneve.

HT Lawless, H Heymann. 1998. Sensory Evaluation of Food: Principles and Practices. New York: Chapman & Hall, pp. 1–26, pp. 358–362, pp. 351–356.

HT Lawless, PB Klein. 1991. Sensory Science Theory and Applications in Foods. Chicago: Marcel Dekker, pp. 1–31.

P Lea, T Næs, M Rødbotten. 1997. Analysis of Variance for Sensory Data. Aas, Norway: John Wiley & Sons Ltd., pp. 23–27.

K Lundström, M Bonneau. 1996. Off-flavour in meat with particular emphasis on boar taint. In: S Taylor, A Raimundo, M Severini and FJM Smulders, eds. Meat Quality and Meat Packaging. Utrecht, The Netherlands: ECCEAMST, pp. 137–143.

H Martens, M Martens. 2001. Multivariate Analysis of Quality. An introduction. Chichester, UK: J Wiley & Sons Ltd., pp. 93–110.

M Meilgaard, GV Civille, BT Carr. 1991. Sensory Evaluation Techniques, 2nd ed. Boca Raton, Florida: CRC Press, pp. 1–5.

MB Mielnik, K Aaby, K Rolfsen, MR Ellekjær, A Nilsson. 2002. Quality of comminuted sausages formulated from mechanically deboned poultry meat. Meat Sci 61:73–84.

E Muguerza, O Gimeno, D Ansorena, JG Bloukas, I Astiasarán. 2001. Effect of replacing pork back fat with pre-emulsified olive oil on lipid fraction and sensory quality of Chorizo de Pamplona—A traditional Spanish fermented sausage. Meat Sci 59: 251–258.

Norsk Institutt for Næringsmiddelforskning (NINF) 1985. Norge, Tecator AB, Sverige.

M Rødbotten, E Kubberød, P Lea, Ø Ueland. 2004. A sensory map of the meat universe. Sensory profile of meat from 15 species. Meat Sci 68:137–144.

LH Stahnke, A Holck, A Jensen, A Nilsen, E Zanardi. 2002. Maturity acceltration of Italian dried sausage by *Staphylococcus carnosus*—Relationship between maturity and flavor compounds. J Food Sci 67: 1914–1921.

H Stone, JL Sidel. 1993. Sensory Evaluation Practices, 2nd ed. San Diego: Academic Press, pp. 1–17.

SS 19102:2004. NCS Color Atlas. SIS Swedish Standards Institute, Stockholm, Sweden.

20
Color

Jens K. S. Møller and Leif H. Skibsted

INTRODUCTION

The manufacturing of fermented meat products relies on centuries of tradition and transference of knowledge over generations. However, these types of meat products also represent a serendipitous utilization of biotechnological principles and specific processing conditions in order to obtain unique attributes of the final product in relation to texture, flavor, and color. During the fermentation or drying process numerous chemical reactions take place including enzymatic and bacterial processes, and this section is dedicated to the pathways involved in color formation in fermented meat products. The color formation and color stability of sausages, hams, and related meat products are, however, closely related to hydrolytic and specifically oxidative processes in the meat matrix.

In raw, cured meat products to which are added either nitrite or nitrate, as is common practice for many bulk products, the nitrosylated form of myoglobin (Mb) primarily determines the color of the final product; in heat-treated products it is a denatured derivative (Killday et al. 1988). The stability of the NO complex of Mb is quite poor, especially when exposed to light in the presence of air. Therefore, a number of systematic investigations have been performed on several interacting factors during production, packaging, storage, and display of nitrite-cured meats in order to identify factors important for quality deterioration. These studies have resulted in formulation of guidelines for meat producers, packers, and retailers in order to ensure optimal production quality and increased shelf life. A schematic outline of parameters that have been found to have an effect on color formation and stability in fermented sausage is shown in Figure 20.1. Here, additional postprocessing aspects are also considered and will be further dealt with in later sections.

COLOR-FORMING COMPOUNDS

The amount of nitrate/nitrite added in the curing process will primarily affect the initial color intensity, as an increasing amount of NO sources will increase the yield of nitrosylmyoglobin, $MbFe^{II}NO$, eventually resulting in higher intensities of reddish or pink color (Froehlich et al. 1983). In relation to color development of fermented meat products, additional aspects have to be considered in relation to activities of specific bacterial cultures each present in very high numbers.

For the meat processing industry, alternative methods to obtain stable color of meat products without the use of nitrite are of continuing interest, also in relation to the growing markets for organically produced foods. Non-nitrite curing can be achieved by various strategies (Shahidi and Pegg 1995), including color formation by adding specific microorganisms or by formation of alternative chromophores in the meat matrix. The latter can be achieved either by addition of specific compounds to the product or even more interestingly through induction of natural endogenous chemical transformations. The chemical identification of Mb derivatives in certain dry-cured products have presented a challenge for the food chemist and is also of specific interest to the industry,

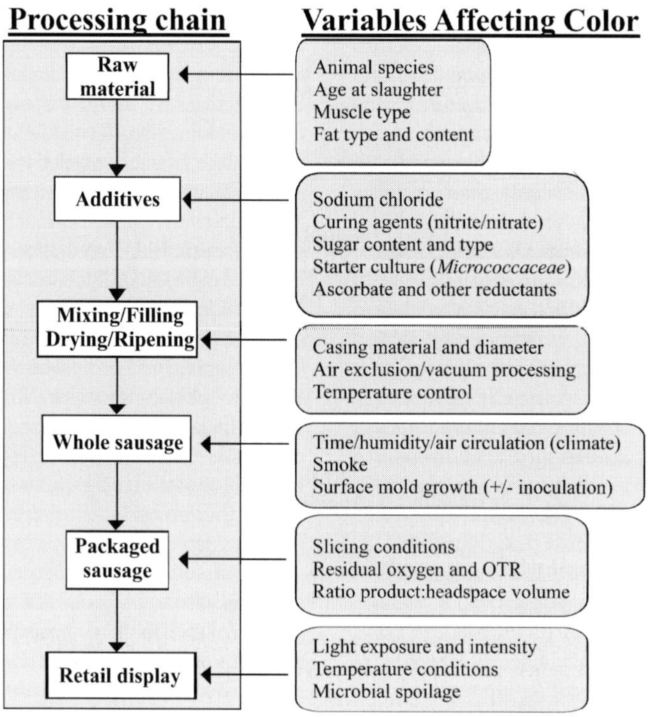

Figure 20.1. Schematic outline of product/ingredient flow and critical process parameters in different steps of fermented sausage manufacturing. OTR is the abbreviation for the oxygen transmission rate of the given packaging material.

because a deep red color is formed in the meat during the maturation process without adding nitrate or nitrite (Parolari 1996).

CHEMISTRY OF MEAT COLOR

Nitrate and nitrite have been used for centuries for curing and preserving meat and fish, and in some types of cheeses (Cammack et al. 1999). For cured, fermented sausages, in which nitrate and/or nitrite are added during the curing process, flavor, appearance, and oxidative stability are directly related to the conversion of the curing agent nitrite to NO and the following reaction with meat pigments, forming $MbFe^{II}NO$. Nitrate is converted to nitrite by endogenous or added reductants prior to formation of the nitrosation agents, which transfer the NO moiety to Mb or other proteins.

$MbFe^{II}NO$ has a bright red color in its native form and is pinker when the protein moiety is denatured by mild heat treatment yielding nitrosylmyochrome, $dMbFe^{II}NO$. However, both forms are very labile toward degradation, which causes formation of metmyoglobin, $MbFe^{III}$, giving the meat surface a dull grayish color. The color formation is affected by the presence of reductants in the meat, where substances such as thiol-groups in peptides and enzymatic cofactors may reduce added nitrite. In fermented products, the initial reduction of added nitrate is influenced by the choice of starter culture (Krotje 1992), as will be discussed in more detail later. Addition of reductants, especially ascorbate or erythorbate, in excess during the curing process facilitate the color formation. The reaction steps involving myoglobins and the curing agents nitrite and nitrate, which is often referred to as the NO generating pool, are summarized in Figure 20.2.

The final products often contain salt in the concentration range of 2–5%. Of the initial nitrite addition, typically ranging between 60–150 ppm, only a small fraction of the nitrite can be detected even shortly after production due to the numerous reactions in which nitrite participate, as also illustrated in Figure 20.2 (Cassens et al. 1979).

Formation and Reactivity of NO

The high reactivity of added nitrite in the meat batter will result in an initial discoloration due to oxidation

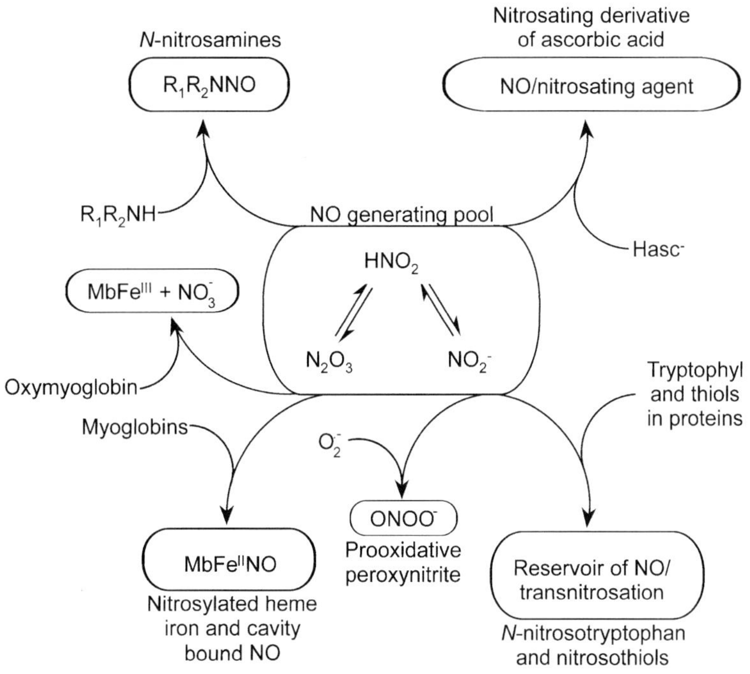

Figure 20.2. The NO generating pool in nitrite-cured meat can participate in numerous reactions with various meat constituents yielding modified proteins, altered pigments, and other nitrosyl products.

of oxymyoglobin ($MbFe^{II}O_2$) and deoxymyoglobin ($MbFe^{II}$), present in the raw meat, to the brown metmyoglobin, $MbFe^{III}$ as indicated in Equations 20.1 and 20.2 (Möhler 1974; Killday et al. 1988). These reactions occur rapidly after meat batter mixing and addition of the additives, and the discoloration persists for hours. Subsequently, and under the influence of added reducing agents, both $MbFe^{III}$ and the formed nitrate will be reduced into species that are active in the color formation.

$$MbFe^{II} + 2H^+ + NO_2^- \rightarrow MbFe^{III} + NO(g) + H_2O \qquad (20.1)$$

$$4MbFe^{II}O_2 + 4NO_2^- + 4H^+ \rightarrow 4MbFe^{III} + 4NO_3^- + O_2 + 2H_2O \qquad (20.2)$$

The complex chemistry occurring during nitrite curing of meat has been studied extensively over the years, and recently reviewed (Møller and Skibsted 2002), and will not be discussed in further detail. Nitrosylated heme proteins were early identified as the primary pigment in nitrite-cured meats, and the exact chemical structure as a mononitrosyl complex was fully elucidated by the use of Electron Spin Resonance (ESR) spectroscopy (Bonnett et al. 1980; Pegg et al. 1996). From kinetic studies, a reaction mechanism behind $MbFe^{II}NO$ formation from nitrite and $MbFe^{III}$ has been suggested, as shown in Figure 20.3 (Bonnett et al. 1980). The coordination of NO to Fe(II) in Mb involves a change from high spin (h.s.) to low spin (l.s.) and an increase in coordination number of Fe(II), which may be monitored by ESR spectroscopy:

$$\text{Heme-Fe}^{II}(\text{h.s.}) + \text{NO(g)}(S={}^1\!/_2) \rightarrow \text{Heme-Fe}^{II}(\text{l.s.})-\text{NO}\ (S={}^1\!/_2) \qquad (20.3)$$

A number of comprehensive reviews based on intensive research during several decades can be consulted for further details concerning various practical and theoretical aspects of meat curing, including the structure of pigments in cured meat (Tarladgis 1962; Bonnett et al. 1980; Pegg and Shahidi 1997), chemical reactions of nitrite and chloride under various pH-conditions (Sebranek and Fox 1985), interaction of nitrite and ascorbic acid leading to nitrosylating agents, and oxidative processes in cured meats affecting pigments and lipids (Cassens et al. 1979; Skibsted 1992; Cornforth 1996). The physicochemical features of both nitrite-cured meat pigment and pigment in dry-cured meat products (Sakata 2000) have also been discussed in detail, together with alternatives to nitrite in meat (Shahidi and Pegg 1995).

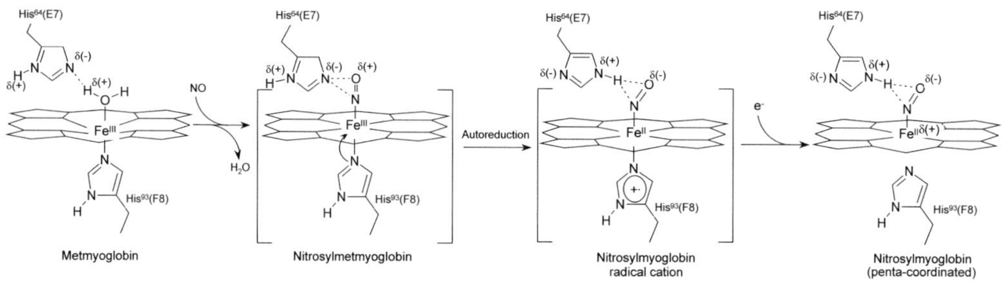

Figure 20.3. Reaction steps involved in the formation of nitrosylmyoglobin in nitrite-cured meat, including a cationic protein radical intermediate (modified from Bonnett et al. 1980).

Formation of Color without NO

Dry-cured ham is a traditional and important meat product in many countries, but it is crucial to distinguish between dried ham variants to which only salt is added, such as Parma ham and most variants of Iberian ham, and other dry-cured hams, such as the Spanish Serrano ham to which are added either nitrate or nitrite together with salt. In dry-cured hams with added nitrate or nitrite, the complex $MbFe^{II}NO$ will be formed (Perez-Alvarez et al. 1999), which, however, seems to have a slightly different structure than the native complex found in brine-cured bacon or the denatured form $dMbFe^{II}NO$ present in cooked ham. Such subtle structural differences are indicated in the ESR spectra (Figure 20.4), which have the same hyperfine splitting pattern due to the paramagnetic ligand NO (Møller et al. 2003a), but which also show varying contribution from the protein moiety as observed from other spectral features, e.g., the anisotropic characteristics. Notably by comparing ESR spectra of raw and heat-treated bacon, similar spectral observations indicate the same type of differences in protein association to the chromophore (Bonnett et al. 1980).

Nitrate and nitrite have been banned from the manufacturing of Parma ham since 1993 (Parolari 1996), and the unique color formation and stability observed for Parma ham is intriguing because the stable color seems to be formed without the action of nitrate/nitrite. Analysis for nitrate and nitrite content in the sodium chloride of marine origin used for processing of Parma ham showed a negligible and insufficient amount to account for the uniform color observed in the final product. Only a few studies are available regarding the nature of the Mb derivatives formed in Parma ham, and spectral changes have been reported to occur gradually during processing (Parolari et al. 2003; Møller et al. 2003a), and some unique features were observed regarding a red-shifted light absorption of the isolated pigment in comparison to the absorption spectra of more well-characterized Mb derivatives. However, recently a Japanese group reported that the main chromophore in Parma ham is a complex of Zn and protoporphyrin IX (Wakamatsu et al. 2004). Subsequently, this Zn complex has been detected in other meat products, but notably addition of nitrite apparently seems to completely inhibit the formation of zinc protoporhyrin IX (Adamsen et al. 2005), possibly through an increased stabilization of the iron binding to porphyrin in heme. Nitrite has previously been suggested to bind at the entrance to the heme cavity protecting Mb against degradation by heat and oxidative processes (Arendt et al. 1997).

Influence of Fermentation Parameters on Color

Optimal color formation in fermented sausages depends strongly on the actual fermentation and resulting pH decrease and oxygen depletion. For one product an initial conversion of 70% of the total pigment into NO derivatives in the meat mince was shown to increase to a 90% conversion in the final product following fermentation (Chasco et al. 1996). Fermented sausages added both nitrate and nitrite together with erythorbate were found to have approx. 80% pigment conversion into $MbFe^{II}NO$ after 8 days of dehydration when the initial fermentation period applied was between 12–21 hours and with final pH around 5.2. A shorter fermentation and a higher final pH provided a lower degree of pigment conversion (Acton and Dick 1977). The nitrite and nitrate have been shown to convert to each other dynamically during dry-fermented sausage production (Stahnke 1995), and nitrate concentration has a tendency to decrease while nitrite concentration increases as higher fermentation temperatures and

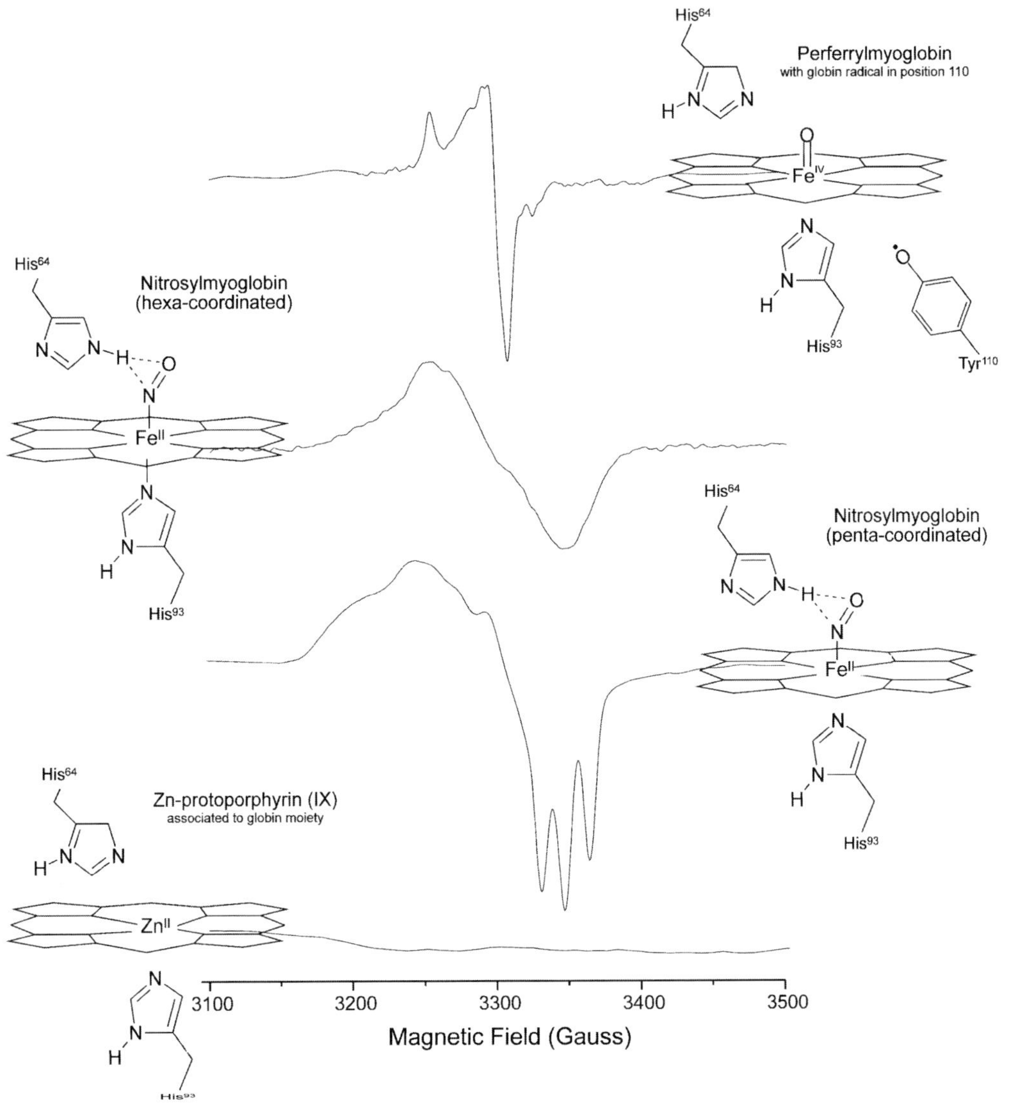

Figure 20.4. Electron spin resonance (ESR) spectra of hypervalente perferrylmyoglobin, two forms of nitrosylated myoglobin and the metal center in the major Parma ham chromophore, Zn porphyrin complex, which is ESR-silent. The detailed chemical structure of various types of metal coordination spheres responsible for the observed ESR spectra are shown in conjunction with the actual spectrum.

higher sugar levels are applied in combination with lower salt concentration. Bacterial activities clearly affect the distribution between nitrate and nitrite and also the dynamics of interconversion and in residual nitrite. Application of backslopping and varying levels of nitrite and nitrate with or without ascorbate addition for fermented sausage production led to the conclusion that optimal color formation was obtained when nitrite was combined with ascorbate, whereas nitrate did not react in this type of product, possibly due to lack of active micrococci as seen as a poor color formation (Alley et al. 1992). Another study in which color formation in fermented and nonfermented sausages was compared showed a higher degree of pigment conversion into $MbFe^{II}NO$ for the fermented product (Demasi et al. 1989). For fermented sausages, the conversion further depends on pH because a more acidic product was found to result in more efficient nitrite utilization. In the processing of Turkish-type fermented sausages that included two fermentations and two ripening stages, the optimal color was present already after the first ripening; no color changes occurred during the second ripening and final production stage (Uren and Babayigit 1997).

BACTERIAL ROLE IN MEAT COLOR

In dry-fermented sausages, flavor and oxidative status depend both on endogenous chemical changes in the meat batter, mainly due to enzymes, and on the nature of the starter cultures added according to the actual recipe. Starter cultures are added in order to ensure a decrease in pH, reduction of nitrate, and flavor formation during product maturation. Some bacterial cultures, however, also seem to affect other chemical processes in the final meat product, because it was found that microbial metabolism led to color changes or prevented oxidative processes (Montel et al. 1998).

NITRATE REDUCTASE

The nitrogen cycle found in many biological systems, including some bacteria, results in numerous N-containing compounds as intermediates in the complex enzymatic metabolism related to nitrification and denitrification (Richardson 2001). Some of these transient nitrogen species can react with meat Mb and thereby affect color. Prokaryotic nitrate reduction serves a number of metabolic functions and depends on a variety of nitrate reductases, as found in nitrogen assimilation and anaerobic respiration (Richardson et al. 2001). The role of bacterial nitrate reductases has been acknowledged and utilized for many decades; however, very little is still known regarding expression or variations in this enzymatic activity for either bacteria present in the environment or commercial starter cultures. The enzymes belong to the class of oxidoreductases, and it is the general reaction of Equation 20.4 that is being catalyzed during the microbial anaerobic respiratory metabolism.

$$\text{Nitrate} + \text{reduced acceptor} \rightarrow \text{nitrite} + \text{acceptor} \qquad (20.4)$$

The assimilatory nitrate reductases can be divided into at least three subgroups depending on the nature and number of cofactors for electron transfer present in the enzymes (Richardson et al. 2001). The enzyme from *Micrococcus halodenitrificans* is an iron protein containing molybdenum (Equation 20.5); the *Pseudomonas* enzyme is a cytochrome enzyme (Equation 20.6). The NADPH dependent variant of nitrate reductase (EC 1.7.1.2) further has FAD or FMN, heme, and molybdenum as cofactors; another variant of the enzyme (EC 1.7.1.3) requires FAD, heme, molybdenum, and an iron-sulfur cluster.

$$NO_3^- + NADPH + H^+ \rightarrow NO_2^- + NADP^+ + H_2O \qquad (20.5)$$

As mentioned, a cytochrome nitrate reductase (EC 1.9.6.1) also exists for which the reaction is the following:

$$2[\text{cyt}]Fe^{II} + 2H^+ + NO_3^- \rightarrow 2[\text{cyt}]Fe^{III} + NO_2^- + H_2O \qquad (20.6)$$

It is generally recognized that most bacterial strains of the *Micrococcaceae* and *Staphylococcaceae* families are positive for nitrate reductase activity in standard biochemical tests, and a study of more than 400 isolates, predominantly *Staphylococcus xylosus*, from chorizo also found 97% to show nitrate reduction (Garcia-Varona et al. 2000). The number of staphylococci has been found to increase only slightly during the fermentation period of sausages from approximately 10^3 up to 10^6 CFU/g, while, for instance, lactic acid bacteria reach numbers as high as 10^8–10^9 CFU/g during processing for more than 35 days (Mauriello et al. 2004). Another study performed to follow ripening of Spanish-type chorizo shows no effect of sodium nitrite level (50 and 150 ppm) and of an added lactobacilli starter culture on the survival of *Micrococcaceae*, although these factors significantly inhibited *Enterobacteriaceae* (Gonzalez and Diez 2002). It is a prerequisite for ripening that these bacteria strains exhibit metabolic activity during sausage fermentation; however, the information about this class of microbial enzymes is quite limited with respect to importance for color formation in fermented meats.

A study including seven staphylococci strains isolated from either sausage or cheese discovered that isolates of *S. xylosus* and *S. carnosus* were able to synthesize nitrate reductase, and a *S. warneri* had no activity of this enzyme under static conditions (Talon et al. 1999). However, during the exponential growth phase, the synthesis was maximal, whereas under shaking condition, the synthesis was maximal at the beginning of the stationary phase of growth. In all cases, nitrate reductase activity was increased in presence of nitrate, which was particularly pronounced for the two *S. carnosus* strains exhibiting the highest activity.

Götz and co-workers have performed a number of extensive investigations with respect to the presence and activity of nitrate/nitrite reductases in a *S. carnosus* strain (Fast et al. 1996; Neubauer and Götz 1996; Pantel et al. 1998; Neubauer et al. 1999; Fedtke et al. 2002). Nitrate reductase is of interest also in other scientific areas such as soil processes, and this enzyme class has been studied in many details relating to anaerobic respiration of several bacterial species. The enzyme contains a molybdenum atom in the active site and several species possess a membrane-bound type of the enzyme, which for this specific isoform has been found to consist of three subunits (Pinho et al. 2005). The *Pseudomonas* nitrate reductase is a cytochrome enzyme, and the enzyme from *Micrococcus halodenitrificans* is an iron protein-containing molybdenum. A critical parameter regarding this type of enzyme and its activity is induction and expression of the enzyme, which is achieved by the presence of nitrate in the growth environment of the bacteria. Figure 20.5 provides a simplified reaction cycle taking into account steps that are either catalyzed by bacterial enzymes or occurring spontaneously.

However, it may be concluded that despite the many studies devoted to bacterial nitrate reductase, our knowledge regarding molecular biological aspects of the nitrate reductase and the activity in bacterial species commonly applied as starter culture for fermented meat products is very limited. More research regarding conditions affecting enzyme activity in starter culture should accordingly be undertaken, because more detailed knowledge would help optimize color formation and the use of additives in meat products such as fermented sausages.

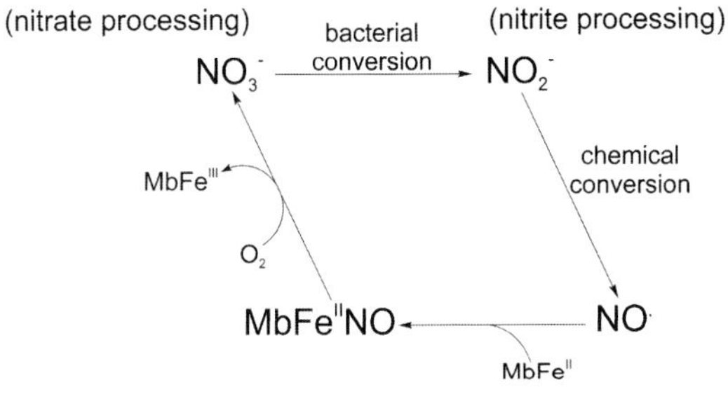

Figure 20.5. Cycle of nitrogen oxides occurring in the meat matrix depending on bacterial and/or chemical conversions resulting in the bright red pigment, nitrosylmyoglobin, $MbFe^{II}NO$, that subsequently can undergo autooxidation into metmyoglobin, $MbFe^{III}$, and nitrate, which may reenter the cycle.

Nitrite Reductase

The nitrite reductases (NIR) and nitrate reductases are active at low oxygen tension using nitrogen oxides as electron acceptors in the metabolism of the bacteria, and the products originating from such metabolism are ammonia, nitrogen oxides, and dinitrogen (NH_3, NO, N_2O, N_2). It has recently been shown that NO exists as a free intermediate in denitrifying bacteria with steady state concentrations in the range 10^{-7}–10^{-9} M (Watmough et al. 1999). In the enzymatic reaction behind NO formation from nitrite in Equation 20.7, two His residues each donate a H^+ to the ligand followed by dehydration and reduction of the nitrosyl complex to yield a ferric NO complex from which NO can escape.

$$[NIR]Fe^{II}NO_2 + 2H^+ \rightarrow [NIR]Fe^{III} + NO(g) + H_2O \quad (20.7)$$

NIR activity has been found in certain *Lactobacillus* species normally employed as starter cultures for raw fermented sausages (Wolf and Hammes 1988), and is found (as well as catalase activity) to depend on the presence of heme-containing compound (hematin, Mb and Hb) (Wolf et al. 1991). Another study showed a significant activity of two types of NIR in various lactic acid bacteria: Type I is heme-dependent and produces ammonia, whereas type II is heme-independent and produces NO and N_2O (Wolf et al. 1990). A periplasmatic cytochrome *c* nitrite reductase from *E. coli* has been studied in detail in relation to the respiratory reduction of nitrite. This enzyme was also found capable of reducing nitric oxide (NO) in vitro, and in vivo assessments also demonstrate that this particular enzyme confers NO detoxification or control in oxygen-limited environments (Poock et al. 2002). A heme protein in *Clostridium botulinum*, found to display femto-molar sensitivity toward NO, has been suggested to be important for a mechanism in which this bacterium recognizes and avoids NO in the growth environment (Nioche et al. 2004). This heme protein may very well be the ultimate target for sodium nitrite inhibition of growth of this heat-resistant and extremely toxic bacteria species responsible for botulism.

NO Synthase (NOS)

Efforts have been made to identify and apply bacterial strains capable of actively changing the color of meat products during fermentation. A number of bacteria are found capable of reducing $MbFe^{III}$ to $MbFe^{II}$, which subsequently binds atmospheric oxygen giving a bright red complex identified as $MbFe^{II}O_2$ (Arihara et al. 1993, 1994), although only a few strains have been identified, which without the presence of nitrite/nitrate can form $MbFe^{II}NO$ (Arihara et al. 1993; Morita et al. 1994, 1998). The identification of bacterial strains forming $MbFe^{II}NO$ has even led to the suggestion that the pigment formed in Parma ham during the lengthy process of ripening was $MbFe^{II}NO$ resulting from microbial activity (Morita et al. 1996), although the involvement of bacteria in the color formation of Parma ham subsequently was questioned (Sakata 2000).

The bacterial generation of $MbFe^{II}NO$ depends on more than a single pathway with respect to the microbial metabolism leading to the nitrosyl derivative of Mb. The interaction between bacteria and NO has mostly been described in relation to enzymatic reductase activity occurring as part of denitrification processes in bacteria (Cutruzzola 1999; Watmough et al. 1999). As already mentioned, recent studies have found NOS (nitric oxide synthetase) activity in bacteria such as *Nocardia* (Chen and Rosazza 1994), *Staphylococcus aureus* (Choi et al. 1997), *Bacillus subtilis* (Pant et al. 2002), and possibly even in *L. fermentum* (IFO3956) (Morita et al. 1997). A recent review dealing with lactic acid bacteria and NO also emphasizes that these bacteria on several occasions have been found to produce NO, a bacterial activity that could be exploited for color development in meat products (Karahan et al. 2005). However, other studies of *L. fermentum* (IFO 3956) species showed less promising results with respect to non-nitrite color formation in salami sausages (Møller et al. 2003). The NOS enzyme found in *S. aureus* has been described in more detail, and it was concluded that it behaved similarly to mammalian NOS with respect to L-arginine binding and dependence on cofactors (Chartier and Couture 2004). The discovery of this particular bacterial NOS, although in a highly pathogenic species, could imply the presence of similar types of enzymes in other non-pathogenic staphylococci species, which could be useful for meat fermentation.

In contrast to the activity of NIR, the enzymatic formation of NO by NOS uses two equivalents of oxygen for every molecule of NO formed from L-arginine, and this reaction has been found to occur in *Nocardia* sp. (Chen and Rosazza 1995) and in *S. aureus* (Choi et al. 1997). Another study has found *L. fermentum* (IFO3956) to incorporate ^{15}N into nitrite/nitrate (the oxidation product of NO) from guanidino-^{15}N labeled L-arginine to increase when various cofactors of the enzyme (NADPH, H_4B, FAD and FMN) are present (Morita et al. 1997).

COLOR STABILITY OF CURED MEAT PRODUCTS

Conditions during storage, transportation, and retail display can jeopardize sound precautions taken earlier in the manufacturing of cured meat products to maintain color stability. Brine-cured products have been studied in most details and only a limited number of studies of packaging conditions for fermented sausages or dry-cured meat products in relation to their chemical stability are available.

INFLUENCE OF OXYGEN

For nitrite-cured meat products, packaging should aim at excluding oxygen completely from the product, and this is under practical conditions obtained either by vacuum or modified atmosphere packaging (MAP) (Eidt 1991). However, the problem of color fading in nitrite-cured meat products has not fully been solved by the industry, which in part is due to the implementation of new and, in other respects, improved packaging technologies that usually cause new problems to emerge. For fermented sausages, a flow-packaging machine is often applied and this results in higher levels of residual oxygen compared to other methods. The use of MAP in relation to fermented sausages has mainly been for presliced sausages in convenient dome packages, which has largely been driven by retailers trying to meet consumer demands for chilled products with high safety and quality. This packaging method may impose problems because the residual oxygen content is difficult to maintain at a sufficient low level, and because the residual oxygen level in the package immediately after closure is clearly a vital parameter in preventing discoloration of MA-packaged cured meat (Muller 1990); even low concentrations of gaseous oxygen present in the package combined with light exposure rapidly cause substantial discoloration of the product (Møller et al. 2003b). Residual oxygen will also be affected by the oxygen transmission rate (OTR) of the packaging material, and a low OTR, which notably is dependent on temperature and relative humidity, is crucial for the stability of the color during storage. A study including 5 packaging materials with OTR ranging from 1–90 $cm^2\ O_2/m^2/atm/24h$ established a similar relationship between color stability and OTR for vacuum-packaged salami (Yen et al. 1988). The critical value for oxygen concentration is mainly affected by

- Headspace to product volume ratio
- Storage period in the dark prior to light exposure
- The amount of cured pigment in the meat product

Increasing the amount of nitrite in the meat product will stabilize the product color, allowing a higher level of residual oxygen in the package headspace without having detrimental effects on color stability, because residual nitrite may regenerate $MbFe^{II}NO$ from $MbFe^{III}$ as long as the meat still has reducing capacity.

Influence of Lipid Oxidation

The flavor of meat products with added nitrite has been described as a pure meat flavor, because no contribution from lipid oxidation during processing, cooking, or storage adds off-flavors, and nitrite-cured meats do not develop Warmed-Over-Flavor (WOF) (Cross and Ziegler 1965). In meat model systems, nitrite alone has been found to act prooxidative at higher levels (>25 ppm), but did inhibit oxidation initiated by Fe ions (Macdonald et al. 1980). Notably, in brine-cured meat products a reducing agent is also added, and this ensures optimal conditions for color formation yielding $MbFe^{II}NO$ including regeneration of $MbFe^{II}$ from $MbFe^{III}$ prior to binding of NO. Likewise, nitrite itself has been shown to act as an antioxidant in model systems with $MbFe^{III}$ as a prooxidant (Arendt et al. 1997). The well-documented antioxidant effect of nitrite in cooked meat may be related to a blocking of release of iron ions from heme pigments (Adamsen et al. 2005).

No significant oxidation, as measured by TBARS, was observed in nitrite (60 ppm) cured beef during 28 days illuminated storage at 8°C, regardless of whether the animal had been fed a vitamin E supplemented diet (Houben and Van Dijk 2001). A concentration-dependent effect of nitrite to inhibit development of rancidity (TBARS or other secondary lipid oxidation products) has been confirmed in meat originating from various species, e.g., fish, chicken, pork, and beef (Morrissey and Tichivangana 1985). Several mechanisms for this well-described antioxidative effect of nitrite have been presented, which are often related to the interaction of nitrite or compounds derived from nitrite with metal ions or heme compounds otherwise acting as prooxidants in meat (Carlsen et al. 2005). Møller et al. (2002) have suggested a possible mechanism for the role of $MbFe^{II}NO$ as a chain-breaking antioxidant where $MbFe^{II}NO$ inactivate peroxyl radicals (Skibsted 1992).

The role of NO in lipid peroxidation is more complex, as seen in Figure 20.6, in which the small radical is described both as a prooxidants and an antioxidant depending on the concentration and other available reagents.

Microbial Effect on Lipid Oxidation

In general, both lipolysis and lipid oxidation contribute to development of the characteristic flavor of dry-cured, fermented sausages (Chizzolini et al. 1998). It is, however, intriguing how certain types of traditional meat products originating from Southern Europe maintain their wholesomeness during lengthy maturation periods and afterward during nonrefrigerated storage. Conditions such as high concentration

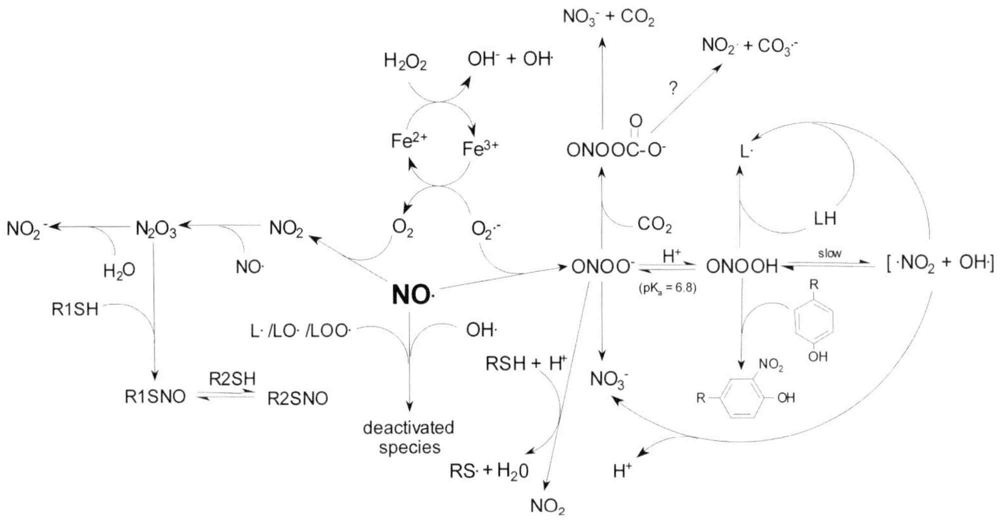

Figure 20.6. Reaction pathways of the small diatomic, radical nitric oxide, NO, and various non-heme biomolecules resulting in nitrosative/oxidative stress, radical scavenging, or stabilization of the antioxidative NO.

of salt and low moisture content help select a beneficial microflora and at the same time retard microbial spoilage in dry-cured meat products. Lipid oxidation in sausages with added starter cultures was found to be lower in comparison to control sausages produced without starter culture. Several of the bacterial strains commonly applied as starter cultures in fermented meat products seem—in addition to acidification, color, and flavor formation—also to be involved in retarding lipid oxidation and development of rancidity. A screening of 19 strains of lactic acid bacteria (species represented: *Lactobacillus acidophilus*; *L. bulgaricus*; *Streptococcus thermophilus* and *Bifidobacterium longum*) for antioxidative activity included effect on rate of ascorbate autoxidation, metal ion chelation, scavenging of reactive oxygen species (ROS), enzyme inhibition, and reducing activity of intracellular cell-free extract (Lin and Yen 1999). All strains investigated chelated metals, scavenged free radicals, and showed general reducing capacities; no activity of superoxide dismutase (SOD) could be detected and was not induced by availability of Mn^{2+}, Fe^{2+}, Cu^{2+}, or Zn^{2+}. Catalase and SOD has been characterized in *S. carnosus* using model systems (Barriere et al. 2001), and the SOD in a *S. xylosus* also has been investigated (Barriere et al. 2001). In a system with linoleic acid as oxidation substrate, more details have been obtained with respect to individual roles of catalase and SOD in *S. xylosus*. It was concluded that the wild-type strain with both enzymes intact had the highest antioxidative potential followed by the mutant lacking catalase but having SOD activity (Barriere et al. 2001).

Certain starter cultures seem to possess antioxidative capacity, which has been proven for strains of staphylococci and lactic acid bacteria when growing on media containing various fatty acids (Talon et al. 2000). This antioxidative effect is improved for most strains when the media is supplemented with manganese, suggesting a mechanism involving SOD activity confirming the results obtained in model systems for the relative importance of catalase and SOD.

The effect of microbial activity on lipid oxidation in dry-fermented sausages is still uncertain, but studies are emerging that point toward an overall antioxidative effect of specific starter cultures. Free fatty acids in dry-fermented sausage was found to increase significantly during 50 days independent of the type of starter cultures added (Kenneally et al. 1998). The role of free fatty acids is interesting because free fatty acids inactivate heme compounds as prooxidants toward unsaturated lipids (Møller et al. 2002). It can be speculated whether a similar mechanism prevails in traditionally processed meat products from Southern Europe in which extensive formation of free fatty acids takes place during maturation, in part due to activities of microbial derived enzymes. However, it is still largely unknown whether these factors also help protect the lipids and preserve the color of dry-cured ham and dry-fermented sausage.

Influence of Other Parameters

At retail display, several conditions such as storage temperature and illumination of the surface of the product will affect the chemical stability during the storage period. Both the spectral characteristics of the illumination and the light intensity are important for the detrimental effect of light on meat quality. Klettner & Terstiege (1999) have examined various types of illumination during display, and they found raw sausage to discolor within 1–3 hours when stored at 5°C and exposed to 800 lux. One thousand lux seems a typical exposure at the package surface in the retail trade in many countries. The wavelength dependence for degradation of the pigment of cured meat is less significant than for $MbFe^{II}O_2$ present in fresh meat. The use of packaging material with incorporated UV filter has accordingly little or no beneficial effect on the color stability of nitrite-cured meat products (Andersen et al. 1988). A level for residual O_2 between 0.2–0.5% (depending on other packaging conditions) at the time of packaging has been identified as critical for color stability upon exposure to light (Møller et al. 2000, 2003b).

In order to control spoilage and pathogenic bacteria in dry-cured ham and other products, high-pressure treatment has been introduced for sliced products. Oxidation of $MbFe^{II}NO$ becomes slower under pressure, and high pressure is accordingly not expected to cause discoloration (Bruun-Jensen and Skibsted 1996). At pressure treatment under 4000 atmospheres, little if any lipid oxidation was found to be induced in Iberian ham (Andres et al. 2004); thus, high hydrostatic pressures is a promising new technology also in relation to the zero-Listeria tolerance enforced in some countries.

CONCLUSION

In fermented meat, color formation depends on two mechanisms. For nitrate-cured products, nitrate reduction mediated through microbial enzyme activity seems to be the bottleneck. In contrast, color formation in nitrite-cured meat is less dependent on microbial activity and may be characterized as "purely chemical." For both of these categories of products, the most important pigment is nitrosylmyoglobin, which is very sensitive to the combined action of light and oxygen. For certain other products, where nitrate and nitrite are not used in curing, other pigments determine the product color. Zinc porphyrin, which is less light-sensitive, is formed in a much slower process requiring longer product maturation.

Future research should be directed toward a better understanding of nitrate reduction by microbial nitrate reductases and especially their expression of relevant genes under the conditions prevailing in meat systems under fermentation. Although the chemistry behind nitrite-induced color formation is rather well documented, the kinetics and mechanism behind formation of zinc porphyrin is largely unknown and deserves further attention also in relation to development of new types of meat products.

ACKNOWLEDGMENT

The Frode and Norma S. Jacobsens foundation is gratefully thanked for its continuing financial support to meat related research carried out within the Food Chemistry group at Department of Food Science, Royal Veterinary and Agricultural University.

REFERENCES

JC Acton, RL Dick. 1977. Cured color development during fermented sausage processing. J Food Sci 42:895–897.

CE Adamsen, JKS Møller, K Laursen, K Olsen, LH Skibsted. 2005. Zn-porphyrin formation in cured meat products. Effect of added salt and nitrite. Meat Sci 72:672–679.

G Alley, D Cours, D Demeyer. 1992. Effect of nitrate, nitrite and ascorbate on color and color stability of dry, fermented sausage prepared using back slopping. Meat Sci 32:279–287.

HJ Andersen, G Bertelsen, G Boegh-Soerensen, CK Shek, LH Skibsted. 1988. Effect of light and packaging conditions on the color stability of sliced ham. Meat Sci 22:283–292.

AI Andres, JKS Møller, CE Adamsen, LH Skibsted. 2004. High pressure treatment of dry-cured Iberian ham. Effect on radical formation, lipid oxidation and colour. Eur Food Res Technol 219:205–210.

B Arendt, LH Skibsted, HJ Andersen. 1997. Antioxidative activity of nitrite in metmyoglobin induced lipid peroxidation. Zeitschr Lebensm Untersuch Forsch 204:7–12.

K Arihara, RG Cassens, JB Luchansky. 1994. Metmyoglobin reduction activity of enterococci. Fleischwirtsch 74:1203–1204.

K Arihara, H Kushida, Y Kondo, M Itoh, JB Luchansky, RG Cassens. 1993. Conversion of metmyoglobin to bright red myoglobin derivatives by *Chromobacterium violaceum*, *Kurthia* sp, and *Lactobacillus fermentum* JCM1173. J Food Sci 58:38–42.

C Barriere, R Bruckner, R Talon. 2001. Characterization of the single superoxide dismutase of Staphylococcus xylosus. Appl Environ Microbiol 67:4096–4104.

C Barriere, D Centeno, A Lebert, S Leroy-Setrin, JL Berdague, R Talon. 2001. Roles of superoxide dismutase and catalase of Staphylococcus xylosus in the inhibition of linoleic acid oxidation. FEMS Microbiol Lett 201:181–185.

C Barriere, S Leroy-Setrin, R Talon. 2001. Characterization of catalase and superoxide dismutase in Staphylococcus carnosus 833 strain. J Appl Microbiol 91:514–519.

R Bonnett, S Chandra, AA Charalambides, KD Sales, PA Scourides. 1980. Nitrosation and nitrosylation of hemoproteins and related Compounds 4. Pentacoordinate nitrosylprotohem as the pigment of cooked cured meat. Direct evidence from Electron Spin Resonance spectroscopy. J Chem Soc—Perkin Trans 18:1706–1710.

L Bruun-Jensen, LH Skibsted. 1996. High-pressure effects on oxidation of nitrosylmyoglobin. Meat Sci 44:145–149.

R Cammack, CL Joannou, XY Cui, CT Martinez, SR Maraj, MN Hughes. 1999. Nitrite and nitrosyl compounds in food preservation. Biochim Biophys Acta Bioenerg 1411:475–488.

CU Carlsen, JKS Møller, LH Skibsted. 2005. Heme-iron in lipid oxidation. Coordination Chem Rev 249: 485–498.

RG Cassens, ML Greaser, T Ito, M Lee. 1979. Reactions of nitrite in meat. Food Technol 33:46–57.

FJM Chartier, M Couture. 2004. Stability of the heme environment of the nitric oxide synthase from Staphylococcus aureus in the absence of pterin cofactor. Biophys J 87:1939–1950.

J Chasco, G Lizaso, MJ Beriain. 1996. Cured colour development during sausage processing. Meat Sci 44:203–211.

Y Chen, JPN Rosazza. 1994. A bacterial nitric-oxide synthase from a *Nocardia* species. Biochem Biophys Res Com 203:1251–1258.

———. 1995. Purification and characterization of nitric-oxide synthase (NOS_{noc}) from a *Nocardia* species. J Bacteriol 177:5122–5128.

R Chizzolini, E Novelli, E Zanardi. 1998. Oxidation in traditional Mediterranean meat products. Meat Sci 49:S87–S99.

WS Choi, MS Chang, JW Han, SY Hong, HW Lee. 1997. Identification of nitric oxide synthase in

Staphylococcus aureus. Biochem Biophys Res Com 237:554–558.

DP Cornforth. 1996. Role of nitric oxide in treatment of foods. In: JR Lancaster, Jr., ed. Nitric Oxide: Principles and Actions. San Diego: Academic Press, pp. 259–287.

CK Cross, P Ziegler. 1965. A comparison of volatile fractions from cured and uncured meat. J Food Sci 30:610–614.

F Cutruzzola. 1999. Bacterial nitric oxide synthesis. Biochim Biophys Acta 1411:231–249.

TW Demasi, LW Grimes, RL Dick, JC Acton. 1989. Nitrosoheme pigment formation and light effects on color properties of semidry, nonfermented and fermented sausages. J Food Prot 52:189–193.

E Eidt. 1991. Films for meat-products in protective gas packs. Fleischwirtsch 71:678–679.

B Fast, PE Lindgren, F Götz. 1996. Cloning, sequencing, and characterization of a gene (narT) encoding a transport protein involved in dissimilatory nitrate reduction in Staphylococcus carnosus. Arch Microbiol 166:361–367.

I Fedtke, A Kamps, B Krismer, F Götz. 2002. The nitrate reductase and nitrite reductase operons and the narT gene of Staphylococcus carnosus are positively controlled by the novel two-component system NreBC. J Bacteriol 184:6624–6634.

DA Froehlich, EA Gullett, WR Usborne. 1983. Effect of nitrite and salt on the color, flavor and overall acceptability of ham. J Food Sci 48:152–154.

M Garcia-Varona, EM Santos, I Jaime, J Rovira. 2000. Characterisation of Micrococcaceae isolated from different varieties of chorizo. Int J Food Microbiol 54:189–195.

B Gonzalez, V Diez. 2002. The effect of nitrite and starter culture on microbiological quality of "chorizo"—A Spanish dry cured sausage. Meat Sci 60:295–298.

JH Houben, A Van Dijk. 2001. Effects of dietary vitamin E supplementation and packaging on the colour stability of sliced pasteurized beef ham. Meat Sci 58:403–407.

AG Karahan, ML Cakmakci, B Cicioglu-Aridogan, A Kart-Gundogdu. 2005. Nitric oxide (NO) and lactic acid bacteria—contributions to health, food quality, and safety. Food Rev Int 21:313–329.

PM Kenneally, G Schwarz, NG Fransen, EK Arendt. 1998. Lipolytic starter culture effects on production of free fatty acids in fermented sausages. J Food Sci 63:538–543.

KB Killday, MS Tempesta, ME Bailey, CJ Metral. 1988. Structural characterization of nitrosylhemochromogen of cooked cured meat: Implications in the meat-curing reaction. J Agric Food Chem 36:909–914.

PG Klettner, H Terstiege. 1999. Importance of illumination on the appearance of meat and meat products. Fleischwirtsch 79:91–94.

D Krotje. 1992. Starter cultures. New developments in meat products. Int Food Ingred 6:14–18.

MY Lin, CL Yen. 1999. Antioxidative ability of lactic acid bacteria. J Agric Food Chem 47:1460–1466.

B Macdonald, JI Gray, LN Gibbins. 1980. Role of nitrite in cured meat flavor—Antioxidant role of nitrite. J Food Sci 45:893–897.

G Mauriello, A Casaburi, G Blaiotta, F Villani. 2004. Isolation and technological properties of coagulase negative staphylococci from fermented sausages of Southern Italy. Meat Sci 67:149–158.

K Möhler. 1974. Formation of curing pigments by chemical, biochemical or enzymatic reactions. In: B Krol, BJ Tinbergen, eds. Proceedings of the international symposium on nitrite in meat products. Zeist, The Netherlands: Wageningen Center for Agricultural Publishing and Documentation, pp. 13–19.

JKS Møller, CE Adamsen, LH Skibsted. 2003a. Spectral characterisation of red pigment in Italian-type dry-cured ham. Increasing lipophilicity during processing and maturation. Eur Food Res Technol 216:290–296.

JKS Møller, M Jakobsen, CJ Weber, T Martinussen, LH Skibsted, G Bertelsen. 2003b. Optimisation of colour stability of cured ham during packaging and retail display by a multifactorial design. Meat Sci 63:169–175.

JKS Møller, JS Jensen, MB Olsen, LH Skibsted, G Bertelsen. 2000. Effect of residual oxygen on colour stability during chill storage of sliced, pasteurised ham packaged in modified atmosphere. Meat Sci 54:399–405.

JKS Møller, JS Jensen, LH Skibsted, S Knöchel. 2003. Microbial formation of nitrite-cured pigment, nitrosylmyoglobin, from metmyoglobin in model systems and smoked fermented sausages by *Lactobacillus fermentum* strains and a commercial starter culture. Eur Food Res Technol 216:463–469.

JKS Møller, LH Skibsted. 2002. Nitric oxide and myoglobins. Chemical Rev 102:1167–1178.

JKS Møller, L Sosniecki, LH Skibsted. 2002. Effect of nitrosylmyoglobin and saturated fatty acid anions on metmyoglobin-catalyzed oxidation of aqueous methyl linoleate emulsions. Biochim Biophys Acta 1570: 129–134.

MC Montel, F Masson, R Talon. 1998. Bacterial role in flavour development. Meat Sci 49:S111–S123.

H Morita, J Niu, R Sakata, Y Nagata. 1996. Red pigment of Parma ham and bacterial influence on its formation. J Food Sci 61:1021–1023.

H Morita, R Sakata, Y Nagata. 1998. Nitric oxide complex of iron(II) myoglobin converted from metmyoglobin by *Staphylococcus xylosus*. J Food Sci 63:352–355.

H Morita, R Sakata, S Sonoki, Y Nagata. 1994. Metmyoglobin conversion to red myoglobin derivatives and citrate utilization by bacteria obtained from meat products and pickles for curing. Anim Sci Technol (Jpn) 65:1026–1033.

H Morita, H Yoshikawa, R Sakata, Y Nagata, H Tanaka. 1997. Synthesis of nitric oxide from the two equivalent guanidino nitrogens of L-arginine by *Lactobacillus fermentum*. J Bacteriol 179:7812–7815.

PA Morrissey, JZ Tichivangana. 1985. The antioxidant activities of nitrite and nitrosylmyoglobin in cooked meats. Meat Sci 14:175–190.

SA Muller. 1990. Packaging and meat quality. Can Inst Food Sci Technol J 23:AT22–AT25.

H Neubauer, F Götz. 1996. Physiology and interaction of nitrate and nitrite reduction in Staphylococcus carnosus. J Bacteriol 178:2005–2009.

H Neubauer, I Pantel, F Götz. 1999. Molecular characterization of the nitrite-reducing system of Staphylococcus carnosus. J Bacteriol 181:1481–1488.

P Nioche, V Berka, J Vipond, N Minton, AL Tsai, CS Raman. 2004. Femtomolar sensitivity of a NO sensor from Clostridium botulinum. Science 306: 1550–1553.

K Pant, AM Bilwes, S Adak, DJ Stuehr, BR Crane. 2002. Structure of nitric oxide synthase heme protein from *Bacillus subtilis*. Biochemistry 41:11071–11079.

I Pantel, PE Lindgren, H Neubauer, F Götz. 1998. Identification and characterization of the Staphylococcus carnosus nitrate reductase operon. Mol Genet Gen 259:105–114.

G Parolari. 1996. Achievements, needs and perspectives in dry-cured ham technology: The example of Parma ham. Food Sci Technol Int 2:69–78.

G Parolari, L Gabba, G Saccani. 2003. Extraction properties and absorption spectra of dry cured hams made with and without nitrate. Meat Sci 64:483–490.

RB Pegg, F Shahidi. 1997. Unravelling the chemical identity of meat pigments. Crit Rev Food Nutr 37:561–589.

RB Pegg, F Shahidi, NJ Gogan, SI DeSilva. 1996. Elucidation of the chemical structure of preformed cooked cured-meat pigment by electron paramagnetic resonance spectroscopy. J Agric Food Chem 44:416–421.

JA Perez-Alvarez, ME Sayas-Barbera, J Fernandez-Lopez, MA Gago-Gago, MJ Pagan-Moreno, V Aranda-Catala. 1999. Chemical and color characteristics of Spanish dry-cured ham at the end of the aging process. J Muscle Foods 10:195–201.

D Pinho, S Besson, PJ Silva, B de Castro, I Moura. 2005. Isolation and spectroscopic characterization of the membrane–bound nitrate reductase from Pseudomonas chlororaphis DSM 50135. Biochim Biophys Acta Gen Subj 1723:151–162.

SR Poock, ER Leach, JWB Moir, JA Cole, DJ Richardson. 2002. Respiratory detoxification of nitric oxide by the cytochrome c nitrite reductase of Escherichia coli. J Biol Chem 277:23664–23669.

DJ Richardson. 2001. Introduction: Nitrate reduction and the nitrogen cycle. Cell Mol Life Sci 58: 163–164.

DJ Richardson, BC Berks, DA Russell, S Spiro, CJ Taylor. 2001. Functional, biochemical and genetic diversity of prokaryotic nitrate reductases. Cell Mol Life Sci 58:165–178.

R Sakata. 2000. Studies on physicochemical characteristics of red pigments in meat products. Anim Sci J 71:1–16.

JG Sebranek, JBJ Fox. 1985. A review of nitrite and chloride chemistry: Interactions and implications for cured meats. J Sci Food Agric 36:1169–1182.

F Shahidi, RB Pegg. 1995. Nitrite alternatives for processed meats. 1223–1241. In: G Charalambous, ed. Food Flavors: Generation, Analysis and Process Influence. Amsterdam: Elsevier Science B.V., pp. 1223–1241.

LH Skibsted. 1992. Cured meat products and their oxidative stability. In: DA Ledward, DE Johnston, MK Knight, eds. The Chemistry of Muscle Based Foods. London: Royal Society of Chemistry, pp. 266–286.

LH Stahnke. 1995. Dried sausages fermented with *Staphylococcus xylosus* at different temperatures and with different ingredient levels. 1. Chemical and Bacteriological Data. Meat Sci 41:179–191.

R Talon, D Walter, S Chartier, C Barriere, MC Montel. 1999. Effect of nitrate and incubation conditions on the production of catalase and nitrate reductase by staphylococci. Int J Food Microbiol 52:47–56.

R Talon, D Walter, MC Montel. 2000. Growth and effect of staphylococci and lactic acid bacteria on unsaturated free fatty acids. Meat Sci 54:41–47.

BG Tarladgis. 1962. Interpretation of the spectra of meat pigments. II. Cured meats. The mechanism of color fading. J Sci Food Agric 13:485–491.

A Uren, D Babayigit. 1997. Colour parameters of Turkish-type fermented sausage during fermentation and ripening. Meat Sci 45:539–549.

J Wakamatsu, T Nishimura, A Hattori. 2004. A Zn-porphyrin complex contributes to bright red color in Parma ham. Meat Sci 67:95–100.

NJ Watmough, G Butland, MR Cheesman, JWB Moir, DJ Richardson, S Spiro. 1999. Nitric oxide in bacteria: Synthesis and consumption. Biochim Biophys Acta 1411:456–474.

G Wolf, EK Arendt, U Pfahle, WP Hammes. 1990. Heme-dependent and heme-independent nitrite reduction by lactic-acid bacteria results in different n-containing products. Int J Food Microbiol 10: 323–329.

G Wolf, WP Hammes. 1988. Effect of hematin on the activities of nitrite reductase and catalase in lactobacilli. Arch Microbiol 149:220–244.

G Wolf, A Strahl, J Meisel, WP Hammes. 1991. Heme-dependent catalase activity of lactobacilli. Int J Food Microbiol 12:133–140.

JR Yen, RB Brown, RL Dick, JC Acton. 1988. Oxygen transmission rate of packaging films and light exposure effects on the color stability of vacuum-packaged dry salami. J Food Sci 53:1043–1046.

21
Texture

Shai Barbut

INTRODUCTION

The development of acceptable texture in fermented sausages is very important for the successful marketing of high-quality products. Obtaining the desired texture is a multiphase process, which is complex and takes anywhere from a few days to several months.

Many factors affect the final texture of fermented meat products. They include the ingredients used, processing parameters, acidification method, possible application of a low/medium heat treatment, drying and ripening conditions, as well as the interactions taking place among these factors over an extended period of time. Overall, texture development takes place in three major steps: extraction of proteins by salt during grinding and mincing, formation of a protein gel during acidification, and strengthening of the protein gel by further denaturation during cooking and/or drying. As shown in other chapters, there are many types of fermented sausages on the market. Describing the unique texture of each one is beyond the scope of this chapter. However, the main factors contributing to texture development will be discussed along with examples of representative products.

TEXTURE OF COMMERCIAL PRODUCTS

Despite the large number of fermented products on the market, there are very few scientific papers describing commercial products. Table 21.1 shows texture variations among five commercial dry-fermented Chorizo de Pamplona produced in Northern Spain. Overall, the authors concluded that there were only small differences in the textural parameters of the products (Gimeno et al. 2000). Some of the observed differences, especially in cohesiveness, were significantly correlated to the pH of the product. It should be noted that a greater variation should be expected when such a Chorizo de Pamplona is produced outside of Spain by non-Spanish or Spanish sausage makers (e.g., immigrants to other countries).

Table 21.2 shows a range of values obtained for 29 commercially produced Italian Felino Salamis, which were produced in the Parma surroundings and marketed by some of the largest Italian manufacturers (products obtained from eight different producers at different times and places). This is a typical dry-cured product with a mean marketing processing time of 73 days; however, as indicated by the authors, time can vary greatly from 17 to 160 days, depending on the product's diameter, moisture content, etc. In this case there were great differences in texture; modulus, elasticity index, and sensory hardness, which varied from 6.27–66.4 N/cm^2, 0.07–0.53, and 2.17–6.51, respectively, for the 29 different products (Dellaglio et al. 1996).

TEXTURE DEVELOPMENT DURING FERMENTATION

Reducing the pH of meat is done for a variety of reasons. Historically, the most important reason was to extend the storage life by producing an acidic environment unfavorable for pathogenic and spoilage microorganisms. Other reasons include the

Table 21.1. Composition, texture, and color (lightness-L* and redness-a*) parameters of five commercial Chorizo de Pamplona dry-fermented Spanish sausages.

	Brand 1	Brand 2	Brand 3	Brand 4	Brand 5
pH	4.65[ab]	4.62[ab]	4.73[b]	4.89[c]	4.55[a]
Moisture (%)	31.62[b]	31.69[b]	33.82[c]	30.02[a]	31.56[b]
Hardness (g)	7154[c]	5438[ab]	5475[ab]	5170[a]	6264[bc]
Springiness (mm)	0.57[ab]	0.61[b]	0.59[b]	0.57[ab]	0.51[a]
Cohesiveness	0.50[b]	0.51[b]	0.49[b]	0.45[a]	0.50[b]
Gumminess (g)	2072[b]	1695[ab]	1640[ab]	1436[a]	1873[ab]
L*	46.8[a]	51.8[bc]	54.3[c]	52.3[bc]	49.2[ab]
a*	20.4[a]	25.9[c]	26.1[c]	25.8[c]	23.5[b]

Source: Gimeno et al. (2000).
[abc] Means followed by a different superscript are significantly different ($P < 0.05$).

Table 21.2. Range of chemical, physical, and sensory values obtained for 29 commercially produced Italian Felino Salamis.

Variable	Mean	Minimum	Maximum	Standard Deviation	CV (%)
Moisture (%)	38.0	28.0	48.9	5.4	14.3
Protein (%)	29.7	23.5	36.7	3.3	11.2
pH	5.92	5.40	6.36	0.26	4.4
Fat (%)	28.0	21.0	34.3	3.3	11.8
NaCl (%)	4.32	3.35	5.20	0.48	11.0
Modulus (N/cm^2)	25.7	6.27	66.4	17.3	67.4
Elasticity index	0.35	0.07	0.53	0.10	28.6
Sensory*-color	5.03	2.89	6.15	0.60	11.9
Saltiness	4.23	3.71	4.91	0.32	7.6
Acidity	4.79	4.05	5.57	0.42	8.8
Hardness	3.76	2.17	6.51	1.06	28.2
Elasticity	4.62	3.16	5.87	0.82	17.8
Overall accept	4.28	2.47	5.56	0.79	18.5

Source: Dellaglio et al. (1996).
* Sensory analysis—13 semitrained panelists, using a 1–7 scale for each attribute (anchor points not mentioned).

production of unique acid/tangy flavors and acidification to assist speeding up the drying process by bringing the pH closer to the iso-electric point of the meat proteins where the binding of water is significantly reduced.

As described in previous chapters, there are different ways to achieve acidification and pH reduction. They can vary from relying on the natural lactic acid bacteria found in the raw meat, inoculation with meat from a previous successful batch (also called backslopping), inoculating the product with selected microorganisms, to using glucono-delta-lacton (GDL) or encapsulated acids. Depending on the method, the time of the acidification process can take a few minutes, using encapsulated acids, to a few days or even weeks, using natural microbial fermentation. The most popular acidification method is microbial fermentation because GDL and encapsulated acids do not produce the unique flavor notes derived from enzymatic activity during microbial fermentation.

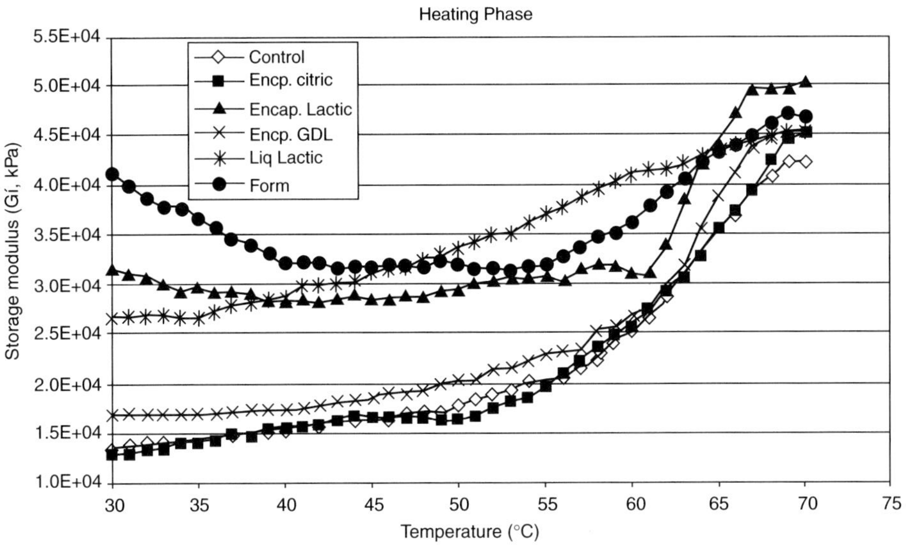

Figure 21.1. Effect of acidification method on storage modulus (G', kPa) of meat batters during heating (1.5°C/min) (from Barbut 2005).

However, various processors around the world do use the fairly fast acidifying GDL and the very fast acidifying encapsulated acids. Inoculation with a specially prepared starter culture is the preferred microbial method now because it minimizes unexpected results, such as failure due to slow/no activity of the lactic acid bacteria, and also help in achieving the safety margins required in terms of reducing the chance of *E. coli* 0157:H7 survival in the product (Pond et al. 2001).

Texture development during the acidification process is primarily the result of meat protein denaturation and coagulation. During the grinding and mincing procedures, the added salt promotes the solubilization and extraction of proteins (primarily myosin and actin) from the myofibrils, forming a sticky protein film around the minced particles. During the succeeding acidification process, the solubilized proteins coagulate and form a strong gel that binds the fat and meat closely together. Fretheim et al. (1985) reported that a 1% myosin solution, kept at 5°C, could spontaneously form a gel if the pH was slowly lowered to below 5.5. The acid-induced denaturation was the basis of the gel formed where the stiffness depended on the pH, with a maximum at pH 4.5 in the presence of 0.6 M KCl. At pH above 5.0, part of the myosin remained dissolved, suggesting an incomplete protein denaturation.

Hermansson et al. (1986) have also demonstrated spontaneous gel formation of myosin solutions dialyzed against pH 4.0 buffers in 0.25 M KCl at 4°C. A fine strand gel was formed, which actually required a lower protein concentration to form the gel than for heated myosin at pH 5.5 or 6.0. Electron microscopy revealed a network formation by parallel alignment of filaments in pairs and end-to-side interactions forming the so-called Y-junctions at low salt level, and a more right-angled network at higher salt level.

Ngapo et al. (1996) showed gel formation, measured as an increase in Young's modulus, due to gradual pH reduction by GDL, of a bovine myofibrillar solution kept at 4°C. Increasing GDL concentration resulted in faster and more extensive disruption of myofibrillar protein, leading to greater stiffness than at high pH values. A similar gelation can be observed in Figure 21.1, which shows the initial storage modulus (G') values of a nonfermented control beef meat batter (pH 5.60), fermented batter (by lactic acid bacteria; pH 4.60), and chemically acidified batter (by liquid lactic acid; pH 4.62) (Barbut 2005). The G' values of the three treatments at 30°C were significantly different (13.1, 41.0, and 26.7 kPa, respectively; $P < 0.05$). The fermented meat batter represents a typical process where the pH is gradually being lowered by lactic acid bacteria. Sausage makers often use such gel formation, or rather the increase in stiffness, to get a rough estimate of the development of the fermentation process; they often squeeze the product and feel the change in texture. The direct addition of liquid lactic acid was used only to demonstrate the devastating effect of such a practice (it is not used in commercial production) due to a very fast protein denaturation. Overall, a sudden acidification results in clumping of the meat particles and a sudden loss of the smooth texture of the raw meat batter. Quinton et al. (1997) tried to add vinegar and liquid citric acid to a meat and vegetable-stick product and made a similar observation. They reported that direct addition of acids "resulted in flaccid, noncohesive raw mix that could not be extruded into sticks, especially at pH 4.6." When they used encapsulated citric and lactic acids, no such problems were encountered because there was no immediate pH drop. Overall, the introduction of microencapsulation in the late 1960s helped address the problem of controlled release of acids in meat acidification.

Roughly, the acidification process used by the meat industry can be compared to other fermented food systems that are produced by lowering the pH. One such popular example is yogurt, where G' increases as fluid milk is fermented to give a gel-like characteristic (Haque et al. 2001). This transformation is attributed to pH reduction (to around 4.5), which results in less suppression of electrostatic repulsion between casein micelles and agglomeration into a three-dimensional casein network structure.

TEXTURE DEVELOPMENT DURING RIPENING

During the ripening process of fermented sausages, which can last from a couple of weeks to a few months (depending on the product, diameter, etc.), there are considerable changes in the meat batter's matrix. The coagulation of proteins that takes place during the fermentation phase is associated with the release of water, and this water is continuously removed during the drying process. As the drying process continues, the more tightly bound water will be released as well, though at a slower rate, and create a more dense and chewy structure due to further protein denaturation and degradation of protein by endogenous or microbial enzymes. Processors sometimes, describe the whole process as the development of a *body* or a typical *bite*. The net result is usually an increase in firmness, but a fine balance is required between forces contributing to hardening and forces contributing to softening or protein breakdown by proteolytic enzymes.

Warnants et al. (1998) followed the development of stiffness in fermented salami dried for 11 weeks. Although the product diameter is not mentioned, the trend reported is typical for dried sausages. Their first measurement of the penetration force after 3 days (the end of fermentation) was 5 ± 1 N (Table 21.3). It more than doubled after 3 weeks of drying at 16°C and 83% relative humidity. Further drying showed a numerical but not statistical increase in the penetration force. Benito et al. (2005) evaluated the changes in myofibrillar proteins, sarcoplasmic proteins, nonprotein nitrogen and amino acid nitrogen in dry-fermented Spanish Salchichon sausage dried for 20 weeks. They also evaluated the effect of adding a pure proteolytic enzyme (EPg 222 from *Penicillium chrysogenum*) to one of the treatments. The results showed a progressive reduction in myofibrillar proteins in both the control and enzyme-added treatment (Table 21.4), with a faster reduction in the EPg 222 sausage. Sarcoplasmic proteins also decreased, and, as expected, nonprotein nitrogen significantly

Table 21.3. Effect of fatty acid composition on oxidation (TBA values), texture, and color of dry salami.

	Ripening Time						Dietary Treatment			
	Day 3	Week 3	Week 5	Week 7	Week 9	Week 11	Diet 1	Diet 2	Diet 3	Diet 4
PUFA:SFA*	0.40	0.42	0.39	0.37	0.40	0.41	0.34	0.38	0.45	0.45
FFA	2.3[a]	5.5[b]	6.5[bc]	8.0[d]	8.4[d]	7.2[cd]	6.3	5.9	6.6	6.6
TBA	–	79	93	207	320	406	134	137	184	429
Penetration	5[a]	13[b]	16[b]	18[b]	17[b]	19[b]	22[a]	17[a]	12[b]	12[b]
L*	55.2[a]	50.8[b]	48.5[b]	48.1[b]	50.6[b]	50.7[b]	49.5	50.4	50.2	50.8
a*	13.0	14.1	14.4	14.4	14.0	13.3	13.9	13.8	14.0	14.2

Source: Warnants et al. (1998).

* PUFA = polyunsaturated fatty acids; SFA = saturated fatty acids; FFA = free fatty acids (in g oleic acid equivalents/100 g fat); TBA = thiobarbituric acid (ug); penetration force to puncture a 1 cm thick slice (N); color L* = lightness value and a* = redness value of the product.

increased over the 20-week period. At the end of drying, the texture profile analysis hardness value was almost twice as high for the control as compared to the EPg 222 treatment (2,925 ± 331 versus 1,593 ± 257 N, respectively; samples: 16 mm thick cm compressed to 8 mm). Other significant differences were reported for gumminess (1,000 ± 299 versus 442 ± 72 N, respectively) and chewiness (610 ± 261 versus 294 ± 70N cm, respectively), but not for springiness (about 0.62) and cohesiveness (about 0.31, ratio).

Severini et al. (2003) reported on the effect of a 55-day drying period (20°C at 80% relative humidity) of a dry Italian salami and showed firmness values (13 mm plunger used for penetration) starting at 7 N for the sausage mixture, 8 N for the sausage after 16 days, 7.8 after 25 days, 4.8 after 42 days (called *half ripening*) and 4.7 N at the end of ripening at 55 days. The product tested was produced by natural fermentation (no started culture added), so a pH of 5.0 was not reached until after 15 days of ripening. The pH stayed at this level until the end, while water activity went down to 0.89. The authors compared the treatment mentioned above to products where 33, 50, and 66% of the pork backfat was replaced with extra virgin olive oil. The replacement resulted in some differences in texture along the ripening period but at the end of the 55-day period the differences were very small.

Table 21.4. Effect of using a proteolytic enzyme (EPg 222) on myofibrillar protein and texture of a dry-fermented sausage.

	Ripening Days	Control	Enzyme Treated
Myofibrillar protein (mg/g DM)	0	123 ± 4[1]	128 ± 3[1]
	3	123 ± 1[a1]	117 ± 2[b2]
	20	45 ± 3[a2]	35 ± 3[b3]
	35	35 ± 2[3]	32 ± 2[3]
	50	27 ± 2[a4]	20 ± 2[b4]
	75	18 ± 1[5]	19 ± 2[4]
	145	17 ± 2[a5]	9 ± 2[b5]
Hardness (N)	145	2925 ± 331[a]	1593 ± 257[b]
Cohesiveness	145	0.35 ± 0.12[a]	0.28 ± 0.05[a]
Gumminess (N)	145	1000 ± 299[a]	442 ± 72[b]
Chewiness (Ncm)	145	610 ± 261[a]	294 ± 70[b]

Source: Benito et al. (2005).
For a given row, values followed by a different letter (a,b) are significantly different ($P < 0.05$). For a given ripening time (column), protein values with different numbers (1–5) are significantly different.

TEXTURE DEVELOPMENT DURING COOKING (NONDRIED/SEMIDRIED)

Some fermented/acidified meat products receive a heat treatment during manufacturing. This includes products such as cooked salami, which is typically not dried, and some products that are only semidried. It is interesting to note that after the mid-1990s, with the introduction of the regulation regarding *E. coli* 0157:H7 reduction/elimination in dry-fermented sausages in North America, a few medium- and large-size processors decided to add a heating step to the production of their traditionally dried noncooked products. The regulation requires demonstrating a 2 or 5 log reduction in *E. coli* 0157:H7, depending on the one of the five options that can be selected by USA and Canadian processors (Pond et al. 2001). This cooking step helps meet the regulation and/or add another level of security; however, it is important to note that a heating step may cause some changes to the traditional products' characteristics.

When a meat product is cooked, the heat causes the proteins to unfold, denature, and form a gel (Barbut 2005). A typical gelation curve of a meat batter is seen in Figure 21.1 (see control treatment), where the beginning of denaturation starts at 45°C (i.e., the slight elevation in G' value), corresponding to myofibrillar proteins denaturation), and the beginning of protein gelation at about 52°C. At this point a stiff protein matrix starts to form and G' values increase by a factor of about four once the temperature reaches 70°C. This basic behavior of a meat protein system has also been shown in beef, pork, and chicken proteins (Montejano et al. 1984). As mentioned above, pH reduction also causes a certain degree of protein denaturation and gelation. However, during heating, preacidified meat batters exhibit different gelation patterns (or G' value curve, Figure 21.1, than the fermented and lactic acid treatments) because of the partial pre-denaturation that occurs prior to cooking. The microbial fermented product (pH 4.60 prior to cooking) had a stiff and coherent texture prior to heating and could be thinly sliced without any problem maintaining its shape. The initial G' value was actually similar to the control treatment after cooking to 70°C (Figure 21.1). During heating of the fermented product a reduction in G' value between 30 and 40°C was probably attributed to fat softening. Later the G' value stayed the same up to 55°C, and then an additional gelation of myofibrillar proteins took place. The final G' value (46 kPa) was significantly higher ($P < 0.05$) than that of the control (42 KPa). The addition of liquid lactic acid (immediate pH drop to 4.60) resulted in very fast protein denaturation, crumbling of the meat particles, and water release. This is clearly seen in the light micrograph (Figure 21.2e), where the denaturation of the meat particles caused an incoherent microstructure of the cooked product compared to the more dense structure of the control and the encapsulated citric and GDL acids treatments (Figure 21.2a,b,d, respectively). The initial high G' value of the liquid lactic acid treatment also indicates that protein denaturation took place prior to cooking, but the texture was different from the LAB fermented treatment. This was reflected in the dense microstructure of the microbial treatment (Figure 21.2f; see also the many tiny black dots of the stained fermenting bacteria). The encapsulated citric and GDL acids (coated

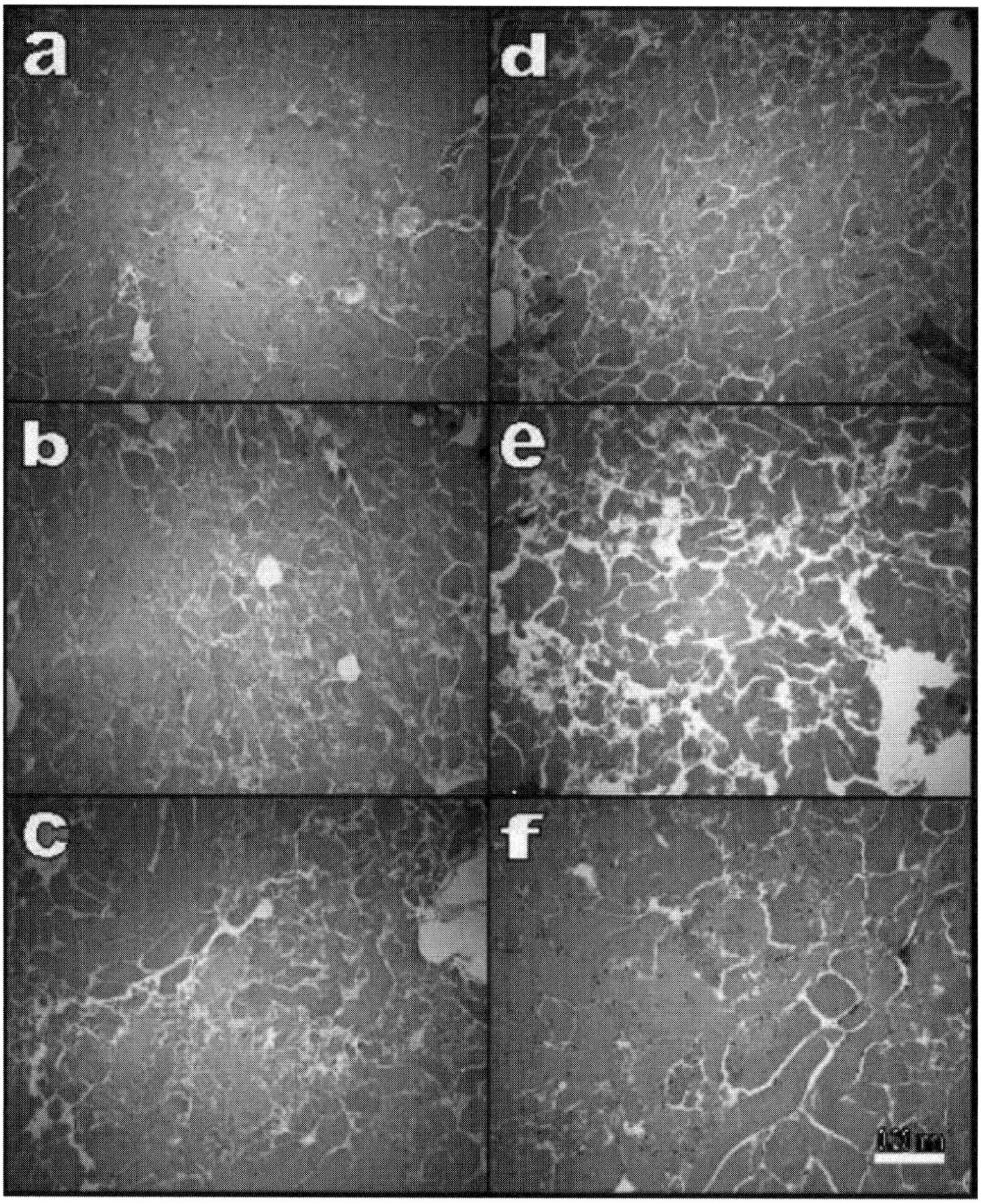

Figure 21.2. Light micrographs of cooked meat batters: (a) control, (b) acidified with encapsulated citric acid, (c) with encapsulated lactic acid, (d) with encapsulated glucono-delta lacton, (e) with liquid lactic acid, (f) fermented with lactic acid bacteria. Bar 0.2 mm (from Barbut 2006).

with hydrogenated vegetable oil, which according to the manufacturer is designed to melt and release the acids at 57–61, and 60–64°C, respectively) showed G' value curves fairly similar to the control, but with a higher final G' values. This was the result of the additional pH reduction and protein denaturation. The deviation, in G' value during cooking, from the control were seen at 62°C for the citric acid and 66°C for the GDL (Figure 21.1). Overall, the advantage of using encapsulated acids is the removal of the adverse effect of crumbling and water release prior to cooking. The microstructures of these two treatments (Figure 21.2b,d) show less dense structure compared to the control, but are fairly similar to the LAB treatment in terms of spacing among meat particles. The encapsulated lactic acid addition showed some water release prior to cooking, and stiffening of the structure prior to cooking (higher initial G', Figure 21.1), probably due to some acid release resulting from damaged/weak capsules prior to heating. In terms of the microstructure, this also resulted in gaps among the meat particles (Figure 21.2c) compared to the control and the other two encapsulated acids. Large deformation testing (fracture force) of the above-mentioned products was also preformed (data not shown), but not always correlated with the small strain values (G'). Other research groups have also reported this phenomenon in different meat systems (Montejano et al. 1984).

EFFECTS OF PROCESSING PARAMETERS

Processing parameters such as particle size of lean meat and fat, temperature of added fat, degree of mixing, etc., have a significant effect on the texture and sensory attributes of any meat product—and fermented products are no exception. However, not much has been reported in the literature in the area of fermented products. van't Hooft (1999) conducted a study in which various factors were compared (Table 21.5). Mathematical models were constructed to investigate the means and variance of the drying rate, true fracture stress and strain, Young's modulus, wall shear stress, firmness, and hardness of the final product (drying to 80% of the original weight). The main factors that were found to influence the binding and texture of the final product were

- Chopping time with salt
- Lean meat temperature
- Fat temperature
- Sharpness of the chopping knives
- Application of vacuum mixing to the raw meat batter prior to stuffing

Overall, the first four factors significantly affected most of the responses to the variables studies, except drying time. However, as indicated by the author, the

Table 21.5. Factors evaluated during the production of semidry-fermented sausage*.

Class	Parameter	Low Level	High Level
Ingredients	Fat hardness (iodine value)	65	70
	Skin, precook time (min)	10	12
	Skin type	Cheek	Back
	Starter culture (addition)	Dry	Suspension
Processing	Fat freezing temp (°C)	–18	–30
	Meat temp (°C)	–2	7
	Knife sharpness	Sharp	Blunt
	Fat adding temp (°C)	–18	–10
	Chopping time (sec)	30	60
	Chopping with salt (sec)	30	60
	Casings restrain	Slack	Tight
	Stuffing rate**	35	60
	Batter vacuum time (sec)	0	90
	Distance knife to wall (mm)	1	2

Source: van't Hooft (1999).

* Batches made from 5 kg pork, 4.4 kg beef, 4.4 kg pork fat, 1.25 kg cooked pork skin, salt, and nitrite. Using a Laska 60 L bowl chopper with six knives. Casing 58 mm cellulose. Fermentation at 25°C and later 18°C for 4 days, cold smoke, and drying at 16°C, 82% relative humidity to 80% of original stuffed weight.

** Machine setting scale 1–100.

relations of the factors to the binding and structure of the semidried products produced are complex because of interactions and because the models used were not decisive enough to optimize all the parameters. In any case, as will be discussed below, these parameters can be used to provide certain guidelines. For example, salt chopping time, which had the greatest effect on the variance of structure parameters, is very important in providing a homogeneous batter. Mixing for 20 seconds was insufficient to obtain a good distribution of the salt and resulted in a large variation in the Young's modulus compared to mixing for 60 seconds. A higher lean meat temperature (−2°C), a higher initial fat temperature (−10°C), blunt knives, longer chopping time with salt, and vacuum mixing after chopping (to remove trapped air bubbles) resulted in improved binding leading to generally higher fracture stress, Young's modulus, and hardness. However, sharp knives resulted in a better visual appearance, and because this is a very important factor in marketing such a product, sharp knives would be preferred. In fact, the parameters leading to improved binding could induce a water-binding capacity of the mince proteins that is too high, impeding the water release during the drying of the product.

In a further study, van't Hooft (1999) concentrated more on the five major factors listed above and added a few more suggestions:

- *Batter chopping time with salt*: 60 to 80 seconds to obtain a sufficient protein exudate to bind lean meat particles. But avoid too much chopping because this may cause high adhesiveness, which leads to greasiness and subsequently a weaker structure as well as too strong a water binding
- *Vacuum mixing*: apply to batters chopped for a short time where it can help binding. However, in appropriately chopped/prolonged mixed batters, such vacuum mixing can result in greasiness and low firmness of the final product
- *Meat temperature*: influenced by the chopping process and is affecting the batter wall shear stress. However, the relationships to the studied processing parameters were not completely clarified
- *Fat temperature*: above −7.5°C resulted in more melted fat and required more salt-soluble proteins exudate to stabilize the fat. Applying a longer chopping time and/or salt chopping time will raise the amount of exudate, but increase water binding and impede the drying process.

These comments illustrate the possible interactions among the different factors, where changing one parameter can influence other parameters. The reader should also keep in mind that processing parameters not found to have a major effect in that study (distance of the knives to the chopping bowl, skin cooking time, and the type of skin used) might be influenced by the type of equipment used in other processing plants (e.g., different ratio of chopping knives speed to bowl chopper speed).

EFFECTS OF PRODUCT MODIFICATION WITH NONMEAT INGREDIENTS

Several modifications, such as changing the fatty acid composition, type of salt, and supplementary dietary fiber, have been reported. The area of substituting/reducing fat and increasing the level of unsaturated fatty acids to obtain a more favorable nutritional profile of fermented sausage commonly produced with pork and/or beef fat has received most attention. Such modifications usually also affect the texture of the final product because of differences in the viscosity of the fats/oils, as well as different interactions with the meat proteins. Warnants et al. (1998) manipulated the diet fed to pigs to increase the level of polyunsaturated fat. When the backfat was used to produce dry salamis (drying up to 11 weeks), they reported significant differences in the penetration force and sensory analysis (Table 21.3). The penetration force values of the first two diets were statistically lower than the last two diets, which contained a higher level of unsaturated fat. The treatments of diets 3 and 4 were rejected by the sensory panel because of an oily texture, and diet 4 was also reported to have a noticeable fishy off-odor due to higher-than-normal oxidation by products (see TBA values; Table 21.3).

Muguerza et al. (2001), replaced pork backfat with 0, 10, 12, 20, 25, and 30% preemulsified olive oil (oil:water:soy protein isolate 10:8:1) in a Chorizo de Pamplona fermented dried Spanish sausage. In terms of nutritional content, cholesterol was lowered by 12–13% in the 20–25% replacement treatments, and about 22% in the 30% treatment. Oleic acid (18:1) was significantly higher ($P < 0.05$) in the 15–30% replacement treatments, and linoleic (18:2) in the 10–25% replacements. All sausages with 10–25% replacements were acceptable by a sensory taste panel. The texture and color were comparable with that of a commercial formulation. However, it should be noted that the hardness, gumminess, and chewiness of the control product (no fat substitution) were higher than in the replacement treatments. The authors concluded that up to 25% replacement with preemulsified olive oil was possible, but a higher level was unacceptable due to considerable fat dripping during ripening.

In a fermented and cooked summer sausage, Shackelford et al. (1990) reported that high content of oleic acid in the pork fat (achieved by replacing the corn and soy bean diet with 10% safflower oil, sunflower oil, or canola oil) resulted in lower sensory panel scores for springiness, cohesiveness, texture, and overall palatability ($P < 0.05$) of the 25% fat level treatment. When 15% fat level was used, the high oleic acid treatments were comparable to the control for all sensory properties. The authors concluded that using high levels of mono unsaturated fat was less desirable in the conventional summer sausage with 25% fat, but produced a highly palatable reduced fat (15%) sausage, which provided a reduction of 13% in the saturated fatty acids.

The use of soluble dietary fibers (inulin; a blend of fructose polymers extracted from plant material) as a fat substitute has also been tried in dry-fermented sausages (Mendoza et al. 2001). The idea was to add an ingredient that can help counteract the increase in hardness due to fat reduction and also help raise dietary fibers in the diet. Adding inulin at 6, 7, 10, and 11.5% to a low-fat (13.5% as compared to a high-fat with 31% fat) naturally fermented sausage (i.e., without a starter culture), resulted in an overall improvement in sensory properties due to the soft texture and tenderness; values were similar to the conventional high-fat control product. Texture profile analysis showed a similar trend. On the other hand, a fat reduction by itself (31.0 to 13.5%) resulted in higher hardness as well as lower springiness and cohesiveness values. Overall, inulin addition helped maintain a softer texture, which is more difficult to maintain in low-fat dried products than in nondried products because of the moisture loss during the long ripening process.

Gimeno et al. (1999) reported on the effect of substituting some of the 2.6% common salt (NaCl) in Chorizo de Pamplona sausage with a mixture of 1.0% NaCl, 0.55% potassium chloride, and 0.74% calcium chloride in order to reach a mixture with the same ionic strength, but with 40% less sodium. They reported that the salt mixture affected the texture parameters, showing significantly lower hardness values (71 versus 62 N for the control and mixture, respectively), cohesiveness (0.54 vs. 0.44), gumminess (38 vs. 27 N), and chewiness (25 vs. 17 Ncm). The sensory quality of sausages with mixed salt treatment was lower, but were classified acceptable. The authors indicated that the lower pH value of the sausage with the salt mixture (pH 4.86 versus 5.04 for the control) could have resulted in greater protein denaturation and decreased binding capacity of the meat proteins, which would explain the results of the salt mixture.

SUMMARY

The complex and long processing procedure of fermented meat products, compared to products such as bologna and cooked ham, requires close attention to details. The steps in the manufacturing of fermented product involve acidification, drying, and ripening that may take anywhere from a few days to a few weeks and even months. The texture of the final products is therefore a combined result of the ingredients and various processing parameters used, as described in this chapter. Furthermore, the interaction among the different ingredients (meat and nonmeat), product's dimension (e.g., stuffing diameter), and processing conditions after stuffing (e.g., drying temperature, relative humidify) also play an important role in the texture and acceptability of the final product. Manufacturing high-quality fermented product requires good understanding of such factors and their interactions.

REFERENCES

S Barbut. 2005. Effects of chemical acidification and microbial fermentation on the rheological properties of meat products. Meat Sci 71:397–401.

———. 2006. Fermentation and chemical acidification of salami-type products—Effect on yield, texture and microstructure. J Muscle Foods 17:34–42.

MJ Benito, M Rodriquez, MG Cordoba, MJ Andrade, JJ Corboda. 2005. Effect of the fungal protease Epg222 on proteolysis and texture in the dry fermented sausage "Salchichon." Sci Food Agric 85: 273–280.

S Dellaglio, E Casiraghi, C Pompei. 1996. Chemical, physical and sensory attributes for the characterization of an Italian dry-cured sausage. Meat Sci 42:25–35.

K Fretheim, B Egelandsdal, O Harbitz, K Samejima. 1985. Slow lowering of pH induces gel formation of myosin. Food Chem 18:169–178.

O Gimeno, D Ansorena, I Astiasarán, J Bello. 2000. Characterization of Chorizo de Pamplona: Instrumental measurements of colour and texture. Food Chem 69:195–200.

O Gimeno, I Astiasarán, J Bello. 1999. Influence of partial replacement of NaCl with KCl and $CaCl_2$ on texture and color of dry fermented sausages. J Agric Food Chem 47:873–977.

A Haque, FK Richardson, ER Morris. 2001. Effect of fermentation temperature on the rheology of set and stirred yogurt. Food Hydrocol 15:593–602.

AM Hermanson, O Harbitz, M Langton. 1986. Formation of two types of gels from bovine myosin. J Sci Food Agric 37:69–84.

E Mendoza, ML Garcia, C Casas, MD Selgas. 2001. Inulin as fat substitute in low fat, dry fermented sausages. Meat Sci 57:387–393.

JG Montejano, DD Hamann, TC Lanier. 1984. Thermally induced gelation of selected comminuted muscle systems—Rheological changes during processing; final strengths and microstructures. J Food Sci 49:146–152.

E Muguerza, O Gimeno, D Ansoreno, JG Bloukas, I Astiasarán. 2001. Effect of replacing pork backfat with pre-emulsified olive oil on lipid fraction and sensory quality of Chorizo de Pamplona—A traditional Spanish fermented sausage. Meat Sci 59: 251–258.

TM Ngapo, BHP Wilkinson, F Chong. 1996. 1, 5-Glucono-delta-lactone induced gelation of myofibrillar protein at chilled temperatures. Meat Sci 42:3–13.

TJ Pond, DS Wood, IM Mumin, S Barbut, MW Griffiths. 2001. Modeling the survival of *Escherichia coli* O157:H7 in uncooked, semi-dry, fermented sausages. J Food Prot 64:759–766.

RD Quinton, DP Cornforth, DG Hendricks, CP Brennand, YK Su. 1997. Acceptability and composition of some acidified meat and vegetable stick products. J Food Sci 62:1250–1254.

C Severini, T De Pilli, A Baiano. 2003. Partial substitution of pork backfat with extra-virgin olive oil in "salami" products: Effects on chemical, physical and sensorial quality. Meat Sci 64:323–331.

SD Shackelford, MF Miller, KD Haydon, JO Reagan. 1990. Evaluation of the physical, chemical and sensory properties of fermented summer sausage made from high-oleate pork. J Food Sci 55:937–941.

B-J van't Hooft. Development of binding and structure in semi-dry fermented sausages. 1999. PhD Thesis, Utrecht University, Utrecht, The Netherlands.

N Warnants, MJ Van Oeckel, Ch V Boucque. 1998. Effect on incorporation of dietary polyunsaturated fatty acids in pork backfat on the quality of salami. Meat Sci 49:435–445.

22
Flavor

Karsten Tjener and Louise H. Stahnke

INTRODUCTION

This chapter deals with flavor and flavor formation in fermented sausage and dry-cured ham: the description of flavor compounds of relevance to the overall sensation of fermented sausage and dry-cured ham, the pathways through which the compounds are formed, and finally how to impact the flavor formation processes in relation to dry-cured ham and fermented sausage production.

Flavor is defined as the total impression of taste, odor, tactile, kinesthetic, temperature, and pain sensations perceived through tasting (Rothe 1972; ISO 5492:1992). However, only taste compounds, defined as molecules with the ability to trigger a response in the taste buds, and aroma (odor) compounds, defined as volatiles with the ability to reach and generate response in the receptors of the olfactory epithelium, will be treated.

IMPORTANT AROMA COMPOUNDS

Identification

Approximately 400 volatiles have been identified in fermented sausage (Stahnke 2002); from dry-cured ham the number is approximately 200 (Ruiz et al. 2002). However, many of those volatiles do not contribute to dry-cured ham or fermented sausage aroma because of high sensory threshold values.

Identification of the more important aroma compounds requires combination of analytical and sensory methods. Gas chromatography-olfactometry (GC-O), combining classical chromatography with the detection system of the human nose, is one of the preferred approaches. The other prevailing approach is correlation of sensory profile and amount of specific volatiles. In principle, it is also possible to compare the amount of a given volatile with the sensory threshold value in a similar matrix, but the approach suffers from the lack of threshold values obtained in relevant matrices as well as quantitative determination of volatiles in fermented sausages and dry-cured hams.

Dry-cured Ham

In a series of studies, GC-O was applied to identify the odor-active compounds of Italian and Spanish dry-cured hams (Flores et al. 1997; Blank et al. 2001; Carrapiso et al. 2002a,b; Carrapiso and Garcia 2004). The studies report from 15 to 44 odorous volatiles belonging to a variety of chemical classes, such as alcohols, aldehydes, acids, esters, ketones, aromatic, sulfur and nitrogen compounds. Table 22.1 summarizes the odorous volatile compounds identified in those five studies.

In Parma ham methional (potatolike), (*E*)-2-nonenal (fatty, leatherlike), and sotolone (seasoninglike) were reported to be the most potent odorants, followed by 3-methylbutanoic acid (sweaty, musty), *p*-cresol (phenolic, musty) and phenylacetic acid (honeylike, spicy) (Blank et al. 2001). The most potent odorants found in three studies of Iberian ham were hexanal (green), (*Z*)-3-hexenal (acornlike),

Table 22.1. Odor-active aroma compounds in Italian and Spanish dry-cured hams.

Compound	Descriptor	Source
Alcohols		
1-Penten-3-ol	Onion, toasted	1
1-Octen-3-ol	Mushroom, rustlike	3,4,5
3-Methyl-1-butanol	Green	1
3-Methyl-2-hexanol	Potato, wheat	1
Aldehydes		
Propanal	Almondlike, green	4
Pentanal	Nutty, toasted, fruity	5
Hexanal	Green	1,3,4,5
Heptanal	Fruity, fatty	4,5
Octanal	Green-fresh	1,3,4,5
Nonanal	Green	1
(E)-2-Hexenal	Olive oil-like	3,4,5
(Z)-3-Hexenal	Acornlike	3,5
(E)-2-Heptenal	Almondlike, fruity, fried foodlike	5
(E)-2-Octenal	Fruity, rancid	4,5
(E)-2-Nonenal	Fatty, leatherlike	2
2-Methylpropanal	Nutty, toasted, pungent	3,4,5
3-Methylbutanal	Cheesy-green, malty, nutty, toasted	1,3,4,5
2-Methylbutanal	Rancid, almondlike, toasted	5
Acids		
Acetic acid	Acidic, vinegar	1,2
Butanoic acid	Sweaty	2
3-Methylbutanoic acid	Sweaty, musty	2
Esters		
Methyl butanoate	Sweet-caramel	1
Methyl hexanoate	Boiled meat	1
Methyl 2-methylpropanoate	Fruity	1
Ethyl butanoate	Fruity	1
Ethyl 2-methylpropanoate	Fruity	4,5
Ethyl 2-methylbutanoate	Fruity	1,3,5
Ethyl 3-methylbutanoate	Fruity	2
Ketones		
2-Pentanone	Green, fruity, tropical	4,5
2-Hexanone	Floral-apple	1
2-Heptanone		4,5
6-Methyl-5-heptan-2-one	Citrus-candy	1
1-Penten-3-one	Rotten, sewerlike, fruity	3,5
1-Octen-3-one	Mushroomlike	2,3,4,5
3-Hydroxy-2-butanone	Fruit-red Jell-o	1
2,3-Butanedione	Buttery, caramellike	1,4,5
Aromatic Compounds		
Phenylacetaldehyde	Honeylike	2
Phenylacetic acid	Honeylike, spicy	2
p-Cresol	Phenolic, musty	2
o-Xylene	Sweet-fruit candy	1

Table 22.1. *(Continued)*

Compound	Descriptor	Source
m-Xylene	Smoked-phenolic	1
p-Xylene	Smoked-phenolic	1
Sulfur Compounds		
Hydrogen sulfide	Boiled or rotten eggs, sewagelike	3,4,5
Methanethiol	Rotten eggs or meat, sewagelike, cheesy	3,4,5
Dimethyldisulfide	Dirty socks	1
Dimethyltrisulfide	Rotten egglike, burnt	5
Methional	Potatolike	2,5
3-Mercapto-2-pentanone	Fruity, sewagelike, fatty, cured hamlike	4,5
2-Methyl-3-furanthiol	Cured hamlike, toasted, nutty	3,4,5
2-Furfurylthiol	Boiled meatlike, cured hamlike	5
Nitrogen Compounds		
2-Acetyl-1-pyrroline	Roasty, overheated meatlike, cured hamlike	2,3,4,5
2-Propionyl-1-pyrroline	Stewlike, boiled meatlike, rancid	3,4,5
Methylpyrazine	Nutty	1
2,6-Dimethylpyrazine	Toasted nut	1
1(H)-Pyrrole	Meaty	1
Miscellaneous		
4-Hydroxy-2,5-dimethyl-3(2H)-furanone	Caramellike, sweet	2
2-Butoxyethanol	Dark, toast-meaty	1
2-Pentylfuran	Hamlike	1
Sotolone	Seasoninglike	2

Sources: (1) Flores et al. (1997); (2) Blank et al. (2001); (3) Carrapiso et al. (2002a); (4) Carrapiso and Garcia (2004); (5) Carrapiso et al. (2002b).

3-methylbutanal (malty, nutty, toasted), 1-octen-3-one (mushroom), 1-octen-3-ol (mushroom, rustlike), hydrogen sulfide (boiled or rotten eggs, sewagelike), methanethiol (rotten eggs or meat, sewagelike) and 2-methyl-3-furanthiol (nutty, dry-cured hamlike, toasted) (Carrapiso et al. 2002a,b; Carrapiso and Garcia 2004). The meaty-smelling compounds 2-methyl-3-furanthiol, 2-furfurylthiol, 3-mercapto-2-pentanone, 2-acetyl-1-pyrroline, and 2-propionyl-1-pyrroline were all detected by Carrapiso et al. (2002b). They all have very low odor thresholds (Rychlik et al. 1998) and may contribute significantly to dry-cured ham aroma although present in low amounts. Meaty aroma compounds found in Serrano dry-cured ham were identified as 2-pentylfuran (hamlike), 1-*H*-pyrrole (meaty) and 2-butoxyethanol (dark toast, meaty), and the latter two further correlated with the pork flavor detected by sensory evaluation (Flores et al. 1997). In studies where instrumental flavor analysis was combined with sensory profiling, cured flavor was correlated to 2-methylbutanal (Ruiz et al. 1999), methyl ketones, and straight-chain alcohols (Buscailhon et al. 1994), respectively. The authors Hinrichsen and Pedersen (1995) found a correlation between nutty, cheesy, and salty flavor notes and the methyl-branched aldehydes, secondary alcohols, methyl ketones, ethyl esters, and dimethyltrisulfide. Finally, aged odor was correlated to methyl-2-methylpropanoate, ethyl-2-methylbutanoate, 2-methyl-3-buten-2-ol, 3-methyl-1-butanol, and 3-hydroxy-2-butanone (Careri et al. 1993).

Fermented Sausage

Table 22.2 compiles the odorous volatiles detected by GC-O in fermented sausages. Compared to dry-cured ham (Table 22.1), more acids, sulfides, terpenes, and phenolic compounds seem to be important in sausages. The presence of sulfides is closely linked to garlic (Mateo and Zumalacárregui

Table 22.2. Odor-active aroma compounds in European dry-fermented sausages, with or without mold coverage, smoked or nonsmoked.

Compound	Descriptor	Source
Alcohols		
1-Octene-3-ol	Mushroom	1,3
Aldehydes		
Acetaldehyde	White glue, green	1,5,7
Pentanal	Synthetic, green leaves	4
Hexanal	Green	1,3,4,5,6,7,8
Heptanal	Pelargonium, potatoes, green leaves	1,3,5
Octanal	Orange, spicy, pelargonium	1,4,5,7
Nonanal	Herb, geranium, dry, spicy, citrus	4,5,8
Decanal	Cucumber, cardboard, dry grass	4
(Z)-x-Hexenal	Old cooked potatoes, nauseous	5
(Z)-4-Heptenal	Unpleasant bread	5
(E)-2-Octenal	Deep-fried, musty	5,8
(E)-2-Nonenal	Fresh, cucumber	4,5
(E)-2-Decenal	Deep-fried	1
(E,E)-2,4-Nonadienal	Fatty	8
(E,E)-2,4-Decadienal	Deep-fried fat	4
(E,Z)-2,4-Decadienal	Rancid, fatty, tallowy	6,8
2-Methylbutanal	Butter, caramel, rancid	3,4,5
3-Methylbutanal	Sour cheese	4,5
3-Methylhexanal	Sour, green, acetone	4,5
Acids		
Acetic acid	Vinegar	1,4,5,6,7
Propanoic acid	Sour, sweaty	5,6
Butanoic acid	Cheesy, sweaty	1,5,6,7
Pentanoic acid	Sweaty	6
Hexanoic acid	Sweaty, putrid	6
Decanoic acid	Pungent, repellent	6
2-Methylpropanoic acid	Cheesy, sweaty	5,6,7
3-Methylbutanoic acid	Cheesy, sweaty socks	1,4,5,6,7
Esters		
Methylbutanoate	Fruity, flowery	6
2-Methylpropylacetate	Candy, fruit	4,5
Ethylacetate	Artificial fruity	6
Ethylpropanoate	Fruity	6
Ethylbutanoate	Fruity, candylike	2,4,5,6,7
Ethylpentanoate	Fruit	4,5
Ethyloctanoate	Fatty, green	8
Ethyl-2-methylpropanoate	Fruity, candylike, pineapple	2,4,5,7,8
Ethyl-2-methylbutanoate	Fruity, candylike, pineapple	2,4,5
Ethyl-3-methylbutanoate	Fruity, candylike, chutney	2,5,6,8
Ethyl lactate (ethyl-2-hydroxypropanoate)	Chutney, sweet	1
Propylacetate	Sourish apple, candy	4
Ketones		
2-Heptanone	Pelargonium, nettle, fruity	5
2-Nonanone	Plant, herb, weak fruity	4,6
3-Octen-2-one	Hot soil, mushroom	5

Table 22.2. *(Continued)*

Compound	Descriptor	Source
1-Octene-3-one	Mushroom	7,8
2,3-Butadione	Butter strong	1,2,4,5,6,7
2,3-Pentadione	Butter weak	5
3-Hydroxy-2-butanone	Butter	6
Aromatic Compounds		
Phenylacetaldehyde	Honey	1
2-Phenylethanol	Flowery, rose	6
2-Methoxyphenol (guiaicol)	Smoked, bandage, phenolic	1,5,7
2-Methoxy-4-methylphenol (methyl guiaicol)	Smoked, car tire	1
p-Cresol	Musty, smoky	6
Naphthalene	Roses, soap	1
Eugenol	Cloves, spicy, smoky	6
Sulfur Compounds		
Methanethiol	Cabbage, feces, putrid	1,2,7
Allyl-1-thiol	Salami, onion, cured meat	1,2,7
Furfuryl thiol	Burned coffee	2
Methylthiirane	Salami, onion	1,2,7
3-Methylthio-1-propene	Gas, onion, unpleasant	3
Methional	Cooked potato	1,7,8
Dimethylsulfide	Sulfur, putrid	7
Dimethyldisulfide	Garlic	6
Dimethyltrisulfide	Cabbage, metallic	1,7
Methylallylsulfide	Strong garlic	1,2,6,8
Methylallyldisulfide	Garlic	6
Diallylsulfide	Garlic	6
Nitrogen Compounds		
x-Pyridineamine	Cooked meat	1
2-Acetyl-1-pyrroline	Popcorn	2,7,8
Dimethylpyrazine	Coffee, toasted	5
Terpenes		
β-Myrcene	Herbal	6
Limonene	Fresh	6
γ-Terpinene	Herbal	6
α-Pinene	Peppery	6,8
β-Pinene	Sharp, dry	6
4-Terpineol	Greenish	6
1,8-Cineol	Menthol	6
Terpinolene	Fruity, eucalyptus	3
Linalool	Flowery, heavy	6,8

Sources: (1) Stahnke (1998); (2) Stahnke (2000); (3) Meynier et al. (1999); (4) Stahnke (1994); (5) Stahnke (1995a); (6) Schmidt and Berger (1998); (7) Stahnke et al. (1999); (8) Blank et al. (2001).

1996), terpenes to pepper (Berger et al. 1990), and the phenolic compounds to smoke (Hollenbeck 1994). Thus, many of the compounds detected in fermented sausage but not in dry-cured ham were "added" to the sausage and not the result of chemical or microbiological processes occurring during ripening.

In a study of French, Italian, Spanish, and German sausages, Schmidt and Berger (1998) found diallyldisulfide (garlic), eugenol (cloves, spicy, smoky),

3-methyl butanoic acid (sweaty), acetic acid (vinegar), linalool (flowery, heavy), methylallyl sulfide (garlic), and diallyl sulfide (garlic) to be the most important odor compounds. In Italian-type salami hexanal (green), methional (cooked potato) and 2-acetyl-1-pyrroline (roasty, popcorn) were identified as the most potent odorants followed by ethyl-2-methyl-propanoate (fruity), ethyl-3-methylbutanoate (fruity, sweet), α-pinene (pepperlike), 1-octen-3-one (mushroom), and (*E*)-2-octenal (musty) (Blank et al. 2001). Based on GC-O studies of two Northern (smoked) and two Mediterranean (nonsmoked, mold-fermented) fermented sausages, Stahnke et al. (1999) concluded that the basic dried sausage odor was composed of the following components: methanethiol (cabbage, putrid), methional (cooked potato), dimethylsulfide (sulfur, putrid), dimethyltrisulfide (cabbage, metallic), diacetyl (butter), ethylbutanoate (fruit), ethyl-2-methylpropanoate (fruit), acetaldehyde (glue, yogurt), acetic acid (vinegar), butanoic acid (cheese), 2-methylpropanoic acid (cheese), 3-methylbutanoic acid (sweaty socks), guaiacol (phenolic), hexanal (green), octanal (orange), 1-octen-3-one (mushroom), and several unidentified compounds (pelargonium, cooked meat, sulfur, onion, dry, rubber, earthy, deep-fried, cress, cucumber). In the same study, the Northern European sausages differed from the Mediterranean molded sausage types by a popcorn note in the latter and coffee/roasted, phenolic, sulfur, and vinegar notes in the former. The coffee/roasted note was ascribed to 2-furfuryl thiol, the phenolic note to guaiacol, the vinegar note to acetic acid, and the popcorn note to 2-acetyl-1-pyrroline. The sulfur note remained unidentified. Another study on 10 different Northern and Southern European sausages confirmed the differences found by Stahnke et al. (1999) and additionally showed that 2-acetyl-pyrroline was mainly present close to the edge of the molded sausages (Stahnke 2000).

The GC-O studies from which Table 22.2 was put together revealed only two components with odor descriptors directly linked to fermented sausage, namely allyl-1-thiol (salami, cured meat) and its ring closure methyl-thiirane (salami, onion) (Stahnke 1998, 2000; Stahnke et al. 1999). However, by combination of volatile and sensory analyses it was shown that salami odor was correlated with the presence of ethyl esters and methyl ketones (Stahnke 1994, 1995a; Stahnke et al. 1999) as well as the methyl-branched aldehydes 2-methyl propanal and 2- and 3-methylbutanal (Stahnke 1994, 1995a; Hagen et al. 1996). Likewise, cured odor correlated with methyl ketones (Berdagué et al. 1993b) and mature flavor with methyl ketones, methyl-branched aldehydes and some allylic sulfur compounds (Stahnke et al. 2002). Volatile analysis combined with sensory assessment was also used to identify volatiles linked to other odor notes. Buttery odor was correlated to the presence of diacetyl (Berdagué et al. 1993b; Viallon et al. 1996), cheesy odor with 2-methyl propanoic acid and 2- and 3-methylbutanoic acid (Stahnke 1995a; Viallon et al. 1996), rancid flavor to straight-chain aldehydes (Stahnke 1994, 1995a; Stahnke et al. 1999), and finally sourish and acid notes to acetic, butanoic, and hexanoic acids (Stahnke 1995a, 1999). The authors Stahnke et al. (1999) also correlated garlic flavor with allylic sulfur compounds, smoked flavor with aromatic and O-heterocyclic compounds, and spicy/piquant flavors with terpenes.

In summary, the main constituents of dry-cured ham and fermented sausage odor are straight- and branched-chain aldehydes; acids; their methyl- and ethyl esters; aromatic, heterocyclic, and sulfur compounds; methyl ketones; and terpenes.

Formation

This section shortly describes the origin of some of the most important groups of aroma compounds. Detailed reaction schemes are outside the scope of this chapter and will not be shown.

Straight-chain alcohols, aldehydes, and to some extent also the acids are formed by autoxidation of unsaturated fatty acids. The complex chemistry of the various autoxidation processes is thoroughly reviewed by several researchers (Love and Pearson 1971; Frankel 1980; Ladikos and Lougovois 1990) and will not be treated further here.

Methyl ketones can be formed by autoxidation of unsaturated fatty acids, but microbial β-oxidation is also possible and would explain the correlation between the presence of staphylococci and elevated levels of methyl ketones observed in several studies (Berdagué et al. 1993b; Stahnke 1995a; Montel et al. 1996). Furthermore, Engelvin et al. (2000) discovered the β-oxidation pathway in *S. carnosus*.

As mentioned above, the straight-chain acids are to some extent formed by autoxidation of unsaturated fatty acids. However, the C2–C6 acids are also derived from microbial degradation of pyruvate. Lactic acid bacteria, staphylococci, and yeasts are all capable of producing short straight-chain acids (Gottschalk 1986; Westall 1998; Olesen and Stahnke 2000; Beck et al. 2002).

Esters are generated by reaction of an alcohol and a carboxylic acid (or acyl-CoA). The amount of ester present in fermented meat is strongly correlated to the amount of precursor alcohol, precursor acid, or both (Stahnke 1994, 1995b, 1999a; Montel et al. 1996; Tjener et al. 2004c). The origin of esters is not well established. LAB, *Staphylococcus*, yeast, and mold can form esters (Stahnke 2002), but chemical formation is also possible, in particular when acyl-CoAs are involved (Yvon and Rijnen 2001).

The methyl-branched aldehydes 2-methylpropanal, 2-methylbutanal, and 3-methylbutanal and the corresponding acids are primarily formed by degradation of the amino acids valine, isoleucine, and leucine, respectively. In fermented sausage the formation of the methyl-branched aldehydes and acids is ascribed to staphylococci. By inactivation of the aminotransferase gene of *S. carnosus* Madsen, Beck et al. (2002) showed that transamination is the first step in the degradation of the branched-chain amino acids. The second step is presumably a decarboxylation reaction whereby the methyl-branched aldehyde is formed, as showed by Beck et al. (2002). Subsequently, the aldehyde may be either reduced or, as in the study by Beck et al. (2002), oxidized into the corresponding acid.

The allylic sulfur compounds listed in Table 22.2 most likely originate from garlic (Schmidt and Berger 1998), whereas methional, methanethiol, dimethyldisulfide, and dimethyltrisulfide are probably derived from methionine through the same pathways as demonstrated in cheese ripening cultures (Stahnke 2002).

The aromatic compounds phenylacetaldehyde, phenylacetic acid, p-cresol, guiaicol, and methylguiaicol are possibly of microbial origin, although the pathways are unrevealed in meat cultures (Stahnke 2002). However, the phenolic compounds listed in Table 22.2 may also arise from smoke.

IMPORTANT TASTE COMPOUNDS

The basic taste sensations salt, sweet, sour, bitter, and umami are triggered by a diverse group of molecules. This group includes inorganic salts, sugars, ATP (adenosine triphosphate) derivatives, organic acids (of which some are volatile), free amino acids, and smaller peptides (Mateo et al. 1996).

Salt

The salty taste in dry-cured ham and fermented sausage is mainly caused by sodium chloride. Typical levels in final products are 4.7–8.7%$^{w}/_{w}$ and 3.5–4.6%$^{w}/_{w}$ for dry-cured ham and fermented sausage, respectively (Wirth 1989; Toldrá 2002).

Sugars

Glucose, sucrose or more complex sugars are typically added to facilitate acid production by lactic acid bacteria during fermented sausage processing. When sugars are added in excess, the residual amount will contribute to sweet taste in the final sausage, given that they are present in levels above their sensory threshold. Threshold values of glucose and sucrose are 1.4 and 0.6%$^{w}/_{w}$, respectively (Plattig 1984). Traditional manufacture of dry-cured hams does not include any addition of sugars, and any sweet taste perceived from this product type is most likely caused by other compounds.

ATP Derivatives

Among the ATP derivatives, IMP (inosine monophosphate) is the dominant component in meat postrigor (Lawrie 1995), but it is gradually converted into inosine and further to hypoxanthine during ripening of fermented sausage and dry-cured ham (Dierick et al. 1974; Kohata et al. 1992; Mateo et al. 1996). In ready-to-eat Chorizo, the level of IMP (0.007–0.13 μmol/g) was below the umami taste threshold, and the level of its predominant derivative, hypoxanthine (6.9–8.7 μmol/g), was considered being negligible to the taste of Chorizo (Mateo et al. 1996). However, in another study, lower levels of hypoxanthine (0.5–1.5 μmol/g) were found to correlate with bitter taste of aged pork meat (Tikk et al. 2005).

Organic Acids

L- and D-lactic and acetic acids are the predominant acids contributing to the sour taste of fermented sausages (Montel et al. 1998; Stahnke 2002). The total lactic acid level varies from 0.4–3.1 g/100 g dry matter among a wide range of Northern European and Mediterranean sausage types (Vandekerckhove and Demeyer 1975; Johansson et al. 1994; Dellaglio et al. 1996; Mateo et al. 1996; Demeyer et al. 2000; Bruna et al. 2000a). Typically, the level of acetic acid is ten- to twentyfold lower than the level of lactic acid (Johansson et al. 1994; Mateo et al. 1996; Demeyer et al. 2000). The presence of C3–C6 straight-chain acids is sometimes reported, but they are unlikely to contribute significantly to the sour taste due to low abundance (100–1000-fold less than acetic acid [Dainty and Blom 1995]). However, the studies by Bruna et al. (2000a, 2001, 2003) showed that propionic acid level of the same magnitude as lactic acid is obtainable by surface inoculation with *Penicillium aurantiogriseum* or *Penicillium camemberti*.

Free Amino Acids and Peptides

The degradation of protein into smaller peptides and free amino acids has been extensively studied in fermented sausages as well as dry-cured hams, and excellent reviews exist for both product types (Toldrá et al. 1997; Toldrá and Flores 1998; Ordonez et al. 1999; Sanz et al. 2002).

The free amino acids and smaller peptides may trigger sweet, sour, bitter, and umami taste when surpassing their respective taste threshold values. Mateo et al. (1996) determined the content of the individual free amino acids in Chorizo and predicted glutamic acid to produce umami response, leucine a bitter response, alanine a sweet response, and valine and lysine both sweet and bitter responses in this product. Henriksen and Stahnke (1997) studied water-soluble fractions from different sausage types and found correlation between sour taste and glutamic acid; between bitterness and lysine, valine, leucine, isoleucine, and proline; and between bouillon taste and a mixture of different amino acids and smaller peptides with high content of methionine, phenylalanine, alanine, glycine, glutamic acid, arginine, and histidine residues.

It is well established that glutamate acts as a flavor enhancer giving umami taste (Kato et al. 1989). In Iberian ham, sodium glutamate together with small, bitter peptides are thought to be the main nonvolatile contributors to the strong characteristic flavor of this product (Ruiz et al. 2002). In other dry-cured hams, the bitter taste was related to high amounts of lipophilic amino acids and lipophilic oligopeptides (Sforza et al. 2001) and to elevated levels of methionine, asparagine, and isoleucine (Virgili et al. 1998). Based on peptide composition, it was predicted that various dipeptides could contribute to sour, bitter, and umami taste of Serrano ham (Sentandreu et al. 2003). However, in the same study it was also stressed that the free amino acids were more important to dry-cured ham flavor as compared to peptides.

HOW TO CHANGE FLAVOR

Dry-cured Ham

In principle, dry-cured ham flavor is determined by meat quality, processing conditions, curing ingredients, and adjuncts. The fatty acid composition and the ratio of various proteolytic, peptidolytic, and lipolytic enzymes in muscle and adipose tissue are some of the important measures of meat quality, and numerous studies report that breed (Berdagué et al. 1993a; Antequera et al. 1994; Gou et al. 1995; Cava et al. 2004) and feed (Manfredini et al. 1992; Isabel et al. 1999; Cava et al. 1999, 2000; Cava and Ventanas 1999; Timon et al. 2002; Virgili and Schivazappa 2002) influence those measures. The combination of the level of curing salts and the process parameters time, temperature, and humidity is decisive for the chemical, enzymatic, and microbiological processes leading to flavor formation in dry-cured ham. The effect of salt, temperature, etc., on enzymatic activity is extensively covered (Sarraga et al. 1989; Toldrá 1992, 2002; Motilva and Toldrá 1993; Toldrá et al. 1997; Toldrá and Flores 1998), and also the effect on chemical processes such as autoxidation is described in some detail (MacDonald et al. 1980; Coutron-Gambotti et al. 1999; Andres et al. 2004). However, microbial activity in dry-cured ham as affected by processing parameters is not well studied. Although meat starter cultures are applied in the dry-cured ham processing industry, the effect of such starter cultures on flavor formation is scarcely researched.

In summary, the desired flavor of dry-cured ham is obtained by choosing the right meat quality (feed, breed, age, sex), the right processing conditions, and curing ingredients to control the enzymatic and microbial activity, and finally to choose the right (if any) adjunct culture.

Fermented Sausage

A fundamental difference between fermented sausages and dry-cured ham is that sausage processing allows the distribution of ingredients, including starter cultures, enzymes, flavor precursors, and cofactors, inside the product and not only on the surface. Consequently, the microflora of fermented sausages has much better opportunities to influence flavor formation compared to the microflora of dry-cured ham. Fermented sausage flavor is probably affected by meat quality in a similar way as described for dry-cured ham. Likewise, the processing conditions (temperature, time, relative humidity) influence the endogenous meat enzyme activities and chemical reactions by the same mechanisms as in hams. However, the processing conditions do, as other environmental factors, also influence microbial activity and thus microbial flavor formation. Before changing process parameters one should therefore consider the effect on chemical, enzymatic, as well as microbiological processes.

The impact of growth parameters on the activity of lactic acid bacteria is described in Chapter 18, "Influence of Processing Parameters on Cultures Performance," and will not be treated further here. However, several studies focus on the aroma formation by staphylococci as affected by ripening time (Mateo and Zumalacárregui 1996; Sunesen et al. 2001; Ansorena et al. 2000; Misharina et al. 2001; Tjener et al. 2003, 2004b; Olesen et al. 2004) and various environmental factors, such as temperature (Stahnke 1995b, 1999b; Tjener et al. 2004c), salt (Stahnke 1995b; Olesen et al. 2004; Tjener et al. 2004c), nitrite, nitrate, and ascorbate (Stahnke 1995b, 1999b; Olesen et al. 2004), and pH (Stahnke 1995b; Tjener et al. 2004c).

The change of growth parameters and ripening time, as described above, is one way to change flavor formation. Another strategy is to change microbial composition within the sausage, e.g., from *S. carnosus* to *S. xylosus*. Generally, *S. carnosus* produces more of the methyl-branched aldehydes and acids than *S. xylosus*, *S. warneri*, *S. saprophyticus*, *S. equorum*, and *K. varians* (Berdagué et al. 1993b; Montel et al. 1996; Stahnke 1999a; Søndergaard and Stahnke 2002; Olesen et al. 2004; Tjener et al. 2004c), whereas *S. xylosus* normally produces more diacetyl than *S. carnosus* (Berdagué et al. 1993b; Montel et al. 1996; Stahnke 1999a; Søndergaard and Stahnke 2002; Olesen et al. 2004; Tjener et al. 2004c).

The use of yeast, streptomyces, or molds is yet another approach reported to influence sausage flavor (Eilberg and Liepe 1977; Bruna et al. 2000b, 2003; Olesen and Stahnke 2000; Lopez et al. 2001; Flores et al. 2004; Hierro et al. 2005).

Another strategy is to change the inoculation level of, e.g., staphylococci. In one study, low inoculum of *S. carnosus* resulted in a low level of, e.g., methyl-branched aldehydes, whereas high inoculum resulted in a high level of methyl-branched aldehydes. The effect was more pronounced in the beginning of the ripening phase and was almost leveled out after 3 weeks of ripening. Thus, the strategy seems most valuable for fast-ripened sausages (Tjener et al. 2004b).

Often staphylococci grow very little or not at all during sausage processing (Lücke 1998). The initial state of a culture, and thereby also the preinoculation conditions, may therefore be very important for the following activity of this culture. Møller et al. (1998) showed that agitation during precultivation increased 3-methyl butanol and 3-methyl butanal formation by *S. xylosus* in minimal medium, and Masson et al. (1999) demonstrated an increase in 3-methyl butanal and 3-methyl butanoic acid production by *S. carnosus*, when grown in complex compared to defined medium prior to inoculation in reaction medium with leucine. Olesen and Stahnke (2003) cultivated *S. carnosus* and *S. xylosus* under various conditions and investigated how this precultivation affected the staphylococci's ability to degrade methyl-branched amino acids in a reaction medium. The authors studied the effect of growth phase, temperature, NaCl concentration, and concentration of leucine, isoleucine and valine and found that all parameters had a significant influence on the production of methyl-branched acids. Optimal production of starter cultures, such as the staphylococci, may thus lead to enhanced flavor formation.

Finally, it is possible to change the flavor profile by adding substrates or cofactors of specific pathways. Some examples are the addition of free amino acids (Herranz et al. 2005) and α-ketoglutarate (Herranz et al. 2003; Tjener et al. 2004a). The addition of proteolytic or lipolytic enzymes with the objective of generating flavor precursors is in this context the same as adding the precursors directly. The addition of enzymes with the purpose of accelerating flavor formation has been extensively studied, and most of the work was reviewed by Fernandez et al. (2000). The authors concluded that the level of flavor precursors by addition of proteolytic or lipolytic enzymes did not necessarily lead to increased flavor formation unless the formation of flavor compounds from their precursors is accelerated as well.

In summary, fermented sausage flavor depends on raw materials, processing parameters, type and amount of starter culture, the conditions at which the starter culture is produced, the use of adjunct cultures as yeast, mold and streptomyces, and addition of substrates or cofactors of specific pathways.

PERSPECTIVES

As shown in the previous sections, considerable knowledge on the identity and the formation of flavor compounds in fermented and dry-cured meat products has been gathered. However, a number of topics still remain to be solved in order to completely understand and control the flavor forming processes. Further research should be focused on two subjects primarily. First of all, after many years of research it is still impossible to make a synthetic dry-cured ham or fermented sausage flavor by mixing the pure components, since *all* of the important components (volatiles as well as nonvolatiles) and the concentrations at which they occur are still not known. Certainly more research is needed within this area in order to know exactly which compounds to focus on. Second, in order to actively manipulate the concentrations of relevant flavor compounds, in this way controlling the flavor development, it is crucial to know their origin and pathway of formation. For several of the already known compounds, the origin is speculated but still not confirmed. Knowing exactly which pathways to target is an excellent starting point for efficient development of new starter cultures or other ingredients with improved flavor-forming capabilities.

REFERENCES

AI Andres, R Cava, J Ventanas, E Muriel, J Ruiz. 2004. Lipid oxidative changes throughout the ripening of dry-cured Iberian hams with different salt contents and processing conditions. Food Chem 84:375–381.

D Ansorena, I Astiasarán, J Bello. 2000. Changes in volatile compounds during ripening of Chorizo de Pamplona elaborated with *Lactobacillus plantarum* and *Staphylococcus carnosus*. Food Sci Technol Int 6:439–447.

T Antequera, C Garcia, C Lopez, J Ventanas, MA Asensio, JJ Cordoba. 1994. Evolution of different physico-chemical parameters during ripening of Iberian ham from Iberian (100%) and Iberian x Duroc pigs (50%). Revista Espan Ciencia Tecnol Alim 34:178–190.

HC Beck, AM Hansen, FR Lauritsen. 2002. Metabolite production and kinetics of branched-chain aldehyde oxidation in *Staphylococcus xylosus*. Enzyme Microb Tech 31:94–101.

JL Berdagué, N Bonnaud, S Rousset, C Touraille. 1993a. Influence of pig crossbreed on the composition, volatile compound content and flavour of dry cured ham. Meat Sci 34:119–129.

JL Berdagué, P Monteil, MC Montel, R Talon. 1993b. Effects of starter cultures on the formation of flavor compounds in dry sausage. Meat Sci 35:275–287.

RG Berger, C Macku, JC German, T Shibamoto. 1990. Isolation and identification of dry salami volatiles. J Food Sci 55:1239–1242.

I Blank, S Devaud, LB Fay, C Cerny, M Steiner. 2001. Odor-active compounds of dry-cured meat: Italian-type salami and Parma ham. In: GR Takeoka, M Guentert, KH Engel, eds. Aroma Active Compounds in Foods: Chemistry and Sensory Properties. Washington: American Chemical Society, pp. 9–20.

JM Bruna, M Fernandez, EM Hierro, JA Ordonez, L de la Hoz. 2000a. Combined use of Pronase E and a fungal extract (*Penicillium aurantiogriseum*) to potentiate the sensory characteristics of dry fermented sausages. Meat Sci 54:135–145.

———. 2000b. Improvement of the sensory properties of dry fermented sausages by the superficial inoculation and/or the addition of intracellular extracts of Mucor racemosus. J Food Sci 65:731–738.

JM Bruna, EM Hierro, L de la Hoz, DS Mottram, M Fernandez, JA Ordonez. 2001. The contribution of *Penicillium auratiogriseum* to the volatile composition and sensory quality of dry fermented sausages. Meat Sci 59:97–107.

———. 2003. Changes in selected biochemical and sensory parameters as affected by superficial inoculation of *Penicillium camemberti* on dry fermented sausages. Int J Food Microbiol 85:111–125.

S Buscailhon, JL Berdagué, J Bousset, M Cornet, G Gandemer, C Touraille, G Monin. 1994. Relations between compositional traits and sensory qualities of French dry-cured ham. Meat Sci 37:229–243.

M Careri, A Mangia, G Barbieri, L Bolzoni, R Virgili, G Parolari. 1993. Sensory property relationships to chemical data of Italian-type dry-cured ham. J Food Sci 58:968–972.

AI Carrapiso, C Garcia. 2004. Iberian ham headspace: Odorants of intermuscular fat and differences with lean. J Sci Food Agric 84:2047–2051.

AI Carrapiso, Á Jurado, ML Timón, C Garcia. 2002a. Odor-active compounds of Iberian hams with different aroma characteristics. J Agric Food Chem 50: 6453–6458.

AI Carrapiso, J Ventanas, C Garcia. 2002b. Characterization of the most odor-active compounds of Iberian ham headspace. J Agric Food Chem 50:1996–2000.

R Cava, J Manuel Ferrer, M Estevez, D Morcuende, F Toldrá. 2004. Composition and proteolytic and lipolytic enzyme activities in muscle Longissimus dorsi from Iberian pigs and industrial genotype pigs. Food Chem 88:25–33.

R Cava, J Ruiz, J Ventanas, T Antequera. 1999. Effect of alpha-tocopheryl acetate supplementation and the extensive feeding of pigs on the volatile aldehydes during the maturation of Iberian ham. Food Sci Technol Int 235–241.

R Cava, J Ventanas. 1999. Oxidative and lipolytic changes during ripening of Iberian hams as affected by feeding regime: Extensive feeding and alpha-tocopheryl acetate supplementation. Meat Sci 52: 165–172.

R Cava, J Ventanas, J Florencio Tejeda, J Ruiz, T Antequera. 2000. Effect of free-range rearing and alpha-tocopherol and copper supplementation on fatty acid profiles and susceptibility to lipid oxidation of fresh meat from Iberian pigs. Food Chem 68:51–59.

C Coutron-Gambotti, G Gandemer, S Rousset, O Maestrini, F Casabianca. 1999. Reducing salt content of dry-cured ham: Effect on lipid composition and sensory attributes. Food Chem 64:13–19.

RH Dainty, H Blom. 1995. Flavour chemistry of fermented sausages. In: G Campbell-Platt, PE Cook, eds. Fermented Meats. Glasgow: Blackie Academic and Professional, pp. 176–193.

S Dellaglio, E Casiraghi, C Pompei. 1996. Chemical, physical and sensory attributes for the characterization of an Italian dry-cured sausage. Meat Sci 42:25–35.

D Demeyer, M Raemaekers, A Rizzo, A Holck, AD Smedt, BT Brink, B Hagen, C Montel, E Zanardi, E Murbrekk, F Leroy, F Vandendriessche, K Lorentsen, K Venema, L Sunesen, L Stahnke, LD Vuyst, R Talon, R Chizzolini, S Eerola. 2000. Control of bioflavour and safety in fermented sausages: First results of a European project. Food Res Int 33:171–180.

N Dierick, P Vandekerckhove, D Demeyer. 1974. Changes in nonprotein nitrogen compounds during dry sausage ripening. J Food Sci 39:301–304.

BL Eilberg, HU Liepe. 1977. Improvements in dry sausage technology by use of *Streptomyces* starter culture. Fleischwirtsch 57:1678–1683.

G Engelvin, G Feron, C Perrin, D Mollé, R Talon. 2000. Identification of β-oxidation and thioesterase activities in *Staphylococcus carnosus* 833 strain. FEMS Microbiol Lett 190:115–120.

M Fernandez, JA Ordonez, JM Bruna, B Herranz, L de la Hoz. 2000. Accelerated ripening of dry fermented sausages. Trends Food Sci Technol 11:201–209.

M Flores, MA Dura, A Marco, F Toldrá. 2004. Effect of *Debaryomyces* spp. on aroma formation and sensory quality of dry-fermented sausages. Meat Sci 68: 439–446.

M Flores, CC Grimm, F Toldrá, AM Spanier. 1997. Correlations of sensory and volatile compounds of Spanish "Serrano" dry-cured ham as a function of two processing times. J Agric Food Chem 45:2178–2186.

EN Frankel. 1980. Lipid oxidation. Prog Lipid Res 19:1–22.

G Gottschalk. 1986. Bacterial Fermentations. Bacterial Metabolism. New York: Springer-Verlag, pp. 208–282.

P Gou, L Guerrero, J Arnau. 1995. Sex and crossbreed effects on the characteristics of dry-cured ham. Meat Sci 40:21–31.

BF Hagen, JL Berdagué, AL Holck, H Næs, H Blom. 1996. Bacterial proteinase reduces maturation time of dry fermented sausages. J Food Sci 61:1024–1029.

AP Henriksen, LH Stahnke. 1997. Sensory and chromatographic evaluations of water soluble fractions from air-dried sausages. J Agric Food Chem 45:2679–2684.

B Herranz, L de la Hoz, E Hierro, M Fernandez, JA Ordonez. 2005. Improvement of the sensory properties of dry-fermented sausages by the addition of free amino acids. Food Chem 91:673–682.

B Herranz, M Fernandez, E Hierro, JM Bruna, JA Ordonez. 2003. Use of *Lactococcus lactis* subsp. *cremoris* NCDO 763 and alfa-ketoglutarate to improve the sensory quality of dry fermented sausages. Meat Sci 66:151–163.

E Hierro, JA Ordonez, JM Bruna, C Pin, M Fernandez, L de la Hoz. 2005. Volatile compound generation in dry fermented sausages by the surface inoculation of selected mould species. Eur Food Res Technol 220:494–501.

L Hinrichsen, SB Pedersen. 1995. Relationship among flavour, volatile compounds, chemical changes, and microflora in Italian-type dry-cured ham during processing. J Agric Food Chem 43:2932–2940.

CM Hollenbeck. 1994. Contribution of smoke flavourings to processed meats. In: F Shahidi, ed. Flavour of Meat and Meat Products. London: Blackie Academic and Professional, pp. 199–209.

B Isabel, M Timon, R Cava, C Garcia, J Ruiz, JM Carmona, M Soares, CJ Lopez-Bote. 1999. Dietary alpha-tocopheryl acetate supplementation modifies volatile aldehyde and sensory properties of dry-cured hams. Ir J Agric Food Res 38:137–142.

ISO 5492:1992. Sensory analysis—Vocabulary. ISO 5492 First ed.

G Johansson, JL Berdagué, M Larsson, N Tran, E Borch. 1994. Lipolysis, proteolysis and formation of volatile components during ripening of a fermented sausage with *Pediococcus pentosaceus* and *Staphylococcus xylosus* as starter cultures. Meat Sci 38:203–218.

H Kato, MR Rhue, T Nishimura. 1989. Role of free amino acids and peptides in food taste. In: R Teranishi, RG Buttery, F Shahidi, eds. Flavor Chemistry: Trends and Developments. Washington, D.C.: ACS, pp. 158–174.

H Kohata, M Numata, M Kawaguchi, T Nakamura, N Arakawa. 1992. The effect of salt composition on taste development in prosciutto. Proceedings of the 38th ICoMST, Clermont-Ferrand, France, 6, 1271–1274.

D Ladikos, V Lougovois. 1990. Lipid oxidation in muscle foods: A review. Food Chem 35:295–314.

R Lawrie. 1995. The structure, composition and preservation of meat. In: G Campbell-Platt, PE Cook, eds. Fermented Meats. Glasgow: Blackie Academic and Professional, pp. 1–38.

TM Lopez, CJ Gonzalez, R Rodriguez, JP Encinas, A Otero. 2001. Use of yeasts and moulds in fermented sausages. Alimentaria 321:25–31.

JD Love, AM Pearson. 1971. Lipid oxidation in meat and meat products—A review. J Am Oil Chem Soc 48:547–549.

FK Lücke. 1998. Fermented sausages. In: BJB Wood, ed. Microbiology of Fermented Foods, vol 2. 2nd ed. London: Blackie Academic & Professional, pp. 441–483.

B MacDonald, JI Gray, Y Kakuda, ML Lee. 1980. Role of nitrite in cured meat flavor: Chemical analysis. J Food Sci 45:889–892.

SM Madsen, HC Beck, P Ravn, A Vrang, AM Hansen, H Israelsen. 2002. Cloning and inactivation of a branched-chain-amino-acid aminotransferase gene from *Staphylococcus carnosus* and characterization of the enzyme. Appl Environ Microbiol 68: 4007–4014.

M Manfredini, A Badiani, N Nanni. 1992. Relationship between dietary fat and quality characteristics of fat and Parma-type ham in heavy pigs. Ital J Food Sci 25–32.

F Masson, L Hinrichsen, R Talon, MC Montel. 1999. Factors influencing leucine catabolism by a strain of *Staphylococcus carnosus*. Int J Food Microbiol 49:173–178.

J Mateo, MC Dominguez, MM Aguirrezábal, JM Zumalácarregui. 1996. Taste compounds in Chorizo and their changes during ripening. Meat Sci 44:245–254.

J Mateo, JM Zumalacárregui. 1996. Volatile compounds in Chorizo and their changes during ripening. Meat Sci 44:255–273.

A Meynier, E Novelli, R Chizzolini, E Zanardi, G Gandemer. 1999. Volatile compounds of commercial Milano salami. Meat Sci 51:175–183.

TA Misharina, VA Andreenkov, EA Vaschuk. 2001. Changes in the composition of volatile compounds during aging of dry-cured sausages. Appl Biochem Microbiol 37:413–418.

JS Møller, L Hinrichsen, HJ Andersen. 1998. Formation of amino acid (L-leucine, L-phenylalanine) derived volatile flavour compounds by *Moraxella phenylpyruvica* and *Staphylococcus xylosus* in cured meat model systems. Int J Food Microbiol 42:101–117.

MC Montel, F Masson, R Talon. 1998. Bacterial role in flavour development. Meat Sci 49:S111–S123.

MC Montel, J Reitz, R Talon, JL Berdagué, AS Rousset. 1996. Biochemical activities of Micrococcaceae and their effects on the aromatic profiles and odours of a dry sausage model. Food Microbiol 13:489–499.

MJ Motilva, F Toldrá. 1993. Effect of curing agents and water activity on pork muscle and adipose subcutaneous tissue lipolytic activity. Z Lebensm Unters Forsch 196:228–232.

PT Olesen, AS Meyer, LH Stahnke. 2004. Generation of flavour compounds in fermented sausages—The influence of curing ingredients, *Staphylococcus* starter culture and ripening time. Meat Sci 66: 675–687.

PT Olesen, LH Stahnke. 2000. The influence of *Debaryomyces hansenii* and *Candida utilis* on the aroma formation in garlic spiced fermented sausages and model minces. Meat Sci 56:357–368.

———. 2003. The influence of precultivation parameters on the catabolism of branched-chain amino acids by *Staphylococcus xylosus* and *Staphylococcus carnosus*. Food Microbiol 20:621–629.

JA Ordonez, EM Hierro, JM Bruna, L de la Hoz. 1999. Changes in the components of dry-fermented sausages during ripening. Crit Rev Food Sci Nutr 39:329–367.

KH Plattig. 1984. The sense of taste. In: JR Piggott, ed. Sensory Analysis of Foods. London: Elsevier Applied Science Publishers, pp. 1–22.

M Rothe. 1972. Zur terminologie der grundbegriffe der aromaforschung. Nahrung 16:473–481.

J Ruiz, E Muriel, J Ventanas. 2002. The flavour of Iberian ham. In: F Toldrá et al., ed. Research Advances in the Quality of Meat and Meat Products. Trivandrum, India: Research Signpost, pp. 289–309.

J Ruiz, J Ventanas, R Cava, A Andrés, C Garcia. 1999. Volatile compounds of dry-cured Iberian ham as affected by the length of the curing process. Meat Sci 52:19–27.

M Rychlik, P Schieberle, W Grosch. 1998. Compilation of Odor Thresholds, Odor Qualities and Retention Indicies of Key Food Odorants. Garching, Germany: Deutsche Forschungsanstalt für Lebensmittelchemie and Institut für Lebensmittelchemie der Technischen Universität München.

Y Sanz, MA Sentandreu, F Toldrá. 2002. Role of muscle and bacterial exo-peptidases in meat fermentation. In: F Toldrá, ed. Research advances in the Quality of Meat and Meat Products. Kerala, India: Research Signpost, pp. 143–155.

C Sarraga, M Gil, J Arnau, JM Monfort. 1989. Effect of curing salt and phosphate on the activity of porcine muscle proteases. Meat Sci 25:241–249.

S Schmidt, RG Berger. 1998. Aroma compounds in fermented sausages of different origins. Lebensm—Wiss Technol 31:559–567.

MA Sentandreu, S Stoeva, MC Aristoy, K Laib, W Voelter, F Toldrá. 2003. Identification of small peptides generated in Spanish dry-cured ham. J Food Sci 68:64–69.

S Sforza, A Pigazzani, M Motti, C Porta, R Virgili, G Galaverna, A Dossena, R Marchelli. 2001. Oligopeptides and free amino acids in Parma ham of known cathepsin B activity. Food Chem 75:267–273.

AK Søndergaard, LH Stahnke. 2002. Growth and aroma production by *Staphylococcus xylosus*, *S. carnosus* and *S. equorum*—A comparative study in model systems. Int J Food Microbiol 75:99–109.

LH Stahnke. 1994. Aroma components from dried sausages fermented with *Staphylococcus xylosus*. Meat Sci 38:39–53.

———. 1995a. Dried sausages fermented with *Staphylococcus xylosus* at different temperatures and with different ingredient levels. Part III. Sensory Evaluation. Meat Sci 41:211–223.

———. 1995b. Dried sausages fermented with *Staphylococcus xylosus* at different temperatures and with different ingredient levels. Part II. Volatile components. Meat Sci 41:193–209.

———. 1998. Character impact aroma compounds in fermented sausages. Proceedings 44th ICoMST, Barcelona, Spain, 2, 786–787.

———. 1999a. Volatiles produced by *Staphylococcus xylosus* and *Staphylococcus carnosus* during growth in sausage minces. Part I. Collection and identification. Lebensm—Wiss Technol 32:357–364.

———. 1999b. Volatiles produced by *Staphylococcus xylosus* and *Staphylococcus carnosus* during growth in sausage minces. Part II. The influence of growth parameters. Lebensm—Wiss Technol 32:365-371.

———. 2000. 2-Acetyl-1-pyrroline—Key aroma compound in Mediterranean dried sausages. In: P Schieberle, K-H Engel, eds. Frontiers of Flavour Science. Garching, Germany: Deutsche Forschungsanstalt für Lebensmittelchemie, pp. 361–365.

———. 2002. Flavour formation in fermented sausage. In: F Toldrá, ed. Research Advances in the Quality of Meat and Meat Products. India: Research Signpost, pp. 193–223.

LH Stahnke, A Holck, A Jensen, A Nilsen, E Zanardi. 2002. Maturity acceleration of Italian dried sausage by *Staphylococcus carnosus*—Relationship between maturity and flavor compounds. J Food Sci 67: 1914–1921.

LH Stahnke, LO Sunesen, AD Smedt. 1999. Sensory characteristics of European, dried, fermented sausages and the correlation to volatile profile. Proceedings of the 13th Forum for Applied Biotechnology, pp. 559–566.

LO Sunesen, V Dorigoni, E Zanardi, LH Stahnke. 2001. Volatile compounds released during ripening in Italian dried sausage. Meat Sci 58:93–97.

M Tikk, K Tikk, MA Tørngren, MD Aaslyng, AH Karlsson, HJ Andersen. 2005. The development of inosine monophosphate and its degradation products during aging of pork of different qualities in relation to basic taste—And retro nasal flavor perception of the meat. Proceedings of 51st ICoMST, Baltimore, USA:97.

ML Timon, L Martin, MJ Petron, A Jurado, C Garcia. 2002. Composition of subcutaneous fat from dry-cured Iberian hams as influenced by pig feeding. J Sci Food Agric 82:186–191.

K Tjener, LH Stahnke, L Andersen, J Martinussen. 2003. A fermented meat model system for studies of microbial aroma formation. Meat Sci 66:211–218.

———. 2004a. Addition of α-ketoglutarate enhances formation of volatiles by *Staphylococcus carnosus* during sausage fermentation. Meat Sci 67:711–719.

———. 2004b. Growth and production of volatiles by *Staphylococcus carnosus* in dry sausages: Influence of inoculation level and ripening time. Meat Sci 67:447–452.

———. 2004c. The pH-unrelated influence of salt, temperature and manganese on aroma formation by *Staphylococcus xylosus* and *Staphylococcus carnosus* in a fermented meat model system. Int J Food Microbiol 97:31–42.

F Toldrá. 1992. The enzymology of dry-curing of meat products. In: FJM Smulders, F Toldrá, J Flores, M Prieto, eds. New technologies for meat and meat products. Nijmegen, The Netherlands: Audet, pp. 209–231.

———. 2002a. Manufacturing of dry-cured ham. In: F Toldrá, ed. Dry-Cured Meat Products. Trumbull, Connecticut: Food & Nutrition Press, Inc., pp. 27–62.

———. 2002b. Effect of raw materials and processing quality. In: F Toldrá, ed. Dry-Cured Meat Products. Trumbull, Connecticut: Food & Nutrition Press, Inc., pp. 189–210.

F Toldrá, M Flores. 1998. The role of muscle proteases and lipases in flavor development during the processing of dry-cured ham. CRC Crit Rev Food Sci Nutr 38:331–352.

F Toldrá, M Flores, Y Sanz. 1997. Dry-cured ham flavor: Enzymatic generation and process influence. Food Chem 4:523–530.

P Vandekerckhove, D Demeyer. 1975. The composition of Belgian dry sausage (salami). Fleischwirtsch 55:680–682.

C Viallon, JL Berdagué, MC Montel, R Talon, JF Martin, N Kondjoyan, C Denoyer. 1996. The effect of stage of ripening and packaging on volatile content and flavor of dry sausages. Food Res Int 29: 667–674.

R Virgili, C Schivazappa. 2002. Muscle traits for long matured dried meats. Meat Sci 62:331–343.

R Virgili, C Schivazappa, G Parolari, CS Bordini, M Degni. 1998. Proteases in fresh pork muscle and their influence on bitter taste formation in dry-cured ham. J Food Biochem 22:53–63.

S Westall. 1998. Characterization of yeast species by their production of volatile metabolites. J Food Mycol 1:187–201.

F Wirth. 1989. Reducing the common salt content of meat products. Possible methods and their limitations. Fleischwirtsch 69:589–593.

M Yvon, L Rijnen. 2001. Cheese flavour formation by amino acid catabolism. Int Dairy J 11:185–201.

Part V

Product Categories: General Considerations

23
Composition and Nutrition

Daniel Demeyer

INTRODUCTION: FOODS FOR NUTRIENT SUPPLY AND/OR HEALTH

The emphasis in any discussion of food composition in relation to nutrient content differs considerably with the overall level of food supply. This is especially true when meat and meat products are discussed in the environment of either developing or industrialized countries. The latter remain characterized by an excessive average per capita supply of food energy, mainly as meat and milk, amounting to 127%, 346%, and 476%, respectively, of the corresponding low values in the former (WHO 2003) (Table 23.1).

Nutritional problems in developing countries therefore continue to relate to the coping of nutrient supply with food shortages and malnutrition, whereas in industrialized countries they mainly reflect the effects of overnutrition. These effects relate to a number of complex societal disturbances, including a convincing and causal relationship between the excessive consumption of energy-dense, micronutrient-poor foods and the incidence of the so-called noncommunicable chronic diseases. The latter involve obesity and the associated risks of type 2 diabetes, cardiovascular diseases (CVD), and cancer as well as dental disease and osteoporosis. Besides sugars and starch, the amounts and types of fatty acids and cholesterol present in animal products are seen as the main culprits in this respect, whereas salt is seen as a main causative factor for CVD (WHO 2003).

For these reasons, the nutrient content of meat is of primary importance in relation to nutrient supply in developing countries, whereas the lack of conformity of meat composition with the goals recently set for healthy nutrition (Table 23.2) (WHO 2003) receive most attention in industrialized countries.

THE NUTRIENT SUPPLY FROM MEAT AND MEAT PRODUCTS

The term *meat* covers all edible tissue of slaughter animals. The live weight of pork and beef, the major species involved in industrial production of meat products, contains on average about 85% of such tissues, involving about 35% red muscle and between 10% (beef) and 20% (pork) adipose tissue (Table 23.3). Pork adipose tissue is almost exclusively used in the manufacture of comminuted meat products, together with predominantly pork muscle tissue, but also involving beef muscle, depending on the type of product.

Protein and Fat

The largest amounts of macronutrients supplied by meat and meat products are by far the protein and fat of animal muscle (19% and 2.5%, respectively) and adipose tissue (90% fat). Actual protein and fat contents, as well as the corresponding amino and fatty acid composition, vary considerably depending on several endogenous (e.g., anatomic location, breeds, sex, species) and exogenous (mainly feeding regime) factors.

Table 23.1. Vegetable and animal sources of dietary energy 1997–1999.

	Industrialized Countries	Developing Countries
kcal/c/d		
Total	3380	2681
Vegetable	2437	2344
Animal	943	337
kg/c/y		
Meat	88.2	25.5
Milk	212.2	44.6

Source: WHO (2003).
kcal/c/d = kcal per capita per day.
kg/c/y = kg per capita per year.

Table 23.2. Ranges of population nutrient intake goals.

Dietary Factor	Goal[a]
Total fat	15–30%
Saturated fatty acids	<10%
Polyunsaturated fatty acids (PUFAs)	6–10%
n-6 PUFAs	5–8%
n-3 PUFAs	1–2%
Trans-fatty acids	<1%
Total carbohydrate	55–75%
Free sugars (Mono- and disaccharides)	<10%
Protein	10–15%
Cholesterol	<300 mg/d
Sodium chloride (sodium)	<2 g/d
Fruits and vegetables	<400 g/d
Dietary fiber and nonstarch polysaccharides (NSP)	From foods[b]

Source: WHO (2003).
[a] % = % of total dietary energy.
[b] >25 and >20 g per day provided by fruits and vegetables.

Table 23.3. Estimated distribution of edible parts of meat animals (% of live weight).

	Beef	Pig
Red muscle tissue	33	44
Removable adipose tissue	11	24
Fat		21
Rind		3[a]
Blood	6	4
Organs	5	4
Stomach and intestines	5	4
Skin	10	–
Gelatin—derived from bones[b]	3	3
All	73	83

Source: from Demeyer (1981).
[a] About 35% in back fat.
[b] 25% in skeleton.

Protein

Muscle protein quality is determined by its content of essential amino acids, a value inversely related to its connective tissue content. Collagen, the major connective tissue protein, is almost completely devoid of methionine and tryptophane, and its proportion in muscle protein may reach values up to about 24% in low-value meat cuts such as those derived from the neck, the head, and the lower limbs (Bender and Zia 1976). The collagen content of the skin covering subcutaneous fat (rind) can be estimated at about 50% (Wood 1990) and, taking into account potential gelatin production from the skeleton, it can be estimated from the data in Table 23.2 that close to 45% of total pork protein consists of collagen. Together with low value meat cuts, rind is used to a variable extent in the production of comminuted meat products. Lee et al. (1978) showed that 95% of the variation in PER (protein efficiency ratio) of meat products is explained by their collagen content, a finding confirmed by Pelczynska and Libelt (1999). In Demeyer (1981), the author could relate the essential amino acid content of meat products to the collagen content (%) in crude protein following $y = -0.28$; $x + 40.06$ ($n = 54$; $R^2 = 0.90$) with y = molar % of [ileu+leu+lys+phe(+tyr)+met(+cys)+thr+val] and x = 100 (8.OH-proline)/crude protein. In view of the high level of protein consumption in industrialized countries, protein quality was already pronounced as being "of no importance whatsoever" by Bender and Zia (1976). According to Lee et al. (1978), an acceptable upper limit for protein quality corresponded to about 30% of collagen.

Fat

It is well known that in contrast to beef adipose tissue, the PUFA content of pork backfat can easily be

Table 23.4. Range of polyunsaturated fatty acid levels in meat (% of total fatty acids).

		Intramuscular Fat[b]	
	Pork Backfat[a]	Beef	Pork
n-6 PUFAs			
Linoleic acid	12–20	12–14	1.5–18
Long chain fatty acids	0.5–1.5	1–3	1–4
n-3 PUFAs		1–3	0.5–4
Linolenic acid	2–11	1–2	<1
Long chain fatty acids	–	<1	<1
cis9trans11 CLA		4–8	1–6
n-6/n-3	1–10		
Total fat (%)	70–82[c]	1–10[bc]	1–6

[a] From Warnants et al. (1998).
[b] From Raes et al. (2004).
[c] From Wood (1990).

increased by dietary means (see, e.g., Warnants et al. 1998) (Table 23.4). As summarized by Raes et al. (2004) and De Smet et al. (2004), the intramuscular fat content may vary between about 1% (high value cuts of very lean pork and beef breeds) and 6% and such variation is also affecting fatty acid composition. The contents of saturated (SFA) and monounsaturated (MUFA) fatty acids increase faster with increasing fatness than does the content of polyunsaturated (PUFA) fatty acids, resulting in a decrease in the relative proportion of PUFA. In beef, fat level and fat composition are to a large extent determined by breed, whereas in pork, nutrition will have a larger impact. The fat level also influences the n-6/n-3 PUFA ratio, due to the difference of this ratio in polar and neutral lipids. However, these effects are much smaller than the effects that can be achieved by dietary means. Although linseed or linseed oil inclusion in animal diet supplies α-linolenic acid (LNA), the conversion of LNA to its longer chain metabolite eicosapentanoic acid (EPA) seems to be limited, resulting in only a small increase in their intramuscular fat. For both ruminants and monogastrics, the only effective way to increase docosahexanoic acid (DHA) levels in intramuscular fat seems to be the feeding of fish oil or fish meal. A desirable increase of c9t11CLA (conjugated linoleic acid) in lamb or beef meat can be achieved to some extent by feeding strategies affecting rumen metabolism. In pigs, however, CLA incorporation is mainly achieved by the inclusion of CLA-rich oils in the diet. Table 23.4 illustrates the type of variability observed for PUFA levels in meat.

In comminuted meat products, protein contents may decrease to values as low as 10% and fat content may increase to levels as high as 40%, as illustrated in Food Composition tables (see, e.g., Anonymous 2005).

Within a total diet, the amounts of protein and fat supplied by meat and meat products are determined by the level of meat consumption within that diet. For years, average per capita consumption of meat has been estimated using carcass production data yielding values in the neighborhood of 90 kg per capita per year (kg/c/y) for the affluent societies of Western Europe and North America. Such values, however, neglect organs but include bone. Correction for the latter lowers such values by about 20%. The removal of other nonedible parts before consumption can again be estimated at 20% (Honikel 1992; Chizzolini et al. 1999). Actual meat consumption can thus be estimated as $0.90 \times 0.8 \times 0.8 = 57.6$ kg/c/y or 158 g per capita per day (g/c/d). Such value is within the range of data collected for Western Europe in dietary survey type of studies, reporting mean value extremes of 85 (low fat consumers in Spain) and 184 (high fat consumers in Belgium) g/c/d for adult males (Williams et al. 1999). More recent data, derived from household budget surveys without correction for inedible parts and other losses (Naska et al. 2006) report an overall average of 191 g/c/d for total meat, covering 63 g/c/d of meat products. Such data are subject to very high variability, with variation coefficients often exceeding 100%. Nevertheless, they suggest that for years now meat and meat products contribute between 34 and 42% to dietary protein and between 13 and 25% to dietary fat supply, respectively (Hermus and Albers 1986; D'Amicis and Turrini 2002; Valsta et al. 2005). In the same period, the percentage contribution of fat to total energy intake has deviated little from 40%, in line with an estimated 10–20% contribution of meat and meat products to total energy intake (Chizzolini et al. 1999). High-fat comminuted meat products such as salami contribute more to fat consumption than fresh meat (D'Amicis and Turrini 2002), whereas the contribution of processed meats in total meat consumption is higher in high-fat than in low-fat consumers (De Henauw and De Backer 1999; Chizzolini et al. 1999).

Other Nutrients

Meat and meat products should not be considered "empty calories" because they are a more important source for several other nutrients than for fat. A selection of data calculated from food composition and food consumption surveys for The Netherlands and Italy, reported between 1986 and 2003 (Table 23.5), illustrates that meat and meat products cover about 20% and between 30 and 40% of dietary fat and protein intake, respectively. Corresponding values for several vitamins (niacin, pyridoxin, thiamine, and riboflavin) and trace elements (zinc, iron, and selenium) exceed 20%, indicating a high energetic density for these nutrients in meat and meat products (Leveille and Cloutier 1986).

It has been reported that within meat and meat products, pork and poultry are the main sources of niacin and selenium, respectively (Lombardi-Boccia et al. 2004), the former being very low in beef because of the specificity of ruminant metabolism. Animal diets are often supplemented with α-tocopherol to control oxidation and color stability of both pork and beef. The amounts clearly exceed that of its use as a food additive and improve the vitamin E status of the meat (McCarthy et al. 2001). As summarized earlier (Demeyer 1981), it has been recognized for years that heme iron is a much more available iron source than vegetable iron and that meat and meat products are major providers of vitamin B_{12}. On the other hand, the very low manganese levels in meat (<0.4 ppm) hamper meat fermentation (Hagen et al. 2000), whereas low values for calcium content (<100 ppm) are reflected in a Ca/P ratio <0.1 and contribute to calcium intakes below recommended levels (Lombardi-Boccia et al. 2003).

Introduction of even low amounts of meat in the diet improves development and/or prevention of anemia, as repeatedly observed in infants (Hallberg et al. 2003; Hadler et al. 2004; Taylor et al. 2004; Tympa-Psirropoulou et al. 2005; Krebs et al. 2006; Lachat et al. 2006). Improvement of available iron and zinc as well as protein supply has been proposed as mechanisms for this finding. The need for dietary heme iron is also suggested by the findings that a high proportion of women of child-bearing age are iron deficient (Hill 2002) and that iron intake of adolescent girls was considerably lower than recent recommendations in Belgium, a typical industrialized country (Pynaert et al. 2005). Recent work from Belgium estimated that beef and pork supplied about 20% of the dietary supply of long chain n-3 PUFA for adolescent boys and girls (De Henauw et al.

Table 23.5. Contribution (%) of meat and meat products to daily nutrient intake.

Reference[a]	(5)	(2)	(4)		(1)	(2)	(3)
Protein		27	34		34	27	42
Thiamine	21		34	Iron	38	20	26
Niacin	40	46		Zinc	24	40	38
Pyridoxin		39	26	Selenium	23	21	
Riboflavin	21		24				
Fat			23				<20
Calcium		2		Sodium		6	
Magnesium		5		Potassium		4	
Phosphorous		7		Copper		12	

[a] (1) Van Dokkum (1987); (2) Lombardi-Boccia et al. (2003); (3) D'Amicis and Turrini (2002); (4) Hermus and Albers (1986); (5) Lombardi-Boccia et al. (2002).

2006). The same authors stated that n-3 PUFA enrichment at the level of animal production would result in a substantial increase of that supply. Recent work from Australia also pointed out that meat and meat products account for more than 40% of long chain n-3 PUFA intake, when docosapentanoic acid (DPA) is included (Howe et al. 2006) and cover about 25–30% of CLA intake in Western populations (Schmid et al. 2006). As pointed out by Razanamahefa et al. (2005) however, more precise analytical data on fatty acid composition of meat and meat products are required to evaluate such findings.

MEAT AND MEAT PRODUCTS IN HEALTHY NUTRITION

Food consumption is clearly related to the rapidly increasing incidence of chronic noncommunicable disease, in both industrialized and developing countries (WHO 2003). Over the past 50 years this association has been related to a steadily increasing consumption of animal fat and, more specifically, to the increased consumption of saturated fat present in dairy products and beef. This has resulted in nutritional guidelines advocating a decrease in fat consumption and in various proposals to decrease fat content and alter fat composition in animal production (Demeyer and Doreau 1999; Givens 2005). However, although available statistics are based on production and trade rather than on actual consumption of foods, they clearly indicate a shift from animal fat to sugar and vegetable oil intake in most industrialized countries. The data also show a total energy intake well above the requirement explaining the failure of population reductions in obesity (Willett 2003). This is illustrated for Belgium in Table 23.6 and also highlights the limited importance of meat and meat products in energy supply.

It is therefore not surprising that in a recent approach, emphasis is shifted from fat intake to intake of "energy-dense foods" as a "convincing increasing risk" of obesity (WHO 2003), in line with the recent suggestion that fat intake has a less important role as a determinant of obesity (Willett 2003). Excessive energy intake is probably associated with excessive generation of ATP by mitochondrial oxidative phosphorylation known to result in free radical production. This then leads to oxidative damage to the lipid, protein, and nucleic acid building blocks of the cell membranes, enzymes, and genes, processes probably involved in the generation of chronic disease and aging (Weidruch 1996; Demeyer et al. 2004). Considerable attention in this respect is directed toward cholesterol. Although dietary goals recommend cholesterol intakes less than 300 mg/d, and meat and meat product intake may provide up to half of that amount, cholesterol intake and serum cholesterol levels are recently considered of less importance for the incidence of CVD (Chizzolini et al. 1999). Other factors relating to genetics and dietary antioxidant intake appear to be very important. The latter could be involved in the protection against oxidation of

Table 23.6. Evolution of food energy intake and distribution in Belgium[a].

	% of 1974 Values			
	1974	1984	1994	2000
Total (kcal/c/d)	3242 (<2300)[b]	106	110	111
Contribution (%) of				
Cereals	21.6	110	97	100
Sugars	11.0	94	133	130
Vegetable oils	10.9	102	127	138
Animal fats	15.4	92	79	75
Meat	8.5	104	97	87
Beef	2.8	78	61	57
Pork	4.5	113	100	84
Poultry	1.2	125	167	167

[a] *Source*: calculated from Anonymous (2006).

[b] ( ) = average energy requirement calculated from the population composition (Anonymous 2006a) and a physical activity level (PAL) = 1.70 (Anonymous 2005).

both dietary and endogenous cholesterol. Cholesterol oxides or oxysterols group about 70–80 compounds, and a number of those have been identified in meats and in atherosclerotic lesions. Their toxicity probably involves interference with the gene expression related to cell viability (Chizzolini et al. 1999).

Another shift in emphasis concerns the substitution of food based for nutrient-based nutritional guidelines. Based on a "probable increasing risk" of colon cancer incidence, "preserved meat (e.g., sausages, salami, bacon, ham)" is one of the rare food groups for which a moderate consumption is recommended (WHO 2003). Since 1975, meat consumption has indeed been associated with colorectal cancer reflected in recommendations for healthy eating from the American Cancer Association and the U.S. National Cancer Institute. These involved less frequent eating of red meat to be substituted with chicken or fish and moderation in the eating of salt-cured, smoked, and nitrite-cured foods (Kim and Mason 1996). Recent European work (Norat et al. 2005) has confirmed the association of red and processed meat consumption with colorectal cancer risk. The hazard ratio significantly increased from 0.96 to 1.35 when increasing average consumption from <20 g/d to >160 g/d. The association was stronger for processed than for unprocessed red meat, and risks were reduced by increased fiber intake. The mechanism underlying the association may relate to endogenous nitrosation of heme iron or colonic formation of toxic heme metabolites (Sesink et al. 1999) and/or to heterocyclic amines (HCAs) and/or polycyclic aromatic hydrocarbons (PAHs) produced during the preparation of meat or production of meat products. The first mechanism is considered to be more likely because chicken is a major contributor to HCA intake and no association was observed between poultry intake and colorectal cancer risk in the study (Norat et al. 2005).

Evidence has been presented for the multifactorial causes for colon cancer (Emmons et al. 2005a) as well as for genetic differences in its association with meat consumption (Chan et al. 2005; Luchtenborg et al. 2005; Brink et al. 2005). Recent reports also highlight the association of red meat and meat products both with the incidence of various diseases as well as with the dietary supply of functional food components. The former suggest association of red meat and/or meat products consumption with type 2 diabetes (Song et al. 2004; Fung et al. 2004), ovarian cancer (Kiani et al. 2006), prostate cancer (Wolk 2005), and degenerative arthritis (Hailu et al. 2006). The latter concerned compounds defined as improving health, beyond the supply of nutrients, having a clearly identified beneficial effect on a target function of the human body and/or lowering the risk of disease (Diplock et al. 1998). Functional food component supply from meat may cover creatine (Pegg et al. 2006) and carnosine (Park et al. 2005) as well as angiotensin converting enzyme inhibitors (Arihara et al. 1999).

As a final note, the role of meat products in sodium intake should be considered. Meat as such is relatively poor in sodium; meat products, however, contain about 2% added salt, values increasing up to 6% in dried products. It can be estimated that meat and meat products provide 20–30% of salt (NaCl) intake (Jiménez-Colmenero et al. 2001). The latter value ranges between 9–12 g/d (3.5 and 5 g/d Na) for industrialized countries, a value well above the dietary goal of <5 g/d (<2 g/d Na) (Table 23.2). The recent finding of a close relationship between consumption of cured meat products and the incidence of hypertension (Paik et al. 2005) highlights the importance of meat processing in sodium overconsumption.

RECOMMENDED MEAT INTAKES

General approaches to healthy nutrition also recommend reductions in meat consumption (Srinivasan et al. 2006; Emmons et al. 2005; Walker et al. 2005). From the arguments presented above and the general agreement on the place of (red) meat as an important food and a major supplier of essential nutrients, it is difficult to rationalize recommended daily meat intakes. In one discussion, Murphy and Allen (2003) recommend a daily consumption between 140 and 200 g of lean meat or meat substitutes for adults. It is evident that a distinction should be made between red (beef, sheep, and pork) and white (poultry) meat (Willett and Stampfer 2003), whereas most moderation should be associated with meat products. In a recent description of a "Mediterranean diet," a moderate meat consumption of a maximum of 100 g/d is proposed (Coene 2001), whereas according to Weisburger (2002), total protein intake can be as low as 10 g/d, equivalent to about 50 g of meat daily. According to the COMA (1998) report, health risks were present only for those consuming over 140 g of red meat daily, whereas Norat et al. (2002) derived the zero-risk level for red meat consumption in relation to colorectal cancer risk at 70 g per week. Bingham (1999) refers to the protective effects of bran, resistant starch, and vegetables, which may not apply, however, at high meat intakes (300–600 g/d).

THE EFFECTS OF FERMENTATION ON THE NUTRITIONAL AND HEALTH PROPERTIES OF MEAT

Both the nutrient content and health properties of fermented meat products (sausages) are affected by

- The mixing of raw materials and additives used.
- The metabolic and chemical changes of the sausage mix components during processing.

EFFECTS OF NATURE OF RAW MATERIALS AND ADDITIVES

The many types of fermented sausages have been narrowed down by industrial production and a major distinction can be made between those produced in shorter and longer ripening periods (Demeyer and Stahnke 2002). In both, pork and/or beef are mixed and comminuted (cuttered) with pork backfat and salt, processes allowing for the incorporation of large amounts of pork fat and low-value meat cuts, including rind. For most fermented sausage types, such practice is reflected in a high calorific value, a high fat content, a high salt content, and a limited nutritional quality of the sausage crude protein. A major action, characteristic for the manufacture of meat products in general, is the addition of 2–3% salt (NaCl) containing about 0.5% of $NaNO_2$. For the longer ripened products, nitrate rather than nitrite may be used. Table 23.7 illustrates the extreme variability of protein quality as well as of fat and salt content for European sausages. It should be noted that the amount of collagen in the sausage protein may reach 35% (Reuterswärd et al. 1985). German legislation differentiates sausages according to collagen content (Anonymous 2006b) (Table 23.7), and this affects price and protein quality but not digestibility as measured in humans (Reuterswärd et al. 1985).

Improvement of the health value of meat products has been tried for years, e.g., by replacement or lowering of fat using collagen (Arganosa et al. 1987) and of salt content using alginates (Schmidt et al. 1986). Fermented sausages with fat contents below 2–3% or where sugar substitutes for salt exist, such as the Thai Nham (Visessanguan et al. 2006) and Chinese products (Savic et al. 1988), respectively. In general, however, sausage levels of fat and salt continue to be considered undesirable in the light of the recent goals for healthy nutrition. Proposals for partial substitution of pork backfat (Mendoza et al. 2001; Muguerza et al. 2004; Valencia et al. 2006), as well as of NaCl (Gelabert et al. 2003) continue to be published. Table 23.8 illustrates to what extent the substitution of pork backfat by soy or fish oil or by inulin can improve the nutrient content and health value of fermented sausages, as reflected in improved PUFA contents, n-6/n-3 ratios, and even dietary fiber content. The interventions referred to affect sensory evaluation to an acceptable extent, provided additional precautions are taken, such as the use of extra antioxidants with fish oil (Valencia et al. 2006). A wider application and development of such technologies involving other meat products is recommended (Fernández-López et al. 2004).

Substitution of salt is the more delicate intervention because of its universal role in meat technology

Table 23.7. Variability in proximate composition of fermented sausages.

	Demeyer et al. (2000)		Salgado et al. (2005)	
	Short Ripening	Long Ripening	Home-Made	Industrial
Dry matter (DM) (%)	57 (4)[c]	67(3)	51–82[d]	46–82
% in DM				
Crude protein (CP)	31 (7)[d]	28 (9)	8–32	9–29
Crude fat	61 (7)	61 (8)	52–85	58–80
NaCl	5.3 (12)	6.1 (11)	0.6–12.2	1.7–3.9
Carbohydrates	0.56 (23)	0.40 (18)	0.3– 3.0	0.05–2.2
Collagen	15–35[a]		1.6– 6.3[b]	2.1–15.9[b]

[a] % of CP: extreme values for types of fermented sausages distinguished in the German legislation (Anonymous 2006b).
[b] % of DM.
[c] ( ) = coefficient of variation.
[d] Minimum-maximum values.

Table 23.8. Effects of fat substitution on nutrient content and health value of fermented sausage.

Reference[b]	(1)		(2)	(3)
% Pork fat Reduction by	Inulin 0 → 75[c]		Fish Oil 0 → 25	Soy Oil 0 → 25
Crude fat[a]	49 → 17	Cholesterol (mg/100 g)	102 → 134	34 → 33
Dietary fiber[a]	0 → 24	% of total fatty acids		
kcal/100 g	392 → 251	Long chain n-3 PUFA	0.02 → 0.82	0.17 → 0.22
		n-6 PUFA	4.33 → 4.20	14 → 22
		PUFA/SFA	0.46 → 0.56	0.40 → 0.73
		n-6/n-3	13.9 → 3.0	11.0 → 11.5

[a] % of dry matter.
[b] (1) Mendoza et al. (2001); (2) Valencia et al. (2006); (3) Muguerza et al. (2004).
[c] → = control → modified formulations.

to improve texture and taste as well as providing a hurdle for bacterial development (Ruusunen and Puolanne 2005). Nevertheless, partial substitution using calcium ascorbate has lowered sausage salt level from 2.3% down to 1.4%, with limited damage to sensory quality while ensuring a simultaneous increase of the sausage calcium and ascorbate content, nutrients known to be very low and absent in meat, respectively (Gimeno et al. 2001).

Arguments for the omission of nitrate and nitrite in meat processing have been formulated for about 20 years now (Wirth 1991), mainly based on the cytotoxic effects of N-nitroso compounds, such as nitrosamines formed by the colon flora and/or during meat processing from meat precursors (Mirvish et al. 2002). Older (Mahoney et al. 1979), although contradictory, results (Lee et al. 1984; Greger et al. 1984) suggest a possible negative effect of nitrite on the bioavailability of iron, and fermented sausages produced without nitrite/nitrate were shown to be acceptable to consumers (Skjelkvåle et al. 1974). It is known now that the need for nitrate/nitrite in color development, flavor, and microbiological safety can be alleviated by use of a purified natural cooked cured-meat pigment (Pegg et al. 2000) and/or an adapted hurdle technology. The latter is mainly based on water activities less than 0.85 (Chawla and Chandler 2004). In the EU, the importance of the cured meat color has, of course, further been diminished by legislation, allowing the use of coloring agents such as Monascus red (Angkak), cochineal, and betanin, derived from yeast, a scale insect, and red beet, respectively (Finkgremmels et al. 1991). These arguments should stimulate both use and further development of alternatives for nitrite. Special attention in this regard should be given to the production of Zn-porphyrin, the red pigment in Mediterranean products such as Parma ham, whose formation is inhibited by nitrite addition (Adamsen et al. 2006). The color of some types of oriental and Mediterranean products, is often also determined by the addition of vegetable components and/or spices such as chili peppers added to Spanish chorizo (Fernández-Fernández et al. 1998). Addition of spices and plant material such as herbs or their extracts does not only affect flavor and color but also may have positive antimicrobial and/or antioxidant effects. This was recently shown for tea catechins (Yilmaz 2006), pepper (Martinez et al. 2006), a lemon peel preparation (Aleson-Carbonell et al. 2004), extracts of grape seed and pine bark (Grün et al. 2006), and paprika (Revilla and Quintana 2005). Furthermore, end products are enriched in desirable plant secondary compounds such as lycopene (Osterlie and Lerfall 2005) and trace elements such as manganese, lacking in meat and essential for microbial fermentation (Demeyer and Stahnke 2002). The "functional compounds" involved may be added to the sausage mix, or raw materials may be used derived from animals fed these compounds. The latter was shown, e.g., to be more effective for the improvement of color stability using vitamin E (Mitsumoto et al. 1993), a finding suggesting preferential use of raw materials derived from animals fed diets enriched in polyphenols (Lopez-Bote 2000; Tang et al. 2000).

Effects of Processing

The main processes occurring after cuttering and affecting the nutritional and health value of fermented sausages are the metabolism of protein, lipid,

Table 23.9. End products of sausage metabolism derived from carbohydrates, protein, and lipid.

	Sausage Type	
	Short Ripening	Long Ripening
DM (%)	57 (4)[a]	67 (3)
From Carbohydrate Metabolism (g/kg DM)		
Lactic acid	19 (12)	15 (17)
Acetic acid	0.6 (14)	0.5 (19)
Residual sugar as glucose	1.0 (23)	0.7 (18)
From Protein Metabolism (% of TN)		
Peptide α-NH_2-N	3.0 (13)	2.7 (16)
Free α-NH_2-N	2.3 (10)	3.7 (13)
Ammonia-N	0.3 (22)	1.0 (18)
Lipid Metabolism		
Free fatty acids (% of total FA)	2.7 (4)	3.7 (11)

Source: calculated from Demeyer et al. (2000).
[a] ( ) = coefficient of variation; DM = dry matter; TN = total nitrogen; FA = fatty acids.

and carbohydrate by both bacterial and meat enzymatic activity and the physicochemical processes of oxidation affecting both the lipid and protein fraction.

Sausage Metabolism

Bacterial fermentation mainly involves the production of lactic acid from carbohydrates added or present. As shown in Table 23.9, the final amount of lactic acid represents about 1–2% of the sausage, being present in about equal amounts as the L- and D-isomer (Demeyer et al. 2000). At these levels, the physiological effects of the slower metabolism of the latter isomer are minor (Ewaschuk et al. 2005).

Besides bacterial fermentation of carbohydrates, however, important hydrolytic changes brought about by the joint action of bacterial and meat enzymes affect the sausage protein and lipid composition. Subject to considerable variation, myofibrillar proteins representing about 5% of the sausage crude protein are hydrolyzed, yielding about equal amounts of shorter peptides and free amino acids, the values increasing with ripening time. Further deamination and decarboxylation yield small amounts of ammonia and amines, respectively. Provided starter organisms do not produce amines, amine production mainly reflects initial bacterial contamination of the raw materials used (Ansorena et al. 2002). Small amounts of the branched-chain amino acids, particularly leucine, are metabolized with production of major aroma compounds. The proteolytic protein changes, together with the acid-induced denaturation of the salt-solubilized myofibrillar protein can explain the improvement of protein digestibility and protein-efficiency ratio after fermentation of sausage mix (Eskeland and Nordal 1980). It was clearly demonstrated that endogenous lipases are by far the main responsible enzymes for the liberation of free fatty acids during ripening. This lipolysis involves preferential release of polyunsaturated fatty acids, both because of the more important phopholipase activity on muscle membrane phospholipids and the specificity of fat cell lipases (Molly et al. 1996). Increasing with ripening time, up to 5% of the total fatty acids present in the sausage may be present as free fatty acids. The nutritional consequences of this change are not obvious but it may increase fatty acid susceptibility to oxidation (Ansorena et al. 1998), although Summo et al. (2005) found evidence for a high degree of primary oxidation forming oxidized triacylglycerols as well as peroxides early in the sausage ripening.

Lipid and Protein Oxidation

Apart from the prooxidant effects of sodium chloride, the PUFAs in sausage lipids are highly susceptible to autooxidation because of the increase in oxygen content due to the high degree of comminution. Also, the added nitrate/nitrite may initially promote oxidation although the later produced nitric oxide may act as an antioxidant. Both lipid autooxidation and microbial betaoxidation of saturated fatty acids are desirable reactions to some extent because the compounds formed are important contributors to flavor. Excessive autooxidation however leads to color defects and ultimately rancidity rendering the product unacceptable for consumption. A concentration of thiobarbituric reactive substances (TBARS) expressed as malondialdehyde <1 à 2mg/kg provides a limit for acceptability (Zanardi et al. 2004). However, before such condition is reached, lipid autooxidation and, in particular, cholesterol oxidation, could generate toxic molecules lowering the health value of the product (Zanardi et al. 2004). For a fast fermented industrial sausage type, the latter authors reported that neither nitrites/nitrates nor ascorbate, added for enhancement of color development, nor added spices significantly affected oxidation evaluated by measurement of TBARS. The extent of cholesterol oxidation (%) was very low in all cases, but values were lowered even more by use of the additives, except for spices. The data suggest that final levels of lipid oxidation are determined to a large extent by the initial level of oxidation in the raw materials. Sammet et al. (2006) recently confirmed the absence of an antioxidant effect of nitrite, but found in contrast that the use of meat from pigs fed α-tocopherol–enriched diets produced fermented sausages clearly less oxidized than those obtained from pigs fed a control diet. Although showing higher contents of PUFAs, sausages prepared from extensively reared pigs were not subject to more oxidation than those prepared from intensively fed animals (Summo et al. 2005). As discussed earlier, addition of antioxidants of plant origin at the start of processing was shown to lower levels of lipid oxidation. However, activity may shift from anti- to prooxidation depending on the concentration used (Lee and Kunz 2005). These findings give little support to suggestions of using plant material as sources of functional food components with additional antimicrobial and antioxidant activity in fermented sausage manufacturing and in meat technology in general (Kondaiah and Pragati 2005). They do reinforce the case, however, for a preferential use of raw materials derived from animals fed diets enriched in polyphenols (Lopez-Bote 2000; Tang et al. 2000). Such practice would not only enrich the meat and sausages with desirable plant secondary compounds but also lower initial levels of oxidation in raw materials as well as limit oxidation during sausage ripening.

Besides the effects of oxidation on the lipid fraction, oxidative lowering of methionine availability during meat processing was recognized years ago (Janitz 1980). Although the associated lipid-protein interactions were mainly studied when affected by heating, it was shown that the process of curing lowered available methionine content from 2.3% to 1.8% in total amino acids. Such change is important because methionine was recognized early on as the limiting amino acid in meat protein (Dvoràk 1972).

Smoking, a technology mainly limited to the shorter and faster production types, merits a final consideration in relation to oxidation. The process is, of course, of paramount importance in relation to both sensory and health properties of the product, as discussed elsewhere. It was also shown, however, that it has antioxidant effects, restoring methionine availability as well as lowering peroxide content (Strange et al. 1980).

REFERENCES

CE Adamsen, JKS Moller, K Laursen, K Olsen, LH Skibsted. 2006. Zn-porphyrin formation in cured meat products: Effect of added salt and nitrite. Meat Sci 72:672–679.

L Aleson-Carbonell, J Fernandez-Lopez, E Sendra, E Sayas-Barbera, JA Perez-Alvarez. 2004. Quality characteristics of a non-fermented dry-cured sausage formulated with lemon albedo. J Sci Food Agric 84:2077–2084.

Anonymous. 2005. Belgische Voedingsmiddelentabel 4th ed. Brussels: Nubel, 92 pp.

Anonymous. 2006. FOD Economie—Algemene Directie Statistiek en Economische Informatie, Belgium. http://www.statbel.fgov.be/port/agr_nl.asp.

Anonymous. 2006a. FOD Economie—Algemene Directie Statistiek en Economische Informatie, Belgium. http://statbel.fgov.be/figures/d21_nl.asp#3.

Anonymous. 2006b. Leitsätze für Fleisch und Fleischerzeugnisse. Germany. http://www.fleischerhandwerk.de/upload/pdf/leitsaetze_fleisch.pdf.

D Ansorena, MC Montel, M Rokka, R Talon, S Eerola, A Rizzo, M Raemaekers, D Demeyer. 2002. Analysis of biogenic amines in northern and southern European sausages. Meat Sci 61:141–147.

D Ansorena, MJ Zapelena, I Astiasarán, J Bello. 1998. Addition of Palatase M (lipase from Rhizomucor miehei) to dry fermented sausages: consequences on lipid fraction and study of the further oxidation process by GC-MS'. J Agric Food Chem 46: 3244–3248.

GC Arganosa, RL Henrickson, BR Rao. 1987. Collagen as a lean or fat replacement in pork sausage. J Food Qual 10:319–333.

K Arihara, T Mukai, M Itoh. 1999. Angiotensin I-converting enzyme inhibitors derived from muscle proteins. Proc 45th ICoMST August 1–6, 1999, Yokohama, Vol 1, pp. 676–677.

AE Bender, M Zia. 1976. Meat quality and protein quality. J Food Technol 11:495–498.
S Bingham. 1999. High-meat diets and cancer risks. Proc Nutr Soc 58:243–248.
M Brink, MP Weijenberg, AFPM de Goeij, GMJM Roemen, MHFM Lentjes, AP de Bruine, RA Goldbohm, PA van den Brandt. 2005. Meat consumption and K-ras mutations in sporadic colon and rectal cancer in The Netherlands Cohort Study. Br J Cancer 92:1310–1320.
AT Chan, GJ Tranah, EL Giovannucci, WC Willett, DJ Hunter, CS Fuchs. 2005. Prospective study of N-acetyltransferase-2 genotypes, meat intake, smoking and risk of colorectal cancer. Int J Cancer 115: 648–652.
SP Chawla, R Chander.2004. Microbiological safety of shelf-stable meat products prepared by employing hurdle technology. Food Control 15:559–563.
R Chizzolini, E Zanardi, V Dorigoni, S Ghidini. 1999. Calorific value and cholesterol content of normal and low-fat meat and meat products. Trends Food Sci & Technol 10:119–128.
I Coene. 2001. De Mediterrane voeding. Nutrinews 9(2):3–9.
COMA. 1998. Chief Medical Officer's Committee on Medical Aspects of Food. Report of the working group on diet and cancer. Nutritional aspects of the development of cancer. Report on health and social subjects no. 48, London: HMSO.
A D'Amicis, A Turrini. 2002. The role of meat in human nutrition: The Italian case. Proc 48th ICoMST, Rome, 25–30 August 2002, vol 1, pp. 117–119.
S De Henauw, G De Backer. 1999. Nutrient and Food intakes in selected subgroups of Belgian adults. Br J Nutr 81:S37–S42 Suppl.
S De Henauw, J Van Camp, G Sturtewagen, C Matthijs, M Bilau, N Warnants, K Raes, M Van Oeckel, S De Smet. 2006. Simulated changes in fatty acid intake in humans through N-3 fatty acid enrichment of foods from animal origin. J Sci Food Agric (in press).
S De Smet, K Raes, D Demeyer. 2004. Meat fatty acid composition as affected by fatness and genetic factors: A review. Anim Res 53:81–98.
D Demeyer. 1981. Technology and nutritional value of meat and meat products (in Dutch). Lecture notes of a Course in Human Nutrition, KVIV, Antwerpen, 27 pp.
D Demeyer, M Doreau. 1999. Targets and procedures for altering ruminant meat and milk lipids. Proc Nutr Soc 58:593–607.
D Demeyer, MA Raemaekers, A Rizzo, A Holck, A De Smedt, B ten Brink, B Hagen, C Montel, E Zanardi, E Murbrekk, F Leroy, F Vandendriessche, K Lorentsen, K Venema, L Sunesen, L Stahnke, L De Vuyst, R Talon, R Chizzolini, S Eerola. 2000. Control of bioflavour and safety in fermented sausages: First results of a European project. Food Res Int 33:171–180.
D Demeyer, K Raes, V Fievez, S De Smet. 2004. Radicals and antioxidants in relation to human and animal health: A case for functional feeding. Comm Appl Biol Sci, Ghent University, 69:75–92.
D Demeyer, L Stahnke. 2002. Quality control of fermented meat products. In J Kerry, J Kerry, D Ledward, eds. Meat Processing, Improving Quality. Cambridge, Woodhead Publishing Ltd., pp. 359–393.
AT Diplock, J-L Charlieux, J Grozier-Willy, FJ Kok, C Rice-Evans, M Roberfroid, W Stahl, J Viña-Ribes. 1998. Functional food science and defence against reactive oxidative species Br J Nutr 80(Suppl.1): 77–112.
Z Dvoràk. 1972. The use of hydroxyproline analyses to predict the nutritional value of the protein in the different animal tissues. Br J Nutr 27:475–481.
KM Emmons, CM McBride, E Puleo, KI Pollak, E Clipp, K Kuntz, BH Marcus, M Napolitano, J Onken, F Farraye, R Fletcher. 2005. Project PREVENT: A randomized trial to reduce multiple behavioral risk factors for colon cancer. Cancer Epidemiol Biomark & Prev 14:1453–1459.
KM Emmons, AM Stoddard, R Fletcher, C Gutheil, EG Suarez, R Lobb, J Weeks, JA Bigby. 2005a. Cancer prevention among working class, multiethnic adults: Results of the healthy directions—Health centers study. Am J Publ Health 95:1200–1205.
B Eskeland, J Nordal. 1980. Nutritional evaluation of protein in dry sausages during the fermentation process. J Food Sci 45:1153–1155.
JB Ewaschuk, JM Naylor, GA Zello. 2005. D-Lactate in human and ruminant metabolism. J Nutr 135: 1619–1625.
E Fernández-Fernández, ML Vázquez-Odériz, MA Romero-Rodríguez. 1998. Colour changes during manufacture of Galician chorizo sausage. Z Lebensm Unters Forsch A 207:18–21.
J Fernandez-Lopez, JM Fernandez-Gines, L Aleson-Carbonell, E Sendra, E Sayas-Barberá, JA Pérez-Alvarez. 2004. Application of functional citrus by-products to meat products. Trends Food Sci & Technol 15 (3-4):176–185.
J Finkgremmels, J Dresel, L Leistner. 1991. Use of monascus extracts as an alternative to nitrite in meat-products Fleischw 71(10):1184–1186.
M Flores, P Nieto, JM Ferrer, J Flores. 2005. Effect of calcium chloride on the volatile pattern and sensory acceptance of dry-fermented sausages Europ Food Res Technol 221:624–630.
TT Fung, M Schulze, JE Manson, WC Willett, FB Hu. 2004. Dietary patterns, meat intake, and the risk of

type 2 diabetes in women. Archiv Intern Medicine 164:2235–2240.

J Gelabert, P Gou, L Guerrero, J Arnau. 2003. Effect of sodium chloride replacement on some characteristics of fermented sausages. Meat Sci 65:833–839.

O Gimeno, I Astiasarán, J Bello. 2001. Calcium ascorbate as a potential partial substitute for NaCl in dry fermented sausages: Effect on colour, texture and hygienic quality at different concentrations. Meat Sci 57:23–29.

DI Givens. 2005. The role of animal nutrition improving the nutritive value of animal-derived foods in relation to chronic disease. Proc Nutr Society 64: 395–402.

JL Greger, K Lee, KL Graham, BL Chinn, JC Liebert. 1984. Iron, zinc, and copper-metabolism of human-subjects fed nitrite and erythorbate cured meats. J Agric Food Chem 32:861–865.

IU Grün, J Ahn, AD Clarke, CL Lorenzen. 2006. Reducing oxidation of meat. Food Technol 60(1): 36–43.

MCCM Hadler, FAB Colugnati, DM Sigulem. 2004. Risks of anemia in infants according to dietary iron density and weight gain rate. Prev Med 39:713–721.

BF Hagen, H Næs, AL Holck. 2000. Meat starters have individual requirements for Mn^{2+} Meat Sci 55: 161–168.

A Hailu, SF Knutsen, GE Fraser. 2006. Associations between meat consumption and the prevalence of degenerative arthritis and soft tissue disorders in the Adventist health study, California USA. J Nutr Health & Aging 10:7–14.

L Hallberg, M Hoppe, M Andersson, L Hulthén. 2003. The role of meat to improve the critical iron balance during weaning Pediatrics 111:864–870.

RJJ Hermus, HFF Albers. 1986. Meat and Meat products in nutrition. Proc 32nd ICoMST, Ghent, 24–29 August 2002, Manuscript.

M Hill. 2002. Meat, cancer and dietary advice to the public. Eur J Clin Nutr 56:S36–S41.

KO Honikel. 1992. Meat and meat-fat ingestion. Fleischw 72(8):1145–1148.

P Howe, B Meyer, S Record, K Baghurst. 2006. Dietary intake of long-chain omega-3 polyunsaturated fatty acids: Contribution of meat sources. Nutr 22:47–53.

W Janitz. 1980. The effect of autolytic changes in meat and of other ingredients added for technological reasons on changes in the content of available lysine, methionine and cystein in sterilized canned meats. Fleischw 60(11):2063–2066.

F Jiménez-Colmenero, J Carballo, S Cofrades. 2001. Healthier meat and meat products: Their role as functional foods. Meat Sci 59:5–13.

F Kiani, S Knutsen, P Singh, G Ursin, G Fraser. 2006. Dietary risk factors for ovarian cancer: The Adventist Health Study. United States. Canc Causes & Contr 17:137–146.

Y-I Kim, JB Mason. 1996. Nutrition and chemoprevention of gastrointestinal cancers: A critical review. Nutr Rev 54:259–279.

N Kondaiah, H Pragati. 2005. Meat sector and its development. Indian J Anim Sci 75:1453–1459.

NF Krebs, JE Westcott, T Butler, C Robinson, M Bell, KM Hambidge. 2006. Meat as a first complementary food for breastfed infants: Feasibility and impact on zinc intake and status J Pediat Gastroenterol and Nutr 42:207–214.

CK Lachat, JH Van Camp, PS Mamiro, FO Wayua, AS Opsomer, DA Roberfroid, PW Kolsteren. 2006. Processing of complementary food does not increase hair zinc levels and growth of infants in Kilosa district, rural Tanzania. Br J Nutr 95:174–180.

K Lee, BL Chinn, JL Greger, L Karen, JE Graham KL, JE Shimaoka, JC Liebert. 1984. Bioavailability of iron to rats from nitrite and erythorbate cured processed meats. J Agric Food Chem 32:856–860.

YB Lee, JG Elliott, DA Rickansud, EC Hagbert. 1978. Predicting protein efficiency ratio by the chemical determination of connective tissue content in meat. J Food Sci 43:1359–1364.

JY Lee, B Kunz. 2005. The antioxidant properties of baechu-kimchi and freeze-dried kimchi-powder in fermented sausages. Meat Sci 69:741–747.

GA Leveille, PF Cloutier. 1986. Role of the food-industry in meeting the nutritional needs of the elderly. Food Technol 40:82–88.

G Lombardi-Boccia, A Aguzzi, M Cappelloni, G Di Lullo, M Lucarino. 2003. Total-diet study: Dietary intakes of macro elements and trace elements in Italy. Br J Nutr 90:1117–1121.

G Lombardi-Boccia, S Lanzi, A Aguzzi. 2002. Contribution of meat and meat products to the daily intakes of some micronutrients. Proc 48th ICoMST, 25–30 August 2002, Rome, Vol 2, p.1004.

G Lombardi-Boccia, Lanzi S, Lucarini M, 2004. Meat and meat products consumption in Italy: Contribution to trace elements, heme iron and selected B vitamins supply. Int J Vitam Nutr Res 74:247–251.

CJ Lopez-Bote. 2000. Dietary treatment and quality characteristics in Mediterranean meat products. In: E Decker, C Faustman, CJ Lopez-Bote, eds. Antioxidants in Muscle Foods. Nutritional Strategies to Improve Quality. New York: Wiley-Interscience, pp. 345–366.

M Luchtenborg, MP Weijenberg, AFPM de Goeij, PA Wark, M Brink, GMJM Roemen, MHFM Lentjes, AP de Bruine, RA Goldbohm, P van't Veer, PA van den Brandt. 2005. Meat and fish consumption, APC gene mutations and hMLH1 expression in colon and rectal cancer: A prospective cohort study. The Netherlands. Canc Causes & Contr 16:1041–1054.

AW Mahoney, DG Hendrickx, TAG Gillett, DR Buck, CG Miller. 1979. Effect of sodium nitrite on the bioavailability of meat iron for the anaemic rat. J Nutr 109:2182–2189.

L Martinez, I Cilla, JA Beltran, P Roncales. 2006. Effect of Capsicum annuum (red sweet and cayenne) and Piper nigrum (black and white) pepper powders on the shelf life of fresh pork sausages packaged in modified atmosphere J Food Sci 71:S48–S53.

TL McCarthy, JP Kerry, JF Kerry, PBLynch, DJ Buckley. 2001. Evaluation of the antioxidant potential of natural food/plant extracts as compared with synthetic antioxidants and vitamin E in raw and cooked pork patties. Meat Sci 58:45–52.

E Mendoza, ML García, C Casas, MD Selgas. 2001. Inulin as fat substitute in low fat, dry fermented sausage. Meat Sci 57:387–393.

SS Mirvish, J Haorah, L Zhou, ML Clapper, KL Harrison, AC Povey. 2002. Total N-nitroso compounds and their precursors in hot dogs and in the gastrointestinal tract and feces of rats and mice: Possible etiologic agents for colon cancer. J Nutr 132:3526S–3529S.

M Mitsumoto, RN Arnold, DM Schaefer, RG Cassens. 1993. Dietary versus postmortem supplementation of vitamin E on pigment and lipid stability in ground beef. J Anim Sci 71:1812–1816.

K Molly, D Demeyer, T Civera, A Verplaetse. 1996. Lipolysis in a Belgian sausage: Relative importance of endogenous and bacterial enzymes. Meat Sci 43:235–244.

E Muguerza, O Gimeno, D Ansorena, I Astiasarán. 2004. New formulations for healthier dry fermented sausages: A review. Trends Food Sci & Technol 15:452–457.

SP Murphy, LH Allen. 2003. Nutritional importance of animal source foods. J Nutr 133:3932S–3935S.

A Naska, D Fouskakis, E Oikonomou, MDV Almeida, MA Berg, K Gedrich, O Moreiras, M Nelson, K Trygg, A Turrini, AM Remaut, JL Volatier, A Trichopoulou. Group Authors.: DAFNE participants. 2006. Dietary patterns and their socio-demographic determinants in 10 European countries: Data from the DAFNE databank. Eur J Clin Nutr 60:181–190.

T Norat, S Bingham, P Ferrari, N Slimani, M Jenab, M Mazuir, K Overvad, A Olsen, A Tjønneland, F Clavel, M-C Boutron-Ruault, E Kesse, H Boeing, MM Bergmann, A Nieters, J Linseisen, A Trichopoulou, D Trichopoulos, Y Tountas, F Berrino, D Palli, S Panico, R Tumino, P Vineis, HB Bueno-de-Mesquita, PHM Peeters, D Engeset, E Lund, G Skeie, E Ardanaz, C González, C Navarro, JR Quirós, M-J Sanchez, G Berglund, I Mattisson, G Hallmans, R Palmqvist, NE Day, K-T Khaw, TJ Key, M San Joaquin, B Hémon, R Saracci, R Kaaks, E Riboli. 2005. Meat, Fish, and Colorectal Cancer Risk: The European Prospective Investigation into Cancer and Nutrition. J Nat Cancer Inst 97:906–16.

T Norat, A Lukanova, P Ferrari, E Riboli. 2002. Meat consumption and colorectal cancer risk: Dose-response meta-analysis of epidemiological studies. Int J Cancer 98:241–256.

M Osterlie, J Lerfall. 2005. Lycopene from tomato products added minced meat: Effect on storage quality and colour. Food Res Int 38:925–929.

DC Paik, TD Wendel, HP Freeman. 2005. Cured meat consumption and hypertension: An analysis from NHANES III (1988–94). Nutr Res 25:1049–1060.

YJ Park, SL Volpe, EA Decker. 2005. Quantitation of carnosine in humans plasma after dietary consumption of beef. J Agric Food Chem 53:4736–4739.

RB Pegg, R Amarowicz, WE Code. 2006. Nutritional characteristics of emu Dromaius novae hollandiae meat and its value-added products. Food Chem 97:193–202.

RB Pegg, KM Fisch, F Shahidi, F Title. 2000. The replacement of conventional meat curing with nitrite-free curing systems. Fleischw 80 (5):86–89.

E Pelczynska, K Libelt. 1999. Nutritional value of sausages as a factor of their connective tissue content. Fleischw 79(7):86–88.

I Pynaert, C Matthys, M Bellemans, M De Maeyer, S De Henauw, G De Backer. 2005. Iron intake and dietary sources of iron in Flemish adolescents. Eur J Clin Nutr 59:826–834.

K Raes, S De Smet, D Demeyer. 2004. Effect of dietary fatty acids on incorporation of long chain polyunsaturated fatty acids and conjugated linoleic acid in lamb, beef and pork meat: A review. Anim Feed Sci Technol 113:199–221.

L Razanamahefa, L Lafay, M Oseredczuk, A Thiebaut, L Laloux, M Gerber, P Astorg, JL Berta. 2005. Dietary fat consumption within French population and quality data on major food contributors composition Bull du Cancer 92(7–8):647–657.

AL Reutersward, H Andersson, NG Asp. 1985. Digestibility of collagenous fermented sausage in man. Meat Sci 14:105–121.

I Revilla, AMV Quintana. 2005. The effect of different paprika types on the ripening process and quality of dry sausages. Int J Food Sci Technol 40(4):411–417.

M Ruusunen, E Puolanne. 2005. Reducing sodium intake from meat products. Meat Sci 70:531–541.

A Salgado, MCG Fontan, I Franco, M Lopez, J Carballo. 2005. Biochemical changes during the ripening of Chorizo de cebolla, a Spanish traditional sausage. Effect of the system of manufacture: Homemade or industrial. Food Chem 92:413–424.

K Sammet, R Duehlmeier, HP Sallmann, C Von Canstein, T Von Mueffling, B Nowak. 2006. Assessment of the antioxidative potential of dietary supplementation with alpha-tocopherol in low-nitrite salami-type sausages. Meat Sci 72:270–279.

Z Savic, ZK Sheng, I Savic. 1988. Wurste nach Chinesischer art. Fleischw 68.5:570–580.

A Schmid, M Collomb, R Sieber, G Bee. 2006. Conjugated linoleic acid in meat and meat products: A review. Meat Sci 73:29–41.

GR Schmidt, WJ Means, AD Clarke. 1986. Using restructuring technology to increase red meat value. J Anim Sci 62:1458–1462.

ALA Sesink, DSML Termont, JH Kleibeuker, R Van der Meer. 1999. Red meat and colon cancer: The cytotoxic and hyperproliferative effects of dietary heme. Cancer Res 59:5704–5709.

R Skjelkvåle, TB Tjaberg, M Valland. 1974. Comparison of salami sausage produced with and without addition of sodium nitrite and sodium nitrate. J Food Sci 39:520–524.

Y Song, JE Buring, JE Manson, SM Liu. 2004. A prospective study of red meat consumption and type 2 diabetes in middle-aged and elderly women. Diabetes Care 27:2108–2115.

CS Srinivasan, X Irz, B Shankar. 2006. An assessment of the potential consumption impacts of WHO dietary norms in OECD countries. Food Policy 31:53–77.

ED Strange, RC Benedict, AJ Miller. 1980. Effect of processing variables on the methionine content of frankfurters. J Food Sci 45:632–634.

C Summo, F Caponio, MT Bilancia. 2005. The oxidative degradation of the lipid fraction of ripened sausages as influenced by the raw material. J Sci Food Agric 85:1171–1176.

SZ Tang, JP Kerry, D Sheehan, DJ Buckley, PA Morrissey. 2000. Dietary tea catechins and iron-induced lipid oxidation in chicken meat, liver and heart. Meat Sci 56:285–290.

A Taylor, EW Redworth, JB Morgan. 2004. Influence of diet on iron, copper, and zinc status in children under 24 months of age. Biol Trace Elem Res 97:97–214.

E Tympa-Psirropoulou, C Vagenas, D Psirropoulos, O Dafni, A Matala, F Skopouli. 2005. Nutritional risk factors for iron-deficiency anaemia in children 12–24 months old in the area of Thess alia in Greece. Intern J Food Sci Nutr 56:1–12.

I Valencia, D Ansorena, I Astiasarán. 2006. Nutritional and sensory properties of dry fermented sausages enriched with n-3 PUFAs. Meat Sci 72:727–733.

LM Valsta, H Tapanainen, S Mannisto. 2005. Meat fats in Nutrition. Meat Sci 70:525–530 Sp. Iss.

W Van Dokkum. 1987. Vlees en Gezondheid. Proc Symposium "Het Vleesvraagstuk," Utrecht, September 1987.

W Visessanguan, S Benjakul, S Riebroy, M Yarchai, W Tapingkae. 2006. Changes in lipid composition and fatty acid profile of Nham, a Thia fermented pork sausage, during fermentation. Food Chem 94: 580–588.

P Walker, P Rhubart-Berg, S McKenzie, K Kelling, RS Lawrence. 2005. Public health implications of meat production and consumption. Public Health Nutr 8:348–356.

N Warnants, MJ Van Oeckel, CV Boucque. 1998. Effect of incorporation of dietary polyunsaturated fatty acids in pork backfat on the quality of salami. Meat Sci 49:435–445.

R Weidruch. 1996. Caloric Restriction and aging. Sci Amer 274(1):32–38.

JH Weisburger. 2002. Lifestyle, health and disease prevention: The underlying mechanisms. Eur J Cancer Prev 11(suppl 2):S1–S7.

WHO. 2003. Diet, Nutrition and the Prevention of Chronic Diseases. Report of a joint WHO/FAO expert consultation, WHO Technical Report Series 919; 148 pp. http://www.who.int/dietphysicalactivity/publications/trs916/summary/en/print.html.

WC Willett. 2003. Dietary fat and obesity: Lack of an important role. Scandin J Nutr 47:58–67.

WC Willett, MJ Stampfer. 2003. Rebuilding the food pyramid. Sci Amer 288.1:64–71.

C Williams, M Wiseman, J Buttriss. 1999. Supplement eds. Food-based Dietary Guidelines—A Staged Approach. Br J Nutr 81(Suppl.2):S29–S153.

F Wirth. 1991. Restricting and dispensing with curing agents in meat-products. Fleischw 71 (9):1051–1054.

A Wolk. 2005. Diet, lifestyle and risk of prostate cancer. Acta Oncologica 44:277–281.

JD Wood. 1990. Consequences for meat quality of reducing carcass fatness. In: JD Wood, AV Fisher, eds. Reducing fat in meat animals. New York: Elsevier Applied Science, pp 344–397.

Y Yilmaz. 2006. Novel uses of catechins in foods. Trends Food Sci & Technol 17:64–71.

E Zanardi, S Ghidini, A Battaglia, R Chizzolini. 2004. Lipolysis and lipid oxidation in fermented sausages depending on different processing conditions and different antioxidants. Meat Sci 66:415–423.

24
Functional Meat Products

Diana Ansorena and Iciar Astiasarán

INTRODUCTION

There is a growing consumer awareness of the relationship between nutrition and health, leading to the relatively new concept of *optimal nutrition*, by which food components have the potential to improve the health and well-being of individuals and maybe to reduce the risk from, or delay the development of, major diseases (Diplock et al. 1999). Advances in food science and technology are now providing the food industry with increasingly sophisticated methods to control and alter the physical structure and the chemical composition of food products. Consumers are also aware of the market potential for functional foods, based on the principle of added value linked to health benefit. In this context, the meat industry is now making a great effort to try to fulfill the expectations of consumers and to offer a wide variety of healthier products, by means of avoiding undesidered substances, reducing them to appropriate limits, and/or increasing the levels of other substances with beneficial properties (Jiménez-Colmenero et al. 2001).

Meat and meat products are significant sources of high biological value protein and they also supply a wide range and amount of micronutrients. Red meat and, consequently, red meat derivatives, are excellent sources of iron, having 50–60% in the heme form, which is absorbed by a more efficient mechanism than the nonheme iron form found in plant foods. Meat is also the major contributor to zinc intakes in developed countries and is one of the richest natural sources of an important reducing agent, glutation (Higgs 2000). The contribution of meat to vitamin B dietary intake is also relevant. However, from a healthy point of view, an excessive presence of these products in the diet can not be recommended, especially for certain population groups, because they are also sources of saturated fatty acids and cholesterol, and in some particular cases, such as occurs with dry-fermented sausages, of significant salt content. These components are known to be implicated in the development in some of the most prevalent diseases today (cardiovascular diseases, hypertension, atherosclerosis, etc.). Health organizations all over the world have promoted the choice of a diet low in saturated fat and cholesterol and moderated in total fat as a mean of preventing cardiovascular heart disease (AHA 2000; USDA 2000). Furthermore, it is known that n-3 fatty acids have important roles in the modulation and prevention of human diseases, particularly coronary heart disease, showing also other well-recognized positive health effects (antitrombogenic, antiinflammatory, hypotrigliceridemic, prevention of arrhythmias, etc.) (Connor 2000). Besides this, hypertension, which is one of the main risk factors of cardiovascular disease, is known to be correlated with an excessive sodium intake from the diet (Law et al. 1991; Truswell 1994).

Those data show the possible target goals for the meat industry to improve their products and supply arguments for the development of healthier dry-fermented sausages that could contribute to minimize the negative effects of some of their constituents. On the other hand, the technological process of elaboration of this type of meat derivatives enables the addition of healthy ingredients, and even the fermentation process that takes place at the beginning of the process leads to study of the potential health benefits of microbial starter cultures selection. In consequence, improvement of nutritional and functional properties of dry-fermented sausages can be tackled by trying different approaches.

MODIFICATION OF THE MINERAL CONTENT IN DRY-FERMENTED SAUSAGES

NaCl has an important influence on the final taste and texture of dry-fermented sausages, and it also plays an important role in the guaranty of the microbiological stability, making it difficult to achieve a significant reduction without affecting the final quality of these products. Gelabert et al. (1995) found a decrease in the overall acceptability of fuet when potassium lactate, glycine, and KCl were used at higher levels than 30%, 20%, and 40%, respectively. These salts were also used by Gou et al. (1996) as sodium chloride substitutes in salami, finding important flavor defects with substitutions of more than 40% for the three substitutes. Metallic taste limits the use of KCl, an excessive sweetness is given by high concentrations of glycine, and an abnormal taste is detected when using K-lactate. Levels of substitution higher than 30% for potassium lactate or higher than 50% for glycine also caused some texture defects. Mixtures of these salts (KCl/glycine and K-lactate/glycine) at levels of 40–70% offered no advantages over those found in the individual components (Gelabert et al. 2003). Ibañez et al. (1997) did not find significant differences between products manufactured with 3% of NaCl (control) and modified products manufactured with 1.5% NaCl and 1% KCl. The sodium reduction in these products was about 25%, in comparison with the traditional formulation, and the Na^+/K^+ ratio decreased from 4.38 to 0.87 (Ibañez et al. 1996). In another work, the same authors found that the nitrosation process and carbohydrate heterofermentative activity of the starter cultures was favoured in dry-fermented sausages manufactured with a mixture containing 1.37% NaCl and 0.92% KCl in comparison with sausages manufactured with 2.73% NaCl (Ibañez et al. 1995).

Other strategies try to formulate healthier dry-fermented sausages not only by the NaCl reduction, but also by the simultaneous addition of interesting minerals such as magnesium and/or calcium, both required for maintaining an adequate bone health. Results of some experiments are shown in Table 24.1. Gimeno et al. (1998), when using 22.5 g/kg of a mixture of 1% NaCl, 0.55% KCl, 0.23% $MgCl_2$, and 0.46% $CaCl_2$ to replace the traditional 26 g/kg salt content in chorizo (16 mm particle size), found that the sodium content decreased from 1.88% in the control sausage to 0.91% in the modified product. However, sensorial acceptability was scored lower, mainly due to the lower salty taste. In chorizo de Pamplona, a traditional Spanish dry-fermented sausage of 3 mm particle size, a mixture of 1% NaCl, 0.55% KCl, and 0.74% $CaCl_2$ (total amount of salts of 22.9 g/kg) was used, decreasing the sodium content from 1.35% in the control to 0.82% in the modified sausage (Gimeno et al. 1999). Significant increases of potassium and calcium (from 0.21 to 0.60% and from 154 mg/100 g to 320 mg/100 g, respectively) were observed in those products, which could be of interest from the nutritional point of view (although no bioavailability studies had been carried out). Concerning sensorial quality aspects, instrumental measures of texture

Table 24.1. Content of Na^+, K^+, and Ca^{2+} of different dry-fermented sausages.

	Chorizo		Chorizo de Pamplona			
	Control	A(*)	Control	B(**)	Control	C(***)
Na^+ (g/100 g sausage)	1.88	0.91	1.35	0.82	1.98	1.07
K^+ (g/100 g sausage)	0.55	1.11	0.21	0.60	–	–
Ca^{2+} (mg/100 g sausage)	84.60	182.90	154.50	319.9	130	400
Mg^{2+} (mg/100 g sausage)	25.6	182.00	41.64	43.16	–	–

A(*) 1% NaCl, 0.55% KCl, 0.23% $MgCl_2$, 0.46% $CaCl_2$. Gimeno et al. (1998).
B(**) 1% NaCl, 0.55% KCl, 0.74% $CaCl_2$. Gimeno et al. (1999).
C(***) 1.4% NaCl, 2.9% sodium ascorbate. Gimeno et al. (2001a).

(TPA analysis) and color (CIEL*a*b*) showed some slight differences with regard to traditional products. The applied mixture of salts did not affect the development of the starter culture and guaranteed the hygienic quality of the products (Gimeno et al. 2001a).

Significant reduction in the sodium content has also been achieved through a partial substitution of NaCl by different percentages of calcium ascorbate. The sodium percentage decreased from 1.98% in the control sausages to 1.07% in the highest ascorbate experimental batch (29.17 g ascorbate/kg and 14 g sodium chloride/kg) (Gimeno et al. 2001b). Calcium content consequently increased from 130 mg/100 g (control) to 400 mg/100 g, which is 50% of the RDAs for this mineral. However, some small differences in the instrumental measures of color and texture were again noticed. The taste of these dry-fermented sausages was generally described as less salty, which is sometimes positively evaluated by consumers.

Finally, addition of $CaCl_2$ at two different concentrations (0.5% and 0.05%) has been assayed in order to formulate calcium-enriched dry-fermented sausages (Flores et al. 2005). The sensory analysis showed that the panel preferred in color ($P < 0.05$), aroma ($P < 0.001$), taste ($P < 0.001$), and overall quality ($P < 0.001$) the control sausage over the Ca 0.5% batch, and no significant differences were detected between the control and Ca 0.05% batch. These differences were associated to the high degree of oxidation suffered by the 0.5% $CaCl_2$ batch, which showed higher TBA values and higher content of typical volatile compounds derived from lipid oxidation. An opposite effect was observed with the 0.05% $CaCl_2$ addition, which showed an important effect on volatile generation during ripening by inhibiting the generation of lipid oxidation products and promoting the generation of methyl branched alcohols and ethyl esters that contribute to proper sausage aroma.

FAT MODIFICATIONS IN DRY-FERMENTED SAUSAGES

Modification of the lipid fraction in meat products can be done by the reduction of the fat content and/or the simultaneous addition of nonlipid fat replacers or substitutes in order to minimize texture defects (Jiménez-Colmenero et al. 2001). Recent studies pointed out the interest of a change in the dietary lipid profile of foodstuffs into a healthier pattern, instead of a reduction in the total dietary fat by using other substances. As a way to reach this objective, different studies try to modify the composition of the animal raw matter (pigs) through a change in the diet (CAST 1991; Myer et al. 1992; Morgan 1992; Irie and Sakimoto 1992; Cherian and Sim 1995; Romans et al. 1995; Mitsuharu et al. 1997; Cava et al. 1997, 1999; de la Hoz et al. 1996; Enser et al. 2000). Concerning dry-fermented sausages Hoz et al. (2004) carried out an experiment in which salchichón, a Spanish dry-fermented sausage, was manufactured using different combinations of backfat and meat obtained from animals fed on diets rich in polyunsaturated n-3 fatty acids (linseed oil) and/or olive oil and α-tocopherol. Enrichment of the animal diets with linseed or linseed and olive oils, significantly increased ($P < 0.05$) the percentage of linolenic acid (5–8 fold), arachidonic acid (3–5fold), EPA (2-fold) and DHA (1.5-fold) in sausages. Products from raw material from pigs fed on diets enriched only with linseed oil (no α-tocopherol) showed high lipid oxidation susceptibility, developing rancid taste.

Other approaches for improving the lipid fraction of dry-fermented sausages introduce modifications in the formulation of products. First trials assayed the incorporation of olive oil, alone or preemulsified with soy protein, as partial substitutes for pork backfat, with the aim of increasing the MUFA content. Bloukas et al. (1997) in Greek dry-fermented sausages and Muguerza et al. (2001) in chorizo de Pamplona, concluded that it was possible to incorporate up to 20% and 25%, respectively, of preemulsified olive oil, developing technologically and sensory acceptable products. When a higher percentage of substitution was used, drip fat was observed. These products showed increments in the MUFA and PUFA fractions, and reduction in the cholesterol content, improving consequently their nutritional quality. A simultaneous study of reduction and incorporation of olive oil was further done in Greek dry-fermented sausages. Reduced- and low-fat dry-fermented sausages can be prepared with 20% and 10% fat in the meat mixture, showing a 37% and 53% lower fat content, respectively, than traditional sausages (30% fat). The reduction in the fat level led to an increase in the weight loss, hardness, and firmness of the modified sausages, which were also darker and with a higher red intensity color (a* value) than control. However, the substitution of 20% of the pork backfat by preemulsified olive oil did not affect weight loss, obtaining lighter products and with a higher yellow intensity (b* value). Muguerza et al. (2002).

Different concentrations of extra-virgin olive oil have also been tested in salami (conventional fat content of 15%). The formulation with 5% olive oil, corresponding to 33.3% substitution of pork backfat was sensorial judged best of all formulations, and it was found to be stable during the shelf life (Severini et al. 2003). Kayaardi and Gök (2004), in a similar fat content fermented product (sucuk), achieved significant cholesterol reductions and increments of oleic and linoleic acid contents in products with a partial substitution of 40% and 60% of beef fat by preemulsified olive oil. Sensorial analysis showed that sucuk prepared by replacing 40% were rated higher in terms of taste, appearance, and texture than the other sausages.

Some other studies modify the lipid fraction by incorporating high PUFA oils from vegetable and animal origin into the dry-fermented sausages formulations. Table 24.2 reports the fatty acids modifications caused in these cases.

Concerning vegetable oils, a 25% substitution of pork backfat by preemulsified soy and linseed oils significantly reduced the content of saturated fatty acids, increasing the PUFA content, mainly at the expense of linoleic acid and α-linolenic acid, respectively (Muguerza et al. 2003; Ansorena and Astiasarán 2004a). The α-linolenic acid content in modified sausages elaborated with linseed oil nearly reached the possibility of including the nutritional claim, still to be approved, "high in omega-3," because it reached 2.5 g/100 g, and 3 g/100 g are needed for that purpose. A significant and positive reduction in the n-6/n-3 ratio was observed when using linseed oil, decreasing from 13.4 in control sausages to 1.9 in modified ones, attributed also to this α-linolenic acid increment. Interesting increments in the PUFA/SFA ratio were also detected in both experiments, from 0.4 to 0.7. The relevance of a high PUFA/SAT ratio was shown in the epidemiological Nurses Health Study, in which it was proved that this ratio was strongly associated with a lower risk of CHD (Hu et al. 1999). Modified sausages were similar to their respective control in sensory properties, and no lipid oxidation was detected in any case.

Concerning animal origin high PUFA oils, a fish oil extract (omega-3 700 Solgar Vitamins) and a deodorised fish oil (Lysi) have been tested, by addition of two different amounts to the formulation (5.3 g/kg-batch A and 10.7 g/kg-batch B) in the first case (Muguerza et al. 2004), and by partial substitution of pork backfat (25%) in the second experiment (Valencia et al. 2006a). It has been postulated that an adequate intake (AI) for adults is 0.22 g/day for both EPA and DHA in 2000 kcal/day (Simopoulos 2002). According to this, 50 g of modified sausages would make up approximately 34% and 30% of the AI for DHA and EPA, respectively, in the case of batch A, and 75% and 59% of the AI in the case of batch B, contributing in a significant way to the dietary intake of EPA and DHA. However, lipid oxidation analysis revealed potential degradation problems in the richest n-3 batch. No sensory analysis was done in these products due to detected off-odor. An experiment carried out by partial substitution of pork backfat by deodorized oil (25% substitution) overcame this inconvenience. The developed products supplied 0.64 g EPA/100 g product and 0.46 g DHA/100 g product, which could mean 85% of the adequate intake in these acids if a portion of 50 g sausage is considered. No signs of oxidation were found by TBA analysis, and none of the dienals and trienals reported as secondary lipid oxidation products typical from fish oil were detected in the modified sausages. A sensory evaluation panel did not find differences in general acceptability, concluding that the modified products can be considered a technologically viable functional food.

One of the potential problems derived from these modifications could be an acceleration of the oxidative processes due to the increment in unsaturated fatty acids, particularly polyunsaturated ones, which are more prone to oxidation. Stability of olive oil or linseed-containing sausages, both products including BHT and BHA as antioxidants, was evaluated after storage during 2 and 5 months under different packaging conditions (aerobic, vacuum, and/or modified atmosphere) (Ansorena and Astiasarán 2004b; Valencia et al. 2006b). No signs of lipid oxidation measured by TBA, peroxides, and volatile aldehydes were detected for modified sausages, regardless of the packaging system used, and only slight changes in the fatty acid profiles were noticed in comparison to those products analyzed just at the end of the ripening process. Control sausages were affected by oxidation when stored aerobically, whereas they were well preserved under vacuum or MAP conditions. These results confirmed the stability of the new formulations and the maintenance of the health benefits during their storage period.

INCORPORATION OF FIBER INTO DRY-FERMENTED SAUSAGE FORMULATION

Well-established physiological properties and health benefits of dietary fiber have been described. It has uniquely significant physical effects in the gut and in addition, through fermentation, is a major determinant of large bowel function (Cummings et al. 2004). Its physical properties in the small bowel affect lipid

Table 24.2. Cholesterol and other lipid-related parameters obtained in dry-fermented sausages elaborated with high PUFA content oils.

	Soy Oil (1)		Linseed +A (2)		Fish Oil (3)			Deodorized Fish Oil +A (4)	
Parameter	Control	25% Substitution	Control	25% Substitution	Control	Modified A (+5.3 g/kg)	Modified B (+10.7 g/kg)	Control	25% Substitution
Cholesterol (mg/g product)	92.96	87.71	nd	nd	94	90	91	102	134
Σ SFA (g/100 g product)	12.82	10.79	14.05	10.33	11.59	10.36	11.57	10.08	9.99
Σ MUFA (g/100 g product)	15.51	13.85	14.25	11.73	14.51	13.28	14.60	12.30	11.98
Σ PUFA 5.16 (g/100 g product)	7.88	5.16	7.22	6.28	5.20	6.77	4.65	5.62	
Σ n-3 (g/100 g product)	0.41	0.66	0.36	2.57	0.36	0.59	1.07	0.31	1.41
Σ n-6 (g/100 g product)	4.75	7.22	4.78	4.96	5.89	4.61	5.70	4.34	4.22
n-6/n-3	11.5	11.0	13.4	1.9	16.1	7.8	5.3	13.9	3.0
MUFA+PUFA/SFA	1.6	2.0	1.4	1.8	1.8	1.8	1.8	1.7	1.7
PUFA/SFA	0.4	0.7	0.4	0.7	0.5	0.5	0.6	0.5	0.6

A: antioxidants (BHT+BHA 200 ppm).
SFA = saturated fatty acids; MUFA = monounsaturated fatty acids; PUFA = polyunsaturated fatty acids. nd = not determined.
1: Muguerza et al. (2003).
2: Ansorena and Astiasarán (2004b).
3: Muguerza et al. (2004).
4: Valencia et al. (2006a).

absorption and the glycemic response. Epidemiological studies point out that many of the diseases of public health significance—obesity, cardiovascular disease, type 2 diabetes—can be prevented or treated by increasing the amounts and varieties of fiber-containing foods. According to the American Dietetic Association, recommendations are established in 20–35 g/day for healthy adults and age plus 5 g/day for children, and in addition, a higher fiber intake provided by foods is likely to be less calorically dense and lower in fat and added sugar (Marlett et al. 2002). However, current intakes of fiber do not cover the recommended values, and incorporating it into frequently consumed foods would be an interesting strategy to reach those values.

Concerning dry-fermented sausages, different types of fiber have been added alone or combined with some other ingredients into various types of formulations (Table 24.3). Mendoza et al. (2001) manufactured low-fat content dry-fermented sausages with different percentages of powdered inulin, which impacts a softer texture and a tenderness, springiness, and adhesiveness very similar to that of conventional sausages. A concentration of 11.5% gave products with the lowest calorific value, and they showed better sensorial results than control low-fat products (no inulin added, 6.3% pork backfat) but still statistically lower than control high-fat products (no inulin added, 25% pork backfat). Also Garcia et al. (2002) manufactured

Table 24.3. Results for % fat reduction, calorific value, and acceptability in relation to the use of inulin, cereals, and fruit fibers as texture modifiers in low-fat dry-fermented sausages.

		% Fat Reduction	Caloric Value (kcal/100 g)	Acceptability
(1) Experiment with inulin	Control high fat		392.2	7[a]
	Control low fat	48.5	305.7	5.9[b]
	R1 (7% inulin)	65.3	257.1	5.5[b]
	R2 (6%inulin)	55.9	271.0	5.4[b]
	R3 (11.5% inulin)	64.6	242.9	6.1[b]
	R4 (10% inulin)	60.1	251.5	5.4[b]
(2.1) Experiment with wheat and oat fibers	Control 25% fat		435.9	6.48
	Control 6% fat	60	275.2	5.71[a]
	Wheat 3%	65	281.9	4.04[b]
	Wheat 1.5%	61	300.6	6.10[a]
	Oat 3%	66	270.8	4.42[b]
	Oat 1.5%	68	264.6	6.12[a]
	Control 10% fat	62	287.9	5.99[a]
	Wheat 1.5%	61	292.3	5.09[a]
	Oat 1.5%	63	289	5.57[a]
(2.2) Experiment with fruit fibers	Control 25% fat		435.9	6.48
	Control 6% fat	60	275.2	5.71[ab]
	Peach 3%	63	277	4.88[b]
	Peach 1.5%	57	290	6.08[a]
	Apple 1.5%	59	289.5	5.39[ab]
	Orange 1.5%	62	287.7	5.94[ab]
	Control 10% fat	62	287.9	5.99[a]
	Peach 1.5%	59	277.5	5.93[a]
	Apple 1.5%	61	269.9	6.00[a]
	Orange 1.5%	57	278.9	6.33[a]

(1) Experiment with inulin. *Source*: Mendoza et al. (2001). Data with the same superscripts did not show significant differences between them at level of significance of $P < 0.05$.

(2) Experiment with dietary fibers (cereal 2.1 and fruit 2.2). *Source*: García et al. (2002). Data with the same superscripts did not show significant differences between them at level of significance of $P < 0.05$. Statistical treatment applied independently.

dry-fermented sausages with 6% and 10% pork backfat with addition of cereal (wheat and oat) and fruit (peach, apple and orange) dietary fibers, at 1.5% and 3% concentration. These authors found that sensorial and textural properties of batches with 3% dietary fiber were the worst, due to their hardness and cohesiveness. The best results were obtained with sausages containing 10% pork backfat and 1.5% fruit fiber especially those with orange fiber. An appreciable decrease of the energetic value was also achieved in these products. Aleson-Carbonell et al. (2003) formulated dry-fermented sausages with the addition of raw and cooked albedo at five concentrations, evaluating the effect on the final chemical composition and on the sensorial properties. Products with 2.5% raw albedo and with 2.5%, 5%, and 7.5% cooked albedo were those with sensory properties similar to control and with improved nutritional benefits due to the fiber increment and to the presence of active biocompounds from albedo, which also contributed to produce a significant ($P < 0.05$) reduction in the residual nitrite levels and to delay oxidation development. The use of by-products of citrus fruits such as albedo is also an environmentally friendly strategy (Fernández-López et al. 2004).

Sugar beet fiber, at a concentration of 2%, has also been used in the production of Turkish-type salami, in combination with different concentrations of three interesterified vegetable oils prepared from either palm, cottonseed, or olive oil, leading to products similar to traditional ones regarding appearance, color, texture and sensory scores, and with a significantly higher fiber content (Javidipour et al. 2005).

USE OF DRY-FERMENTED SAUSAGES AS PROBIOTICS

According to the conclusions of the EC Concerted Action on functional foods, the gut is an obvious target for the development of functional foods because it acts as an interface between the diet and the events that sustain life. Among the functions related to the physiology of the gut are those that involve the colonic microflora, the optimal intestinal function and stool formation, those that are associated with the gut-associated lymphoid tissue (GALT), and those that depend on the products of nutrient fermentation. All these functions can be modulated by different food components, such as probiotics, prebiotics, and symbiotics (Bellisle et al. 1998). As already known, probiotics are "foods that contain live bacteria which are beneficial to health," whereas prebiotics are "nondigestible food components that beneficially affect the host by selectively stimulating the growth and/or activity of one or a limited number of bacteria in the colon, which have the potential to improve host health." Some of the claimed benefits for probiotics are to alleviate lactose maldigestion, increase resistance to invasion by pathogenic species of bacteria in the gut, shorten of rotavirus diarrhoea, stimulate the immune system, and possibly protect against cancer. These microorganisms should fulfill other properties to be considered suitable as prebiotics: resistance to gastric acidity and the action of bile salts, the adhesion to intestinal epithelial cells, and the growth capability in the presence of prebiotic carbohydrates (Bezkorovainy 2001; Ouwehand et al. 2002).

Most probiotic bacteria are lactic acid bacteria, mainly lactobacilli and bifidobacteria, although enterococci have also been used in the past as probiotics (Foulquié Moreno et al. 2006). In consequence, food products fermented by lactic acid bacteria and not receiving heating treatment have been investigated in order to evaluate their potential health benefits. In this context, the meat industry has made an effort to find novel starter cultures with additional value.

Concerning meat products, different probiotic and potential probiotic cultures isolated from other sources than meat have been applied as starter cultures for the production of dry-fermented sausages. However, more interesting strategies that combine commercial probiotic culture and meat starter cultures have been also successfully used. Andersen (1998) used *Bifidobacterium lactis* Bb-12, *Staphylococcus xylosus* DD-34, and *Pediococcus pentosaceus* PC-1 to produce dry-fermented sausages. Even the possibility of using only starter cultures with probiotic properties has been assayed. Erkkilä et al. (2001) examined the technological properties of three selected probiotic *Lactobacillus rhamnosus* strains concluding that particularly *Lactobacillus rhamnosus* GG, LC-705, and E-97800 seem to be suitable for the production of dry sausages, whereas *L. rhamnosus* LC-705 was not highly adapted to the meat environment, resulting in slower growth rates and acidification of the dry sausage.

Other approaches in this field identify different starter cultures or isolated strains from different types of dry-fermented sausages, with the aim of exploring their probiotic potential by means of in vitro and in vivo tests. For that purpose, Klingberg et al. (2005) isolated 22 strains of nonstarter *Lactobacillus* from 15 Scandinavian-type fermented sausages. Nine of them were able to grow at 37°C and lowered pH below 5.1 in a meat model. These strains and another 19 from a culture collection were evaluated by in vivo methods, including survival upon exposure to pH 2.5, bile tolerance, adhesion capacity to the human colon adenocarcinoma cell lineCaco-2, and antimicrobial activity against potential pathogens. Results obtained in this study confirmed that 3 strains isolated from sausages, 2 identified as *Lactobacillus plantarum* and another as *Lactobacillus pentosus*, were successfully applied as starter cultures for production of fermented sausage. No sensorial differences were detected with regard to commercial sausages, and good technological properties were described for those products, reaching high cell numbers ($4.7*10^7$—$2.9*10^8$ cfu/g) and good pH decrease (to 4.8–4.9) at the end of ripening.

In a similar work, Pennacchia et al. (2004) performed a screening procedure, by isolation at low pH and in presence of bile salts, to select potentially probiotic *Lactobacillus* strains directly from 10 fermented meat products. Twenty-eight *Lactobacillus* strains, from the initially 150 isolated ones, were able to resist low pH and to grow in conditions simulating the intestinal environment, suggesting their value as probiotics. Twenty-five of them, identified by 16S rDNA sequencing as 13 *L. plantarum*, 7 *L. brevis*, and 6 *L. paracasei*, were selected for a further study, in which a complete characterization was carried out, together with adhesion tests and analysis of growth capability in the presence of different carbohydrates (Pennacchia et al. 2006). The results of this paper demonstrated that several *Lactobacillus* strains isolated from sausages possess these two properties, pointing at the potential in future development of novel dry-fermented sausages by using potential functional starter cultures. They also show that it would be useful to take into account the specific utilization patterns of a probiotic strain before incorporating it into a product containing prebiotic substances, such as novel symbiotic fermented sausages. Nevertheless, these authors also conclude that the particular probiotic effect on human health can be given based only on in vivo experiments.

This is in agreement with the European Consensus Document on functional foods (Diplock et al. 1999), where it was concluded that the most pertinent aspect in communicating health-related benefits of functional foods is that any claim of their functionality must be scientifically based.

CONCLUSIONS

The following are conclusions based on the research cited by this chapter:

- Partial substitution of NaCl in dry-fermented sausages by other salts as calcium ascorbate could imply interesting health benefits, because of both the sodium reduction and calcium supply.
- The use of highly unsaturated oils as partial substitutes for pork backfat is an interesting way to change the fatty acid profile of dry-fermented sausages. Reduction of saturated fatty acids and better PUFA/SFA and n-6/n-3 ratios can be achieved by using olive oil, linseed oil, and deodorized fish oil, among others. Addition of antioxidants and proper packaging conditions are effective against potential lipid oxidation processes.
- Partial substitution for pork backfat by inulin and other dietary fibers such as albedo, seems to be a viable strategy to develop low-fat dry-fermented sausages and to incorporate active biocompounds into formulations.
- Research into functional properties of starter cultures point out the possible development of novel dry-fermented sausages with health benefits.

REFERENCES

AHA. 2000. Dietary guidelines revision. A statement for Healthcare Professionals from the Nutrition Committee of the American Heart Association. Circulation 102, pp. 2284–2299.

L Aleson-Carbonell, J Fernández-López, E Sayas-Barberá, E Sendra, JA Pérez-Alvarez. 2003. Utilization of lemon albedo in dry-cured sausages. J Food Sci 68(5):1826–1830.

L Andersen. 1998. Fermented sausages produced with the admixture of probiotic cultures. 44th International Congress of Meat Science and Technology, Barcelona, pp. 826–827.

D Ansorena, I Astiasarán. 2004a. The use of linseed oil improves nutritional quality of the lipid fraction of dry-fermented sausages. Food Chem 87:69–74.

———. 2004b. Effect of storage and packaging on fatty acid composition and oxidation in dry fermented sausages made with added olive oil and antioxidants. Meat Sci 67:237–244.

F Bellisle, AT Diplock, G Hornstra, B Koletzko, M Roberfroid, S Salminen, WHM Saris. 1998. Functional Food Science in Europe. Br J Nutr 80(Suppl 1):S1–S193.

A Bezkorovainy. 2001. Probiotics: Determinants of survival and growth in the gut. Am J Clin Nutr 73(2): 399S–405S.

JG Bloukas, ED Paneras, GC Fournitzis. 1997a. Effect of replacing pork backfat with olive oil on processing and quality characteristics of fermented sausages. Meat Sci 45:133–144.

CAST (Council for Agricultural Science and Technology). 1991. Foods, Fats and Health. Task Force Report No. 118.

R Cava, J Ruiz, C López-Bote, L Martín, C García, J Ventanas, T Antequera. 1997. Influence of finishing diet on fatty acid profiles of intramuscular lipids, triglycerides and phospholipids in muscles of the Iberian pig. Meat Sci 45:263–270.

R Cava, J Ruiz, J Ventanas, T Antequera. 1999. Oxidative and lipolytic changes during ripening of Iberian hams as affected by feeding regime: Extensive feeding and alpha-tocopheryl acetate supplementation. Meat Sci 52:165–172.

G Cherian, JS Sim. 1995. Dietary α-linolenic acid alters the fatty acid composition of lipid classes in swine tissues. J Agric Food Chem 43:2911–2916.

WE Connor. 2000. Importance of n-3 fatty acids in health and disease. Am J Clin Nutr 71(1 SUPPL): 171S–175S.

JH Cummings, LM Edmond, EA Magee. 2004. Dietary carbohydrates and health: Do we still need the fibre concept? Clin Nutr Suppl 1(2):5–17.

L De la Hoz, MO López, E Hierro, JA Ordoñez, MI Cambero. 1996. Effect of diet on the fatty acid composition of intramuscular and intermuscular fat in Iberian pig cured hams. Food SciTech Int 2(6):391–397.

AT Diplock, PJ Agget, M Ashwell, F Bornet, EB Fern, MB Roberfroid. 1999. Scientific concepts of functional foods in Europe: Consensus document. Br J Nutr 81: Suppl.

M Enser, RI Richardson, JD Wood, BP Gill, PR Sheard. 2000. Feeding linseed to increase the n-3 polyunsaturated fatty acids of pork: Fatty acid composition of muscle, adipose tissue, liver and sausages. Meat Sci 55:201–212.

S Erkkilä, ML Suihko, S Eerola, E Petäjä, T Mattila-Sandholm. 2001. Dry sausage fermented by *Lactobacillus rhamnosus* strains. Int J Food Microbiol 64(1–2):205–210.

J Fernández-López, JM Fernández-Ginés, L Aleson-Carbonell, E Sendra, E Sayas-Barberá, JA Pérez-Alvarez. 2004. Application of functional citrus by-products to meat products. Trends Food Sci Tech 15(3–4):176–185.

M Flores, P Nieto, JM Ferrer, J Flores. 2005. Effect of calcium chloride on the volatile pattern and sensory acceptance of dry-fermented sausages. Eur Food Res Tech 221(5):624–630.

MR Foulquié Moreno, P Sarantinopoulos, E Tsakalidou, L De Vuyst. 2006. The role and application of enterococci in food and health. Int J Food Microbiol 106(1):1–24.

ML García, R Domínguez, MD Gálvez, C Casas, MD Selgas. 2002. Utilization of cereal and fruit fibres in low fat dry fermented sausages. Meat Sci 60: 227–236.

J Gelabert, P Gou, L Guerrero, J Arnau. 1995. Efecto de la sustitución del NaCl por Lactato Potásico, Glicina o KCl sobre las características del "fuet." AIDA, Zaragoza.

———. 2003. Effect of sodium chloride replacement on some characteristics of fermented sausages. Meat Sci 65:833–839.

O Gimeno, I Astiasarán, J Bello. 1998. A mixture of potassium, magnesium and calcium chlorides as a partial replacement of sodium chloride in dry fermented sausages. J Agric Food Chem 46:4372–4375.

———. 1999. Influence of partial replacement of NaCl with KCl and $CaCl_2$ on texture and colour of dry fermented sausages. J Agric Food Chem 47:873–877.

———. 2001a. Influence of partial replacement of NaCl with KCl and $CaCl_2$ on microbiological evolution of dry fermented sausages. Food Microbiol 18: 329–334.

———. 2001b. Calcium ascorbate as a potential partial substitute for NaCl in dry fermented sausages: Effect on colour, texture and hygienic quality at different concentrations. Meat Sci 57:23–29.

P Gou, L Guerrero, J Gelabert, J Arnau. 1996. Potassium chloride, potassium lactate and glycine as sodium chloride substitutes in fermented sausages and in dry-cured pork loin. Meat Sci 42:37–48.

JD Higgs. 2000. The changing nature of redmeat: 20 years of improving nutritional quality. Trends Food Sci Tech 11:85–95.

L Hoz, M D'Arrigo, I Cambero, JA Ordóñez. 2004. Development of an n-3 fatty acid and α-tocopherol enriched dry fermented sausage. Meat Sci 67(3): 485–495.

FB Hu, MJ Stampfer, JE Manson, A Ascherio, GA Colditz, FE Speizer, CH Hennekens, WC Willett. 1999. Dietary saturated fats and their food sources in relation to the risk of coronary heart disease in women. Am J Clin Nutr 70(6):1001–1008.

C Ibáñez, L Quintanilla, I Astiasarán, J Bello. 1997. Dry fermented sausages elaborated with *Lactobacillus plantarum-Staphylococcus carnosus*. Part II. Effect of partial replacement of NaCl with KCl on the proteolytic and insolubilization processes. Meat Sci 46:277–284.

C Ibáñez, L Quintanilla, C Cid, I Astiasarán, J Bello. 1996. Part I. Effect of partial replacement of NaCl with KCl on the stability and the nitrosation process. Meat Sci 44:227–234.

C Ibáñez, L Quintanilla, A Irigoyen, I García-Jalón, C Cid, I Astiasarán, J Bello. 1995. Partial replacement of sodium chloride with potassium chloride in dry fermented sausages: Influence on carbohydrate fermentation and the nitrosation process. Meat Sci 40:45–53.

M Irie, M Sakimoto. 1992. Fat characteristics of pigs fed fish oil containing eicosapentaenoic and docosahexanoic acids. J Animal Sci 70:470–477.

I Javidipour, H Vural, OO Özbaş A Tekin. 2005. Effects of interesterified vegetable oils and sugar beet fibre on the quality of Turkish-type salami. Int J Food Sci Tech 40(2):177–185.

F Jiménez-Colmenero, J Carballo, S Cofrades. 2001. Healthier meat and meat products: Their role as functional foods. Meat Sci 59:5–13.

S Kayaardı, V Gök. 2004. Effect of replacing beef fat with olive oil on quality characteristics of Turkish soudjouk (sucuk). Meat Sci 66(1):249–257.

TD Klingberg, L Axelsson, K Naterstad, D Elsser, BB Budde. 2005. Identification of potential probiotic starter cultures for Scandinavian-type fermented sausages. Int J Food Microbiol 105(3):419–431.

MR Law, CD Frost, NJ Wald. 1991. By how much does dietary salt reduction lower blood presure?. I-Analysis of observational data among population. BMJ 302: 811–815.

JA Marlett, MI McBurney, JL Slavin. 2002. Position of the American Dietetic Association Health Implications of Dietary Fiber. J Am Diet Assoc 102 (7):993–1000.

E Mendoza, ML García, C Casas, MD Selgas. 2001. Inulin as fat substitute in low fat, dry fermented sausages. Meat Sci 57:387–393.

I Mitsuharu, K Yoshihiro, S Keiichi, O Yuuko, A Hiroyuki. 1997. The effects of fish oil enriched with n-3 polyunsaturated fatty acids on lipids and tasty compounds of pork loin. Nippon Shokuhin Kagaku Kogaku Kaishi 43:1219–1226.

CA Morgan. 1992. Manipulation of the fatty acid composition of pig meat lipids by dietary means. J Sci Food Agric 58:357–368.

E Muguerza, D Ansorena, I Astiasarán. 2004. Functional dry fermented sausages manufactured with high levels of n-3 fatty acids: Nutritional benefits and evaluation of oxidation. J Sci Food Agric 84(9): 1061–1068.

E Muguerza, D Ansorena, JG Bloukas, I Astiasarán. 2003. Effect of fat level and partial replacement of pork backfat with olive oil on the lipid oxidation and volatile compounds of Greek dry fermented sausages. J Food Sci 68(4):1531–1536.

E Muguerza, G Fista, D Ansorena, I Astiasarán, JG Bloukas. 2002. Effect of fat level and partial replacement of pork backfat with olive oil on processing and quality characteristics of fermented sausages. Meat Sci 6:397–404.

E Muguerza, O Gimeno, D Ansorena, JG Bloukas, I Astiasarán. 2001. Effect of replacing pork backfat with pre-emulsified olive oil on lipid fraction and sensory quality of chorizo de Pamplona, a traditional Spanish fermented sausage. Meat Sci 59:251–258.

RO Myer, JW Lamkey, WR Walker, JH Brendemuhl, GE Combs. 1992. Performance and carcass characteristics of swine when fed diets containing canola oil and added copper to alter the unsaturated:saturated ratio of pork fat. J Anim Sci 70:1147–1423.

AC Ouwehand, S Salminen, E Isolauri. 2002. Probiotics: an overview of beneficial effects. Antonie Van Leeuwenhoek 82(1–4):279–89.

C Pennacchia, D Ercolini, G Blaiotta, O Pepe, G Mauriello, F Villani. 2004. Selection of *Lactobacillus* strains from fermented sausages for their potential use as probiotics. Meat Sci 67:309–317.

C Pennacchia, EE Vaughan, F. Villani. Potential probiotic *Lactobacillus* strains from fermented sausages: Further investigations on their probiotic properties. Meat Sci 73:90–91.

JR Romans, DM Wulf, RC Johnson, GW Libal, WJ Costello. 1995. Effects of ground flaxseed in swine diets on pig performance and on physical and sensory characteristics and omega-3 fatty acid content of pork II. Duration of 15% dietary flaxseed. J Anim Sci 73:1987–1999.

C Severini, T De Pilli, A Baiano. 2003. Partial substitution of pork backfat with extra-virgin olive oil in "salami" products: Effects on chemical, physical and sensorial quality. Meat Sci 64(3):323–331.

AP Simopoulos. 2002. The importance of the ratio o omega-6/omega-3 essential fatty acids. Biomed Pharmacother 56:365–379.

AS Truswell. 1994. The evolution of diets for western diseases. In: Food. Multidisciplinary Perspectives. B Harris-While, R Hoffenderg, eds. Oxford, UK: Basil Blackwell.

USDA. 2000. US Department of Agriculture-US Department of Health and Human Services: "Nutrition and your health: Dietary guidelines for Americans." Homes and Garden Bulletin No. 232. Washington, D.C. U.S. Government Printing Office.

I Valencia, D Ansorena, I Astiasarán. 2006a. Nutritional and sensory properties of dry fermented sausages enriched with n-3 PUFAs. Meat Sci 72:727–733.

———. Stability of linseed oil and antioxidants containing dry fermented sausages: A study of the lipid fraction during different storage conditions. Meat Sci 73:269–277.

25
International Standards: USA

Melvin C. Hunt and Elizabeth Boyle

INTRODUCTION

Fermented products made from muscle foods have been produced for centuries. Demeyer and Toldrá (2004), Incze (2004), and Toldrá (2004) provide excellent descriptions of the art and science of the manufacture of these products. Globally, considerable variation exists in the sensory, chemical, and physical properties of fermented (and often dried) meat products. Because of this variation, it is challenging for regulatory agencies to categorize and create standards for these unique products. The purpose of this chapter is to present the U.S. "standards" for these products. Although there is no one document that describes in detail the regulatory issues for fermented meat and poultry products, a description of United States Department of Agriculture (USDA) regulatory documents that relate to this product category are presented below.

U.S. REGULATORY PROCESS

In the U.S., a total of nine government agencies are involved with ensuring that meat and poultry are safe and wholesome (Hale 1994). The USDA Food Safety and Inspection Service (FSIS) is the primary government agency that develops and administers regulations, standards, instructions, and guidelines for the meat and poultry industry. The following is a description of official documents that are the primary means for delivering rules, policies, and guidelines to processors.

Federal Register Publications and Related Documents

All official FSIS policy documents are published in the Federal Register, including notices and proposed rules, and interim and final rules. Final rules are incorporated into the U.S. Code of Federal Regulations, commonly referred to as the *CFR*, the year after a rule becomes final. Definitions and standards of identity or composition are found in 9 CFR, with Subparts I and J reserved for semidry-fermented sausage and dry-fermented sausage, respectively. At this time, there are no standards of identity mandated for these products. There are, however, other regulations that pertain to these product categories that are described later in this chapter.

FSIS Directives: Directives

Directives contain instructions of an indefinite duration. Table 25.1 lists the series of directives that provide instructions for processors.

FSIS Notices

Notices are temporary instructions that are scheduled to expire no later than 1 year from the issuance date.

Table 25.1. Series of FSIS directives providing instructions to processors.

Series Number	Category
5000	Program services
6000	Slaughter inspection
7000	Processed products
8000	Compliance evaluation and enforcement
9000	Exports and imports
10000	Laboratory services
11000	Facilities, equipment, and sanitation
12000	Voluntary inspection

Proposed or Final Rule

A proposed or final rule typically includes the following:

1. Preamble discussion of the provisions of the rule, and the need for and the statutory basis for the provisions.
2. A regulatory impact analysis estimating the future costs and benefits of the rule and discussion of regulatory alternatives considered. If the rule is designated "nonsignificant" or "exempt," a detailed regulatory impact analysis is not required.
3. A discussion of the estimated economic impact of the regulations on small businesses.
4. Statements addressing relevant Executive orders.
5. Proposed regulatory provisions or final regulatory provisions that will be codified in the CFR.
6. A request for comment on a proposed rule, or, if a final rule, a summary of comments received to the proposed rule and FSIS response to the comments.
7. A risk assessment of the rule for a "major" rule that affects human health, human safety, or the environment.

The Food Standards and Labeling Policy Book (USDA 2005)

This book provides definitions and descriptions that aid processors with preparation of product labels. To be in compliance with all regulatory issues, the Policy Book must be used in conjunction with other regulatory documents, including those found in the CFR and the Meat and Poultry Inspection Regulations, Directives, Notices, and Guidelines.

Navigating these documents can be challenging and requires some diligence; nevertheless, becoming familiar with their contents is essential for inspection and product approval.

REGULATORY DEFINITIONS AND SPECIFICATIONS

USDA FSIS requires that shelf-stable sausages meet specific moisture protein ratio and pH requirements. The following is a description of selected items that pertain to fermented meat and poultry products.

Moisture Protein Ratio (MPR) and pH

Nonrefrigerated or shelf-stable sausages must have an MPR of 3.1:1 or less and a pH of 5.0 or less, unless commercially sterilized. This does not apply to products containing more than 3.5% binders or 2% isolated soy protein. Specific MPR for dry, semidry, and other sausages are shown in Table 25.2.

Shelf-Stable Sausage

Dry sausage must have an MPR of 1.9:1, unless there is a specified MPR. Semidry, shelf-stable sausage must have an MPR of 3.1:1 or less and pH of 5.0 or less, unless commercially sterilized or unless MPR is specified for a particular type of sausage. Alternatively, nonrefrigerated, semidry, shelf-stable sausages are those that meet the following description:

1. Are fermented to a pH of 4.5 or lower (or the pH may be as high as 4.6 if combined with product water activity no higher than 0.91).
2. Are in an intact form or, if sliced, are vacuum packed.
3. Have internal brine concentration no less than 5%.
4. Are cured with nitrite or nitrate.
5. Are smoked with wood.

Dry and Semidry Sausages

Dry sausages may or may not be characterized by a bacterial fermentation. When fermented, the intentional encouragement of a lactic acid bacterial growth is useful as a meat preservative, as well as producing the typical tangy flavor. They may be smoked, unsmoked, or cooked. Dry sausages that are "medium" dry generally lose about 70% of their original weight, but the sausages typically range

Table 25.2. Specific moisture protein ratio (MPR) for dry, semidry, and other sausages.

Type	Classification	Maximum MPR
Blockwurst	Semidry	3.7:1
Calabrese	Salami	
Cervelat	Semidry or dry	No MPR requirement
Chorizo	Semidry or dry	
Easter Nola	Salami	
Farmer Sausage Cervelat	Semidry	
Farmer Summer Sausage	Dry	1.9:1
Frizzes	Dry	1.6:1
Genoa or Genoa Salami	Dry	2.3:1
Goteborg	Dry	
Gothaer Cervelat	Semidry	
Landjaeger Cervelat	Semidry	
Lebanon Bologna	Semidry	3.1:1 or less and pH 5 or less
Lyons Sausage	Dry	
Metz Sausage	Semidry	
Milan or Milano Salami	Dry	1.9:1
Mortadella	Semidry or dry	
Pepperoni	Dry	1.6:1
Poultry Salami	Dry	1.9:1
Salami	Dry	1.9:1
Sicilian Sausage	Dry	2.3:1
Hard Salami	Dry	1.9:1
Italian Sausage	Dry	1.9:1
Sarno	Dry	
Soppresate	Dry	1.9:1
Soujouk	Dry	2.04:1
Summer Sausage	Semidry	
Thuringer	Semidry	3.7:1
Touristen Wurst	Semidry	3.7:1
Ukranian Sausage	Dry	2.0:1

Adapted from USDA (2005).

from 60–80% of the original weight. Semidry sausages are usually partially dried and fully cooked in a smokehouse. They are semisoft sausages with good keeping qualities due to their lower pH.

Trichinae

Because fermented sausages (with no or minimal thermal processing) often include pork, the following is applicable to using this species. Products that contain pork muscle tissue (not including pork hearts, pork stomachs, and pork livers) must be effectively heated, refrigerated, or cured to destroy any possible live trichinae. USDA FSIS describes in 9 CFR 318.10 various methods for control of trichinae in fermented meat and poultry products (CFR 2006).

Many other terms listed in the Policy Book (USDA 2005) may be essential knowledge for meeting USDA regulations, depending on the product, claims of regionality, or labeling terms that may be used with the product.

HACCP OPTIONS

On July 25, 1996, the USDA FSIS mandated requirements designed to reduce the occurrence and numbers of pathogens on meat and poultry products, reduce the incidence of foodborne illness associated with consuming these products, and provide a framework for modernization of the meat and poultry inspection system. The new regulations required establishment of four new programs, three as pathogen reduction measures and one for HACCP.

USDA FSIS requires that a written HACCP plan include all decision-making documents and supporting documentation used in the development of an HACCP plan. HACCP records for slaughter activities and refrigerated product must be maintained for 1 year; records for frozen, preserved, or shelf-stable products must be stored for 2 years. All HACCP records must be maintained on-site for a minimum of 6 months. After 6 months, records may be stored off-site, providing they can be retrieved within 24 hours if requested by inspection authorities.

Fermented meat and poultry products are shelf-stable and ready-to-eat and they must meet USDA HACCP regulations. Products such as dry salami, summer sausage, and pepperoni would be considered to fall under the HACCP category of not heat-treated, shelf-stable meat and poultry products. If a fermented sausage was fully cooked, it would fall into the heat-treated, shelf-stable meat and poultry product category.

USDA FSIS has mandated a rule that requires producers to employ effective measures to control *Listeria monocytogenes* in ready-to-eat products. Most FSIS verification activities are concentrated in those establishments that rely solely on sanitation for preventing *Listeria* contamination, and in which there is no limitation on the growth of the pathogen if present in the product. Establishments would need to select the most effective strategies to control *Listeria* and would select one of the three following alternatives:

- *Alternative 1*: Processors employ both a postlethality treatment and a growth inhibitor for *Listeria* on RTE products. Establishments opting for this alternative are subject to FSIS verification activities that focus on the postlethality treatment effectiveness. Sanitation is important, but it is built into the degree of lethality necessary for safety as delivered by the postlethality treatment.
- *Alternative 2*: Processors employ either a postlethality treatment or a growth inhibitor for *Listeria* on RTE products. Establishments opting for this alternative are subject to more frequent FSIS verification activity than for Alternative 1.
- *Alternative 3*: Processors employ sanitation measures only. Establishments opting for this alternative are targeted with the most frequent level of FSIS verification activity. Within this alternative, FSIS places increased scrutiny on operations that produce hot dogs and deli meats. In a 2001 risk ranking, FSIS and the Food and Drug Administration identified these products as posing relative high-risk for illness and death.

Published guidelines (FSIS 2003) will assist processors with implementing alternatives 1, 2, or 3.

VALIDATION

From a regulatory standpoint, USDA FSIS requires that manufacturers of dry and semidry sausage products either: (1) demonstrate, through a process validation study, the ability to reduce a 7 log inoculum of *E. coli* O157:H7 by $\geq$5 logs, or (2) institute a statistically based sampling program for finished product that affirms the absence of *E. coli* O157:H7 (Appendix B). In 1997, the American Meat Institute released a white paper titled "Good Manufacturing Practices for Fermented Dry & Semi-Dry Sausage Products" (AMIF 1997). This paper presents pathogens of concern in fermented products and discusses strategies for controlling pathogens, including *Staphylococcus aureus*, *E. coli* O157:H7, *Listeria monocytogenes*, *Salmonella*, and *Trichinella spiralis*. In addition, good manufacturing practices for different stages throughout production, cleaning, and sanitation are discussed.

ACKNOWLEDGMENT

Contribution No. 07-9-B from the Kansas Agricultural Experiment Station, Manhattan, KS 66506 U.S.A.

REFERENCES

AMIF (American Meat Institute Foundation). 1997. Good Manufacturing Processes for Fermented Dry & Semi-dry Sausage Products. http://www.amif.org/FactsandFigures/SAUSAGE.pdf.

CFR (Code of Federal Regulations). 2006. Code of Federal Regulations (CFR): Main Page. http://www.gpoaccess.gov/cfr/index.html.

Demeyer D and F Toldrá. 2004. Fermentation. In: Jensen, Devine, Dikeman, eds. Encyclopedia of Meat Sciences. Oxford: Elsevier, Academic Press, pp. 467–474.

FSIS (Food Safety and Inspection Service). 2003. FSIS Rule Designed to Reduce *Listeria monocytogenes* in Ready-to-Eat Meat & Poultry. http://www.fsis.usda.gov/Fact_Sheets/FSIS_Rule_Designed_to_Reduce_Listeria/index.asp.

DS Hale. 1994. Inspection. In: Kinsman, Kotula, Breidenstein, eds. Muscle Foods: Meat, Poultry and Seafood Technology. New York: Chapman and Hall, pp. 163–185.

K Incze. 2004. Dry and semi-dry sausage. In: Jensen, Devine Dikeman, eds. Encyclopedia of Meat Sciences. Oxford: Elsevier, Academic Press, pp. 1207–1216.

F Toldrá. 2004. Dry Curing. In: Jensen, Devine, Dikeman, eds. Encyclopedia of Meat Sciences. Oxford: Elsevier, Academic Press, pp. 360–366.

USDA (United States Department of Agriculture). 2005. Food Standards and Labeling Policy Book. http://www.fsis.usda.gov/OPPDE/larc/Policies/Labeling_Policy_Book_082005.pdf.

USDA-FSIS. 1999. Compliance Guidelines for Cooling Heat-Treated Meat and Poultry Products (Stabilization). Appendix B. http://www.fsis.usda.gov/OPPDE/rdad/frpubs/95-033F/95-033F_Appendix%20B.htm.

26
International Standards: Europe

Reinhard Fries

INTRODUCTION

Food technology has grown historically, reflecting summarized experience from the past, giving also tribute to changes of nutritional habits of a population. Today's scientific insight (technological as well as from the hygienic point of view) potentially provides a high level of food technology and hygiene.

THE TERM OF QUALITY

General Remarks

A food chain consists of several different operators—sellers, customers, and finally consumers—all paying attention to different aspects of quality. So, a general agreement on quality does not exist: Quality is dependent on the point of view—legal or internal use, producer's or customer's—as well as the given circumstances. Factors of quality also reflect the period of history or a particular geography and the sociocultural situation.

In Figure 26.1, each circle represents the view of quality of particular groups, with one point of view being separate from the others, but overlapping, and simultaneously covering a different scope of the total of quality factors. However, there are some common and basic items: In each food commodity, main issues are the health factor of the product as well as its nutritional and technical value (composition and shelf life).

Food Chain versus End Product

For appropriate testing, sometimes the chain is observed, sometimes the final product should be observed. The end product with its chemical composition is defined by recipes characterizing raw incoming materials as well as the manufacturing procedure. This is true also for meat quality factors (e.g., tenderness). Meanwhile, some factors can be determined only in the food chain itself, i.e., ecological balance in the food chain, or the observation of bioethical factors during the keeping of food animals.

Disciplines, Parameters, and Methods

For comparison of results, an agreement on the technical performance (the method that is finally used to characterize the quality factor) is necessary. Disciplines and parameters of analysis may be:

- Biochemistry (glycogen, lactate, content of ATP/ADP/AMP, IMP or Inosine and further decomposition products thereof).
- Chemistry (representing the basic composition of a sample, e.g., protein or adipose-tissue content).
- Physical examination (using pH, water binding capacity, or temperature).
- Bacteriology, representing hygiene relevant agents (spoilage), agents of technological relevance (lactobacilli), or zoonotic agents (health factors). The bacteriological composition is commanded by the ecology (intrinsic and extrinsic factors) of a food as well as the initial status of incoming raw materials.

- Organoleptic (e.g., pH, color, structure, consistency of the sample). Spoilage, i.e., decomposition, would be recognized at a given time. So the sensory of foods or esthetical quality (Latin *qualitas*: "how is it") is an important issue, too (Figure 26.2).

Poultry Meat

Poultry meat is accepted worldwide without any religious or sociocultural restrictions. Accordingly, consumption has greatly increased in the last decades.

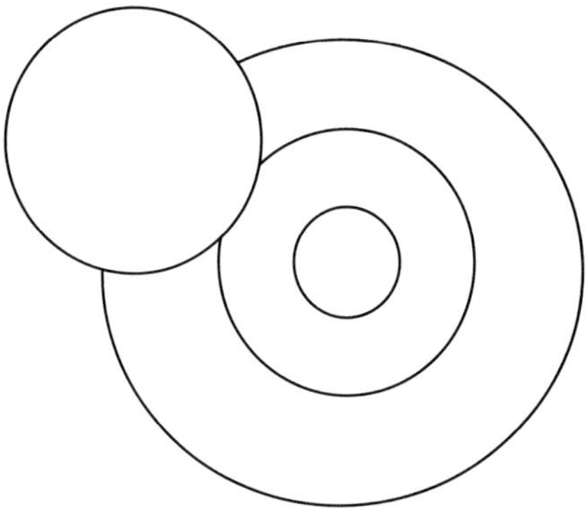

Figure 26.1. Overlapping circles of quality adoption.

Regarding the nutrition value, the best food conversion rate of the food animal species provides a good relationship of effort and efficacy.

Changing Ideas of Quality in the Past

In the EU, the concern with poultry meat was different at different times (Petersen 1997), beginning with the need for supply of high-quality animal protein, followed by the technical ability to stock high numbers of birds independent of the climate in combination with keeping the animals healthy in large numbers. Consecutively, human concerns (consumer protection, amount of workload) were raised, and finally the environmental impact and animal well being was a concern.

Food Chain Technology and Hygienic Consequences

In low-cost countries, one will find abattoirs with a lower grade of mechanization, whereas in the EU, poultry meat production is characterized by high technological standards in slaughter and processing, including further processing. More sophistication is to be expected.

To run an abattoir at high speed (up to 12,000 birds/h) requires sophisticated technology. Accordingly, in primary production, large herds supply

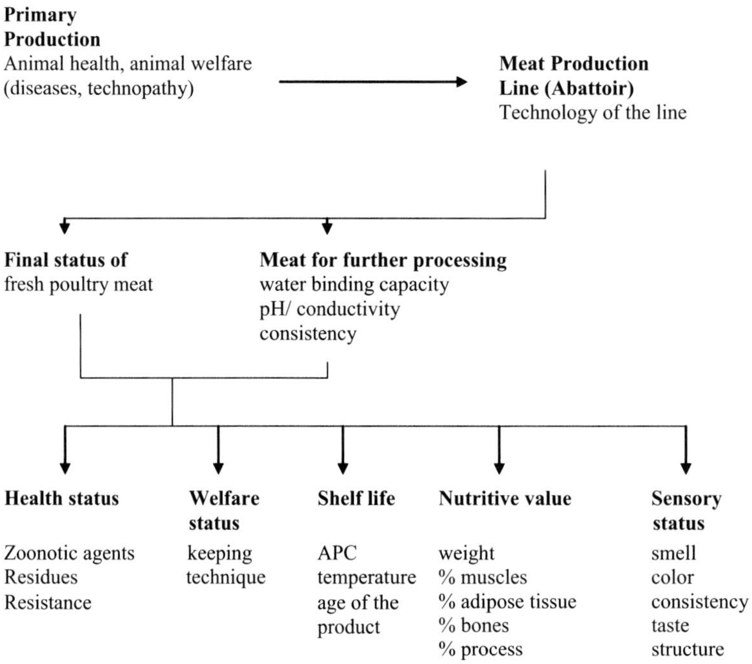

Figure 26.2. Different quality aspects are caused from different stages.

these abattoirs, with vertical flux from breeding via several steps of husbandry, transport, lairage to slaughter and processing, with a transport-technology already fitting into the slaughter line.

In such chains exists a permanent transfer of (zoonotic) agents from primary production into the abattoir. Identical strains of salmonellae or Campylobacter might be found over days and throughout the line, vertically as well as horizontally, and at several positions, the bird brings in its burden from prior stages. Such events are—at present—either an unavoidable part of the technology, or they happen because of accidents.

So, even an elaborate and reliable hygiene program in the abattoir cannot prevent that transfer, because effective barriers, which could be able to control zoonotic agents, do not exist at this stage of the line (Fries 2002).

Over the years, *Salmonella* was found most frequently in low numbers. Recently in the Netherlands, Dufrenne et al. (2001) detected in most samples *Salmonella* in numbers of <10/carcass, which is still comparable to the quantitative data from broiler carcasses (end product) some 20 years ago, where most frequently positive cases did not exceed a contamination rate of <10/total skin (Fries 2002).

Animal Welfare and Well Being

Animal welfare in the holdings is an issue in intensive keeping. Circumstances of holding technology and management lead to losses because of condemnation and downgrading, which touches the ethics of animal keeping, another and only recently raised term of *quality*. These days, such animal rights would be discussed. In 2002, Germany was the first country to grant the animals rights in the rank of the constitution (Cherney 2004).

Poultry Meat as Such

Organoleptic aspects (taste, smell, consistency, appearance) or the usability of the food (shelf life, storability under extreme climates, or the meat, from its composition, fitting into further processing lines) are important. With regard to shelf stability, the general spoilage microflora and their pattern throughout the processing line should be considered, too. Having used nonselective media for isolation, Table 26.1 gives a picture of microbiological colonization of poultry and (scalding- and chilling-) water samples during processing.

Environment

Finally, impact on the environment (release of NH_3, energy costs of processing, quantity, and quality of waste) also are of growing importance (Table 26.1).

Table 26.1. Isolates found during poultry meat processing: Carcasses and water (Fries 1988, 2005).

Micrococcaceae	463
Gram-positive irregular rods	395
Streptococcus and related	158
Enterobacteriaceae	87
Bacillus	159
Lactobacillus	99
Listeria	45
Flavobacterium	11
Acinetobacter	7
Moraxella	5
Others	11
Total	1,440 isolates

Legal Requirements

Poultry meat processing has to observe human and animal health and must be aware of animal well being. Worker's protection might be an additional task.

So, in primary production, foreign substances, zoonotic agents and resistance, zoonoses, other diseases, as well as technical performance of the procedures (reflecting animal welfare) are important factors of the process. In the end, a wholesome meat or meat product with acceptable quality and shelf stability should be available.

For that, the animals must remain under control during their whole lifespan ("from stable to table"). It might be done on a mandatory basis ("control of control" or direct supervision) as well as based on an internal control, which is a question of organization, personnel, and political decisions, and which is not to be discussed here.

According to Regulation (EC) 178/2002, the farmer as a food entrepreneur bears responsibility for the animal intended to produce food for human consumption (product responsibility). Consequently, the food chain character of animal production has been brought into Regulation (EC) 854/2004 on the surveillance of food animals. Moreover, for poultry, Regulation (EC) 2160/2003 must be followed, requiring freedom of *Salmonella*, e.g., for layers from November 17, 2007, for broilers from November 17, 2008, or for turkeys from November 17, 2009.

In anticipation of the philosophy of the food chain, earlier legislation of the EU (by then EEC) reflects particular details of the whole production line:

- *Regulation (EEC) 1538/91 ("quality")*: Classification of carcasses (carcass weight, skin condition, fat content), state of evisceration (e.g., New York dressed), trade conditions
- *Directive 92/117/EEC for "Zoonoses": Concealment of zoonotic agents*
- *Directive 71/118/EEC for Poultry Meat Hygiene and Inspection Procedures*: Flock health, residues, antemortem inspection, post mortem inspection with macroscopic lesions to be observed, processing procedures (water uptake, evisceration efficacy).

FERMENTED MEAT PRODUCTS

The term of (poultry) *meat products* implies that the material has been processed beyond the point of fresh meat. Such lines are located at the very end of the food chain. Here, characteristics of the incoming raw material are needed, because it is the initial "raw" material for the next, and qualitatively different, processing line.

Food Preservation as Such

For food technology in general, one should expect the product to cover specific requirements:

- No health impacts to consumers (safety)
- Storable within a certain and product-typical time frame (shelf stability)
- Acceptable in the sense of sensory value
- No loss of nutrients or losses remaining in an unavoidable low range (nutritive value)
- Production costs in an acceptable range
- No rendering problems or remaining residues from the applied technology
- No risk in the application of the technique for workers

Ingredients, recipes, and technology result in a particular commodity with a given composition and with given quality items as well as a product-specific microbiological composition—i.e., the respective technical sequence has a specific impact on microorganisms, leading to no restriction, to the cease of metabolism, or even to the destruction of the biological agent. Such a preservation potential of actions is based on chemical, physical, or biological action; these three main disciplines convey different ways of antimicrobiological pressure:

- **Chemical factors**
  - Alcohols
  - Aldehydes (glutare-di-aldehyde)
  - Organic acids and related compounds
  - Anorganic acids and related compounds
    - Phosphorus (sodium phosphates, trisodiumphosphate TSP, tripotassiumphosphate TPP, sodiumhypophosphite)
    - Sulphur (sulfur dioxide, sodium disulfite)
    - Anorganic nitrogen compounds such as nitrite.
- **Physical factors**
  - Application of different wave lengths
    - X-rays
    - Ultraviolet
    - Microwaves
    - Ultrasonic
  - Application of temperature
    - High temperature
    - Low temperature
  - Use of mechanical devices
    - Washing and sanitizing
    - Desiccation of surfaces
    - Pressure
    - Prohibition of atmosphere (vacuum, modified atmosphere)
- **Biological factors**
  - Competing flora (starters, protective cultures)
  - Metabolites (bacteriocines, colicine, natamycine)
  - Enzymes (lysozyme, lactoperoxidase)

In the end, factors affecting microbial associations and metabolism are (Mossel 1979, modified).

- **Intrinsic Factors:** Nutrient composition, structural barriers, antimicrobial constituents (naturally occurring or added), pH, a_W, E_h, amount of initial bacteria
- **Extrinsic Factors:** Temperature, type of packaging, duration of exposure
- **Implicit Factors:** Antagonisms and synergisms, generation time of bacteria.

From their microbiological background, particular technological procedures can correspond with a particular impact on microbiological colonization of the material (Table 26.2).

The Product

In Europe, many fermented meats are cured using nitrites or nitrates. In that case, preservative measures and factors are chemical (nitrite or nitrate), physical (a_W), and (micro-) biological (lactobacilli, as well as micrococci, if nitrate is used as a preservative), which results in the characteristic sensory of that product.

Table 26.2. Impact of technological commodity on microbiological background.

Technology	Microbiological Background	Commodity
Organic acids, lactobacilli	Impact on pH (lower)	Fermented sausage
Wrapping (folia), MAP	E_h (atmosphere change)	All products
Drying, salting, sugar treatment	a_w (lowering)	Fermented sausage, dried products
Chilling/deep freezing	Slowdown/cease of metabolism	Fresh meat, convenience products
Preserves/pasteurization	Heat/pressure (destroying of the structure)	Canned foods, milk, egg products
Starter cultures	Microbiological competition	Fermented sausage
Nitrite/nitrate, smoking	Adverse (toxic) chemical agents	Salami-type sausage
Combinations	pH/a_W/nitrite	Shelf-stable products

Fermented products need some time to reach the final pH. In some areas such as Germany, some products are marketed within 3 to 5 days, after which the product would be sold as a fresh fermented sausage. That environment would allow salmonellae to survive. Consequently, "fresh" fermented products must be considered still to harbor still salmonellae, if such organisms have been initially present.

Recently a new type of product has emerged ("meat preparations"), in which meat is only salted and some spices are added. Products like this do not belong to the category of meat products, because they are not preserved by means of reduced a_W/pH, and because they are only refrigerated. Therefore the risk of salmonellae in such "meat preparations" remains high, especially in products stemming from raw material with high prevalence of salmonellae, such as poultry. These products have not been fermented.

For a fermented product, preservation techniques and sensory state are typical. A product of European origin (e.g., salami) would be characterized as follows:

- **Chemical composition:** In Germany, the German Food Collection (Deutsches Lebensmittelbuch, N.N. 2002) refers to the percentage of animal-originated (muscle) protein as well as a product of typical sensory and appearance.
- **Product typical sensory:** The characteristic taste and odor, consistency, and mincing state depends on the product characteristics. So, some fermented sausages have been minced very intensively, and others only to a lesser extent (the original product of "salami") or not at all (cured and fermented "raw ham").

Besides that, important results of examination depend on the technology used (Table 26.3).

Table 26.3. Comparison of examination results between fast-ripening and slow-ripening products.

	Fast-ripening Products	Slow-ripening Products
pH	4.8–5.2 (use of nitrite)	5.3–.8 (use of nitrate)
a_W	<0.90	>0.90
lactobacilli	+++	+++
micrococci	+	++
APC	log 8	log 8

For health-related microbiology, the presence/absence of zoonotic agents, in particular in very recently prepared "fresh" fermented products, are important items. However, the risk of a positive Salmonella result is low because of the low pH and low a_w.

MICROBIOLOGICAL SAFEGUARDING IN FOOD CHAINS

Risk Analysis (RA)

Risk analysis comprises the scientific assessment of hazards in a given line as well as—separately—drawing consequences as managerial steps (Figure 26.3). So, RA is an analytical instrument directed backward toward the given line, first to find out possible hazards and second to determine possible risks.

The *risk profile* shortens this procedure in a more practical way. However, the approach is basically the same. A risk profile provides an initial evaluation of the food safety issue in relation to the scope of

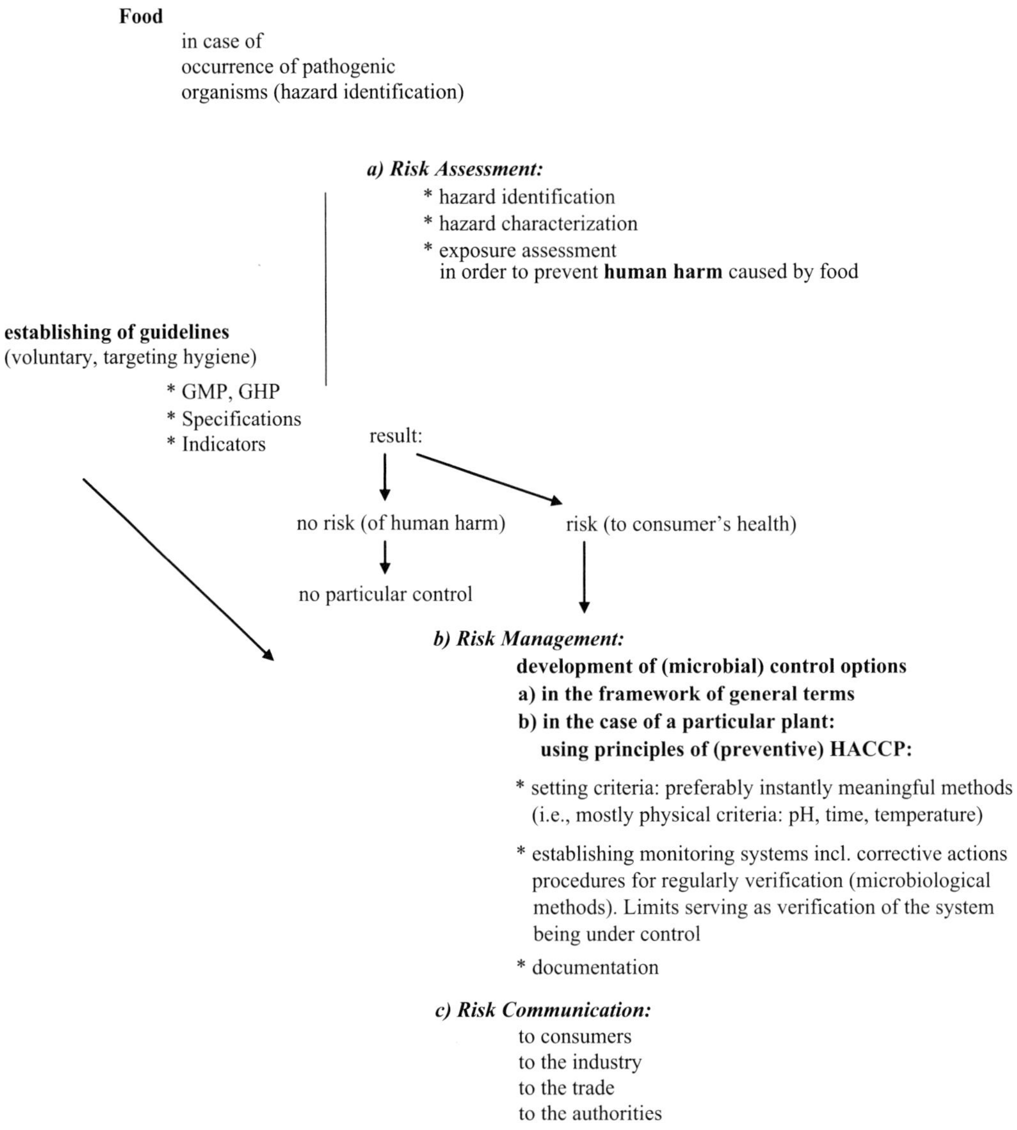

Figure 26.3. Interrelations between Risk Assessment and Risk Management, as well as HACCP (based on European Commission 1998; Lammerding 1997).

public health concerns, based on available information and surveying available control measures.

The terms *Appropriate Level of Protection* (ALOP) and *Food Safety Objectives* (FSO) may serve as sort of practical and achievable blueprints for what could be realistically reached at the present state. The SCVPH (2003 I) stated:

Appropriate Level of Protection (ALOP) is a quantification of the disease burden within a country linked to the implementation of food safety systems. ALOP is derived from a risk assessment and is expressed as, e.g., the likelihood to suffer a food-related illness from a food serving, or the number of cases per 100,000 consumer years. The setting of an ALOP is a risk management decision.

Food Safety Objectives (FSO) may be an important element in guidance on options to be taken for the future safety of foods. The concept is still evolving and no definition has yet been agreed upon. The SCVPH (2003 III) quotes Codex document CX/FH (2001) as follows: "The maximum frequency and/or concentration of a microbiological hazard in a food at the time of consumption that provides the appropriate level of health protection (ALOP)."

Food Safety Concepts in Practice

After a period of end product control, today the whole sequence of the technical procedure gets under control, using hygienic elements (e.g., GHP) or recording methods (e.g., observation sheets). Such elements are needed for the construction of more sophisticated practical solutions, such as HACCP. In that sense, Good Hygiene Practice (GHP in primary as well as secondary production) or Good Management Practice (GMP) are prerequisites for the final HACCP performance.

The SCVPH (1999 I) stated that the use of GHP and other elements and the implementation of Hazard Analysis Critical Control Point (HACCP) systems ensure that undesirable microorganisms are eliminated or minimized to an extent that they cannot cause harm to human health.

HACCP means knowledge of a line and regular checking and recording its control. The concept looks forward and determines Critical Control Points in that line and, as far as possible, their elimination (Figure 26.3). If the critical point cannot be deleted, at least it must be kept under technical control by regular checking and recording.

HACCP systems need also to be rapid. So, fast and reliable results are requisites. In most cases, instruments based on physical factors such as time, temperature, volume (e.g., water consumption), weight, or also magnetism (metal detectors) or pH would be used. In general, results of direct measurements can be data from the technical procedure; observations provided by human beings such as macroscopic parameters during meat inspection; other processing records (observation sheets); or results based on microbiological tests such as recording, cleaning, and disinfection efficacy in the premises.

From its origin, HACCP is an instrument for aspects of human health concern, having been developed for safeguarding the manufacturing of space food. Using the term *hazard* (and *HACCP*), this approach is limited to human safety (NACMCF 1998).

From the legal point of view, control measures, based on the principles of HACCP, must be implemented in the food chain, with the exception of primary production (Reg. 852/2004, Art. 5 and Annex 1) (see also Figure 26.3).

CRITERIA—REGULATION (EC) 2073/2005

The implementation of limits into food legislation has been frequently discussed between the member states of the EEC or later on of the EU. Already years ago, in the case of special foods being estimated as important, legal quality parameters and microbiological limits were laid down. Generally, in cases of legal background, the estimation of results of an analysis is easily done.

As an example, with Decision 2001/471/EC, the examination of carcasses (not for poultry) was based on legal parameters; however, with limited legal consequences (improvement of hygiene in an abattoir after checking the procedure and appropriate measures to be taken).

Moreover, in several other legislative documents we find already microbiological criteria, which have been recently repealed with the Directive 2004/ 41/EC.

Most of such criteria had a national template. As an example, from the German Society of Hygiene and Microbiology, criteria expressed as guidelines and as limits have been published, but not on a legal base (N.N. 2006). These criteria serve as guidance (m) values or warning limits (M). For finally fermented meat products (fully fermented and shelf-stable), the values read as shown in Table 26.4).

Table 26.4. Guidance (m) values or warning limits (M) for finally fermented meat products.

	Product	m	M
Enterobacteriaceae	For spreading	log 2/g	log 3/g
	For slicing (firm)	log 3/g	log 4/g
Coagulase-pos.staphylococci		log 3/g	log 4/g
E. coli		log 1/g	log 2/g
Salmonella			n.d. in 25 g
L. monocytogenes			log 2/g

n.d.: not detectable.

Since the days of the Fresh Meat Directive 64/433/EEC from the 1960s, community food legislation has proceeded very far. Having been confronted with the BSE disaster in the EU, and triggered also by the recognition of a need for a basic readjustment of legislation philosophy, EU hygiene legislation experienced a new arrangement. Parallel to, and after the implementation of the European Food Safety Administration (EFSA), now located in Parma, Italy, the Regulations of the Hygiene Package went into force. Among them, Regulation (EC) 2073/2005 is based on several opinions in particular from the Scientific Committee on Foods (SCF) and the Scientific Committee on Veterinary Measures Related to Public Health (SCVPH), later also from the newly founded scientific panel on biological hazards (BIOHAZ). Opinions are scientific-based statements answering to questions from the commission (Terms of Reference). Here, opinions were produced on:

- The setting of criteria as such (result: Criteria should be relevant and effective in relation to consumer health protection): *SCVPH (1999I)*
- Listeria monocytogenes in foodstuffs (result: Concentration in foodstuffs should be <100 cfu/g): *SCVPH (1999 II)*
- Vibrio vulnificus and Vibrio parahaemolyticus in seafood (result: Codes of practice should be implemented): *SCVPH (2001)*
- Norwalklike viruses (result: No use of fecal bacteria as indicators, but E. coli should be used in case of application of bacteriological indicators): *SCVPH (2002)*
- Gelatin specifications (result: In terms of consumer health, the *Salmonella* criterium would be sufficient): *SCF (2002)*
- Verotoxigenic E. coli in foodstuffs (result: Guidelines to reduce fecal contamination along the food chain should be established for special food category lines): *SCVPH (2003 I)*
- Staphylococcus enterotoxins in milk and cheeses (result: The amount of staphylococcal enterotoxins should be laid down in particular specific milk products): *SCVPH (2003 II)*
- Salmonella in foodstuffs (result: Criteria for Salmonella should be set only if they are practical and able to protect consumers; otherwise, guidelines should be established): *SCVPH (2003 III)*
- Salmonella and Enterobacter sakazakii in infant formula (result: E. sakazakii seems to be more difficult to control than Salmonella; the family of Enterobacteriaceae should be used as an indicator): *SP BIOHAZ (2004)*

In the end, scientific information from these opinions served as a template for Regulation (EC) 2073/2005, which has been in force in all member states since January 1, 2006. With this Regulation, food safety and process hygiene criteria were introduced into EU food business, simultaneously repealing Decision 93/51/EEC. Moreover, Decision 2001/471/EC as amended had also been taken out of force. In particular, the regulation contains (see Annexes 2 and 3):

- Food safety criteria and process hygiene criteria
- The commodity of concern
- The parameter for each commodity to be tested for
- The number to be sampled
- The limit for c (accepted number of n exceeding the limit m)
- Limits for m and M
- Amount of sample material to be taken and to be given into analysis
- Sampling days shall be changed each week to ensure coverage of each working day

From its basic philosophy, the Regulation states the following:

Foods should not contain microorganisms/toxins/metabolites in quantities that present an unacceptable risk for human health.

With this, the idea of zero tolerance was abandoned. On the contrary, here the idea of ALOP and of FSO is taken over as a schedule for practice, which has already been mentioned in the opinion of the SCVPH on Salmonella.

Food business operators have an obligation to withdraw unsafe food from the market.

Here, the idea of responsibility, of transparency, and of the obligation for action from the basic Hygiene Regulation (Reg. (EC) 178/2002) has been taken over.

The producer of the food decides on the commodity (ready to be consumed or with a need for cooking or other procedure in order to provide its safety and compliance with the said criteria, which should be labelled).

Here, the responsibility of the food producer is stated again, but also the freedom to decide upon the product category. However, the operator remains responsible, in particular, in case of an accident.

Sampling plans, analytic methods should be comparable for the comparability of results. However, methods other than reference methods should be allowed.

Here, the choice of the method has been left to the analyst. However, analysts should be able to make their applied method transparent and they should be able to make it comparable with the reference method.

For a certain transitorial time, a derogation should be granted. In such a case, the product should be legal only on the domestic market.

With this, a transitorial time for the adoption was granted out of practical reasons; however, with the product to be marketed only on the limited domestic market, a fast implementation of the content of this regulation should be expected.

Finally, studies may be performed on compliance with the criteria during the shelf life of a product.

Here, and with other comments also, the content is open for continuous further development, which should enable the commission to keep in contact with the users of such a regulation, to improve it and to find other solutions in case of impracticability. It should be kept in mind that in the EU, a historically grown and different picture of nutritional records and traditions exists.

GENERATING MICROBIOLOGICAL DATA IN PRACTICE

Basically, criteria must not be understood exclusively in the sense of the microbiological discipline. There exist also technical (e.g., temperature or time) or chemical criteria (e.g., residue limits or lower or upper limits of other—spoilage—substances). In any case, if using criteria, they should be clearly elaborated and located at the appropriate place in the line. More decisions should be made:

- The character of the criterion
- The stage of setting the criterion
- The laboratory aspects (prelaboratory, in-laboratory, postlaboratory)

Several aspects influence the outcome of a test. Much work has already been done on comparison of methods (in-laboratory), however, in particular preanalytic decisions have a huge influence on the practical use of an analytic result.

PRELABORATORY

Stage of processing, respectively, points of sampling:

Samples might be taken from very different places and stages: within/during the food chain or production line, at its end (end-product-control), or in the environment of a holding or a factory with a given machinery and processing line. It should also be considered whether the sampling should be done prior to or after packaging the original bulk.

Number of samples to be taken in relation to the lot for sampling:

In larger bulks, the level of aggregation of bacteria is of importance for the probability of recovery, and the sensitivity of a result depends (beside others) on the amount of samples taken. The more samples examined, the higher the probability of finding, e.g., Salmonella. If the prevalence of an environmental pathogen such as Salmonella in food commodities is regarded as low, the probability of detecting the pathogen might also be low, resulting in a perhaps false feeling of security when the samples are to be taken.

However, for purposes of international trade, the sampling plan has to be standardized, and two- or three-class plans are widely in use. The following are their advantages:

- Small and fixed amounts of samples (e.g., $n = 5$)
- One or two limits for the specific parameter to be examined (e.g., $m = \lg 3$; $M = \lg 5$)
- Fixing the number of samples allowed to exceed the limit of m (e.g., $c = 2$).

Two-class plans are performed as presence-absence tests.

The amount of materials to be sampled:

Especially in large bulks such as powders, the amount of material for the actual analysis must be considered. Here, different directives have requested different sampling size. Simultaneously, the frequency of sampling is a point of concern. In the Regulation (EC) 2073/2005, different frequencies have been established for specific parameters.

IN-LABORATORY

Qualitative (yes/no) approach versus quantitative testing:

Testing for Salmonella is hampered by the most frequently used yes/no procedure. In these cases, only the percentage of infected animals or contaminated samples can be recovered. Quantitative data are lacking as background information for the severeness of Salmonella as a prevalence in processing lines. Using a qualitative approach, pooling of samples is another option. Regulation (EC) 2073/2005 refers to this technique with fishery products (food safety criteria) and with poultry carcasses (process hygiene criteria), too.

The microbiological nature of a criterion as such:

The use of a criterion depends on the quality of information that it is intended to obtain. So, in the case of Salmonella, the particular serotype and phagetype should be determined. Moreover, molecular-biological criteria are appropriate in order to find out sources of a contamination and the epidemiological transfer pathways.

Methods and internal laboratory quality assurance:

Methods should follow particular criteria, which reflect the "truth" of a result, i.e., specificity, accuracy, precision, lower limit of detection, or sensitivity. Practicability and applicability also may be of interest (see also Figure 26.4):

- *Accuracy* (N.N. 1982): Ability of a test to give a true measure of the item being tested.
- *Repeatability* (Precision) (N.N. 1982): Ability of a test to give consistent results in repeated tests.
- *Sensitivity* (N.N. 1982): Ability of a test to give a positive result when an animal is diseased (or a sample is positive). It is measured as the proportion of samples (animals) that gives a positive test result. Too-low sensitivity may lead to false negative results.
- *Specificity* (N.N. 1982): Grade of exclusiveness of a reaction. The ability of a test to give a negative result when the sample is negative. Lack of specificity may lead to false positive results.

Methods applicable uniformly to various groups of foods should be given preference over methods that apply only to individual commodities. Methods to be used for legal purposes require validated and/or ISO or EN methods. National reference laboratories have been established for cases of specific problems.

Postlaboratory

Interpretation of the Test Results

For the assessment of a result, limits necessarily should be set. As can be demonstrated with the class plans, the number of c determines the number of batches to fall within the unwanted range. So, criteria can be set as a quantitative limit or as a yes/no result. This question remains to be solved with respect to each food commodity. For food categories referred to in this Regulation, this has been done.

It is also of importance to know which outcome is expected from this testing. The SCVPH (1999 I) as a scientific body of the EU has offered as characterization: "A microbiological criterion for foodstuffs defines the acceptability of a process, product or food lot based on the absence or presence, or number of microorganisms and/or quantity of their toxins/ metabolites, per unit of mass, volume or area."

As food safety items, the purpose of microbiological criteria is to protect the health of the consumer by providing safe, sound, and wholesome products, and to meet the requirements of fair practices in trade.

The mere existence of criteria does not protect consumer health per se. So, such criteria should be established and applied only where there is a need and where they can be shown to be effective, practical and realistic (SCVPH 1999 I).

The Character of Microbiological Criteria

Microbiological criteria currently used can be understood differently. Basically, we separate the terms of *standard*, *guideline*, and *specification*.

Use in the sense of standard:

The term *microbiological standard* has a legal meaning; it is based on legislation and a failure to comply with the limits, which results in actions such as rejection, withdrawal, or other.

Standards should be used only if the aim can be achieved. Otherwise, the tool becomes weak and counterproductive.

Use in the sense of guideline:

Microbiological guidelines are used to assess whether GMP has been followed in a food processing line. They may represent limits, which are possible

Type of Mistake		Accidental Mistake	Systematic Mistake	Bad Mistake
	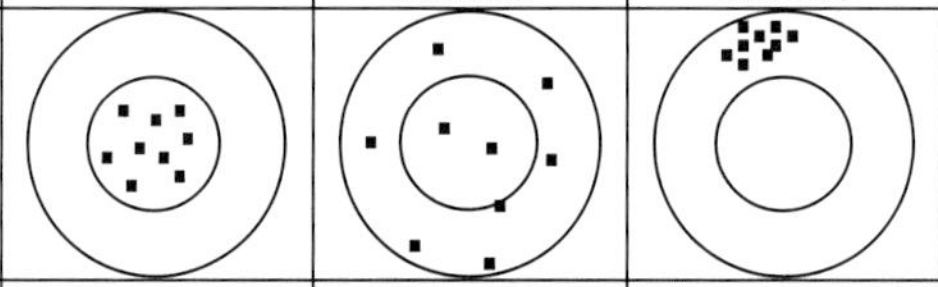			No Result
Precision	Optimal	Poor	Good	-
Accuracy	Optimal	Good	Poor	-

Figure 26.4. Precision and accuracy (Henniger 1980).

from the technological capacity, and they should not be exceeded.

In this sense, a test result reflects the appropriateness of a procedure within a given line. Data beyond the limit indicate possible failures, a need for correcting actions to be taken in technology in order to improve the hygiene at that specific point of the line. Such criteria are intended to guide the manufacturer and help ensure good hygiene practice.

In such a case, only the collecting of data is mandatory: For statistical purposes or in order to obtain more information about suspect circumstances and environments—at present without market or any other legal consequences—data are to be gathered (in the sense of a continuous sampling in a given lot) on a mandatory basis in order to get the requested information.

For poultry farming, for example, it would make sense to identify holdings with a high prevalence of Salmonella in order to focus on such holdings for reduction measures. Such a system has been shown to be useful over the years in pig production. This way the situation might improve stepwise, using Salmonella testing before setting a criterion in the sense of a standard.

Use in the sense of specifications:

Such criteria are used only for contract purposes between producer and customer. Such particular specifications are not binding for other contractors.

REFERENCES

DJR Cherney. 2004. Western Coordinating Committee—204 Goals and why they are important to the future of animal production systems. Poult Sci 83:307–309.

Decision 2001/471/EC of 8 June 2001 laying down rules for the regular checks on the general hygiene carried out by the operators in establishments according to Directive 64/433/EEC on health conditions for the production and marketing of fresh meat and Directive 71/118/EEC on health problems affecting the production and placing on the market of fresh poultry meat. Off. J. EC L165/48 of 21.6.2001.

Decision 2004/379/EC of 26 April 2004 amending Decision 2001/471/EC . Off. J. EC L144/1 of 30.4.2004.

Decision 93/51/EEC of 15 December 1992 on the microbiological criteria applicable to the production of cooked crustaceans and molluscan shellfish. Off. J. EC, L13/11 of 21.1.1993.

Directive 2004/41/EC of the European Parliament and of the Council of 21 April 2004 repealing certain Directives concerning food hygiene and health conditions for the production and placing on the market of certain products of animal origin intended for human consumption and amending Council Directives 89/662/EEC and 92/118/EEC and Council Decision 95/408/EC. Off. J. EC L195/12 of 2.6.2004.

Directive 71/118/EEC of 15 February 1971. Off. J. EC L055/23 of 8.3.1971.

Directive 91/497/EEC of 29 July 1991 Amending and consolidating Directive 64/433/EEC. Off. J. EC L268/69 of 24.9.1991.

Directive 92/117/EEC of 17 December 1992. Off. J. EC L62/38 of 15.3.1993.

J Dufrenne, W Ritmeester, E Delfgou-Van Asch, F Van Leusden, R de Jorge. 2001. Quantification of the contamination of chicken and chicken products in The Netherlands with Salmonella and Campylobacter. J Food Prot 64:538–541.

European Commission. 1998. Microbiological criteria: Collation of scientific and methodological information with a view to the assessment of microbiological risk for certain foodstuffs. Food—Science and techniques; Directorate—Gen. Industry, Brussels, EUR 17638 EN, 107 pp.

R Fries. 1988. Bakteriologische Prozeβkontrolle in der Geflügelfleischgewinnung.Habilitationsschrift, Tierärztliche Hochschule Hannover.

———. 2002. Reducing Salmonella transfer during industrial poultry meat production. World's Poult Sci J 58:527–540.

———. 2005. Spoilage Microorganisms in the Course of Poultry Processing. Feedinfo News Scientific Reviews. October 2005. Available from URL. http://www.feedinfo.com.

G Henniger. 1980. Enzymatische Analyse: Empfindlichkeit, Richtigkeit, Präzision, Fehlerquellen. Vortrag GDCh—Kursus: "Enzymatische Analyse" am 16.9.1980 in München.

AM Lammerding. 1997. An overview of microbial food safety risk assessment. J Food Prot 60:1420–1425.

DAA Mossel. 1979. The microbial associations of foods of animal origin. Archiv für Lebensmittelhygiene 30:82–83.

N.N. 1982. Livestock Disease Surveys: A Field Manual for Veterinarians. Australian Bureau of Animal Health, 35 pp.

———. 2002. Deutsches Lebensmittelbuch—Leitsätze 2002, Verkehrsbezeichnung, Qualität und Zusammensetzung. Bundesanzeiger Verlagsges. mbH, Köln, 398 S.

———. 2006. Veröffentlichte mikrobiologische Richt und Warnwerte zur Beurteilung von Lebensmitteln (Stand Juni 2005).(http: www.lm-mibi.uni-bonn.de/DGHM.html).

National Advisory Committee on Microbiological Criteria for Foods (NACMCF). 1998. Hazard Analysis and Critical Control Point Principles and Application Guidelines. J Food Prot 61:762–775.

J Petersen. 1997. Trends und Entwicklungen in der Zucht und haltung von Geflügel und Kaninchen. Vorträge der 49. Hochschultagung der Landwirtschaftlichen Fakultät der Universitüt Bonn vom 18.2.1997 in Bonn, pp 165–75. Landwirtschaftsverlag Münster-Hiltrup.

Regulation (EC) No 2073/2005 of 15 November 2005 on Microbiological Criteria for Foodstuffs. Off. J. EU L338/1 of 22.12.2005.

Regulation (EC) No 854/2004 of the European Parliament and of the Council of 29 April 2004. Off. J. EC L226/83 of 25.6.2004.

Regulation (EC) No. 178/2002 of the European Parliament and the Council of 28 January 2002. Off. J. EC L31/1 of 1.2.2002.

Regulation (EC) No. 2160/2003 of the European Parliament and the Council of 17 November 2003. Off. J. EC L325/1 of 12.12.2003.

Regulation (EC) No. 852/2004 of the European Parliament and the Council of 29 April 2004. Off. J. EC L226/3 of 25.6.2004.

Regulation (EEC) No 1538/91 of 5 June 1991 introducing detailed rules for implementing Regulation (EEC) No 1906/90 on certain marketing standards for poultry. Off. J. EC L143/11 of 7.6.1991.

Scientific Committee on Foods (SCF). 2002. Opinion of the SCF on specifications for gelatine in terms of consumer health. Adopted 27.2.2002. http://europa.eu.int/comm/food/fc/sc/scf/out122_en.pdf.

Scientific Committee on Veterinary Measures relating to Public Health (SCVPH). 1999 I. Opinion of the SCVPH on "The evaluation of microbiological criteria for food products of animal origin for human consumption." (http://europa.eu.int/comm/food/fs/sc/scv/outcome_en. html# opinions).

Scientific Committee on Veterinary Measures relating to Public Health (SCVPH). 1999 II. Opinion of the SCVPH on *Listeria monocytogenes* (adopted 23.9.1999). http://europa.eu.int/comm/food/fs/sc/scv/out25_en.pdf.

Scientific Committee on Veterinary Measures relating to Public Health (SCVPH). 2001. Opinion of the SCVPH on *Vibrio vulnificus* and *Vibrio parahaemolyticus* (in raw and undercooked seafood) (adopted 19–20 September 2001). http://europa.eu.int/comm/food/fs/sc/scv/out45_en.pdf.

Scientific Committee on Veterinary Measures relating to Public Health (SCVPH). 2002. Opinion of the SCVPH on Norwalk-like Viruses (adopted 30–31 January 2002). http://europa.eu.int/comm/food/fs/sc/scv/out49_en.pdf.

Scientific Committee on Veterinary Measures relating to Public Health (SCVPH). 2003 I. Opinion of the SCVPH on Verotoxogenic *E. coli* (VTEC) in foodstuffs (adopted 21–22 January 2003). http://europa.eu.int/comm/food/fs/sc/scv/out58_en.pdf.

Scientific Committee on Veterinary Measures relating to Public Health (SCVPH). 2003 II. Opinion of the SCVPH on Staphylococcal enterotoxins in milk products, particularly cheeses (adopted 26–27 March 2003). http://europa.eu.int/comm/food/fs/sc/scv/out61_en.pdf.

Scientific Committee on Veterinary Measures relating to Public Health (SCVPH). 2003 III. Opinion of the SCVPH on Salmonellae in Foodstuffs (adopted on 14–15 April 2003). http://europa.eu.int/comm/food/fs/sc/scv/onl66_en.pdf.

Scientific Panel on Biological Hazards. 2004. Opinion of the SPBIOHAZ on a request from the commission related to the microbiological risks in infant formulae and follow-on formulae. EFSA J 113:1–35.

ANNEXES

Annex 1: Microbiological Criteria for Foodstuffs in the Reg. (EC) 2073/2005

- Food Safety Criteria
  - 26 food categories
  - 6 parameters (microorganisms/toxins/metabolites)
    - Salmonella
    - Listeria monocytogenes
    - Staphylococcal enterotoxins
    - Enterobacter sakazakii
    - E. coli
    - Histamines
  - Different limits
  - Different sample numbers
  - Different stages of sampling
  - Different interpretation support
- Process Hygiene Criteria
  - Meat and meat products
  - Milk and dairy products
  - Egg products
  - Fishery products
- Vegetables, fruits, products thereof
- Rules for sampling

Annex 2 (Chapter 1 of Annex I of Reg. (EC) 2073/2005): Food safety criteria

Food category	Microorganisms/their toxins, metabolites	Sampling-plan (1) n	c	Limits (2) m	M	Analytical reference method (3)	Stage where the criterion applies
1.2. Ready-to-eat foods able to support the growth of *L. monocytogenes*, other than those intended for infants and for special medical purposes	*Listeria monocytogenes*	5	0	100 cfu/g (5)		EN/ISO 11290-2 (6)	Products placed on the market during their shelf-life
		5	0	Absence in 25 g (7)		EN/ISO 1129-0-1	Before the food has left the immediate control of the food business operator, who has produced it
1.3. Ready-to-eat foods unable to support the growth of *L.monocytogenes*, other than those intended for infants and for special medical purposes	*Listeria monocytogenes*	5	0	100 cfu/g		EN/ISO 11290-2 (6)	Products placed on the market during their shelf-life
1.5. Minced meat and meat preparations made from poultry meat intended to be eaten cooked	*Salmonella*	5	0	From 1.1.2006 Absence in 10 g From 1.1.2010 Absence in 25 g		EN/ISO 6579	Products placed on the market during their shelf-life
1.8. Meat products intended to be eaten raw, excluding products where the manufacturing process or the composition of the product will eliminate the salmonella risk	*Salmonella*	5	0	Absence in 25 g		EN/ISO 6579	Products placed on the market during shelf-life
1.9. Meat products made from poultry meat intended to be eaten cooked	*Salmonella*	5	0	From 1.1.2006 Absence in 10 g From 1.1.2010 Absence in 25 g		EN/ISO 6579	Products placed on the market during their shelf-life

[1] n = number of units comprising the sample; c = number of sample units giving values over m or between m and M.
[2] For points 1.1–1.24 m = M.
[3] The most recent edition of the standard shall be used.
[5] This criterion applies if the manufacturer is able to demonstrate, to the satisfaction of the competent authority, that the product will not exceed the limit 100 cfu/g throughout the shelf-life. The operator may fix intermediate limits during the process that should be low enough to guarantee that the limit of 100 cfu/g is not exceeded at the end of the shelf-life.
[6] 1 ml of inoculum is plated on a Petri dish of 140 mm diameter or on three Petri dishes of 90 mm diameter.
[7] This criterion applies to products before they have left the immediate control of the producing food business operator, when he is not able to demonstrate, to the satisfaction of the competent authority, that the product will not exceed the limit of 100 cfu/g throughout the shelf-life.
[8] Products with pH $\leq$ 4.4 or $a_w \leq 0.92$, products with pH $\leq$ 5.0 and $a_w \leq 0.94$, products with a shelf-life of less than 5 days are automatically considered to belong to this category. Other categories of products can also belong to this category, subject to scientific justification.

Interpretation of the test results

The limits given refer to each sample unit tested.
The test results demonstrate the microbiological quality of the batch tested. However, they can be used also for demonstrating the effectiveness of the HACCP or good hygiene practice of the process.
L. monocytogenes in ready-to-eat foods able to support the growth of *L. monocytogenes* before the food has left the immediate control of the producing food business operator when he is not able to demonstrate that the product will not exceed the limit of 100 cfu/g throughout the shelf-life:

- satisfactory, if all the values observed indicate the absence of the bacterium,
- unsatisfactory, if the presence of the bacterium is detected in any of the sample units.

L. monocytogenes in other ready-to-eat foods:

- satisfactory, if all the values observed are < the limit,
- unsatisfactory, if any of the values are > the limit.

Salmonella in different food categories:

- Satisfactorily, if all the values observed indicate the absence of the bacterium.
- Unsatisfactorily, if the presence of the bacterium is detected in any of the sample units.

Annex 3 (Chapter 2 of Annex I of Reg. (EC) 2073/2005): Process hygiene criteria

2.1. Meat and products thereof

Food category	Microorganisms	Sampling-plan ([1])		Limits ([2])		Analytical reference method ([3])	Stage where the criterion applies	Action in case of unsatisfactory results
		n	c	m	M			
2.1.5. Poultry carcases of broilers and turkeys	Salmonella	50 ([5])	7 ([6])	Absence in 25 g of a pooled sample of neck skin		EN/ISO 6579	carcases after chilling	Improvements in slaughter hygiene and review of process controls, origin of animals and biosecurity measures in the farm of origin

[1]n = number of units comprising the sample; c = number of sample units giving values between m and M.
[2]For points 2.1.3–2.1.5 m = M.
[3]The most recent edition of the standard shall be used.
[5]The 50 samples are derived from 10 consecutive sampling sessions in accordance with the sampling rules and frequencies laid down in Regulation (EC) 2073/2005.
[6]The number of samples where the presence of salmonella is detected. The c value is subject to review in order to take into account the progress made in reducing the salmonella prevalence. Member States or regions having low salmonella prevalence may use lower c values even before the review.

Interpretation of the test results

The limits given refer to each sample unit tested, excluding testing of carcases where the limits refer to pooled samples.
The test results demonstrate the microbiological quality of the process tested.
Salmonella in carcases:

- satisfactory, if the presence of *Salmonella* is detected in a maximum of c/n samples,
- unsatisfactory, if the presence of *Salmonella* is detected in more than c/n samples.

After each sampling session, the results of the last ten sampling sessions are assessed in order to obtain the n number of samples.

27
Packaging and Storage

Dong U. Ahn and Byungrok Min

INTRODUCTION

The primary function of food packaging is to prevent or retard the deterioration of nutritional, organoleptic, and aesthetic quality in foods and to provide the protection against environmental contamination. Packaging of meat products is defined as

> . . . the enclosure of products in a wrap, pouch, bag, box, cup, tray, can, tube, bottle or other container form to perform one or more of the following functions: 1) containment for handling, transportation and use; 2) preservation and protection of the contents for required shelf and use life; 3) identification of contents, quantity, quality and manufacturer; 4) facilitate dispensing and use [FSIS 2000].

Packaging of meat products can be practically classified into three levels: primary, secondary, and tertiary level (Dawson 2001). The primary level is the package that directly contacts with meat products, and consumers usually purchase the products with the primary package at retail. Therefore, it should provide most of the functions of food packaging, such as protection, information, etc. The most common packaging material for the primary packages of meat products is plastic film. The secondary level of packaging contains several single primary packages and protects the primary packages from environmental hazards during distribution. The tertiary packages hold many secondary packages for loading, unloading, and shipping. The most common example is a pallet with stretched wrap. This chapter focuses on the primary packaging.

Most fermented meat products consumed today are processed sausage-type products. They undergo several key processing steps, such as fermentation, drying, smoking, and cooking after the meat batter is stuffed into casings. At the end of the process, the whole sausages with casing or sliced products are packaged for storage and commercial distribution. The characteristics of final products and their distribution conditions for retail are very important factors that should be considered for the choice of packaging material and systems. This chapter describes the function of packaging, packaging materials, and packaging systems that can be applied to fermented meat products.

FUNCTIONS OF FOOD PACKAGING

The functions of food packaging can be categorized into four areas: containment, protection, convenience, and communication (Robertson 1993).

CONTAINMENT

Containment is the basic function of packaging. The products must be contained in appropriate packages before they are transferred. The products should be well contained in order for the packages to achieve their functions successfully.

Protection

Protection is considered as the major function of food packaging. Moreover, the packaging is regarded as one of the major preservation methods of foods and retards or prevents quality deterioration during handling, transportation, and storage. Packages should:

- Protect food products from the contamination of microorganisms and vermin to ensure safety and prevent spoilage, which is one of the hottest issues in foods.
- Minimize the transfer of heat and light energy, which can activate or accelerate chemical reactions such as lipid oxidation and discoloration and microbiological process.
- Prevent gas transfer from outside to inside and vice versa. Therefore, packages should serve as a barrier against oxygen, water vapor, and volatile compounds. The permeability of packages to oxygen is very important because oxygen is the most important factor that causes oxidative changes in meat products during storage. Packaging as a barrier for water and volatiles transfer prevents dehydration and weight loss, loss of good flavor from food products, and contamination of volatile compounds that influence the flavor of food products negatively.
- Prevent the physical damage by external force or pressure and the contamination of dust and foreign elements. However, it should be reminded that packaging cannot improve the original quality but retard the rate of quality loss.

Convenience

Packaging should be designed for the convenient use of products not only from the consumer's standpoint but also from the handler's. One of the consumer's great needs in a modern lifestyle is to eat and handle foods conveniently. Two concepts, the *apportionment* of products and the *shape* of the primary packaging in package design, have contributed to the improvement of the convenience in food packaging (Robertson 1993). The apportionment function of food packaging allows us to apportion industrial-sized products into serving-sized products, which can be consumed conveniently. The effective methods and package design for apportioning should be considered for individual food products because consumers' needs and manufacturing conditions for each food product vary. In addition, the proper shape of the primary package can improve not only the convenience in opening, storing, reclosing, and reusing by consumers but also improves the efficiency in making secondary and tertiary packages for easy handling, shipping, and storing, which are directly related to cost.

Communication

The package carries information that is required by law, wants to be known by consumers, and is claimed by the manufacturer. It provides an interface for the communications between consumers and manufacturers. The information that *must* be shown on the label of meat products is product name, USDA inspection legend, net weight, handling statement, ingredient statement, nutrition facts, date (related to shelf life), name and address of manufacturer, and safe handling instructions. In addition, packaging can serve as an excellent marketing tool and is called the "silent salesman." The purchasing desire of consumers can be boosted by package design, information provided, etc. For example, the use of transparent packaging enables consumers to observe the whole product and improve their sense of reliability about the product. Unique shape, colors, and graphics of packaging help consumers spot and identify the products at retail. In addition, providing information such as instructions for use, nutrients content, ingredients list, and expiration date is critical for customers purchasing the product.

PACKAGING MATERIALS

The most important factors that should be considered for the selection of suitable packaging materials are the identification of the most limiting factors for the preservation of food quality and safety, such as sensitivity to oxygen and water, shelf life, potential of microbial contamination, storage conditions, and development of requirements for the packaging materials on the basis of limiting factors. Other packaging features are strength factors such as tensile, elongation, tear strength, and resistance to puncture; permeability to gas, moisture, grease, and volatiles; machinability factors such as stiffness, static accumulation, and slip; marketing factors such as transparency, gloss, and film color; anticlouding characteristics; and cost (Kropf 2004).

The packaging materials can be classified into three categories on the basis of their properties: barrier, strength, and sealing properties (Dawson 2001; Lundquist 1987). For example, aluminum foil is usually used as a layer because of its barrier properties for light and gases. Polyester (polyethylene terephthalate, PET) is added for strength, and polyethylene as a sealing agent. The classification of major packaging materials for each property is as follows:

- *Barriers*: aluminum foil, ethylene vinyl alcohol (EVOH), polyvinyl chloride (PVC), polyvinylidene dichloride (PVDC, Saran®), and acrilonitriles (Barex®)
- *Strength properties*: polyester (PET), nylon, and polypropylene (PP)
- *Sealing properties*: polyethylene (PE), ionomers (Surlyn®), and polystyrene (PS)

No individual packaging material can satisfy all desired requirements for packaging materials. Thus, in most cases, a number of individual packaging materials are combined and arranged as a layer to produce a multilayer-structured material called a *film* that has the desired characteristics (Dawson 2000). The selection of specific individuals for a multilayer film to use for the packaging depends on the specific requirements of package for a specific product. The thickness of a film commonly ranges from 0.0254 mm (1 mil or 0.001 in) to 0.3054 mm (12 mil or 0.012 in) (Kropf 2004). Generally, the film consists of three or more layers. Each layer, arranged from outside to inside, contributes specific properties to overall structure: the outside layer should provide scuff/abrasion-resistance ability and printability. The middle layer(s) generally has (have) barrier properties, especially to oxygen and water vapor, and mechanical structure. The inner layer next to the product (called the *sealant layer*) should have heat sealability and compatibility with the product. Another important consideration in the selection of packaging materials, especially for the inner layer of a film, is the *migration*, which refers to the transfer of compounds from the packaging materials to food products by leaching or diffusion, resulting in food safety issues and flavor deterioration. Slip and antifogging agents, heat-degradable products from film, and ink components are considered major causes for the migration problem in plastic package (Driscoll and Paterson 1999). Therefore, those should be chemically inert.

Major quality deteriorations in fermented meat products are discoloration; development of lipid oxidation, which causes flavor deterioration; weight loss; and microbial spoilage. The pigment compounds that are responsible for the color of fermented meat products are nitric oxide myoglobin for uncooked and nitrosyl hemochromogen for cooked meat (Aberle et al. 2001). The color stability in fermented meat products is considerably influenced by light, oxygen, and microbial spoilage. Nitrosyl hemochromogen is very sensitive to light-induced discoloration in the presence of oxygen. In the absence of oxygen or under anaerobic conditions, nitric oxide separated from the heme group is not oxidized and can be reattached to the heme. Also, oxygen can accelerate lipid oxidation, which is closely related to the flavor deterioration and discoloration in meat. In addition, microorganisms are directly and indirectly involved in discoloration. Under aerobic conditions, some aerobes can produce hydrogen peroxide, which can oxidize heme pigments. Also, because most of the fermented meat products undergo a drying process, rehydration by residual moisture or inflow of outside water vapor can provide favorable environments for microbial growth, resulting in microbial spoilage. Therefore, the presence of moisture, oxygen, and/or air is the primary quality limiting factor for the fermented meat products. In addition, loss of moisture, loss of flavor compounds, and influx of undesirable volatiles can be other quality limiting factors during storage. Therefore, low permeability to oxygen, water vapor, and other volatiles; strength and durability throughout the distribution; machinability factors such as flexibility, heat sealability, shrinkability, and printability; marketing factors; and cost are important in selecting packaging materials for fermented meat products. Oxygen transmission rate (OTR) and water vapor permeability (WVP) are defined as "gas or vapor exchange rates, at a relative humidity and temperature conditions stipulated for a one mil thick (0.0254 mm or 0.001 in) film at one atmospheric pressure" (Dawson 2000).

Numerous packaging materials are available for fermented meat products: paper and coated-paper products; cellulose products; cellophane; metals such as tin, aluminum, and stainless steel; ceramics; glass; chemically treated rubber; and plastics. However, multilayer films made of various plastics are the most common packaging materials for the fermented meat products because of its versatile functions, cost, and convenience.

The following section is a brief description of individual packaging materials (Table 27.1) that can be used as a part of the multilayer film for the packaging of fermented meats.

Barrier Properties

Aluminum Foil

Aluminum foil provides an ultimate barrier for oxygen, water vapor, and light, and is heat- and cold-resistant. In addition, it offers a great surface for the application of graphics on the package. However, aluminum foil should be used with plastic polymers

Table 27.1. Properties of plastics for packaging of fermented meat products.

Plastics	Primary Functions	Oxygen Transmission Rate[2]	Water Vapor Transmission Rate[3]	References
Barrier Properties				
EVOH[1]	Oxygen barrier	0.2–1.6[4]	24–120[4]	Day (1995)
PVC[1] (unplasticized)	Oxygen and water vapor barrier; rigidity	120–160	22–35	Day (1995)
PVC[1] (plasticized)	High oxygen and water vapor permeability; high flexibility; good heat sealability	2000–10000[5]	200[5]	Day (1995)
PVDC[1] (Saran®)	Oxygen and water vapor barrier	0.8–9.2	0.3–3.2	Day (1995)
Acrilonitriles (Barex®)	Oxygen and water vapor barrier	15.5	77.5	Soroka (1999)
Strength Properties				
PET[1] (Polyester)	Strength; good heat and cold stability	50–100	20–30	Day (1995)
Nylon (nylon6)	Strength, thermostability; thermoformability	80[4]	200	Day (1995)
PP[1] (Oriented)	Strength; water vapor barrier; heat resistance	2000–2500	7	Day (1995)
Sealing Properties				
PE[1]	Heat sealing agent	2100–7100	7–24	Day (1995)
Ionomer (Surlyn®)	Excellent heat sealing agent	4650–6975 at 25°C	23–31	Dawson (2001)
PS[1]	Heat sealing agent; thermoformability	2500–5000	170	Day (1995)

[1]EVOH = ethylene vinyl alcohol; PVC = polyvinyl chloride; PVDC = polyvinylidene chloride; PET = Polyethylene terephthalate; PP = Polypropylene; PE = Polyethylene; PS = Polystyrene.

[2]$cm^3 \cdot mil \cdot m^{-2} \cdot day^{-1} \cdot atm^{-1}$ at 23°C, 0% RH unless noted. ORT: <50, barrier; 50–200, semi-; 200–5000, medium-; >5000, high-.

[3]$g \cdot mil \cdot m^{-2} \cdot day^{-1}$ at 38°C, 90% RH. WVTR: <10, barrier; 10–30, semi-; 30–100, medium; >100, high-.

[4]Depending on moisture.

[5]Depending on moisture and level of plasticity.

with sealing properties, such as PE, because the foil is not heat-sealable. Its common thickness in the multilayer film is 0.0003 to 0.0005 in. Although aluminum foil has been primarily used in flexible retort pouches, it is applicable for the packaging of fermented meat products because of its great barrier properties for oxygen, water vapor, and light.

Ethylene Vinyl Alcohol (EVOH)

Ethylene vinyl alcohol is produced by the saponification of ethylene, vinyl alcohol, and vinyl acetate polymers. The superior properties of EVOH are its impermeability to oxygen and other gases such as CO_2, N_2, and other volatiles. However, EVOH is hydrophilic and can absorb moisture due to the presence of hydroxy group in its backbone structure, and thus adversely affects oxygen barrier properties of EVOH under high humidity conditions. The gas barrier properties of EVOH are determined by the ratio between vinyl alcohol, which has gas barrier properties, and ethylene polymer, which has water-resisting properties. EVOH is often used as a core layer in multilayer film between layers of other polymers such as PE, PP, and PET with high moisture barrier properties. In addition, EVOH has high mechanical strength, elasticity and surface hardness, transparency, high gloss, good abrasion resistance, great barrier to oil and organic solvents, and heat stability. EVOH has been extensively used for the packages of various meat products, which need protection from oxygen.

Polyvinyl Chloride (PVC)

Polyvinyl chloride is produced by the polymerization of vinyl chloride monomers, and is simply called *vinyl*. Two major forms of PVC are used in food packages: unplasticized and plasticized. The unplasticized PVC is very clear and glossy, and it has high tensile strength and rigidity, shows low water vapor and oxygen permeability, and is resistant to oil and grease as well as acid and alkali. The plasticized PVC has excellent transparency and gloss, high flexibility, good heat sealability and high oxygen and water vapor permeability, and has been widely used for wrapping fresh red meat cuts at supermarkets in order to keep fresh meat color.

Polyvinylidene Dichloride (PVDC, Saran®)

Polyvinylidene dichloride is produced by the copolymerization of polyvinylidene and polyvinyl chloride. The characteristics of PVDC include excellent barrier properties to water vapor, oxygen, and other gases; good clarity; good resistance to grease; and great abrasion-resistance. It can be heat-sealed, and it is printable and heat-resistant. PVDC has been commonly used as a layer of multilayer pouches, bags, and thermoformed packages for various meat products such as frankfurters and ham. Furthermore, PVDC is preferably used for modified atmosphere packaging (MAP).

Acrilonitriles (Barex®)

Acrilonitriles provides excellent barrier properties to water vapor, oxygen, and other gases; good clarity; good strength; and good chemical resistance to most organic solvents. Acrilonitriles are commonly used for thermoformed plastic containers due to its rigidity.

Strength Properties

Polyester (Polyethylene Terephthalate, PET)

Polyester is produced by condensation of polyhydric alcohol such as ethyleneglycol and polyfunctional acid such as dimethyl terephthalic acid. PET provides the excellent tensile strength, great stability over a wide range of temperature (–60 to 220°C), good chemical resistance, outstanding mechanical properties, good clarity, excellent printability, and moderate barrier properties to oxygen, water vapor, and other gases for the food packaging film. PET is applicable to vacuum packaging, sterilizable pouches, and cook-in-bag applications for meat products because of its strength and heat stability.

Nylon

Nylon, polyamide, is produced by the condensation of diamines and diacids. Nylon provides excellent heat resistance, thermal stability, thermoformability, and low temperature flexibility; it is strong and abrasion-, alkali-, acid-resistant, and offers good barrier properties to oxygen and other gaseous phases. However, nylon can absorb moisture and is well permeated by water vapor, resulting in weakening tensile strength and lowering oxygen barrier properties. Therefore, it is often used as a layer of multiple layer film with other layers such as PE and PP, which can provide moisture barrier properties and heat sealability. It can be used for vacuum packaging, pouches, cook-in-bag, and ovenproof trays for meat products because of its strength, thermal resistance, and stability.

Polypropylene (PP)

Polypropylene is a polymerized product from propylene monomers. Two types of PP depending on its manufacturing processes have been generally used as packaging materials: disoriented (or cast, CPP) and oriented (OPP). CPP is very weak, so that it is not recommended for the layer with strong properties. CPP is primarily used as a sealing layer in retort pouches due to its relatively high temperature-resistance. OPP has very good tensile strength, great water vapor barrier properties, resistance to high temperature, clarity, and rigidity. For meat products, OPP can be applied to cook-in-steam or boil processed products and cook-in-bag products because of its high heat resistance and water vapor impermeability in boiling or steam cooking.

Sealing Properties

Polyethylene (PE)

Polyethylene is polymerized from polyethylene monomers, and is generally divided into three major types with different molecular structure, properties, and manufacturing processes: low density (LDPE), linear low density (LLDPE), and high density PE (HDPE). LDPE is a good sealing agent at relatively low temperature, which can be fused into itself in the opposite layer to make good liquid-tight seals. LDPE is slightly semitransparent; has good tensile, impact, and tear strength. It also has excellent resistance to acids, alkalis, and inorganic solvents, but is sensitive to oil, grease, and organic solvents. In addition, LDPE provides good barrier properties to water vapor, but poor to oxygen and other gases, which makes it appropriate for packaging of fresh fruits and vegetables. LLDPE has greater chemical resistance, higher strength, and higher resistance to stress cracking than LDPE. HDPE shows greater heat resistance, tensile strength, rigidity, and chemical resistance than LDPE. In addition, HDPE provides improved barrier properties to water vapor as well as oxygen and other gases. However, HDPE has less translucence, lower impact and tear strength, and considerably worse sealability than LDPE. PE is a relatively low-cost packaging material.

Ionomer (Surlyn®)

Ionomer (Surlyn®) is produced by polymerizing ethylene with a small amount of unsaturated carboxylic acid, such as methacrylic acid, followed by neutralization of the polymers with derivatives of metals, such as sodium or zinc, which results in the polymers with low levels of covalently attached sodium or metal ions that form ionic cross-links to provide enhanced rigidity. Ionomer is an excellent sealing agent and is used over a wide range of conditions. Ionomer is primarily used for the meat-contact and heat-sealing layer in multiple layer films because of its broad range of heat-sealing temperature and great adhesion capability to other packaging materials including aluminum foil. In addition, ionomer has good oil and grease resistance, good clarity, and resistance to stress cracking. It is commonly used as a sealing agent of vacuum packages for processed meat products, but is expensive.

Polystyrene (PS)

Polystyrene (PS) is produced by the polymerization of styrene. PS is transparent, hard, fragile, and low-strength material, which can be thermoformed into disposable containers or clear trays. PS can be foamed to form expanded polystyrene (EPS) (Styrofoam), which is thermoformed into trays for fresh meats. Both PS and EPS have poor oxygen and water vapor barrier properties as well as good resistance to acids and alkalis.

PACKAGING SYSTEMS

The major quality changes that limit the shelf life of fermented meat products include discoloration, lipid oxidation, rehydration, dehydration, and microbial spoilage. In addition to selecting suitable packaging materials, appropriate packaging systems should be applied in order to retard or prevent those unfavorable quality changes in the products during storage and distribution. Because barrier properties to oxygen, water vapor, and other gases are crucial for packaging materials, the capability of removing or minimizing headspace inside the packages or changing headspace to low oxygen conditions is important for packaging systems. Among available packaging systems, vacuum packaging and modified atmosphere packaging (MAP) have been widely used for fermented meat products. In recent years, active packaging has been introduced and intensively developed in response to consumer and market demands. Major packaging systems including vacuum packaging, modified atmosphere packaging (MAP), active packaging, and aseptic packaging are described below.

Vacuum Packaging

Vacuum packaging is the primary packaging system used in fermented meat products to delay or prevent the quality deterioration by oxygen. Vacuum packaging is generally described as a packaging system in which all air in a gas-impermeable container is substantially evacuated by vacuum pump before sealing the container. Vacuum packaging provides oxygen-free or, strictly speaking, extremely low levels of oxygen environment for the products to prevent or delay quality deterioration such as discoloration, lipid oxydation flavor deterioration, weight loss, rehydration, and microbial spoilage in packaged products. The extremely low level of oxygen as well as production of carbon dioxide (CO_2) from the products in the container not only prevents the growth of aerobic microorganisms but also provides a favorable condition for facultative and/or anaerobic microorganisms, such as lactic acid bacteria, which can produce lactic acids (Kropf 2004). These conditions are also very helpful for protecting color fading in cured meat products because oxygen plays a critical role in the oxidation of heme pigments such as nitrosyl hemochromogen, the major heme pigments in cooked, cured meat products.

The most critical properties of packaging materials required for vacuum packaging are excellent barrier properties to gases, especially oxygen and water vapor, in order to maintain vacuum condition during storage and distribution. The flexibility of material to be tightly attached to the entire surface of products and good puncture resistance are also important properties. Because no plastic material can meet all requirements, a multilayer film of several plastics with different properties has been usually used for vacuum packaging. Among the materials, nylon and PE have been commonly used for vacuum packaging (Toldrá et al. 2004). Nylon (polyamide) is an ideal material for vacuum packaging because of its high tensile strength, good puncture and abrasive resistance, and good barrier properties to oxygen and other gases. However, nylon is hydrophilic and has poor heat sealability, and is generally laminated with PE. In nylon/PE multilayer film, nylon serves as a barrier to oxygen and other gases and a mechanical supporter, and PE is a barrier to water vapor and a heat sealing agent. The number of layers in a multilayer film is dependent upon designated shelf life and storage and distribution conditions. For the products that need long shelf life and are stored and distributed at room temperature, as in most of the fermented meat products, plastic layers with more powerful barrier properties to oxygen and other gases such as EVOH and PVDC should be added to the film (Toldrá et al. 2004). However, products with short shelf life and stored at refrigerated temperature can be packaged with a film with reduced barrier properties or even single-layer HDPE.

Several vacuum systems have been developed on the basis of how to make air-deficient environment within the package: vacuum clip, snorkel, Pi-vac, and vacuum chamber systems (Kropf 2004). In vacuum chamber systems, a flexible pouch or bag containing product is placed in the vacuum chamber; this is followed by vacuuming out. The open end of the bag is heat-sealed, and then the chamber is exposed to atmosphere so that the package is collapsed onto products by higher external pressure. The package collapse may cause deformation if the product does not have certain extent of resistance. Therefore, vacuum packaging may not be appropriate for less-firm products. Use of a thermoformed plastic tray as a container can be a good approach for soft products. PP, PS, nylon, or PET can be used to provide structural properties in laminates for thermoformed base trays, where those are combined with an EVOH as a gas barrier and PE as a heat sealing agent (e.g., PP/EVOH/PE, PS/EVOH/PE, nylon/EVOH/PE, or PET/EVOH/PE structures) (Toldrá et al. 2004).

Modified Atmosphere Packaging

Modified atmosphere packaging (MAP) is one of the most common preservative technologies wherein air in the package is replaced by a single gas or a mixture of inert gases, such as carbon dioxide and nitrogen to prevent quality deterioration and microbial spoilage of food products. The gas composition of within-package atmosphere is not controlled after incorporation of gas mixtures and sealing, but can be changed by metabolism within the products itself and/or microorganisms present, diffusion of certain gases into or out of the products, and transfer of gases through the packaging film into or out of the package (Kropf 2004).

In practice, the products are retained within a flexible pouch or bag, or on the thermoformed base tray, and then a single gas or gas mixtures are flushed to modify the atmosphere, followed by sealing. There are two methods to modify the atmosphere within the package: gas flushing and compensated vacuum gas flushing (Mullan and McDowell 2003). The gas flushing method is a simple method wherein air within the package is removed by flushing with a continuous gas stream before sealing. It is less effective for removing air and consumes large amounts of flushing gas although the operation rate is very fast. Thus, the gas flushing method is not suitable for oxygen-sensitive products. On the other hand, the compensated vacuum gas flushing method consists of two steps: evacuation and gas flushing steps. The air within a package is evacuated to achieve a vacuum state, and, subsequently, the package is flushed with modified atmosphere. The consumption of flushing gas for this method is low, but the operation rate is slow. This is a more efficient method in decreasing residual air (especially oxygen) than the gas flushing method, and thus is a better method for oxygen-sensitive products such as fermented meat products. Zanardi et al. (2002) reported that MAP filled with 100% nitrogen is more efficient than vacuum packaging in preventing lipid oxidation and discoloration in Milano-type fermented sausage under commercial conditions because of lower residual oxygen content in MAP.

Three major gases have been commonly used in MAP: oxygen, carbon dioxide, and nitrogen. Others, such as carbon monoxide, sulfur dioxide, nitrous oxide, ozone, chlorine, and argon also have been investigated for use in MAP (Davies 2003). Gas mixtures with various compositions can be used in MAP depending on the features of products. As mentioned previously, it is essential to minimize the amount of residual oxygen within a package atmosphere for fermented meat products. Therefore, modified atmosphere within a package is generally composed of carbon dioxide and nitrogen: commonly, 20–30% carbon dioxide and 70–80% nitrogen (Toldrá et al. 2004). Carbon dioxide is well known to have inhibitory effects on the growth of aerobic microorganisms, especially gram-negative aerobes such as pseudomonads, resulting in delaying the initiation of microbial spoilage. It also has shown to have inhibitory effect on enzymatic activities. Increasing the proportion of CO_2 within the atmosphere of a package up to about 25% improves the antimicrobial effect (Kropf 2004). In addition, the antimicrobial effect of CO_2 is significantly greater at temperatures below 10°C than at 15°C or higher (Mullan and McDowell 2003). On the other hand, carbon dioxide promotes the growth of psychotropic lactic acid bacteria and increases the production of lactic acid (Kropf 2004). Carbon dioxide is soluble in water, lipids, and some other organic compounds. Carbon dioxide, therefore, can be dissolved in meat products, and its solubility increases with decreasing temperature. Due to its solubility in meat products and loss during storage, CO_2 should be sufficiently introduced into the package atmosphere to maintain its antimicrobial effect. In addition, the loss of CO_2 can cause reduction of headspace volume in the package, which results in package collapse onto products. Another important gas consisting of modified atmosphere for fermented meat products is nitrogen. Nitrogen is relatively inert and has a low solubility in water and other food components. Therefore, it is generally used as a filler gas (about 70–80% for meat products) to avoid the possibility of package collapse when a high concentration of carbon dioxide is used. Nitrogen as a filler gas plays a significant role in the dilution of residual oxygen to a minimal level, resulting in prevention or delay of oxidative changes in products. Nitrogen does not have antimicrobial effects, but provides an anaerobic condition where the growth of aerobic microorganisms responsible for microbial spoilage is inhibited.

Various types of containers (e.g., flexible pouches or bags and thermoformed trays) can be used for MAP. Use of flexible pouches is cost effective because they can be applied for fast gas-flushing methods. However, without adequate information, consumers may consider the MAP flexible pouches as vacuum-failed, impaired products because of the baggy appearance, resulting in rejection of the products (Toldrá et al. 2004). Therefore, semirigid or rigid thermoformed trays have been generally utilized as the main container for MAP in fermented meat products. The requirements of packaging materials for MAP are similar to those of vacuum packaging, and thus the same approaches can be used for the selection of packaging materials and design of multilayer films for MAP.

Active Packaging

In recent decades, huge attention has been paid to active packaging as a promising technology, capable of satisfying increased consumer demand for safer, more convenient, and better quality foods as well as changes in business environments such as diversification of markets (globalization, internet-shopping, etc.), increased complexity of logistics and distribution, and more tightened law. *Active packaging* is defined as an innovative technology where certain additives are introduced into a packaging film or within a packaging container and provide desired roles, by the interaction with headspace or foods, to extend shelf life and maintain and/or improve nutritional and organoleptic quality and microbial safety of foods (Labuza and Breene 1989; Day 2003). Packaging may be termed *active* when it provides desired roles for food preservation other than an inert barrier to external environments (Day 2003). The prerequisites for the development and application of active packaging for specific food products are to understand the mechanisms of quality deterioration and identify the key factors limiting shelf life. The development and application of an active packaging system is based on a fine match of packaging properties to the requirements of packaging for specific food products (Rooney 1995). As shown in Table 27.2, various types of active packaging techniques have been introduced and developed: oxygen scavengers, carbon dioxide scavengers/emitters, moisture absorbers, ethanol emitters, antimicrobial

Table 27.2. Various active packaging systems applicable for fermented meat products.[1]

Type of Active Packaging System	Core Substances Applied and Mechanisms
Oxygen scavengers	Chemical system Oxidation of iron-based powder with catalysts Sulfite salt/copper sulfate Oxidation of ascorbic acid Oxidation of photosensitive dyes/singlet oxygen acceptor Oxidation of oxidizable polymers Oxidation of unsaturated fatty acids Consumption by immobilized yeast on solid materials Conversion of oxygen to water vapor by platinum catalysts Enzymatic system Reaction of glucose oxidase and alcohol oxidase with substrates
Carbon dioxide	Ferrous carbonate/metal halide
Scavengers/emitters	Ascorbic acid/sodium bicarbonate Iron powder-calcium hydroxide Calcium oxide/hydrating agents (e.g., silica gel, activated charcoal)
Ethanol emitters	Absorbed or encapsulated ethanol Ethanol spraying
Antimicrobial releasers	Organic acids (e.g., acetic acid, citric acid, benzoic acid, sorbic acid) Salts and acid anhydrides of organic acids Bacteriocins (e.g., nisin, pediosin, lacticin) Enzymes (e.g., lysozyme, lactoperoxidase, chitinase) Metals (e.g., silver zeolite, copper) Antibiotics (e.g., natamycin) Fungicides (e.g., sulfur dioxide, benomyl, imazail) Metal chelators (e.g., EDTA, lactoferrin, conalbumin) Natural phenols (e.g., catechin, hydroquinone) Oligo and polysaccharides (e.g., chitosan and its oligomer) Plant volatile compounds (e.g., allylisothiocyanate) Antimicrobial peptides (e.g., attacin, cecropin, magainin) Extracts (e.g., grapefruit seed extract) and essential oils Amines and ammonium compounds
Antioxidant releasers	BHA, BHT, TBHQ, tocopherol
Moisture absorbers	Silica gel, natural clays (e.g., montmorilonite), calcium oxide, molecular sieves Superabsorbent polymers (e.g., polyacrylate salts, graft copolymers of starch, propylene glycol)
Flavor/odor absorbers	Baking soda, activated carbon/clays, active cellulose acetate, citric acid, ferrous salt/organic acids
Flavor/odor releasers	Desired food flavors
Light absorbers/regulators	UV blocking agents, hydroxybenzophenone
Time-temperature indicators	Color change or development on packages

[1] *Sources*: Vermeiren et al. (1999); Suppakul et al. (2003); Day (2003); Ozdemir and Floros (2004).

and antioxidant releasers, flavor/odor absorbers/releasers, temperature control packaging, and selective gas permeable films (Ozdemir and Floros 2004).

The major factor for quality deterioration in fermented meat products is presence of oxygen, which facilitates oxidative rancidity, flavor deterioration, discoloration, loss of nutrition, and growth of aerobic bacteria, mold, and fungi. Therefore, oxygen scavengers are highly applicable to fermented meat products, which are sensitive to oxygen. Various techniques are currently available for oxygen scavenging systems. The majority of commercial oxygen scavenging systems is based on the oxidation of iron powder with catalysts. The most well-known technique is incorporation of a small size of sachet containing oxygen scavengers such as iron-based powders along with catalysts into the package (Ozdemir and Floros 2004). The major disadvantage of this method is the possibility of accidental ingestion of the contents in the sachet by consumers. Thus, as alternatives to sachet, other new approaches have been developed: incorporation of oxygen scavenging layers into laminated trays and plastic films or inserts in the form of flat packets, cards, or sheets; and oxygen scavenging adhesive labels, which can be applied to the inside of packages (Vermeiren et al. 1999). An enzymatic oxygen scavenging system, which uses oxidation enzymes such as glucose oxidase and alcohol oxidase, is also available, either in packaging structure–incorporated or sachet-inserted form (Vermeiren et al. 1999). In addition, several other technologies such as oxidation of ascorbic acid, photosensitive dyes coupled with singlet oxygen acceptors, oxidizable polymers and unsaturated fatty acids, immobilized yeast on solid materials, and catalytic conversion of oxygen to water vapor by platinum in the presence of hydrogen gas are also available as oxygen scavenging systems (Labuza and Breene 1989; Vermeiren et al. 1999; Ozdemir and Floros 2004). Oxygen scavengers are commonly used in association with vacuum packaging or MAP to remove residual oxygen within a food package. The use of oxygen scavengers in combination with a film that has intermediate oxygen barrier properties is more cost effective than using a film with very high oxygen barrier properties alone (Vermeiren et al. 1999).

Carbon dioxide emitters are useful active packaging materials for fermented meat products because CO_2 is effective in preventing the growth of surface microorganisms. Because CO_2 is more permeable than oxygen through most films used in food packaging, it is very difficult to control the concentration of CO_2 by film alone (Labuza and Breene 1989). The CO_2 emitters on the basis of either ferrous carbonate or a mixture of ascorbic acid and sodium bicarbonate can dually function as both CO_2 emitter and possible oxygen scavenger (Vermeiren et al. 1999). The use of this dual-functioned system in MAP with carbon dioxide, which is soluble to water and lipid, is highly useful in extending shelf life.

Other useful packaging for fermented meat products is antimicrobial and antioxidant active packaging systems. Ethanol is widely recognized as a microbial growth suppressor. Ethanol emitters are generally used in sachets where ethanol is absorbed or encapsulated in carrier materials and is released by exchange with water vapor (Ozdemir and Floros 2004). Therefore, the rate of ethanol release from sachet to head space within a package depends upon the permeability of sachet barrier to water vapor. Ethano-emitting films with an adhesive-backed film that can be attached on the inside of a package is also available (Suppakul et al. 2003).

In recent years, however, most attention has been paid to the development of antimicrobial active packaging systems because of food safety concerns. In general, the package for antimicrobial active packaging systems is made by incorporating and immobilizing antimicrobial agents into a film or by surface medication and surface coating of the agents (Suppakul et al. 2003). The mode of actions of antimicrobial agents in or on a film can be divided into three categories: (1) direct diffusion to the products in the packaging system without headspace (e.g., vacuum packaging); (2) in addition to (1), equilibrium evaporation from the film into the headspace followed by equilibrium sorption into the products; and (3) surface inhibition of microbials by surface-immobilized antimicrobial agents without migration (Quintavalla and Vicini 2002). Antimicrobial agents that can be applied to active packaging are countless: for example, organic acids (e.g., acetate, citrate, sodium benzoate, and sodium sorbate and their acid anhydrides), bacteriocins (e.g., nisin, pediosin, lacticin), enzymes (e.g., lysozymes), metals (e.g., silver and copper), fungicides (e.g., sulfur dioxide, benomyl, and imazalil), oligo and polysaccharides (e.g., chitosan), antimicrobial peptides, amines, plant volatiles (e.g., allylisothiocyanate), metal chelators (e.g., ethylenediamine tetra acetic acid (EDTA), lactoferrin, and conalbumin), and natural phenols (e.g., catechin, extracts of grapefruit seed, essential oils) (Suppakul 2003; Cha and Chinnan 2004). Factors that should be taken into consideration in designing and manufacturing antimicrobial films include the chemical nature of films/coatings, casting process conditions, residual antimicrobial activity, chemical interaction of additives with film matrix, characteristics of antimicrobial agents and foods, storage temperature, mass transfer coefficient, physical properties of packaging materials, food contact approval of antimicrobial agents, and cost (Quintavalla and Vicini 2002; Suppakul 2003).

Antioxidants such as BHT and BHA can be incorporated into plastic films in order to stabilize polymers and thereby prevent film degradation. These antioxidants are able to migrate into headspace or products, resulting in an extended shelf life (Vermeiren et al. 1999). Other active packaging materials such as moisture absorbers, flavor/odor absorbers/releasers, time-temperature indicators, and light absorbers/regulators are applicable for fermented meat products.

Aseptic Packaging

Aseptic packaging is defined as the processes wherein products, which are commercially sterilized or not, are placed in sterile packages under aseptic conditions and then the packages are hermetically sealed (Robertson 1993). The main objective of aseptic packaging is to extend the shelf life of products at normal temperatures by reducing the initial microbial contamination and avoiding recontamination by microbes. There are two specific fields of application for aseptic packaging depending on whether the products are sterilized or not before filling. The requirements for aseptic packaging are eradication of microorganisms and protection of the packaging atmosphere for preventing any recontamination. In order to sterilize the food contact surface of packaging materials, numerous methods have been applied: irradiation, such as ultraviolet, infrared, and ionizing-radiation; steam; hot air; extrusion; and chemical treatment with hydrogen peroxide, peracetic acid, ethylene oxide, and so on (Robertson 1993).

STORAGE

The major characteristics of fermented meat products are low pH and water activity, which can prevent the growth of food poisoning microbes during storage. The presence of salt, nitrite, and/or smoking components added during processing inhibits bacterial growth during storage. Therefore, if whole processes are well monitored by a well-designed HACCP system, microbiological risk would not be the major concerns for fermented products. However, it is recommended to keep the products below 15°C for semidry sausages and below 25°C for dry and mold-ripened sausages during storage in order to control the growth of major pathogens such as *Salmonella enteritica*, *Staphylococcus aureus*, and *Listeria monocytogenes* (Lücke 1998).

For fermented meat products, chemical and physical changes during storage are more critical than microbial concerns. The major quality deteriorations in fermented meat products during storage are discoloration, development of off-odors and off-flavors, dehydration, and weight loss. Vacuum packaging or MAP with a high barrier plastic film to oxygen and moisture can prevent or retard those primary quality deteriorations during storage. However, vacuum packaging and MAP is not suitable for all fermented meat products, especially mold-ripened sausages (Incze 2004). The surface of mold-ripened sausages should be dry to maintain ideal appearance, but no film can meet the properties of high moisture and low oxygen permeability. Therefore, the quality of mold-ripened sausages is more affected by storage conditions. In this case, MAP combined with an active packaging system such as moisture absorbers may be one of the solutions to maintain the product's appearance and prevent oxidative reactions and weight loss at even room temperature during storage. In addition, during cold storage of mold-ripened sausages, phosphate crystals can appear on the products.

An important food safety risk in fermented meat products is production of biogenic amine, toxic compounds for human health. The amounts of biogenic amines can be increased to hazardous levels during storage at room temperature (Komprda et al. 2001, 2004). Therefore, it is recommended that fermented meat products be stored and distributed at refrigerated temperature—not only because the rates of chemical reactions such as lipid oxidation and discoloration, as well as transmission of oxygen, through all plastic films are generally reduced as temperature is lowered (Romans et al. 2001), but also because the production rate of biogenic amines during storage can be decreased at refrigerated conditions (Komprda et al. 2001, 2004).

REFERENCES

ED Aberle, JC Forrest, DE Gerrard, and EW Mills. 2001. Principle of Meat Science. 4th Ed. Dubuque, Iowa: Kendall/Hunt Publishing Co. DS Cha and MS Chinnan. 2004. Biopolymer-based antimicrobial packaging: A review. Crit Rev Food Sci Nutr 44:223–237.

AR Davies. 2003. Modified atmosphere and vacuum packaging. Ch. 11 in: Food Preservatives, 2nd ed. NJ Russell and GW Gould, eds. New York: Kluwer Academic/Plenum Publishers, pp. 218–239.

PL Dawson. 2001. Packaging. Ch. 6 in: Poultry Meat Processing. AR Sams, ed. New York: CRC Press, pp. 73–95.

BPF Day. 1995. Active packaging for fresh produce. In: Foods and Packaging Materials—Chemical Interactions. P Ackermann, M Jägerstad, and T Ohlsson, eds. Cambridge, UK: The Royal Society of Chemistry, pp. 189–200.

———. 2003. Active packaging. Ch. 9 in: Food Packaging Technology. R Coles, D McDowell, and MJ Kirwan, eds. Boca Raton, Florida: CRC Press LLC., pp. 303–339.

RH Driscoll and JL Paterson. 1999. Packaging and food preservation. Ch. 23 in: Handbook of Food Preservation. MS Rahman, ed. New York: Marcel Dekker, Inc., pp. 687–733.

Food Safety and Inspection Service (FSIS). 2000. Meat packaging materials. http://www.fsis.usda.gov/OA/pubs/meatpack.htm.

K Incze. 2004. Mold-ripened sausages. Ch. 24 in: Handbook of Food and Beverage Fermentation Technology. YH Hui, L Meunier-Goddik, AS Hansen, J Josephsen, W Nip, PS Stanfield, and F Toldrá, eds. New York: Marcel Dekker, Inc., pp. 445–458.

T Komprda, J Neznalovà, S Standara, and S Bover-Cid. 2001. Effect of starter culture and storage temperature on the content of biogenic amines in dry fermented sausage poličan. Meat Sci 59:267–276.

T Komprda, D Smělá, P Pechová, L Kalhotka, J Štencl, and B Klejdus. 2004. Effect of starter culture, spice mix and storage time and temperature on biogenic amine content of dry fermented sausages. Meat Sci 67:607–616.

DH Kropf. 2004. Packaging in Encyclopedia of meat science, v. 3. WK Jensen, C Devine, and M Dikeman, eds. New York: Elsevier Academic Press, pp. 943–976.

TP Labuza and WM Breene. 1989. Application of "active packaging" for improvement of shelf-life and nutritional quality of fresh and extended shelf-life foods. J Food Process Preserv 13:1–69.

FK Lücke. 1998. Fermented sausages. Ch. 14 in: Microbiology of Fermented Foods, vol. 2, 2nd Ed. BJB Wood, ed. New York: Blackie Academic & Professional, pp. 441–483.

BR Lundquist. 1987. Protective packaging of meat and meat products. Ch. 14 in: The Science of Meat and Meat Products, 3rd Ed. JF Price and BS Schweigert, eds. Westport, Connecticut: Food & Nutrition Press, Inc., pp. 487–505.

M Mullan and D McDowell. 2003. Modified atmosphere packaging. Ch. 10 in: Food Packaging Technology. R Coles, D McDowell, and MJ Kirwan, eds. Boca Raton, Florida: CRC Press LLC., pp. 303–339.

M Ozdemir and JD Floros. 2004. Active food packaging technologies. Crit Rev Food Sci Nutr. 44: 185–193.

S Quintavalla and L Vicini. 2002. Antimicrobial food packaging in meat industry. Meat Sci 62:373–380.

G L Robertson. 1993. Food Packaging: Principle and Practice. New York: Marcel Dekker, Inc.

JR Romans, WJ Costello, CW Carlson, ML Greaser, and KW Jones. 2001. The Meat We Eat, 14th ed. Danville, Illinois: Interstate Publishers, Inc.

ML Rooney. 1995. Overview of active food packaging. Ch. 1 in: Active Food Packaging. ML Rooney, rd. New York: Blackie Academic & Professional, pp. 1–37.

W Soroka. 1999. Fundamentals of Packaging Technology. Herndon, Virginia: Institute of Packaging Professionals.

P Suppakul, J Miltz, K Sonneveld and SW Bigger. 2003. Active packaging technologies with an emphasis on antimicrobial packaging and its applications. J Food Sci 68:408–420.

F Toldrá, R Gavara and JM Lagarón. 2004. Fermented and dry-cured meat: Packaging and quality control. Ch. 26 in: Handbook of Food and Beverage Fermentation Technology. Y H Hui, L Meunier-Goddik, AS Hansen, J Josephsen, W Nip, PS Stanfield, and F Toldrá, eds. New York: Marcel Dekker, Inc., pp. 445–458.

L Vermeiren, F Devlieghere, M van Beest, N de Kruijf, and J Debevere. 1999. Developments in the active packaging in foods. Trends Food Sci Technol 10:77–86.

E Zanardi, V Dorigoni, A Badiani, and R Chizzolini. 2002. Lipid and colour stability of Milano-type sausages: Effect of packing conditions. Meat Sci 61:7–14.

Part VI

Semidry-fermented Sausages

28
U.S. Products

Robert E. Rust

INTRODUCTION

Semidry sausages in the United States are generally considered to be acidified processed meat products with a Moisture/Protein Ratio (MPR) of 3.7:1 or less. Once an MPR of 2.3:1 is reached, the product would be considered a dry sausage. For the purposes of this discussion, we will concentrate on those products that fall into the MPR range of 3.7:1. With semidry sausages, shrinkage of about 15% would be expected during processing. The U.S. Department of Agriculture Food Safety Inspection Service (USDA-FSIS) requires a "Keep Refrigerated" label on sausage products unless these are validated as being shelf-stable. Previously they were considered to be shelf-stable when they reached an MPR of 3.1:1 and a pH of 5.0 or less. Because the requirements for a shelf-stable semidry sausage are constantly changing, it is recommended that the processor consult the current regulatory standards.

METHODS OF ACIDIFICATION

Most U.S. semidry products are fermented to a pH of approximately 4.8, although in some regions of the U.S. a pH of as low as 4.5 is considered desirable. Rapid fermentation is considered desirable, not only for safety considerations but for maximizing production as well. It is important that the product reach a pH of 5.3 or less before coagulase-positive *Staphylococcus* can grow and produce toxin.

Chemical Acidulants

Acidification is generally accomplished through fermentation, although it can be also accomplished through the use of chemical acidulants such as encapsulated lactic or citric acid or glucono delta lactone (GDL). Although these chemical acidulants can achieve the desired acidity, there is some question as to whether the same subtleties of flavor can be achieved as with microbial fermentation. The products containing encapsulated acidulants require heating to ~140°F (60°C) to release the acid from the vegetable fat coating. Acidulants are sometimes combined with lactic acid starter cultures to achieve rapid pH drop.

Microbial Fermentation

For semidry sausages, a rapid fermentation cycle is usually used. The fermenting organisms of preference are *Pediococcus sp.*, often combined with *Lactobacillus*, which have an optimum growth temperature of ~100°F (38°C). USDA-FSIS requires that "lactic acid starter culture" be included in the ingredient statement where a commercial starter culture is used. There are still some processors that depend on natural fermentation or the *backslop* method of inoculation. It is difficult to justify these practices in light of Hazard Analysis Critical Control Point (HACCP) and Good Manufacturing Practice (GMP) regulations.

FOOD SAFETY

Because these products are considered "ready to eat" (RTE), several foodborne pathogens must be addressed during manufacturing. USDA-FSIS considers the presence of *Salmonella*, *Listeria monocytogenes*, and *E. coli* O157:H7 (products containing beef) in an RTE product as an adulterant. The processor must be able to document, to the satisfaction of the regulatory agency (in this case USDA-FSIS), that the process used in production renders the product safe from these foodborne pathogens. Appropriate HACCP plans and GMPs must be in place. Various researchers are working toward the development of computer models for the lethality of foodborne pathogens. This computer modeling might eventually supplant the necessity of lengthy and expensive challenge studies.

To detail the specific steps needed in an HACCP plan and the processes needed to justify an acceptable GMP for any specific semidry sausage would require several lengthy chapters in themselves. Again, as previously mentioned, regulations are in a constant state of change, so a processor would be advised to consult the up-to-date references on regulations.

Although *Trichinae* are no longer a serious threat in the U.S., detailed regulations on the processing of any RTE product containing pork are still parts of the USDA-FSIS regulations in force and need to be followed.

MANUFACTURING PROCESSES

In discussing manufacturing processes it is common in the U.S. to express the level of ingredients added as a proportion of the meat block rather than as a proportion of the total formulation.

Meat Selection

Most of the sausages discussed here are either beef, pork, or a mixture of the two. Poultry may sometimes be included, but chicken or turkey semidry sausages are rare.

It is important to select meat ingredients that have minimum microbial loads to begin with. High levels of bacteria can compete with the lactic acid starter culture and often introduce unwanted flavors and textures as a result of their growth during the fermentation process. Obviously, selection of meat that is as free as possible from foodborne pathogens is critical.

Both lean and fat colors are important. Lean should have a distinct red color and the fat should be white. Yellow fat detracts from the appearance of a product where discrete particle distinction is important. In the case of pork, meat from the shoulder area carries a higher level of pigmentation and is preferable.

Some processors like to include a limited amount of high-collagen meats to provide the distinct collagen "eyes" that are often deemed desirable. For this reason beef or pork cheeks are often included in the meat ingredient list. These two meat ingredients, however, often carry high microbial loads. Meats with high amounts of connective tissue—shank meat, for example—should be avoided because of the negative effect on texture.

In any case, meats with abnormal conditions such as soft oily fat or the PSE condition in pork and "dark cutting" beef should be avoided if a quality end product is desired.

For most of the semidry sausage products, a meat block with a fat content of 20–25% is typical. Leaner products tend to be rubbery in texture unless a texture modifier is added. To be labeled as a *fat reduced* product, the fat content must be at least 25% less than the manufacturer's regular product.

Nonmeat Ingredients

Salt, NaCl, is normally added at levels of 2–3% depending on market demands. Recently there has been a trend toward reducing salt levels because consumer's tastes have changed.

Sodium nitrite is added at 156 ppm. Traditionally, sodium nitrate was used as the curing ingredient because there was sufficient time for nitrate to be reduced to nitrite by microbial action. With today's faster processing, the inclusion of nitrate has been dropped. The only exception might be Lebanon Bologna, where a mixture of nitrate and nitrite is still often used.

Sugar, particularly the simple sugar, dextrose, is critical to support fermentation. Dextrose is usually added at levels of 0.5–0.75% depending on the pH drop required. Some products like Lebanon Bologna that are fermented to a very low pH may actually require levels of as high as 1.25%. Some of the more complex sugars may also be added, but generally these are limited unless a very low pH product is desired, because they may also support some fermentation.

Spices are what differentiate the semidry products. Because of the long storage life, most processors prefer the flavor of whole or ground spices over the spice extractives. With whole or ground spices, sterilized spices should be used so as not to introduce undesirable microorganisms into the product. Garlic, in the form of garlic powder, may or may not be added.

Natural smoke flavorings may be added to intensify smoke flavor. These flavorings can be added to the meat mixture in the form of an oil-based product, a maltodextrin-based product or a water-based solution, provided the latter is appropriately buffered and designed for direct inclusion in the product.

Grinding/Mixing

Grinding and mixing are critical steps to ensure good particle distinction and to prevent fat smearing. The meats should be kept as cold as possible and temperatures of above 40°F (4°C) should be avoided. Depending on the grinder, it is often preferable to work with partially frozen meat to maintain good particle distinction. Close attention should be paid to grinder maintenance. Knives and plates should be sharp and carefully matched. A bone removal system should be used on the final grind.

A bowl chopper may be used for particle reduction. Although producing excellent particle distinction, it has the disadvantage of being highly operator-dependent for uniform particle size. Further, it cannot be fitted with any type of bone removal system as can a grinder.

A coarse grind (1/2–3/8 in; 9–13 mm) of the meat ingredients is the first step. This may be followed by final grind to desired particle size or a mixing step with the final grind following. The grind-grind-mix is preferable because it reduces the chance of fat smearing and allows the use of a vacuum mixer prior to stuffing. It is also mandatory if whole spice or other inclusions are to be added. In any event, a vacuum mix prior to stuffing is desirable.

For mixing, a paddle mixer with vacuum capabilities is preferred. Mixing should be sufficient just to uniformly distribute the ingredients. Overmixing should be avoided. The accumulation of fat on the paddles or sides of the mixer is an indication of overmixing.

Casings

Fibrous, collagen, or natural casings are used for these semidry sausages as occasion demands. When using fibrous casings, if the casing is to shrink with the sausage as it loses moisture, a special coated casing designed for dry and semidry sausage should be used. For a product that will be peeled and sliced, a regular fibrous casing would be the casing of choice. As with all sausage casings, observing the Recommended Stuffing Diameter (RSD) is critical.

Laminated natural casings provide another alternative for summer sausage as do fabric-reinforced collagen casings. Many of these have imbedded netting or sting patterns for decorative purposes. There is also a variety of novelty casings made from fibrous, collagen, or fabric materials. These come in a wide number of shapes and printed designs ranging from beer bottles to little pigs to pumpkins to Christmas stockings, and they are useful when supplying summer sausage to a niche market.

Stuffing

A stuffer that will minimize smear is mandatory. Most processors seem to lean toward a vane-type vacuum stuffer. Keeping the meat below 40°F (4°C) is critical to minimize smearing and prevent fatting-out during the cooking step. The largest possible stuffing horn should be used, and this horn should be free of nicks and dents.

Fermentation, Smoking, and Cooking

Some type of conditioning step to bring the surface of the product up to the initial fermentation temperature in order to avoid moisture condensation on the product as it is introduced to a high humidity atmosphere normally precedes the fermentation step in the thermal process. For these products, a culture that ferments at a temperature of about 100°F (38°C) is the lactic acid culture of choice. The fermentation step is carried at 95% relative humidity (RH) and the desired pH can be attained in 24 hours or less.

Following fermentation, a brief drying step precedes the smoking step. Smoke may be applied as a vapor from a smoke generator or in the form of atomized liquid smoke. If liquid smoke is to be applied as a drench it can be done at this point. With a liquid smoke drench, no predrying step is needed. If liquid smoke is atomized, it is necessary to follow the smoke supplier's recommendation for application, which usually involves atomization followed by a dwell step.

Because of microbial concerns, summer sausage and similar semidry products are usually fully cooked, reaching a final internal temperature of 160°F (71°C). Lebanon Bologna is an exception to this, but a lengthy process including cold smoking and holding at specified temperatures is involved. If the product is not fully cooked, the process must be validated to ensure foodborne pathogen reduction.

Packaging and Storage

Vacuum packaging in a high-barrier plastic film is the most common packaging employed. With a sliced product, if compaction of the slices is to be avoided, an inert gas Modified Atmosphere package should be used. When held in inventory, even the shelf-stable products should be stored under refrigeration 32–36°F (0–2°C) to obtain maximum storage life.

DIFFERENT TYPES OF U.S. SEMIDRY SAUSAGES

In this section the characteristics of some of the most traditional products are explained.

Summer Sausage

The term *summer sausage* is applied to a loosely defined variety of U.S. semidry sausages similar to some of the European cervelats. They generally have a rather low pH, sometimes as low as pH 4.6. Summer sausage is usually a mixture of beef and pork, although sausages made of beef alone are common.

The final grind size for this product ranges from 1/8–3/16 in (3–5 mm), and the predominant spice is black pepper along with coriander and mustard. Sometimes whole peppercorns and whole mustard seed are added. The product may contain garlic and is smoked, or natural smoke flavoring is added. Some processors add nonmeat inclusions such as diced cheese. When cheese is added, it is probably at the level of about 10% of the formulation.

Fibrous, collagen, natural, or laminated casings are used and casing size ranges from 40–120 mm. Some of these sausages may be stuffed in a variety of novelty casings ranging from those shaped like American footballs to beer bottles. Summer sausage is often a component of food gift boxes in the U.S. because of its shelf stability. A shelf life of up to a year is not uncommon.

Lebanon Bologna

This sausage is unique to the area around the town of Lebanon, Pennsylvania. It is an all-beef product with a final grind size of 3/32–3/16 in (2–5 mm). In addition to a prolonged cold smoking process, liquid smoke flavoring may also be added. The lactic acid starter culture used is often one that has been developed specifically for this product in order to enhance the pathogen reduction in light of the fact that no "cooking step" is involved. Casings would be the same as those used for summer sausage. It is fermented to a very low pH (approximately pH 4.2–4.4). Typically, it is not heated to a temperature much above 135°F (57°C). The predominant spice is black pepper along with cinnamon, cloves, allspice, and ginger. It is heavily smoked and has a distinctly "tangy" and smoky flavor. In the traditional process, the coarse ground meat is presalted and held under refrigeration for a period of up to 1 week.

Sweet Bologna

Sweet bologna is a variation of Lebanon Bologna that is made with 10–12% sugar as opposed to 2–4% in the Lebanon Bologna.

Snack Sticks

This is a shelf-stable semidry sausage with a finished diameter of approximately 19–15 mm. It is normally vacuum packaged in individual serving sizes and used as a snack item. The formulation is similar to summer sausage. Sometimes jalapeno, chili, or other spicy peppers are added to produce a "hot" finished product. These are normally stuffed in an edible collagen casing that is designed specifically for this type of product.

Sausage for Pizza

This is a derivative of summer sausage with a spice flavoring similar to the dry sausage, pepperoni. It can be used as a substitute for true pepperoni on price-competitive pizzas if appropriate labeling restrictions are observed. Sizing would be the same as for the true pepperoni.

REFERENCES

S Barbut. 2006. Validating Lethality for Processing Dry Fermented Meat Products. Proc Thermal Processing of RTE Meat Products. The Ohio State University, March 28–30, 2006, The Ohio State University, Columbus, Ohio.

JF De Holl. 1993. Encyclopedia of Labeling Meat and Poultry Products. 10th Edition. St. Louis: Meat Business Magazine.

SA Palumbo, JL Smith, SA Ackerman. 1973. Lebanon Bologna I. Manufacture and processing. J Milk and Food Technol 36:497–503.

RE Rust. 2004. Ethnic Meat Products/North America. Encyclopedia of Meat Sciences Vol. 1 Oxford, UK: Elsevier Ltd, pp. 455–456.

———. 2006. Good Manufacturing Processes Dry and Semi Dry Sausages. Proc Dry and Semi Dry Sausage Short Course, Iowa State University April 11–13, 2006, Iowa State University, Ames, Iowa.

Standards and Labeling Policy, Book Change #10. 1995. United States Department of Agriculture (USDA)—Food Safety and Inspection Service Washington, D.C.

29
European Products

Kálmán Incze

INTRODUCTION

When humans began to gather and hunt food in excessive amounts, they must have experienced that, sooner or later, food becomes inedible, either because of its offending taste or odor (spoilage) or because it caused illness (toxicity). They must also have learned that, with the help of methods that affect the food (cooking, frying, cooling, drying, salting), it remained edible for a longer time, with an added benefit of improvement in palatability. Drying food in the sun or above fire in caves or in tents, where smoking contributed additionally to taste and length of storage, was later combined with salting and, if the salt was contaminated with nitrate, preservative and palatability effect must have been experienced noticed too. Because not only harmful microorganisms were present in the ambient world of mankind, and because salting and drying can cause a natural selection of the useful microorganisms, this combination of methods can be considered as an ancient way of fermentation. (In this chapter, the word *fermentation* is used in a broader sense: metabolic activity of microorganisms as well as enzymic activity of tissue enzymes, bringing about all the chemical, biochemical, physical, and sensoric changes considered as fermentation.)

Fermentation of vegetables has been known for many thousands of years; fermented meat products have a much younger known history, although salting and drying (not fermentation) of meat was also known by Egyptians. Fermented salami (a dry sausage type) was apparently "invented" 300 years ago in Italy (Leistner 2005), and its method of manufacturing was learned by other European butchers in Germany, Hungary, France, etc. Its history goes back close to 200 years in Hungary.

This process of manufacturing salami was based on experience without knowing what microbial activities take place, and the success of the manufacture depended highly on the knowledge of the esteemed salami masters. Research and inventions of Louis Pasteur revolutionized the fermentation industries; yet such a revolution in meat industry was witnessed much later, when thanks to the works of Niven et al. (1959) and Niinivara (1993), strains with favorable effects were isolated, tested, and selected.

The results of this research were introduced in the meat industry about half a century ago and became step-by-step a generally accepted technology, yet with basic differences among countries. In the United States, strong acidifying starter cultures (Pediococci) were used, whereas in Europe strains with good nitrate reducing and aroma formation (Micrococci) were introduced originally. Now, several different bacterial, yeast, and mold strains are available as starter cultures—less often as single strains, more often in combinations. Part III of this book, "Microbiology and Starter Cultures for Meat Fermentation," deals with starter cultures in detail.

SEMIDRY SAUSAGES, DEFINITION

The term *semidry sausage* is etimologically unequivocal; these products are dryer than water-added

Table 29.1. Different types of semidry sausages.

Types	Typical pH*	Typical a_w*	Storage Temperature
Raw fermented and dried	<5.0–5.3	0.91–0.95	Ambient
Raw fermented and cooked	<5.0–5.5	0.92–0.96	Ambient
Raw fermented cooked and dried	5.0–5.5	0.91–0.95	Ambient
Raw fermented dried and cooked	5.0–5.5	0.91–0.95	Ambient
Cooked and dried	5.6–5.7	0.93–0.96	<18°C**
Hot smoked and dried	5.6–5.7	0.93–0.97	<18°C**
Raw fermented, slightly dried (spreadable)	<5.0–5.3	<0,97	Chilled**

*Serves only as guidance.
**Limited shelf life.

cooked meat products, but have higher moisture content than dry sausages. According to the definition in the United States, semidry sausages are fermented and cooked but usually not dried, with the weight loss only during cooking (Sebranek 2004).

In Europe, the term *semidry sausage* is used in a broader sense, because it may involve the types listed in Table 29.1.

As shown in the table, one may distinguish among seven different types of semidry sausage with the following common features:

- Most of these sausages are dried or at least lose weight after fermentation because of cooking or hot smoking; consequently they have a lower a_w value than cooked sausages or cooked hams to which water is added.
- The typical a_w values are higher than characteristic a_w values of dry sausages, which are below 0.90.
- As a consequence of a higher a_w value, texture is softer than that of dry sausages.
- There are also cooked or hot smoked and dried products without previous fermentation that can also be considered as semidry sausages.
- The taste of semidry sausages, with the exception of the only cooked and dried sausages (where neither starter culture nor chemical acidulant is used), is more acidic than that of dry sausages because of the higher moisture content.
- Semidry sausages are sliceable due to coagulation caused by acidification and/or heat treatment, with the exception of spreadable sausages.
- Spreadable semidry sausages are always raw and are dried to a lesser extent; for this reason, their shelf life is limited and chilled storage may be needed.

DIFFERENT TYPES OF SEMIDRY SAUSAGES

The most common semidry sausages in Europe are the *raw fermented and dried* products. Although earlier natural microflora supported by carbohydrate addition had the task to complete fermentation in a desired way, ensuring both good sensory value and food safety, and backslopping was used only sometimes, real breakthrough came with the introduction of starter cultures or chemical acidulants or both. This process contributed not only to safety and sensory value but also to the more consistent quality. The combination of lower pH value caused by lactic acid fermentation or by direct addition of acidulant and the lower water activity ensures the storage at room temperature without spoilage or growth of pathogens. Some of these types are manufactured with mold growth on the surface, but the majority are processed without it. Spontaneous mold growth has been changed to mold starter application in most cases. For safety reasons, molded semidry sausages have to be dried to a lower a_w value than the other products because molds can metabolize lactic acid, which causes elevated pH. This pH increase making possible the growth of staphylococci, among other undesired microorganisms.

It is valid for the raw fermented and dried sausages just as for the other groups that—within the ranges given in Table 29.1—if a_w is higher, a lower pH value is needed to give a satisfactory protection against undesired growth of microorganisms and vice versa. Final moisture content is in general below 40%.

Raw fermented and cooked sausages were also manufactured earlier without starter culture or chemical acidulant, since the "age of starter cultures" now these products are cooked after a starter culture–controlled fermentation. This heat treatment, inactivating most vegetative forms of microbes, lowered pH, and a_w values ensure a fair safety record even with a somewhat higher a_w value. Final moisture content is, in general, above 40%, because usually weight loss during cooking serves as the only drying effect. These types and spreadable sausages sometimes are called *moist sausages*, referring to the high moisture content (Campbell-Platt 1995).

Some *raw fermented sausages* are further *dried after cooking*. This evidently makes them more stable and their sensory value is also upgraded. Heat treatment is, of course, advantageous because pathogenic microorganisms (e.g., *E. coli* O157:H7, *Listeria monocytogenes*, etc.) are inactivated with much higher probability than if the a_w and pH were decreased only. It has to be mentioned, however, that raw semidry sausages in comparison with cooked ones show a higher sensory value because bacterial and tissue enzymes will contribute to the flavor and aroma.

Although not too common, there exists a technology in which *cooking is applied* only *after fermentation and drying*, with the aim of inactivating harmful microbes and ensuring a better palatability than if heat treatment is applied immediately after fermentation, thus inhibiting extended metabolic activity of microorganisms.

There are also types of semidry sausages when neither *spontaneous nor controlled* (by starter culture) *fermentation* is applied, but the sausages are *cooked and dri*ed or *hot smoked and dried*, until the a_w value is decreased. In these cases the a_w value itself is not inhibitory enough, but cooking or hot smoking inactivates most vegetative forms and nitrite has also some inhibitory effect, so these products can be stored without refrigeration but usually for a shorter period of time (some weeks).

All the semidry sausages discussed until now are considered sliceable products, but a large group of *spreadable sausages* should be listed among semidry sausages, too. These types of products are manufactured mostly in Germany and Austria. Spreadable sausages are raw and are dried only for some days after smoking.

Semidry and dry sausages can be distinguished on the basis of water activity value (Incze 2004) or on the basis of moisture:protein (M:P) ratio (Sebranek 2004). According to Sebranek, the M:P ratio is mainly used in the United States, whereas water activity (a_w) is more common in Europe, even if the M:P is also given sometimes in recipes (Wahl 2004).

Moisture content alone gives no sufficient information in terms of distinction or shelf life and safety; a_w and pH values are more informative. Initial moisture content, initial salt (low molecular weight substances in general) content, and extent of drying, or initial moisture content, initial protein content, and extent of drying define final a_w and the final M:P ratio, respectively. As a matter of fact, the M:P ratio gives information on the extent of drying of the lean meat part. It is worth mentioning that a borderline between dry and semidry sausages is not always set similarly. In Europe, an a_w value below 0.89–0.90, which equals an M:P ratio of about 1.2–1.3:1, is considered as a criterion for dry sausages (Incze 2004), whereas the U.S. uses an M:P ratio of 1.9:1 for the same criterion (Sebranek 2004) corresponding to an a_w of cca 0.93.

One of the main reasons of this difference may be that low acidic sausages are not uncommon in Europe, where a lower a_w value is needed for good shelf life and safety reasons. It is also interesting to note that the initial moisture content defines initial protein content as well if the sausage is manufactured only of meat and fat without another protein source; *different a_w values* can belong to *the same M:P ratios* (Table 29.2).

TECHNOLOGY OF SEMIDRY SAUSAGES

Unlike in the dairy industry where milk is (can be) pasteurized, thus the majority of pathogenic as well as spoilage microorganisms are inactivated before starter cultures are added, there is practically no way in the meat industry to support the growth of starter cultures by destroying competing natural microflora of meat batter. Consequently, good manufacturing practice is even more important than with other types of meat products in order that starter cultures can overcome undesired microbes and become dominating.

Referring to Table 29.1, technologies differ from one another in fermentation, heat treatment, and drying. Differences also can be in raw materials (pork, beef, nonmeat proteins), in mincing, in smoking, and evidently in seasoning.

PRODUCTION OF MEAT BATTER

Selection of Ingredients and Additives

Semidry sausages are manufactured mainly of pork or beef or a mixture of them, and less often from poultry and mostly with pork fat. (As ethnic food, pure beef or lamb with tallow is also a possible choice for manufacturing semidry sausages.) Chilled meat and fat are used for grinding technology, frozen

Table 29.2. Differences between final moisture contents of dry and semidry sausages and their relation to moisture:protein ratio depending on initial moisture content and drying loss.

Initial Moisture	Drying Loss	Moisture:Protein Ratio	Moisture Content	a_w
40%	30%	0.8:1	14.3%	0.836
50%	40%	0.7:1	16.7%	0.839
50%	38%	0.8:1	19.4%	0.864
40%	10%	2.5:1	33.3%	0.938
50%	15%	2.5:1	41.2%	0.946
60%	20%	2.5:1	50.0%	0.951
50%	30%	1.4:1	28.6%	0.913
60%	37%	1.4:1	36.5%	0.923
40%	5%	1.9:1	36.8%	0.946
50%	5%	3.2:1	47.4%	0.957
50%	10%	2.9:1	44.4%	0.951
50%	20%	2.1:1	37.5%	0.939
60%	5%	3.4:1	57.9%	0.962
60%	10%	3.1:1	55.6%	0.959

Data were calculated on the basis that 40%, 50%, or 60% initial moisture content was chosen; initial protein content was accordingly 12%, 14%, or 16% respectively; initial salt content was 2.5% uniformly.

meat and fat (–4 to 8°C) are used for bowl chopper technology. Firm fat is necessary instead of soft fat for a nice, clear-cut surface. Nonmeat ingredients, plant proteins and skin (rind) can also be used. For safety reasons it is preferable to precook or to handle the skin with acidic treatment; these treatments are advantageous not only from a hygienic but also from a sensory point of view.

Semidry sausages usually are produced in a short time; therefore, chiefly nitrite curing salt is applied (0.4–0.6% $NaNO_2$ blended with 99.4–99.6% NaCl). Bacterial breakdown of nitrate to nitrite would need longer time, and only nitrite is effective for color formation and color stability, for microbial inhibition, and as an antioxidant.

Finely or coarsely ground black pepper is the most common spice used both in southern and northern Europe, whereas red pepper and garlic are more typical for Hungarian and southern European (Spanish, Italian) sausages. Laurel is a typical seasoning for several spreadable sausages, and common spices of other meat products find their way to semidry sausage products, too. It is worth mentioning that the relatively high carbohydrate content of red pepper can cause an overacidification if not calculated with added carbohydrate. Another extreme is the excessive amount of red pepper, which makes good bind and sliceability questionable.

As a carbohydrate source for lactic acid formation, dextrose is used the most, but other types of sugars can also be used in a percentage of usually between 0.3–0.7%. Dextrose supports the fastest, lactose slowest, acid formation that can be used, e.g., in retarding pH drop and making metabolism of micrococci possible. Sodium or potassium ascorbate improves color and retards oxidation of lipids (which acts also against nitrosamine formation); therefore, it is used extensively, being most effective in the case of cooked semidry sausages. With these latter types of products, the addition of phosphates helps prevent fat separation during heat treatment. For pH drop chemical acidulants also can be used alone or in combination with starter cultures. The most well-known substances are glucono-delta-lactone and encapsulated organic acids. The advantage of their use is the almost immediate pH drop; sausages with starter cultures are nevertheless of better quality and have longer shelf life.

Comminution

Meat and fat are chilled (for grinding) or frozen (bowl chopper technology); comminuted to the desired particle size; and mixed with carbohydrate, a starter culture if applicable, curing salt, and seasonings. A low temperature of meat and fat is necessary for avoiding fat smearing, which causes not only esthetic failure (individual meat and fat particles are not seen) but may retard or even inhibit drying if a fat layer on the surface stops evaporation, and as a result water activity is not reduced. Pathogenic and spoilage microorganisms have a good chance for growth when this happens, and safety cannot be ensured.

Comminution of ingredients and blending with additives takes place in one step in a bowl chopper, and yet extra blending is necessary if grinder technology is applied. In both cases, chopping, grinding, blending, and stuffing should be carried out under vacuum for microbiological and color formation purposes.

Development of Binding, Structure, and Texture

Structure and texture of meat products are very important characteristics perceived and judged by consumers. The pH value and pH changes have a definite influence on binding and structure. In high acidic sausages, major structural changes take place during the first 48 hours (van't Hooft 1999) when pH decreases to 5.3 or below. The solubilized myofibrillar proteins gelify and ensure the fixation of meat and fat particles affecting binding. Myosin is the main gelating protein, but collagen fibrils also take part in cohesion and sliceability; at the same time sodium chloride plays an important role in binding. van't Hooft (1999) also quantified the amount of meat exudates necessary for good binding: exudates below 15% gave poor cohesion, adhesion, and fat binding.

Selection of Starter Cultures

Northern European countries introduced starter cultures earlier than the Mediterranean meat industry, and semidry sausages are manufactured in most cases with starter cultures, but there are still some products that are fermented and dried, cooked and dried, or hot smoked and dried without starter cultures.

One of the reasons that manufacture of high acidic semidry sausages with higher a_w value have been characteristic in northern countries for long time while low acidic dry sausages with a low a_w value were more common in southern European countries is possibly because of safety reasons: Long-lasting drying at low temperature gives a product with a low a_w value, ensuring good stability and safety at room temperature even without the acidification that would perform at higher temperature, a risky process if no starter culture is applied. This difference in praxis and in consumers' preference can still be witnessed within one country, e.g., Germany. Sausages in Northern Germany are more acidic than those produced in the southern part.

Fast acidification and lower pH can be ensured by pediococci (*P. acidilactici*) that grow and metabolize carbohydrates at higher temperatures (37–40°C) or by lactobacilli (*L. sakei*, *L. curvatus*) growing at lower temperatures (20–24°C).

If higher final pH and more richness in aroma is required, staphylococci (*S. carnosus*, *S. xylosus*) or *Kocuria varians* are used alone or in combination with lactobacilli. In the latter case, only slow acidification makes growth and the metabolic activity of staphylococci—which do not tolerate low pH—possible. This combination of starter cultures gives mutual benefit: better safety and more aroma richness. To enhance aroma formation, *Debaryomyces hansenii* also is generally used together with *Lactobacillus* and *Staphylococcus* starter. Although these microorganisms are blended in the meat batter, molds are inoculated onto the surface of sausages with the aim of meeting consumers' demand, to render specific aroma and protect the sausage to some extent from the effect of light (rancidity) and from drying too rapidly. The most common mold starters are *Penicillium nalgiovense*, *P. camemberti*, *and P. chrysogenum*.

Casings

Several decades ago, natural casings (casing of pig, cattle, sheep, or horse) were used, and as a shortage of natural casings occurred, research and development intensified for replacing them with artificial ones. All the artificial casings used for dried sausages are common in their water vapor and gas permeability, a necessity for making evaporation (drying) and smoking possible. The main types are *cellulose casings* (nonfibrous and fibrous) and *collagen casings*. These latter types are resistant to molds that can solubilize cellulose on molded sausages, causing plaques and fattening, through.

Basically two types of cellulose casings exist, the nonfibrous and the fibrous types (Savic and Savic 2002b). Both are permeable to water vapor and gases; therefore, they can be used for raw fermented sausages, too. Nonfibrous cellulose casings have excellent eye appeal with their bright surface and transparency. In fibrous casings, regenerated cellulose fibers reinforce a cellulose hydrate matrix; thus they possess higher strength and can be used for manufacture of semidry sausages, both cooked and raw. Depending on the end users' need, the adhesion properties of fibrous cellulose casings can be adjusted to facilitate casing removal (slicing operations) or to ensure a better adhesion.

The other biodegradable polymer casing that is used in the meat industry is collagen casing. This type is also suitable for stuffing raw or cooked (hot smoked) semidry sausages because of its good vapor, gas, and smoke permeability and sufficient strength, as well as good adhesion (Savic and Savic 2002a).

The advantage of cellulose and collagen casing is the uniform caliber and, in the case of small-diameter collagen casings, there are edible varieties, too. The mechanical strength of these artificial casings is higher than that of natural casings (Effenberger 1991). More on casings can be found in Part II, "Raw Materials."

Smoking

Smoking is also an ancient way of meat preservation; archeological findings prove that it was used 90,000 years ago (Sikorski 2004).

Depending on the temperature of smoke, we may distinguish between cold smoking (15–25°C), warm smoking (approx. 30–50°C) and hot smoking (approx. 60–90°C). For raw semidry products, cold smoking is applied. Warm smoking is common with cooked semidry sausages, but hot smoking also is used with some semidry products, when the internal temperature (in the core) can reach a temperature of 65°C or above. In order to produce a safe smoke low in polycyclic hydrocarbons, the combustion temperature of the wood (mostly beech and other hardwoods) is strictly controlled.

Smoking has several advantages in addition to sensory value (making air-dried and smoked products entirely different in taste), both chemically and microbiologically. If applied sufficiently, possibly more than once for several hours intensively, smoking has a fair antioxidative as well as a bacteriostatic and fungistatic effect. These latter effects make protection against surface slime formation and mold growth easier and more reliable. Diffusion of smoke constituents is restricted mainly to the outer layer of sausage and cannot be considered as an efficient antimicrobial factor inside the sausage.

In order to avoid greyish discoloration, smoking is applied onto the dry surface of sausages and always at lower relative humidity (70–80%).

In south Austria, Germany, and Hungary, smoking is not as typical as in these and in northern countries, whereas in Mediterranean countries air drying instead of smoking is the common technology for semidry products. As mentioned, this brings about different sensory values.

For several decades, not only smoke directly generated in the meat plant by controlled combustion of wooden chips or by friction of wooden logs has been applied, but liquid smoke also has been used, wherein an atomized liquid is deposited on the sausage.

The food laws of different countries regarding liquid smoke regulate labeling differently. As an advantage of liquid smoke, a low concentration of polycyclic hydrocarbons is claimed, which can also be guaranteed by well-controlled smoke generation. A meeting of environmental requirements is more costly, however. More on smoking can be found in Part I, "Meat Fermentation Worldwide: History and Principles."

FERMENTATION, RIPENING, AND DRYING

In cases when a starter culture is applied, optimum incubation temperature for microorganisms suggested by the supplier is adjusted so that starter microorganisms can perform expected metabolic activity ensuring desired changes in pH drop, resulting in acid coagulation and texture formation, enhancing microbial stability and safety, breaking down of lipids and proteins, forming color and stability as well as safety, and forming good aroma. On these changes many papers are published where all the details can be found (Montel et al. 1998; Toldrá 2002; Demeyer and Toldrá 2004; Talon et al. 2004; Incze 2004; Part IV of this book). During incubation (fermentation), sausages can be smoked only if their surface is dried. Depending on the type of starter, the incubation temperature, and the technological aim, fermentation lasts shorter if product is cooked afterward, or longer if no heat treatment is applied. As mentioned earlier, some semidry sausages are dried after heat treatment, some are not. Fermentation with raw sliceable semidry products lasts longer than that of cooked, with reduced intensity parallel with drying in order to increase salt concentration.

With products that are heat treated right after stuffing and eventual smoking, practically no fermentation takes place because bacterial and tissue enzymes are inactivated. As a result of drying, the a_w value decreases in all types of semidry sausages, resulting in an extended shelf life and safety. The initial moisture and salt content and the degree of drying determine the extent of shelf life and safety. Sausages with lower pH and/or being heat treated can lose moisture easier, i.e., the drying rate can be higher than in the case of raw sausage with higher pH (5.7 and above). Even then, sausages with lower pH can be dried at a rate not higher than the moisture-losing ability of the sausage. The driving force of moisture diffusion (the difference between the relative humidity of air and equilibrium relative humidity of product) has to be selected in such a way that neither case hardening nor insufficient drying will result. The range between 5–10% can be chosen with acidic and/or heat treated sausages, but air velocity must be taken into consideration.

Reduction of ripening and drying time has always been a thoroughly investigated research topic. The main methods in achieving this objective were the addition of enzymes such as proteinases (Blom et al. 1996), reduction of initial moisture content by addition of freeze dried meat or a remarkable volume of meat batter of already dried sausage (Ulmer et al. 2006). Although drying time could be reduced in several cases, sensory characteristics were in most cases not equal to products manufactured, ripened, and dried for the normal length of time. Sensory improvement has nevertheless been found (Bolumar et al. 2006) with the addition of cell-free extracts of *Debaryomyces hansenii* and *L. sakei*.

Physical Changes

During fermentation, ripening, smoking (if applicable), cooking, or hot smoking (if applicable) and drying, sausages lose weight. Because of lower pH (near to the isoelectric point) or because of heat treatment (coagulation), muscle protein loses moisture more rapidly than traditional, low acidic sausages. As a result of this and because these products are of higher a_w than dry sausages, there are huge differences between the whole processing time of these two groups: 2–3 days versus 100 days or more. As for the drying rate of semidry sausages, meat batter with higher fat content loses weight slower and vice versa. Mainly muscle is capable of losing water, because the moisture content of fat is much lower and it takes more time to evaporating the same amount of moisture. In practice, it means that the lower initial a_w (higher fat content) of a sausage does not result in a lower final a_w than a sausage of higher initial a_w (lower fat content) after the same drying period, unless a significantly higher driving force of moisture is applied, which then means a high risk of case hardening.

SAFETY AND STABILITY OF SEMIDRY SAUSAGES

Safe and stable products can evidently be produced only from ingredients and additives of high hygienic quality and applying GHP and GMP during processing. Combinations of (1) pH and a_w drop; (2) pH drop, heat treatment, and a_w drop; or (3) heat treatment and a_w drop ensures safety and stability of such products. Although publications report on higher resistance of *Clostridium botulinum* (Lund and Peck 2000; Setlow and Johnson 2001; Austin 2001) against pH, a_w, and other inhibiting factors, with pH = 4.6 and a_w = 0.94–0.96 as limiting values for growth and similar growth-limiting values mentioned concerning other pathogenic bacteria (*E. coli*, EHEC, *Listeria monocytogenes*, *Salmonella*, *Staphylococcus*, *B. cereus*); these are nevertheless measured under otherwise optimal conditions. In semidry sausages, a combination of several growth-limiting factors act simultaneously, ensuring safety at values that individually would not be sufficient for inhibition. This effect of hurdles (Leistner 2000) works well at pH values of $\leq$5.3 and parallel a_w values at $\leq$0.95. A combination of reduced pH and a_w as well as other hurdles ensure inhibition or inactivation of not only bacteria (*Salmonella*, *Staphylococcus*, *Listeria*, EHEC, etc.) but also parasites like *Toxoplasma* (Anon 2006). Further safety-improving factors include heat treatment, in special cases, heat treatment in combination with a_w reduction. More details on the safety of fermented sausages, with special regard to enterohaemorrhagic *E. coli* are discussed in publications (Leistner 1995; Incze 1998, 2003; Toldrá 2002). On biogenic amines, polycyclic hydrocarbons, and nitrosamines, see publications of Bauer and Paulsen (2001) and Bauer (2004). Mycotoxins are less a problem because semidry sausages are in most cases produced without molds, and in the few cases when mold growth is desired, nontoxigenic mold starters are used.

Treatments ensuring safety are usually sufficient for stability of semidry sausages: growth of spoilage microorganisms is also suppressed or they are inactivated. For this reason, shelf life is mainly limited by physical changes such as fat melting, rancidity, discoloration, extreme drying, and weight loss. Most of these changes can be prevented by MAP or vacuum packaging. For more details on spoilage and pathogenic microorganisms, see Kröckel (1995) and Chapters 41, "Spoilage Microorganisms: Risks and Control," and 42, "Pathogens: Risks and Controls," of this book.

QUALITY DEFECTS OF SEMIDRY SAUSAGES

Some of the quality defects can be detected from outside, others only after slicing. The reasons for defects can be manifold; the following are some more common failures:

- Contamination of raw materials (spoilage or safety risk)
- Poor technological quality of ingredients (soft fat, offensive odor)
- Comminution of meat and fat not uniform
- Temperature of fat during comminution not low enough (fat smearing, failing of structure)
- Stuffing not suitable (fat smearing, air holes)
- Inadequate temperature during smoking, ripening, and drying (too low: not suitable for starters; too high: spoilage or safety risk; fat melting, not uniform in the room: remarkable differences among sausages)
- Inadequate relative humidity and/or air velocity during smoking, ripening, and drying (too slow drying: spoilage risk; undesired growth of Micrococci, yeasts, and molds on the surface; too rapid drying: case hardening risk; too high RH during smoking: color defects)
- Inadequate pH drop (too low pH: sensory failure; too high pH: slow moisture loss; insufficient inhibition of microbial growth: spoilage and safety risk, poor sliceability)
- Inadequate a_w reduction (spoilage and safety risk)
- Selection of wrong packaging film or gaseous atmosphere (color defect, rancidity, eventual mold growth if film has high oxygen permeability or residual oxygen percent is not sufficiently low ($<1\%$) in modified atmosphere packages)
- Storage temperature is not suitable (too high: fat melting, separation, rancidity, spoilage possibility; too low: phosphate crystallization inside and outside the sausage)

TYPES OF SEMIDRY SAUSAGES IN DIFFERENT COUNTRIES

Although different types of semidry sausages are characteristic for different countries, alterations can be detected between products under the same name even in the same country, depending on smaller or greater changes in technology. A difference can be found in the a_w value because sausages of the same recipe and diameter can be dried for a shorter or longer period of time. Drafting somewhat arbitrarily, we might say that a raw fermented dried sausage becomes a semidry product if dried for a shorter time and will be a dry sausage if dried for a longer period. Evidently this is true only for high acidic sausages to avoid safety and/or spoilage risk. As a result of this different tenure, the a_w values will differ from one another and the products will count, accordingly, as dry or semidry (see definitions earlier in the section "Different Types of Semidry Sausages").

Typical Semidry Sausages of Different Countries

Austria

Manufacture and consumption of dry and semidry sausages have a long tradition in Austria. Before starter culture application, mainly dry sausages with or without a mold layer on the surface were produced. But besides keeping the tradition of producing these types, the combination of milder drying with the use of a starter culture made the production of raw sausages with higher M:P ratio (semidry sausages) possible, while ensuring microbial stability and safety. Known types are fresh salami (*Frische Salami*); *Dekorsalami*, which are peeled after drying, dipped in gelatin, and then covered with spices or grated cheese and called, accordingly, pepper, garlic, and parmesan-decorated salami; *Debreziner Rohwurst* (Debreziner raw sausage); *Pusstawürstel* (Puszta sausage); *Knoblauchwurst* (garlic sausage); *Haus würstel* (home sausage); and *frische Rohwurst* (fresh raw sausage). Garlic sausage and home sausage are eaten usually after cooking at home, because of their moderate drying (15%). All these are sliceable products made of beef and/or pork, pork fat, starter culture, spices, and curing salt, and dried to a weight loss of 15–30%. Paprika is not a typical spice of these products, but in some regions in Eastern Austria, paprika seasoning can also be characteristic (Puszta sausage). The ratio of main ingredients is approximately 70% meat and 30% fat, and the M:P ratio varies between 1.6:1 and 3.0:1.

Production technology is similar to German processing, and comminution and blending of raw materials is done with a bowl chopper. Spreadable sausages are always raw fermented. "Officially" they are not dried, but during fermentation and smoking they lose weight about 5 max. 10%. As a result of pH drop and slight drying, these products are relatively stable and rather safe with a shelf life of 1–3 weeks depending on the storage temperature. Characteristic products are different types of *Teewurst* and *Mettwurst*, with M:P ratio ranging between 2.4:1 and

3.2:1; nevertheless, their share is definitely lower than that of the sliceable products. Depending on product and starter culture, fermentation (incubation) temperature can vary between 19 and 28°C with 90–95% relative humidity (RH) that are decreased step by step to 14–18°C and 72–78% relative humidity. Smoking temperature is in general less than 20°C, and relative humidity during smoking is less than 85%. Austrian semidry sausages are less acidic (pH 5.2–5.3) than (North) German products.

Denmark

There are two Danish dried products that fulfill the definition of semidry sausages mentioned earlier: Danish salami (*Dansk spegepølse*) and South Jutlandic salami (*Sønderjysk spegepølse*).

Danish salami is produced from pork, and has a caliber of 90–120 mm, a final pH of 4.7–4.9, and an a_w of about 0.90. Seldom a starter culture, but more often glucono-delta-lactone, helps in pH drop. The salami has a bright red color due to cochineal (E 120); nitrite salt is also used. Two production methods are known: *direct curing*, wherein all curing additives containing the meat-fat batter is stuffed in fibrous cellulose casing, and *tank curing*, wherein sausages are put in a tank with brine causing salt enrichment in the sausage and moisture removal from it at the same time. In the latter method, the original sausage mix contains less salt. Because of the high initial fat, high salt, and low moisture content (49%, 4.5% and 36%, respectively), a rather moderate drying loss of 13% causes a low final a_w value, which in combination with a pH value of 4.7–4.9 ensures a good microbial stability and safety.

South Jutlandic salami has a caliber of 55–60 mm, the final pH is 4.7, the a_w value is 0.90–0.95. Produced with pork, beef, and pork fat and 3–3.5% salt, sugar, spices, and nitrite, the meat is pre-cured for 24 hours in salt. Generally used starter cultures are *L. plantarum*, *P. cerevisiae*, and Micrococci, often in combination. Fermentation and drying starts similarly at 24°C and 94–95% RH, which are decreased gradually. After the pH drop, the products are smoked. Being a traditional Danish product, this type is manufactured both by small and large enterprises.

France

If we go southward, sausages with higher moisture content (semidry products) become less and less known and less popular; fully dried products (dry sausages) on the other hand represent the definite majority of such types. French semidry sausages are only moderately dried, 10% weight loss being an average; in some of them a starter culture is used, others are manufactured without it. Traditionally these sausages are eaten raw, but sometimes heat treated (cooked or fried) before consumption, which contributes to safety. Some known sausages are *saucisse de Montbeliard*, smoked, also eaten raw, produced along the Swiss border between Geneva and Basle; *saucisse de Morteau*, smoked and dried to some extent; and cabbage sausage, also smoked and cooked as in Switzerland.

Further types are *longeoles*, *pormoniers*, and *diots* (can be smoked), produced in the region of the French Alps to Italy. The typical Corsican products with pork liver are *figatelles-figatelli*. *Saucisson lorrain* is the well-known, although produced only in small quantity, smoked, dried (20–25% weight loss) sausage of Northern France along the Belgian-German border. Also along the German border a *Teewurst-Mettwurst*–type smoked spreadable sausage is manufactured in Alsace with a starter culture.

Germany

Germany can be considered as "the home of sausages" in general, but also as "the home of semidry sausages." Similarly to Austrian products, sliceable and spreadable sausages are produced. A wide range of production includes manufacture both in small and medium enterprises (SMEs) and at a large industrial scale, but in a huge variety. It is interesting to note that in a book of recipes (Wahl 2004), analytical data, M:P ratio, etc., are given for weight (drying) losses of 5, 10, and 15% (spreadable) and 10, 20, and 30% (sliceable) to the same named products supporting the view mentioned earlier concerning a wide range of semidry sausages depending only on the extent of drying. Because of text restrictions, only some of the most typical products are discussed below.

Sliceable Sausages

Schinkenplockwurst, coarse granulation, caliber 70–90 mm, smoked, dried (M:P ratio if 20% drying loss = 2.3:1, with 30%=1.71).

Schlackwurst, finely chopped, stuffed in bung, smoked, dried (M:P ratio with 20% drying loss = 2.2:1, with 30% = 1.6:1).

Schinkenmettwurst, coarse granulation, pork meat only, natural casing, air-dried (M:P ratio with 20% drying loss = 2.5:1, with 30% = 1.9:1).

Air-dried *Mettwurst*, coarse granulation (M:P ratio with 20% drying loss = 2.4:1, with 30% = 1.9:1). Mettwurst is a collective designation including several types under different names.

Mettenden, medium-size granulation, pork intestine or collagen casing smoked, dried (M:P ratio with 10% drying loss = 3:1, with 20% = 2.5:1, with 30% = 1.9:1).

In addition to a type-specific seasoning mixture, a starter culture, garlic, and MSG are added to practically all these sausages.

Spreadable Sausages

Teewurst, finely chopped, caliber up to 45 mm, smoked, moderately dried (M:P ratio if 5% drying loss = 3.2:1, with 10% = 2.8:1).

Mettwurst, finely chopped, caliber same as above, smoked, moderately dried (M:P ratio similar to above).

Coarse Mettwurst, caliber same as above, smoked, moderately dried, with garlic (M:P ratio if 5% drying loss = 2.5:1, with 10% = 2.3:1).

Schmierwurst, finely chopped, caliber as above, smoked and moderately dried, with paprika (M:P ratio with 5% drying loss = 3.5:1, with 10% = 3.0:1).

All these spreadable sausages are produced with starter cultures in order to decrease the pH value, an absolute necessity because of a short drying period and relatively high a_w. Spreadable sausages are stored usually chilled in retail shops, though it is also suggested (Leistner, personal communication) that storage at ambient temperature helps in reducing or eliminating pathogenic bacteria (*Salmonella*) through metabolic exhaustion, a phenomenon that does not work at low temperatures.

Hungary

Semidry sausages are manufactured mainly from pork and pork fat. The classic Hungarian semidry sausage, summer tourist (*nyári túrista*), has been produced for more than 60 years similarly to the original technology. The coarsely chopped pork and pork fat, blended with curing salt, now nitrite salt, and seasonings, is stuffed into collagen casings of 60–70 mm, warm smoked and cooked right after smoking, and then dried to a weight loss of about 10–20%. Because starter culture or chemical acidulant has never been used, and the pH of the product (5.6–5.7) would have no inhibitive effect against spoilage and pathogenic microbes, cooking ensures a fair microbial stability and safety even at a_w values of 0.95–0.96 by inactivating these microorganisms. A similar product, though with paprika and smaller diameter, Transylvanian tourist (*erdélyi túrista*), was produced earlier in larger volume. Farmer sausage (*Farmer kolbász*) was originally manufactured for export purposes; it is a cooked and dried product that now is also produced for the domestic market.

Three decades ago, starter culture technology was introduced in Hungary, too, and many different types have been developed, practically all of them are raw fermented and dried. Semidry varieties fulfill the safety requirements: $a_w < 0.95$ and pH < 5.3 at the same time. Natural as well as artificial casings (small intestine of pork, fibrous cellulose and collagen casing, respectively) are used with 30–65 mm.

Black and white pepper, garlic, and caraway are common spices used. Most of these sausages contain mild or hot or mixed paprika up to 1–2%. Starter cultures and nitrite salt (2.4–2.5%) are added to the meat batter, and all the sausages are smoked 2–4 hrs/day after drying the surface for some days. Incubation temperature: 20–22°C, RH starting with 92–94%, decreasing parallel with temperature to about 80% and 13–14°C, respectively. Mold growth on the surface is undesired.

Some typical products that are (were) produced: sausage of Kunság (kunsági kolbász), vine dresser sausage (*vincellér kolbász*), Hercules salami, "thick" paprika sausage (*paprikás vastagkolbász*), Goliath sausage, Bajaer deli sausage (*Bajai csemegekolbász*), Bácskaer hot sausage (*Bácskai csìpös kolbász*), *Kulen*, *Stifolder*, etc. All these are sliceable products in small intestine of pork, in cecum (*Kulen*), or in artificial casings of 30–80 mm with a final moisture of 32–38%. Spreadable sausage is not manufactured in Hungary, even though earlier several trials with good products have been launched; because of lack of consumers' acceptance they failed.

Hot smoked (*lángolt kolbász*) products, on the other hand, are very popular and are produced with nitrite curing salt and spices, paprika dominating. After hot smoking, they may be dried or not, and because neither starter culture nor GDL is used, these products with 10–15% weight loss are usually consumed in cooked form.

Italy

In the birthplace of dried sausages, classic dry sausages are still most popular, even if, unlike originally, starter cultures now are extensively used inside the meat batter and outside (mold starter). If the pH drops below 5.3, drying can be stopped earlier than if dry sausage is aimed at, before the a_w reaches 0.90: such can be the case, e.g., with *Salame friulano*. In addition to salamis and raw dry sausages, several of them with mold cover—and in addition to raw sausages that are dried to less weight loss and meet the requirements of semidry products—a very famous Italian sausage *Mortadella* can be considered as a semidry product that contains visible fat cubes in finely ground meat batter.

Stuffed in large diameter casing of several hundred mm, mortadella has to be heat-treated for many hours, during which a remarkable cooking loss takes place. This way, *Mortadella* has a low initial microbial count, having killed most undesired microorganisms, and a good shelf life at ambient temperature, proven by the fact that these products are kept without refrigeration in the shops. These sausages have nothing to do with so-called "mortadella" that are produced in other countries and have a similar technology to Bologna-type sausages.

Italian dry and semidry sausages are made mostly of pork, in some cases pork and beef (*Salame Napoli*), and pork fat with salt and black pepper—some with garlic, red pepper, or dill seed. They are stuffed in natural casings 40–70 mm, with some exceptions (Napoli) not smoked, and ripened-dried for 25, 40, or 45 days (*Salciccia*, *Soppressata*, *and Friulano*, *resp*ectively). If they are ripened longer, weight loss is higher and they can be considered dry sausages.

The Netherlands

Dutch semidry sausages are manufactured from pork and/or beef, and in some products with cooked pork rind. Pork fat, starter culture, (mixed lactobacilli and Micrococci) carbohydrate, nitrite salt, and spices like pepper, cardamom, coriander, and garlic are comminuted and blended in a bowl chopper at −1 to −5°C to a particle size between 1 and 10 mm, stuffed mostly in cellulose or collagen casing, incubated (fermented) until pH drop, smoked during fermentation and dried afterward. During fermentation, color formation, acidification, and gelation take place. During fermentation and drying temperature, RH and air velocity is controlled, initial temperature of 25°C is gradually decreased to 18°C, and RH is reduced from 95% to below 85%. A detailed description of the technology is given by van't Hooft (1999). The most popular Dutch products are *Salami*, finely chopped; *Cervelat*, finely chopped, with higher fat content (35–40%); *Snijworst*, finely chopped, with higher fat content (35–45%) and rind added; Farmersmetwurst, coarsely chopped; and *Chorizo* (less spicy than the Spanish product). These products are ripened-dried for 2–3 weeks resulting in a weight loss of about 12–20%. Water activity drops to 0.92–0.95, pH drops to 4.5–5.2.

Semidry sausages, also with reduced fat content (18% fat), are produced. Snack size *Farmersmetwurst* is dried to a lower a_w and therefore considered dry sausage.

Poland

Semidry sausages are cooked and dried, made from pork or mixed with beef, with starter culture, carbohydrate, and nitrite salt. After stuffing in a natural or collagen casing, they are smoked and dried to a total weight loss of 12–17%. The following are typical products.

Polish smoked sausage contain spices black pepper, marjoram, and garlic. Different-quality pork is ground, stuffed in pork small intestine, paired, smoked at 25–30°C to brown color, and dried at 18–20°C, 70–80% RH until a 12% weight loss.

Smoked frankfurters are made from pork and beef, nitrite salt, nutmeg and black pepper, and starter culture; stuffed in sheep or edible collagen casing (22 mm), and smoked and dried at 15–18°C, 80–85% RH until a 15% weight loss.

Semidry cervolat is made of pork and beef, paprika in addition to black pepper, garlic, and coriander, and curing salt; it is stuffed in 60 mm collagen casing, and smoked and dried to a yield of 83%.

ACKNOWLEDGEMENT

Sincere thanks are due to F. Bauer, L. Cocolin, B.-J. van't Hooft, W. K. Jensen, A. Pisula, E. Puolanne, and J.-L. Vendeuvre for the information on national semidry product types.

REFERENCES

Anonymous. 2006. Technologie spielt Schlüsselrolle. Fleischwirtschaft 1:43–44.

JW Austin. 2001. Clostridium botulinum. In: MP Doyle, LR Beuchat, TJ Montville, eds. Food Microbiology. Fundamentals and Frontiers. Washington, D.C.: ASM Press, pp. 329–349.

F Bauer. 2004. Residues in meat and meat products. Residues associated with meat production. In: WK Jensen, C Devine, M Dikeman, eds. Encyclopedia of Meat Sciences. London: Elsevier Ltd. Academic Press, pp. 1187–1192.

F Bauer, P Paulsen. 2001. Biogenic amines in meat and meat products. In: DML Morgan, V Milovic, M Krizek, A White, eds. COST 917—Biogenically Active Amines in Food. Luxembourg: Office for Official Publications of the European Commission, pp. 88–94.

H Blom, BF Hagen, BO Pedersen, AL Holck, L Ayelsson, H Naes. 1996. Accelerated production of dry fermented sausage. Meat Sci 43:229–242.

T Bolumar, Y Sanz, M Flores, M-C Aristoy, F Toldrá, J Flores. 2006. Sensory improvement of dry-fermented sausages by the addition of cell-free extracts from Debaryomyces hansenii and Lactobacillus sakei. Meat Sci 72:457–466.

G Campbell-Platt. 1995. Fermented meats—A world perspective. In: G Campbell–Platt, PE Cook, eds.

Fermented Meats. London: Blackie Academic & Professional, pp. 46–47.

D Demeyer, F Toldrá. 2004. Fermentation. In: WK Jensen, C Devine, M Dikeman, eds. Encyclopedia of Meat Sciences. London: Elsevier Ltd. Academic Press. pp. 467–474.

G Effenberger. 1991. Wursthüllen. Kunstdarm. Bad Wörishofen: Hans Holzmann Verlag, pp. 33–34.

K Incze. 1998. Dry fermented sausages. Meat Sci 49(Suppl 1):169–177.

———. 2003. Ungarische Roh und Dauerwürste. Fleischwirtschaft 11:30–34.

———. 2004. Dry and semi-dry sausages. In: WK Jensen, C Devine, M Dikeman, eds. Encyclopedia of Meat Sciences. London: Elsevier Ltd. Academic Press, pp. 1207–1216.

L Kröckel. 1995. Bacterial fermentation of meats. In: G Campbell-Platt, PE Cook, eds. Fermented meats. London: Blackie Academic & Professional, pp. 69–109.

LF Leistner. 1995. Stable and safe fermented sausages world-wide. In: G Campbell–Platt, PE Cook, eds. Fermented Meats. London: Blackie Academic & Professional, pp. 160–175.

———. 2000. Basic aspects of food preservation by hurdle technology. Inter J Food Microbiol 55: 181–186.

———. 2005. European Raw Fermented Sausage: Fundamental Principles of Technology and Latest Developments. Proc 2005 International Symposium on Fermented Foods. Chonbuk Natl Univ Jeonju, South Korea, pp. 29–39.

BM Lund, MW Peck. 2000. Clostridium botulinum. In: BM Lund, TC Baird-Parker, G Gould, eds. The Microbiological Safety and Quality of Food. Gaithersburg MD: Aspen Publishers, Inc., pp. 1057–1109.

MC Montel, F Masson, R Talon. 1998. Bacterial role in flavour development. Meat Sci 49(Suppl 1):111–123.

FP Niinivaara. 1993. Geschichtliche Entwicklung des Einsatzes von Starterkulturen in der Fleischwirtschaft. 1. Stuttgarter Rohwurst Forum. Gewürzmüller, pp. 9–20.

CF Niven, RH Deibel, GD Wilson. 1959. Production of fermented sausages. US Patent 2, 907, 661.

Z Savic, I Savic. 2002a. Collagen casings. In: Sausage Casings. Vienna: Victus Lebensmittelbedarf Vertriebsgesellschaft, pp. 183–200.

———. 2002b. Cellulose casings. In: Sausage Casings. Vienna: Victus Lebensmittelbedarf Vertriebsgesellschaft, pp. 215–238.

JG Sebranek. 2004. Semidry fermented sausages. In: YH Hui, L Meunier-Goddik, ÅS Hansen, J Josephsen, W-K Nip, PS Stanfield, F Toldrá, eds. Handbook of Food and Beverage Fermentation Technology. New York: Marcel Dekker, Inc., pp. 385–396.

P Setlow, EA Johnson. 2001. Spores and their significance. In: MP Doyle, LR Beuchat, TJ Montville, eds. Food Microbiology. Fundamentals and Frontiers. Washington, D.C.: ASM Press, pp. 33–70.

ZE Sikorski. 2004. Traditional smoking. In: WK Jensen, C Devine, M Dikeman, eds. Encyclopedia of Meat Sciences. London: Elsevier Ltd. Academic Press, pp. 1265–1272.

R Talon, S Leroy-Sétrin, S Fadda. 2004. Dry fermented sausages. In: YH Hui, L Meunier-Goddik, ÅS Hansen, J Josephsen, W-K Nip, PS Stanfield, F Toldrá, eds. Handbook of Food and Beverage Fermentation Technology. New York: Marcel Dekker, Inc., pp. 397–416.

F Toldrá. 2002. Dry-cured meat products. Trumbull, Connecticut: Food & Nutrition Press Inc., pp. 113–172, 221–231.

K Ulmer, J Stankeviciute, M Gibis, A Fischer. 2006. Reifezeitverkürzung bei schnittfester Rohwurst 2. Teil. Fleischwirtschaft 2:96–99.

B-J van't Hooft. 1999. Development of binding and structure in semi-dry fermented sausages. A multifactorial approach. Thesis. Universiteit Utrecht: Waageningen, Ponsen & Looyen, pp. 3–5, 109–132.

W Wahl. 2004. Excellent German sausage recipes. Bad Wörishofen: Holzmann Verlag GmbH & Co KG, pp. 1.1–1.18, 2.1–2.10.

Part VII

Dry-fermented Sausages

30
Dry-fermented Sausages: An Overview

Fidel Toldrá, Wai-Kit Nip, and Y. H. Hui

INTRODUCTION

Historically, the manufacturing procedures for fermented sausages have been adapted to the climatic conditions of the production area. For instance, Mediterranean sausages have dried to low water activity values, taking advantage of long dry and sunny days, whereas in Northern Europe fermented sausages require smoking for further preservation (Toldrá 2006). Summer sausage is another example traditionally produced in the summer and heated for safety reasons (Zeuthen 1995).

Preservation is the result of a successive and specific series of factors known as *hurdle effects* (Leistner 1992). These hurdles follow a chain of events:

- Addition of nitrite, salt, and/or sugar
- Reduction of the redox potential
- Introduction of lactic acid bacteria
- Lowering pH
- Decreasing water activity
- Smoking

The sausage remains stable after this sequence of hurdles (Leistner 1995). Lactic acid bacteria play important roles in safety, nutrition, and sensory quality (Toldrá et al. 2001). These microorganisms develop important reactions essential for the development of adequate color, texture, and flavor (Demeyer and Toldrá 2004). Details have been presented in previous chapters.

Some of the most important fermented sausages produced worldwide are listed in this chapter, with a short description for each type. However, more details on sausages produced in North America, the Mediterranean area, and Northern Europe are provided in three separate chapters in this part. Space limitation prohibits an in-depth discussion of fermented sausages from other parts of the world. If interested, one can refer to the references provided at the end of the chapter.

FERMENTED SAUSAGES WORLDWIDE

North American

Manufacturing practices were brought to the United States by the first European settlers. Today, many typical European fermented sausages can be found in northern states, such as Wisconsin (Toldrá and Reig 2006). Lebanon bologna is a semidry-fermented sausage originally from Lebanon (Pennsylvania). It is produced entirely from beef, with the addition of black pepper, fermented to a very high pH and heavily smoked (Rust 2004). Pepperoni is produced from pork and/or beef and seasoned with red pepper, ground cayenne pepper, pimento, anise seed, and garlic. This sausage has a small diameter and is smoked.

South and Central American

There is a general Spanish and Italian influence on different typical fermented meat products in many Latin American countries. This is the case of Italian Milano and Cacciaturi, which are consumed in many countries, such as Uruguay, Brazil, and Mexico.

Mediterranean

There are many types of dry-fermented sausages that have been produced for centuries in the Mediterranean area. They are usually dried due to the particular climate and are rarely smoked. These types of sausages depend on the diameter, shape, size, spices and seasonings, and sensory characteristics. They receive different names according to the geographic origin; sometimes the names differ even between very close and small areas (Toldrá 2006). Pork is the main meat used, and fungi starters may be used for development on the external surface (Talon et al. 2004). Salamis of medium diameter (around 6 cm) include French Menage, French Saucisson d'Alsace, Italian Turista, and Spanish Salchichón; those of larger diameter such as French Varzi, Italian Milano, and Italian Crespone may be ripened for more than 60 days (Toldrá 2004). Spanish chorizo, which has a strong red color, is seasoned with garlic, pepper, and oregano (Toldrá 2002).

Northern European

Many fermented meats are produced in North Europe, for example, the Greuβner salami is produced in Thuringia (Germany). It is a sliceable sausage produced from beef and some pork, and fat and flavored with garlic, pepper, and other spices. Production includes a long-term fermentation process and cold smoking. Another example is the Rügenwalder teewurst, a semidry-fermented sausage, also produced from beef and pork, which is fermented and cold-smoked (Gibis and Fischer 2004). The Austrian katwurst is another example of long-dried sausage. The Swedish metwursk contains some potato in addition to spices and seasonings. Other meats can be added to the formulation of sausages in Scandinavian countries: for instance, horse meat in farepolse, toppen, trondermorr, stabbur, and sognekorr in Norway and kotimainen meetwurst in Finland; lamb meat in lambaspaeipylsa in Iceland; reindeer meat in poro meetwurst in Finland and rallersnabb gilde in Norway (Campbell-Platt 1995).

Several fermented sausages are produced in Poland from pork and some beef, game, or poultry (Pisula 2004). Krakowscha sucha is produced from pork and beef with the addition of black pepper, nutmeg, and garlic. These sausages are dry-cured, smoked, cooked, and dried for about 3 weeks. Kabanosy is produced from pork with addition of black pepper, nutmeg, and caraway. It is smoked and dried for 3–5 days. Jalowcowa is produced from pork and a little beef with the addition of pepper and juniper. It is dry-cured, smoked, cooked, and dried for 3–5 days (Pisula 2004).

Eastern European

The Hungarian salami is a good example of a typical salami, which is intensively smoked and then the surface is inoculated with mold starters or spontaneous mold growth covering the surface (Incze 2004). This salami is seasoned with white pepper, garlic, red wine, and paprika. Similar sausages are Winter salami, also produced in Hungary, and Hermannstädler produced in Romania (Roca and Incze 1990). Russian salami and Moscow salami are produced in Russia from pork, and, optionally, some beef.

Middle Eastern

Fermented sausages are produced from many varieties of animals, such as beef, buffalo, mutton, lamb, goat, camel, and horse meat in Middle Eastern countries. Pork meat is not used because of religious prohibition. Sausages, which can contain rice, wheat, corn, and rice flour, are cured and smoked. Different flavors are given depending on the addition of olive oil, garlic, cinnamon, onion, paprika, black pepper, rosemary, etc. Fermented and strongly smoked beef sausages were produced in Lebanon and expanded to other countries (El Magoli and Abd-Allach 2004). Soudjouk sausages, using only beef, buffalo, and/or mutton and fat-tailed sheep, are produced in Turkey (Gökalp and Ockerman 1985). These sausages may be heavily seasoned with spices and other ingredients, including garlic, red and black pepper, cumin, pimento, and olive oil.

African

Biltong is a typical South African meat product. It is produced from young and lean carcasses from either cattle or game, especially from areas like the round, loin, or tenderloin. The meat is ripened and dried until losses exceed 50%. Salt, sugar, pepper, and roasted ground coriander are added. Vinegar and saltpeter can also be used (Strydom 2004). Other typical fermented and sun-dried products have been summarized by Campbell-Platt (1995). Most of them are produced in Northeastern Africa. Miriss and Mussran are made from fat surrounding a lamb's stomach and small intestine, respectively. Other similar products are Twini-digla and Um-tibay. Beirta is made in the Sudan from goat meat with offal; Kaidu-digla is made from chopped bones; and Dodery, Mulaa el-sebit, and Aki-el-Muluk are made from crushed bones, marrow, and fat.

EAST ASIAN

Most information on fermented sausages for East Asian countries has been derived from recipe books in both Chinese and English. The English references include Aidells et al. (2000), Campbell-Platt (1995), Inglis et al. (1998), Leistner (1995), Rogers (2003), Solomon (2002), and Trang (2006). Although their descriptions here do not detail the fermentation stage, most East Asian sausages require short- or long-term fermentation during the manufacturing processes.

Chinese Sausage

Lap Cheong (La chang): The name *Lap Cheong* is a general term for Chinese sausages, but it can also be used to describe Chinese pork sausages. Literally, it means winter (lap) intestines (cheong or sausage), which can be interpreted as intestines stuffed in the winter. Traditionally, Lap Cheong is made in the winter months to take advantage of the lower temperatures, which reduce the chance of spoilage during curing after the sausages are stuffed. The ingredients used to make Lap Cheong in China vary from place to place, but basically they are cut-up pork and pork fat (now they are ground-up pork and fat), sugar, and salt, with optional ingredients such as soy sauce, alcoholic beverages, spices, and others. The amount of pork fat used also varies depending on the types of Lap Cheong, with regular and low-fat types more common now. In the old days, when the intestines or casings were stuffed at the beginning of the winter season, they were hung with a string in a ventilated area to gradually dry up the ingredients and the surface, and also to produce the typical flavor (odor, color) and texture. With the decrease in water activity and moisture content, the product finally hardened and could be kept edible (after cooking) for the winter and spring months. It was not uncommon to have mold and yeast development on the surface of the dried product when the relative humidity was high. To extend the shelf life, some people also stored small amounts of the dried products in oil in a sealed container in a cool place. This practice could keep the product available for the summer and even the autumn months. For several decades, industrial production of Lap Cheong has been modified by curing or drying the green sausage in temperature- and humidity-controlled dryers to speed up the process. The products are also now packed in vacuum-sealed pouches with recommendations to store them in the refrigerators to maintain quality all year round. However, the traditional procedure is still practiced in the rural areas of China.

Aap Gon Cheong: Aap (duck) Gon (liver) Cheong (sausage) is a specialty Cantonese product made in a similar manner as the Lap Cheong, with the cut-up duck liver replacing the pork. The amount produced is small because of less availability of the duck liver. Traditionally, soy sauce is one of the main ingredients in the making of duck liver sausage and contributes to a special flavor.

Gam Ngan Cheong: Gam (gold) Ngan (silver) Cheong (sausage) is a very special Cantonese product, which does not use intestine or casing as the "container." A chunk of pork fat is cut into a wedge shape and wrapped with thin slices of pork liver, which has been marinated with salt and sugar. The Gam Ngan Cheong is then dried by the natural process the same way Lap Cheong is cured. Because of concern about cholesterol and fat intakes, this product is less popular today.

Chicken liver sausage w/wo pig liver: This is a modification of the Aap Gon Cheong, with chicken and/or pig liver replacing the duck liver.

Singaporean Sausage

Singaporean sausages are similar to Chinese sausages. This is understandable because the majority of the population in Singapore is of Chinese ethnic origin. They produce sausages such as Special grade (less fat) pork sausage, Chicken sausage, and Pig liver sausage.

Thai Sausage

Sai Ua (a dried Northern Thai sausage): This Thai sausage is made by stuffing pork with Thai curry paste (onion, galangal, lemon grass, parsley's root, curcuma, chili, and salt mashed in shrimp paste) into a pork casing. It is dried and roasted before consumption.

Northeastern sour Thai-style sausage: This sour Northeastern Thai sausage is made with ground pork, cooked rice, nitrite, erythorbate, pepper, salt, and sugar. After stuffing the mixture into a pork casing, it is kept at room temperature for about 24 hours to allow lactic acid fermentation. The sour sausage requires thorough cooking, such as roasting or frying before consumption.

Nham (Thai fermented sausage): This Thai sausage is made similarly to the making of Northeastern Thai sour sausage except that chili and pork skin are also added and they are packed in a bamboo leaf or plastic film. After keeping (fermenting) for 3–4 days at room temperature, they are then ready for cooking and consumption. Currently, some manufacturers apply irradiation treatment to kill parasites and ensure safety.

Goon Chiang: This Thai sausage is made first by marinating the pork with nitrite at refrigerating temperature for 24 hours, followed by grinding and mixing with sugar and erythrobate before stuffing into a pork casing. It is dried at 60°C to appropriate dryness. This sausage requires cooking before consumption.

Filipino Sausage

Longamisa: Longamisa is a sweet sour sausage made at the rural level using lean pork, pork fat, white vinegar, soy sauce, and sugar as ingredients. After stuffing, it can be smoked or cooked fresh.

Korean Sausage

Sundae: This popular street-vendor Korean stuffed sausage is made with pig's blood, rice, green onions, garlic, minced pork, and sweet potato vermicelli, stuffed into small and large pig's intestines. It is steamed before consumption.

Soonday: This is also a popular Korean stuffed sausage sold at public markets. The stuffing consists of firmly cooked rice, crushed garlic and cloves, crushed fresh ginger, black or white pepper, Korean sesame oil, crushed sesame seeds, crushed scallions, and either beef or pork blood. The mixture is stuffed into small beef intestines. The sausages are cooked in water before consumption.

Nepalese Sausage

Nepalese sausage is similar to German sausage except chicken is used as the main meat ingredient.

Sri Lankan Sausage

Sri Lankan sausage is made with lean pork, pork fat, toasted and ground coriander seeds, ground cinnamon, ground clove, ground black pepper, finely grated nutmeg, salt, and vinegar. After stuffing, they are cold smoked at temperature not higher than 30°C for a few hours to appropriate dryness. They require proper cooking before consumption.

Countries such as India, Indonesia, Japan, and Malaysia do not have specialties such as fermented sausages, although ethnic groups within these countries have their ethnic heritage sausages, such as Chinese, Thais, and so on. However, European-style sausages are now produced in Japan and India.

Pacific Rim

Pepperoni is produced in Australia. Vento salami is also produced in Australia from beef with peppercorns and red wine (Campbell-Platt 1995).

IMPORTANCE OF FERMENTED SAUSAGES

Fermented sausages are very popular with most population groups—such as American, Chinese, Spanish, German, and so on—who consume meat. Science and technology have played an important role in better quality and storage time for fermented sausages, especially in Europe, and Central, South, and North America. However, the development and production processes for fermented sausages in countries such as China, Hong Kong, and Taiwan are still an art more than a science.

Most of us like fermented sausages, and this type of processed meat products has been and will continue to be significant in our diets.

ACKNOWLEDGMENT

The authors thank their personal friends and colleagues in East Asian countries for contributing significant information on sausages associated with their native countries.

REFERENCES

B Aidells. 2000. Bruce Aidells's Complete Sausage Book. Berkeley, California: Ten Speed Press, 314 pp.

G Campbell-Platt. 1995. Fermented meats—A world perspective. In: Fermented Meats. G Campbell-Platt, PE Cook, eds. London: Blackie Academic & Professional, pp. 39–51.

D Demeyer, F Toldrá. 2004. Fermentation. In: Encyclopedia of Meat Sciences. W Jensen, C Devine, M Dikemann, eds. London: Elsevier Science, pp. 467–474.

SB El-Magoli, MA Abd-Allach. 2004. Ethnic meat products: Middle East. In: Encyclopedia of Meat Sciences. W Jensen, C Devine, M Dikemann, eds. London: Elsevier Science, pp. 453–455.

M Gibis, A Fischer. 2004. Ethnic meat products: Germany. In: Encyclopedia of Meat Sciences. W Jensen, C Devine, M Dikemann, eds. London: Elsevier Science, pp. 444–451.

HY Gökalp, HW Ockerman. 1985. Turkish-style fermented sausage (soudjouk) manufactured by adding different starter cultures and using different ripening temperatures. Fleischwirtschaft 65:1235–1240.

HF Ho. 2004. Ethnic meat products: China. In: Encyclopedia of Meat Sciences. W Jensen, C Devine, M Dikemann, eds. London: Elsevier Science, pp. 441–444.

K Incze. 2004. Mold-ripened sausages. In: Handbook of Food and Beverage Fermentation Technology.

YH Hui, LM Goddik, J Josephsen, PS Stanfield, AS Hansen, WK Nip, F Toldrá, eds.New York: Marcel Dekker Inc., pp. 417–428.

K Inglis, G Francione, L Invernizzi. 1998.Tropical Asian Style. North Clarendon, Vermont: Periplus Editions, 223 pages.

L Leistner. 1992. The essentials of producing stable and safe raw fermented sausages. In: New Technologies for Meat and Meat Products. JM Smulders, F Toldrá, J Flores, M Prieto, eds. Nijmegen, The Netherlands: Audet, pp. 1–19.

———. 1995. Stable and safe sausages world-wide. In: Fermented Meats. G Campbell-Platt, PE Cook, eds. London: Blackie Academic & Professional, pp. 161–175.

FK Lücke. 1985. Fermented sausages. In: Microbiology of Fermented Foods. BJB Wood, ed. London: Elsevier Applied Science, pp. 41–83.

A Pisula. 2004. Ethnic meat products: Poland. In: Encyclopedia of Meat Sciences. W Jensen, C Devine, M Dikemann, eds. London: Elsevier Science, pp. 456–458.

M Roca, K Incze. 1990. Fermented sausages. Food Reviews International 6:91–118.

J Rogers. The Essential Asian Cookbook, 2003. Berkeley, California: Thunder Bay Press, 304 pp.

RE Rust. 2004. Ethnic meat products: North America. In: Encyclopedia of Meat Sciences. W Jensen, C Devine, M Dikemann, eds.. London: Elsevier Science, pp. 455–456.

C Solomon. The Complete Asian Cookbook. 2002. North Clarendon, Vermont: Tuttle Publishing, 512 pp.

PE Strydom. 2004. Ethnic meat products: Africa. In: Encyclopedia of Meat Sciences. W Jensen, C Devine, M Dikemann, eds. London: Elsevier Science, pp. 440–441.

R Talon, S Leroy-Satrin, S Fadda. 2004. Dry fermented sausages. In: Handbook of Food and Beverage Fermentation Technology. YH Hui, LM Goddik, J Josephsen, PS Stanfield, AS Hansen, WK Nip, F Toldrá, eds. New York: Marcel Dekker, Inc., pp. 397–416.

F Toldrá. 2002. Dry-cured Meat Products. Trumbull, Connecticut: Food & Nutrition Press, pp. 63–88.

———. 2004. Ethnic meat products: Mediterranean. In: Encyclopedia of Meat Sciences. W Jensen, C Devine, M Dikemann, eds. London: Elsevier Science, pp. 451–453.

———. 2006. Meat fermentation. In: Handbook of Food Science, Technology and Engineering. YH Hui, E Castell-Perez, LM Cunha, I Guerrero-Legarreta, HH Liang, YM Lo, DL Marshall, WK Nip, F Shahidi, F Sherkat, RJ Winger, KL Yam, eds. Boca Raton, Florida: CRC Press, volume 4, pp. 181-1–181-12.

F Toldrá, M Reig. 2006. Sausages. In: Handbook of Food Product Manufacturing. YH Hui, ed. In press. John Wiley & Sons.

F Toldrá, Y Sanz, M Flores. 2001. Meat fermentation technology. In: Meat Science and Applications. YH Hui, WK Nip, RW Rogers, OA Young, eds. New York: Marcel Dekker, Inc., pp. 537–561.

C Trang. The Asian Grill: Great Recipes, Bold Flavors. 2006. San Francisco: Chronicle Books, 168 pp.

P Zeuthen 1995. Historical aspects of meat fermentations. In: Fermented Meats. G Campbell-Platt, PE Cook, eds. London: Blackie Academic & Professional, pp. 53–67.

31
U.S. Products

Robert Maddock

INTRODUCTION

As with most meat processing endeavors, United States dry sausage production borrows heavily from European knowledge and background, especially German and Italian (Pearson and Gillet 1996). In addition, true dried sausages are not especially common in the United States marketplace, and are generally produced by smaller, specialized manufacturers for specific ethnic markets (Price and Schweigert 1987)—for example, Italian markets in the Northeast and German in the Midwest. The annual production of dried sausage in the United States is hard to estimate, but is likely less than 5% of total sausage production. Essentially no large United States sausage manufacturers produce a true dried sausage. In fact, in the United States it is uncommon to find dried sausage in a typical supermarket or grocery store. Some sausages commonly thought of as dried, such as pepperoni or salami, are easy to find, but the typical United States manufacturing process for these sausages involves a cooking and smoking step, making them more similar to a semidried sausage than to a dried. Adding a cooking step to dried sausage manufacturing has been done in response to the many United States government regulations and definitions associated with dried sausage production. These regulations have resulted in many processors either discontinuing their production of dried sausages or adding a cooking step to their traditional sausage production, resulting in semidried sausage rather than a true dried sausage.

DEFINITIONS

In order to understand dry sausage production in the United States, it is important to know the definitions associated with American dry sausage production. The United States Department of Agriculture (USDA) is responsible for the regulatory naming and labeling of dried sausage. Some commonly used terms in United States sausage production and some of the actual USDA definitions of all sausage types follow (in alphabetical order):

Acidulation: The process of reducing the pH by the direct addition of organic acids. The majority of acidulants in common use require the protection of an encapsulation process. Because many different encapsulating agents are in use, adherence to the supplier's specific recommendations is required.

Approved source: A source that has been determined to conform to principles, practices, and standards that protect public health.

Casings: The natural animal stomachs, intestines, or bladders, or manufactured casings of cellulose or collagen, which are used to contain comminuted meat or poultry mixtures for sausages.

Cooked sausages and/or smoked sausages: Sausages that are chopped or ground, seasoned, cooked, and/or smoked and come in various shapes and sizes. The most common of all United States sausage products, they include sausages such as cotto salami, liver sausage, wieners, bologna, knockwurst, and others.

Degree-hours: Degrees are measured as the excess over 60°F (15.6°C) (the critical temperature at which staphylococcal growth begins). Degree-hours are the product of time at a particular temperature and the "degrees." Degree-hours are calculated for each temperature used during fermentation. The limitation of the number of degree-hours depends upon the highest temperature in the fermentation process prior to the time that a pH of 5.3 or less is attained. Processes attaining less than 90°F (32.2°C) prior to reaching pH 5.3 are limited to 1200 degree-hours; processes exceeding 89°F (31.7°C) prior to reaching pH 5.3 are limited to 1000 degree-hours, and processes exceeding 100°F (37.8°C) prior to reaching pH 5.3 are limited to 900 degree-hours.

Dry sausage: A sausage made of chopped or ground meat that may or may not be characterized by a bacterial fermentation. Dried sausages are always stuffed into casings and started on a carefully controlled air-drying process. Some dry sausage is given a light preliminary smoke, but the key production step is a relatively long, continuous air-drying process. When fermented, the intentional encouragement of a lactic acid bacteria growth is useful as a meat preservative as well as producing the typical tangy flavor. The meat ingredients, after being mixed with spices and curing materials, are generally held for several days in a curing cooler. As a result of bacterial action, a pH of 5.3 or less is obtained and the sausage is then dried to remove 25–50% of the moisture to yield a moisture/protein ratio in compliance with USDA Food Safety Inspection Service (FSIS) requirements. Medium-dry sausage is about 70% of its "green" weight when sold. Less dry and fully dried sausage ranges from 80–60% of original weight at completion. According to the USDA, dry sausage must have a Moisture Protein Ratio (MPR) of 1.9:1 or less. Nonrefrigerated, semidry, shelf-stable sausage must have an MPR of 3.1:1 or less and a pH of 5.0 or less, unless commercially sterilized. Some of the dry sausages produced in the United States are chorizo (Spanish, smoked, highly spiced), frizzes (similar to pepperoni but not smoked), pepperoni (not cooked, air-dried), Lola or Lolita and Lyons sausage (mildly seasoned pork with garlic), salami (Italian; many different types of "salami" are made; most are usually made from pork but may have a small amount of beef and are seasoned with garlic), summer sausages including cervelats (may be fully cooked and dried after cooking or fermented and dried), and landjäger (American landjäger is often fully cooked).

Fermentation: That part of the process in which lactic acid producing bacteria (from the starter culture or "mother" culture) convert fermentable carbohydrate (i.e., dextrose or sucrose) in the meat mixture to lactic acid, and thus lower the pH. Once the pH reaches 5.3 or less, the environment for *Staphylococcus aureus* and other pathogenic microbes growth is effectively controlled if the process continues lowering the pH to a more stable value or the process begins drying the product at a low temperature. However, reducing pH to 5.3 is insufficient to destroy *Escherichia coli* O157:H7 without a cooking step in the sausage-making process (Muthukumarasamy and Holley 2007). Fermentation to a pH of ≤4.6 at 90°F (32.2°C) or 110°F (43.3°C) requires further holding and potentially a heat process in order to destroy *E. coli* O157:H7 if it were present. During fermentation of sausages to a pH of 5.3, it is necessary to limit the time during which the sausage meat is exposed to temperatures exceeding 60°F (15.6°C).

Fermentation culture: An active culture of one or more bacteria, which affects the rapid development of a lower pH in dry and semidry-fermented sausages.

Fresh sausage: A sausage made of fresh, uncured meat, generally cuts of fresh pork, and sometimes beef. When ice or water is used to facilitate chopping and mixing, it is limited to a maximum of 3% of the total formula. Product must be kept refrigerated and is cooked before consumption. Bratwurst, "Italian" sausage, whole-hog, and most pork sausages, are common examples.

Identifiable source: Can include the name and address of the immediate supplier and the actual source or location of the supplies.

Moisture/Protein Ratio (MPR): The ratio of moisture content to protein content in the sausage.

Potentially hazardous food (PHF): A food that is natural or synthetic and that requires temperature control because it is in a form capable of supporting the rapid and progressive growth of infectious or toxigenic microorganisms; the growth and toxin production of *Clostridium botulinum*.

Semidry-fermented sausage: A sausage made of chopped or ground meat that, as a result of fermentation, reaches a pH of 5.3 or less and undergoes up to 15% removal of moisture during the fermentation/heating process. Examples of semidry-fermented sausages made in the United States include summer sausage, thuringer, cervelat, and Lebanon bologna.

Trichina treatment: A defined method to render food products, i.e., pork and game meat products, free of the *Trichinella spiralis* parasite.

Uncooked smoked sausage: A unique product that has characteristics of fresh sausage, but is smoked without the addition of heat (cooking), resulting in a product with a different flavor and color. This sausage must be thoroughly cooked before serving, and requires prominent labeling for consumers, because the product can look fully cooked. "Smoked Pork Sausage" is included in this class.

UNITED STATES MANUFACTURING PROCESSES FOR DRIED SAUSAGES

The properties of most American dry sausages depend upon the products of bacterial fermentation and by biochemical and physical changes occurring during the long drying or aging process. Many producers will use starter cultures in the manufacturing of dry sausages; however, it is still common to use bacteria existing in the meat as a source for fermentation, even though this practice is discouraged by many experts and inspectors. Production time depends upon many factors including possible smoking, drying period, diameter and physical properties of casings, sausage formulation, choice and methods of preparing meat, conditions of drying, etc. Often, overall processing time may require up to 90 days. Also, the final pH of dry sausages is usually somewhat higher (5.0–5.5) than that of semidry sausages, and can increase during the aging process.

American dry sausages are mainly coarsely or moderately chopped (the majority of small-diameter dry sausages) and very occasionally finely chopped. Some varieties of dry sausages are cold smoked (54–64°F, 12–18°C) but sometimes they are not; in some countries, dried sausage may be heavily spiced, however, United States dried sausages tend to be milder, and they usually have less smoked flavor and salt than found in European sausages. In principle, American dried sausages are processed by long, continuous air-drying, sometimes after a comparatively short period of smoking. It is important to note that processing of dried sausages can be as varied as the production of any other sausage type. Some production systems utilize a long period of moderate temperatures (90–120°F, 32–49°C) and light smoke to promote fermentation and flavor development. These extremely smoky sausages are more common in the upper Midwest of the United States and are usually marketed as "summer sausages" versus the heavier-spiced dried sausages found in many ethnic communities that are often referred to collectively as "salamis."

American dry sausages are stuffed in both natural and artificial casings of varying diameters, with 2.5 in (60 mm) and larger being common. Natural casings are preferred by many because they adhere closely to the sausages as sausages shrink. However, modern collagen and fibrous casing will provide similar performance as natural casing. The shelf life of dry sausages is excellent, attributed to the high salt: moisture ratio and a lower than optimal pH for bacterial growth. Dried sausages can be kept without refrigeration, but are commonly refrigerated by consumers after purchase. Raw sausages, which are not submitted to the smoking process, are known as air-dried sausages and often have a highly attractive appearance and a yeasty-cheesy flavor. Air-dried sausages are marked with or without mold overlay and are more commonly produced in the northeastern United States.

American dry sausages can be sold as new dry sausages (about 20% weight loss from original weight), moderately dry sausages (about 30% weight loss) and dry sausages (about 40% weight loss), depending upon the market. Drying or aging is a key operation, especially in dry sausage production. The most critical point in drying is to avoid the pronounced surface coagulation of proteins and the formation of a surface skin. If the sausages lose moisture too rapidly during the initial stages of the drying period, the surface becomes hardened and a crust or ring develops immediately adjacent to the casing. This hardened ring inhibits further loss of moisture and the sausage has an excessively moist center. Only a sufficiently wet and soft casing, a high relative humidity at the outset of the drying operation, and the use of a lower relative humidity in the advanced stages of the process will permit moisture to migrate from the interior of the sausage into the outer layer. Thus, the sausage should dry from the inside outward. If the outer layers of the sausage become hard, the diffusion of water will be inhibited and the sausage tends to spoil.

Unlike much American sausage manufacturing, the production of dried sausage is still somewhat "low tech." It is not uncommon for much variation in the steps, conditions, and drying of the sausage. At the start of the drying or the drying and smoking process, relative humidity can be as high as 98%, but can vary depending upon the sausage being produced. In the following 2–4 days the relative humidity must gradually but slowly be reduced. Too much humidity in the drying room favors the development of mold and sliming of products. Dried sausage from the midwestern and western United States will usually not be displayed with visible mold. However, other areas do not object to a light white surface mold at the beginning of the drying process. It is believed that mold contributes to the specific flavor of some products.

In the traditional production process of dried sausage, fermentation is accomplished by natural flora or a starter culture. In order to achieve safe fermentation of the raw sausage, it is important to give the microflora the proper growth conditions. Types of bacteria used can sometimes be controlled by the incubation times and temperatures used. For example, the raw sausage mixture, containing meats, curing salts, and sugar, can be placed in 15–18 cm deep pans and kept for 2–4 days at 37–39°F (3–4°C). After remixing, the mix is stuffed into casings and the drying process continued at 54–59°F (12–15°C) with or without simultaneous smoking. A number of alternative procedures are found in practice.

American dried sausage makers will use frozen concentrated starter cultures rather than natural microflora to improve the speed and consistency of production. The addition of starter culture allows for a shorter "greening" or microbial growth time. When using starter cultures in current production practices, desired acidity can be achieved within 24 hours at a high (95–106°F, 35–41°C) incubation temperature. This is in contrast to traditional manufacturing processes, which often require up to 72 hours to achieve the required microbial growth to lower pH.

In general, American dried sausage manufacturers will add sugar from 0.3–2.0% of the weight of the raw product. If dextrose or other easily degradable sugars are used, acidification is fast and the amount of sugar added should be somewhat lower. In opposition, corn syrup solids, which are common in many sausage recipes, must be added at higher levels.

Basic Formulations and Processes for Selected Large-Diameter Dried Sausages

Table 31.1 is an example of a formulation for a hard salami made in the United States, naturally fermented (no starter culture). The procedure follows.

Table 31.1. Formulation for U.S. hard salami, naturally fermented.

Ingredients	kg	g
Beef chuck	18	
Pork jowl	18	
Pork trimmings	9	
Salt	1.5	
Sugar	0.7	
White pepper		85
Sodium nitrate		60
Garlic powder		10

Procedure:
Grind lean meat through 30 mm plate and fat pork through a 60 mm plate. Mix all ingredients for about 5 minutes. Store in 25 cm deep trays for 48–96 hours. Stuff into suitable-sized casings (sewed bungs for natural, or No. 5 fibrous). Hold for 9–11 days at 39–43°F (4–6°C) and 60% relative humidity. Allow to dry until desired MPR is reached.

Table 31.2 shows an example of a formulation for a pepperoni made in the United States, starter culture added. The procedure follows:

Table 31.2. Formulation for U.S. pepperoni, starter culture added.

Ingredients	kg	g
Beef chuck	14	
Pork heart	7	
Pork trimmings	20	
Pork cheek	4.5	
Salt	1.5	
Sugar	0.5	
Sweet paprika		340
Ground pepper		170
Whole fennel		100
Red pepper		100
Sodium nitrite		7
Commercial starter culture		

Procedure:
Grind lean meat through 30 mm plate and fat pork through a 60 mm plate. Mix all ingredients for about 5 minutes. Stuff into suitable sized casings (around 40 mm for natural, or No. 1 fibrous). Hold for 9–11 days at 39–43°F (4–6°C) and 60% relative humidity. Transfer to green room at 50°F (10°C) and 70% relative humidity for 48 hours. Smoke for 24–60 hours at 100°F (38°C). Hold for approximately 21 days at 54°F (12°C) and 70% relative humidity.

SAFE PRODUCTION OF DRIED SAUSAGES IN THE UNITED STATES

Dried sausages were generally produced without any special USDA inspection protocols prior to 1994, when an outbreak of *E. coli* O157:H7 poisoning was attributed to dry sausages. According to reports from USDA, from November 16 through December 21, 1994, a total of 20 laboratory-confirmed cases of diarrhea caused by *E. coli* O157:H7 were reported to the Seattle-King County Department of Public Health, in the Pacific Northwest of the United States. Epidemiologic investigation linked *E. coli* O157:H7 infection with consumption of a commercial dry-cured salami product distributed in several western states. Three additional cases subsequently were identified in northern California. These illnesses raised questions about the effectiveness of controlling *E. coli* during the production of dry-fermented sausage. In response, the USDA's Food Safety and Inspection Service (FSIS) developed specific protocols to identify problems, which encompass options to correct them. These protocols must be followed or the product must be heat treated to control potential *E. coli* contamination. Specifically, USDA requires commercial producers of dry and semidry-fermented sausages in the United States to follow 1 of 5 safety options:

- Achieve a 5-log kill using a heat process (145°F, 63°C for 4 minutes).
- Develop and validate individual 5-log inactivation treatment plans.
- Conduct a hold-and-test program for finished product. Depending on type of product, 15–30 individual chubs must be subsampled per lot.
- Propose combinations that demonstrate a collective 5-log kill.
- Initiate a hazard analysis critical-control point system that includes raw batter testing and a 2-log inactivation in fermentation and drying.

Process Control Points for Dried Sausage Manufacturing

pH Control: Fermented and acidulated sausages shall attain a pH of 5.3 or lower through the action of lactic acid forming bacteria or by direct acidulation within the time frame defined in the GMP. It is important to reach a pH below 5.3 to control the growth of pathogenic microorganisms, including staphylococci and pathogenic *E. coli*. To assure that the pH decreases normally, pH readings must be taken from each lot. It is important to record all pH measurements before the product surface temperature reaches 110°F (43°C) or before the degree-hour limitation has been reached and before any final heat treatment, if used, is initiated.

Fermentation: There are two general means recommended for use by which lactic acid–forming bacteria used for fermentation may be incorporated into the chopped or ground meat to produce fermented products safely:

1. The preferred and most reliable method is to use a commercially prepared culture, which is handled and used as prescribed by the manufacturer.
2. A less acceptable procedure is the use of a portion of a previously fermented and controlled "mother" batch. Because this is less precise than using a commercial culture, it is important that the inoculum derived from the mother batch be composed of a vigorous culture capable of producing a rapid pH decline.

A third method, used historically, relied on lactic acid bacteria, which naturally occur in fresh meat, to initiate the fermentation. Although this practice had been used in the past and was the original art of making fermented sausage, the method is highly unreliable and should be used by only experienced sausage manufacturers and may require more rigorous examination of the safety of the final product.

Acidulation: An alternative to commercial starter cultures for reducing pH in sausage batters is direct acidulation by USDA-approved acidulants such as citric acid, lactic acid or glucono delta lactone (GDL). These ingredients should be incorporated into the sausage batter following procedures recommended by the manufacturer.

Because dry sausages are not usually cooked, people "at risk" (the elderly, very young children, pregnant women, and those with weakened immune systems) might want to avoid eating them. The following are recommendations made for the safe consumption of dried sausages (MacDonald et al. 2001):

- Dry-fermented sausages produced should be subject to heat treatment or equivalent processes that will result in acceptable reductions of pathogenic *E. coli* and other pathogens.
- The effectiveness of the hold-and-test option should be reevaluated.
- Individuals who are at high risk of serious outcomes following infection with *E. coli*, including children, the elderly, and individuals with immuno-compromising conditions, should not consume uncooked, dry-fermented sausage products.
- This warning should be noted in the HACCP plan and on the label.
- Questions about exposure to dry-fermented sausages should routinely be asked of all reported cases of *E. coli* O157:H7 and other infections.

REFERENCES

D MacDonald, M Fyfe, A Paccagnella, J Fung, J Harb, K Louie. 2001. *Escherichia coli* O157:H7 outbreak linked to salami, British Columbia, Canada, 1999. Health Canada Field Epidemiology Training Program Abstracts.

P Muthukumarasamy, RA Holley. 2007. Survival of *Escherichia coli* O157:H7 in dry fermented sausages containing micro-encapsulated probiotic lactic acid bacteria. Food Microb 24:82–88.

AM Pearson, TA Gillet. 1996. Processed Meats, 3rd ed. New York: Chapman & Hall.

JF Price, BS Schweigert. 1987. The Science of Meat and Meat Products, 3rd ed. Westport, Connecticut: Food & Nutrition Press.

32
Mediterranean Products

Juan A. Ordóñez and Lorenzo de la Hoz

INTRODUCTION

The production of foods using microorganisms is one of humanity's most ancient activities. Throughout history, man soon realized the beneficial effects of some fermentations and, in fact, archaeological remains and historical documents show that man produced wine, beer, bread, cheeses, and sausages, among other products, many years ago. All these products are the result of fermentations, some accompanied by ripening phenomena, which are, essentially, processes of breakdown and synthesis due to the action of enzymes of different origin: some naturally present in the raw material, others intentionally added for this purpose, and those provided by the microorganism. The result is a product with a longer shelf life and the accumulation in it of aromatic and sapid compounds responsible for the taste and odor of the final product.

Drying and fermentation of meat is, probably, the oldest form of meat conservation (Roca and Incze 1990). The two processes are mentioned together because, in practice, they cannot be separated. Curing of meat probably first occurred by chance as a way to conserve meat by adding salt in countries of the Mediterranean area (Adams 1986). Around 1500 B.C., it was found that the shelf life of meat was considerably extended if, after mincing finely, it was mixed with salt and aromatic herbs, introduced into casings and then dried (Palumbo and Smith 1977), which also resulted in a product highly appreciated for its pleasant taste. The first documented evidence for this process was found in book XVIII of *The Odyssey*, written by Homer 900 years B.C., which referred to "goats casings filled with blood and fat" (Liepe 1982). The Romans inherited the custom of eating these food products from the Greeks, perfecting the techniques for their preparation and adding other ingredients. In pre-Christian Rome, during the flower festivals and lupercalias, large amounts of fermented sausages were eaten ("termicina," "circeli," and "botuli"), made from pork, pork fat, pepper, and cumin seeds. Given the pagan nature of these festivals, during the first few centuries A.D. the church banned the festivals and also prohibited the consumption of fermented sausages because of its close associations with these celebrations. This prohibition was maintained until 494 A.D., when these festivities were Christianized (Sanz 1988). Since the Roman Empire, these products have spread worldwide.

Today, the importance of curing as a method of preserving meat has declined owing to the development of other preservation systems (refrigeration, freezing, heat treatments, etc). However, in contrast, other aspects of the curing process have become more important, such as the flavor and color of the foods, and today, fermented sausage production can be considered more as a method of transformation and diversification than preservation. In any case, the Mediterranean dry-fermented sausages have an exceptional hygienic background, which was guaranteed by both low pH ($<$4.5–5) and low water activity ($<$0.90) of the end product, when it is delivered to the market.

Raw fermented sausages, in general, can be defined as products made by selecting, chopping, and mincing meat and fat, with or without offal, mixed with seasoning, spices, and authorized additives, that are then ripened and dried (cured) and, sometimes, smoked. However, regional customs, environmental variations, family recipes and other factors have given rise to a wide range of fermented sausages, and it can be said that there are almost as many types of sausages as there are geographical regions or even manufacturers, although their production process always requires the combination of fermentation and dehydration. In the specific case of Mediterranean countries, and also Portugal, air-dried sausages made with spices are the most common type, with relatively long ripening times compared to central and northern European countries, where fermentation is usually accompanied by a smoking process or a less intense drying process (Flores 1997). In the United States, the most common fermented sausages are semicured, fermented rapidly at relatively high temperatures, with a short drying period (Lücke 1998).

Dry-fermented sausages can be classified according to a range of criteria, such as the acidity, the mincing size of the ingredients, the addition or absence of starter cultures, the presence or absence of molds on the surface, the addition of some ingredients and of seasoning, the diameter and type of casing used, etc. Mediterranean sausages can be classified as fermented meat products with a high acidity because, usually, the fermentation and ripening temperature, concentration of curing salts used, pH, and addition of carbohydrates most commonly used in these products are similar to those described by Incze (1992) for that type of fermented sausage.

PRODUCTION OF MEDITERRANEAN DRY-FERMENTED SAUSAGES

There is a very wide range of Mediterranean fermented sausages because, as mentioned above, it depends on several factors, although its manufacturing always is based on a combination of fermentation and drying. For example, in Spain, more than 50 varieties have been described, although many are handmade on a small-scale (MAPA 1997). Table 32.1 shows the ingredients for quite a few Mediterranean fermented sausages from several countries. The manufacture of dry-fermented sausages consists, essentially, of a number of sequential steps that are shown in the flow chart in Figure 32.1. The meat used depends on eating habits, customs, and the preferences availing in the geographical region where the fermented sausage is produced. This is usually pork, sometimes mixed with beef, although, for religious reasons, in Turkey lamb or cow is used. The meat also has a variable composition, and its fat content depends on the species used and the anatomical region of the animal from which it is taken. Generally, meat from adult animals is preferred, owing to its higher myoglobin content, which favors a more appropriate color (Lücke 1998). Likewise, the fat should be firm, with a high melting point and a low content of polyunsaturated fatty acids, because this causes the fermented sausage to turn rancid more quickly and to exude fats (Frey 1985). One exception to this is found in Spain, where some regions manufacture sausages with adipogenic pork (Iberian pig), which contains a high level of unsaturated fats (Ordoñez et al. 1996), giving rise to products highly appreciated by the consumer.

Mincing of the meat and fat is done at low temperatures (between –5 and 0°C) to achieved a clean cut and to avoid the release of intramuscular fat from fatty meats, which could cause changes in the color and the drying process during ripening (Frey 1985). Once the meat and fat have been comminuted, the starter culture (lactic acid bacteria, belonging to the *Lactobacillus* genus and the nitrate and nitrite reducer bacteria *Micrococcaceae*), curing salts (salt, nitrates, and nitrites), additives (colorants, ascorbic acid, etc.), and other ingredients (sugars glutamate, aromatic herbs, and spices) are added. All the ingredients are placed in a kneader, after which the sausages are made by filling natural or synthetic casings. This operation is carried out in a vacuum to exclude, as far as possible, oxygen from the matrix and to prevent abnormal fermentations and the development of undesirable colors and flavors.

Once the casings have been filled, the sausages are placed in ripening rooms under conditions of controlled temperature (18–24°C), relative humidity (~90%) and air movement (0.5–0.8 m/s). The sausages are kept here for 1 to 2 days during which the fermentation is produced. In this phase, two basic microbiological reactions proceed simultaneously, and they influence one another. These consist of the formation of nitric oxide by nitrate and nitrite-reducing bacteria (*Micrococcaceae*) and the generation of lactic acid by lactic acid bacteria (mainly *Lactobacillus* spp. in Mediterranean sausages). These two phenomena are well known, and several authoritative reviews have been published in which the mechanisms and interaction of the reactions taking place are thoroughly discussed (Palumbo and Smith 1977; Liepe 1982; Lücke 1998; Nychas and Arkoudelos 1990; Demeyer et al. 1992; Ordoñez et al. 1999; Toldrá 2002). Although the nitrite reactions occurring during curing have been studied extensively, the actual chemical reactions are not conclusive due to the high reactivity of nitrite, the complexity of the meat

Table 32.1. Typical formulation (g/100 g) of selected Mediterranean dry-fermented sausages.

	Salchichón/Saucisson	Salami	Fuet	Chorizo	Lukanka
	(Spain/France) (1)	(Italy) (1)	(Spain) (1)	(Spain, Portugal) (1)	(Bulgaria) (2)
Lean pork	35–70	45–84	60–70	80–65	25
Pork fat	10–25	14–25	30–40	20–40	20.0
Lean beef	0–50	0–37	0–20	0–20	55.0
Fermentable sugars*	0.2–0.5	0.3–0.7	0.1–0.4	0.6–0.8	–
Curing salts**	2.0–2.4	1.8–2.5	2.0–2.4	1.8–2.1	2.24
Flavoring or spices:					
Whole/ground black pepper	0/0.2–0.2/0.4	0/0.2–0.1/0.14	0/0.3–0/0.2		0.30
White pepper	–	0–0.2		0–0.3	–
Red pepper	–	–	–	–	0.20
Sweet/hot paprika	–	–	–	1.5/0–2.5/1.5	–
Cumin	–	–	–	–	0.20
Red pepper paste	–	–	–	–	–
Garlic	–	0–0.2	–	0.2–1.2	–
Other spices	***	***	***	–	–
Others****:					
Sodium glutamate	0.25	–	0–0.15	–	–
Powdered milk	0–0.6	0–2.5	–	0–2.5	–
Caseinate	0–0.6	–	0–1.0	–	–
White wine	–	–	–	–	–
Liquid smoke	–	–		–	0.20

Sources: (1) Several sources; (2) Balev et al. (2005).

*Glucose is commonly used and, sometimes, glucose plus lactose.

**Curing salts are composed by NaCl, nitrate, nitrite, and ascorbic acid. A typical mix is, respectively, NaCl/NO_3Na/NO_2Na/ascorbic acid. 2.5/0.095/0.0065/0.094)(w/w/w/w) (Díaz et al. 1997).

***The spices, flavorings, and aromatic herbs used are very variable cinnamon, oregano, aniseseed, dill, posemary, etc.

****Coloring is used in some products, (e.g., cochineal, ponceau 4R).

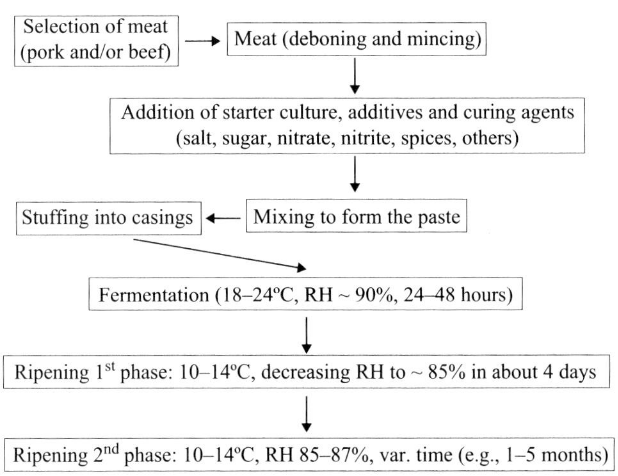

Figure 32.1. Steps in the manufacture of dry-fermented sausages.

substrate and the different processing types of the product (Cassens 1995). Nevertheless, several functions have been attributed to nitrites: it is involved in both color (Eakes et al. 1975; Giddings 1977) and odor (Cross and Ziegler 1965; Bailey and Swain 1973) development; it prevents the growth of several undesirable microorganisms (Hauschild et al. 1982; Pierson and Smoot, 1982) and it has an antioxidant activity (Sato and Hegarty 1971; MacDonald et al. 1980).

The sausage ingredients play an important role in the selection of the proper microbiota of the sausages. The initial microbial population of meat is always varied and is similar to that found in fresh meat. It includes gram-negative (*Pseudomonas* spp., *Achromobacter* spp., *Moraxella* spp., *Enterobacteriaceae*, etc.) and gram-positive (*Bacillus* spp., *Lactobacillus* spp., *Enterococcus* spp., *Brochotrix thermosphacta*, *Micrococcaceae*, etc.) bacteria (Gill 1986; Gill and Newton 1977). On the contrary, the water activity (a_w) has passed from 0.98 (in fresh meat) to 0.96 (in fresh sausage) due to the addition of curing salts and other solutes (e.g., sugars), which, together with the specific inhibitory effects of nitrate and nitrite and the low oxygen tension, favors the selection of both lactic acid and *Micrococcaceae* bacteria. These circumstances are normally brought about in naturally produced dry-fermented sausages (without starters), which causes the inhibition of organisms responsible for the spoilage of fresh meat (gram-negative bacteria, especially pseudomonads) and allows the rapid growth of the former lactic acid bacteria and *Micrococcaceae*, which are the dominant bacteria until the end of the ripening period (Figure 32.2). Similarly, these conditions stimulate the growth of the starter cultures when these are added. The sugar metabolism of lactic acid bacteria naturally present, or those included in the starter, in the Mediterranean dry-fermented sausages is mainly homolactic, yielding (Gottschalk 1986) approximately 1.8 mol of lactic acid per mole of metabolized hexose and around 10% of other substances (formic and acetic acids and ethanol) ready to act as substrates for the synthesis of some flavor compounds. This phenomenon produces a decrease in the pH from an initial value of 5.8–6.2 close to 5.0 or below (Figure 32.2), which has several beneficial effects on both the manufacturing process and the shelf life of the product, such as the control of microbiota and enzymic reactions, the reduction of water retention capacity of proteins favoring the drying, acceleration of the gelation of myofibrillar proteins, and regulation of color formation reactions.

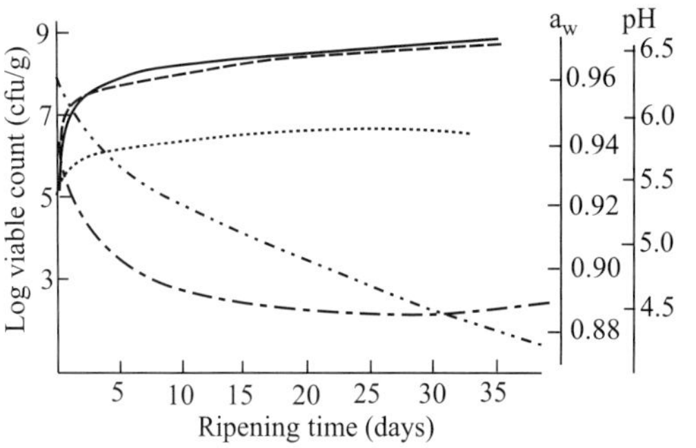

Figure 32.2. Microbial (total viable count [______]), lactobacilli (----------) *Micrococcaceae* (...........)] and chemical [pH (-·-·-·-·-) and a_w (-·-·-·-·-)] changes during ripening of dry fermented sausages.

The presence of molds in Mediterranean fermented sausages is usually attractive for the consumers and, in fact, spores now are often inoculated to enhance growth of a given species. Owing to its aerobic character, its growth is limited to the surface of the sausage (Geisen 1993). The count is very slow in the first few days (on the order of 10^2–10^3 cfu/cm^2) but rapidly increases, reaching values of 10^6–10^7 cfu/cm^2 in around 25 days (Roncales et al. 1991). The molds most frequently isolated belong to the genera *Penicillium* and *Aspergillus*, although sometimes molds of the genera *Cladosporium*, *Scopularopsis*, *Alternaria*, and *Rhizopus* (Leistner and Eckardt 1979; Pestka 1986) are identified. The presence of molds on the sausage surface can lead to both desirable and detrimental effects. The main positive actions are the antioxidant effect by oxygen consumption, peroxide degradation, and protection against light, which leads to the inhibition of oxidative phenomena and color stabilization (Bruna et al. 2001; Sunesen and Stanhke 2003); the protection against spontaneous colonization with undesirable molds, including toxinogenic ones (Lücke and Hechelmann 1987); the favorable characteristic white or greyish appearance of the mycelium and conidia (Lücke 1998); and the contribution to the development of flavor compounds, which has been recently reviewed by Sunensen and Stahnke (2003). The adverse effects are usually related to the growth of unwanted species, and the main effects correspond to the production of mycotoxins and antibiotics. The inoculation of nontoxinogenic strains is a very useful approach to avoid colonization by undesirable molds. *Penicillium nalgiovense* is a much-used strain with this goal because it is nontoxinogenic, easily establishes on the surface, and provides an attractive appearance.

After fermentation, the temperature and relative humidity are reduced to reach values of 10–14°C and 85–87%, respectively, in 4–6 days, which are maintained until the end of the ripening period, during which many flavor compounds develop.

RIPENING OF MEDITERRANEAN DRY-FERMENTED SAUSAGES

A ripening sausage is a highly complex biochemical system in which numerous reactions are taking place, many equilibria are established, and several metabolic and chemical pathways cross. The resulting product of these reactions often becomes the substrate for other ones. The composition of a fermented sausage, therefore, is constantly changing during the ripening process. Small extrinsic or endogenous modifications can produce significant shifts in the equilibrium of the reactions involved, which gives rise to some of the actions taking place during the manufacturing process. One example, is the repercussions of the casing diameter of salchichon (>3 cm) versus fuet (<3 cm); in the first case, the volatile fraction is dominated by compounds derived from microbial activities (e.g., esters, alcohols, etc.) and, in the second case, volatile compounds produced by autooxidative phenomena are more prevalent (e.g., aldehydes and ketones) due, in this case, to the mixture inside being more exposed to oxygen (Edwards et al. 1999). In any case, during ripening a wide range of compounds are generated (peptides, free amino acids, free fatty acids, amines, ketones, aldehydes, etc). Some of these are accumulated in the matrix, contributing directly to the taste (e.g., amino acids and amines and long chain fatty acids, which are nonvolatile sapid compounds) and/or odor (e.g., short chain fatty acids released by the action of esterases) and others that become aromatic substances (e.g., amino acids and amines, which are transformed into ketoacids, aldehydes, etc., by oxidative deaminations or free fatty acids, which, mainly unsaturated, can undergo chemical oxidations and yield, among other compounds, ketones or fungal catabolism that give rise to the generation of methyl ketones and secondary alcohols). All these compounds accumulate in the medium at different concentrations, and they are responsible for the flavor and odor of the final product. These compounds can be of diverse origin: some are the result of microbial phenomena (glycolysis and numerous other reactions derived from the enzymatic luggage of the microorganisms present); some arise from biochemical reactions of endogenous origin (lipases and proteases) and others from chemical phenomena (mainly nitric oxide reactions and lipid autooxidation); and, finally, others accumulate from the addition of certain ingredients (e.g., spices and seasonings). The phenomena that produce these low molecular weight compounds, generally absent or only present in small amounts in the initial substrate, correspond to processes of degradation or synthesis. The former are metabolic and chemical transformations of major compounds in the raw material (proteins and triglycerides and other less-abundant lipids) and the added compounds (sugars). The metabolic reactions are catalyzed by enzymes of different origin: some proceed from the microorganisms which, at one time or another, participate in the ripening process; and others come to the sausage directly from the meat and fat that form part of its normal composition. Undoubtedly, the first reactions to come into play are degradative phenomena: hydrolytic reactions (glycolysis, proteolysis and lipolysis) catalyzed by microbial or endogenous enzymes, and chemical reactions (mainly lipid autooxidation) catalyzed by the oxygen from the air that penetrates the fermented sausage.

The first phases of the hydrolytic and chemical processes are well known and, essentially, consist, respectively, of the degradation of sugar by lactic acid bacteria during the aforementioned process of fermentation; the rupture of the triglyceride-ester bond by the action of lipases, with the subsequent formation of free fatty acids and diglycerides, monoglycerides, or glycerol, depending on the extent and position of the bond cleavage in the lipid molecule; the fragmentation of proteins due to the action of proteinases, yielding polypeptides of different molecular size that are, later, attacked by other proteases producing nonprotein nitrogen fractions (small peptides, amino acids, amines); and, finally, the formation of hydroperoxides from unsaturated fatty acids, which, in the more advanced stages, result in the appearance of low molecular weight organic compounds. All these reactions involved in the mentioned degradative phenomena are well known and have been analyzed in detail in several articles (Dainty and Blom 1995; Ordoñez et al. 1999; Toldrá 2002) and an update of chemistry of reactive oxygen in food is described by Choe and Min (2006).

In this context, it is well known that the enzymatic luggage of the LAB is responsible for the metabolism of sugars but, perhaps, the origin of enzymes predominant in the lipolytic and proteolytic processes occurring in the sausage mixture is less clear.

Traditionally, the microbiota of fermented sausages has been considered to be responsible for the breakdown of lipids that occurs during ripening (Alford et al. 1971; Cantoni et al. 1967; Demeyer et al. 1974; Selgas et al. 1988; Talon et al. 1992). In fact, many studies in vitro have been focused on the extracellular and intracellular activity of lactic acid bacteria (Nordal and Slinde 1980; Sanz et al. 1988; Papon and Talon 1988, 1989; Samelis and Metaxopoulus 1997) and *Micrococcaceae* (Selgas et al. 1988, 1993;

Talon et al. 1992; Coppola et al. 1997) isolated from Mediterranean fermented sausages (Italian, French, Greek, and Spanish). In general, all these authors have detected some activity and, in accordance with the results, speculations have been made about their participation in lipolytic phenomena during ripening. However, Garcia et al. (1992) observed a similar content of free fatty acids in fermented sausages manufactured aseptically and in others inoculated with lactobacilli and/or micrococci. They concluded that the release of fatty acids could be due, at least in part, to the endogenous lipases in the meat. Montel et al. (1993) also studied the lipolysis of dry-fermented sausages. They demonstrated that noninoculated sausages presented slightly lower levels of fatty acids than batches inoculated with different starter cultures, and, therefore, also considered the endogenous lipases as main responsible of lipolysis. Also, Hierro et al. (1997), with aseptic fermented sausages, concluded that more than 60% of the free fatty acids released proceeded from the action of these enzymes. They also described a greater release of linoleic and oleic fatty acids than in batches inoculated with starter cultures. Kenneally et al. (1998) manufactured batches of fermented sausages with different starter cultures using, as a control, a batch to which glucone-δ-lactone and several antibiotics (penicillin, streptomycin, and amphotericin) were added to prevent microbial growth. The results showed that there were no significant differences in the levels of free fatty acids among the batches made with or without the starter culture. From this, they also concluded that lipolysis is mainly due to endogenous enzymes. It, therefore, seems that endogenous lipases play an important role in the breakdown of lipids in fermented sausages.

Cured ham is a good model in which to study the activity of endogenous enzymes because it usually presents bacterial loads below 10^4 cfu/g inside it (Molina and Toldrá 1992). Hence, Motilva and Toldrá (1993), in samples of *Biceps femoris* of pork, studied the effect of curing salts and a_w on muscular and adipose lipase activity and they reported that acid lipase could participate actively in muscular lipolysis throughout the curing process because it is strongly activated as the concentration of NaCl is increased and a_w is reduced. However, the basic and neutral lipases are strongly inactivated as the a_w diminishes. It was confirmed in dry hams, where the activity of these two enzymes increases during the first 2 months of curing to diminish rapidly afterward, at the start of the drying phase, maintaining some degree of activity until the end of the process. On the other hand, acid lipase remains active during the whole ripening period (Toldrá et al. 1991; Motilva et al. 1993). Taking into account that the pH of the fermented sausages varies from 4.8 to 6.0 along the ripening, acid lipase could develop an important activity in these products. Also, in studies conducted on dry hams, lipases of subcutaneous adipose tissue presented less stability than muscular lipases (Toldrá et al. 1991). The neutral lipase would be the main agent responsible for lipolysis in adipose tissue because it would not be affected by variations in the NaCl concentration between 1 and 10 g/l and would, also, be slightly reduced as the a_w dropped from 0.98 to 0.62 (Motilva and Toldrá 1993). The role played by this enzyme in the lipolysis of fermented sausages during ripening may not be highly important, because its optimum pH is almost neutral and, as mentioned previously, fermented sausages have an acid pH. Moreover, the NaCl concentration of these products is approximately 22–25 g/Kg (Toldrá et al. 1992), which could also adversely affect the activity of neutral lipase.

Finally, there is some doubt about the possible participation of muscular and adipose esterases in the ripening process of fermented sausages and cured hams. Esterases are more sensitive to variations in NaCl concentrations and a_w than lipases (Motilva and Toldrá 1993). However, in experiments carried out in dry hams, both muscle tissue and adipose tissue esterases presented a good activity during the whole process (Toldrá et al. 1991). Nonetheless, it seems that they do not play an important role in lipolytic phenomena, due, among other reasons, to the lack of a suitable substrate (Motilva et al. 1993).

Similarly, the fragmentation of protein macromolecules in dry-fermented sausages has been traditionally attributed to the activity of major microorganisms present during the ripening process. Many studies have been conducted in the past to investigate the proteolytic activity of *Micrococcaceae* and lactic acid bacteria in isolated strains from fermented sausages. In the case of *Micrococcaceae*, several authors have reported that these bacteria actively participate in the proteolytic phenomena that take place during the ripening of fermented sausages (Bacus 1986; Selgas et al. 1993), causing an important increase in the free amino acid content (Bacus 1986). In fact, endopeptidase, aminopeptidase, and dipeptidase activities have been detected in these bacteria (Bhowmik and Marth 1989). However, studies conducted on *Staphylococcus* sp. (Montel et al. 1992), *S. carnosus* (Hammes et al. 1995), and *S. xylosus* (Bermell et al. 1992) show that *Micrococcaceae* do not present a very important proteolytic activity. Hierro et al. (1999) did not found differences in myofibrillar proteins or in the amino acid profile after inoculating sausages with a strain of *Staphylococcus* sp.

Similar considerations can be made about the lactic acid bacteria. Hence, some authors (Cantoni et al. 1975; Montel et al. 1992; Fadda et al. 1998) have observed in vitro proteolytic activity of lactic acid bacteria strains isolated from fermented sausages, and they concluded that these bacteria played a very important role in the protein breakdown. However, other authors (Nordal and Slinde 1980; Lücke and Hechelmann 1987; Bermell et al. 1992; Hierro et al. 1999) did not detect proteolytic activity, or only negligible activity, in these bacteria, and Law and Kolstad (1983) indicated that there was no direct evidence that these microorganisms contribute significantly to the flavor and odor of fermented meat products, and that the role of their enzymes in meat proteolysis is uncertain. Later, Kröckel (1995) considered that the extent to which lactic acid bacteria and/or *Micrococcaceae* participate in the proteolytic changes during the ripening of dry-fermented sausages and cured hams is unknown.

Data supporting the proteolytic activity due to endogenous enzymes, however, are more consistent. To demonstrate their importance in sausage ripening, some authors conducted experiments in which microbial growth was inhibited by making the sausages in aseptic conditions (Garcia et al. 1992; Montel et al. 1993; Hierro et al. 1999) or by adding antibiotics (Verplaetse et al. 1992; Molly et al. 1996). No differences were observed between the nitrogen fractions in batches without microorganisms and batches to which different starter cultures were added. Moreover, the addition of pepsatin and leupeptin to inhibit the proteinases produced a sharp drop in the rate of actin and myosin degradation (Verplaetse et al. 1992). On the basis of these results, these authors deduced that endogenous proteases play a more important role than bacterial ones in the development of proteolytic phenomena, and Hierro et al. (1999) even express that these bacteria have a very low participation, if any, in the process.

The importance of muscular proteases in protein fragmentation is also supported, as in the case of lipolysis, by studies conducted in cured ham (Toldrá et al. 1993; Buscailhon et al. 1994; Martin et al. 1998). Catepsins are the proteases that develop most activity during ripening, whereas the calpains have a neglible action because they are relatively sensitive to acidity because calpain I loses 82% of its maximum activity at pH 5.5, and calpain II is even more sensitive (Dransfield 1993). This, together with the addition of curing salts (Sarraga et al. 1989), makes it highly unlikely that calpains participate in these processes. With regard to catepsins, research has been carried out in model systems that simulate the conditions of the ripening of fermented sausages (Toldrá et al. 1992), to determine which of these could be the most relevant in proteolytic phenomena. They concluded that catepsins B, L, and D are mainly active in the first stages of the production process (mixing and fermentation), whereas only catepsin L could present a significant activity during the drying step. On the other hand, they also agreed with Demeyer et al. (1992) that muscular proteins (endopeptidases) of the type of catepsin D, are mainly active during the fermentation phase, whereas bacterial exopeptidases are mainly active in the drying phase. Nonetheless, other muscular enzymes, such as type II dipeptidyl aminopeptidases would collaborate with the latter because the pH of the dry-fermented sausages is adequate for their activity (Bird and Carter 1980) and also that of the carboxypeptidases (Toldrá et al. 1992).

FLAVOR DEVELOPMENT IN MEDITERRANEAN DRY-FERMENTED SAUSAGES

Conversion of Free Amino Acids and Fatty Acids

One of the most important phenomena that occurs during the ripening of fermented sausages is the transformation of amino acids that appear as a consequence of the breakdown of proteins. These amino acids can accumulate in the matrix, contributing to the overall taste of the product, but play a much more important role as a substrate for other reactions that can produce a wide range of aromatic and sapid compounds, some of which are shown in Figure 32.3. As in cheeses (Engels and Visser 1996; Yvon et al. 1997), these transformations are mainly due to the activity of microbial enzymes (Hinrichsen and Pedersen 1995) although some chemical reactions (Strecker degradation) could also occur (Yvon et al. 1997; Ordoñez et al. 1999). Among these reactions, transamination is a very important event because it is the first step in amino acid catabolism, which requires the presence of an α-keto acid acceptor for the amino group, commonly α-ketoglutarate (Yvon et al. 1998). This reaction is catalyzed by aminotransferases and these enzymes have been demonstrated to initiate the catabolism of aromatic, branched-chain, and sulphur amino acids (Dias and Weimer 1998; Yvon et al. 1997; Engels et al. 2000), yielding, among other products, branched aldehydes, which play a central role in the flavor of the final

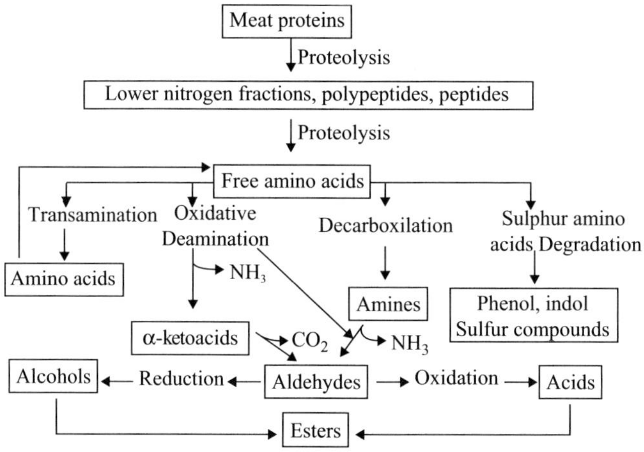

Figure 32.3. Formation of aromatic and sapid compounds from the protein degradation.

product because they have been associated with the ripened flavor of both dry ham (Careri et al. 1993; Hinrichsen and Pedersen 1995) and dry-fermented sausages (Stahnke et al. 2000). In fact, they have been detected in several cured products, such as bacon (Andersen and Hinrichsen 1995); Italian dry ham (Barbieri et al. 1992; Hinrichsen and Pedersen 1995); Iberian dry ham (Garcia et al. 1991; Lopez et al. 1992); and Italian (Stahnke and Zeuthen 1992), French (Berdague et al. 1991), and Spanish (Edwards et al. 1999; Mateo and Zumalacarregui 1996) dry-fermented sausages. These compounds may be generated by certain microorganisms from amino acids (Hinrichsen and Pedersen 1995) via transamination-decarboxylation, as demonstrated in *Streptococcus lactis*, var. *maltigenesis* for leucin, which is converted in 3-methylbutanal (MacLeod and Morgan 1955) and several other lactic acid bacteria, mainly from cheeses (Alting et al. 1995; Christensen et al. 1999; Smit et al. 2000; Yvon et al. 1998). Other pathways may also proceed, such as the deamination-decarboxylation of amino acids by Strecker degradation (Ordoñez et al. 1999), which has been demonstrated in dry-cured Iberian ham (Garcia et al. 1991; Barbieri et al. 1992; Ventanas et al. 1992). This reaction involves the interaction of α-dicarbonyl compounds and α-amino acids, yielding an aldehyde with one less carbon atom than the initial amino acid, which results in the formation of 3-methylbutanal, 2-methylbutanal, and phenylacetaldehyde from, respectively, leucine, isoleucine, and phenylalanine. A third possibility for the generation of branched aldehydes is through the biosynthesis of amino acids. In this pathway, 2-methylpropanal and 3-methylbutanal are formed as by-products of valine and leucine synthesis and, similarly, butanal and 2-methylbutanal are produced during isoleucine formation.

Like amino acids, free fatty acids are the source of many volatile compounds, which are integrated in the flavor compounds. It has been mentioned above that lipolytic activity has been demonstrated in several lactic acid bacteria, but these organisms are not able to degrade the resultant free fatty acids. Therefore, the contribution of these substances to the flavor directly derives from their accumulation in the sausage. However, the molds may play an important role in lipid degradation, because these microorganisms can degrade, via partial β-oxidation, the free fatty acids released by lipases, yielding methyl-ketones, which may be reduced to secondary alcohols, as it has been described in cheeses for *Penicillium camemberti* and *Penicillium roqueforti* (McSweeney 2004).

Role of Molds in the Flavor Compounds Generation

Given the aerobic character of molds, these cannot be present in the fermented sausage interior, but are often found on its surface. Maybe this is why little attention has been paid to their potential role in the flavor and odor of fermented sausages. Several authors (Mateo and Zumalacarregui 1996; Meynier et al. 1999; Schmidt and Berger 1998; Stahnke et al. 2002) have detected a number of volatiles compounds in dry-fermented sausages with molds in the fungal casing coat, but few authors have analyzed the origin of these substances and the effect of different molds on the overall quality and volatile composition in Mediterranean dry-fermented sausages.

Singh and Dincho (1994) studied the protective effect of molds in Bulgarian dry sausages Sredna Gora and observed, in relation to the flavor, that the three strains (two belonging to *Penicillium camemberti* and one to *P. nalgiovensis*) used slightly improved the overall acceptability compared to the controls (without molds), although there were not differences between the strains. Garcia et al. (2001) compared the sensory properties of Spanish dry-fermented sausages (salchichon), using one strain of *Mucor* spp., a commercial strain of *P. nalgiovensis*, and two of *Penicillium* spp. A correlation between proteolytic and lipolytic activities of molds and overall acceptability was observed. *P. nalgiovensis* had a better external aspect than the other strains, but it presented the lowest enzyme activity, giving rise to the lowest sensory parameters. A recent study (Hierro et al. 2005) deals with volatile generation in dry-fermented sausages by three mold species (*Mucor racemosus*, *Penicillium aurantiogriseum*, and *Penicillium camemberti*) inoculated onto the surface of sausages. A starter culture composed of *Lactobacillus plantarum*, *Staphylococcus carnosus*, and *Staphylococcus xylosus* was used. As a control, a noninoculated batch was prepared, which was sprayed on the surface with a 25% potassium sorbate solution. Four patterns were distinguished. Each one was dominated by different volatile classes, i.e., aldehydes from lipid degradation, esters, branched aldehydes from amino acid catabolism and alcohol, and branched aldehydes, which, respectively, corresponded to control, *M. racemosus*, *P. camemberti*, and *P. aurantiogriseum* batches. The four batches were perfectly discriminated by a principal component analysis. The authors even concluded that "it is possible to generate aroma profiles in mold-ripened sausages depending on the fungal strain used" . . . and, at the same time, . . . "control the level of lipid autooxidation."

Role of Herbs and Spices in the Flavor Compounds Formation

Other factors providing volatiles substances to dry-fermented sausages are the herbs and spices and the physical state (whole or ground) in which these are added to the sausage. Black pepper, for example, provides terpenes, and garlic and paprika provide sulphurous compounds and 3-hexenol, respectively (Edwards et al. 1999).

In summary, flavor development is a very complex phenomenon, in which many degradation and synthesis reactions take place. These are modulated by different internal and external agents. The work of Edwards et al. (1999) clearly demonstrates this. These authors analyzed Spanish commercial dry sausages with different times of ripening, diameters (salchichon versus fuet), and spices (salchichon versus chorizo), which had different mincing degrees (whole versus ground). Sulphurous compounds and 3-hexanol were detected only in chorizo, derived, respectively, from garlic and paprika. In sausages made with ground pepper, the terpenes were the major components and reached much higher levels than in dry-fermented sausages that were manufactured with whole peppercorns. The narrower the diameter of sausages, the higher the levels of compounds from lipid oxidation, and, finally, the longer the ripening times, the greater the ester contents.

ATTEMPTS TO ACCELERATE THE RIPENING OF MEDITERRANEAN DRY-FERMENTED SAUSAGES

Since, on the one hand, in the Mediterranean regions a longer ripening time is preferred to produce dry meat products (Flores 1997) and, also, since protein and fat degradation are two central phenomena during dry-fermented sausage ripening, several attempts have been made to either accelerate the ripening period or enhance the flavor. This would permit storage time to be shortened and would increase the profit margin and competitiveness of the final product. To do this, enzymes of different origins have been used: bacterial (Næs et al. 1992, 1994; Hagen et al. 1996; Zapelena et al. 1997, 1998), fungal (Diaz et al. 1992, 1997; Zapalena et al. 1998, 1999; Ansorena et al. 1998b, 2000) and vegetal (Diaz et al. 1996, 1997; Melendo et al. 1996) proteinases, mixture of bacterial proteases (Diaz et al. 1993, 1997), and bacterial (Næs et al. 1992), fungal (Zalacain et al. 1997a,b,c; Ansorena et al. 1998a,b, 2000) yeast (Zalacain et al. 1995, 1996), and animal (Paleari 1991; Fernandez et al. 1991, 1995a,b) lipases. All these authors obtained similar results, i.e., an increase in the amounts of substances derived from proteolysis (nonprotein nitrogen and free amino acids) and lipolysis (free fatty acids) was observed, but no, or few, effects were obtained in relation to an improvement in the sensory properties and overall acceptability. Probably the most reasonable conclusion has been described by Diaz et al. (1997). In brief, when proteinases are added to the sausages, proteolysis is enhanced, with an increase in the different nitrogen fractions (water soluble, nonprotein, and total volatile basic nitrogens) and quantitative and qualitative changes in free amino acids. Similarly, the addition of lipases accentuates the hydrolysis of triglycerides, yielding variable amounts of free fatty acids. However, the sensory analysis showed that solely, and only in some cases, a slight increase in the flavor may be obtained. Therefore, the addition of lipases and proteinases alone is not useful for shortening the ripening period, but this approach may be used to provide precursors (free amino acids or fatty acids) to be transformed into flavor compounds (aldehydes, ketones, alcohols, etc.). After this statement, the same authors (Diaz et al. 1997; Ordoñez et al. 1999; Fernandez et al. 2000) hypothesized that to shorten the ripening, or to improve the sensory quality of fermented sausages, in addition to adding proteinases or lipases, it is necessary to create conditions or to add either an efficient starter or another kind of enzyme, to convert the free amino acid and fatty acids into volatile flavor compounds in a shorter time than usual.

According to the former hypothesis, several attempts have been made to increase the transaminase and oxidative deaminase activities during production of dry-fermented sausages, by either surface inoculation of molds or by adding microbial (fungal or bacterial) crude extracts as an ingredient, together with proteases, to provide increased amounts of amino acids. These experiments showed that the addition of mold intracellular extracts from *Mucor racemosus* (Bruna et al. 1999, 2000a), *Penicillium aurantiogriseum* (Bruna et al. 2000b, 2001a,b), and *Penicillium camemberti* (Bruna et al. 2002) produce, in comparison with the control, an increase in both the free amino acid and ammoniacal fractions. Likewise, higher levels of branched aldehydes were observed, which led to an improvement in the flavor. The fungal superficial inoculation of these species produced similar results (Bruna et al. 2000a, 2001a,b, 2003). However, the combination of both approaches was much more effective (Bruna et al. 2000a, 2001a,b). In similar experiments with bacterial extract, the same trend was observed but a weaker effect, with negligible effects on the sensory quality (Herranz et al. 2006). When commercial free amino acids were added during the mixing of ingredients, both a higher total volatile content and a slight improvement in flavor was observed, without addition of any enzyme extract (Herranz et al. 2005).

Another approach to enhance the flavor of dry sausages has been based on the fact that α-ketoglutarate is, as mentioned previously, the first step in amino acid catabolism. In this sense, it has been described (Herranz et al. 2004) as the effect of the addition of α-ketoglutarate on both volatile substance generation and flavor. An increase in the glutamic acid content was observed and, at the same time, an increase in 2-methyl propanal and 3-methyl butanal, which means that valine and leucine were the donors of the amino group. However, the best results were obtained when addition of α-ketoglutarate was combined with a lactic acid bacteria strain of cheese origin (*Lactococcus lactis*, subsp. *cremoris* NCDO 763) highly active in glutamic acid metabolism (Yvon et al. 2000). It has also been described (Tjener et al. 2004) that addition of α-ketoglutarate enhances the formation of volatile compounds by *Staphylococcus carnosus* during the fermentation of sausages.

CONCLUSIONS

Mediterranean dry-fermented sausages have been produced and consumed for many centuries. Traditionally, they were made by artisanal methods in both industrial and craft level. However, the developments in both food engineering and meat product technology, the consumer demands, and the globalization trend in the latter twentieth century have pushed the meat industry to produce dry sausages on a grand scale. As a consequence, it has been necessary to manufacture a normalized end product. To reach this goal, the conditions of fermentation/ripening (temperature, humidity, and air flow) are narrowly controlled and starter cultures are generally used. The production of Mediterranean dry-fermented sausages is based, as in other varieties from Northern and Central Europe, on a combination of fermentation and drying, but in Mediterranean countries the smoking treatment is not generally applied, and air-drying and the addition of spices are prevalent circumstances, together with a longer ripening time than that of Northern and Central European products.

In relation to the scientific advances, many studies have been made to clear up both the source of agents responsible for the phenomena occurring during ripening and the precursors and reactions giving rise to flavor compounds. Likewise, several attempts have been made to either accelerate the ripening or enhance the flavor of dry-fermented sausages. For that, the addition of exogenous enzymes (lipases and proteinases) from bacterial, fungal, vegetal, or animal origin has been assayed. The results showed that higher amounts of compounds from hydrolytic processes (e.g., amino acids and free fatty acids) were formed, but sensory analysis demonstrated that solely, and only in some cases, a slight increase in flavor was obtained. Accordingly, an approach based on the amino acid transformation has been also investigated, which involves the enrichment of the matrix with aminotransferase and deaminase activities by adding bacterial or fungal extracts or inoculating molds at the surface. It is expected that the results of these investigations could be transferred to the industrial production of dry-fermented sausages to optimize and standardize the typical attributes of each variety.

REFERENCES

MR Adams. 1986. Fermented flesh foods. Prog Ind Microbiol 23:159–198.

JA Alford, JL Smith, HD Lilly. 1971. Relationship of microbial activity to changes in lipids of foods. J Appl Bacteriol 34:133–146.

AC Alting, WJM Engels, S Vanschalkwijk, FA Exterkate. 1995. Purification and characterization of cystathionine beta-lyase from *Lactococcus lactis* subsp *cremoris* B78 and its possible role in flavor development in cheese. Appl Environ Microbiol 61: 4037–4042.

HJ Andersen, LL Hinrichsen. 1995. Changes in curing agents, microbial counts and volatile compounds during processing of green bacon using 2 different production technologies. J Sci Food Agric 68:477–487.

D Ansorena, I Astiasarán, J Bello. 2000. Influence of the simultaneous addition of the protease flavourzyme and the lipase novozym 677BG on dry fermented sausage compounds extracted by SDE and analyzed by GC-MS. J Agric Food Chem 48:2395–2400.

D Ansorena, MJ Zapelena, I Astiasarán, J Bello. 1998a. Addition of palatase M (Lipase from *Rhizomucor miehei*) to dry fermented sausages: Effect over lipolysis and study of the further oxidation process by GC-MS. J Agric Food Chem 46:3244–3248.

———. 1998b. Simultaneous addition of palatase M and protease P to a dry fermented sausage (Chorizo de Pamplona) elaboration: Effect over peptidic and lipid fractions. Meat Sci 50:37–44.

JN Bacus. 1986. Fermented meats and poultry products. In: AM Pearson and TR Dutson, eds. Advances in Meat Research. Meat and Poultry Microbiology. London: AVI Publishing, pp. 123–164.

ME Bailey, JW Swain. 1973. Influence of nitrite in meat flavor. Proc of Meat Industries and Research Conference. Chicago, p. 29.

D Balev, T Vulkova, S Dragoev, M Zlatanov, S Bahtchevanska. 2005. A comparative study on the effect of some antioxidants on the lipid and pigment oxidation in dry-fermented sausages. Int J Food Sci Technol 40:977–983.

G Barbieri, L Bolzoni, G Parolari, R Virgili, R Buttini, M Careri, A Mangia. 1992. Flavor compounds of dry-cured ham. J Agric Food Chem 40:2389–2394.

JL Berdagué, C Denoyer, JL Le Quéré, E Semon. 1991. Volatile componentes of dry cured ham. J Agric Food Chem 39:1257–1261.

S Bermell, I Molina, C Miralles, J Flores. 1992. Study of the microbial flora in dry-cured ham .6. Proteolytic activity. Fleischwirtschaft 72:1703–1705.

T Bhowmik, EH Marth. 1989. Esterolytic activities of *Pediococcus* species. J Dairy Sci 72:2869–2872.

JWC Bird, JH Carter. 1980. Proteolytic enzyme in striated and non-striated muscle. In: K Widenthal, ed. Degrative Process in Heart and Skeletal Muscle. Amsterdam: Elsevier North Holland, pp. 51–85.

H Bozkurt, O Erkmen. 2004. Effects of temperature, humidity and additives on the formation of biogenic amines in sucuk during ripening and storage periods. Food Sci Tech Int 10:21–28.

JM Bruna, M Fernandez, E Hierro, L de la Hoz, JA Ordoñez. 1999. Effect of the combined use of Pronase E and a fungal extract (*Mucor racemosus* forma *sphaerosporus*) on the ripening of dry fermented sausages. Food Sci Technol Inter 5:327–337.

JM Bruna, M Fernandez, EM Hierro, JA Ordoñez, L de la Hoz. 2000a. Improvement of the sensory properties of dry fermented sausages by the superficial inoculation and/or the addition of intracellular extracts of *Mucor racemosus*. J Food Sci 65: 731–738.

JM Bruna, M Fernandez, E Hierro, L de la Hoz, JA Ordoñez. 2000b. Combined use of Pronase E and a fungal extract (*Penicillium aurantiogriseum*) to potentiate the sensory characteristics of dry fermented sausages. Meat Sci 54:135–145.

JM Bruna, M Fernandez, JA Ordoñez, L de la Hoz. 2001. Papel de la flora fúngica en la maduración de los embutidos crudos curados. Alim, Equip Tecnol. Septiembre:79–84.

———. 2002. Enhancement of the flavour development of dry fermented sausages by using a protease (Pronase E) and a cell-free extract of *Penicillium camemberti*. J Sci Food Agric 82:526–533.

JM Bruna, EM Hierro, L de la Hoz, DS Mottram, M Fernandez, JA Ordoñez. 2001b. The contribution of *Penicillium aurantiogriseum* to the volatile composition and sensory quality of dry fermented sausages. Meat Sci 59:97–107.

———. 2003. Changes in selected biochemical and sensory parameters as affected by the superficial inoculation of *Penicillium camemberti* on dry fermented sausages. Inter J Food Microbiol 85:111–125.

JM Bruna, JA Ordoñez, M Fernandez, B Herranz, L de la Hoz. 2001a. Microbial and physico-chemical changes during the ripening of dry fermented sausages superficially inoculated with or having added an intracellular cell-free extract of *Penicillium aurantiogriseum*. Meat Sci 59:87–96.

S Buscailhon, G Monin, M Cornert, J Bousset. 1994. Time-related changes in nitrogen fractions and free amino acids of lean tissue of french dry-cured ham. Meat Sci 37:449–456.

C Cantoni, S d'Aubert, MA Bianchi, G Beretta. 1975. Metabolic aspects of lactobacilli during ripening of sausage (salami). Ind Alim 14:88–92.

C Cantoni, MR Molnar, P Renon, G Giolitti. 1967. Lipolytic micrococci in pork fat. J Appl Bacteriol 30:190–196.

M Careri, A Mangia, G Barbieri, L Bolzoni, R Virgili, G Parolari. 1993. Sensory property relationships to chemical data of Italian type dry-cured ham. J Food Sci 58:968–972.

RG Cassens. 1995. Use of sodium nitrite in cured meats today. Food Technol 49:72–80.

E Choe, DB Min. 2006. Chemistry and reactions of reactive oxygen species in foods. CRC Crit Rev Food Sci Nutr 46:1–22.

JE Christensen, EG Dudley, JA Pederson, JL Steele. 1999. Peptidases and amino acid catabolism in lactic acid bacteria. Antonie Van Leeuwenhoek Internat J Gen Mol Microbiol 76:217–246.

R Coppola, M Iorizzo, R Saotta, E Sorrentino, L Grazia. 1997. Characterization of micrococci and staphylococci isolated from soppressata molisana, a Southern Italy fermented sausage. Food Microbiol 14:47–53.

CK Cross, P Ziegler. 1965. A comparison of the volatiles fractions of cured and uncured meats. J Food Sci 30:610–614.

R Dainty, H Blom. 1995. Flavour chemistry of fermented sausages. In: G Campbell-Platt and PE Cook, eds. Fermented Meat. London: Blackie Academic & Professional, pp. 176–194.

D Demeyer, EY Claeys, S Ötles, L Caron, A Verplaetse. 1992. Effect of meat species on proteolysis during dry sausage fermentation. Proc 38th Inter Cong Meat Sci Technol, Clermond Ferrand, pp. 775–778.

D Demeyer, J Hoozee, H Mesdom. 1974. Specificity of lipolysis during dry sausage ripening. J Food Sci 39:293–296.

B Dias, B Weimer. 1998. Conversion of methionine to thiols by Lactococci, Lactobacilli, and Brevibacteria. Appl Environ Microbiol 64:3320–3326.

O Diaz, M Fernandez, GD García de Fernando, L de la Hoz, JA Ordoñez. 1992. Effect of the addition of the aspartyl proteinase from *Aspergillus oryzae* on dry fermented sausage during ripening. Proc 38th Inter Cong Meat Sci Technol, Clermond Ferrand, pp. 779–782.

———. 1993. Effect of the addition of Pronase E on the proteolysis in dry fermented sausages. Meat Sci 34:205–216.

———. 1996. Effect of the addition of papain on the dry fermented sausage proteolysis. J Sci Food Agric 71:13–21.

———. 1997. Proteolysis in dry fermented sausages: The effect of selected exogenous proteases. Meat Sci 46:115–128.

E Dransfield. 1993. Modeling postmortem tenderization .4. Role of calpains and calpastatin in conditioning. Meat Sci 34:217–234.

BD Eakes, TN Blumer, RJ Monroe. 1975. Effect of nitrate and nitrite on color and flavor of country style hams. J Food Sci 40:973–976.

RA Edwards, JA Ordoñez, RH Dainty, EM Hierro, L de la Hoz. 1999. Characterization of the headspace volatile compounds of selected Spanish dry fermented sausages. Food Chem 64:461–465.

WJM Engels, AC Alting, M Arntz, H Gruppen, AGJ Voragen, G Smit, S Visser. 2000. Partial purification and characterization of two aminotransferases from *Lactococcus lactis* subsp *cremoris* B78 involved in the catabolism of methionine and branched-chain amino acids. Inter Dairy J 10:443–452.

WJM Engels, S Visser. 1996. Development of cheese flavour from peptides and amino acids by cell-free extracts of *Lactococcus lactis* subsp *cremoris* B78 in a model system. Neth Milk Dairy J 50:3–17.

S Fadda, G Vignolo, APR Holgado, G Oliver. 1998. Proteolytic activity of *Lactobacillus* strains isolated from dry-fermented sausages on muscle sarcoplasmic proteins. Meat Sci 49:11–18.

M Fernandez, O Diaz, I Cambero, L de la Hoz, JA Ordoñez. 1991. Effect of the addition of pancreatic lipase on lipolysis during the ripening of dry fermented sausage. Proc of the 37th Inter Cong Meat Sci Technol, Kulmbach, pp. 867–870.

M Fernandez, L de la Hoz, O Diaz, MI Cambero, JA Ordoñez. 1995a. Effect of the addition of pancreatic lipase on the ripening of dry fermented sausages. I Microbial, physicochemical and lipolytic changes. Meat Sci 40:159–170.

———. 1995b. Effect of the addition of pancreatic lipase on the ripening of dry fermented sausages. II Free fatty acids, short chain fatty acids, carbonyls and sensorial quality. Meat Sci 40:351–362.

M Fernandez, JA Ordoñez, JM Bruna, B Herranz, L de la Hoz. 2000. Accelerated ripening of dry fermented sausages. Trends Food Sci Technol 11:201–209.

J Flores. 1997. Mediterranean *vs* northern European meat products. Processing technologies and main differences. Food Chem 59:505–510.

W Frey. 1985. Fabricación fiable de embutidos. Zaragoza: Acribia, pp. 1–39.

C Garcia, JJ Berdague, T Antequera, C Lopez-Bote, JJ Cordoba, J Ventanas. 1991. Volatile components of dry cured Iberian ham. Food Chem 41:23–32.

ML Garcia, C Casas, VM Toledo, MD Selgas. 2001. Effect of selected mould strains on the sensory properties of dry fermented sausages. Eur Food Res Technol 212:287–291.

ML Garcia, MD Selgas, M Fernandez, JA Ordoñez. 1992. Microorganisms and lipolysis in the ripening of dry fermented sausages. Inter J Food Sci Technol 27:675–682.

R Geisen. 1993. Fungal starter cultures for fermented foods: Molecular aspects. Trend Food Sci Technol 4:251–256.

GG Giddings. 1977. Basis of color in muscle foods. CRC Crit Rev Food Sci Nutr 9:81–114.

CO Gill. 1986. The control of microbial spoilage in fresh meats. In: AM Pearson and TR Dutson, eds. Advances in Meat Research. Meat and Poultry Microbiology. Westport, Connecticut: AVI Publishing Co., pp. 49–88.

CO Gill, KG Newton. 1977. The development of aerobic spoilage flora on meat stored at chill temperatures. J Appl Bacteriol 43:189–195.

G Gottschalk. 1986. Bacterial Metabolism. New York: Springer-Verlag, pp. 214–224.

BF Hagen, JL Berdague, AL Holck, H Næs, H Blom. 1996. Bacterial proteinase reduces maturation time of dry fermented sausages. J Food Sci 61:1024–1029.

WP Hammes, I Bosch, G Wolf. 1995. Contribution of *Staphylococcus carnosus* and *Staphylococcus piscifermentans* to the fermentation of protein foods. J Appl Bacteriol 79:S76–S83.

AHW Hauschild, R Hilsheimer, G Jarvis, DP Raymond. 1982. Contribution of nitrite to the control of *Clostridium botulinum* in liver sausage. J Food Prot 45:500–506.

B Herranz, M Fernandez, L de la Hoz, JA Ordoñez. 2006. Use of bacterial extracts to enhance amino acid breakdown in dry fermented sausages. Meat Sci 72:318–325.

B Herranz, M Fernandez, E Hierro, JM Bruna, JA Ordoñez, L de la Hoz. 2004. Use of *Lactococcus lactis* subsp *cremoris* NCDO 763 and alpha-ketoglutarate to improve the sensory quality of dry fermented sausages. Meat Sci 66:151–163.

B Herranz, L de la Hoz, E Hierro, M Fernandez, JA Ordoñez. 2005. Improvement of the sensory properties of dry-fermented sausages by the addition of free amino acids. Food Chem 91:673–682.

E Hierro, L de la Hoz, JA Ordoñez. 1999. Contribution of the microbial and meat endogenous enzymes to the free amino acid and amine contents of dry fermented sausages. J Agric Food Chem 47:1156–1161.

———. 1997. Contribution of microbial and meat endogenous enzymes to the lipolysis of dry fermented sausages. J Agric Food Chem 45:2989–2995.

E Hierro, J Ordoñez, JM Bruna, C Pin, M Fernandez, L de la Hoz. 2005. Volatile compound generation in dry fermented sausages by the surface inoculation of selected mould species. Eur Food Res Technol 220: 494–501.

LL Hinrichsen, SB Pedersen. 1995. Relationship among flavor, volatile compounds, chemical changes, and microflora in Italian type dry-cured ham during processing. J Agric Food Chem 43:2932–2940.

K Incze. 1992. Raw fermented and dried meat products. Fleischwirtschaft 72:1–5.

PM Kenneally, G Schwarz, NG Fransen, EK Arendt. 1998. Lipolytic starter culture effects on production of free fatty acids in fermented sausages. J Food Sci 63:538–543.

L Kröckel. 1995. Bacterial fermentation of meat. In: G Campbell-Platt and PE Cook, eds. Fermented Meats. Glasgow: Blackie Academic and Professional, pp. 69–109.

BA Law, J Kolstad. 1983. Proteolytic systems in lactic-acid bacteria. Antonie Van Leeuwenhoek. J Microbiol 49:225–245.

L Leistner, C Eckardt. 1979. Occurrence of toxinogenic Penicillia in meat-products. Fleischwirtschaft 59: 1892–1896.

HU Liepe. 1982. Starter cultures in meat production. In: HJ Rhem and G Reed, eds. Biotechnology. Basilea: Verlag Chemie, pp. 169–174.

MO López, L Hoz, MI Cambero, E Gallardo, G Reglero, JA Ordoñez. 1992. Volatile compounds of dry hams from Iberian pig. Meat Sci 31:267–277.

FK Lücke. 1985. The microbiology of fermented meats. J Sci Food Agric 36:1342–1343.

FK Lücke. 1998. Fermented sausages. In: BJB Wood, ed. Microbiology of Food Fermentation. London: Appl Sci Publ pp. 441–483.

FK Lücke, H Hechelmann. 1987. Starter cultures for dry sausages and raw ham composition and effect. Fleischwirtschaft 67:307–314.

B MacDonald, JI Gray, LN Gibbins. 1980. Role of nitrite in cured meat flavor. Antioxidant role of nitrite. J Food Sci 45:893–897.

P MacLeod, ME Morgan. 1955. Leucine metabolism of *Streptococcus lactis* var *maltigenes* .1. Conversion of alpha-ketoisocaproic acid to leucine and 3-methylbutanal. J Dairy Sci 38:1208–1214.

MAPA (Ministerio de Agricultura, Pesca y Alimentación). 1997. Inventario Español de productos tradicionales. Madrid: Servicio de Publicaciones del MAPA.

L Martin, JJ Cordoba, T Antequera, ML Timon, J Ventanas. 1998. Effects of salt and temperature on proteolysis during ripening of Iberian ham. Meat Sci 49:145–153.

J Mateo, JM Zumalacarregui. 1996. Volatile compounds in chorizo and their changes during ripening. Meat Sci 44:255–273.

PLH McSweeney. 2004. Biochemistry of cheese ripening. Inter J Dairy Technol 57:127–144.

JA Melendo, JA Beltran, I Jaime, R Sancho, P Roncales. 1996. Limited proteolysis of myofibrillar proteins by bromelain decreases toughness of coarse dry sausage. Food Chem 57:429–433.

A Meynier, E Novelli, R Chizzolini, E Zanardi, G Gandemer. 1999. Volatile compounds of commercial Milano salami. Meat Sci 51:175–183.

I Molina, F Toldrá. 1992. Detection of proteolytic activity in microorganisms isolated from dry-cured ham. J Food Sci 57:1308–1310.

K Molly, D Demeyer, T Civera, A Verplaetse. 1996. Lipolysis in a Belgian sausage: Relative importance of endogenous and bacterial enzymes. Meat Sci 43:235–240.

MC Montel, R Talon, JL Berdague, M Cantonnet. 1993. Effects of starter cultures on the biochemical characteristics of French dry sausages. Meat Sci 35: 229–240.

MC Montel, R Talon, M Cantonnet, J Fournaud. 1992. Identification of *Staphylococcus* from French dry sausage. Lett Appl Microbiol 15:73–77.

MJ Motilva, F Toldrá. 1993. Effect of curing agents and water activity on pork muscle and adipose subcutaneous tissue lipolytic activity. Z Lebens Unte Forsch 196:228–232.

MJ Motilva, F Toldrá, P Nieto, J Flores. 1993. Muscle lipolysis phenomena in the processing of dry cured ham. Food Chem 48:121–125.

H Næs, AL Holck, L Axelsson, HJ Andersen, H Blom. 1994. Accelerated ripening of dry fermented sausage by addition of a *Lactobacillus* proteinase. Inter J Food Sci Technol 29:651–659.

H Næs, BO Pedersen, AL Holck, L Axelsson, V Holten, H Blom. 1992. Fermentation of dry sausage. The effect of added proteinase and lipase from lactobacilli. Proc 38th Inter Cong Meat Sci Technol, Clermond Ferrand, pp. 815–818.

J Nordal, E Slinde. 1980. Characteristics of some lactic acid bacteria used as starter cultures in dry sausage production. Appl Environ Microbiol 40:472–475.

GJE Nychas, JS Arkoudelos. 1990. Staphylococci. Their role in fermented sausages. J Appl Bacteriol 69:S167–S188.

JA Ordoñez, EM Hierro, JM Bruna, L de la Hoz. 1999. Changes in the components of dry-fermented sausages during ripening. CRC Crit Rev Food Sci Nutr 39:329–367.

JA Ordoñez, MO Lopez, E Hierro, MI Cambero, L de la Hoz. 1996. Efecto de la dieta de cerdos ibéricos en la composición en ácidos grasos del tejido adiposo y muscular. Food Sci Technol Inter 2:383–390.

MA Paleari, L Piantoni, F Caloni. 1991. Prove sull impiego della lipasi nella preparazione dei salumi. Ind Alim 30:1072–1074.

SA Palumbo, JL Smith. 1977. Chemical and microbiological changes during sausage fermentation and ripening. In: L Orly and J St. Angelo, eds. ACS Symposium Series no. 47, pp. 279–294.

M Papon, R Talon. 1988. Factors affecting growth and lipase production by meat lactobacilli strains and *Brochothrix thermosphacta*. J Appl Bacteriol 64: 107–115.

———. M Papon, R Talon. 1989. Cell location and partial characterization of *Brochothrix thermosphacta* and *Lactobacillus curvatus* lipases. J Appl Bacteriol 66:235–242.

JJ Pestka. 1986. Fungi and mycotoxins in meats. In: AM Pearson and TR Dutson, eds. G Advances in Meat Research. Meat and Poultry Microbiology. Westport: AVI Publishing, Vol 2, pp. 277–309.

MD Pierson, LA Smoot. 1982. Nitrite, nitrite alternatives, and the control of *Clostridium botulinum* in cured meats. CRC Crit Rev Food Sci Nutr 17: 141–187.

M Roca, K Incze. 1990. Fermented sausages. Food Rev Inter 6:91–118.

J Samelis, J Metaxopoulus. 1997. Lipolytic activity of meat lactobacilli isolated from naturally fermented Greek dry sausage. Fleischwirtschaft 77:165–168.

B Sanz. 1988. El ayer, hoy y mañana de la Bromatología. Madrid: Instituto de España. Real Academia de Farmacia, pp. 1–92.

B Sanz, D Selgas, I Parejo, JA Ordoñez. 1988. Characteristics of Lactobacilli isolated from dry fermented sausages. Inter J Food Microbiol 6:199–205.

C Sarraga, M Gil, J Arnau, JM Monfort, R Cusso. 1989. Effect of curing salt and phosphate on the activity of porcine muscle proteases. Meat Sci 25:241–249.

K Sato, GR Hegarty. 1971. Warmed-over flavor in cooked meat. J Food Sci 36:1098–1102.

S Schmidt, RG Berger. 1998. Aroma compounds in fermented sausages of different origins. Food Sci Technol Lebens Wissen Technol 31:559–567.

MD Selgas, ML Garcia, GD Garcia de Fernando, JA Ordoñez. 1993. Lipolytic and proteolytic activities of micrococci isolated from dry sausages. Fleischwirtschaft 73:1175–1176, 1179.

MD Selgas, B Sanz, JA Ordoñez. 1988. Selected characteristics of micrococci isolated from Spanish dry fermented sausages. Food Microbiol 5:185–193.

BJ Singh, D Dincho. 1994. Molds as protective cultures for raw dry sausages. J Food Prot 57:928–930.

G Smit, A Verheul, R van Kranenburg, E Ayad, R Siezen, W Engels. 2000. Cheese flavour development by enzymatic conversions of peptides and amino acids. Food Res Inter 33:153–160.

LH Stahnke, A Holck, A Jensen, A Nilsen, E Zanardi. 2000. Flavour compounds related to maturity of dried fermented sausage. Proc 46th Inter Cong Meat Sci Technol, Buenos Aires, pp. 236–237.

———. 2002. Maturity acceleration of Italian dried sausage by *Staphylococcus carnosus*. Relationship between maturity and flavor compounds. J Food Sci 67:1914–1921.

LH Stahnke, P Zeuthen, 1992. Identification of volatiles from Italian dried salami. Proc 38th Inter Cong Meat Sci Technol, Clermond Ferrand, pp. 835–838.

LO Sunesen, LH Stahnke. 2003. Mould starter cultures for dry sausages—Selection, application and effects. Meat Sci 65:935–948.

R Talon, MC Montel, G Gandemer, M Viau, M Cantonnet. 1992. Lipolysis of pork fat by *Staphylococcus warneri*, *S. saprophyticus* and *Micrococcus varians*. Appl Microbiol Biotechnol 38:606–609.

K Tjener, LH Stahnke, L Andersen, J Martinussen. 2004. Addition of alpha-ketoglutarate enhances formation of volatiles by *Staphylococcus carnosus* during sausage fermentation. Meat Sci 67:711–719.

F Toldrá. 2002. Dry-cured Meat Products. Trumbull, Co: Food & Nutr Press, Inc., pp. 89–148.

F Toldrá, MJ Motilva, E Rico, J Flores. 1991. Enzyme activities in the processing of dry-cured ham. Proc 37th Inter Cong Meat Sci Technol. 1–6. Kulmbach, Germany, pp. 954–957.

F Toldrá, E Rico, J Flores. 1992. Activities of pork muscle proteases in model cured meat systems. Biochimie 74:291–296.

———. 1993. Cathepsin-B, cathepsin-D, cathepsin-H and cathepsin-L activities in the processing of dry cured ham. J Sci Food Agric 62:157–161.

SO Tomek, A Bulgay. 1991. Influence of environmental RH on some physical and chemical properties of Turkish sukuk. Proc 37th Inter Cong Meat Sci Technol 1–6. Kulmbach, Germany, pp. 958–961.

J Ventanas, JJ Cordoba, T Antequera, C Garcia, C Lopezbote, MA Asensio. 1992. Hydrolysis and maillard reactions during ripening of Iberian ham. J Food Sci 57:813–815.

A Verplaetse, D Demeyer, S Gerard, E Buys. Endogenous and bacterial proteolysis in dry sausage fermentation. Pro 38th Inter Cong Meat Sci Technol, Clermond Ferrand, 1992, pp. 851–854.

M Yvon, S Berthelot, JC Gripon. 1998. Adding alphaketoglutarate to semi-hard cheese curd highly enhances the conversion of amino acids to aroma compounds. Inter Dairy J 8:889–898.

M Yvon, E Chambellon, A Bolotin, F Roudot-Algaron. 2000. Characterization and role of the branched chain aminotransferase (BcaT) isolated from *Lactococcus lactis* subsp. *cremoris* NCDO 763. Appl Environ Microbiol 66:571–577.

M Yvon, S Thirouin, L Rijnen, D Fromentier, JC Gripon. 1997. An aminotransferase from *Lactococcus lactis* initiates conversion of amino acids to cheese flavor compounds. Appl Environ Microbiol 63:414–419.

I Zalacain, MJ Zapelena, I Astiasarán, J Bello. 1995. Dry fermented sausages elaborated with Lipase from *Candida cylindracea*. Comparison with traditional formulations. Meat Sci 40:55–61.

———. 1996. Addition of lipase from *Candida cylindracea* to a traditional formulation of a dry fermented sausage. Meat Sci 42:155–163.

I Zalacain, MJ Zapelena, MP de Peña, I Astiasarán, J Bello. 1997a. Application of lipozyme 10,000 L (from *Rhizomucor miehei*) in dry fermented sausage technology: Study in a pilot plant and at the industrial level. J Agric Food Chem 45:1972–1976.

———. 1997b. Lipid fractions of dry fermented sausages change when starter culture and/or *Aspergillus* lipase are added. J Food Sci 62:1076–1079.

———. 1997c. Use of lipase from *Rhizomucor miehei* in dry fermented sausages elaboration: Microbial, chemical and sensory analysis. Meat Sci 5:99–105.

MJ Zapelena, D Ansorena, I Zalacain, I Astiasarán, J Bello. 1998. Dry fermented sausages manufactured with different amounts of commercial proteinases: Evolution of total free alpha-NH2-N groups and sensory evaluation of the texture. Meat Sci 49:213–221.

MJ Zapelena, I Astiasarán, J Bello. 1999. Dry fermented sausages made with a protease from *Aspergillus oryzae* and or a starter culture. Meat Sci 52:403–409.

MJ Zapelena, I Zalacain, MP de Peña, I Astiasarán, J Bello. 1997. Effect of the addition of a neutral proteinase from *Bacillus subtilis* (Neutrase) on nitrogen fractions and texture of Spanish fermented sausage. J Agric Food Chem 45:2798–2801.

33
North European Products

Jürgen Schwing and Ralf Neidhardt

INTRODUCTION

This chapter covers the dry sausage traditions of northern Europe; the quality aspects of dry-fermented sausages, the ingredients and processing technologies commonly used, and the characteristic recipes of various countries. Of course, the great number of products makes it impossible to give an exhaustive overview, also it is generally acknowledged that a great number of names and designation very often cover the same type of product or a very similar product. This makes it particularly difficult to describe differences.

Northern Europe represents a huge number of sausages and production traditions in the following countries: Benelux, Scandinavia, Germany, Switzerland, Hungary, Poland, Russia, Baltic countries, Bulgaria, Romania, Czech Republic, Slovakia, Austria, Slovenia, Ukraine, Serbia, Croatia, England, Ireland, etc. Some of these countries produce fermented sausages, which can be classified either as north European or as south European depending on the consumers' appreciation of the quality characteristics: however, this chapter will deal with only the typical northern types.

CHARACTERISTICS OF NORTH EUROPEAN SAUSAGES

In general, north European sausages are smoked and lower in pH than south European sausages. The majority of the sausages are prepared with a finely minced meat batter, which is fermented and dried in a relatively short time and which leads to a distinct acidic flavor, though astringency and stingy mouth-feel should be avoided. This meat batter resembles the meat batter of Milan salami very much, with its very fine grain size (Löbel 1989), hence the habit of the English-speaking world to call all fermented sausages "salami." The same is done in this chapter. Molded sausages are still of relatively little importance in northern Europe, even if the market trend seems to point toward a milder, more rounded Mediterranean sausage flavor.

Quality Criteria

The definition of quality is quite difficult, because it depends not only on objective criteria laid down in a written form and measurable, but depends as well on subjective perceptions linked to cultural background and history. Hofmann (1974) defined quality as the summary of all quality-relevant parameters. For administrative safety, measurable parameters like pH, water activity, and microbiology aspects are relevant quality parameters. In addition, protein/fat ratio, connective tissue, water binding capacity, and type of fat of the raw materials are important for processors. Of course the consumer expects good flavor, good mouth-feel, and consistency on top of the safety aspects.

A long tradition for producing dry-fermented sausages in northern Europe exists with a great number of product names. Some are very similar in appearance; others are very different in their features (BAFF 1985). The following characteristics are used in order to describe the quality of a dry-fermented sausage:

- General appearance (color, molded, smoked, coated, etc.)
- Caliber of casing/type of casing
- Grain size of meat and fat particles
- Meat composition (lean, fat, connective tissue, others)
- Degree of fattiness and quality of fat (color, taste)
- Salt contents (type of salt, level, and health aspects)
- Color (added colorant, natural color formation)
- Aroma (smell of smoke, molds, yeasts, fermentation, etc.)
- Taste features (acidity level, fermentation flavor, maturity)
- Texture (firmness, elasticity, crumbliness, etc.)

Sausage Ingredients

Raw Materials

Dry, red meat from older animals is preferred, but sometimes cooked pork skin is used as well. The raw material is usually used fresh and/or frozen. Pork meat must fulfill the following criteria: $pH_1>5.8$ and $pH_{24}<5.8$; if chilled at 0°C, meat must be used within 3 to 5 days postmortem; if frozen at –30°C to 18°C, meat must be stored for a maximum of 90 days. For beef, the following criteria must be met: $pH_{24}<5.8$; if chilled at 0°C, meat must be used within 3 to 14 days postmortem; if frozen at –30°C to 18°C, frozen storage should be a maximum of 180 days (Fischer 1988).

The fatty part is usually from pork, which has to be fresh, white, and firm to avoid strong "oiling out" and rancidity. The shelf life of a fermented sausage is often limited by changes in the fat. The best fat is from neck or back of the pork carcass. The amount of polyunsaturated fatty acids should be below 12%. The fat must fulfill the following: well chilled before freezing; chilled <2°C, stored for a maximum of 3 days; frozen at –30°C to –18°C, storage for a maximum of 90 days (Fischer 1988).

Meat from poultry should be dry and dark; therefore, meat from the leg is preferred. Poultry fat is problematic for its softness and may lead to strong oiling out. Horse, reindeer, and venison meat is used for specialties. These meats need, in general, an addition of backfat in order to obtain good finished fermented products. The fermentation profile has to be adjusted as compared to products with pork and beef because these meats have initially high pH well over pH 6.2.

Dry Ingredients

The addition of nitrite is important for the color formation and for the microbiological stability, especially in the first days of ripening (Leistner and Gould 2002). Furthermore, nitrite participates in the flavor formation. Nitrite is added to nearly all north European fermented sausages in a mixture with common salt; only a few specialties are made with nitrate.

To achieve a proper acidification and an excellent flavor and to improve and stabilize color, the use of bacterial starter cultures is widely spread in northern Europe (Hansen 2003). As compared to southern Europe, processors in northern Europe mostly use starter cultures, which results in a fast pH drop. Due to the fast starter cultures, the use of GDL (glucono-delta-lactone) is becoming less common.

Sugar is primarily added as an energy source for the starter cultures. Glucose, maltose, sucrose, and derivatives from starch hydrolysis are predominantly used. In the past, sucrose was the most commonly used sugar because it was easy to get hold of. Now it is mostly substituted by glucose to obtain a fast pH drop. Lactose becomes more and more unpopular for allergenic reasons, in particular in Germany. It is important that the sugar type corresponds to the needs of the specific starter culture, especially when fast fermentation is targeted (Hansen 2003). Carbohydrates like potatoes, potato starch, or potato flakes are added in specific sausage types (see Swedish Mettwurst) and are responsible for the characteristic acidification profile, flavor, and texture of those products.

The most important spice for north European salami is pepper (*Piper Nigrum L.*). White pepper is usually used as fine ground pepper and black as coarsely ground pepper. Other spices are ginger, cardamom, garlic, coriander, cumin, mace, paprika, rosemary, nutmeg, and clove (especially in northern Holland); sweet peppers or mustard seeds are used in Hungarian products (see below).

Processing Technology

The microbiological stability of a north European sausage is based on low pH, low water activity, and smoke application externally (Potthast and Eigner 1988). Traditionally, fermented sausage processors did not use cultures, and they produced under climatic conditions, which were strongly influenced by the weather of the moment. Therefore, sausages were often dried extensively to make the sausages safe. Now the use of starter cultures and controlled fermentation and drying conditions result in a controlled pH drop and an optimization of water loss. The hurdle principle ensures that a sausage technology can be considered safe even when drying is less extensive than in the past (Stiebing 1993).

Meat Batter Preparation

The most important comminution technology in northern Europe in fermented sausage production is the combined chopping and mixing technology of the bowl chopper. The standard procedure is the following: Frozen lean meat is chopped coarsely in the bowl chopper. All ingredients except salt are added. Chopping continues to desired particle size (1–4 mm), frozen fat (−18°C, 0°F) is added, chopping continues to the final particle size of the end product, and fresh, minced meat is supplied and mixed in homogeneously. At the very end, salt is added and minimal blending is applied to ensure a homogeneous distribution. The final batter must have a temperature not higher than about −4°C (25°F). Smearing has to be avoided as much as possible during the subsequent stuffing procedure (Stiebing 1993).

Fermentation and Drying

After stuffing, the sausages are fermented and dried/ripened by applying the appropriate temperature, humidity, and speed of air (circulation) with respect to the wanted end product (Stiebing and Rödel 1987). Roughly, the fermentation pattern may be divided into three schemes:

- *Traditional fermentation*: fermentation temperature usually 18–24°C, more than 40 hours to achieve pH 5.3, final pH often >5.0, final water activity <0.90, total production time >3 weeks (BAFF 1990).
- *Fast fermentation*: fermentation temperature 22–26°C, less than 30 hours to archive pH 5.3, final pH usually 4.5–4.9, final water activity >0.90, total production time 2–3 weeks (Neuhäuser 1988).
- *Very fast American-style fermentation*: fermentation temperature 32–43°C, less than 15 hours to archive pH 5.3, final pH <4.8, final water activity >0.90, total production time 1–3 weeks, cooking often applied to stop fermentation and/or increase safety (Bacus 1984).

The boundaries between the three technologies overlap, but in northern Europe the fast fermentation is the most used technology now. Due to less technical possibilities and a smaller range of available starter cultures, the slower traditional fermentation was used in the past. Other technologies, such as ripening in brine or ripening under vacuum and/or pulsating atmospheric pressure, disappeared mostly when the use of industrial fermentation equipment became more widespread (Neuhäuser 1988). In fact, the machinery plays a crucial role in modern sausage making, and this role is very often underestimated. This is especially true in the area of fermentation equipment (Stiebing 1993). The ripening scheme in Table 33.1 shows a standard process for fermentation and drying in northern Europe.

SAUSAGE TYPES OF CENTRAL EUROPE

Germany

German sausage quality is mainly characterized by the usage of valuable parts of the carcass (ham, belly, trimmings, etc.); hence, the consumer does not consider the stage of drying as an obligingly important

Table 33.1. Standard ripening program for north European technology.

Time (Hours)	Temperature (°C/°F)	RH* (%)	Smoking
4	25/77	–	
24	24/75	94	
24	22/72	92	Moderate 1–2 h
24	20/68	90	
24	20/68	88	Moderate 1–2 h
24	18/64	86	
24	18/64	84	Moderate 1–2 h
	14–16/57–61	78–82	Postripening

* RH = relative humidity.

feature of quality assessment. Yet in some areas and in some products, this is the case, e.g., for landjäger sausage in south Germany. Traditionally, the German butcher chooses good raw materials as a basis for successful production. Many recipes are based on pork, beef, and backfat of excellent quality (Löbel 1989). The typical characteristics are:

- *Northwestern Germany*: strongly smoked, relatively soft sliceable or spreadable products having soft fats and a "fresh meat" characteristic with mild, low acid features.
- *Westphalia*: smoked, firm, and sliceable products with a distinct sour feature; ideally there is a distinct fermentation flavor; very often a fast technology is used and sausages are stuffed in big diameter casings with limited drying.
- *Southern Germany*: mildly smoked, well-dried products with only very mild acidic characteristics; sausages may be molded instead of smoked.
- *Eastern Germany*: mildly smoked, well-dried products with low acidic features and special characteristic spice combinations (cumin).

German specialties are Teewurst, Thuringer Bratwurst (raw fermented type of semidry coarse grained sausage), Cervelatwurst, Feldkieker, Eichsfelder (based on freshly slaughtered meat), Katenrauchwurst, etc.

The most-often sold German salami product is a sliced large salami type called Stapelpack salami, which is based on Westphalian salami (Figure 33.1). Both the meat block and the seasoning of the standard Westphalian salami vary between producers. The following recipe represents the standard recipe. The meat block: 40% frozen, lean sow meat, 40% chilled, lean pork meat, and 20% frozen backfat. Further ingredients are nitrite salt, pepper, garlic, sugars, starter cultures, and sometimes, mustard seeds. The standard production procedure is as follows: Sow meat is chopped coarsely in the bowl chopper and dry ingredients except salt are added. Further comminution continues to a particle size of approximately 2 mm. Fat is put in and minced to a particle size of 3–4 mm. Pork meat, processed through a 2–3 mm plate, is added, and the whole mixture is chopped to the desired particle size. During the last rounds of chopping, nitrite salt is added and mixed in homogenously. The mince is stuffed into casings and ripened at 24°C at 95% relative humidity (RH) in the beginning. The temperature is lowered to 12–14°C and the humidity to about 80% (±2) step by step, until a water loss of approximately 25% is obtained.

Schlackwurst has a very fine particle size and its name originates from the casing type used (colon or Schlacke in German). The basic meat block is 24% lean beef, 10% lean pork, and 66% pork belly. Typical ingredients are nitrite salt, pepper, and honey. There are several regional differences. Braunschweiger Schlackwurst contains only pork meat and backfat instead of pork belly. The spices are pepper, ginger, and red wine. Thüringer Schlackwurst contains pepper and cardamom. Westfälische Schlackwurst is made of 20% beef, 20% pork belly, 20% backfat, and 40% very lean pork. The seasonings are based on pepper, ginger, raspberry syrup, rum, and mustard seeds.

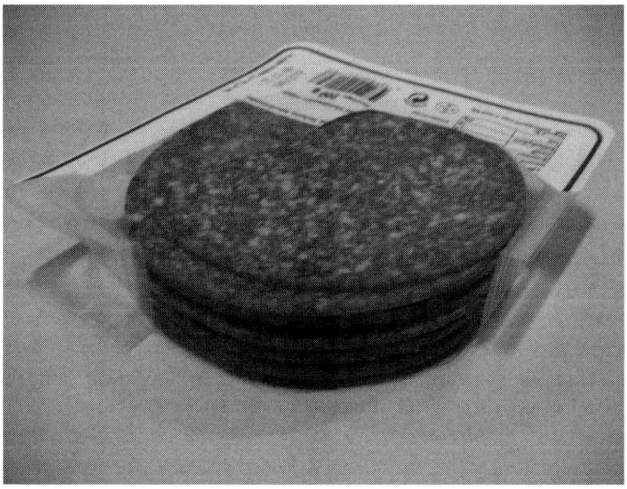

Figure 33.1. Stapelpack salami from Germany.

Cervelatwurst has as fine a particle size as Schlackwurst or even finer, but it is stuffed in different casings. The origin of Cervelatwurst is a cooked Milan-type of salami from Italy. The meat block for this high-quality product is 30% very lean beef, 46% lean pork, and 24% backfat. Typical seasonings are ground pepper, pepper seeds, and juniper brandy, but different recipes exist in each region of Germany, such as Westphalia, Thüringen, or Holstein. One main recipe variation stems from the use of different brandy types like rum, whiskey, or Steinhäger, which is based on herbal extracts.

Feldkieker is a specialty from a region around the town of Göttingen. The special feature of this product is the use of just-slaughtered warm meat (derived from the traditional sausage making technique when cold stores were not available and meat had to be preserved immediately after slaughtering). The 70% lean pork and 30% backfat are minced through a 2 mm plate and mixed with salt, pepper, raspberry syrup, honey, and red wine or rum. The mince is stuffed into a calf's bladder and ripened for 8–12 months. The sausage is either air-dried or smoked.

Landjäger is a sausage from the south of Germany. The origin is from the region of Voralberg in Switzerland, Elsaβ in France, and Schwarzwald in Germany. Landjäger is cased in small pork intestine and gets its characteristic rectangular shape from a special pressing step during fermentation. The meat block is based on beef rich in collagen tissue, pork, and backfat. Typical spices are pepper and cumin. In some recipes, parts of the beef are substituted with smoked cured ham, which results in a very strong smoke flavor. A typical ripening scheme is the following: Press in a rectangular shape for 24–48 hours at 24°C, 94% RH, the meat hanging on sticks, followed by continued ripening for 24 hours at 20°C, 88% RH and 48 hours at 18°C, 82% RH with occasional smoking; and further dry at 14–16°C, 76–78% RH to a weight loss of 35%.

Teewurst is a high-quality, spreadable raw sausage (Figure 33.2). It can be divided into two groups: coarsely grained Teewurst (3–4 mm particle size) and finely minced Teewurst (no visual particles, creamy appearance). Teewurst must not have a firm structure, and therefore the meat particles should be surrounded by fat particles or an oily fat phase to avoid protein coagulation. In addition, pH should not be lowered below 5.3. The use of about 3% sunflower oil helps maintain the soft texture. The water loss is usually 5–10%. A typical processing scheme is the following: The meat block (2 mm) contains 20% lean beef, 60% lean pork, 17% backfat, and 3% sunflower oil; meat temperature is −2°C to +2°C. Backfat is chopped to a cream in the bowl chopper. Meat, sunflower oil, and all other ingredients except salt are added. The blend is chopped with high speed until 15°C to desired consistency. Salt is added while chopping at low speed in order to reach a homogenous distribution of salt. The end temperature should be approximately 18°C. After filling, the ripening scheme is the following: 2 days at 20°C/85–90% RH; smoking at 20°C, 85% RH until the desired golden color is obtained, followed by 6 h at 26°C, 80% RH to obtain a slight film of fat and further ripening at 14°C, 78–80% RH.

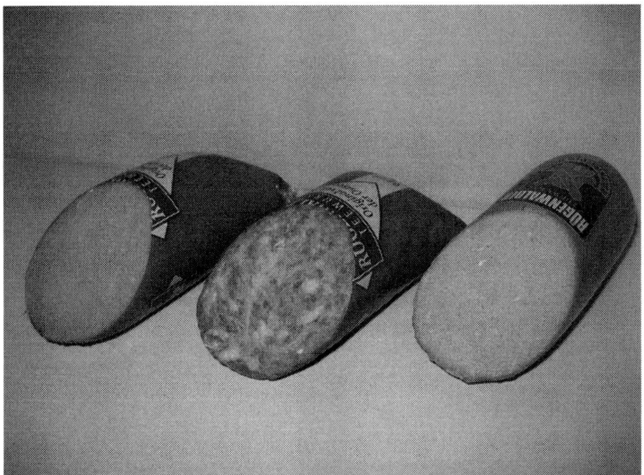

Figure 33.2. German Teewurst.

Austria

Austrian sausages are very similar to the German salamis; however, a typical Austrian product is Kantwurst. Kantwurst gets its characteristic square shape and unique taste from a special pressing and curing process close to traditional Bavarian technology. Kantwurst is cured for 7 weeks and reaches a weight loss of about 35%. Traditionally, Kantwurst could be kept for a long time when stored in a cool and dry place.

The Netherlands (and Flemish-speaking Belgium)

Dutch dry sausages can be divided into two major groups: spreadable and nonspreadable sausages, the latter having about two-thirds of the market. The traditional process is slow, aiming at a very aromatic fermentation flavor. Typical Dutch dry sausages are Boerenmetworst, salami and cervelat (Dutch: snijworst).

The most common product of Boerenmetworst is without added garlic (Figure 33.3). The meat block contains 20% lean beef, 50% pork meat, and 30% backfat. Normally, the meat is 80% frozen and 20% fresh. The frozen meat is ground to 3 mm in the bowl chopper. Halfway through this process, the backfat is added and while comminuting further the fresh meat part is put on top of the blend. Salt is added close to the end. The final mince has a meat particle size of about 3 mm with larger particles of backfat (3–5 mm). The mince is stuffed in a natural pork casing formed into a ring of approximately 400 g. Drying normally takes 15–20 days and results in a weight loss of 25–30%. The standard product is treated with smoke. Seasonings are pepper, coriander, mustard, and salt. In the northern part of the Netherlands, ginger and clove are added as a specialty and the product is dried to about 40%.

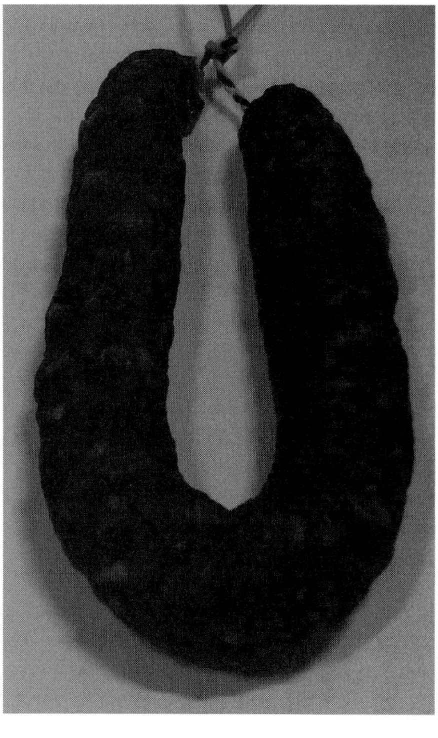

Figure 33.3. Dutch Boerenmetworst.

The Dutch salami always has added garlic. The meat block contains 10% lean beef, 40% pork meat, 10% belly, and 30% backfat. Normally, the meat is 80% frozen and 20% fresh. Sometimes about 10% ground pork skin is added for a softer texture. The frozen meat is ground to 3 mm in the bowl chopper. Halfway through this process, the backfat is added and while comminuting further the fresh meat part is put on top of the blend. Salt is added close to the end. The final mince has a uniform particle size of 3–5 mm. The grain size could be smaller in the western part of the country. Sometimes the particle size is as fine as 0.5–1 mm. The sausages are normally stuffed into a beef casing of 90–110 mm diameter to a total sausage weight of 4.5–5 kg. The drying and ripening takes around 30 days, and a weight loss of 25–30% is common. The normal product is nonsmoked. Seasonings are pepper, garlic, red wine, and green pepper (sometimes visible). In the southern part of the Netherlands, the sausages are more of the French type with respect to spices and application of molds.

Snijworst (Dutch cervelat) is always smoked and very close to some German products. The spices are pepper, nutmeg, mace, and coriander. In the south of the country, the product is more coarse-grained and beef replaces part of the pork. The meat block contains 10% lean beef, 40% pork meat, 10% belly, and 30% backfat. Normally, the meat and fat is added frozen to the bowl chopper and ground to a uniform size of 3–5 mm. The sausages are normally stuffed into a beef casing of 90–110 mm diameter to a total sausage weight of 4.5–5 kg. The drying and ripening takes around 30 days, and a weight loss of 25–30% is common.

Belgium and Luxembourg

Belgium salami is a smoked sausage similar to the Dutch salami. The special Baguette salami in

Belgium is a nonsmoked sausage, which resembles a Mediterranean product.

Ireland and England

Ireland and England do not have a real fermented sausage tradition, but today American-style pepperoni and continental-style products are produced in both countries in small amounts.

SAUSAGE TYPES OF EASTERN EUROPE

Hungary

Hungarian salami is one of the two world tradenames for salami (Figure 33.4). The other is Milano salami. Hungarian company tradenames like Pick from Szeged or Herz from Budapest have carried the fame of Hungarian sausage making all over the world. The Hungarian salami is one of the rare products combining smoke and mold application. In original Hungarian salami, sugars and starter cultures are still not allowed. The traditional technology was based on the Italian predrying technique: After a coarse grinding (6 mm), the meat was stored at low temperature for 1 day, followed by a second and final grinding (2.6–3.2 mm). The sausages were stuffed and hung for smoking over open fires. The pH did not drop below 5.5 during processing (Incze 1986). The seasonings used for Hungarian salami comprise white pepper, hot and sweet peppers, Tokay wine, garlic, etc. The flavor profile of traditional Hungarian salami should not contain any acidic notes at all and it is a typical feature of the product that droplets of free oil form on the slice surface during slicing (Incze 1986).

Szegediner salami is made of beef, pork, and backfat, but sometimes the recipe can be on a pork-only basis. The traditional casing was the big horse intestine; today the stuffing is done in a 60 mm fibrous casing. After the classical cold smoking period for 1 day, the sausages are fermented for 3–4 days. The target of 30% weight loss is reached in 3–4 months. The salamis acquire a nice whitish grey mold cover over time and have a firm to very firm texture and excellent keeping qualities. Usually, processors inoculate only the ripening chamber with molds and not the sausages directly (dipping or spraying).

Budapest salami is a sausage based on 66% lean well-trimmed pork and 34% backfat. Its production is very similar to Szegediner salami.

Hot Kolbász is a pure pork or a pork/beef product with hot spices, coarsely ground, stuffed into 28–30 mm casings, fermented for 3–4 days, smoked for 1 day, and dried afterward according to desired weight loss. This can take up to 6 weeks; the normal time is 2–3 weeks. The end products weigh 200 to 250 g. Typical spices used in Kolbász are white pepper, hot and sweet paprika, cumin, coriander, and garlic. Kolbász is not mold covered.

Bulgaria

Bulgaria has a long-standing tradition for making fermented meat products under naturally well-suited climate conditions. One of the outstanding products is Loukanka, with its traditionally flat shape, very

Figure 33.4. Hungarian salami.

dry and slightly crumbly texture, and extremely mild fermentation flavor (total absence of acidic note). Loukanka is a finely grained lean sausage with good definition of the particles and with a typical cumin-pepper spice background in addition to the normally strong garlic flavor. It is mainly made from pork only and stuffed into casings of 28–30 mm. The flat shape is obtained by pressing the sausage during the initial mild fermentation cycle; a typical pH curve should not go below 5.3. When pressing is done too late, the sausage forms crevices inside and the end product is very often spoiled. Its strong drying leads to extremely good keeping. Other Bulgarian specialties are Ambaritsa, Babek, and Kalofer.

Czech Republic and Slovakia

In Slovakia, German-type salamis with mild acidity and good fermentation flavor are common. A lightly smoked, well-rounded flavor profile is typical. On the other hand, in the Czech Republic, sausages are traditionally German-type salamis with very strong smoke features and relatively fatty meat batters (Figure 33.5). Furthermore, the use of clove is very typical for Czech salamis. Hercules salami, Polican, and Paprikas, are the most important ones. A specialty is Lovecky salami, which has a characteristic rectangular shape from a special pressing.

Poland

Most Polish salamis are more or less German-type salamis, but for safety reasons the sausages are very often cooked to an internal temperature of 68–70°C after production. A typical Polish sausage type is Polska. Traditionally, this sausage was not fermented, but now manufacturers use starter cultures to achieve a pH drop. Pork meat is minced through an 8 mm plate and mixed with all other ingredients. Afterward, the mince is stuffed into natural casing (30–32 mm) longer than 35 cm, ripened at 6°C for 2 days and smoked at 22°C for about 2 days. The finished Polska is still quite soft due to the limited drying.

Russia

Typical Russian fermented sausages with a long tradition are Salami Moscow-type and Salami Russian-type. In Russia, the product must lose 40% weight, and the pH must be 5.0 (±0.1). Salami Moscow-type is made of 25% backfat and 75% lean beef. The particularity is the big size of the fat (7–8 mm). The mince is stuffed in 45 mm casings, and the product is finished when a weight loss of 45% is obtained. The surface of the sausage is quite rough due to the big fat pieces, which have less shrinkage than the lean part.

Russian Salami is made in two different qualities. The lean type is made of 30% backfat and 70% lean meat, which can be pork, beef, or a combination of the two (50/50). The standard type is made of 50% backfat, 25% lean pork, and 25% lean beef. Both types have a particle size of 3 mm, are stuffed in 45 mm casings, and have a weight loss of 40%.

Figure 33.5. Czech dry sausages.

SAUSAGE TYPES OF SCANDINAVIA

Denmark

Danish dry-fermented sausages basically cover two sausage types: the red-colored salami and the spegepølse that exist in many different qualities.

The typical Danish salami sausage is smoked and is characterized by added color and finely minced meat with a high fat content of up to 55% fat in the finished product. The meat block contains 55% pork/beef and 45% backfat, and typical ingredients are nitrite salt, potato starch, maltodextrin, glucose, white pepper, garlic powder, sodium ascorbate, carmine, and starter culture. The meat is chopped coarsely in a bowl chopper, dry ingredients except salt are mixed thoroughly into the minced meat, and the coarsely chopped fat, nitrite salt, and salt are mixed in at the end. The mince is stuffed into 60–120 mm collagen casings and the sausages fermented and dried until the desired water loss is achieved.

The second type of Danish fermented sausage is similar to German Westphalian salami in production technology, but is often more fatty, more acidic, and less aromatic. The sausages are smoked and the degree of comminution may vary from coarse (6 mm) to fine (2 mm). Often, juniper berries are used for smoking to obtain a special flavor. A special smoked taste called *eel-smoked* is obtained when the backfat is smoked before freezing. The meat block contains, typically, 1/3 pork, 1/3 beef, and 1/3 backfat.

Sweden

Swedish Medwurst is characterized by its content of boiled potatoes (usually 10–20%) or potato flakes. Quite often a high fermentation temperature, such as 30–35°C, is used, and sometimes the Medwurst is also heat-treated after fermentation. The meat block contains 50% pork, 22% fat, 20% boiled potatoes, and 3% water; other ingredients are salt, saccharose, pepper, onion, ascorbate, and starter culture. Meat and fat are chopped in the bowl chopper. The dry ingredients are mixed thoroughly into the chopped meat and the mince stuffed into casings. The sausages are allowed to equilibrate to room temperature over night. The sausages are fermented at 29°C, 80% RH for 5 days and smoked shortly each day.

Norway

Many Norwegian salamis are similar to the German-style sausages, but typical for Norwegian fermented sausages is the use of "unusual" meats, such as mutton, lamb, goat, horse, offal (liver, heart), blood, reindeer, moose, etc. A large range of so-called Morr sausages are produced based on old traditional recipes. As an example Fårepølse contains mutton meat, goat, and beef, and Stabbur contains different meats that could be pork, beef, horse, mutton, etc.

Traditionally dry sausages from sheep and salamis with meat from wild animals are produced all over Scandinavia. The use of meat from free-ranging animals and the addition sometimes of blood to the sausages gives a special strong flavor to the end products. Variation in quality is quite frequent because there is a natural spread in the quality of raw materials under these conditions.

Finland

The two famous and most important Finish fermented sausages are Kotimainen and Venäläinen. In general, these salamis are quite sour and acidic (pH 4.6–4.9), and they have a strong smoke taste. The particle size and the fat contents are of major importance. Products with a lot of visual fat are unpopular in Finland. The fat contents of salamis are usually lower than 20% and the particle size is about 2 mm. Technology and ripening is according to the German procedure. Dominant spices are pepper and garlic.

REFERENCES

J Bacus. 1984. Utilization of microorganisms in meat processing. A handbook for meat plant operators. Letchworth, UK: Research Studies Press Ltd, pp. 1–170.

BAFF. 1985. Kulmbacher Reihe Band 5. Mikobiologie und Qualität von Rohwurst und Rohschinken. Kulmbach, DE: Bundesanstalt für Fleischforschung, pp. 2–15, 34–39.

———. 1990. Kulmbacher Reihe Band 10. Sichere Produkte bei Fleisch und fleischerzeugnissen. Kulmbach, DE: Bundesanstalt für Fleischforschung, pp. 70–80, 96–97, 102–103.

Chr. Hansen. 2003. Bactoferm™ Meat Manual, vol 1. Production of Fermented Sausages with Chr. Hansen Starter Cultures, 1st ed. Hørsholm, DK: Chr. Hansen A/S, pp. 1–31.

A Fischer. 1988. Handbuch der Lebensmitteltechnologie—Fleisch, Produktbezogene Technologie, Rohwurst. Stuttgart, DE: Ulmer Verlag, pp. 518–546.

K Hofmann. 1974. Notwendigkeit und Vorschlag einer einheitlichen Definition des Begriffes "Fleischqualität." Fleischwirtsch 54:1607–1609.

K Incze. 1986. Technologie und Mikrobiologie derUngarisches Salami. Fleischwirtsch 66:1305–1311.

L Leistner, GW Gould. 2002. Hurdle Technologies: Combination Treatments for Food Stability, Safety and Quality. New York: Kluwer Academic/Plenum Publishers, pp. 1–208.

J Löbel. 1989. Mailänder salami & Co. Düsseldorf, DE: Econ Taschenbuchverlag, pp. 50–80.

S Neuhäuser. 1988. Fleischwaren-Handbuch, Technik und technologie. Hamburg, DE: Behr's Verlag, pp. 129–197.

K Potthast, G Eigner. 1988. Neuere Ergebnisse über die Zusammensetzung von Räucherrauch. Fleischwirtsch 68:651–655.

A Stiebing. 1993. Vorlesungsscripe im Fach—Technologie der fermentierten Fleischwaren, Studiengang Fleischtechnologie. Lemgo, DE: University of Applied Sciences.

A Stiebing, W Rödel. 1987. Einfluβ der Luftgeschwindigkeit auf den Reifungsverlauf von Rohwurst. Fleischwirtsch 67:236–240, 1020–1030.

Part VIII

Other Fermented Meats and Poultry

34
Fermented Poultry Sausages

Sunita J. Santchurn and Antoine Collignan

INTRODUCTION

Poultry meat is very popular among Western consumers (Bruhn 1994), while also offering an excellent source of animal protein for consumers in developing countries (Gascoyne 1989). According to FAO (FAOSTAT data 2004), poultry meat ranks second in meat produced worldwide, after pork, and represents more than 30% of the total world meat production with 77 million tons/year. It also ranks first in exported meat with 8 million tons/year. Raw-poultry-meat-based production (i.e., salted, dried, smoked, etc,) is fairly recent (20 years old) and, in today's world market, remains very marginal compared to pork or beef delicatessen meat. This explains the poor indexing of these products.

As a matter of fact, traditionally, poultry animals are consumed just after slaughtering. This is due to the family size of the animals, which allowed total consumption of the animal without the need to preserve parts of its carcass. In contrast, other farm animals (pig, cattle), once slaughtered, could not be eaten at once and a large proportion of the carcass needed to be preserved for later use. In this way, people started to salt, dry, and, unknowingly, ferment pork or beef portions.

Nevertheless, a narrow range of poultry delicatessen meats have appeared on the market, chiefly cooked products such as cured thigh meat (Bater et al. 1993), patties (Reddy and Rao 1997), sausages (Beggs et al. 1997), and raw products to be cooked, e.g. bacon (Walters et al. 1992). More recently, there has been a renewed interest in poultry and poultry products, especially 100% pure poultry products including sausages. This can be explained by several factors. First, the increasing price of red meat and the facility to set up poultry production (compared with cattle or pigs), particularly in developing countries, have played a determining role in the worldwide spread of the poultry industry. Poultry meat also benefits from a healthier image than red meats and, as such, appeals to the ever-increasing health-conscious consumers. Last but not least, poultry and its derived products are not snubbed by any religious or cultural group (Santos and Booth 1996; Jackson 2000; Kubberod et al. 2002), contrary to pork and beef. In fact, it is highly appreciated by Muslims, Hindus, and Tamils. Moreover, halal practice is commonly implemented in poultry processing plants in countries where a Muslim community forms a non-negligible part of the population demographics. Demand from these groups for new poultry products has led to the development of poultry bologna, franks, burgers, nuggets, and sausages.

In the particular case of fermented poultry products, there is no real tradition of consumption as with red meat. Nevertheless, some raw cured poultry-based products have been developed, and a few are even available on the market, such as smoked or dried duck fillet (Lesimple et al. 1995), poultry jerky (Carr et al. 1997), and dry sausage with partial or complete poultry meat content (Holley et al. 1988). These fermented delicatessen products and, in particular, fermented sausages do not yet have a broad market distribution. This may be explained by problems associated with the raw material characteristics. Indeed, the first development of fermented poultry sausages, dating back to the 1970s, met with two main obstacles. The first one concerns sanitary hazards arising from the presence of pathogens on the poultry carcass. In fact, *Salmonella* spp. and *Campylobacter jejuni* are commonly present in poultry meats (Bryan and Doyle 1995). These pathogens constitute a serious public health hazard accentuated by the high raw poultry meat pH. Moreover, prevalent contamination by *Listeria monocytogenes* in slaughterhouses, poultry processing plants (Franco et al. 1995), and all raw poultry meats (Glass and Doyle 1989) has also limited the development of these products.

The second limitation is linked to the poor technological properties of poultry fat tissue. The latter is weakly structured and is characterized by a low melting point of fat leading to greasiness defects (Arnaud et al. 2004). As such, this fat cannot be used as lard and is not suitable for products like dry-fermented sausages requiring a structured fat.

This chapter sums up the different existing poultry fermented products with particular reference to fermented sausages, the technological constraints associated to their sanitary and sensorial qualities, and promising alternatives to overcome them.

FERMENTED POULTRY SAUSAGE

Specific Characteristics of Raw Poultry Meat and Fat in Relation to Process and Quality

The characteristics of poultry meat and fat, together with those of other meats and fats, are presented in Tables 34.1 and 34.2, respectively.

Apart from the fact that poultry meat is generally lower in fat than pork or beef, it is also less fibrous and more tender. Moreover, its ultimate pH tends to be higher than that of pork and beef, with high-activity muscles (thigh and leg) generally showing a higher ultimate pH than low-activity muscles (breast, wing) (Table 34.1) (Jones and Grey 1989). In addition, poultry meat has on average, a lower myoglobin content than beef. This explains the lighter color of poultry compared to beef. Furthermore, color varies with the type of muscle within a carcass: in poultry, thigh meat is darker than breast meat (Miller 2002).

The composition of poultry fat (abdominal and skin) is significantly different from that of beef and pork fats (Table 34.2). First, its water content is higher and fat content lower, with chicken skin showing more marked differences. Like beef and pork fat, chicken abdominal fat is an association of fat cells held in a loose framework of connective tissue. However, it contains more water and less

Table 34.1. Characteristics of poultry, beef, and pork lean.

	Poultry	Beef	Pork	References
Postrigor pH	Breast: 5.6–5.8 Thigh: 6.1–6.4	5.5–5.7	5.7–5.9	Jones and Grey 1989; Varnam and Sutherland 1995; Ricke et al. 2001
Myoglobin content (mg/g); visual color	8-week-old poultry: dark meat 0.40, dull red; white meat 0.01, grayish white 14-week-old turkey: dark meat 0.37, dull red; white meat 0.12, dull red	3-year-old: 4.60; bright red	5-month-old: 0.30; grayish pink	Miller 2002
Water content (wt %)	75.4	74.0	71.5	Kirk and Sawyer 1991
Protein (wt %)	23.4	20.3	20.7	
Fat (wt %)	1.0	4.6	7.1	

Table 34.2. Characteristics of poultry, beef, and pork fat.

	Poultry Fat				
	Abdominal	Skin	Beef Fat	Pork Fat	References
Water (wt %)	33.8	57.1	24.0	21.1	Kirk and Sawyer 1991
Protein (wt %)	5.4	14.0	8.8	6.8	
Fat (wt %)	63.3	31.2	66.9	71.4	
Cholesterol (mg/100 g)	40–63		75–140	50–120	Viau and Gandemer 1991; Foures 1992
Fat melting point (°C)	26.6 ± 0.2		42–45	32–38	Foures 1992; Arnaud et al. 2004
Consistency at room temperature	semi liquid		solid	solid	
MUFA* (g/100 g)	44.0			46.6	
Oleic acid C18:1(g/100 g)	38.2		36.5–43.0	39.0–45.0	
PUFA* (g/100 g fat)	25.7			11.4	
Linoleic acid C18:2 (g/100 g)	23.8		1.4–3.9	8.5–12.0	
SFA* (g/100 g)	30.3			42.0	
Stearic acid C18:0 (g/100 g)	5.8		15.5–23.0	12.8–17.7	

*MUFA = monounsaturated fatty acids; PUFA = polyunsaturated fatty acids; SFA = saturated fatty acids.

collagen, accounting for its lower protein content and very weak structure. In contrast, chicken skin is composed of cellular layers (a superficial epidermal and inner dermal layers) that hold the subcutaneous adipose tissue (Dingle 2001). Hence its higher protein content and greater level of structure compared to pork or beef fat. Poultry fat also contains less saturated fatty acids and more monounsaturated and polyunsaturated fatty acids than pork or beef fat. This justifies the lower melting point of chicken fat and its semiliquid consistency at ambient temperature (Foures 1992; Arnaud et al. 2004).

The distinctive features of poultry lean and fat have a direct bearing on formulations, processes, and end-product organoleptic quality, as will be discussed later in the chapter. Not to be neglected are their repercussions on the nutritional quality of processed products.

Indeed, from a nutritional standpoint, poultry meat is leaner than red meats and thus constitutes an excellent source of high-quality proteins for weight control diets (Mountney and Parkhurst 1995). Furthermore, compared to pork and beef fat, poultry fat contains higher levels of mono- (MUFA) and polyunsaturated (PUFA) fatty acids and lower levels of saturated fatty acids (SFA) (Foures 1992; Arnaud et al. 2004). More than 82% of fatty acids of chicken fat in the internal position (position 2) on the triglyceride (TGA) are unsaturated fatty acids and 70% of SFA are in the external position (positions 1 and 3) (compared to 30% in lard) (Arnaud et al. 2004). According to Entressangles (1992), this is of significance to intestinal absorption of fatty acids because most of the unsaturated fatty acids will be assimilated by the organism, and most of the SFA will be eliminated. Considering that SFA, unlike MUFA and PUFA, raise blood cholesterol levels and cause atherosclerosis (Higgs 2002), chicken fat thus represents a nutritionally better fat than pork or beef fats.

The Fermented Poultry Sausage-making Process

The fermented poultry sausage making process follows the same basic technology used for the processing of fermented red meat sausages (Figure 34.1). The main process steps are grinding/blending of ingredients, stuffing into casings, fermentation, and drying. Nevertheless, due consideration must be given to the specific nature of poultry raw materials when designing formulations and monitoring processes.

Grinding/Blending of Ingredients

Formulations include standard ingredients, namely meat (80–90%), fat (7–15%), curing agents (sodium

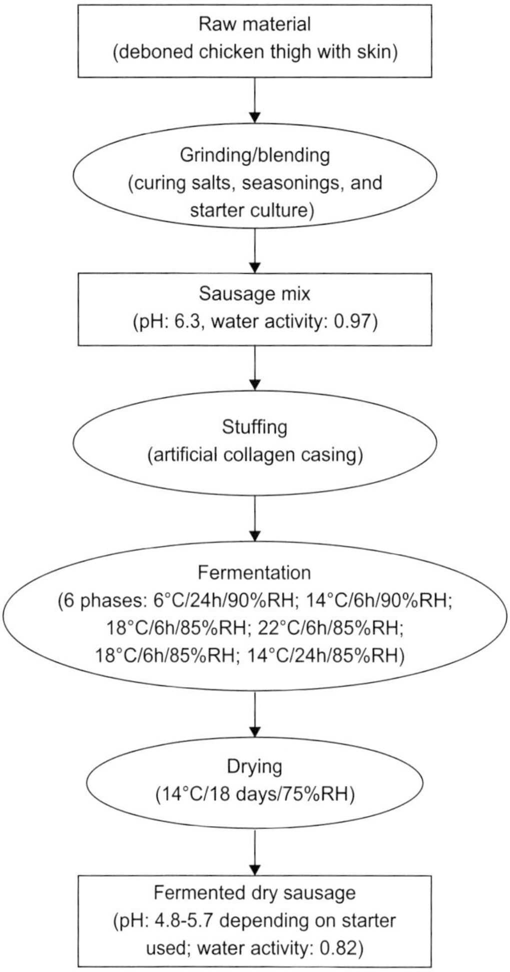

Figure 34.1. Fermented dry poultry sausage-making process.

nitrite 0.008%, sodium nitrate 0.008%, salt 1.5–3%), seasonings (e.g., ground pepper, ground mustard, paprika, garlic powder, allspice, and/or coriander), and carbohydrates (e.g., glucose, sucrose, lactose, 0.7–1.5%) (Keller and Acton 1974; Deumier and Collignan 2003). A grinder equipped with a metal plate bearing holes of specific diameter (3–8 mm) or a bowl cutter may be used. Although skin-on meat is more conveniently ground as such, it is advisable to grind/chop lean and fat separately for better particle definition in the end product. Mechanically deboned poultry meat (MDPM), still cheaper than hand-deboned poultry meat, has also been successfully incorporated in sausage formulations (Holley et al. 1988; Ricke et al. 2001). However, the maximum proportion of MDPM should stand at 15% of the meat block, beyond which excessive softening of the final sausage is noted. Use of chilled material and sharp knives ensures a cleaner cut and better particle definition. Once the meat is chopped, the other ingredients are mixed in. It is to be borne in mind that comminution and any other mechanical working tend to cause more physical disruption to poultry meat that red meat, poultry being less fibrous and more tender. In addition, the presence of NaCl, at ionic strength 0.6 or higher, generally causes extraction of myofibrils from meat fibers into exudate. Chicken, and especially its white muscle, produces exudates easily (Varnam and Sutherland 1995). Although exudate is important for binding and maintenance of product structure, high levels of myofibrillar protein solubilization into exudates lead to a decrease in fibrous texture, hence less chewiness in the final sausage. Furthermore, the water-holding capacity of solubilized myofibrillar proteins is higher, which tends to retard subsequent drying rate. Use of dark meat (thigh and leg), coarse grinding of lean, and addition of salt in the last instance are measures that minimize the above-mentioned problems.

Another point important to consider is the nature of the fat component. Fat sources in pure poultry sausages are either abdominal or skin fat. As has been pointed out earlier, poultry fat is more fluid at room temperature. As a result, during ingredient blending, and contrary to pork and beef fats, which remain discrete particles, poultry fat liquefies easily and tends to coat the lean meat particles. In the case of dry sausages, this fat layer constitutes a barrier to moisture removal during the drying stage. Consequently, the amount of poultry fat or skin is usually restricted to 15%, which is low compared to red meat formulations (23–32%) (Bacus 1986). Also, fat is better incorporated toward the end of the blending cycle. In addition, and of particular relevance for pure poultry sausage, mixing time should be just sufficient for the purpose and temperature maintained low (e.g., use of frozen ingredients, cooling of mixer/blender, cooling of processing room).

Traditionally, fermented red meat sausages have a typical red color arising from the reactions between myoglobin and added nitrite. Poultry meat, especially white muscle, is poorly pigmented and develops a less intense red color than pork or beef. The incorporation of poultry fat in formulations tends to further lighten the overall sausage color due to fat-coating of the lean. However, processors may have recourse to permitted colorings, such as Monascus red (Angkak), cochineal, and betanin, to achieve specified color intensity (Miller 2002).

As mentioned before, poultry meat and fat have a high average moisture content. Furthermore, the amount of fat in poultry sausage formulation is generally low. As a result, the sausage mix has a relatively high initial moisture content and water activity (0.97–0.98) (Deumier and Collignan 2003; Alter et al. 2006). In addition, its pH value is also fairly high, ranging from 5.9 to 6.53 (Holley et al. 1988; Deumier and Collignan 2003; Alter et al. 2006). These two factors make the sausage mix very favorable to microbial growth. Use of poultry skin, a major source of microbiological contamination, may further magnify the hazard. The presence of nitrite and salt is important to inhibit the growth of *Clostridium botulinum* and gram-negative pathogens such as *Pseudomonas*, *Escherichia coli,* and *Salmonella* (Bacus 1986; Ricke et al. 2001). Still, the high level of raw poultry meat and a high salt content may selectively promote the growth of *Staphylococcus aureus*. In this regard, it is essential to limit the time the sausage mix is left at temperatures above 15.6°C in the initial stages of processing while the pH is still above 5.3 (limit value for toxin production by *S. aureus*) (Bacus 1986).

As a pH-control measure, commercially available starter cultures, most commonly composed of lactic acid bacteria (LAB), are used to rapidly acidify the sausage mix (Baran et al. 1973; Acton and Dick 1975; Deumier and Collignan 2003). When nitrates are used instead of nitrites, starter cultures containing nitrate-reducing strains (*Kocuria* or *Staphylococcus*) are used. The inoculum is more often and conveniently added at the end of the blending cycle, although it may sometimes be manually mixed after a short holding time to allow proper binding of the meat (Baran et al. 1973).

Stuffing

Following inoculation, the mix is stuffed into a casing using a mechanical stuffer/clipper. The only appropriate option for pure poultry sausages is artificial casing such as fibrous collagen casings. Again, care must be taken to minimize temperature increase to prevent fat smearing (Varnam and Sutherland 1995). In some cases, namely for dry sausages, sausages are then dipped in a suspension of mold culture (*Penicillium candidum* and *P. nalgiovencis*), which eventually grows into a white or gray coat on the outer surface of the casing (Deumier and Collignan 2003).

Fermentation

Fermentation is the next stage, during which incubation conditions are set to favor active growth and metabolism of the lactic acid-producing bacteria. This is of particular relevance to poultry sausage mix, which presents a high initial pH. Rapid pH decrease is essential to curb the growth of pathogenic bacteria that may be present in poultry material (meat and skin). At the end of the fermentation cycle, the pH value should decrease to at least 5.3 to prevent staphylococcal toxin production (Bacus 1986). Besides, so long as the pH remains above the isoelectric point of meat proteins (pH 5) (Keller and Acton 1974), the water-holding capacity of poultry lean stays high, which in turn may retard the subsequent rate of drying. Instead of relying solely upon fermentation-produced acidulants (lactic acid), processors may employ chemical acidulants, in the form of gluco-delta-lactone or organic acids, to stimulate pH decrease (Bacus 1986).

Drying

Depending on the extent of drying, two types of sausages may be distinguished: semidry and dry sausages, showing a product water activity of 0.92–0.94 and 0.85–0.86, respectively (Roca and Incze 1990). As mentioned earlier, poultry sausage formulations are generally high in moisture. Consequently, measures must be taken to prevent too rapid release of surface moisture and to ensure uniform drying and product integrity (Bacus 1986). This involves, in the initial stages of processing, the control of the degree of meat and fat comminution, temperature to limit fat liquefaction, level and time of salting, and the rate of acidification. Uneven drying may also be minimized through the use of casings of suitable water permeability and appropriate time/temperature/relative humidity drying schedules. Drying process temperatures must also be controlled to minimize fat loss, thus maximizing process yield.

End Product Quality

The most noticeable defect in fermented pure poultry sausages is associated with the nature of poultry fat. Despite the fact that pure poultry sausages are lower in fat than pork or beef sausages, an unappealing oily surface invariably appears due to fat exudation even at room temperature (Keller and Acton 1974; Deumier and Collignan 2002), thus giving a false impression of a high-fat product to consumers. Using beef or pork fat in place of poultry fat usually corrects the greasiness problem (Acton and Dick 1975; Bacus 1986). However, sausages so produced are not acceptable for no-pork and no-beef consumer groups.

In addition, the higher level of unsaturated fatty acids in poultry fat makes sausages more susceptible to rapid fat oxidation, thereby increasing the risks of rancidity during drying and storage. The use of permitted antioxidants may, however, inhibit such oxidative defects.

Moreover, abdominal fat is soft and lacks structure. This leads to poor particle definition in the finished product, which significantly reduces its marketability (Deumier and Collignan 2002). For this reason, poultry skin fat is preferred because it confers more chewiness and may simulate the white flecks in salami-like products (Bacus 1986). However, fat exudation still prevails. In addition, poultry skin is a notorious source of microbiological contaminants and its incorporation into sausage formulations, especially raw sausages, implies accrued risks of pathogens.

Indeed, poultry and poultry products have been the cause of many foodborne disease outbreaks most frequently involving *Salmonella* spp., *Clostridium perfringens*, and *Staphylococcus aureus* (Mead 1989). *Campylobacter jejuni* is another pathogen associated mainly with poultry-meat consumption (Bryan and Doyle 1995). The high incidence of *Listeria monocytogenes* in raw and even processed poultry products is also a major concern (Franco et al. 1995). To counteract these sanitary hazards, at the end of the fermentation cycle and prior to drying of fermented poultry sausages, a mild heating treatment may be applied to kill heat-sensitive pathogenic bacteria (Keller and Acton 1974; Raccach and Baker 1979). Such types of sausages fall into the semidry group and have a significantly different organoleptic profile from that of dry sausages. In the case of more traditionally processed dry sausage, no thermal treatment for preservation or safety purposes is practiced. As a result, these sausages are in essence raw meats that have been salt-cured, incubated, and dried. Moreover, dry sausage is usually eaten by the consumer without further cooking. For these reasons, in addition to good manufacturing practices during production and slaughtering, and better sanitization of carcasses at the primary levels of production, poultry sausage processors must also abide by strict hygiene regulations (sanitation, employee hygiene, and good manufacturing practices).

Recent Advances

Successful commercialization of fermented dry pure chicken sausage was recently made possible through the use of thigh meat and skin fat originating from a selected slow-growing breed of chicken (Guignol) slaughtered at 63 days (Viénot 2003). The chicken fat has a higher melting point and therefore does not cause any of the previously mentioned problems associated with standard poultry fat. However, sanitary hazards associated with poultry skin still persist.

Another promising avenue for pure poultry delicatessen meats, including fermented sausages, rests on the improvement of the technological properties of poultry fat. Recently, pilot-scale dry fractionation of chicken fat succeeded in producing a fat fraction, stearin, with a higher melting point (Arnaud et al. 2004) and with good sanitary quality due to the heat treatment of fat before fractionation. Stearin is solid at room temperature and closely resembles lard, and as such would eliminate greasiness defects in all-poultry sausages. However, unlike pork or beef fat, the stearin fraction is not sufficiently structured for use in poultry formulations. And ongoing research by the same authors aims at constructing a stearin with sufficient body for good particle definition in the end product.

OTHER FERMENTED PRODUCTS

Although dry sausage is the most common fermented poultry meat, a few other fermented poultry products may also be found.

Among these, the French dry-cured magret is probably the most famous. It is made with the breast of Moulard duck, a breed used for foie gras (fattened duck liver). The magret refers specifically to the lobe, or half breast of the Moulard duck, well known for its excellent quality. In particular, the flavor of this processed fillet is related to exogenous and endogenous compounds and their distribution between the fatty panicle and the lean (Lesimple et al. 1995). The process is well described by Poma (1998). After cutting carcasses, raw duck fillets (with subcutaneous fatty panicle) are subjected to salting and drying. Salting consists of a dry rub of the fillet with salt (20 g/kg) and pepper (2 g/kg) followed by curing at 4°C for 12 h. After curing, the magret is hung in a thermostated chamber at 25°C and 80% of relative humidity during 48 h for fermentation. Next, the magret is dried in the same chamber with an initial step at 16–18°C for 24 h followed by a second one at 12–13°C for 1–2 weeks. The dry-cured fillets are commercialized either entire or thinly sliced. Often vacuum-packed, they keep for 2 months under refrigeration. Dry magret makes a great appetizer. It may be served with fruit slices (pear, melon, or fig depending on the season) as part of an antipasto platter, or sliced thinly on top of a green salad.

Other dried and cured duck meats that are encountered may be classified as fermented products. It is the case of NAN-AN, a traditional dried duck meat (more than 100 years old) from China (Jiangxi province) processed from Big Gunny ducks. The process mainly includes cutting of the carcass into five pieces, salting in brine and sun drying (Yongchang 1989).

Prosciutto is originally a traditional Italian cured pork meat which is cured and further matured by air drying. However, it can also be made with magret, Moulard duck breast (Smith 2005). In the particular case of Moulard magret prosciutto, a mixture of salt and spices (juniper berries, bay leaf, coriander, peppercorns, and garlic) is used for curing under refrigeration (4°C) for 24 hours. Without rinsing, the magret is then hung in a dry chamber at 10°C for 2 weeks.

For many years, jerky and biltong meats have been favorite snack foods for Americans and southern Africans, respectively. They are produced by using a combination of curing, drying, and sometimes smoking procedures. People generally associate jerky and biltong with beef. However, the meat sources are unlimited and it is not uncommon to use meat from turkey, ostrich, and emu (Carr et al. 1997). Depending on the drying and/or storage conditions, spontaneous lactic acid fermentation may be favored, thus leading to a fermented end-product (unpublished work).

REFERENCES

JC Acton, RL Dick. 1975. Improved characteristics for dry, fermented turkey sausage. Food Prod Devel 9:91–94.

T Alter, A Bori, A Hamedi, L Ellerbroek, K Fehlhaber. 2006. Influence of inoculation levels and processing parameters on the survival of Campylobacter jejuni in German style fermented turkey sausages. Food Microbiol 23:701–707.

E Arnaud, P Relkin, M Pina, A Collignan. 2004. Characterisation of chicken fat dry fractionation at the pilot scale. European J Lipid Sci Technol 106:591–598.

JN Bacus. 1986. Fermented meat and poultry products. In: Advances in Meat Research, vol. 2. Meat and poultry microbiology, AM Pearson, ed. Westport: Avi Publishing Company, pp. 123–164.

WL Baran, LE Dawson, KE Stevenson. 1973. Production of a dry fermented turkey sausage. Poultry Sci 52:2358–2359.

B Bater, O Descamps, AJ Maurer. 1993. Quality characteristics of cured turkey thigh meat with added hydrocolloids. Poult Sci 72:349–354.

KLH Beggs, JA Bowers, D Brown. 1997. Sensory and physical characteristics of reduced-fat turkey frankfurters with modified corn starch and water. J Food Sci 62:1240–1244.

CM Bruhn. 1994. Consumer perceptions of quality. In: Minimal Processing of Foods and Process Optimization: An Interface RP Singh, FAR Oliveira, eds. London: CRC Press, pp. 493–504.

FL Bryan, MP Doyle. 1995. Health risks and consequences of Salmonella and Campylobacter jejuni in raw poultry. J Food Protect 58:326–344.

MA Carr, MF Miller, DR Daniel, CE Yarbrough, JD Petrosky, LD Thompson. 1997. Evaluation of the physical, chemical and sensory properties of jerky processed from emu, beef, and turkey. J Food Qual 20:419–425.

F Deumier, A Collignan. 2002. Evaluation des risques sanitaires au cours du procédé de fabrication d'un saucisson sec de volaille. 14èmes Rencontres Scientifiques et Technologiques des Industries Alimentaires, AGORAL 2002, Nancy, France, 2002, pp. 443–449.

———. 2003. The effects of sodium lactate and starter cultures on pH, lactic acid bacteria, Listeria monocytogenes and Salmonella spp. levels in pure chicken dry fermented sausage. Meat Sci 65:1165–1174.

J Dingle. 2001. Manufacturing leather from chicken skin. Kingston, Australia: Rural Industries Research and Development Corporation 1–16.

B Entressangles. 1992. Digestion et absorption des lipides. In: Les Lipides. Institut Français Pour la Nutrition, ed. Paris, France, pp. 17–33.

FAOSTAT data. 2004. http://faostat.fao.org/faostat (24 May 2006).

C Foures. 1992. Tissus animaux. In: Manuel des Corps Gras. I. A Karleskind, ed. Paris: Lavoisier Tec & Doc, pp. 242–260.

CM Franco, EJ Quinto, C Fente, L Rodriguez-Otero, A Cepeda. 1995. Determination of the principal sources of Listeria spp. contamination in poultry meat and a poultry processing plant. J Food Protect 58: 1320–1325.

J Gascoyne. 1989. The world turkey industry, structure and production. In: Recent Advances in Turkey Science. C Nixey, TC Grey, eds. London: Butterworths, pp. 3–9.

KA Glass, MP Doyle. 1989. Fate of Listeria monocytogenes in processed meat products during refrigerated storage. Appl Envir Microbiol 55:1565–1569.

J Higgs. 2002. The nutritional quality of meat. In: Meat Processing—Improving Quality. J Kerry, J Kerry, DA Ledward DA, eds. Cambridge, England: Woodhead Publishing Ltd., pp. 64–104.

AR Holley, PA Jui, M Wittman, P Kwan. 1988. Survival of *S. aureus* and *S. typhimurium* in raw ripened dry sausages formulated with mechanically separated chicken meat. Fleischwirtschaft 68:194–201.

MA Jackson. 2000. Getting religion—For your products that is. Food Technol 54:60–66.

JM Jones, TC Grey. 1989. Influence of processing on product quality and yield. In: Processing of Poultry. GC Mead, ed. Essex, England: Elsevier Science Publishers Ltd., pp. 127–181.

JE Keller, JC Acton. 1974. Properties of a fermented, semidry turkey sausage during production with lyophilized and frozen concentrates of *Pediococcus cerevisiae*. J Food Sci 39:836–840.

RS Kirk, R Sawyer. 1991. Pearson's Composition and Analysis of Foods. Essex, England: Longman Scientific & Technical.

E Kubberod, O Ueland, A Tronstad, E Risvik. 2002. Attitudes towards meat and meat-eating among adolescents in Norway: A qualitative study. Appetite 38:53–62.

S Lesimple, L Torres, S Mitjavila, Y Fernandez, L Durand. 1995. Volatile compounds in processed duck fillet. J Food Sci 60:615–618.

GC Mead. 1989. Hygiene problems and control of process contamination. In: Processing of Poultry. GC Mead, ed. Essex, England: Elsevier Science Publishers Ltd., pp. 183–220.

RK Miller. 2002. Factors affecting the quality of raw meat. In: Meat Processing. J Kerry, J Kerry, DA Ledward, eds. Cambridge: Woodhead Publishing Ltd., pp. 27–63.

GJ Mountney, CR Parkhurst. 1995. Chemical and nutritive characteristics. In: Poultry Products Technology. New York: Food Products Press, pp. 67–82.

JP Poma. 1998. Le jambon sec et les petites salaisons. Paris: Editions Erti.

M Raccach, RC Baker. 1979. Fermented mechanically deboned poultry meat and survival of Staphylococcus aureus. J Food Protect 42:214–217.

KP Reddy, BE Rao. 1997. Influence of binders and refrigerated storage on certain quality characteristics of chicken and duck meat patties. J Food Sci Technol 34:446–449.

SC Ricke, I Zabala Diaz, JT Keeton. 2001. Fermented meat, poultry, and fish products. In: Food Microbiology: Fundamentals and Frontiers MP Doyle, LR Beuchat, TJ Montville, eds. Washington D.C.: ASM Press, pp. 681–700.

M Roca, K Incze. 1990. Fermented sausages. Food Rev Inter 6:91–118.

MLS Santos, DA Booth. 1996. Influences on meat avoidance among British students. Appetite 27: 197–205.

P Smith. 2005. Duck Breast Prosciutto. www.articledashboard.com (30/05/06).

AH Varnam, JP Sutherland. 1995. Meat and Meat Products—Technology, Chemistry and Microbiology. London, UK: Chapman & Hall.

M Viau, G Gandemer. 1991. Principales caractéristiques de composition des graisses de volaille. Revue française des corps gras 38:171–177.

E Viénot. 2003. Saroja volailles lance le saucisson sec "pur poulet." Filières Avicoles 43.

BS Walters, MM Lourigan, SR Racek, JA Schwartz, AJ Maurer. 1992. The fabrication of turkey bacon. Poultry Sci 71:383–387.

Z Yongchang. 1989. Preservation of the NAN-AN pressed salted ducks. International Congress of Meat Science and Technology, Copenhagen, Denmark, 1989, pp. 507–510.

35
Fermented Sausages from Other Meats

Halil Vural and Emin Burçin Özvural

INTRODUCTION

In fermented sausage production, utilization of other animals' meat rather than pork, beef, and poultry is not widely spread. Nevertheless, there are several studies investigating the use of mutton, lamb, camel, ostrich, horse, buffalo, and game meats for sausage production. In this chapter, each section includes information on the raw materials from the mentioned animals, related studies on their manufacture into fermented sausages, and some details on local products.

MUTTON AND LAMB

Mutton and lamb are consumed in almost all countries around the world. Moslem, Jewish, and Hindu people do not eat pork and/or beef due to religious and cultural taboos; however, there is not any handicap for mutton and lamb. Nevertheless, the production and consumption of sheep meat is lower than pork and beef because of its odor, flavor, and difficulty in deboning (Baliga and Madaiah 1971; Anderson and Gillett 1974; Shaikh et al. 1991; Young et al. 1994).

Fat in mutton and lamb contains higher proportions of saturated fatty acids than that of pork, and supplies a number of flavor components (Bartholomew and Osuala 1986; Beriain et al. 1997). Lamb meat flavor depends on age, diet, sex, breed, and body weight of the animal (Bartholomew and Osuala 1986; Beriain et al. 1997; Rousset-Akrim et al. 1997; Young et al. 1997; Sanudo et al. 2000). Reduction of mutton flavor has been achieved by decreasing mutton fat to a level of 10% or less in meat products (Bartholomew and Osuala 1986; Beriain et al. 1997). Also, undesirable mutton characteristics may be prevented by application of spices, curing, fermentation, and smoking (Bartholomew and Osuala 1986; Wu et al. 1991; Beriain et al. 1997). Limited work has been done on fermented mutton sausages.

Bartholomew and Osuala (1986) used mutton in the manufacture of summer sausages by incorporating lean beef (10% fat), beef fat (87% fat), lean mutton (5.5% fat), untrimmed mutton (16.5% fat), or fat mutton trims (45% fat) into the formulation. They showed that mutton products containing 10% or less mutton fat with added high levels of spices were acceptable. Pepperoni is a heavily seasoned, smoked, dried fermented sausage generally made from pork and/or beef and sometimes from other meats such as lamb. It is produced in central, northern, and eastern Europe; North America; and Australia (Campbell-Platt 1995). Bushway et al. (1988) examined chemical, physical, and sensory properties of pepperoni produced with mutton (50%), fowl meat (50%), or beef fat. Their work indicated that spice mix, beef fat, and fermentation at 28°C for 48 hours, or until the pH was reduced to 5.0 or less, prevented mutton flavor in the product. Wu et al. (1991) produced fermented mutton sausages with incorporation of *Pediococcus acidilactici H*, *Lactobacillus plantarum 27*, and *Lactacel 75* as a starter culture. The mince was stuffed into 20 mm collagen casings and fermented at 30°C and 90–92% relative humidity. When the pH was 4.9–5.0, the temperature was raised to 68°C for cooking. After cooking, the sausages were dried until achieving a moisture loss of 40–45%. According to the results, *P. acidilactici H* and *Lb. plantarum 27* were the more suitable cultures for fermented mutton sausage. Cured dry mutton sausages formulated with 76% mutton and 24% fat were compared to pork sausages prepared with the same meat and fat ratio. Sausages were fermented at 24°C with 60% and 90% relative humidity until pH was 4.9, and then dried in a drying chamber at 12–15°C while reducing the relative humidity gradually from 85% to 75%. It was reported that both mutton and pork sausages showed similar technological characteristics during processing, whereas the flavor and texture of mutton sausages were not desirable (Beriain et al. 1997).

In Norway, fermented lamb sausage is a common sausage type that is prepared differently in various parts of the country (Helgesen et al. 1998). Fjellmorr in Norway and lambaspaeipylsa in Iceland are dried fermented sausage types that contain lamb in addition to beef and pork. In warmer countries of southern Europe (France) and North Africa, such as Algeria, Tunisia, and Morocco, Merguez-type fermented sausages, which contain pork, beef, lamb, or mixed meats, are consumed. Those sausage types are usually stuffed in medium-large casings (50–70 mm), fermented at 30–35°C for 3–6 days and dried for a couple of days (Campbell-Platt 1995). Utilization of mutton has also been tested in South African dry sausages (Mellett 1991).

Turkish fermented sausage, which is called *soudjouk* or *sucuk*, is a popular fermented meat product in Turkey. This product is produced and consumed all over the country. Soudjouk or sucuk is made from beef and/or water buffalo meat and/or mutton. Soudjouk mixture also contains beef fat and sheep tail fat. The mixture is stuffed into air-dried bovine small intestines (Gökalp and Ockerman 1985; Gökalp et al. 1988; Vural 1998; Bozkurt and Erkmen 2002; Soyer et al. 2005). In traditional soudjouk production, starter cultures are not added and the fermentation and drying stages are applied under natural climatic conditions. The ripening periods, temperature, and relative humidity applied in the production vary depending upon the traditional method used. Due to variable conditions and long processing time, manufacturers have started to use starter cultures (Gökalp 1986; Soyer et al. 2005).

Vural (1998) produced Turkish semidry-fermented sausages using *P. acidilactici*, *Staphylococcus xylosus* plus *P. pentosaceus* and *S. carnosus* plus *Lb. pentosus*. The sausages were prepared with 74.5% beef, 9.1% mutton, 10% beef fat, and 6.4% tail fat. Fermentation was carried out for 36 hours at 26°C, 96% relative humidity, and 1 m/s air velocity and then heated for 30 min at 55°C, followed by 70°C. The products were cooled to 20°C and placed in a drying room at 15°C (relative humidity 59%) for 3 days. Vural (1998) reported that sausages with *P. acidilactici* had the best color, appearance, flavor, and general acceptability.

Sheep tail fat is an important constituent of Turkish fermented sausage due to its strong influence on the sensory characteristics. Gökalp and Ockerman (1985) and Gökalp (1986) used 10% tail fat and Ayhan et al. (1999) added 15% tallow for the soudjouk manufacture. Soyer (2005) and Soyer et al. (2005) used 10, 20, and 30% sheep tail fat content and ripened the products at 20–22°C and 24–26°C. Soyer et al. (2005) showed that fat level affected water activity, color, and total viable and lactic acid bacteria counts of the sausages. Soyer (2005) found that total acidity, free fatty acids, and thiobarbituric acid (TBA) values were affected by fat level, ripening temperature, and time. It was shown that high fat level and temperature increased the rancid flavor and decreased the overall acceptability. In another study, Vural (2003) manufactured Turkish semidry-fermented sausages by substituting 0, 20, 60 or 100% of the beef fat and tail fat with interesterified palm and cottonseed oils. Control groups had the highest sensory scores, whereas vegetable oils reduced the ratio of saturated to unsaturated fatty acids and increased the ratio of polyunsaturated to saturated fatty acids.

GOAT

Goat meat consumption varies globally due to cultural, social, and economic conditions of consumers (Norman 1991; Webb et al. 2005). Goat meat is generally preferred by people of local regions and sold for domestic consumption. The world production of goat meat was 1.1 million tons in 1961 and 3.7 million tons in 2001. However, the increase in goat meat production was less than other meats (Dubeuf et al. 2004).

Goat meat is also called *chevon*; Henry Wallace gave this name in 1924 (Norman 1991). Although goat meat (chevon) contains lower fat than other types of red meats, its intense and natural flavor is undesirable to most consumers (Cosenza et al. 2003; Kannan et al. 2006). However, Nassu et al. (2000) claimed that fermented goat meat sausages could be a valuable alternative for fresh goat meat because fermented sausages produced with goat meat had good consumer acceptability and suitable physicochemical and microbiological characteristics during storage. The sausages were fermented at 23–12°C and 95–70% relative humidity during 14 days with the addition of the starter cultures *Lb. farciminis*, *S. xylosus*, and *S. carnosus*. In a similar study it was shown that rosemary extract (0.050%) gave an effective protection against oxidation in fermented goat meat sausages (Nassu et al.2003). In a different study, Nassu et al. (2002) investigated the effect of different fat contents (5, 10, and 20%) in the processing, chemical characteristics, and sensory acceptance of fermented goat meat sausages. To the sausages were added 0.02% *S. xylosus* and *P. pentosaceus*, and they were matured according to Nassu et al. (2003). The study indicated that the fat level did not significantly affect the processing and sensory acceptance of the sausages. The authors also reported that a fat level of 10–20% was considered the most adequate due to the low fat content in goat meat.

The Angora goat originated in the district of Angora in Asia Minor (Anatolia) from where it spread to the rest of the world. It is primary bred for mohair (wool) production. Kolsarıcı et al. (1993) investigated the utilization of Angora goat meat at different ratios (25, 50, 75 and 100%) in soudjouk production. Physical, chemical, and sensory characteristics of the soudjouks were examined during fermentation and storage periods that took place at 24 ± 2°C, 90–95% relative humidity on the first day, followed by 22 ± 2°C, 80–85% relative humidity for the last 6 days. The authors stated that goat meat addition up to 50% level in soudjouk production did not cause any differences on physical, chemical, and sensory properties of soudjouks, in general, when compared with the control (100% beef). For sausages with 75% and 100% goat meat, physical and chemical properties were good, but sensory properties not acceptable. Goat meat has also been tested in fermented cabrito snack stick products (Cosenza et al. 2003).

CAMEL

Camel meat is not consumed all over the world, but it is eaten in Sudan, Libya, Egypt, Saudi Arabia, and India (El-Faer et al. 1991; FAO 2005). It is an alternative product in arid areas where the climate conditions adversely affects other animals. In some societies, camel meat is the food of ritual celebrations (FAO 2005). Both young and mature animals are used in camel meat production. The carcass of animals younger than 1 year of age weighs about 100 kg and the carcasses of the older animals, 200–300 kg (Dawood and Alkanhal 1995). El-Faer et al. (1991) examined the minerals and proximate composition of muscle tissues from shoulders, thighs, ribs, necks, and humps of seven young (1–3 years) male camels. The mineral elements, protein, and ash contents in the various muscle tissues of the camels were generally similar to those in the same tissues of beef, but camel meat contained significantly less lipids (1.2–1.8% versus 4.0–8.0%) and 5–8% higher moisture content than beef. Dawood and Alkanhal (1995) determined the proximate composition of some Najdi-male camels. They found that skeletal muscles had 68.8–76% moisture, 19.4–20.5% protein, 4.1–10.6% fat, and 1.0–1.1% ash.

Limited work has been done on the use of camel meat in fermented sausages. Kalalou et al. (2004) produced fermented camel meat sausages with added *Lb. plantarum* to extend the shelf life of ground camel meat. They mixed minced camel lean meat with hump fat in the proportion of 80/20, stuffed the mixture into natural casings and dried the sausages at 15–18°C at 70–80% relative humidity. The results showed that in 21 days pH dropped to 4.5; water activity to 0.7; and coliforms, *Enterococci*, and yeasts to nondetectable levels. However, the authors concluded that the processing conditions should be better monitored for improving the organoleptic quality of the product.

OSTRICH

The ostrich meat industry is largely developed in South Africa and has also been extended to Australia, North and South America, Canada, Israel, and recently to European countries such as Spain and Italy (Kolsarıcı and Candoğan 2002; Fernandez-Lopez et al. 2003). Ostrich meat can be used as a healthy alternative to other meats (Fisher et al. 2000; Kolsarıcı and Candoğan 2002; Fernandez-Lopez et al. 2003). It contains a favorable fatty acid profile due to the high amount of polyunsaturated ω-3 fatty acids (Sales 1998; Fisher et al. 2000; Kolsarıcı and Candoğan 2002). In addition, ostrich meat has higher pH value, darker color, lower collagen solubility, and higher pigment content than beef. Its cooking loss and cholesterol content are similar to those of beef (Sales et al. 1996; Kolsarıcı and Candoğan 2002; Fernandez-Lopez et al. 2003).

The published data on fermented ostrich sausages are very few. Böhme et al. (1996) produced Italian-type salami with ostrich meat by a combination of *Lb. sake*, *Lb. curvatus*, and *Micrococcus* spp. Sensory characteristics of these products were compared with the salami produced with glucono-delta-lactone (GdL). Ostrich meat was fermented for 4 days at 20–22°C, 97–99% relative humidity and ripened for a further 11 days at 16–18°C, 40–60% relative humidity. The study indicated that the combination of *Lb. curvatus* and *Micrococcus* spp. resulted in salami with acceptable texture and sensory quality. Dicks et al. (2004) produced ostrich meat salami by using the cultures *Lb. plantarum* 423 and *Lb. curvatus* DF126 to prevent the growth of *monocytogenes* during production. The results indicated that bacteriocins produced by the lactic acid bacteria, inhibited the growth of *monocytogenes* in salami.

HORSE

The application of horsemeat in fermented sausages is not common but is popular in Norway and Kazakhstan. Horsemeat consumption shows differences in historic times according to economy and cultural societies. Milk and meat of horse are still an important diet for the nomadic people of Eurasia. However, in Western Europe, its consumption was partly restricted because of its relation with pagan rites and festivals. There is a contamination risk of *Salmonella* and *Yersinia enterocolitica* and an infection possibility of *Trichinella* from horsemeat. However, in some countries, particularly France, Italy, and Japan, some people believe that raw consumption of horsemeat is safer than other meats and eat it raw or after only minimal cooking (Gill 2005).

Kazy and Shuzhuk are popular sausages in Kazakhstan. In Kazy production, raw sausage is dried, boiled, or smoked. Hanging it outside for a week in a sunny, aired place dries the sausage. Smoking is applied with dense smoke at 50–60°C for 12–18 hours followed by drying at 12°C for 4–6 hours. Shuzhuk is made from horsemeat and suet at equal amounts. The meat is rubbed with salt and kept for 1–2 days in a cool place at 3–4°C. The meat and fat are cut into small pieces and stuffed into the gut with salt, pepper, and greens. The chubs are hung in a cool place for 3–4 hours, smoked at 50–60°C for 12–18 hours, and dried for 2–3 days at 12°C (Chenciner 2005).

Trøndermorr, Fjellmorr, Sognemorr, and Stabbur are dried fermented sausages produced in Norway. In some of these products, horsemeat is added into the formulation. Horsemeat may also be used in Kotimainen sausage in Finland (Campbell-Platt 1995).

BUFFALO

Because marketing of buffalo meat does not have any specific regulations and the meat is generally sold as if beef, its consumption is not known precisely. The tenderness of buffalo meat resembles that of beef. It has lower cholesterol content, and better hardness and resistance than beef (Paleari et al. 2000). Sachindra et al. (2005) claimed that buffalo meat is cheaper when compared to other meats and that products made from buffalo meat would be a profitable foreign income. Buffalo meat produced in India is largely exported in frozen condition. Because there is not any taboo against the consumption of buffalo meat in India, popularity of this meat has been rising, and buffalo meat covers now 35.7% of the total meat consumption (Sekar et al. 2006).

In Turkey, water buffalo meat is generally used together with beef and/or mutton in fermented sausages (soudjouks) (Gökalp and Ockerman 1985; Gökalp 1986; Gökalp et al. 1988). Gökalp and Ockerman (1985) manufactured Turkish-style fermented sausage (soudjouk) with 45% beef, 45% water buffalo meat, and 10% sheep tail fat and studied the effects of *Lb. plantarum*, *Lb. plantarum* + *M. aurantiacus*, *Lb. plantarum* + *M. aurantiacus* + *Debaryomyces hansenii* as starter cultures at different ripening temperatures (12–14°C, 16–18°C, and 20–22°C). The authors showed that a mixed culture of *Lb. plantarum* and *M. aurantiacus* should be used and that the addition of *D. hansenii* had beneficial effects on the decline of proteolytic and lipolytic microorganisms. It was reported that the lowest proteolytic and lipolytic counts were found by using the mixed culture of starters with ripening at 20–22°C. Gökalp (1986) showed that using mixed cultures significantly reduced the ripening period; dropped the pH faster; and in addition to these effects, provided better development of color, appearance, firmness, texture, flavor, odor and general acceptability. *D. hansenii* in the starter mixture had positive effects in the development of soudjouk characteristics.

Shehata (1997) prepared salamis with only buffalo meat and buffalo fat and evaluated the effects of *Lb. plantarum* and white mold coating on the quality of the products. The white molds provided the best organoleptic scores, whereas sausages with starter cultures had the best color. In another study, Abou-Arab (2002) used buffalo and lamb meats in a sausage mixture and studied the effect of the fermentation stage by a meat starter on the pesticides DDT and lindane in fermented sausage. Fermentation was accomplished at 30°C for 72 hours. The fermentation process decreased the pesticide residues in the products as a result of the activity of the meat starter.

GAME MEAT

Game meat includes deer, reindeer, wild boar, rabbit, gazelle, and more. Generally, game meat has a dark color, strong taste, and tough texture, but these characteristics may vary according to the type and age of the animal (Soriano et al. 2006). Processing of game meat is not common due to the difficulties of its preparation and because most of the domestic consumption is unprocessed. Also, the marketing of game meat products is limited due to the hunting season (Soriano et al. 2006). However, today there is a tendency for incorporating game meat to processed meat products (Paleari et al. 2003; Soriano et al. 2006). The literature on fermented sausages is still limited, though.

In Sudan, whole wild rabbit is skinned and then fermented for 3 days. After fermentation, the carcass is boiled in water and pounded to give a thick meaty sauce (Campbell-Platt 1995). Reindeer meat is added into some kinds of fermented sausages in Scandinavian countries. In Finland Poro sausage and in Norway Reinsdyr pølse include reindeer meat. Venison (deer meat) can also be utilized in summer sausage (semidried fermented sausage) production. A summer sausage of venison has three types: regular, garlic, or cheddar (includes cheddar cheese) (VIC 2005).

REFERENCES

AAK Abou-Arab. 2002. Degradation of organochlorine pesticides by meat starter in liquid media and fermented sausage. Food Chem Toxicol 40:33–41.

JR Anderson, TA Gillett. 1974. Organoleptic acceptability of various cooked mutton salami formulations. J Food Sci 39:1150–1152.

K Ayhan, N Kolsarici, GA Özkan. 1999. The effects of a starter culture on the formation of biogenic amines in Turkish soudjoucks. Meat Sci 53:183–188.

BR Baliga, N Madaiah. 1971. Preparation of mutton sausages. J Food Sci 36:607–610.

DT Bartholomew, CI Osuala. 1986. Acceptability of flavor, texture and appearance in mutton processed meat products made by smoking, curing, spicing, adding starter cultures and modifying fat source. J Food Sci 51:1560–1562.

MJ Beriain, J Iriarte, C Gorraiz, J Chasko, G Lizaso. 1997. Technological suitability of mutton for meat cured products. Meat Sci 47:259–266.

HM Böhme, FD Mellett, LMT Dicks, DS Basson. 1996. Production of salami from ostrich meat with strains of L. sake, L. curvatus and Micrococcus sp. Meat Sci 44:173–180.

H Bozkurt, O Erkmen. 2002. Effects of starter cultures and additives on the quality of Turkish style sausage (sucuk). Meat Sci 61:149–156.

AA Bushway, MR Stickney, D Bergeron, RH True, TM Work, GK Criner. 1988. Formulation and characteristics of a fermented pepperoni using mutton and fowl. Can Inst Food Sci Technol J 21: 415–418.

G Campbell-Platt. 1995. Fermented meats—A world perspective. In: G Campbell-Platt, PE Cook, eds. Fermented Meats. London: Blackie Academic and Professional, pp. 39–52.

R Chenciner. 2005. http://www.polosbastards.com/artman/publish/horsemeat.shtml.

GH Cosenza, SK Williams, DD Johnson, C Sims, CH McGowan. 2003. Development and evaluation of a fermented cabrito snack stick product. Meat Sci 64:51–57.

AA Dawood, MA Alkanhal. 1995. Nutrient composition of Najdi-camel meat. Meat Sci 39:71–78.

LMT Dicks, FD Mellett, LC Hoffman. 2004. Use of bacteriocin-producing starter cultures of L. plantarum and L. curvatus in production of ostrich meat salami. Meat Sci 66:703–708.

JP Dubeuf, P Morand-Fehr, R Rubino. 2004. Situation, changes and future of goat industry around the world. Small Ruminant Res 51:165–173.

MZ El-Faer, TN Rawdah, KM Atar, MV Dawson. 1991. Mineral and proximate composition of the meat of the one-humped camel (Camelus dromedarius). Food Chem 42:139–143.

FAO. 2005. Available at http://www.fao.org/docrep/x0560e/x0560e06.htm.

J Fernandez-Lopez, E Sayas-Barbera, C Navarro, E Sendra, JA Perez-Alvarez. 2003. Physical, chemical and sensory properties of Bologna sausage made with ostrich meat. J Food Sci 68:1511–1515.

P Fisher, LC Hoffman, FD Mellett. 2000. Processing and nutritional characteristics of value added ostrich products. Meat Sci 55:251–254.

CO Gill. 2005. Safety and storage stability of horse meat for human consumption. Meat Sci 71:506–513.

HY Gökalp. 1986. Turkish style fermented sausage (soudjouk) manufactured by adding different starter cultures and using different ripening temperatures II. Ripening period, some chemical analysis, pH values, weight loss, color values and organoleptic evaluations. Fleischwirtschaft 66:573–575.

HY Gökalp, HW Ockerman. 1985. Turkish-style fermented sausage (soudjouk) manufactured by adding different starter cultures and using different ripening temperatures I. Growth of total, psychrophilic, proteolytic and lipolytic micro-organisms. Fleischwirtschaft 65:1235–1240.

HY Gökalp, H Yetim, M Kaya, HW Ockerman. 1988. Saprophytic and pathogenic bacteria levels in Turkish soudjouks manufactured in Erzurum. J Food Protect 51:121–125.

H Helgesen, R Solheim, T Naes. 1998. Consumer purchase probability of dry fermented lamb sausages. Food Qual Prefer 9:295–301.

I Kalalou, M Faid, TA Ahami. 2004. Improving the quality of fermented camel sausage by controlling undesirable microorganisms with selected lactic acid bacteria. Int J Agri Biol 6:447–451.

G Kannan, KM Gadiyaram, S Galipalli, A Carmichael, B Kouakou, TD Pringle, KW McMillen, S Gelaye. 2006. Meat quality in goats as influenced by dietary protein and energy levels, and postmortem aging. Small Ruminant Res 61:45–52.

N Kolsarıcı, K Candoğan. 2002. Devekuşu eti. Standart 4:35–42.

N Kolsarıcı, A Soyer, K Turhan. 1993. Tiftik keçisi etinin sucuk üretiminde kullanılabilme olanakları üzerine araştırma, Ankara Üniversitesi Ziraat Fakültesi Yayınları, pp. 1–15.

FD Mellett. 1991. South African type dried sausage Addition of caudal fat from sheep carcasses. Fleischwirtschaft 71:680–681.

RT Nassu, FJ Beserra, LAG Gonçalves, T Feitosa. 2000. Manufacturing of a goat meat fermented sausage. 7th International Conference on Goats, France, pp. 672.

RT Nassu, LAG Gonçalves, FJ Beserra. 2002. Effect of fat level in chemical and sensory characteristics of goat meat fermented sausage. Pesq Agropec Bras 37:1169–1173.

RT Nassu, LAG Gonçalves, MAAP da Silva, FJ Baserra. 2003. Oxidative stability of fermented goat meat sausage with different levels of natural antioxidant. Meat Sci 63:43–49.

GA Norman. 1991. The potential of meat from the goat. In: RA Lawrie, ed. Developments in Meat Science. Essex: Elsevier Science Publishers Ltd, pp. 89–157.

MA Paleari, G Beretta, F Colombo, S Foschini, G Bertolo, S Camisasca. 2000. Buffalo meat as a salted and cured product. Meat Sci 54:365–367.

MA Paleari, VM Moretti, G Beretta, T Mentasti, C Bersani. 2003. Cured products from different animal species. Meat Sci 63:485–489.

S Rousset-Akrim, OA Young, J-L Berdague. 1997. Diet and growth effects in panel assessment of sheepmeat odour and flavor. Meat Sci 45:169–181.

NM Sachindra, PZ Sakhare, KP Yashoda, D Narasimha Rao. 2005. Microbial profile of buffalo sausage during processing and storage. Food Control 16: 31–35.

J Sales. 1998. Fatty acid composition and cholesterol content of different ostrich muscles. Meat Sci 49:489–492.

J Sales, D Marais, M Kruger. 1996. Fat content, caloric value, cholesterol content, and fatty acid composition of raw and cooked ostrich meat. J Food Comp Anal 9:85–89.

C Sanudo, ME Enser, MM Campo, GR Nute, G Maria, I Sierra, JD Wood. 2000. Fatty acid composition and sensory characteristics of lamb carcasses from Britain and Spain. Meat Sci 54:339–346.

A Sekar, K Dushyanthan, KT Radhakrishnan, R Narendra Babu. 2006. Effect of modified atmosphere packaging on structural and physical changes in buffalo meat. Meat Sci 72:211–215.

NS Shaikh, AT Sherikar, KN Bhilegaonkar, UD Karkare. 1991. Studies on preparation of mutton sausages. Journal Food Sci Technol 28:323–325.

H Shehata. 1997. Production of buffalo salami. Effects of bacterial starter cultures, gelatine coat with pasterma spices and mold coat on product quality. Fleischwirtschaft 77:1116–1118.

A Soriano, B Cruz, L Gómez, C Mariscal, AG Ruiz. 2006. Proteolysis, physicochemical characteristics and free fatty acid composition of dry sausages made with deer (Cervus elaphus) or wild boar (Sus scrofa) meat: A preliminary study. Food Chem 96:173–184.

A Soyer. 2005. Effect of fat level and ripening temperature on biochemical and sensory characteristics of naturally fermented Turkish sausages (sucuk). Eur Food Res Technol 221:412–415.

A Soyer, AH Ertaş, Ü Üzümcüoğlu. 2005. Effect of processing conditions on the quality of naturally fermented Turkish sausages (sucuks). Meat Sci 69: 135–141.

VIC, Venison Information Centre. 2005. Available at http://www.shopsaskatchewan.com/NorthBattleford/venison_information_centre.htm.

H Vural. 1998. The use of commercial starter cultures in the production of Turkish semi-dry fermented sausages. Z Lebensm Unters Forsch A 207:410–412.

———. 2003. Effect of replacing beef fat and tail fat with interesterified plant oil on quality characteristics of Turkish semi-dry fermented sausages. Eur Food Res Technol 217:100–103.

EC Webb, NH Casey, L Simela. 2005. Goat meat quality. Small Ruminant Res 60:153–166.

WH Wu, DC Rule, JR Busboom, RA Field, B Ray. 1991. Starter culture and time/temperature of storage influences on quality of fermented mutton sausage J Food Sci 56:916–919, 925.

OA Young, J-L Berdague, C Viallon, S Rousset-Akrim, M Theriez. 1997. Fat-borne volatiles and sheepmeat odour. Meat Sci 45:183–200.

OA Young, DH Reid, ME Smith, TJ Braggins. 1994. Sheepmeat odour and flavor. In: F Shahidi, ed. Flavor of Meat and Meat Products. London: Blackie Academic & Professional, pp. 71–97.

Part IX

Ripened Meat Products

36
U.S. Products

Kenneth J. Stalder, Nicholas L. Berry, Dana J. Hanson, and William Mikel

INTRODUCTION

There is no doubt that the dry-curing of meat, or as we in the United States call it, *country curing*, specifically of pork, followed from European traditions as the New World was beginning to be settled as early as the 1500s (Voltz and Harvell 1999). Explorers and settlers arrived in what eventually became the east coast of the United States of America and often brought with them pigs to provide a source of food. The pigs were preferred because they had many offspring per female every year. Additionally, the pigs were allowed to roam freely and forage for their own food. As the pigs gained weight and became full grown, they were harvested. Settlers found it desirable to preserve meat in order to provide food for their families over a longer period of time. Of course, in those days there was no refrigeration and, hence, other methods of preservation were required. The use of dry-cure methods had long been utilized in Europe, and the early U.S. settlers applied the knowledge they brought with them. Consequently, a new industry was born that still survives, mainly in the southeastern U.S., albeit dramatically changed.

COUNTRY HAM STANDARDS

The United States Department of Agriculture (USDA) Food Safety and Inspection Service (FSIS) has developed standards (Code of Federal regulations 2001) of identity in order to assure that consumers are provided products they expect, based on label claims. These standards can also serve to protect a market by not allowing a competing product to use the same name—for example, *country ham* versus a ham and water product. These are both *ham*, but they are produced in a very different manner.

Title 9, Section 319.106 of the USDA-FSIS code classifies *Country Ham*, *Country Style Ham*, *Dry Cured Ham*, *Country Pork Shoulder*, *Country Style Pork Shoulder*, and *Dry Cured Pork Shoulder* as follows:

(a) *Country Ham*, *Country Style Ham*, or *Dry Cured Ham*, and *Country Pork Shoulder*, *Country Style Pork Shoulder* or *Dry Cured Pork Shoulder* are the uncooked, cured, dried, smoked or unsmoked meat food products made respectively from a single piece of meat conforming to the definition of ham, as specified in Sec. 317.8(b) (13), or from a single piece of meat from a pork shoulder. They are prepared in accordance with paragraph (c) of this section by the dry application of salt (NaCl), or by the dry application of salt (NaCl) and one or more of the optional ingredients as specified in paragraph (d) of this section. They may not be injected with curing solutions nor placed in curing solutions.

(b) The product must be treated for the destruction of possible live trichinae in accordance with such methods as may be approved by the Administrator upon request in specific instances and none of the provisions of this standard can be interpreted as discharging trichinae treatment requirements.

(c)

1. The entire exterior of the ham or pork shoulder shall be coated by the dry application of salt or by the dry application of salt combined with other ingredients as permitted in paragraph (d) of this section.
2. Additional salt, or salt mixed with other permitted ingredients, may be reapplied to the product as necessary to insure complete penetration.
3. When sodium or potassium nitrate, or sodium or potassium nitrite, or a combination thereof, is used, the application of salt shall be in sufficient quantity to insure that the finished product has an internal salt content of at least 4%.
4. When no sodium nitrate, potassium nitrate, sodium nitrite, potassium nitrite or a combination thereof is used, the application of salt shall be in sufficient quantity to insure that the finished product has a brine concentration of not less than 10% and a water activity of not more than 0.92.
5. For hams or pork shoulders labeled country or country style, the combined period for curing and salt equalization shall not be less than 45 days for hams, and shall not be less than 25 days for pork shoulders; the total time for curing salt equalization, and drying shall not be less than 70 days for hams, and shall not be less than 50 days for pork shoulders. During the drying and smoking period, the internal temperature of the product must not exceed 95°F (35°C), provided that such temperature requirement shall not apply to product dried or smoked under natural climatic conditions.
6. For hams or pork shoulders labeled dry cured, the combined period for curing and salt equalization shall not be less than 45 days for hams, and shall not be less than 25 days for pork shoulders; and the total time for curing, salt equalization, and drying shall not be less than 55 days for hams and shall not be less than 40 days for pork shoulder.
7. The weight of the finished hams and pork shoulders covered in this section shall be at least 18% less than the fresh uncured weight of the article. The total time for curing, salt equalization, and drying shall not be less than 55 days for hams and shall not be less than 40 days for pork shoulders.

(d) The optional ingredients for products covered in this section are

1. Nutritive sweeteners, spices, seasonings and flavorings.
2. Sodium or potassium nitrate and sodium or potassium nitrite if used as prescribed in this section and in accordance with a regulation permitting that use in this subchapter or 9 CFR Chapter 111, Subchapter E, or in 21 CFR Chapter 1, Subchapter A or Subchapter B.

COMMERCIAL DRY-CURE HAM PRODUCTION IN THE U.S.

History

The country ham industry has its roots in the southeastern U.S. Early U.S. colonial history suggests that the ham industry began around the time European settlers arrived in what is now Virginia. As early as the 1600s, colonists realized that hogs were a critical valuable food resource if they were to survive in the New World.

In 1779, Captain Mallory Todd of Smithfield, Virginia, is often credited with starting the commercial country ham industry in the United States, after he shipped cured hams to the West Indies. The country ham industry grew into a viable industry in this region, and by the early 1900s Smithfield, Virginia boasted that it was the "ham capital of the world" (Evans-Hylton 2004).

Another factor that contributed to the success of the industry was the fact that peanuts were a readily available feed source for the hogs raised in the area. Peanut-fed hogs quickly became synonymous with a Smithfield Ham. In 1926, the Virginia General Assembly passed legislation to protect the identity of ham produced in Smithfield. A genuine Smithfield Ham was defined as a dry-cured ham that was produced within the city limits of Smithfield, Virginia from peanut-fed hogs from the "peanut belt" of Virginia or North Carolina (Evans-Hylton 2004). This law was altered in 1966, when the peanut-fed hog distinction was removed because peanut-fed hogs became more difficult to procure (Evans-Hylton 2004).

Commercial country ham production has followed the evolution of the U.S. from a totally agrarian society to a more industrialized nation (D. Pilkington, personal communication, 2006). Beginning after World War II, many farmers or their children began to move to the city to look for work, and they desired familiar food including country ham. This started commercialization of the country ham industry in the late 1940s and early 1950s (D. Pilkington, personal communication, 2006).

Organization

In 1992, commercial dry ham producers in the U.S. organized themselves in order to have the interests of country ham producers represented nationwide (National Country Ham Association 2006). The National Country Ham Association headquarters is located in the heart of country ham production in Conover, North Carolina. The majority of, but not all, dry-cure or country ham producers are members of this organization. Currently, there are 21 members of the National country Ham Association (National Country Ham Association 2006). The organization is funded through a self-imposed check-off currently set at $0.0125 per ham, with minimum dues of $250.00 and a maximum of $6,250.00 annually (National Country Ham Association 2006). Additionally, many producers have organized state county ham associations.

The organization maintains a website providing contact information, products produced, and websites of its membership (National Country Ham Association 2006). Additionally, the association holds an annual meeting for its membership. This meeting not only includes an annual business meeting, but typically also includes educational seminars that address the latest dry-cure research and regulatory issues. The association also publishes a quarterly newsletter addressing emerging issues.

The organization primarily functions to perform duties that individual country ham producers cannot, or would find difficult to do individually (National Country Ham Association 2006). Some of the defined responsibilities include "encourage the sale and use of country cured meats through cooperative methods of production, promotion, education, advertisement, and trade shows" (National Country Ham Association 2006). Some of the issues addressed by country ham producers through actions of this organization include Hazard Analysis Critical Control Points (HACCP), nutritional labeling, various research projects. Ultimately, the goals of the organization are designed to increase consumer demand for country ham and other dry-cured meat products, thereby improving the profitability of its members. Country ham enthusiasts can find out where to buy product and find new and traditional country ham recipes at the www.countryham.org website.

County Ham Production

U.S. commercial country ham production is concentrated in the southeastern portion of the country (Figure 36.1). This region of the country is where country ham originated. This area also has the desirable climate for producing country hams utilizing traditional methods. Today, country ham could be produced in any area of the U.S. because commercial ham processing facilities have the ability to control the curing environment.

Members of the National Country Ham Association cured approximately 3.4 million hams in 2004

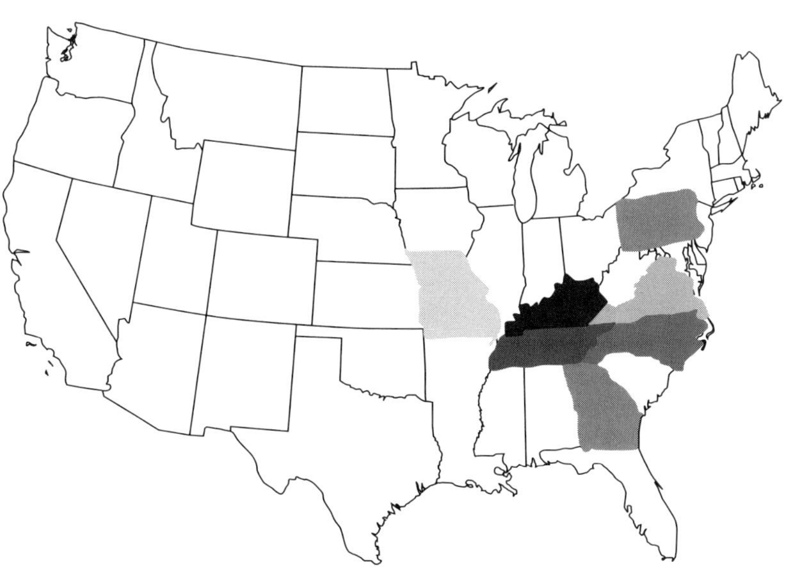

Figure 36.1. U.S. states (Georgia, Kentucky, Missouri, Pennsylvania, North Carolina, Tennessee, and Virginia) where country-cure meat products are produced.

(C. Cansler, personal communication, 2006). Assuming a 9.1 kg average weight for each ham and a retail value of $11.00 per kg., the value of U.S. produced country hams produced by member of the National Country Ham Association was approximately 340 million dollars in 2005. Nationwide, it is estimated that 6.5 million hams were produced in 2004 (C. Cansler, personal communication, 2006) by both nonmembers and members of the National Country Ham Association. These hams had a retail value in excess of 600 million dollars. Although this represents a large quantity of hams, it is still a relatively minor portion of pork meat produced in the U.S annually. Approximately 110 million market pigs are processed annually in the U.S. These carcasses would produce 220 million hams annually (National Pork Board 2005). The numbers of country hams processed annually represent approximately 3% of the total fresh hams produced in the U.S.

The estimated production of country ham in the U.S. by state is shown in Table 36.1. North Carolina led the U.S. in the estimated production of country ham, with over 2.7 million hams cured in 2004. These hams had a retail value in excess of 273 million dollars. North Carolina was followed by Tennessee, Missouri, Kentucky, Virginia, Georgia, and Pennsylvania in the production of country hams. As more country ham producers and even pork producers recognize the opportunity to enter a value-added market chain, country ham production is likely to continue to rise. Additionally, country ham is moving from a product that is regionally produced and consumed to one that is being served in more areas of the country. This is particularly true of chain restaurants that have their origins in the southeastern U.S. and have branched out into other areas of the country. Moreover, country ham producers have recognized an opportunity to add value when competing against imported prosciutto and other types of ham produced in other areas of the world. Imported hams have traditionally demanded a relatively high retail price, particularly on both the East and West Coasts of the U.S. U.S. country ham producers are taking their product to these regions of the country where opportunity exists to expand their market and sell product for premium prices. A good example is a new dry-cure ham facility that was recently completed in Iowa, not considered the traditional dry-cure ham production area of the U.S. This new facility is making a prosciutto-like dry-cured ham that is being marketed to high-end value-added markets on the east and west coasts of the U.S. The company is using hams from pigs raised without the use of antibiotics and in an outdoor or "natural" environment. Furthermore, country ham producers are exploring a variety of marketing methods, ranging from bubble packages of thinly sliced country ham, center-cut slices, biscuit-sized portions, etc., in addition to the traditional method of marketing whole, bone-in cured hams. This is being done to capture new markets and to serve more traditional markets that demand product in a manner that meets the consumer's desire for meal components that are both easy and quick to prepare.

Commercial Curing Process

The curing process begins with the purchase of quality fresh or what the industry terms as *green hams*. Country ham processors begin by purchasing green hams from a harvest facility. These harvest facilities follow USDA guidelines for humanely and safely harvesting food animals. As the harvest process begins, the animal, and subsequently the carcass, and all internal organs are thoroughly inspected. A veterinarian or someone supervised by a veterinarian closely examines every carcass for signs of disease or abnormalities. After harvest, the carcass is chilled and then broken into primal cuts. One of these primal cuts, the ham, serves as the raw ingredient for the country ham industry. In order for any portion of the harvested carcass to be sold commercially, it must have a stamp signifying that it has passed required inspection. This includes all primal cuts sold for further processing. Hence, all commercial country ham processors purchase their green hams from USDA-inspected harvest facilities so both the processor and later the consumer can be assured that the country curing process began with fresh hams that were wholesome and safe.

Once green hams have been purchased, the first step in the dry-curing process occurs. The hand application of the dry-cure ingredients has not changed since the process was first identified. The ingredients in the curing mixture include (1) salt, (2) sodium nitrite/nitrate, (3) sugar, and (4) other spices.

Salt is really the only ingredient needed to perform the dry-curing process. Salt provides flavor to dry-cured meat and acts as a preservative by inhibiting growth and destroying microorganisms (Marriott and Graham 2000). Additionally, it helps other dry-cure components migrate through the raw product and dehydrates the meat, which also limits bacterial growth (Marriott and Graham 2000).

Sodium nitrite/nitrate helps give the product the reddish color that country ham is known for. Additionally, the sodium nitrate/nitrite acts as a preservative and inhibits the growth of *Clostridium*

Table 36.1. U.S. country ham producers, estimated production, percentage produced, and retail value by state.[1]

State	Number of Producers	Estimated Number of Hams Produced[2]	Weight of Hams Produced, kg	Percentage of Total Hams Produced, %	Retail Value, $
North Carolina	20–25	2,730,000	24,843,000	42	273,270,000
Tennessee	10	1,690,000	15,379,000	26	125,170,000
Missouri	5	910,000	8,281,000	14	91,100,000
Kentucky	10	780,000	7,098,000	12	78,050,000
Virginia	7	260,000	2,366,000	4	26,030,000
Georgia	2	65,000	591,500	1	6,500,000
Pennsylvania	2	65,000	591,500	1	6,500,000
Total	54–59	6,500,000	59,150,000	100	606,620,000

[1] Data from members of National Country Ham Association only. Information was obtained from C. Cansler, National Country Ham Association, personal communication, 2006.

[2] Estimated 6.5 million hams produced by nonmembers and members of the National Country Ham Association in 2004. Assumes average weight of dry-cure hams is 9.1 kg and a retail value of $11.00 dollars per kg.

botulinum, the organism that causes botulism. Nitrates and nitrites contribute the flavor distinctive of cured meats and reduce the development of oxidative rancidity and the associated rancid taste (Marriott and Graham 2000).

Sugar is included to sweeten the taste of the ham and reduce the salty flavor of the ham to some degree. Sugar also provides an energy source for beneficial microorganisms that convert nitrate to nitrite during the curing process (Marriott and Graham 2000).

Other spices are added to provide flavor differences, and their use varied from processor to processor. The amount of each of these ingredients varied slightly from processor to processor and is what contributes to flavor differences between individual processor's products. Consumers usually favor the flavor of ham from a given country ham processor. The slight differences in the curing ingredients contribute to these consumer flavor preferences.

The dry-cure mixture is spread evenly or rubbed onto the ham, paying particular attention to the lean surfaces of the ham. Typically, the dry-cure mixture is packed into the hock to prevent spoilage that occurs around the bone of the ham. Approximately, one-half of the dry-cure mixture is applied in the initial application of the dry-cure ingredients. After the application of the dry-cure mixture has occurred, the hams are placed fat side down into large containers or plastic vats (Figure 36.2) and refrigerated. Plastic vats are used because they are easily cleaned. After 5 to 7 days of refrigeration, the hams are resalted; the industry calls this process *overhauling*. During this process the remainder of the dry-cure is applied to the partially cured meat. The ham is then refrigerated for an additional 35 days at approximately 2.5–3.0°C.

After the initial curing process, excess salt is washed from the surface of the ham. After washing, the hams are placed in a mesh stockinette made of plastic or cotton with the *hock*, the small portion of the ham, down. The hams are hung on ham trees (Figure 36.3). The next step is called *equalization* or *drying*. In this process, the hams are placed in environmentally controlled chambers for a minimum of 13 days and up to 30 days. The temperature of the chambers is maintained at 10–12.5°C, and the relative humidity is maintained between 55% and 70%. This process dries the hams. Controlling the relative humidity helps prevent mold growth on the surface of the ham. Additionally, the temperature is used to kill *Trichina* spiralis.

After equalization has occurred, hams can be smoked for 7 to 8 days at a temperature of 41°C if desired. Smoking of the hams is not necessary, but it gives the surface of the hams a desirable deep golden brown color (Figure 36.4). Smoking is not only for aesthetic purposes, but for purposes of adding flavor and aroma, which may be important when the final product is marketed as whole ham. Generally, smoking occurs by exposing the hams to smoke from sawdust made from oak, hickory, apple, or some other desirable wood material. Trichina control can occur by manipulating the temperature and duration during the smoking and drying periods. Smoking is a process that is traditional to a Virginia Style Country

Figure 36.2. Salted hams in large plastic vats after dry-cure mixture has been placed on the surface.

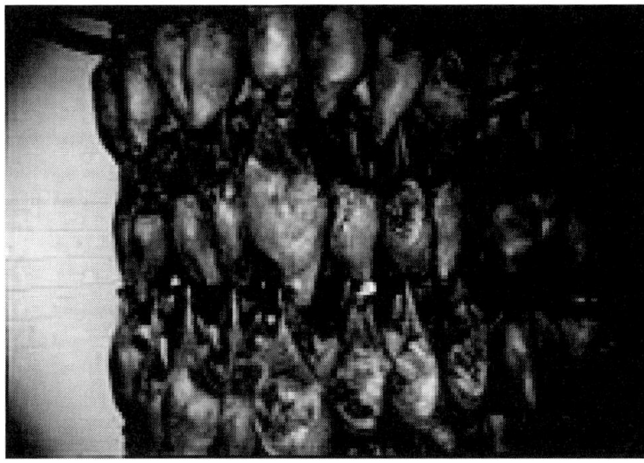

Figure 36.3. Hams are placed in stockings and hung on "trees," with the hock portion of the ham at the bottom.

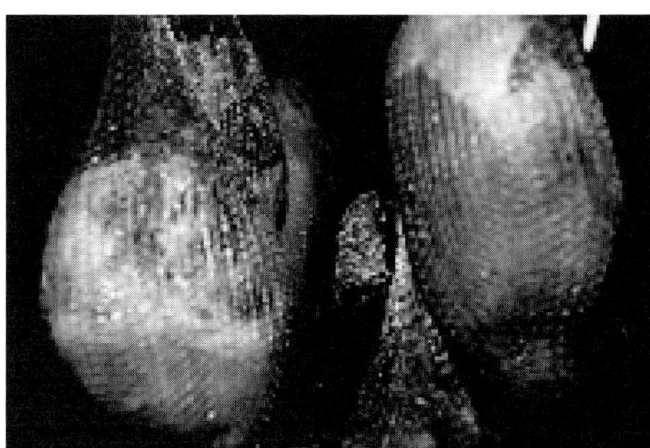

Figure 36.4. Country-style hams that have been smoked have a golden brown color.

Ham. This processing step is applied to most hams produced in the U.S. The exception to this rule is found in North Carolina, where almost all country hams are not smoked. This unsmoked country ham is often referred to as a North Carolina Style Country Ham.

The next step in the process is referred to as *aging*. The aging period can last 30 days or longer. Many specialty hams and dry-cure hams from other areas of the world are aged 18 to 24 months. The longer the ham is allowed to age, the bolder and more complex the flavor will become. Drying temperatures normally range from 21–25°C, with the relative humidity (55–70% similar to that utilized during the equalization step.

HAM CURING AT HOME

Long before dry-cure ham production became commercialized or industrialized, dry-cure hams were produced at home on the farm. The reason that country ham is produced and became popular in the southeastern U.S. was that it had the climate required to properly dry-cure ham. Settlers and farmers of the ham-producing states commonly experience temperatures between 0° and 5.5°C during the winter. These temperatures allowed farmers, curing their own hams, to avoid spoilage and freezing during these months (American Museum of Food 2006). The mild springs in these areas of the U.S. were ideal for equalization of the salt content in the ham

(American Museum of Food 2006). The hot and humid summers experienced in this area were ideal for the aging process (American Museum of Food 2006). The hams were ready to eat around Thanksgiving time the following year. Thus, a tradition was born whereby country hams were produced and consumed during the holiday time of the year (American Museum of Food 2006).

Today, many country-cured hams are still produced at home, passing the tradition of country ham production on to future generations. Many of the home curers of country-cured ham remember days gone by where many family members were involved with the entire process of slaughtering hogs on the farm and producing country ham. These producers continuously try to keep the tradition alive for future generations.

If producers want to raise and harvest their own pigs on the farm, slaughtering should occur in the winter months when the temperature drops below 40°F (4.5°C) (Cole and Houston 1961). This is the most extreme temperature that will still allow the carcass to go through the rigor process without spoiling (USDA 1967). The average weight of market hogs in 2004 was 121.1 kg (267 lb) in the U.S. (National Pork Board 2005). Carcasses produced from a pig weighing 121 kg will weigh approximately 91 kg (200 lb). This large carcass will produce a raw ham in the 9 to 10 kg range. Both hams represent approximately 25% (22.7 kg or 50 lbs.) of the total carcass weight (National Pork Board 2005). Because average market weight continues to rise in the U.S., ham weight will continue to increase. These heavier hams produced from today's market pig make curing hams at home much different for current generations of curing enthusiasts when compared with their predecessors. Heavy weight hams can affect the rate of cure penetration, the amount of cure ingredients needed, and overall aging time. Individuals that want to cure a smaller weight ham may consider raising the pigs themselves, or make arrangements with a pork producer to purchase a pig(s) that is lighter than normal marketing weight.

After the carcasses have gone through proper chilling, the hams are dissected (USDA 1967) from the carcass and are ready to have the dry-cure mixture applied or rubbed on the hams. Rather than harvesting and processing the pigs at home, many people who want to produce their own country ham at home purchase fresh hams from a local supplier. High-quality fresh hams can be purchased from grocery stores, local butcher shops, or locker plants. After the fresh hams have been prepared or purchased, the dry-cure mixture is applied to the hams. The dry-cure mixture ingredients include: salt, sugar, and saltpeter. The exact mixture can vary slightly. Table 36.2 shows examples of dry-cure mixtures commonly used for the home curing process. Only salt is really needed to cure the ham. The other ingredients provide the roles previously described.

After the dry-cure mixture has been applied to the surface, the ham is placed in a container of some sort; a clean barrel or crock works well (USDA 1967). If many hams are being cured, small wooden boxes made of oak or other hardwood or a small building with shelving works well (Winfree and

Table 36.2. Dry-cure mixture ingredients commonly used when curing hams at home.

Formula	Ingredient	Amount, %	Reference
1.	Salt	79.0	Winfree (1975)
	Sugar	19.8	United States Department of Agriculture (1967)
	Saltpeter	1.2	Marriott and Ockerman (2004)
2.	Salt	71.5	Cole and Houston (1961)
	Sugar	27.2	
	Saltpeter	1.7	
3.	Salt	76.1	Skelley and Ackerman (1965)
	Sugar	21.8	
	Saltpeter	2.1	
4.	Salt	83.0	Varney (1968)
	Sugar	8.9	
	Saltpeter	1.8	

Melton 1990). The temperature should be maintained at 2–4°C (Cole and Houston 1961). The hams should remain in the cured mixture for a minimum of 30 days. After the salt curing, the hams should be placed in cold water to remove excess salt and then allowed to dry (Cole and Houston 1961). The hams "equalize" approximately 30 days, which allows the salt to spread evenly throughout the meat. If smoking is desired, a smoldering smoke made from hardwoods (again oak, hickory, etc., are commonly used) is utilized for a period of 2 days in order to obtain the desired color and flavor.

Cured meat should be stored in a cool, dry, well-ventilated area that is free of insects and rodents (Cole and Houston 1961). There are many things that have been done in an attempt to control pests when storing whole cured hams. The first attempt is processing hams in cooler weather when freezing temperatures will kill insects. Also, wrapping hams in meat wrapping paper ((USDA 1970) and inserting the wrapped ham into a closely woven bag and hanging it in a cool dry place will help eliminate pest damage. Coating the outside of a cured ham in lard or coating with a heavy layer of pepper is also thought to reduce the incidence of insect infestation if curing at home (Cole and Houston 1961). A good summary of the required curing times, temperatures, and relative humidity can be found in a fact sheet by Varney (1968).

SAFETY

Country ham in the U.S. is produced commercially under very controlled conditions using only the safest raw ingredients. A plant must adhere to many regulations before USDA inspectors will allow a label (indicating inspected and passed) to be placed on the product (Code of Federal Regulations 2001). Additionally, all country ham processors must have a HACCP program in place. The HACCP program has seven principles including (1) conduct a hazard analysis, (2) determine the critical control points (CCPs), (3) establish critical limits, (4) establish monitoring procedures, (5) establish corrective actions, (6) establish verification procedures, and (7) establish record-keeping and documentation procedures (USDA 201).

The HACCP program focuses on (1) identification of hazards and preventing them from contaminating the food supply, (2) basing the plan on sound science, (3) allowing more efficient and effective government oversight (record-keeping allows inspectors to determine whether a firm is complying with food safety laws over a period of time rather than how well it is doing on any given day), (4) placing responsibility for ensuring food safety appropriately on the food manufacturer or distributor, (5) helping food companies compete more effectively in the world market, and (6) reducing barriers to international trade. These factors go a long way toward ensuring that food, including country ham produced in the U.S., is safe and wholesome (USDA 2001).

COOKING

As with the European prosciutto or Serrano dry-cured ham counterparts, properly cured country ham does not need to be cooked to be safely eaten. As more and more country ham is distributed around the U.S., more of it is eaten in the more traditional European way, uncooked. Some think that country ham or dry-cured ham has the best flavor this way (Marriott and Graham 2000).

Traditions in the U.S. usually dictate that country ham is cooked prior to consumption. One of the most common ways to cook country ham is by pan-frying it. Thin pieces of pan-fried country ham can be served with a variety of side dishes or put in the middle of a homemade biscuit and eaten for breakfast. Other methods of cooking country ham include oven-steaming, slow-cooking in the oven, and roasting in an oven bag (Voltz and Harvell 1999; Marriott and Ockerman 2004).

Today's consumer might feel that a large bone-in country ham is difficult to prepare. However, with relative minimal effort a great meal can be prepared with this American delicacy. The following steps are an easy way to cook a whole country ham:

1. Remove the whole ham from its package. Country hams will often develop mold on the surface of the skin during storage. This mold does not detract from the quality or safety and is easily removed by washing it off with tap water and a stiff kitchen brush. In order to facilitate easier handling during cooking the shank may be removed at this time.
2. The harshness of the salt may be reduced by soaking the ham for 24 to 48 hours prior to cooking. An easy way to accomplish this task is to soak the ham in a plastic cooler large enough to hold the ham and 10 to 15 liters of water. For best results, change the water 2 to 3 times during the soaking process.
3. The ham now is ready to begin the cooking process. Place the ham in a pot large enough that it will hold the ham yet allow for the entire ham to be covered with water. Simmer the ham over

medium-low heat for 3–4 hours or until the bone is nearly lose enough to remove by hand.
4. Allow the ham to cool long enough that it may be safely removed from the cooking water. At this time the ham can be transferred to an oven-safe cooking pan. Prior to oven-baking, the ham can be dressed with pineapple rings, cherries, and clove or brushed with the cook's favorite glaze. Cook the ham in a 175°C oven for 20 to 30 minutes. This final cooking step is really needed only to brown the outer surface of the ham and give it a pleasing appearance.
5. Carve thin slices (5 to 6 mm) of the cooked ham prior to serving.

RESEARCH

Because the country ham industry represents such a small portion of the U.S. pork industry, relatively little funding opportunities exist to evaluate ways to improve it. Recent work has explored breed differences (Stalder et al. 2003) and molecular marker (Stalder et al. 2005) effects on various quality and yield evaluations in country ham. These studies have reported that hams from purebred Duroc may have a yield and quality advantage when compared to hams from pigs of unknown genetic background (Stalder et al. 2003). Additionally, two molecular markers, CAST and PRKAG3, have been shown to impact processing yield and moisture content of country-style cured hams. Other current studies under way include a project designed to identify methyl bromide alternatives for use as pest infestation deterrents. These studies are being conducted as a joint project between Mississippi State, North Carolina State, and the University of Kentucky in conjunction with the National Country Ham Association.

REFERENCES

American Museum of Food. 2006. Raw Talent: America's ham artisans take on Europe's masters. Available at http://www.americanmuseumoffood.org/. Accessed on April 30, 2006.

Code of Federal Regulations. 2001. Title 9, Volume 2. Parts 200 to end. Revised as of January 1, 2001. 9CFR319.106, pp. 300–301.

JW Cole, JW Houston. 1961. Country Style Pork. University of Tennessee Agriculture Extension Service Publication 441. Knoxville, Tennessee: The University of Tennessee.

P Evans-Hylton. 2004. Images of America: Smithfield, Ham Capital of the World. Charleston, South Carolina: Arcadia Publishing.

NG Marriott, PP Graham. 2000. Some solutions to difficulties of home-curing pork. Virginia Cooperative Extension. Publication Number 458–872. Blacksburg, Virginia: Virginia Polytechnic Institute and State University.

NG Marriott, HW Ockerman. 2004. The ultimate guide to country ham an American delicacy. Radford, Virginia: Brightside Press.

National Country Ham Association. 2006. National Country Ham Association web site. Available at http://www.countryham.org. Accessed on April 30, 2006.

National Pork Board. 2005. Quick Facts. Clive, Iowa: National Pork Board. Available at http://www.pork.org/NewsAndInformation/QuickFacts/porkFactPDFS/pg68-93.pdf. Accessed on April 20, 2006.

GC Skelley, Jr., CW Ackerman. 1965. Pork Processing and Curing. Clemson Agriculture Extension Service. Bulletin 77. Clemson, South Carolina: Clemson University.

KJ Stalder, CC Melton, GE Conatser, SL Melton, MJ Penfield, JR Mount, D Murphey, KJ Goddard. 2003. Fresh and processed qualities of dry-cured hams from purebred Duroc and genetically undefined market hogs. J Muscle Foods 14:253–263.

KJ Stalder, MF Rothschild, SM Lonergan. 2005. Association between two gene markers and indicator traits affecting fresh and dry-cured ham processing quality. Meat Sci 69:451–457.

United States Department of Agriculture (USDA). 1967. Slaughtering, cutting, and processing pork on the farm. Farmers' Bulletin No. 2138. Washington, D.C.: United States Department of Agriculture.

———. 1970. Protecting home-cured meat from insects. Home and Garden Bulletin No. 109. Washington, D.C.: United States Department of Agriculture.

United States Food and Drug Administration. 2001. HACCP: A state-of-the-art approach to food safety. Rockville, Maryland: U.S. Food and Drug Administration. Available at http://www.cfsan.fda.gov/~lrd/bghaccp.html. Accessed on April 30, 2006.

WY Varney. 1968. Curing and Aging Country-Style Hams Under Controlled Conditions. University of Kentucky Cooperative Extension Service. Circular 617. Lexington, Kentucky: University of Kentucky.

J Voltz, EJ Harvell. 1999. The Country Ham Book. Chapel Hill, North Carolina: The University of North Carolina Press.

SK Winfree. 1975. Preserving Pork at Home. The University of Tennessee Agriculture Extension Service. Knoxville, Tennessee: The University of Tennessee.

SK Winfree, CC Melton. 1990. Home Guide to Meat Curing. The University of Tennessee Agricultural Extension Service. Publication PB700-500-4/90. Knoxville, Tennessee: The University of Tennessee.

37
Central and South American Products

Silvina Fadda and Graciela Vignolo

INTRODUCTION

Food habits in Central and South America are a blend of cultural backgrounds, available foods, cooking styles, and the foods of colonial Europeans. Although Spanish and Portuguese influence is largely documented in Latin American countries, eating habits could be divided into two main characteristic groups: the highland foods of the Andes with a maize-based diet and the lowland dishes of southern countries that are major beef producers. The cuisine in the mountain areas is unique in South America, preserving many dishes of the Incas.

Regional Characteristics

The diets of people in Mexico and Central America (Guatemala, Nicaragua, Honduras, El Salvador, Belize, and Costa Rica) have several common features, though within the region great differences in methods of preparation and local recipes exist. The basis of the traditional diet in this part of the world is corn (maize), beans, and rice, with the addition of meat, animal products, local fruits, and vegetables. Historically, major changes in the traditional diet occurred during colonial times, when the Spaniards and other European settlers introduced the region to wheat bread, dairy products, and sugar. The consumption of meat and animal products (pork, beef, chicken, fish, and eggs), although popular, is often limited due to cost. However, an Americanized version of popular foods as a result of culture intermingling has been produced, with the increasing use of processed foods contributing to obesity, diabetes, and other chronic conditions in this region (Romieu et al. 1997). The balance between improving access to variety and maintaining dietary quality poses a challenge for public health. In Mexico and Central America, as elsewhere, this transition has been fueled by globalization and urbanization.

The cuisine of South America differs among countries and regions. Some regions have a largely maize-based diet, and others use rice as a meal base. Meat is also very popular and traditionally sides of beef, hogs, lamb, and goats are grilled slowly for hours. Another cooking method consists of steaming foods in a pit oven, as with *pachamanca* in Perú, which typically includes a young pig or goat (as well as chicken, guinea pig, tamales, potatoes, and corn) cooked under layers of hot stones, leaves, and herbs. Venezuelan and Colombian foods have strong Spanish influences, the food being prepared and cooked with olive oil, onions, parsley, cilantro, garlic, and cheese. Hot chili peppers as well as local fruits and vegetables are used as a garnish in most dishes. Chicken stew and *sancocho* (a meat stew with starchy vegetables) are popular in Colombia; *pabellón caraqueño*, consisting of flank steak served on rice with black beans, topped with fried eggs and garnished with plantain chips, is a typical meat dish in Venezuela. The cuisine of Guyana and Suriname is a culinary hybrid with African, Indian, Portuguese, Dutch, Creole, and Chinese influences. A favorite meat dish is *pepper pot*, a stew made with bitter cassava juice, meat, hot pepper, and seasoning. In Brazil, the Portuguese contributed with *linguiça* (Portuguese sausage) and spicy meat stews. The national dish of Brazil is *feijoada*, which consists of black beans cooked with smoked meats and sausages served with rice, sliced oranges, boiled greens, and hot sauce. In Perú and Ecuador, food features an abundant use of potatoes and chili peppers. Due to the scarce availability of other sources of meat as well as to geographical isolation, communities in the Andean plateaus consume meat from *Camelidae* (llama, guanaco, and alpaca), whereas in the tropical coastal regions *ceviche*, prepared from fish and seafood marinated in sour lemon, is a typical dish. In Central and South America, along the Andes highland regions, meat pies called *empanadas* as well as *tamales* containing a mixture of chopped meat (beef and pork), onions, potato, and different seasonings are traditionally consumed.

The southern countries (Argentina, Chile, Uruguay, and Paraguay), including Brazil, are major beef producers. Argentineans eat more beef per capita than people from any other country in the world (67 kg/person/year). Argentina is famous for *asados*, grilled meat dishes, mainly beef, but also pork, lamb, and chicken. Meat production in South America is concentrated in Argentina, Uruguay, and Brazil, where meat exportation constitutes an important source of income. Nevertheless, a great diversity of meat products is produced and demanded by consumers. Poultry meat consumption has increased worldwide together with the increase in production. Brazil and Mexico are among the major Latin American poultry producers; broiler meat production is increasing steadily in response to strengthened domestic demand and record exports. Due to high beef price and consumer nutritional concerns, the demand for poultry meat and poultry processed products has increased among consumers. Pork is the world's major supplier of animal proteins as the principal meat staple in Central America. Domestic consumption of pork meat as well as processed pork products is of great importance. The consumption of meat and meat products in some Latin American countries is summarized in Table 37.1. In this chapter, a contribution is presented on the meat foods currently prepared and consumed in Latin America as well as their safety status.

TYPICAL MEAT PRODUCTS, MICROBIAL ECOLOGY, AND SAFETY RISKS

Processed Meat Products

Even though in Central America and the western areas of South America the consumption of meat and animal products (pork, beef, chicken, fish, and eggs) is limited due to their high cost, a wide variety of meat- and poultry-based products is consumed. Meat is a perishable product unless some methods to restrict bacterial proliferation are applied. Together with drying and salting, fermentation is one of the oldest forms of food preservation technology in the

Table 37.1. Meat and meat products consumption (kg/person/year) in Latin America (2002–2005).

Country	Processed Meat Products	Beef	Pork	Poultry
Argentina	8	67	5.7	23
Brazil	5.7	38	11	38
Chile	12	25.3	19.3	31.2
Uruguay		78	8.9	
Mexico	4.4	15.4	17.7	26.6
Colombia		14.8	2.4	16
Costa Rica	6	16.4	3	21
Peru		5.5	5.2	23.9
Ecuador		14.9		16.4
Venezuela	4.6	17	4.3	31
Bolivia		19	8	16.24
Paraguay		37.6	28	17

world. For centuries, drying and salting have been used to keep meats available for consumption. Processed meat refers to those products that have been subjected to a technological process such as heat treatment or decrease of water activity. Some examples are cooked meat products such as hams, pastrami, roast beef, corned beef, mortadella, wiener sausages, turkey breast, and turkey sausages. Mexico, Costa Rica, Brazil, Chile, and Argentina are among the major processed meat producers. The microbiological status of these products is highly dependent on raw materials and hygiene of handling procedures. Cooked meat and poultry products are generally either prepackaged in a permeable overwrapping film or cut from a bulk block and may become contaminated during production or processing in homes or food establishments. Even when processing has been adequate and the product has been correctly handled after cooking, the most relevant microorganisms of public health significance traditionally implicated are *Staphylococcus aureus*, *Listeria monocytogenes*, *Salmonellae*, and shiga toxin-producing *Escherichia coli*. They are likely to be present as a result of postprocessing contamination due principally to poor cooling procedures and mishandling practices. Staphylococcal food poisoning is generally associated with highly manipulated foods, the etiological agent being *S. aureus* and its related enterotoxins (Balaban and Rasooly 2000). Even when statistics for foodborne diseases due to contaminated meat products are scarcely available in Latin American countries, contamination of processed meat and poultry products by *L. monocytogenes* has resulted in outbreaks of listeriosis. This pathogen can survive on processing equipment such as meat slicers as well as under different technological and storage conditions (Lin et al. 2006). On the other hand, it is well known that cattle are a major reservoir of enterotoxigenic *E. coli*, an organism associated with food illness that is often due to consumption of undercooked meat (Hussein and Bolliger 2005). Although ground beef products are among the more commonly contaminated foods, the heat treatment usually applied in processed meat products would decrease the prevalence of this meatborne pathogen. Other major types of spoilage of these products results from the growth of yeast and, more commonly, molds on the surface of the product, which may be exacerbated by environmental conditions (type of packaging, temperature changes).

Comminuted Meat Products

In Latin America, the high consumption of food specialties containing comminuted beef, pork, and poultry meat is strongly linked to culture and tradition, especially in rural households and village communities, making a significant contribution to the diet of millions of individuals. Although wide local variations between types of products and, indeed, names were found, the differences usually amounted to little more than changes in the relative amounts of the filling ingredients or in the type of pastry used. In addition, the bulk of such products is marketed on a national or at least a supraregional scale, although a relatively large number of small manufacturers distributing on a local basis or even from a single outlet do exist. Raw comminuted meat products are those intended for sale in their natural raw state, such as fresh sausages and burger products. These are made from chopped (minced, ground) pork, beef, and poultry meat and fat and usually include meat from various parts of the animal, including trimmings. In specific products, offal is also used. A major type of comminuted raw meat product is the burger-type product made from beef but also from poultry meats. These, often circular in shape, may also contain filler and spices. Fresh pork sausages as well as blood sausages, which are very popular in Central and South American countries, are often implicated in foodborne diseases. Even when it has been a long tradition in these countries, the existence of street food sell-points constitutes a major health risk because hygienic conditions for food preparation cannot always be assured (Carrascal et al. 2002). The raw comminuted meat products are among the foods more commonly contaminated, and the strains isolated from them have been frequently found to carry virulence factors. Enterotoxin-producing *E. coli*, *Yersinia enterocolitica*, and *Salmonellae* associated with raw beef, pork, and poultry meat, respectively, are causative agents of food poisoning. Basic hygienic precautions should be included during slaughtering, dressing, and conditioning of meat. On the other hand, comminuted meats are used as fillings in traditional Central and South American staples such as *tacos* and *enchiladas*, generally made with beef, pork, and poultry meat and served on corn *tortillas* or *tamales*, in which meat mixed with corn meal is wrapped in corn and banana leaves and *empanadas*, a tasty meat filled pastry. Because these meat fillings are cooked as a stew containing onions, potatoes, and seasonings, the most important organism compromising public health that may be present is *S. aureus*, although Salmonella and coliforms (as fecal contamination markers) could also be encountered (Carrascal et al. 2002). This is likely to occur principally as a result of poor cooling procedures and contamination by handling. It is estimated that 25% of the population carry *S. aureus* on their skin and hands and, consequently, the organism may be transferred to the cooked product. Due to the heat stability of Staphylococcal enterotoxins, once formed in the food they cannot be eliminated. Furthermore, if the storage temperature exceeds 8°C, *S. aureus* may then multiply to significant numbers. Slow cooling of these products also constitutes a serious risk for heat-resistant spores of the genera *Clostridium* and *Bacillus*, it being particularly favorable for *Cl. perfringens*.

Fermented and Meat Products

Fermentation is a relatively efficient, low-energy preservation process, which increases shelf life and decreases the need for refrigeration or other forms of food preservation technology. It is, therefore, a highly appropriate technique for use in developing countries and remote areas where access to sophisticated equipment is limited. Fermented foods should be recognized as part of each country's heritage and culture and efforts should be made to preserve the production methods. The most common types of fermented meat products are salamis and salami-type sausages. These products are microbiologically extremely stable by virtue of their low pH (usually 5.0) and water activity levels (0.7–0.8). Dry-cured sausages have traditionally been produced by fermentation combined with drying without heat treatment. Bacterial food-poisoning hazards are minimal with dry sausages because the manufacturing processes used usually inhibit pathogen growth. Fermentation and drying have been reported to reduce the number of *E. coli* O157:H7 and *L. monocytogenes* mainly through lactic acid and bacteriocin production by lactic acid bacteria (LAB) naturally present in the meat batter or added as starter cultures (Lahti et al. 2001). Although fermented dry sausages are produced in different Latin American countries, their manufacture has a long history in Argentina, mainly due to Italian and Spanish traditions as well as to the well-known quality of meat. Fermented dry sausages (*salame*) are produced in different regions of Argentina using local artisanal techniques for their preparation, which includes beef and/or pork meat, pork fat, salt, and different spices. The meat, fat and other ingredients are mixed together and stuffed in pieces that can have different sizes according to the type of sausage to be produced in each region. In the northern region of Argentina, a traditional fermented sausage is produced without microbial starters from fresh pork meat and lard that are mixed with sugar, NaCl, nitrate/nitrite, and spices. LAB, in particular lactobacilli and gram-positive, coagulase-negative cocci (*Staphylococcus* and *Kocuria*), are the two major groups involved in the fermentation of dry-cured sausages; these bacterial groups account for texture, color, and sensorial changes. The ability of LAB to lower the pH and produce bacteriocins prevents the growth of pathogenic and spoilage microorganisms, improving hygienic safety. LAB also contribute to color and texture development, mainly through their acidification capacity. On the other hand, *Staphylococcus* and *Kocuria* reduce nitrate to nitrite, leading to the formation of the typical cured red color. By means of the application of molecular techniques (RAPD analysis, DGGE and 16SrDNA sequencing), the identification and intraspecific differentiation of strains responsible for artisanal Argentinean sausage fermentation was carried out (Fontana et al. 2005). *Lactobacillus plantarum* and *Lactobacillus sakei* were found to be the dominant LAB during fermentation, and *Staphylococcus saprophyticus* represented the dominant species among the *Staphylococcus* strains (Fontana et al. 2005). During sausage fermentation, a rapid increase in LAB numbers within the first 5 days of ripening was observed, and these organisms were the dominant population at 14 days. As a result of acid production, total coliforms decreased after 5 days whereas yeast and molds were detected throughout the fermentation period.

Dried, Cured, and Salted Meat Products

The preservation of foods in a sound and safe condition has long been an ongoing challenge for humans, drying and salting being ancient methods of keeping meat available for consumption when refrigeration was not yet available. It is well known that salted beef was the very first "industry" in southern South America, especially in Argentina and Uruguay, from colonial times to the advent of refrigeration around the end of the 19th century. Beef from *criollo* cattle naturally raised was salted and shipped to Europe as *tasajo*. The major South American meat producers have also traditionally been important manufacturers of a wide variety of cured and salted meat products, such as dry-cured ham and *bondiola* made from pork hindlegs and neck muscle, respectively. In these products, the increased concentration of salt and the progressive reduction in water activity are the most important factors contributing to shelf stability. Even when molds can grow and develop on the outer surface of these meat products, they can be considered as safe regarding public health concerns. Due to the higher cost of these cured and salted products when compared with other processed ones, they are not massively consumed.

Another dried meat product is jerky or *charqui*, which traditionally was made by exposure to sun, wind, and smoke as a way to preserve and extend the shelf life of meat. *Jerky* refers to dried meat strips (beef, pork, lamb, venison, poultry) or ground and formed meat. In South America, jerky is also made from llama, alpaca, and guanaco, these being the major protein source for many Amerindians groups in the Andes and in southern South America (Franklin 1982). The fresh meat is cut into small pieces, salted, pressed for several days, and dried, with the moisture allowed to drain freely from the product. Because most of the water is removed, jerky is shelf-stable and can be stored for months without refrigeration. The standards for jerky require a water content no higher than 45% and at least 15% salt in the lean. However, illnesses due to *Salmonella* and *E. coli* O157:H7 from homemade jerky raise questions about the safety of traditional drying methods. The USDA recommends meat to be heated at 71–74°C before the dehydrating process in order to destroy pathogenic organisms. After drying, bacteria become much more heat-resistant and likely to survive and, if pathogenic, they can cause foodborne illness. Comparison between heated and unheated meat samples inoculated with *E. coli* O157:H7 before further processing into jerky strips and dehydration showed a higher destruction rate of bacteria when meat was precooked to 71°C prior to drying (Harrison et al. 1998; Kauer et al. 1998). Although jerky has traditionally been a product eaten by low-income people, recently it has become fashionable to use as snacks and ready-to-eat foods.

CONCLUSIONS

With many of the traditional products in Africa, Asia, and Latin America, knowledge of the processes involved is poor and the production conditions vary enormously from region to region, thus giving rise to numerous variations of the basic product. The main idea is not to standardize the process and thereby lose this huge diversity, but to harness the tremendous potential of these methods have to increase not only the quantity but also the quality of food available to the world's population. Developing countries need to build their resources of trained, knowledgeable individuals able to apply the basic microbiological and technological principles to the production of traditional foods. The implementation of the HACCP system and Good Manufacturing Practices (GMPs) during food preparation is an essential complement to improve hygienic quality and minimize safety risks of traditional meat products.

REFERENCES

N Balaban, A Rasooly. 2000. Staphylococcal enterotoxins. Int J Food Microbiol 61:1–10.

A Carrascal, G Arrieta, S Máttar. 2002. Estudio preliminar de la calidad microbiológica de los alimentos en la Costa Atlántica Colombiana. Inf Quinc Epidemiol Nac 7:161–176.

C Fontana, P Cocconcelli, G Vignolo. 2005. Monitoring the bacterial population dynamics during fermentation of artisanal Argentinean sausages. Int J Food Microbiol 103:131–142.

C Fontana, G Vignolo, P Cocconcelli. 2005. PCR-DGGE analysis for the identification of microbial populations from Argentinean dry fermented sausages. J Microbiol Meth 63:254–263.

W Franklin. 1982. Biology, ecology, and relationship to man of the South American camelids. In: M Mares, H Genoways, eds. Mammalian Biology in South America. University of Pittsburgh, pp. 457–489.

J Harrison, M Harrison, B Rose. 1998. Survival of E. coli O157:H7 in ground beef jerky assessed on two plating media. J Food Prot 61:11–13.

H Hussein, L Bolliger. 2005. Prevalence of Shiga toxin-producing *Escherichia coli* in beef cattle. J Food Prot 68:2224–2241.

J Kauer, D Ledward, R Park, R Robson. 1998. Factors affecting the heat resistance of *Escherichia coli* O157:H7. Lett Appl Microbiol 26:325–330.

E Lahti, T Johansson, T Honkanen-Buzalski, P Hill, E Nurmi. 2001. Survival and detection of *Escherichia coli* O157:H7 and *Listeria monocytogenes* during the manufacture of dry sausage. Food Microbiol 18, 21:75–85.

C Lin, K Takeuchi, L Zhang, C Dohm, J Meyer, P Hall, M Doyle. 2006. Cross-contamination between processing equipment and deli meats by *Listeria monocytogenes*. J Food Prot 69:71–79.

I Romieu, M Hernández-Ávila, J Rivera, M Ruel, S Parra. 1997. Dietary studies in countries experiencing a health transition: Mexico and Central America. Am J Clin Nut 65(Suppl):1159S–1165S.

38
Mediterranean Products

Mario Estévez, David Morcuende, Jesús Ventanas, and Sonia Ventanas

INTRODUCTION

The necessity for meat preservation in the Mediterranean countries has been fulfilled through the production of ripened meats since ancient times. The traditional ripening process includes the addition of curing salt and the subsequent drying process aimed at reducing the water activity of the product, which results in a shelf-stable muscle food. Among the large variety of dry-cured meats produced in the Mediterranean area, the dry-cured ham is the most outstanding example of a traditional ripened product, highly appreciated by European consumers. The manufacture of dry-cured meats using raw hams from autochthonous pigs has been done in these countries throughout centuries. However, the introduction of improved pig breeds during the 20th century led to a significant decrease of interest in these traditional products. Today, the production of dry-cured hams is mainly carried out in Spain, France, and Italy, in which two different types of dry-cured hams can be found, depending on the origin of raw material:

- Dry-cured hams from free-range–reared pigs derived from rustic genotypes, slaughtered at high weights, and producing fatty carcasses (i.e., Iberian hams)
- Those hams from intensively reared pigs derived from industrial genotypes, with leaner carcasses, and slaughtered at lower weights (Parma or Serrano hams).

The latter show, however, higher levels of fat than industrial genotype pigs raised for other purposes (i.e., meat for fresh consumption) and are slaughtered at higher weights, because pigs reared for the production of Parma hams are slaughtered at 130–160 kg. The characteristics of the raw material largely determine the procedure for the ham processing and the sensory features of the products. For instance, the former require a weak salting stage and a long ripening process (over 20 months), whereas the latter are moderately salted and ripened during a shorter period of time (around 7–14 months). In general, hams produced from free-range–reared rustic pigs are considerably more appreciated and reach higher prices than those from intensively reared industrial genotypes. The great acceptability of these products by consumers allows both ham versions to coexist in each country, keeping their peculiar characteristics. For cultural and geographical reasons, consumers prefer the appearance, taste, and flavor of the dry-cured hams produced in their own countries according to the consumption data and the preference and choice studies carried out. In the present chapter, the most relevant aspects regarding the production of dry-cured hams will be presented. After that, the biological and technological factors determining the particular features of dry-cured hams produced in the Mediterranean area will be described, with special attention to those protected by the European policies 2081/92 and 2082/92 (P.D.O., P.G.I., T.S.G.).

PRODUCTION OF DRY-CURED HAMS

Although the times and procedures for the manufacture of dry-cured hams are highly variable depending on costumes and the type of product, there are common stages for all of them. In general, two different phases are clearly discerned:

- The *cold phase*, which comprises the selection of the raw material and the salting and postsalting stages, aimed to reduce the water activity of the green ham down to 0.95 through the addition of marine salt and the loss of water
- The *hot phase*, which includes the drying and cellar stages, intended to smooth the progress of the dehydration process through the increase of the temperature under certain relative humidity conditions and the development of the peculiar sensory features of a ripened ham, caused by the onset of chemical and biochemical reactions

Whereas the traditional procedures followed for the production of ripened hams in Mediterranean countries were perfectly adapted to the natural environmental conditions during the seasons, today, in some cases, the development of artificial salting, drying, and cellar rooms with controlled environmental conditions of temperature and relative humidity allow the development of standardized ripening processes. Figure 38.1 summarizes the length of each phase for the production of the most representative dry-cured hams in the Mediterranean countries.

Selection of the Raw Material

Green hams are selected principally by considering the genetic and feeding backgrounds of the pigs because these two factors have been highlighted as the more influential on the sensory quality of the final product (Ventanas et al. 2005). Particularly in hams from rustic breeds reared in traditional systems and protected by EU legislation, the origin of the raw material is specially taken into consideration, because only certain pig breeds and finishing feeds are accepted in order to fit established quality policies. Among the most relevant quality traits, the fat content and fatty acid composition, myoglobin, and antioxidants content decisively influence the appearance, flavor, and taste of the hams affecting the consumer's preference and acceptability (Ventanas et al. 2005). From a technological point of view, some other parameters such as the muscle pH, the temperature, and the weight of the green ham are also considered. The preparation of the green hams prior to the technological process includes the removal of remaining blood from the core of the ham and the adjustment of the shape and appearance of the ham according to the quality policy of each type of product.

Salting/Curing Stage

The raw ham is treated with marine salt leading to the penetration of the sodium chloride into the core of the ham. Depending on the type of product, the marine salt can be used together with potassium nitrate, sodium nitrite, and reducing sugars. For this purpose, different techniques can be used: (1) direct salting in piles with alternative layers of hams and salting mixture (i.e., Serrano and Iberian ham), and (2) mechanical or manual massage of the hams with the salting mixture at fixed periods of time (i.e., Parma hams). The length of this stage generally depends on the weight of the ham, ranging from

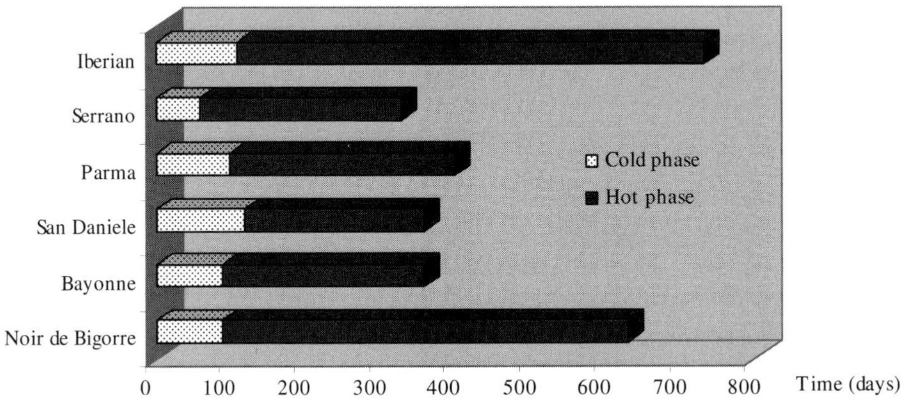

Figure 38.1. Length of the main phases of the production process of the most representative European ripened hams.

17 hours to 48 hours per kg of weight. The salting chamber is kept at refrigeration temperatures (below 5°C) in order to avoid microbial spoilage and at high relative humidity (RH) (around 85–90%) to facilitate the penetration of the salt and to avoid an excessive dehydration of the muscles' surface.

Postsalting Stage

The salt is homogeneously distributed throughout the lean of the ham muscles according to osmosis and diffusion mechanisms. Consistent with the previous stage, the conditions of the environment are still controlled in terms of temperature (3–5°C) and RH (85–95%). At the end of this stage (30–90 days), the water activity of the lean tissues decreases (below 0.96) as a result of the addition of salt and the dehydration of the hams, with weight shrinkages up to 17–18%. The low water activity reached at the end of this stage allows the increase of the temperature and continues the gradual dehydration of the hams.

Drying Stage

The environmental conditions in the Mediterranean area during late spring and the summer (relatively high temperatures and decreasing relative humidity) allowed the dehydration of the hams following traditional procedures in drying rooms in which the temperature and the RH were controlled by opening or closing the windows. Nowadays, the drying process is carried out in artificial drying chambers, making the production of dry-cured hams season-independent and a more homogeneous practice. The length of the process, the temperature and relative humidity program, and the number of drying cycles are designed depending on the type of dry-cured ham. Longer drying periods are commonly used for high-quality dry-cured hams from free-range–reared rustic pigs (i.e., Iberian hams) following traditional procedures, whereas a short drying process at higher temperatures is usually chosen for hams from industrial genotype pigs (i.e., Serrano hams).

The degradation of lipids and proteins mainly occur during this stage, with those having a decisive effect on the sensory properties of the final product. The lipolytic degradation of muscle lipids leads to the generation of free fatty acids, which are subsequently degraded through oxidative reactions (Gandemer 1998; Toldrá 1998). Muscle proteins are denatured and degraded by proteases leading to the generation of low-weight nitrogen containing molecules (amino acids and peptides) and causing loss of protein functionality and solubility (Toldrá 1998). The intensity of the oxidative processes during the ripening of dry-cured hams is dependent on the balance between anti- and prooxidant factors, mainly the fatty acid composition and antioxidant status of the muscles (Cava et al. 2000), whereas the lipolytic changes are caused by endogenous (muscle) and exogenous (microbial) enzymes (Toldrá 1998).

Cellar Stage

After the drying stage, the dry-cured hams are kept in cellars under lower temperatures (12–20°C) during a highly variable period of time, depending on the type of product. The length of this stage is generally related to the sensory quality of the final product because the development of greatly appreciated sensory features (*bouquet*) in dry-cured hams occurs during this last stage. The length of this stage is particularly remarkable for dry-cured hams derived from rustic breeds with high fat content, such as the Iberian ham (up to 24 months).

SPANISH DRY-CURED HAMS

The production of dry-cured meat products in Spain has been done since ancient times. Currently, Spain is the world's largest producer of dry-cured ham, with approximately 39 million cured hams per year. In the Iberian Peninsula, especially in mountainous regions, with climates that are mild and dry in the summer and cold in the winter, the environmental conditions throughout the year allowed the production of such shelf-stable meat products as sausages, shoulders, loins, and especially dry-cured hams. Among dry-cured hams, the most representative products are those coming from intensively reared pig breeds such as the Serrano ham, and those obtained from extensively reared rustic pigs, such as the Iberian pigs (Iberian dry-cured ham) and the Alentejano pigs (*Barrancos* ham in Portugal).

Serrano Ham

At the present time, Serrano ham is protected by the European Community as a Traditional Speciality Guaranteed, and the production is carried out in modern installations called *secaderos* equipped with modern technology, keeping a constant high-quality level. Serrano hams are produced using raw material from intensively reared industrial genotype pigs (Landrance, Large White, and other commercial crossbred pigs including Duroc). Serrano ham is produced throughout the whole country and, therefore, certain differences or nuances regarding production procedures might exist depending on the zone and according to the producers. The fundamental stages in the Serrano ham manufacture are the following:

The ham is first salted for a period of between 0.8 and 2 days per kg in order to avoid microbial spoilage. It is then left to stand for at least 40 days to allow the salt to diffuse through the ham. This is followed by a period of maturing, at least 110 days, and, finally, by a period of aging during the time necessary to complete a minimum of 7 months. During the last stage, biochemical processes caused by oxidative reactions and enzymes from microorganisms and from porcine tissues take place leading to the traditional quality and its peculiar taste and aroma. A Serrano ham slice has a characteristic pink to purple color, and the fat has a shiny appearance, with an aromatic and pleasant taste.

Serrano ham can be bought in three different forms: (1) with bone, the way that it is usually eaten in Spain, which requires a good knife in order to obtain very thin, almost transparent, slices; (2) boneless hams, processed the same way as the Serrano ham with a bone, although at the end of the maturing process and before packaging, the bone is removed; and (3) sliced and vacuum packaged.

Particular Serrano-type Hams

Protected Designation of Origin (PDO) Teruel ham: The entire province of Teruel (northeast Spain) is included in the production of Teruel ham, the first Spanish P.O.D. for meat products. Teruel ham is obtained from 50% Duroc x (Landrace/Large White) pigs slaughtered with a live-weight of 120 to 130 kg. Fresh ham with hoof and weight no less than 11.5 kg are cured during 12 to 14 months in the highlands of Teruel, more than 800 miles above sea level.

Protected Geographical Indication Trevélez ham: Trevélez ham is elaborated in the rocky area of Alpujarras (Granada, Andalucia region), in factories that are higher than 1200 meters above sea level. Cured hams from Trevélez have a rounded shape, keep their rind and hoof, and have a final weight of approximately 8 kg.

Iberian Ham

The Iberian dry-cured ham is the most representative traditional Spanish meat product, and it represents an outstanding example of a high-quality meat product, comparable to the most exquisite food products in the world. Iberian ham production, which represents only about 8% of the total hams produced in Spain, has largely contributed to the development of the regional meat industry in the southwest of the Iberian Peninsula. Furthermore, it supports and protects the exceptional environmental system, *Dehesa*, in which Iberian pigs have been traditionally free-range–reared. As a consequence of their exceptional characteristics, the European Union protects these traditional products with five Protected Designations of Origin, one in Portugal—the *Presunto de Barrancos* in the Alentejo region—and four in Spain—*Jamón de Guijuelo* in Castilla y León region, *Dehesa de Extremadura* in the homonymous region, and *Jamón de Huelva* and *Jamón Valle de los Pedroches* in the Andalucia region. The Iberian pig has been a constant inspiration source for researchers who have profusely studied, during more than 15 years, different aspects related to their genetic traits, their traditional rearing system, and the muscle foods produced with their tissues, with special attention to the Iberian dry-cured ham (Ventanas et al. 2005).

Elaboration of the Iberian Dry-cured Ham

The final product quality depends on the quality of the raw matter, clearly influenced by the genotype and the feeding background, which also affects the technological process.

Raw Matter

The Iberian pig is a rustic and highly lipogenic breed. The lipogenic ability of Iberian pigs together with the feeding strategy used during its fattening period lead to green hams with a thick subcutaneous fat layer (above 5 cm) and high intramuscular fat (8–10% in *biceps femoris*) content (Ruiz-Carrascal et al. 2000; Ventanas 2001; Carrapiso and García 2005), which is appropriate from a technological point of view. In addition, the feeding background influences the quality of the final product as it affects the fatty acid composition and antioxidant level of the porcine tissues leading to Iberian dry-cured hams with different nutritional and sensory characteristics. With the purpose of improving reproductive and productive parameters, the Iberian pig has been commonly crossed with other breeds. Currently, only Iberian x Duroc (up to 50%) crosses are allowed for the production of Iberian dry-cured hams, and in such case, only the pure Iberian breed is allowed as a maternal line (BOE 2001). The productive cycle of Iberian pigs involves an extensive system of breeding and comprises three stages:

- *Weaning*, for around 50 days
- *Growing*, from postweaning until 90–100 kg
- *Fattening*, the last stage, which has a great importance regarding the classification of the animal and their products

There are three models of fattening allowed (López-Bote 1998): *Montanera, Recebo*, and *Cebo.* With Montanera, pigs are fed during the final fattening period ad libitum with natural resources (grass and acorns) in the Dehesa, a sylvopastoral system with a variety of Quercus species and grass. The length of this period is variable, but it is generally carried out from October to February, when maturation of acorns takes place. Feeding pigs on acorns and pasture has two main consequences on the compositional traits of meat from Iberian pigs: high oleic acid and antioxidant (tocopherol) content (Cava et al. 2000; Daza et al. 2005). The high amount of oleic acid in both subcutaneous and intramuscular fat (IMF) leads to a very bright slice surface and very soft fat. Moreover, together with the high oleic acid content, the high amount of natural antioxidant make muscles from Iberian pigs reared in the Montanera system highly stable against lipid oxidation. When the fattening period finishes, the animals are slaughtered at around 160 kg live weight and are 14–18 months old.

The Recebo fattening model is employed when the resources of the Dehesa are insufficient for the pigs to reach the optimum slaughtering weight, so it is necessary to add a complementary diet based on commercial feeding. With the Pienso model, the pigs are mainly fed commercial feed.

Processing

The traditional manufacture process of Iberian hams is perfectly adapted to the natural environmental conditions of the mountainous areas located in southwestern Spain, with cold winters and warm and dry summers. Although the basic steps followed for dry-cured ham production are similar in all types of ham, temperatures reached during ripening of Iberian hams are higher, and the length of the process is much longer than those used for the production of other Mediterranean dry-cured hams. The following are the fundamental stages in the Iberian ham production (Ventanas 2001):

Green hams preparation: This step includes the slaughtering operations, pig quartering, and V-shape cutting. The hoofs are kept for the manufacture of Iberian hams (see Figure 38.2).

Salting: This step is carried out by rubbing the legs with salt containing nitrites and nitrates. Then the pieces are piled up in salting chambers or in containers. The chambers must be at 0–4°C and 90–95% relative humidity. The pieces are buried in salt from 0.70 to 2 days per kg (see Figure 38.3). The trend is to employ shorter times than 1 day per kg.

Postsalting: After salting, hams are gently brushed in order to remove remaining salt from their surfaces and are transferred to the postsalting chamber. The initial temperature is 5°C, and it is slowly increased to 13°C; the relative humidity decreases from 90% to 80%. This step lasts around 80–100 days, longer than in the processing of other hams. The high intramuscular fat content in Iberian hams hinders the water and salt diffusion through the whole piece, which explains the length of the postsalting stage.

Drying: In this step, the hams are exposed to higher temperatures and lower relative humidity, in order to

Figure 38.2. Green Iberian hams.

Figure 38.3. Salting stage: piles of alternative layers of hams and marine salt.

Figure 38.4. Drying stage: dry-cured hams hanging in the drying chambers.

accelerate the drying process. This phase is partially carried out in summertime, and, thus, length and temperature varies with the geographical location, reaching almost 30°C in some areas (see Figure 38.4). During this stage, the fat melts and it impregnates the muscular fibers, acting as a real solvent of volatile compounds.

Cellar: Finally, hams are kept in a cellar for at least 12–24 additional months at a temperature of 10 to 20°C and relative humidity from 65% to 82%. The length of this step is critical for quality attributes to fully develop, especially the generation of volatile compounds (Ruiz et al. 2002).

Sensory Features of Iberian Dry-cured Hams

Sensory characteristics of Iberian dry-cured hams are greatly appreciated by consumers, explaining the considerably high prices reached in the market. Among these features, marbling, color, and flavor seem to be largely influential for Iberian ham acceptability (Ruiz-Carrascal et al. 2000). The marbling of Iberian dry-cured ham is typically abundant and evident, much more intense than in other dry-cured hams (see Figure 38.5). This feature, together with the high levels of oleic acid, supports the formation of pleasant flavor compounds, affects in a positive way the sensorial perception (e.g., juiciness), and provides a healthy nutritional quality (Ventanas 2001). Iberian dry-cured ham shows a more intense color than other hams due to the high myoglobin concentration of the muscles of these pigs, as a consequence of their age at slaughter and the exercise the pigs should accomplish to gain the food, especially in free-range systems.

An intense and pleasant flavor perception has been highlighted as one of the most peculiar features of Iberian dry-cured hams. The characterisation of certain odor-active compounds (Carrapiso et al. 2002) among the total volatile components isolated from the headspace (HS) of Iberian dry-cured hams (Andrés et al. 2002) has been one of the most outstanding approaches to understanding the mechanisms of generation of volatile compounds in dry-cured hams and the role of particular volatile compounds on the aroma perception of such dry-cured products. The main routes involved in the formation of relevant volatile compounds include proteolysis, lipid

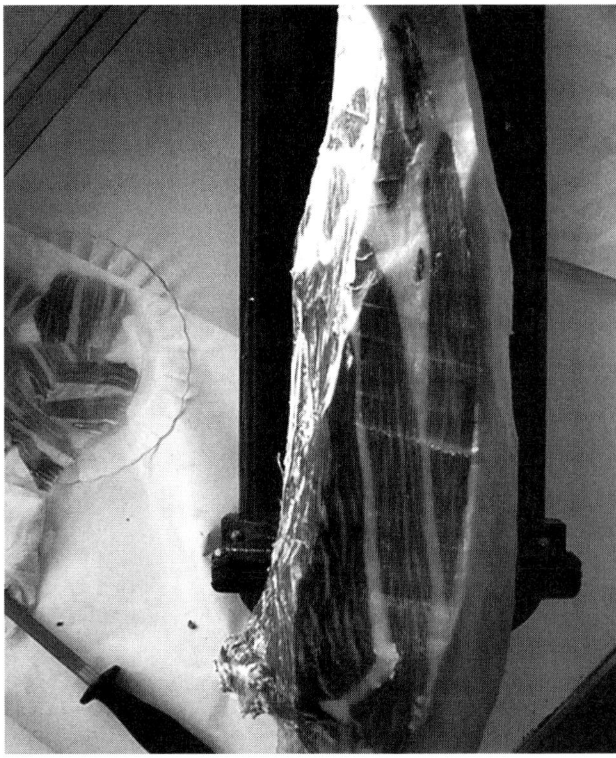

Figure 38.5. Surface of an Iberian dry-cured ham.

oxidation, and amino acid degradation through Maillard reactions and Strecker degradation (Ventanas et al. 1992; Ruiz et al. 2002). Table 38.1 summarizes the major volatile compounds isolated from the HS of dry-cured hams produced in the Mediterranean area.

ITALIAN DRY-CURED HAMS

Parma Ham

The Parma hams are the most representative Italian dry-cured hams because of their mass production and rate consumption. The Parma ham owns a Protected Designation of Origin (PDO) since 1970. These hams are produced using raw material from Landrace and Large White pigs, although Duroc genotypes are accepted for crossing purposes. Animals are usually slaughtered at 160 kg, and are at least 9 months of age. All Parma ham producers must be located within the geographical boundaries of the Parma production area, 5 km south of the via Emilia, limited to the east by the river Enza and on the west by the river Stirone, and up to an altitude of 900 m. Pigs are intensively reared and slaughtered in the Emilia-Romagna, Lombardy, Piedmont, Veneto, Tuscany, Umbría, Marche, Abruzzi, Lazzio and Molise areas, under regulated conditions and fixed feeding arrangements.

For the manufacture of Parma hams, the green hams are first placed in special cold storage rooms for 24 hours in order to reduce the leg temperature from 40°C to 0°C and harden the meat, thereby facilitating the trimming. During this phase, the weight losses are around 1%. The trimming of the green hams involves removing external fat and rind from the leg. The green hams acquire their peculiar "chicken drumstick" shape, with those showing the slightest imperfection being discarded. In addition to the aesthetic purpose of the trimming phase, it is also aimed to facilitate the salting step. After trimming, hams lose up to 24% of their weight in fat and muscle.

Parma hams are salted using wet and dry salt: wet salt for treating the skin, and sprinkled dry salt for the lean parts. Using any other type of additive such as nitrates and nitrites is not allowed. Hams are stored in cold storage rooms at a temperature ranging between 1°C and 4°C at about 80% relative

Table 38.1. Major volatile components of different types of dry-cured hams.

	Iberian	Serrano	Parma	Bayonne
Aldehydes				
2-Methylpropanal	***	***	**	**
3-Methylbutanal	****	****	***	***
2-Methylbutanal	****	***	***	***
Pentanal	****	***	***	***
Hexanal	****	****	****	****
Heptanal	***	***	***	***
2,4-Nonadienal	**	**	**	**
Octanal	***	**	***	**
Phenylacetaldehyde	***	**	**	**
Nonanal	***	***	***	***
Decanal	**	*	*	*
Alcohols				
2-Methylpropanol	**	**	**	**
1-Penten-3-ol	***	***	***	***
2-Pentanol	***	***	*	*
3-Methylbutanol	***	***	**	**
1-Pentanol	***	***	***	***
1-Octen-3-ol	**	**	**	**
Ketones				
Butanone	***	***	***	***
2-Pentanone	***	***	***	***
2-Hexanone	***	**	*	*
2-Heptanone	***	***	*	*
2-Octanone	***	**	*	*
n-Alkanes				
Hexane	***	**	**	**
Octane	***	***	**	***
Sulphur Compounds				
Dimethyl disulphide	***	***	**	**
Dimethyl trisulphide	**	**	*	*

Adapted from Sabio et al. (1998).
Amount of volatile compound in Area Units (AU): * ($AU<10^5$), ** ($10^5<AU<10^6$), *** ($10^6<AU<10^7$), **** ($10^7<AU<10^8$).

humidity. After 6–7 days stored in these rooms, called *preliminary salting rooms*, the hams are cleaned of residual salt and sprinkled with a second thin coating of salt. Hams are then placed into a new cold stored room, known as the *final salting room*, where they remain for 15–18 days depending on their weight. A highly trained *maestro salatore* ensures that the hams absorb just enough salt to guarantee the safety and technological stability, preserving simultaneously the peculiar "sweetness" of Parma hams. At the end of this period, weight losses are around 3.5–4%.

After removing the residual salt, hams are stored in the so-called *resting rooms* for 60–90 days at 1–5°C and 75% relative humidity (this is known as the *resting stage*). During this stage, hams are allowed to "breathe" and should not become too dry or too wet. The air in the resting rooms is changed at frequent intervals. The absorbed salt penetrates deeply, becoming evenly distributed in the muscular mass. During this phase, the ham weight reduces around 8–10%. Then, hams are washed with warm water and afterward the rind is brushed to remove any residual salt or impurity. Hams are set out to dry in the air when the weather conditions are favorable (dry, windy, and sunny) or in special drying rooms.

The precuring phase is carried out in large rooms with windows where hams are hung on the traditional *scalere* (ladders). Windows are opened, considering the internal/external humidity as well as that of the product. After precuring, hams are beaten to give them their typical rounded shape. The hollow surrounding the best end is sprinkled with pepper to keep the contact dry. Between 8–10% of the ham weight is lost during this stage.

During the "greasing stage," the hollow surrounding the best end, exposed muscular parts, and any cracks are covered with a layer of ground pork fat, mixed with a pinch of slay and ground pepper and, if necessary, rice flour. The greasing has the purpose of softening the superficial muscular layers, preventing them from drying up too quickly compared to the inner layers, as well as causing additional moisture loss. According to the Italian legislation, this grease mixture is not considered an ingredient.

After greasing and upon reaching the seventh month, hams are moved to the cellars, which are colder and less ventilated than precuring rooms. Before being removed, hams are subjected to sampling, in which a special needle, made of horse bone and having the feature of being able to quickly absorb the aroma and subsequently lose it again, is inserted into various points of the muscular mass and then sniffed by experts. During the curing phase, chemical and biochemical reactions take place, leading to the development on the particular scent and taste of Parma hams. Hams lose about 5% of their weight during this phase.

After a period of 10 months for hams with a final weight of 7–9 kg and 12 months for hams weighing more than 9 kg, the Ducal Crown fire brand is affixed as a sign of a Parma ham elaborated following the established policies.

The Parma ham is a seasoned dry-cured ham with a curved shape, without the distal part of the ham (the hoof is removed) or any other external blemishes likely to impair the product's image. The hams weigh between 8 kg and 10 kg, but never less than 7 kg. The ham has a mild, delicate flavor, is slightly salty, and has a fragrant and distinctive aroma. Among the variety of factors influencing the sensory and technological quality of Parma dry-cured hams, the compositional characteristics of the fat from the carcass has been highlighted as the most relevant (Bosi et al. 2000). Although the fatty acid composition of Parma hams is highly dependent on the feeding background of the pigs, the levels of saturated (38%), monounsaturated (51%), and polyunsaturated fatty acids (11%) are within the expected range for the fatty acid compositions of other commercial Italian, French, or Spanish dry-cured hams (Bosi et al. 2000; Gandemer 2002). Compared to the fatty acid profiles of dry-cured hams produced from free-range–reared pigs fed on acorns or chestnuts (i.e., Iberian ham, Corsican ham) Parma hams generally contain lower levels of MUFA and higher of SFA and PUFA (Gandemer 2002) (Table 38.2). Several recent studies have reported the possibilities of enhancing the fatty acid composition, oxidative stability, and aromatic profile of Parma hams through dietary means (Bosi et al. 2000; Pastorelli et al. 2003).

San Daniele Ham (Prosciutto di San Daniele)

This ham has been produced in the northeastern region of Friuli Venezia-Giulia for centuries, although it has owned a Protected Designation of Origin (PDO) since 1996. As with the Parma hams, San Daniele hams are another clear example of Italian hams produced using raw material from industrial genotype pigs (Landrace and Large White) and crossbred pigs using Duroc genotypes. San Daniele hams are manufactured in the municipality of San Daniele del Friuli, in the province of Udine (Italy). Pig rearing and slaughter take place in Friuli-Venezia Giulia, Veneto, Lombardy, Piedmont, Emilia Romagna, Umbria, Tuscany, Marche, Abruzzi, and Lazio. At slaughter, pigs must have average weights of no less than 160 kg and be at least 9 months of age.

The whole production process lasts 12 months at least, of which a minimum of 8 months is used for air curing. Pig thighs are selected and suspended in well-ventilated or refrigerated rooms for 24 to 36 hours. The fat and hide is then trimmed, and the salt curing process begins. The salt is applied through a manual massage, repeating the procedure once a week for a month. The hams are then washed, brushed, and dried either in sunlight or indoors, where temperatures never exceed 15°C. Once the hams are dried, they are inspected for flaws and are coated with a mixture of flour, lard, water, and pepper. After that, San Daniele hams must be aged for at least 12 months, but some are aged for up to 2 years. During this time, the hams lose up to 30% of their original weight. Whereas the ingredients and conditions involved in the production of San Daniele hams are identical to other types of Italian dry-cured hams, the unique climatic conditions of the Friuli region, with its higher altitudes and drier air, has been considered as one of the key factors influencing

Table 38.2. Intramuscular fat content and fatty acid composition of m. *Biceps femoris* from different types of hams.

	Iberian	Serrano	Parma	Nero Siciliano	Bayonne
IMF (g/100 g)	11.3	4.78	3.57	7.07	3.53
Fatty Acids (%)					
C12:0	0.07	0.07	–	0.08	–
C14:0	1.27	1.37	1.18	1.09	1.08
C16:0	22.92	24.48	21.65	22.55	22.91
C18:0	7.45	10.98	12.67	11.08	12.53
C20:0	0.22	–	0.14	0.14	–
Saturated	31.93	37	35.99	34.94	36.52
C16:1	3.39	3.41	3.05	2.94	3.32
C18:1	54.51	47.99	49.99	39.53	43.6
C20:1	–	0.97	0.86	0.8	0.57
Monounsaturated	57.9	52.37	54.04	43.29	47.49
C18:2	9.41	9.62	7.77	16.75	11.7
C18:3	0.65	0.53	0.21	0.93	0.5
C20:4	0.11	0.97	0.61	2.35	3.1
Polyunsaturated	10.17	11.01	8.59	20.03	15.3

Adapted from Timón et al. (2001), Lo Fiego et al. (2005), and unpublished data from the authors.

the unique flavor and texture of these hams. It is a matured, dry-cured ham with a guitar-shaped exterior and intact distal part (the hoof remains in the ham). The most peculiar external feature of San Daniele hams is the presence of the bottom part of the leg bone, providing a very rustic-looking appearance. It is appreciated for its pink meat, its creamy, smooth texture, and salty-sweet flavor. Whole San Daniele hams weight from 8 to 10 kg, but never less than 7.5 kg.

Cinta Senese Ham

These hams are known as *Toscano* dry-cured ham according to the policy of the Protected Designation of Origin. In contrast to the Parma and San Daniele hams, the Cinta Senese ham is manufactured using green hams from an ancient pig breed very common in Tuscany since the twelfth century. Pigs are characterized by a black skin and white stripe all over the front legs, the chest and the shoulders. Cinta Senese is one of the pig breeds allowed by the PDO for the production of the Toscano dry-cured ham.

Toscano ham processing takes place in the traditional production area including the entire territory of the region of Tuscany. Production and slaughter of animals for raw material are carried out in the regions of Emilia-Romagna, Lombardy, Marche, Umbria, Latium, and Tuscany.

These are salted, naturally cured, uncooked hams with a rounded shape and bowed at the top. The hams normally weight around 8 to 9 kg and have a light to bright red color when sliced, with little inter- and intramuscular fat.

Recently, Pugliese et al. (2005) have studied the fatty acid composition and sensory characteristics of Toscano dry-cured hams from the Cinta Senese breed pig compared to those from crossbreeding Cinta Senese with a commercial breed pig (Large White). Fat of hams from Cinta Senese purebred pigs had lower SFA and PUFA but higher MUFA proportions than those from crossbred pigs. Moreover, Toscano dry-cured hams from Cinta Senese purebred pigs scored higher for juiciness than hams from crossbred pigs.

Nero Siciliano Ham

The Nero Siciliano pigs derive from a population of pigs whose presence in Sicily is well documented at least since the nineteenth century (Moretti et al. 2004). This breed, reared on wood pasture, is characterized by high roughness and strength. The population of Nero Siciliano pigs is currently declining. The population size is probably below 900 sows, distributed in a large number of small herds and including a free-ranging population of about 200 sows. However, there are conditions for developing successful conservation programs. Traditionally, Nero Siciliano pigs have used both for fresh meat consumption and for the production dry-cured sausages (Moretti et al. 2004) and most recently for the production of dry-cured hams. According to recent studies (Pugliese et al. 2004) the fat and meat compositional traits of extensively reared Nero Siciliano pigs are similar to those found in other free-range–reared pigs (i.e., Iberian or Corsican pigs) fed on natural resources such as acorns and chestnuts.

FRENCH DRY-CURED HAMS

The tradition of producing dry-cured hams from free-range–reared autochthonous pigs fed on acorns and chestnuts in France arises from the Roman Empire. The coins carved into a ham shape found in Nîmes in the first century B.C. support that hypothesis.

However, the population of rustic breeds in France is now decreasing and the production of dry-cured hams is relegated to southern France, whereas northern producers are devoted to the production of cooked (Paris) or smoked (Alsace) hams. Lately, several initiatives aimed to retrieve particular rustic breeds (Noir de Bigorre, Corsican, Vasque, Limousine) and traditional production systems have been started in order to fulfill increasing consumer demand. The production of improved industrial genotype pigs (Large White, Landrace, and Duroc) specifically for the manufacture of certain dry-cured products such as Bayonne and Lacaunne hams has been recently increased as well.

Bayonne Ham

The Bayonne hams own a Protected Geographical Indication and are the most representative French hams due to the large amount of hams produced per year (around 500,000). These hams are manufactured using raw material from industrial genotype pigs slaughtered at around 120 kg. Green Bayonne hams show intermediate characteristics between those from rustic pig breeds (i.e., Noir de Bigorre) and other commercial pigs in terms of subcutaneous fat thickness (around 1.4 cm) and intramuscular fat content (2.8%), which largely determine the process followed for their ripening. Bayonne hams are ripened during an intermediate long process (9–12 months) leading to dry-cured hams of commercial quality.

The production of French dry-cured hams is clearly influenced by the procedures followed in Spain and Italy. Bayonne hams are nitrified, salted using dry salt and then kept in postsalting chambers around 10 weeks. The ripening process lasts between 9–10 and 12 months with the temperatures going from 20–22°C (at the first stages) to 14–18°C following the procedures followed for the production of Parma hams.

Dry-cured Bayonne hams exhibit a rounded shape and their external appearance reminds of that of Parma hams: the hoof is removed and the head of the femur can be superficially seen. Compared to other types of dry-cured hams, Bayonne hams contain relatively high moisture content and relatively low pigment concentration. Buscailhon et al. (1994) reported the relationship between the concentration of heme pigments in Bayonne hams, their appearance, and their quality, highlighting these parameters as decisive for consumer acceptability. The lean of Bayonne hams has a pink, mild-ripened appearance. In order to improve the quality of Bayonne hams in terms of texture and flavor, the aforementioned authors suggested extending the duration of the ripening process up to the length currently used. According to Buscailhon et al. (1994), a longer ripening process would increase the generation of proteolysis-derived compounds and aroma volatiles. Additionally, a higher drying course would lead to more firm hams and an extended chewing quality, which would increase the perception of taste and flavor compounds.

Noir de Bigorre Ham

The Noir de Bigorre ham is produced using raw material from the Noir de Bigorre (traditionally called Gascon) rustic and vigorous pig breed traditionally reared in southwest France. The Noir de Bigorre pigs are reared in a semiextensive system with availability of natural resources (grass, acorns, and chestnuts), and they are slaughtered at high weights (around 160 kg) with at least 12 months of age. Currently, the Noir de Bigorre ham is in the process of obtaining a Protected Designation of Origin. In opposition to the commercial pigs used for the production of Bayonne hams and similarly to other rustic breeds (i.e., Iberian pigs), Noir de Bigorre pigs generally have thick subcutaneous fat (above 5 cm), and their muscles contain high amounts of IMF (around 4%) and tocopherols. Therefore, green hams from Noir de Bigorre pigs show a high suitability for being subjected to a lengthy ripening process. In fact, Noir de Bigorre hams are processed following traditional procedures similar to those carried out for the production of the Iberian ham: a long drying course (above 18 months) under natural environmental conditions with severe temperature changes.

The appearance of the Noir the Bigorre ham is considerably different from that of Bayonne ham. The coxal bone and the hoof are kept, giving to this ham a similar appearance to that of the Spanish Serrano-type ham. Compared to the lean of Bayonne hams, that of Noir the Bigorre hams exhibits a redder and more intense color as a result of the higher concentration of heme pigments (0.32 versus 0.12 mg/100 g) and a more severe drying process (Ventanas unpublished data; Robert et al. 2005). Large differences between Bayonne and Noir de Bigorre hams also have been described regarding other chemical components. Compared to other commercial dry-cured hams (i.e., Parma or Bayonne hams), higher oleic acid (~46%) and tocopherol (~7 mg/kg) levels have been reported in Noir de Bigorre hams as a likely consequence of the feeding background of the pigs (pasture, acorns) in outdoor systems (Ventanas unpublished data). Recent sensory studies carried out in different French dry-cured hams established a certain relationship between compositional traits and sensory quality, revealing large differences between types of hams. As a likely consequence of the higher levels of IMF, Noir de Bigorre hams showed a higher juiciness than Bayonne hams (Tovar et al. 2005; Robert et al. 2005), with that parameter highlighted as decisive for the overall acceptability of the dry-cured products. Hedonic tests carried out by Tovar et al. (2005) in collaboration with the Centre Technique de la Salaison, de la Charcuterie et des Conserves de Viandes (CTSCCV) revealed that amongst French dry-cured hams Noir de Bigorre hams received higher scores by French consumers than Bayonne hams. The aroma perceived from Bayonne hams was described as "mild and slightly rancid," whereas that from Noir de Bigorre hams showed a more intense and complex aromatic profile. The authors considered that the high level of IMF and the high levels of oleic acid and tocopherols caused the generation of high levels of certain lipid-derived volatiles during the long ripening process of the Noir de Bigorre hams, which likely contributed pleasant aromatic notes and enhanced their flavor traits. Consistently, volatile compounds described as odor-active compounds in Iberian dry-cured hams (Carrapiso et al. 2002) and contributors of pleasant flavor notes, have been described as volatile components of Noir de Bigorre hams (Sans et al. 2004). The Noir de Bigorre hams are currently considered a promising option to French consumers looking for dry-cured hams of higher quality than that of Bayonne hams, with the latter being, nevertheless, generally appreciated by consumers.

REFERENCES

AI Andres, R Cava, J Ruiz. 2002. Monitoring volatile compounds during dry-cured ham ripening by solid-phase microextraction coupled to a new direct-extraction device. J Chrom A 963:83–88.

BOE. 2001. Real Decreto 1083/2001, de 5 de octubre, por el que se aprueba la norma de calidad para el jamón ibérico, paleta ibérica y caña de lomo ibérico elaborados en España. Boletín Oficial del Estado, 247, 15 de octubre, 37830–37833.

P Bosi, JA Cacciavillani, L Casini, DP Lo Fiego, M Marchetti, S Mattuzzi. 2000. Effects of dietary high-oleic acid sunflower oil, copper and vitamin E levels on the fatty acid composition and the quality of dry-cured Parma ham. Meat Sci 54:119–126.

S Buscailhon, JL Berdagué, J Bousset, M Cornet, G Gandemer, C Touraille, G Monin. 1994. Relations between compositional traits and sensory qualities of French dry-cured ham. Meat Sci 37:229–243.

S Buscailhon, G Gandemer, G Monin. 1994. Time-related changes in intramuscular lipids of French dry-cured ham. Meat Sci 37:245–255.

AI Carrapiso, C García. 2005. Instrumental colour of Iberian ham subcutaneous fat and lean (biceps femoris): Influence of crossbreeding and rearing system. Meat Sci 71:284–290.

AI Carrapiso, A Jurado, M Timón, C García. 2002. Odor-active compounds of Iberian hams with different aroma characteristics. J Agric Food Chem 50: 6453–6458.

R Cava, J Ventanas, J F Tejeda, J Ruiz, T Antequera. 2000. Effect of free-range rearing and α-tocopherol and copper supplementation on fatty acid profiles and susceptibility to lipid oxidation of fresh meat from Iberian pigs. Food Chem 68:51–59.

A Daza, AI Rey, J Ruiz, CJ Lopez-Bote. 2005. Effects of feeding in free-range conditions or in confinement with different dietary MUFA/PUFA ratios and α-tocopheryl acetate, on antioxidants accumulation and oxidative stability in Iberian pigs. Meat Sci 69:151–163.

G Gandemer. 1998. Lipids and meat quality, lipolisis, oxidation and flavour. In: Proc of the 44th ICoMST, Barcelona, Spain, pp. 106–119.

———. 2002. Lipids in muscles and adipose tissues, changes during processing and sensory properties of meat products. Meat Sci 62:309–321.

DP Lo Fiego, P Macchioni, P Santoro, G Pastorelli, C Corino. 2005. Effect of dietary conjugated linoleic acid (CLA) supplementation on CLA isomers content and fatty acid composition of dry-cured Parma ham. Meat Sci 70:285–291.

C López-Bote. 1998. Sustained utilization of the Iberian pig breed. Meat Sci 49:S17–S27.

VM Moretti, G Madonia, C Diaferia, T Mentasti, MA Paleari, S Panseri, G Pirone, G Gandini. 2004. Chemical and microbiological parameters and sensory attributes of a typical Sicilian salami ripened in different conditions. Meat Sci 66:845–854.

G Pastorelli, S Magni, R Rossi, E Pagliarini, P Baldini, P Dirinck, F Van Opstaele, C Corino. 2003. Influence of dietary fat on fatty acid composition and sensory properties of dry-cured Parma ham. Meat Sci 65: 571–580.

C Pugliese, R Bozzi, G Campodoni, A Acciaioli, O Franci, G Gandini. 2005. Performance of Cinta Senese pigs reared outdoors and indoors: 1. Meat and subcutaneous fat characteristics Meat Sci 69: 459–464.

C Pugliese, G Calagna, V Chiofalo, VM Moretti, S Margiotta, O Franci, G Gandini. 2004. Comparison of the performances of Nero Siciliano pigs reared indoors and outdoors: 2. Joints composition, meat and fat traits. Meat Sci 68:523–528.

N Robert, S Basly, C Dutertre. 2005. Efecto del proceso de maduración en la textura del Jamón IGP Bayonne. Proceedings of the 3th Mondial Congress of Dry-cured Ham, Teruel, 2005, pp. 411–412.

J Ruiz, E Muriel, CJ Lopez-Bote. 2002. The flavour of Iberian ham. In: F Toldrá, ed. Research Advances in the Quality of Meat and Meat Products. Trivandrum, India: Research Signpost, pp. 289–309.

J Ruiz-Carrascal, J Ventanas, R Cava, A I Andrés, C García. 2000. Texture and appearance of dry cured ham as affected by fat content and fatty acid composition. Food Res Int 33:91–95.

E Sabio, MC Vidal-Aragón, MJ Bernalte, JL Gata. 1998. Volatile compounds present in six types of dry-cured ham from south European countries. Food Chem 61:493–503.

P Sans, MJ Andrade, E Muriel, J Ruiz. 2004. Etude du profil des composes volatils de jambons secs issus de porcs de race Gasconne. Proceedings of the 5th International Symposium of the Mediterranean Pig, Tarbes, France, 2004, 66–67.

ML Timón, J Ventanas, AI Carrapiso, A Jurado, C García. 2001. Subcutaneous and intermuscular fat characterisation of dry-cured Iberian hams. Meat Sci 58:85–91.

F Toldrá. 1998. Proteolysis and lipolysis in flavour development of dry-cured meat products Meat Sci 49:S101–S110.

J Tovar, E Gilbert, C García, J Ventanas. 2005. Establecimiento de un perfil sensorial del jamón francés Noir de Bigorre. Proceedings of the 3th Mondial Congress of Dry-cured Ham, Teruel, 2005, pp. 419–420.

J Ventanas. 2001. Tecnología del Jamón Ibérico. Madrid: Ediciones Mundi-Prensa, pp. 15–45.

J Ventanas, JJ Cordoba, T Antequera, C García, C Lopez-Bote, MA Asensio. 1992. Hydrolysis and Maillard reactions during ripening of Iberian ham. J Food Sci 57:813–815.

S Ventanas, J Ventanas, J Ruiz, M Estévez. 2005. Iberian pigs for the development of high-quality cured products. In: SG Pandalai, ed. Recent Research Developments in Agricultural and Food Chemistry. Trivandrum, India: Research Signpost (6), pp. 27–53.

39
North European Products

Torunn T. Håseth, Gudjon Thorkelsson, and Maan S. Sidhu

INTRODUCTION

Salting and drying are considered to be of the oldest and most important methods for preservation of meat (1). The main focus of this chapter is on the typical and traditional dry-cured meat products from Norway, Iceland, and the Faeroe Islands. Among these dry-cured products, some are generally unknown and different from dry-cured products in the rest of Europe.

NORWAY

Several types of dry-cured meats are produced in Norway, and the annual production is about 1,700 tons (ACNielsen Norway AS), in addition to an import of about 300 tons. The most important dry-cured products of whole meat in Norway are *fenalår* (fenalaar), which is dry-cured leg of lamb or mutton, and spekeskinke, which is dry-cured ham. The processes described in this chapter are for the dry-cured meat products with bone, but processing of deboned tumble-salted products are widespread. Dry-cured meat has a long tradition in Norway, and the consumption is about 0.5 kg per capita. Sweden, Finland, and Denmark do not have the same tradition in producing and eating dry-cured meat products, and therefore the consumption here is much lower. The industrial "bulk type" dry-cured products are low-priced and have a high salt content. More expensive high-quality products with low salt content and more pronounced dry-cured aroma and taste are produced at a small, but increasing, scale.

Fenalår: Norwegian Dry-cured Leg of Lamb or Mutton

Fenalår is to a large extent a Norwegian specialty, although there is a small production in some other European countries such as Iceland (2). The name *fenalår* comes from the word *fenad*, which is an old Norwegian word for *mutton*, and the word *lår* meaning *leg*. Curing of mutton and lamb legs has a long tradition in Norway, back to the time of the Vikings more than a thousand years ago. Today, fenalår is a popular dry-cured product. The classic way of eating it is with very thin and unfermented crisp bread or with normal bread, followed by scrambled eggs and beer. Traditionally, a sheath knife was used to cut off slices of the dry-cured meat, and many people still do it this way at home. Modern use includes a variety of dishes where dry-cured ham can be used.

The most common raw material for fenalår is leg of lamb or mutton. Norway has the largest sheep production among the Nordic countries (3), mostly the long-tailed crossbreed Norwegian White Sheep (T. Kvame, personal communication, 2006). Norway is a long country with differing climate and differing geography from the sea to the mountains; therefore there is a wide variety in pasture, but as much as 80% of the sheep graze in the highland or mountains during the summer (Norwegian Sheep Recording System 2005). In September the sheep and lambs are brought to the farms a short time before slaughtering for cultivated pasture and grain feeding if necessary

(S. Vatn, personal communication, 2006). The animal welfare is generally good and strongly regulated by legislation. The animal transportation has a high standard, and the drivers have comprehensive coursing in animal welfare and transportation (E. Tolo, personal communication, 2006).

Most sheep and lamb are slaughtered in the autumn, and use of frozen and thawed legs is common to ensure constant meat supply to the production during all seasons. In addition, salting time is reduced (4). The legs should preferably be at least 3 kg after cutting and covered by an evenly distributed fat layer. The leg is traditionally cut similarly to Spanish hams, with the aitch bone (ischium) kept on, but the foot is removed. There is a new trend among some producers to remove the aitch bone and form the leg to a rounded shape, similar to Parma ham.

Traditionally dry-cured leg of mutton and lamb were produced on farms holding sheep. The legs were pickle- or brine-cured in bins. Today, there are numerous small- and medium-size production plants in Norway, in addition to butchers, specialty shops, and some private production. Brine- or pickle-curing is common (Figure 39.1), but there seems to be a trend to convert to dry salting, especially among producers of higher-quality products. They also tend to reduce the amount of salt in the product because the salt amount in industrial bulk products tends to be very high, often as much as 7–10 %. Combinations of finely ground vacuum salt, nitrite salt, and coarse marine salt has long been preferred, but coarse marine salt used alone gains ground as dry salting becomes more common. Some producers use herbs, spices, honey, syrup, or other natural ingredients along with the salt, mainly for taste contribution.

Figure 39.1. Salting Norwegian fenalår in a Norwegian meat processing plant. The legs are soaked in saturated brine before being pickle-salted (private photo, M. S. Sidhu).

Two important conditions separate the Norwegian technology from the South European. First, the only climatic control of the salting rooms is temperature control. The relative humidity (RH) varies widely, depending on conditions such as the outdoor climate and which other products are stored in the salting room. Second, keeping the cured legs cold for salt equalization (postsalting) was, until recently, rarely used in Norway. The extensive salting allows the salt to reach the center of the leg during salting, and the temperature can safely be raised immediately after salting. Focus on salt reduction has led to extended use of salt equalization after salting (5). A nondestructive method for measuring salt is a large advantage in both research and product development of expensive dry-cured products, and computed tomography (CT) is such a method (6). The CT scanner measures density as the attenuation of X-rays sent through an object and visualizes this in a gray-scale image (7). The density is related to the salt content in the meat. The distribution of salt through the meat is also visualized (8).

Historically, farmers dried the salted hams in traditional Norwegian storehouses on pillars (*stabbur*), each cured leg hanging in a separate bag of cloth. A wire prevented contact between cloth and meat, which was important to prevent flies from placing their eggs in the meat. Today, modern production plants have conditioned rooms where temperature, relative humidity, and air velocity are controlled. Small producers control only the temperature. Bulk products are dried until a weight loss of 26–33%, with little time for ripening processes, like proteolysis and lipolysis, to occur. Fortunately, the industry has increased its development activity the last several years and an extended ripening time is increasingly common. Smoking was traditionally used for additional preservation in the western part of Norway, where the climate is very humid. Today, many producers smoke part of their production. The smoking is very slight compared to the heavily smoked German Schwarzwald hams. The purpose of the smoking is to add taste and flavor to the product and to prevent growth of mold.

Spekeskinke: Dry-cured Ham with Bone

Hams used for Norwegian bulk dry-cured hams are 9–10 kg, usually with less than 10 mm subcutaneous fat. The small size, leanness, and high enzymatic activity of these fresh hams make them best suited for shorter production times (9). A common industrial curing method for bulk bone-in ham is pickle-curing, where dry salt and meat extract turns into brine, but dry salting and brine salting are also used. Bulk hams are heavily salted for 2–4 weeks and given none or a short salt equalization. Some hams are smoked, but not all, in contrast to what may be the general idea (10). The hams are dried for 2.5–3 months at roughly 15°C and 75% RH. Bulk hams have a weight loss of 25–30% and the salt content can be as high as 10%. Salt distribution, proteolytic index, and protein degradation is shown to be variable among Norwegian dry-cured hams (11).

High-quality hams are generally produced from older, heavier, and fattier pigs. The hams are dry-salted with coarse marine salt, the curing time being considerably shorter than for bulk ham (12). The equalization period lasts from a few weeks up to 4 months. Climatic conditions during drying are similar to those of bulk ham, but small scale producers may also use stabbur for drying, where the climatic conditions are dependent on the outdoor conditions. During drying, the meat surface is covered with a thin layer of fat or a mixture of fat and rice flour to prevent excessive drying and to keep the surface smooth. The total production time is from 7 to 24 months, depending upon initial size and weight of the fresh hams and the desired degree of ripening. These hams have a pronounced dry-cured ham flavor and aroma, due to proteolysis and lipolysis (13, 14). The final salt content is usually between 5 and 7%.

Food safety is an increasingly important public health issue. Dry-cured meat is more or less shelf-stable, depending on the ingredients and process used, but is frequently contaminated with molds. All together, mold contamination on dry-cured meat is a significant quality problem leading to a major economic impact for the producers due to cleaning and loss of products. Molds may also produce mycotoxins that are toxic compounds with diverse effects on health.

Pinnekjøtt: Norwegian Dry-cured Side of Lamb or Mutton

The dry-cured specialty *pinnekjøtt* is the traditional and popular Christmas Eve dinner in the western part of Norway, and its popularity is increasing in the rest of the country. The name *pinnekjøtt* comes from *pinne* meaning *wooden stick*, and *kjøtt*, which is Norwegian for *meat*. The dry-cured meat is placed on wooden sticks during cooking. Pinnekjøtt is produced from the side of lamb or mutton. It is mainly produced in the slaughtering season for lamb, and for this reason, the raw material is mainly used fresh. The meat is placed in bins and coarse salt between the layers ensures that all meat surfaces are in contact with salt. The meat is dry-, pickle-, or brine-cured, or combinations of these methods. There are two salting strategies, either restricted or complete salting. Restricted salting lasts from 1.5 days (15) and up to 3–4 days, whereas complete salting lasts until the meat is more or less saturated with salt after 4–10 days. Some producers rinse the salted meat in fresh water for some hours to a couple of days, to prevent salt precipitation during drying.

The salted meat is dried at 10–15°C, usually in rooms without humidity control, but often under high air circulation from fans. If controlled, the humidity is usually quite low, about 65–75% RH. The drying time varies widely; 10–15 days is common, but as much as 6–7 weeks can be found (15). About 25% are cold-smoked at room temperature, either during or after drying. The variation in total water loss is considerable. This is due to large differences in processing conditions such as time and temperature, and differences in the raw material, such as the size of the meat pieces, bone mass, and amount of fat (T. C. Johannessen, personal communication, 2006).

ICELAND AND THE FAEROE ISLANDS

Iceland and the Faeroe Islands were settled from Norway during the 8th and 9th centuries. Sheep farming has been the most important part of agriculture, and the production systems were adapted according to local environmental conditions. The sheep breeds are directly descended from the first sheep brought to the islands, belonging to the North European short-tailed group of sheep (16). The sheep live outdoors, grazing on wild pasture and cultivated hayfields, but in Iceland they are kept in sheds and fed on dried hay or silage during the coldest part of the winter.

Hangikjöt: Cold-smoked Icelandic Lamb Meat

The traditional preservation methods of lamb meat in Iceland are drying, smoking, and whey pickle fermentation (17). *Hangikjöt* is a traditional product in Iceland, which is eaten as a main part of festive dinners like Christmas and Easter, but also as cold cut slices on bread and as a starter.

Hangikjöt was traditionally produced from fresh meat after slaughtering in the autumn. Today, hangikjöt is produced throughout the year from frozen lamb carcasses of 17–19 kg. The carcass is cut into legs, forequarters, rack, and flanks. The legs and forequarters are thawed overnight in cold running water before salting. Different salting methods are used. At one time, nitrate salt was common for dry-curing, but it is now fully replaced by nitrite salt. Dry-curing is also combined with, or replaced by, brine immersion. Multineedle injection was introduced during the 1970s and is used by some factories in combination with brine-curing. Brine strengths (9–15 %), salting times (1–6 days), and use of other ingredients (ascorbates, phosphates) vary between producers or brands of hangikjöt.

Until the late 19th century, smoking took place over open fireplaces (18) for 2–3 weeks before the meat was moved to other parts of the kitchen or in storage huts (*hjallur*) to dry. The fuel in the old kitchens was dried peat, dried sheep dung, and in some places birchwood. Today, specially designed cold smoke ovens are used in the meat processing plants. Dried sheep dung is always needed because the smoke from it gives the meat its distinct flavor. The smoking time is 1–5 days, depending on how many times the fuel is lit.

Hangikjöt is still an important product in Iceland. In 1975, the consumption was about 4.5 kg/person (17), but it has declined to about 2.5 kg/person in 2005, calculated as meat with bone. Traditionally produced hangikjöt is eaten both raw and cooked. But today, most smoked lamb meat in Iceland can be consumed only if cooked thoroughly. Some products are also produced in order to be eaten raw. They are always smoked with dried sheep dung and the total weight loss is 30–35%. There are different preferences in Iceland for the various types of hangikjöt. Less salt and less smoke is the main trend.

The consumption of smoked foods in Iceland was very high before the introduction of modern preservation methods like freezing. It has been suggested that it might have been a factor of importance in the high incidence of stomach cancer in Iceland, among other factors due to considerable amounts of polycyclic aromatic hydrocarbons (17). A high amount of nitrosamines has been found; most are in homemade products, but they are also in commercially produced products (19). The meat industry has shortened the smoking times, cut down on the use of nitrates, and introduced ascorbates as brining aids in order to reduce the formation of N-nitrosamines (20).

Skerpikjøt: Air-dried Faeroese Sheep Meat

The local sheep production system and the ancient slaughtering and processing traditions for sheep meat in the Faeroe Islands are unique, compared to both the Nordic countries and to Europe. Most lambs are home-slaughtered at the farms or in the villages and used for production of the traditional air-dried lamb meat called *skerpikjøt*. The Faeroese have preserved a Viking food culture and it is part of daily life in the islands, and skerpikjøt is a national dish. *Skerpikjøt* means meat sharpened by cold and wind. It is cut into thin slices and eaten, or placed on bread—preferably buttered rye bread.

Newly slaughtered, whole carcasses of lamb, ewe, and ram are cut directly after slaughtering. The meat is neither salted nor smoked but dried directly. To facilitate and speed up the drying, the carcass is opened up and flattened by cutting through the ribs on one side close to the backbone. Skerpikjøt is dried in hjallur, mostly placed along rivers or brooks on low banks or at places with good air circulation, where drying of meat, by experience, is good.

Drying takes place during winter from October to March. Early winter the temperature is around 5–10°C and the humidity is rather high; later, the temperature and humidity decreases. A weight loss of about 35% during drying has been reported (21). Water activity decreases from 0.97% to 0.90% during the first month but changes little after that. Protein degradation due to proteolysis leads to an increase in the pH of the meat during the first 4 months, before it decreases again—perhaps due to yeast fermentation of carbohydrates in the meat. The chemical composition of three cuts of skerpikjøt is shown in Table 39.1 (22).

The quality of the final product is very much dependent on weather conditions, especially in the first period when temperature, humidity, and wind have a great influence. The quality of the fresh meat—that is freshness, carcass conformation, fat condition, and fat composition—is also critical for processing and quality of the end product. Further studies are needed to explain the influence of raw material and processing conditions on the quality of skerpikjøt. But parallels can be drawn to studies on dry-cured ham, e.g., the age at slaughter can affect the color, amount of intramuscular fat, and activity of proteases (23), whereas raised temperature at the

Table 39.1. Chemical composition (in percent) of three cuts of Faeroese air-dried lamb meat, skerpikjøt.

Types	Protein	Fat	Dry Matter	Salt
Shoulder	27.6	27.9	57.0	0.54
Leg	28.5	19.0	59.5	0.41
Flank	22.2	65.6	89.0	0.39

Adapted and modified from Marita Poulsen (22).

end of the process can increase the risk of faults, as with soft hams and color changes (24).

Health authorities have questioned the safety of skerpikjøt because the slaughtering and drying, which is the only hurdle, are not carried out under controlled conditions. Four outbreaks of botulism were reported in the Faeroe Islands in 1979, 1988, and 1989 (25). Today, the health authorities in the islands are strict on accomplishing good hygienic practices to avoid microbial contamination.

LAPP TRADITIONS

Lapp traditions of dry-curing reindeer meat are widespread in northernmost parts of Norway, Sweden, and Finland. The production is generally for private consumption. Most parts of the carcass, such as side, ham, and shoulder, are dry-cured. In the Norwegian regions, there are two main types of products: *tørka* (dried) meat and *bokna* meat (M. Trumf, personal communication, 2006). Both types are dry salted using coarse marine salt, but the amount and length of salting varies. After salting, the meat is dried outdoors, hanging on wooden racks. The drying time depends on both the climatic conditions and the type of commodity produced. Tørka meat is dried for roughly 1.5 months, sometimes 2 months, before it is then ready for consumption. The final water content is generally lower than in products such as dry-cured ham. On the other hand, Bokna meat is dried for only a couple of weeks. This product is roasted in the oven or in a pan before consumption.

In northern Sweden, there is a long tradition for drying reindeer meat. The meat is sometimes smoked, and many different types and pieces of muscles are used, including tongue and heart. Production is usually on a small scale for home consumption or for local distribution and sale (S. Berg, personal communication, 2006).

Reindeer, mainly loin and steak, is also dry-cured in Lapland in the north of Finland (J. Hallikas, personal communication, 2006). The production is traditional and on a small scale. The meat is heavily salted and dried to give a long shelf life. The drying takes place outdoors during the winter and spring.

GERMANY

Various types of dry-cured meats are widely used in Germany, such as smoked ham like *Westphalia* ham, air-dried ham, cured lard, and various cured muscles (26). Smoked hams are usually pickle-cured, filling up with brine after some days. The meat is commonly left in slow running water for 12–24 hours and smoked for 4 to 20 days depending upon product and the desired degree of smoking (27). Air-dried hams are produced from round-shaped cooled hams. They are rubbed with nitrite salt and dry salted in sea salt for 25–30 days at 0–2°C and 80% RH (27). The hams are slightly dried at 20–23°C to get a dry rind, before being kept at 8–12°C and 60–70% RH for salt equalization. A mixture of salt and pepper is rubbed into the ham before drying, initially at 8–12°C at 60–70% RH and later at 12°C at 70–80% RH for several months.

Perhaps the most famous and exclusive dry-cured meat from Germany is *Schwarzwälder Schinken*, Black Forest hams. These hams can be produced only in Schwarzwald (the Black Forest) in the southwest of Germany following strict production regulations, due to their EU origin protection and production protection (28). The hams must come from specially raised pigs. To be accepted, the fresh hams must have the correct pH (usually below 6.0), the right amount of fat, and a good hygienic quality including a sufficiently low core temperature (29, 30).

The hams are deboned and trimmed before salting. A mixture of salt, herbs, and spices, such as garlic, pepper, coriander, and juniper berries is rubbed onto the hams by hand (31). This mixture is considered important for the products' final characteristics, and the exact blend is each producer's secret. Brine is formed after 3–4 days in the salt/spice mixture and the hams are left in the brine until they are considered salty enough, approximately after 2 weeks (32). Excessive salt is removed and the hams are left in cold rooms for salt equalization for 2–7 weeks (29, 33). Smoking is an important part of the tradition. The hams are cold-smoked over branches and sawdust of pine or fir from the Schwarzwald district, at about 25°C for 2–3 weeks (33). The hams are kept at about 15°C from 2 to several weeks for ripening.

OTHER NORTH EUROPEAN COUNTRIES

The production volume of whole dry-cured meat in Denmark and Sweden is small, though that of dry-fermented sausages is considerable. During the last years, local production of dry-cured meat has been initiated in Sweden to compete with an increasing import from Germany, Italy, Spain, and Norway. Whole hams are not produced, but, instead, cold-smoked products of whole muscles are preferred. These products are sold in pieces or sliced. Cold-smoked whole meat products are also produced at farms and small enterprises, but the drying process and final water activity are hardly controlled (S. Berg, personal communication, 2006).

Historically, dry-cured hams were produced at Danish farms for home consumption, but there is no industrial production of whole dry-cured meat in Denmark to date (K. Teglmand, personal communication, 2006). However, fresh hams are exported to Italy for production of dry-cured ham, whereas Parma ham and other hams are imported to Denmark. Local butchers and gourmet shops produce some dry-cured hams, the most famous being *Skagenskinke* (C. Vestergaard, personal communication, 2006). Finland has a small production of dry-cured reindeer meat (J. Hallikas, personal communication, 2006). The meat is salted and dried, but not as heavily as the traditional Lapp production. The products are usually smoked.

Poland has a tradition for dry-curing different whole muscles of pork (J. Mielnik, personal communication, 2006). Typical muscles are loin, tenderloin/filet, neck, clod (outside chuck; *Triceps brachii*), and ham muscles. The muscles are slightly salted, usually without any other additives, but lactic acid bacteria are sometimes used. Dry salting is most common, but brine injection is also commonly used in larger muscles. The salted muscles are predried, usually for about 10 days, before smoking for 4–12 hours. Predrying allows smoke to fasten to the meat surface. The drying is continued after smoking until a total weight loss of 20–30% is achieved. The final salt content is usually about 3–3.5%.

Smoked meat products of pork and beef are popular in Lithuania, whereas sheep and game are hardly used for dry-curing. People in cities use industrially produced dry-cured products. Small farms make different products for their own consumption, like smoked fat, ham, and sausages (S. Nominaitis, personal communication, 2006). The production of dry-cured meat in England is limited; bacon and cooked cured ham are much more widespread. Dry-cured meat is mainly sold presliced among the delicacies in larger stores (U. Væreth, personal communication, 2006).

ACKNOWLEDGMENTS

The authors wish to acknowledge the Norwegian dry-cured meat producers who anonymously shared their production processes for the documentation purpose of the chapter. We also wish to thank Mohamed Abdella and Terje Frøystein at the Norwegian Meat Research Centre for useful comments. Special thanks to Per Berg at the Norwegian Meat Research Centre for general discussion about the production of fenalår in Norway and for valuable comments.

REFERENCES

1. A Riddervold. 2004. Matkonserveringsmetodene. In: A Riddervold, P Berg, et al.: Spekemat. 2nd ed. Norway: H.W. Damm & Søn AS, pp. 35–65.
2. P Berg. 2004. Fenalår—fra konserveringsmetode til smak av kultur. In: Kjøttets Tilstand 2004. Norwegian Meat Research Centre, pp. 4–8.
3. ÓR Dýrmundsson. 2004. Sustainability of sheep and goat production in North European countries—From the Arctic to the Alps. 55th Annual Meeting of the European Association for Animal Production, Bled, Slovenia, p. 4.
4. F-S Wang. 2001. Lipolytic and proteolytic properties of dry-cured boneless hams ripened in modified atmospheres. Meat Sci 59:15–22.
5. T Thauland. 2003. The effect of resting on the distribution of salt during the production of dry-cured ham. Proceedings of the 49th International Congress of Meat Science and Technology, Brazil, pp. 479–480.
6. A Dobrowolski, R Romvári, P Allen, W Branscheid, P Horn. 2003. X-ray computed tomography as an objective method of measuring the lean content of a pig carcass. Proceedings of the 49th International Congress of Meat Science and Technology, Brazil, pp. 371–372.
7. CS Vestergaard, SG Erbou, T Thauland, J Adler-Nissen, P Berg. 2005. Salt distribution in dry-cured ham measured by computed tomography and image analysis. Meat Sci 69:9–15.
8. T Thauland. 2005. Accuracy of CT scanning for prediction of salt content in salted meat. Proceedings of the International Skjervold-symposium, Hamar, Norway, p. 44.
9. F Toldrá. 2002. Dry-cured meat products. Trumbull, Connecticut: Food & Nutrition Press, pp. 193–194.

10. J Flores. 1997. Mediterranean vs. northern European meat products. Processing technologies and main differences. Food Chem 59:505–510.
11. MS Sidhu, K Hollung, P Berg. 2005. Proteolysis in Norwegian dry-cured hams: Preliminary results. Proceedings of the 51st International Congress of Meat Science and Technology, Maryland, 2005, pp. 50–51.
12. P Berg, T Thauland. 2004. Den norske, italienske og spanske spekeskinken. In: A Riddervold, P Berg, et al.: Spekemat. 2nd ed. Norway: H.W. Damm & Søn AS, pp. 96–133.
13. F Toldrá. 2006. The role of muscle enzymes in dry-cured meat products with different drying conditions. Trends Food Sci Technol 17:164–168.
14. F Toldrá, M Flores. 1998. The role of muscle proteases and lipases in flavor development during the processing of dry-cured ham. Crit Rev Food Sci Nutr 38:331–352.
15. A Brimi, B Wilson. 1987. Frå Lom til Lyon. Kokebok frå naturens kjøken. 2nd ed. Norway: Universitetsforlaget AS, p. 132.
16. I Tapio, L Grigaliunaite, E Holm, S Jeppsson, J Kantanen, I Miceikiene, I Olsaker, H Viinalass, E Eythorsdottir. 2002. Mitochondrial differentiation in Northern European sheep. The 7th World Congress on Genetics Applied to Livestock Production, Montpellier, France. Session 26. Management of genetic diversity, pp. 26–38.
17. T Thorsteinsson, G Thordarsson, G Hallgrimsdottir. 1976. The Development of Smoking of Food in Iceland. In: Advances in Smoking of Foods. International Joint IUPAC/IUFoST Symposium in Warszawa, Poland, pp. 185–189.
18. H Gísladóttir 2004. Hangikjöt í rót upp rís: um reykhús og önnur reykingarými. In: Árbók Hins Islenzka Fornleifafélags 2002–2003, pp. 151–162.
19. MJ Dennis, GS Cripps, AR Tricker, RC Massey, DJ McWeeny. 1984. N-nitroso compounds and polycyclic aromatic hydrocarbons in Icelandic smoked cured mutton. Food Chem Toxicol 22:305–306.
20. G Thorkelsson. 1989. The effect of processing on the content of polycyclic aromatic hydrocarbons and volatile N-Nitrosoamines in cured and smoked lamb meat. In: Bibliotheca "Nutritio et Dieta." JC Somogyi, HR Muller, eds. Basel: Karger, pp. 188–198.
21. Annual Report of Heilsufrødiliga Starvstovan. 1990.
22. Marita Poulsen. 1995. Føroyskar Føðslutavlar. (Faeroese Food Composition tables). Heilsufrøð iliga Starvsstovan, pp. 16–18.
23. C Sárraga, M Gil, JA Farcia-Regueiro.1993. Comparison of calpain and cathepsin (B, L and D) activities during dry-cured ham processing from heavy and light large white pigs. J Sci Food and Agric 62:71–75.
24. L Martin, JJ Córdoba, T Antequera, ML Timón, J Ventanas. 1998. Effects of salt and temperature on proteolysis during ripening of Iberian ham. Meat Sci 49:45–153.
25. EP Wandall, T Videro. 1976. Botulism in the Faeroe Islands. Ugeskr Laeger 153:833–835.
26. H Koch. 1966. Die Fabrikation feiner Fleisch—und Wurstwaren. 15th ed. Frankfurt am Main: Verlagshaus Sponholz, pp. 124–130.
27. H Koch, H Fuchs, Gemmer. 1992. Die Fabrikation feiner Fleisch—und Wurstwaren. 20th ed. Frankfurt am Main: Deutscher Fachverlag, pp. 69–76.
28. http://www.tannenhof-schinken.de/english/schutzverband.html, printed 060214.
29. http://www.adler-schinken.de/start.htm, printed 60214.
30. http://www.schinken-wein.de/content/index.html, printed 060214.
31. http://www.pfau-schinken.de/komplett-rahmen.html, printed 060214.
32. http://www.tannenhof-schinken.de/english/produktion.html, printed 060214.
33. http://www.schwarzwaelder-schinken-verband.de/herstellung/index.html, printed 060214.

40
Asian Products

Guang-Hong Zhou and Gai-Ming Zhao

INTRODUCTION

Asia has a long and glorious history with splendid dietary culture. Many traditional ripened meat products were developed in Asia, among which Jinhua ham is the most famous in the world. Jinhua ham has attractive color, unique flavor, and bamboo-leaflike shape. Its roselike muscle, golden-yellow skin, and pure white fat make Jinhua ham one of the most preferred items in Chinese cuisine. The processing technology of Jinhua ham was introduced to European countries by the famous Italian traveler Marco Polo in the 13th to 14th centuries and had an important impact on dry-cured ham processing technology in the world, especially on Parma ham of Italy. In fact, the processing technologies of most dry-cured hams in China are evolved from Jinhua ham. In this chapter, the traits and traditional processing technology of Jinhua ham are introduced.

HISTORY AND TRAITS OF JINHUA HAM

Jinhua ham is formed and produced in Jinhua District, Zhejiang Province of China. The earliest legend regarding the Jinhua ham processing method may be traced back to the Tang Dynasty (A.D. 618–907). However, *Jinhua ham* was formally named by the first emperor of the South Song Dynasty about 800 years ago (Wu and Shun 1959).

Typical Jinhua ham processing generally takes 8–10 months, starting from winter and finishing in autumn of the next year. After a long ripening process, muscle protein and fat are hydrolyzed to some extent by internal enzymes, and many small peptides, free amino acids, free fatty acids, and volatiles are produced, which eventually contribute to the unique flavor of Jinhua ham (Zhao et al. 2005a,b,c; Du and Ahn 2001; Huan et al. 2005). For example, the content of the free amino acids in final Jinhua ham products is 14 to 16 times that in green ham, among which, Leu, Ile, Phe, Glu, Ser, and Lys are 30 times higher (Zhao et al. 2005a). There are 168 volatile compounds identified in the biceps femoris from Jinhua ham using SPME coupled with GC-MS, including hydrocarbons, aldehydes, ketones, esters, alcohols, carboxylic acid, heterocyclic compounds, lactones, and sulphur compounds, etc. (Zhao 2004; Huan 2005).

PROCESSING OF JINHUA HAM

Processing of Jinhua ham consists of six stages: green ham preparing, salting, washing, sun drying and shaping, ripening, and postripening. There are more than 90 steps. The flow route of Jinhua ham processing is shown in Figure 40.1.

Jinhua ham is traditionally processed under natural conditions. Its unique quality is due not only to elaborate processing technology, but also related to the unique local geographic terrain and climate. About 70% of Jinhua District is a mountainous region with four distinct seasons; air temperature goes up and down regularly without tremendous

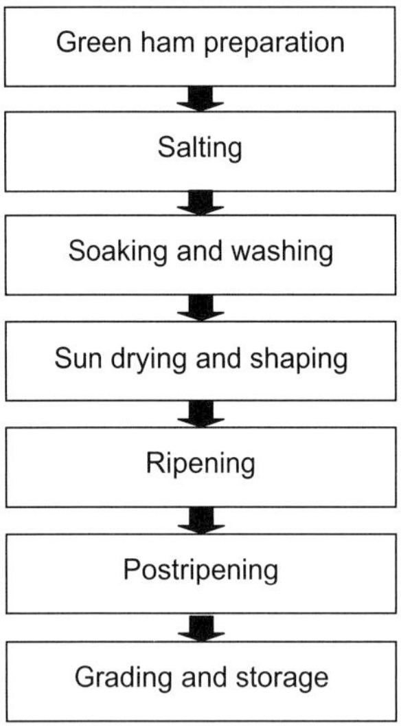

Figure 40.1. Flow route of Jinhua ham processing.

fluctuation, which is desirable for processing dry-cured meat products. Usually, the room temperature in winter is from 0–10°C, favorable for ham salting. In spring, air temperature goes up to 20°C or higher, suitable for sun drying. The summer in Jinhua District is usually very hot. The temperature can go up to 40°C, ideal for ham ripening. Postripening begins in autumn when temperature falls. The whole processing of Jinhua ham, beginning in winter and finishing in the next autumn, takes 8–10 months.

Green Ham Preparation

Traditionally, only hindlegs from the Jinhua "Liangtouwu" pig or its cross-offspring could be used for producing Jinhua ham. Desirable legs should be fresh, with thin skin, slim shank-bone, and well-developed muscle; and covered by a thin layer of white fat. Broken bone should be particularly avoided. The exposed part of bones, as well as the fat, tendon, and muscle membrane on the meat surface of selected legs are cut off and then trimmed into a shape like a bamboo leaf. The remaining blood should be squeezed out. A leg weight between 5.5 and 7.5 kg after trimming is preferred.

Salting

Salting is a critical stage in Jinhua ham processing, and inappropriate salting may cause ham spoilage. The ambient temperature and humidity have great effects on the salting process. With regard to the temperature, salt has difficulty penetrating meat when it is below 0°C, whereas the fast growth of microbes will occur when temperature is above 15°C. According to the experience, undesirable water loss will happen, which causes insufficient salt penetration when the ambient humidity is below 70% RH. Otherwise, much of the salt will flow away in the form of brine, causing pastiness on the ham surface when the humidity is above 90%. So, the desirable ambient temperature and humidity for salting is 5–10°C and 75%–85% RH, respectively.

Average salting duration is about 30 days, varying from 25 days for small hams (less than 5 kg) and 35 days for big hams (more than 8 kg), during which each ham is salted 5–7 times. Usually no nitrate or nitrite is used.

The first salting is also called *drain salt*. The aim is to remove the remaining blood and superabundant water from the muscle and restrain microbe intrusion. The first salting requires the entire meat surface to be covered with a thin layer of salt. Where bones and main blood vessels are located, a little more salt should be used. As a result of salting, part of the water-soluble proteins, the remaining blood, and some water seep out of the meat surface, which is favorable for growth of microorganisms. So the second salting should start no later than 24 hours after the first salting. It requires the entire meat surface to be covered with a thick layer of salt. The quantity of salt used this time is the largest among all the salting procedures. The following salting processes are carried out at 5-day intervals for the purpose of supplementary salt in the ham. After 5–7 times of salting, the muscle normally becomes firm and the color of the meat surface becomes dark red, indicating that the ham has been well salted.

Soaking and Washing

The purpose of this procedure is to remove superfluous salt and wash off dirty substances on the meat surface. Usually, hams are initially soaked in water for 4–6 hours and then washed with bamboo brushes. Water in the pools is changed after washing, and hams are soaked again in water for another 16–18 hours.

Sun Drying and Shaping

The objective of sun drying is to achieve appropriate dehydration because insufficient dehydration may cause ham spoilage. To keep balance, a pair of hams of similar weight are tied by a rope and hung from a rack. Positioning is such that each ham gets enough sunshine and sound ventilation. When hams are hung, the husk of the hoof is removed, water and dirt on the skin is razed off, and the brand is sealed on the skin. Then hams can be dismounted from racks for shaping. First, the ankle joint is straightened, and then the claw is pressed to curve inside; finally, the two sides of the ham are pressed toward the middle to make a bamboo leaflike shape. After shaping, hams are hung up in racks again. Sun drying can be terminated when hams start to drop oil. It generally requires about 7 sunny days. If several continuous rainy days are incurred during this period, a layer of yellow and sticky substances formed on the meat surface should be washed off with water on a sunny day. The remaining hair on the skin is usually burnt off with flame before dismounting the hams from racks.

Ripening

Ripening is the key process for the generation of ham flavor substances. During this period, muscle proteins and fat are hydrolyzed to some degree, mainly by an endogenous enzyme, which results in a great amount of peptides, free amino acids, and free fatty acids. The products constitute the main part of ham taste substances and may continue to react with each other or be hydrolyzed to produce volatiles that contribute to the unique aroma of Jinhua ham.

Ripening rooms are usually on the second floor. When transferred to the ripening room, ham pairs are fastened to the centipede rack (named for its centipedelike shape) with the meat surface toward the windows. Normally, various molds begin to grow 20 to 30 days later and the meat surfaces of the hams are gradually covered by several dominant molds. It was reported (Toldrá and Etherington 1988; Molina and Toldrá 1992; Sánchez and Crespo 2000; Zhao 2004; Huan et al. 2005) that mold growth is not the main contributor to ham flavor. However, the growth of some dominant molds may reflect a ham-ripening condition in Jinhua ham processing. Normally, when gray and greenish molds dominate, it indicates desirable muscle salt content and water activity (a_w). White mold domination indicates higher and undesirable water content and/or insufficient salt concentration in the muscle. No mold growth indicates too much salt in the muscle, which will make it difficult for the hams to develop sufficient aroma and easy to grow worms (Wu and Shun 1959; Gong 1987).

Ham quality is susceptible to ripening room microclimate. High temperature with low humidity stimulates weight loss and fat oxidation, and with high humidity may result in ham spoilage. On the other hand, low temperature, especially in the later phase of ripening, slows aroma formation in hams. Usually, the ripening room should be well ventilated, room temperature should increase from 15°C to 37°C gradually, and humidity should be controlled within 55% and 75%.

During ripening, skin and muscle shrink to some extent because of loss of water, and bones around joints protrude out of the meat surface. Therefore, hams are usually dismounted from the centipede rack and retrimmed in mid-April. This is normally the last shaping, and it requires cutting off protruding bones, superfluous skin, and fat. After reshaping, hams are retied to the centipede rack for ripening. Usually, the ripening process terminates in mid-August when the temperature begins to drop.

Postripening

After ripening, the meat surface becomes very dry and often covered with a thin layer of mold spores and dirt. So hams are first brushed clean and daubed with a thin layer of vegetable oil to soften the muscle and prevent excessive fat oxidation; then they are piled with skin side up for postripening. The postripening process is a process to stabilize and intensify the ham flavor; it is carried out in a warehouse and usually takes 2 months. During postripening, ham piles need turning over from time to time in case of unexpected fermentation that affects ham quality.

Grading and Storage

It's a common practice to grade Jinhua ham into different categories according to its quality, mainly depending on its intensity of aroma. The government authorized panel judges a ham's aroma by inserting a bamboo prod into a ham and smelling it when it's pulled out. There are three fixed locations: namely the "up," "middle," and "lower" positions on a ham for this special purpose. Jinhua ham can be stored for years, and it has the best flavor when stored for around 12 months.

REFERENCES

M Du, DU Ahn. 2001. Volatile substances of Chinese traditional Jinhua ham and Cantonese sausage. J Food Sci 66:827–831.

YL Gong. 1987. Jinhua ham Processing Technology. 1st ed. Beijing: Popular Science Press, pp. 4–22, 73–75.

YJ Huan. 2005. Studies on the forming mechanism of flavor compounds originated from lipid during processing of Jinhua ham. PhD dissertation, Nanjing Agricultural University, Nanjing, China.

YJ Huan, GH Zhou, GM Zhao, XL Xu, ZQ Peng. 2005. Changes in flavor compounds of dry-cured Jinhua ham during processing. Meat Sci 71:291–299.

I Molina, F Toldrá. 1992. Detection of proteolytic activity in microorganisms isolated from dry-cured ham. J Food Sci 57:1308–1310.

FJC Sánchez, FL Crespo. 2000. Influence of molds on flavor quality of Spanish ham. J Muscle Foods 11:247–259.

F Toldrá, DJ Etherington. 1988. Examination of cathepsins B, D, H and L activities in dry-cures hams. Meat Sci 59:531–538.

AF Wu, CY Shun. 1959. Jinhua Ham. 1st ed. Beijing: China Light Industry, pp. 3.

GM Zhao. 2004. Studies on the effects of muscle proteolytic enzymes in the processing of Traditional Jinhua ham. 2004. PhD dissertation, Nanjing Agricultural University, Nanjing, China.

GM Zhao, GH Zhou, W Tian, XL Xu, YL Wang, X Luo. 2005a. Changes of alanyl amino peptidase activity and free amino acid contents in biceps femoris during processing of Jinhua ham. Meat Sci 71: 612–619.

GM Zhao, GH Zhou, YL Wang, XL Xu, YJ Huan, JQ Wu. 2005c. Time-related changes in cathepsin B and L activities during processing of Jinhua ham as a function of pH, salt and temperature. Meat Sci 70: 381–388.

GM Zhao, GH Zhou, XL Xu, ZQ Peng, YJ Huan, ZM Jing, MW Chen. 2005b. Studies on time-related changes of dipeptidyl peptidase during processing of Jinhua ham using response surface methodology. Meat Sci 69:165–174.

SW Zhu. 1993. Study on Taste and Taste Substances of Jinhua Ham. Food Sci (Chinese) 159:8–11.

Part X

Biological and Chemical Safety of Fermented Meat Products

41
Spoilage Microorganisms: Risks and Control

Jean Labadie

INTRODUCTION

In Europe, the most popular dry-cured sausages are salamis and salami-type sausages. Although such products are, according to the customs of the countries, highly variable in terms of flavors, texture, color and acidity, they are always the result of fermentations by two groups of bacterial species, which are included in the *Lactobacillus* and *Staphylococcus* genera. These species are mainly *Lactobacillus sakei*, *Lactobacillus pantarum*, *Staphylococcus saprophyticus*, *Staphylococcus xylosus*, *Staphylococcus warneri*, and *Staphylococcus equorum* (Gevers et al. 2003; Fontana et al. 2005; Corbière et al. 2006). All these bacteria, which are now well identified by many taxonomical approaches, including the most up-to-date molecular techniques, principally degrade the sugars (glucose or saccharose) added to the raw sausage mixture at the beginning of the processing. Such fermentations largely contribute to the pH shift observed early in the batter and to the more-or-less acidic pH of the finished products. In terms of microbiology, the consequence of a rapid acidification inside a meat mixture is the selection of a few microbial species well fitted to such conditions that are able to survive and/or to grow. Additionally, because salamis are always progressively dried during the course of their ripening, the water activities (a_w) rapidly reach levels (0.7–0.8), which constitute insuperable hurdles for most of the microorganisms not involved in fermentations (Leistner 2000). Hence, spoilages are fortunately rare and they are always the consequence of defective manufacturing practices. This chapter describes these defects, the spoilages, their consequences, and in each case the solutions proposed to limit their risk.

STICKY OR "SLIMY" SAUSAGES

These sausages result from an overgrowth of microorganisms on casings. The surface microflora is clearly visible very early, noticeably during the course of initial fermentation when the relative humidity of the fermentation chamber is high and favorable to the microbial growth. Two factors explain the defect: the high temperature inside the fermentation room and a film of liquid water clearly visible on the surface of the sausages. The microorganisms involved include yeasts, molds, and many bacteria that grow heavily and form thick and sticky biofilms. Ammonia flavors are almost always thoroughly produced and easily noticed. The aspect of the biofilms are reminiscent of those observed inside water pipes; they are grey, slimy, and have a strong smell. *Pseudomonas* spp. constitute the dominant part of the flora isolated from these biofilms (Fournaud 1976; Migaud and Frentz 1978). The thick biofilms grown on the surface of casings threaten sausage manufacturers because they slow down and limit the dehydration inside the ripening chambers. The color of these sausages rapidly turned to yellow, green, or a combination of these colors. The formation of the biofilms could also be observed when processing is correctly managed, but in this case it is observed later, noticeably during the ripening phase. The defect is generally the result of a malfunctioning of the drying room, but in this condition the biofilms are not so pronounced and more progressively formed, because of a lower temperature. The microorganisms involved are resistant to lower a_w values; they are different from those isolated from biofilms observed at the beginning of fermentations. They are mostly yeasts and *Staphylococci*.

The odors of slimy sausages are so pronounced that they could be compared to some French soft cheese like Epoisse or Livarot. These odors result from the activity of numerous microbial enzymes, proteases, peptidases, and desaminases. Many bacteria are involved, among which *Pseudomonas* spp., *Brevibacterium* spp., and *Corynebacterium* spp. are the most important.

Slimy sausages are also produced when the following occurs:

- Products are dried at too low temperatures.
- Water pellicles are observed on casings. They result of rapid temperature and RH (relative humidity) shifts.
- Water condensations exist inside packaging.

Solutions that could be used to avoid the production of sticky sausage are easy to set up: drying temperature higher than 12°C, increasing the air velocity inside the drying chamber, and packaging the finished products in a room of similar temperature.

MOLDY SAUSAGES

Moldy sausages are not spoiled inside the meat and fat mixture, because molds are present only on the casings. Moreover, as their growth on casings is demanded by consumers of Romanic (France, Italy) type sausages (Rödel et al. 1994), it is a difficult task for the producers to avoid an overgrowth of molds on sausages that are stored for several weeks before their sale. In France, to get this "moldy" aspect, the most often inoculated species is *Penicillium nalgiovense*. This mold gives the sausages a white attractive color. However, sometimes other mold and yeast species, also present on casings, grow so heavily that they give the sausages unwanted colors (green or brown, or a mixture of both) which are a cause of rejection by the consumers.

Several solutions could be applied to avoid or to limit the growth of molds. They are briefly listed in the next sections (Sirami 1998).

Elimination of the Molds by Cleaning and Brushing

Eliminating the molds is only a part of the solution because the aspect of the sausages always demands the use of another surface treatment, for instance by using flour, to give the sausages a white color and a smoothness indicating the moldy surfaces of normally ripened products. Unfortunately, these treatments are costly and they almost systematically demand a reintroduction of the sausages in packages that often favor the mold regrowth.

In fact, practically, the best solution is a combination of different treatments.

Inoculation of the Casings with White Molds and Yeasts

This treatment can be efficient, but results are hazardous because microbial growth can be long before a visible, acceptable result. It means that sometimes other wild molds already present in the ripening chambers are able to grow in place of the added strains. In other cases, no mold growth is observed for unexplained reasons.

Increasing the Air Velocity

This solution is difficult to use because intense drying often leads to "crusting" of the sausages. Additionally, air fluxes will have to be avoided from the brushing chamber to the fermentation chamber in order to limit the airborne contaminations.

Fumigation of the Working Plant

This cleaning and disinfection approach is not easy to set up (hazardous for humans) and often inefficient because many places are not cleaned and disinfected, such as crevices, corners, pipes, wood surfaces, etc. (Migaud and Frentz 1978). It is important to understand that disinfection is to be carried out in the first part of the plants, where RH (relative humidity) is still high. The materials used in the drying room are to be cleaned by the same approach or another specific one.

Treatment with Natamycin (Piramicin)

This treatment inhibits the growth of any molds on many food products, including cheese and sausages. It is not convenient for all kind of sausages.

Controlling Relative Humidity

As indicated above, stabilizing sausages with a mold flora biofilm that remains acceptable is a difficult task, because it is always impossible to give all sausages the same aspect—that is, controlling the mold growth in the same way on every sausage (Leistner 1995; Rödel et al. 1994). This is mainly the consequence of the conditions chosen during the ripening phase. These conditions are optimal for fermentations to get tasty, colorful products, but they are also optimal for mold growth on casings. This means that some sausages are inevitably covered with a thick mold biofilm.

Water is the most important parameter for the growth of bacterial biofilm (Migaud and Frentz 1978). It is also true for mold biofilms, the growth of which is favored by increased water quantities in the vicinity of the sausage surface during the course of drying. This water comes from the meat particles of the batter and goes through the air of the drying room by passing across the casings. By speeding up the air velocity inside fermentation rooms it is possible to eliminate the surface water of sausages more rapidly; however, crusting frequently results from such operations. That is why other approaches are preferable.

Among the molds sometimes isolated, several *Penicillium* spp. strains producing toxins could grow on the casings (Leistner 1995). Their growth, however, never or rarely leads to toxin production, at least inside the sausages. Moreover, because generally most of the sausages are peeled off before being eaten, the intoxication hazard is limited.

As already outlined, inside the ripening chambers, RH and temperatures are the most important factor to be controlled and monitored (Sirami 1998). During the ripening time, RH must remain below 75% on the sausage casings; molds grow slowly up to an RH of 80%. Avoiding the growth of molds means that the air of the ripening chamber is to be renewed regularly, and that the RH value is below 80% anywhere in the ripening room. Getting homogeneous RH inside the chamber is a difficult task, and great care in the renewing of the air is necessary. The growth of the molds on only one sausage is detrimental to all the sausages present in the ripening chamber. To get good results, manufacturers are generally using very dry air fluxes, which are highly favorable to the production of crusty surfaces. Sausages must also be well separated inside the chamber to avoid air pockets where RH could remain high.

Smoking

A slight smoking every day prevents, delays, or inhibits the mold flora; this explains why smoked sausages of northern Europe are contaminated by molds lately (Leistner 1987).

Antimold Products

Among the efficient products that could be used on sausages, pimaricin, a natural product produced by *Streptomyces natalensis*, is often recommended. During the first 2 weeks, the pimaricin inhibits the growth of molds on the casings, but the efficiency of this antimicrobial molecule decreases rapidly (Sirami 1998). This time is, however, sufficient for products with a short ripening period of not longer than 2 weeks. In case of longer ripening, generally 4 weeks, it seems impossible to avoid the growth of molds because dehydration of the casings considerably limits the action of the piramicin. On German raw sausages, potassium sorbate solutions (20%) are recommended and approved. These solutions are used to avoid the "sausage bloom" sometimes observed on some batches (Leistner 1985).

SOURING

This defect is generally detected in romanic (French, Italian) salamis (Fournaud 1976; Flores and Bermell 1996; Leistner 1985), but it could also be observed sometimes in Hungarian and German salamis heavily contaminated by heterofermentative lactobacilli (Leistner 1995). Generally, it occurs at the end of ripening, and the problem is discovered mostly when the sausages are eaten. Acidic sausages are often observed in the largest pieces of the product and are practically not visible from the outside. Acidic sausages display similar aspects of normal sausages, although some of them could be paler than the others. Sometimes, the absence of molds, which are normally present on casings, is an indication that the sausages could be soured, but many other causes could explain the less abundant growth of molds. However, a combination of properties observed on big sausages, i.e, nude casings and soft texture at the end of ripening, is a good indication that the batters of these big pieces are soured (Migaud and Frentz 1978). After slicing, the batters of soured sausages never show any transition to a coagulated state; moreover, off-flavors reminiscent of lactic acid are detectable. The taste of these products is clearly acidic and even strongly acidic. The color is brown or pale pink on the outside of the sausage, close to the casing. Surprisingly, the surface is a little "crusty," tough, and very dry. The casings are tightly attached to the batter. When analyzed, the water content of these sausages is too high and the pH of the batter is higher in the middle than on the outside of the sausages (Sirami 1998). Additionally, nitrate contents are high in the center of the sliced sausages, whereas no nitrite is observed in the same location. The acidic taste, which is obvious in such products, is generally well correlated to the pH. Because the D lactate is clearly involved in the acidic taste, it is likely that heterofermentative bacteria, at least *Leuconostoc* spp. (*L. mesenteroides*, *L. carnosum*, *L. gelidum*), which only produce D lactate from sugars, are involved in the manufacturing defect. As a matter of fact, the defect is easily reproduced when these bacteria are experimentally introduced in batters (Sirami 1998).

Avoiding the defect is generally difficult because it could arise for unpredictable reasons. However, it is possible to limit the souring by different ways. One of the most efficient approaches is to slow down the fermentation process in order to get batters with higher pH at the end of the ripening time. In order to achieve such pH values, decreasing quantities of sugars should be introduced into the batter, and conditions favoring rapid degradation of these carbohydrates must be favored. Moreover, because the final pH of the product is reached earlier in the fermentation process, a slight increase of its value is generally observed during the ripening phase. To be efficient, this approach demands the use of high microbial quality batters in order to avoid the growth of *Enterobacteriaceae*, which could be less inhibited by the higher pH values achieved at the end of ripening.

Selecting the Starters

Some selected strains degrade the sugars more rapidly than the heterofermentative *Leuconostoc* spp., which are thus inhibited.

Increasing the Salt Content of the Batter up to 35 g/kg

It is first necessary to predry the fats and chop up meat at low temperatures. Such operations are carried out in cold rooms at −1/−3°C for 24 hours. They are, however, responsible for weight losses that could be as large as 10% of the total weight of the products.

Decreasing the Fermentation Temperatures

Low temperatures improve the biochemical activities involved in aroma production, particularly in the *Staphylococci* (*S. xylosus*, *S. warneri*, *S. carnosus*). These species thus produce less acid and more aromatic compounds.

Optimizing the Processing to Avoid Crusting and Biofilm Growth

It is difficult to decrease simultaneously the sugar content of batters and the fermentation temperature without favoring some spoilage microorganisms, for instance, *Enterobacteriaceae*. In this case, it is necessary to work with meat of high microbiological standards.

ROTTEN SAUSAGES

This defect is fortunately rare in industrial productions. It is, however, not so uncommon in small workshops of southern Europe, which are producing sausages without starters. In these workshops, the lactic microflora responsible for fermentations are only those present inside the processed chopped meat and fat mixture and on the natural casings generally used (Fournaud 1976; Migaud and Frentz 1978; Sirami 1998). Two parameters are responsible for the defect. First, acidification inside fermentation chambers is weak. The final pH of the finished products is too high (pH 6.2). Second, the raw products that are used to prepare the batters are sometimes of low microbiological quality—that is, high numbers of *Pseudomonas* spp. and *Enterobacteriaceae* are observed ($>10^5$/g) on meat and fat as well. The sausages produced with such quantities of spoilage organisms become rapidly full of crevices that contain different glazes. They look like balloons. Chopping liberates spoiled, putrid odors. These sausages are obviously not consumable. As indicated above, this defect is caused by the use of low-quality raw materials, but also by a lack of hygiene during the course of processing. The solutions to this problem are clear, i.e., the choice of high-quality meat and fat, and the respect of good manufacturing practices on the processing lines.

DISCOLORED SAUSAGES

This problem is basically due to defective fermentation processes. Very often, the origin is a lack of favorable bacterial species and particularly those producing the microbial enzymes involved in color formation (Hechelman 1985; Leistner 1985; Lücke and Hechelman 1987; Sirami 1998). These enzymes are the nitrate reductase of *Staphylococci* or sometimes *Enterobacteriaceae*. However, as *Enterobacteriaceae* are often involved in the production of biogenic amines, meat containing too many of these microorganisms should be eliminated. A selection of starters with efficient nitrate-reductase activities, among which staphylococci are the most interesting, is certainly a better solution.

No, or poor, reductase activities could also result in a very rapid acidification of the batter, which often inhibits the growth of staphylococci. Lactic acid bacteria (mainly *L sakei*, *L. curvatus*, and *L. plantarum*) which are responsible of the pH reduction, are not inhibited and could also participate to the inhibition observed by other mechanisms. The sausages with this defect are grey-brown or brown after chopping. The discoloration is often observed when the quantities of sugars added to the raw sausage mix are large (15 g/kg or more). Generally, the sugars are thoroughly degraded so that the pH of the batter drops almost immediately; it reaches, generally within 48 hours, values that are lower than pH 5.

Discoloration of sausages could also be the result of peroxide production by greening lactic acid bacteria, almost always present in raw sausage mixes. Peroxides, which are normally present in low quantities at the end of the ripening, are still present in these sausages, because the *Staphylococci* which normally produce the enzymes (catalases and peroxydases) involved in their destruction are inhibited by the lactic acid bacteria. Such fermentations are generally observed when high a_w (0.96) values and high temperature (25°C) are combined. The solution to this problem is simple, although not always easy to apply: decreasing the a_w of the batter before stuffing casings and decreasing the temperature inside fermentation chambers.

HOLLOW SAUSAGES

Hollow sausages are generally observed in large-diameter, heavy sausages, although the problem could be observed sometimes inside small pieces (Fournaud 1976; Migaud and Frentz 1978; Sirami 1998). The hollows are located inside the upper part of the sausage, close to the hanging string. The result of such hollows is that fat rapidly becomes rancid, with fat and lean pieces of the batter not tightly cemented in the same coagulated structure. The pieces of meat are less colored and more-or-less large crevices are observed between fat and meat, inside the sausage mix. Inside crevices, meat pieces appear dark-red or red-brown, and fats are yellow-orange. These crevices grow with time because fermentations due to spoilage bacteria inevitably occur in the increasing hollows. Molds like *Mucor* or *Penicillium*, heterofermentative lactic acid bacteria like *Leuconostoc* spp., and spoilage bacteria such as *Pseudomonas* spp. and *Enterobacteriaceae* (Hechelman 1985) often form slimy biofilms inside the hollows.

Sometimes hollows are observed inside sausages of small sizes. They are due to a defective coagulation of proteins by salt, or to the use of partly degraded muscle proteins in the sausage mix. The microorganisms growing inside these sausages are the same as observed in bigger pieces.

This problem is difficult to detect, but it could be definitely eliminated by hanging the big sausages inside nets.

CRUSTY SAUSAGES

This defect is one of the most frequently observed in industrial workshops. The sliced sausages display discolored zones. From the outside, close to the casings, a dark zone of several mm thickness is formed. This crusty surface, which is generally very tough, slows down (Rödel et al. 1994) the dehydration of sausages during the ripening. The phenomenon could have two main origins:

- Dehydration is too fast; the water that, from the inner parts of the sausages reaches the surface and evaporates, remains inside the core of the batter.
- An overgrowth of molds increases the speed of dehydration. As a result, a crust is formed under the mold biofilm.

Molds consume nutrients and water on the surface of the sausages, which become crusty. These sausages remain soft, and they dry very slowly. Acidification is slowed and the batter is destructured. As molds are consuming the lactic acid produced by fermentation, the taste of these sausages is neutral and the slices rapidly lose their color in the air. The sausages are circled by a dry surface. If the problem is identified during the course of ripening, it is possible to wash the surfaces with salted water in order to make possible a further ripening.

CONCLUSION

Most of the batters used to process dry-cured sausages contain high counts of microorganisms, which are numerous species of bacteria, molds, and yeasts. Fortunately in most cases, these organisms are not harmful and they generally grow very slowly or are progressively eliminated inside the sausages during fermentations and ripening. Thus, spoiled fermented sausages are very uncommon, but they could be observed in peculiar conditions described in this chapter. As shown, spoilage generally results from processing defects in which the normally two insuperable hurdles, i.e., acidification and drying, are either too pronounced or insufficient. Finally, to avoid the defects due to spoilage in dry-fermented sausages, it is important to set up rapid and efficient corrective measures during the processing and good manufacturing practices along the fermentation and drying process.

REFERENCES

S Corbière Morot-Bizot, S Leroy, R Talon R. 2006. Staphylococcal community of a small unit manufacturing traditional dry fermented sausages. International J Food Microbiol 108(2):210–217.

J Flores, S Bermell. 1996. Dry-cured sausages factors influencing souring and their consequences. Fleischwirtschaft 76 (2):163–165.

C Fontana, P Sandro Coconcelli, G Vignolo. 2005. Monitoring the bacterial population dynamics during fermentation of artisanal Argentinean sausages.Inter J Food Microbiol 103(2):131–42.

J Fournaud. 1976. La microbiologie du saucisson sec. L'Alimentation et la Vie 64:2–3.

D Gevers, M Danielsen, G Huys, J Swings. 2003. Molecular characterization of tet(M) genes in *Lactobacillus* isolates from different types of fermented dry sausage. Appl Environ Microbiol 69(2): 1270–1275.

H Hechelman. 1985. Mikrobell verursachte fehlfabrikate bei Rohwurst und rohschinken. In: Mikrobiologie und Qualitât von Rohwurst und Roschinken. Herausgegeben vom Institût fûr Mikrobiologie, Toxikologie und Histologie der Bundesanstadt fûr Fleishforshung, Kulmbach, pp. 103–127.

L Leistner. 1985. Empfehlungen fûr sichere produkte. In: Mikrobiologie und Qualitât von Rohwurst und Roschinken. Herausgegeben vom Institût fûr Mikrobiologie, Toxikologie und Histologie der Bundesanstadt fûr Fleishforshung, Kulmbach, pp. 219–244.

———. 1987. Fermented meat products. Proc 33rd Int Cong Meat Sci Technol August 2–7, Helsinki. Finland, pp. 323–326.

———. 1995. Stable and safe fermented sausages worldwide. In: Fermented Meats. G. Campbell-Platt, P. Cook, eds. England: Blackie Academic and Professional, pp. 160–175

———. 2000. Basic aspects of Food preservation by hurdle technology. Inter J Food Microbiol 55: 181–186.

FK Lücke, H Hechelman. 1987. Starter cultures for dry sausage and raw ham: Composition and effect. Fleischwirtschaft 67:307–314.

M Migaud, JC Frentz. 1978. In: La Charcuterie Crue et les Produits Saumurés, Ch 47, Les défauts de fabrication. Orly France: Soussana 12 rue des lances, pp. 224–239.

W Rödel, A Stiebing, L Krockel. 1994. Ripening parameters for traditionnal dry sausages with a mould covering. Fleishwirtschaft 73:848–853.

J Sirami. 1998. Saucisson sec: Défauts de fabrication. In:L'Encyclopédie de la Charcuterie, Ch 28. Défauts de Fabrication. Frentz et al., eds. Orly, France: Soussana 12 rue des lances, pp. 1122–1124.

42
Pathogens: Risks and Control

Panagiotis Skandamis and George-John E. Nychas

INTRODUCTION

Preservation of meat by fermentation and drying has been used for hundreds of years. For a long time, the technology of fermentation has been considered an art; however, more recently the process has been studied, and as a result, products of quality can now be repeatedly produced under controlled conditions. Dry or semidry-fermented sausages are prepared by mixing ground meat with various combinations of spices, flavorings, salt, sugar, additives, and bacterial cultures. The mixtures, in bulk or after stuffing, are allowed to ferment at different temperatures for varying periods of time. Following fermentation, the product may be smoked and/or dried under controlled conditions of temperature and relative humidity.

Today, these meat products are part of the daily diet in many countries all over the world; recently they have become fashionable food products, and in urban centers their market has been increasing significantly. In general, fermented products are considered safe foods due to the pH drop and, in the case of meat fermentation, simultaneous reduction of water activity during the transformation of raw tissue (muscle) to final (fermented) product. Although both these factors are considered to restrict detrimental parameters that cause growth of pathogenics, their presence has been previously reported in different stages of production as well as at the retail level (Ananou et al. 2005; Samelis and Metaxopoulos 1999; Thévenot et al. 2005a). Different surveys have revealed the presence of *Listeria monocytogenes*, *Escherichia coli*, *Clostridium* spp., and *Staphylococcus aureus* not only in the final products, e.g., sausages, but also in different production levels. This can be due either to frequently contaminated raw material, where there is a probability that some of the pathogenic organisms could cross the antimicrobial barriers imposed during processing, or to cross-contamination at any stage of retailing. In both cases, they may be present in the final product, causing a concern for the producers and for those responsible for public health, which has been the topic of study of several research groups (Moore 2004).

The risk analysis assessment related to pathogenic bacteria in sausages is based on the following steps:

1. *Hazard identification*, where all pathogenic agents capable of causing adverse health effects and which may be present in sausages are identified.
2. *Hazard characterization*, the qualitative and/or quantitative evaluation of the nature of the adverse health effects associated with the above-mentioned bacteria present in sausages. This step requires a dose-response assessment for chemical agents, and if the data are obtainable, this should also be attempted for biological or physical agents.
3. *Exposure assessment*, the qualitative and/or quantitative evaluation of the likely intake of pathogenic agents via sausages, as well as exposures to other sources if relevant.
4. *Risk characterization*, the qualitative and/or quantitative estimation, including uncertainties around the model parameters, of the probability

of occurrence and severity of known or potential adverse health effects in a given population based on the previous three steps.

HAZARD IDENTIFICATION

It is generally accepted (Nychas et al. 2007) that, with the exception of external surfaces of the animal and the gastrointestinal and respiratory tracts, the tissue of healthy living food animals is essentially free of microorganisms—or at least they are undetectable or at extremely low populations (Mackey and Derrick 1979). Once the animal is slaughtered, however, the internal defense mechanisms (e.g., lysozyme, antimicrobial peptides) that combat infectious agents in the living body are lost. Thus the resulting meat becomes exposed to increasing levels of contamination and, depending on various extrinsic parameters (temperature, packaging, processing method, etc.), may undergo rapid microbial decay. Unless effectively controlled, the slaughtering process may lead to extensive contamination of the exposed cut surfaces of muscle tissue with a vast array of gram-negative and gram-positive bacteria as well as fungi (Nychas et al. 2007). Sources of these microorganisms include the external surfaces of the animal, including the gastrointestinal tract, as well as the environment (e.g., floors, contact surfaces, knives, hands, etc.) with which the animal had contact at some time before or during slaughter (Nychas et al. 2007). Many slaughter procedures contribute to the contamination as well, i.e., during removal of the hide/fleece of cattle/sheep, during dehairing of pigskin, and during evisceration (Grau 1986). As a consequence, pathogenic bacteria are present on the meat products thereof. For example, testing of the microbiological quality of ready-to-eat dried and fermented meat and meat products performed on 2,981 samples sold at retail in the U.K., Wales, and Northern Ireland revealed that *Staphylococcus aureus* and Salmonella were present at 1.18% and 0.06%, respectively (Little et al. 1998). In processed meat products, including ready-to-eat food, the contamination ranges from 8–92% (Borges et al. 1999). In a similar survey in Italy, 5.3% of samples were found to be positive for Salmonella; in the U.S., it was found that the prevalence in small-diameter cooked sausages was 0.20% and the cumulative 3-year *Salmonella* prevalence for dry and semidry-fermented sausages was 1.43% (Levine et al. 2001).

For *L. monocytogenes* in dry and semidry-fermented sausages, in the same study, the prevalence was 3.25%; none of the RTE products tested positive for *E. coli* O157:H7 or staphylococcal enterotoxins. However, a higher incidence of *L. innocua* in meat products, compared to *L. monocytogenes*, was reported (Borges et al. 1999). In salami, the occurrence of *L. monocytogenes* varies from 5–23% (Borges et al. 1999). Salami does not undergo heat treatment and are fermented under variable temperatures (mostly between 25–30°C, according to the processing method adopted by the manufacturer), having a final pH between 4.8 and 5.2 and water activity around 0.85–0.90 (Borges et al. 1999). The fermentation process is also variable and can be conducted by the meat natural flora or by the addition of lactic acid starter cultures. The characteristics of these products make them susceptible to the survival of *L. monocytogenes* and, therefore, there is a potential risk to the consumer's health. Despite the importance of *L. monocytogenes* in meat products, there are only a few studies on the occurrence of this microorganism in fermented sausage.

In addition to *Salmonella*, *Staphylococcus aureus*, and *Clostridium botulinum*, the microorganisms mainly studied in these products in the past decades (Tompkin et al. 1974), *L. monocytogenes* and *Escherichia coli* O157:H7, are the predominant pathogens currently under study (Hew et al. 2005; Naim et al. 2004; Pond et al. 2001; Samelis et al. 2005).

HAZARD CHARACTERIZATION—DEFENSE MECHANISM

Although raw meat is an ideal medium for the growth of many organisms due to richness in low molecular compounds—e.g., fermentable carbohydrates, proteins, peptides, amino acids, and other plentiful growth factors such as vitamins and minerals—meat fermented products have a less favorable matrix for bacterial growth, due to dynamic changes of pH and a_w as a consequence of the development of a microbial association dominated by lactic acid bacteria. Indeed, the pH drop, reduction in a_w values, and bacterial activity (e.g., competition and production of specific metabolites) are the important factors in the control of the development or die-off of pathogenic organisms. However, many pathogens such as *L. monocytogenes*, *E. coli* O157:H7, *Salmonella* spp., and *Staphylococcus aureus* have been reported to survive or even grow in such unfavorable conditions (Table 42.1). The ability of these pathogens to survive in acidic foods has been recently studied. It needs to be noted, however, that the effect of structure and the physicochemical characteristics (buffering capacity, local pH gradients, and nutrient availability or diffusion etc.) of the fermented sausages have not

Table 42.1. Pathogenic bacteria commonly found on meats.

		Type of Muscle Meat	
Bacteria	Gram Rxn	Fresh	Processed
Bacillus	+	X	X
Campylobacter	–	X	
Clostridium	+	X	
Escherichia	–	X	
Klebsiella	–	X	
Kluyvera	–	X	
Listeria	+	X	X
Pseudomonas	–	XX	X
Salmonella	–	X	
Staphylococcus	+	X	X
Vibrio	–	X	
Yersinia	–	X	

X = known to occur; XX = most frequently isolated. Based on Nychas et al. 2007.

been taken into account in microbial safety risk assessment of these products.

Escherichia coli

Strains of the bacterium *E. coli* capable of producing certain cytotoxins are reported as Verotoxigenic *Escherichia coli* (VTEC). Enterohemorrhagic *E. coli*, commonly referred to as EHEC, are a subset of the VTEC, harboring additional pathogenic factors. More than 150 different serotypes of VTEC have been associated with human illness; however, the majority of reported outbreaks and sporadic cases of VTEC infections have been attributed to serotype O157. There is a wide spectrum of symptoms associated with VTEC infections, ranging from mild to bloody diarrhea, often accompanied by severe abdominal cramps but usually without fever. VTEC infection can also result in hemolytic uremic syndrome (HUS). HUS is characterized by acute renal failure, anemia, and lowered platelet counts. HUS develops in up to 10% of patients infected with VTEC O157 and is the leading cause of acute renal failure in young children.

Listeria monocytogenes

The genus *Listeriae* comprises six species, but human cases are almost exclusively caused by the species *Listeria monocytogenes*. *Listeriae* are ubiquitous organisms, which are widely distributed in the environment. In humans, infections most often affect the pregnant uterus, the central nervous system, or the bloodstream. Symptoms vary, ranging from mild flulike symptoms and diarrhea to life-threatening infections characterized by septicemia and meningencephalitis. In pregnant women, the infection spreads to the fetus, which will either be born severely ill or die in the uterus, resulting in abortion. Although human infections are rare, it is worth to note that is considered to be of the high mortality type.

Listeriosis is an important disease in Europe due to high morbidity and mortality in vulnerable populations, although it remains a relatively rare disease in the EU. Meat and meat products (e.g., sausages) that are contaminated with more than 100 *L. monocytogenes* bacteria per gram, and that are to be consumed without further heat treatment, are considered to form a direct risk to human health. These food categories (ready-to-eat meat products) have been typically identified as risk products for contamination with *Listeria*. Member states in the European Community have been instructed to report the presence of *Listeria*. So far comparison between data from different countries is difficult due to differences of sample sizes and testing protocols. However, it is generally considered that concentrations of *L. monocytogenes* greater than 100 cfu/g are required to cause human disease in healthy populations; therefore, qualitative results alone are not necessarily an indicator of risk (EU Regulation 2073/2005).

Salmonella spp.

Salmonella has long been recognized as the most important zoonotic pathogen of economic significance in animals and humans.

Human salmonellosis is usually characterized by acute onset of fever, abdominal pain, nausea, and sometimes vomiting. Symptoms are usually mild and most infections are self-limiting, lasting a few days. However, in some patients, the infection may be more serious and the associated dehydration can become life threatening. In these cases, as well as when *Salmonella* causes bloodstream infection, effective antimicrobials are essential for treatment. Salmonellosis has also been associated with long-term and sometimes chronic sequelae, e.g., reactive arthritis. There are numerous foodborne sources of *Salmonella*, including a wide range of domestic and wild animals and variety of foodstuffs (meat and meat products, including sausages). Transmission often occurs when organisms are introduced in food preparation areas and are allowed to multiply in food, e.g., due to inadequate storage temperatures or because of inadequate cooking or cross-contamination of cooked food. The organism may also be transmitted through direct contact with infected animals and fecally contaminated environments.

Staphylococcus aureus

Staphylococcus aureus is of significant practical importance in meat. Not only can this bacterium cause a variety of infections in meat animals, as well as in humans, but it is also the causative agent of a major form of food poisoning. For example, in the U.S. during the period 1975–1979, 540 food-poisoning outbreaks were reported, with *Staph. aureus* responsible for 28% (153 outbreaks). Misuse of foods in food service operations seems to be the major cause of outbreaks, followed by mishandling in the home.

Bad practices during manufacturing, especially poor conditions of fermentation and poor conditions of storage, have led to food-poisoning outbreaks from products, especially those that are consumed raw (Nychas and Arkoudelos 1990; Pullen Genigeorgis 1977). The occasional outbreaks of *Staph. aureus* food poisoning, which appear to be more common in the USA and Canada than in Europe, have led to studies of the ability of *Staph. aureus* to grow in fermented sausages (Nychas and Arkoudelos 1990; Pullen and Genigeorgis 1977; Metaxopoulos et al. 1981a,b).

Illness results from the ingestion of water-soluble, heat-stable enterotoxins (Minor and Marth 1972a,b) secreted by the staphylococcal cells. Secretion of these enterotoxins occurs at different periods of the growth phase, either as primary or secondary metabolites (Minor and Marth 1972a).

Among these many ubiquitous organisms, *Clostridium botulinum* will inevitably be present from time to time (Table 42.1). However in this overview the hazard of *C. botulinum* will not be discussed.

EXPOSURE ASSESSMENT

Although it is well established that the majority of pathogens may gain access to the foods as a result of contamination during processing and storage, raw materials constitute the principal source of hazards (Adams and Mitchell 2002). Meat fermentation may be viewed as a process that takes potentially hazardous raw materials and, through a two-stage process of fermentation and ripening, develops an inhibitory ecosystem for most bacterial pathogens, thereby transforming the perishable and hazardous raw materials into products with extended keeping quality and reduced risk of causing illness. The overall antimicrobial effect elicited in fermented meats is the sum of multiple hurdles of microbial (organic acids, ethanol, and bacteriocins) or nonmicrobial (reduced a_w, nitrites, nitrates, spices, phenolics, and heating temperature) origin. However, the manufacturing of traditional fermented products is characterized by considerable variability in raw materials, operation units, fermentation, and/or ripening conditions. This in turn is reflected in the final pH and a_w of the produced sausages, which have been reported to vary from 4 to 7 and from <0.60 to >0.95, respectively (Little et al. 1998).

Starting in 1950, the outcome of numerous studies was the establishment of a new fermentation technology involving the application of starter cultures comprised of lactic acid bacteria (Incze 1998). Since then, a variety of starter cultures has been evaluated and used in practice, including acidifying cultures, aroma-forming cultures, nitrate-reducing starters and even mold starters (Incze 1998). The application of starter cultures was a "triumphal" step in the manufacturing of fermented products, because it assured safer products, especially those with short ripening time. Among the antimicrobial factors associated with LAB, such as bacteriocins, hydrogen peroxide, CO_2, diacetyl, low redox potential, crowding low pH, and organic acids (Adams and Nicolaides 1997), the main factors are organic acids and low pH. Commonly, the level of organic acids, mainly lactic acid, produced by lactic acid bacteria during fermentation exceeds the 100 mM (1% v/v), whereas the ultimate pH is often in the range of 3.5–4.5. Another feature that is required so that LAB may elicit their antimicrobial effectiveness is their numerical superiority over any pathogens present.

However, despite the increase in safety level conferred by starter cultures and the manufacturing process itself, there are still emerged or "developed" pathogens that are capable of combating the inhibitory environment of fermented meats ecosystems. The dominant pathogens that have been reported to be presently capable of surviving during storage of fermented sausages include *Salmonella*, enterohemorrhagic *Escherichia coli* (EHEC), *Staphylococcus aureus*, and *Listeria monocytogenes* (Incze 1998; Adams and Nikolaides 1997). Therefore, knowledge of the ability of pathogens to survive and proliferate in fermented meats is an important prerequisite in order to achieve the food safety objectives related to these specific food ecosystems in a reliable and consistent manner. Such knowledge may derive from challenge tests of pre- and post-process inoculation of pathogens in the meat batter and final products, respectively. These approaches will allow elucidation of the effect of the manufacturing process as well as the combined effect of storage and product pH and a_w on the inactivation of pathogens (Incze 1998). Moreover, these data allow for the development of predictive models that mathematically describe the inactivation of pathogens in fermented sausages (Pond et al. 2001). However, it is expected that survival responses of pathogens will vary depending not only on the manufacturing process and product characteristics, but also on whether inoculation is carried out in the meat batter or the final products, because the pathogens will be exposed to different stresses. By way of example, Shadbolt et al. (2001) demonstrated that exposure of *E. coli* to lethal pH (pH 3.5) followed by exposure to lethal NaCl (10%, a_w 0.90), a sequence of stresses simulating the order of stresses encountered during fermentation and ripening, is more deleterious than the reversed sequence, or the simultaneous exposure to both stresses, a situation that simulates postfermentation inoculation. Thus, in order to conduct reliable exposure assessment reports related to fermented products, the need for challenge testing is intensive. A review of the reports on the survival/growth of the major pathogens in a variety of fermented dry or semidry sausages is presented in the next paragraphs. It needs to be noted that literature reports have focused considerably on EHEC and *L. monocytogenes* due to the severity of these pathogens and their ability to survive at extreme environmental conditions.

Escherichia coli O157:H7

Enterohemorrhagic *E. coli* is considered highly sensitive to the low pH and a_w of dry-fermented sausages. Moreover, the presence of nitrite further contributes to the reduction of this bacterium. Therefore, once good manufacturing practices are in effect, presence of *E. coli* O157:H7 in final products should be avoided (Incze 1998). However, *E. coli* O157:H7 was implicated in an outbreak involving consumption of presliced dry-fermented salami (CDC 1995a) with an estimated dose of less than 50 cells (Tilden et al. 1996). A year later, another *E. coli* outbreak was linked to mettwurst, an uncooked semidry-fermented sausage in Australia (CDC 1995b). These two outbreaks prompted U.S. food authorities to develop guidelines through which fermented sausage manufacturers were required to exhibit the 5-log unit reduction in the enterohemorrhagic *E. coli* during processing (Reed 1995; Hinkens et al. 1996), which were further established as lethality performance standards (USDA 2001). Following the release of these guidelines, research was focused on finding alternatives in the manufacturing process that will ensure 5-log reductions (Freeman 1996). Among the predominant alternatives was a heating step at 63°C for 4 minutes as well as combined antimicrobial treatments (Hinkens et al. 1996). Ellajosyula et al. (1998) showed that fermentation of Lebanon bologna to pH 4.7 followed by gradual heating to 48.9°C, in 10.5 hours is sufficient to destroy 5-$\log_{10}$ CFU of *E. coli* O157:H7. However, despite the indisputable effectiveness of cooking in reducing pathogens, it is against the traditional character of some dry-fermented sausages, which are known as raw products, whereas in some cases cooking to an internal temperature of 63°C leads to unacceptable products (Calicioglu et al. 1997, 2002). Thus, in order to evaluate the eligibility of the target of 5-log reduction, many challenge studies were performed to validate various manufacturing processes, as well as to ensure that storage of the final products may also contribute to the inactivation of a pre- or postprocess contamination (Table 42.2). Calicioglu et al. (2001) evaluated the survival of *E. coli* O157:H7 during manufacturing of semidry Turkish soudjouk fermented with a commercial *Pediococcus* and *Lactobacillus* starter culture (pH 4.48–4.55, a_w 0.88), or naturally fermented (pH 5.28–5.44, a_w 0.90–0.91), without cooking or cooked at 54°C for 0–60 minutes and subsequently stored under vacuum at 4 or 21°C. They found that cooking reduced *E. coli* O157:H7 to undetectable levels in sausages fermented with starters, while it caused only 2–3 logs in naturally fermented products. At both temperatures, the population of *E. coli* O157:H7 that had survived the manufacturing process (without cooking) was further reduced by 1–4 logs within 28 days of storage, while higher reduction was achieved at 21 compared to 4°C. Higher log reductions of *E. coli* O157:H7 during storage occur in sausages fermented with starter cultures compared to naturally fermented sausages, apparently due to the lower pH and a_w sustained with starters (Calicioglu et al. 2001, 2002). Note, however, that the potential of starters to produce bacteriocin does not influence the gram-negative EHEC, as is the case with gram-positive pathogens, such as L. monocytogenes (Lahti et al. 2001). The majority of studies on prefermentation inoculation of *E. coli* O157:H7 have concluded that the reduction of the pathogen caused by the manufacturing process of fermented sausages, including summer sausage, Turkish soudjouk, and local dry-fermented sausages, with pH 4.5–5.0 and

Table 42.2. Data related to prevalence, survival, and growth of *E. coli* O157:H7 in traditional meat products.

			Process Conditions							Final Product				
a/a	Type of Sausages	Type of Study[§]	Pre-valence	Concen-tration Data	Starter Cultures (Y/N)	Sugars (Y/N)	NO_3^-/ NO_2^- (Y/N)	Fermen-tation T°C	Drying T°C	pH	a_w/ RH %	Conditions of Storage T°C/	Packaging	Pathogen Response
1	Salami (matrix)	C[b]	nd[¥]	6–7 Log_{10} CFU/g	nd	nd	nd	nd	nd	4.7–5.1	nd	nd	nd	D_{50} = 91.14–116.93 min strain dependent, D_{55} = 17–21.9 min strain dependent, D_{60} = .18–2.21 min strain dependent
2	Dry sausage	C[a]	nd	7.3 Log_{10} CFU/g	Y	N	Y	24	14	4.9 ± 0.06	0.90 ± 0.01	nd	nd	2.13 ± 0.08 Log_{10} CFU/g reduction
3	Salami	C[a]	nd	6 Log_{10} CFU/g	Y	Y	Y	18–23	15	4.5–4.7	24.7 ± 1.36	nd	nd	Present at <1 Log_{10} CFU/g after processing
4	Fermented sausages (sliced)	C[b]	nd	10^3–10^6 CFU/ 25 g	nd	nd	nd	nd	nd	4.5	nd	7,22	A	HI: 2 Log_{10} CFU/g reduction after 35d, LI: not detected after 7 or 22d

5	Lebanon bologna luncheon slices	C[b]	nd	7 Log_{10} CFU/g	nd	nd	nd	nd	nd	4.4–4.6	nd	3.6, 13	V	$D_{3.6}$= 21.5d, D_{13}= 12.05d
6	Lebanon bologna (model sausages)	C[a/b]	nd	7.5 Log_{10} CFU/g	Y	nd	nd	37.7	nd	4.4–4.6	nd	3.6	V	$D_{3.6}$= 21.65d
7	Dry sausage	C[a]	nd	4.5 Log_{10} CFU/gg	Y	Y	Y	24	17	5.0–5.2	0.90	nd	nd	~2.5–3.0 Log_{10} CFU/g reduction
8	Turkish soudjouk	C[a/b]	nd	7.65 ± 0.04 Log_{10} CFU/g	Y	Y	Y	24	22	4.55 ± 0.026	0.88 ± 0.1	4,21	V	4.0–7.0 Log_{10} CFU/g reduction depending on cooking
9	Turkish soudjouk	C[a/b]	nd	7.65 ± 0.04 Log_{10} CFU/g	N	N	Y	24	22	5.48 ± 0.011	0.88 ± 0.1	4,22	V	1.7–4.7 Log_{10} CFU/g reduction depending on cooking
10	Dry-fermented sausage	C[a/b]	nd	4.68 ± 0.02 log CFU/g	Y/N	Y	Y	nd	nd	4.4 ± 0.01	nd	4	V	~2 Log_{10} CFU/g reduction
11	Dry-fermented sausage	C[a/b]	nd	2.30–5.68 log CFU/g	Y	Y	N	20–23	17	4.4–4.6	0.76–0.87	15–17	A	HI: 2.79–4.84 Log_{10} CFU/g reduction depending on starter culture, LI: bellow 1 Log_{10} CFU/g after 14 or 21d

(*Continues*)

Table 42.2. (*Continued*)

				Process Conditions						Final Product			
a/a	Type of Sausages	Type of Study§	Pre-valence	Concen-tration Data	Starter Cultures (Y/N)	Sugars (Y/N)	NO_3^-/ NO_2^- (Y/N)	Fermen-tation T°C	Drying T°C	pH	a_w/ RH %	Conditions of Storage T°C/ Packaging	Pathogen Response
12	Turkish soudjouk	C[a/b]	nd	10^5 CFU/g	N	N	N	22–24	22–24	4.86–4.9	31–32.5%	4 A/V	Not detected after 60, 90d
13	Salami (lumberjack beef roll)	C[b]	nd	10^2–10^5 CFU/ml	Y	Y	Y	nd	nd	4.63–4.86	0.90–0.95	5,20 A	Not detected at 20°C, ~3.5 Log_{10} CFU/g reduction depending on inoculum history at 5°C
14	Dry/ semidry sausage	P	0%	Absence in 25g	nd	nd	nd	nd	nd	4.0–6.0	0.85–0.94	nd nd	nd
15	Spreadable fermented sausage	P	0%	Absence in 25g	nd	nd	nd	nd	nd	4.0–6.0	0.85–0.94	nd nd	nd
16	Fermented minced meat, fermented meat cubes	C[a]	nd	10^5 CFU/g	Y	Y	nd	30,37	nd	4.0–4.2	nd	nd nd	Not detected

17	Dry sausage	$C^{a/b}$	nd	10^3–10^7 CFU/g	Y	nd	nd	27	14	nd	nd	4,20	V	HI: detected at 4°C, not detected at 20°C, LI: not detected at 4, 20°C
18	Fermented sausage	S-NP	nd	<10 CFU/g–1.25 Log_{10} CFU/g	N	Y	Y	nd	0–14	5.62–5.70	nd	nd	nd	<10 CFU/g

§P: prevalence study; C^a: preprocessing inoculation (challenge) without storage; C^b: postprocessing inoculation (challenge); $C^{a/b}$: preprocessing inoculation (challenge) and storage; S-NP: survival study on naturally contaminated products.

¥nd: no data, HI: high inoculum; LI: low inoculum.

A: aerobically packaged; V: vacuum packaged.

Based on Duffy et al. (1999); Naim et al. (2006); Samelis et al. (2005); Uyttendaele et al. (2001); Chikthimmah and Knabel (2001); Erkkila et al. (available at 2000); Calicioglou et al. (2001); Glass et al. (1992); Lahti et al. (2001); Cosansu and Ayhan (2000); Rocelle et al. (1996); Little et al. (1998); Sakhare and Rao (2003); Nissen and Holck (1998); Rantsiou et al. (2005).

a_w 0.80–0.90 does not exceed the 2–3-log units (Glass et al. 1992; Faith et al. 1998a,b; Calicioglu et al. 1997, 2001, 2002; Nissen et al. 1998; Cosansu and Ayhan 2000; Erkkilä et al. 2000; Chiktimmah and Knabel 2001; Lahti et al. 2001; Naim et al. 2003). In contrast, all the above studies agree that significant reductions of *E. coli* O157:H7 are sustained during subsequent storage of products for 1 to 6 months, especially at higher temperatures. In general, the reduction in pathogen numbers achieved by storage at ambient temperatures is greater than that achieved by storage at cooler temperatures.

Furthermore, given the high acid resistance of some EHEC strains at pH 2.5–4.0 (Buchanan and Endelson 1996; Zhao et al. 1993), the survival of *E. coli* inoculated postfermentation has been an important area of investigation. Chiktimmah and Knabel (2001) noted a 7-log reduction of *E. coli* inoculated postfermentation on sliced vacuum-packed Lebanon bologna (pH 5.0), fermented with starters (pH 4.4–4.6), after 55 and 65 days of storage at 13 and 3.6°C, respectively. Conversely, when *E. coli* O157:H7 was inoculated in the meat batter (prefermentation inoculation) the total reductions obtained after fermentation and storage for 90 days did not exceed the 4-log units, suggesting that the gradual reduction of pH during fermentation induced acid resistance, which enhanced survival of pathogen during subsequent storage (Buchanan and Edelson 1998). Uyttendaele et al. (2001) illustrated that the lower are the levels of *E. coli* O157:H7 inoculated postfermentation on fermented sausage stored at 7 and 22°C the higher is the inactivation. This conclusion is also consistent with the findings of Nissen and Holck (1998) and Lahti et al. (2001). It is notable however, that Nissen and Holck (1998) observed an initial growth of the low inoculation level of *E. coli* O157:H7 at the early stages of fermentation followed by rapid decrease to undetectable levels.

Of high concern is also the survival of stress-adapted EHEC cultures on the final products, considering potential cross-contamination scenarios by cells that are habituated in acidic or osmotically stressful niches within a plant producing dry-fermented sausages. Leyer et al. (1995) inoculated (10^4–10^5 CFU/g) meat batter at pH 5.8 before fermentation for 15 hours at 32°C and shredded dry salami (pH 5.0) stored at 5°C for 5 days with a composite of five acid-adapted or nonadapted *E. coli* O157:H7 strains. Acid-adapted cultures were capable of surviving longer than nonadapted cultures in both foods. On the other hand, Riordan et al. (2000) found that acid-adapted *E. coli* O157:H7 were more sensitive than nonadapted cells when heated (55–62°C) in pepperoni, whereas higher susceptibility to heat was observed in postfermentation inoculation (pH 4.8) compared to prefermentation inoculation (pH 6.7). Clavero and Beuchat (1996) investigated the survival of heat-stressed (52°C, 30 min) and nonstressed *E. coli* O157:H7 populations onto salami slices stored at 5 and 20°C for 32 days. They found that heat stressed *E. coli* O157:H7 reduced by 4 logs after 8 days at 20°C and by 3 logs after 32 days at 5°C, while the nonstressed cells reduced by 4 and 2 logs at 20°C and 5°C, respectively, after 32 days.

Another useful insight was provided by Naim et al. (2004) who studied the survival of *E. coli* O157:H7 during an in vitro challenge in a model system, which comprised of synthetic saliva (pH 6.7), synthetic gastric fluid (pH 2.5), and synthetic pancreatic juice (pH 8.0). *E. coli* O157:H7 was inoculated during meat batter preparation, survived the manufacturing process, and then was exposed to the GI tract model under static or dynamic pH conditions. It was found that the population of pathogen reduced by 2 logs during fermentation and ripening (final pH 5.0 and a_w 0.90), it remained stable throughout 120 minutes of exposure to gastric acidity (pH 2.5), and showed hundred- or thousand-fold increase during subsequent exposure to pancreatic juice (pH 8.0), depending on the duration of previous exposure to gastric fluid. Finally, Duffy et al. (1999) evaluated the thermo-tolerance of three *E. coli* O157:H7 meat product isolates, injected (10^9 CFU/g) in the core of dry salami. The inoculated salami samples were heated at 50, 55, and 60°C. The D values of the three isolates ranged from 91 to 110 minutes at 50°C, 17–22 minutes at 55°C, and 1.1 to 2.2 minutes at 60°C.

Listeria monocytogenes

Although *L. monocytogenes* is among the most frequently detected pathogens in fermented sausages reaching prevalence levels of up to 30% (Levine et al. 2001; Thévenot et al. 2005a), critical listeriosis cases have yet to be linked to the consumption of fermented sausages contaminated with *L. monocytogenes* (Nightingale et al. 2006). So far, there is only one epidemiological association of *L. monocytogenes* with salami, in Philadelphia, in the U.S, but no confirmed outbreak due to the consumption of such product with *Listeria* is reported (Moore 2004). However, *L. monocytogenes* is ambiguous in nature (Tompkin 2002) and is thus considered a common contaminant, which is highly difficult to control in food-processing environments due to its high tolerance to low pH and high salt concentration (Farber and Peterkin 1991). Therefore, it is likely that this bacterium may also survive the common manufacturing technologies of dry-fermented products, as has been the case with fermented dairy products (Gahan et al. 1996). In this respect, legislation authorities, such as the U.S. Department of Agriculture (USDA) have published rules and guidelines to control the specific pathogen in fermented meats regarded as ready-to-eat foods (USDA 2003, 2004; Ingham et al. 2004). By way of example, the reduction of pH due to the fermentation process, the reduction of water activity *via* drying, and the deposition of inhibitory compounds *via* smoking are considered antimicrobial processes. These processes, according to the compliance guidelines (USDA 2004), should not allow more than 1-log increase of *L. monocytogenes* throughout the shelf life of the treated products. Moreover, these guidelines state that a_w <0.92, a relatively high water activity compared to that of dry sausages, or pH <4.39, a marginal pH achieved after fermentation depending on the use of fermentable carbohydrates and starter cultures, are the lowest levels permitting growth of *L. monocytogenes* based on scientific evidence. Moore (2004) summarized the guidelines for microbiological quality of fermented meats and in the case of *L. monocytogenes*, acceptable fermented meat products are considered those that contain equal to or less than 100 CFU/g of *L. monocytogenes*, consistent to some extent with the new EC regulation 2073/2005 for microbiological criteria in foods. However, in contrast to *E. coli* O157:H7, there are no lethality performance standards proposed for this bacterium in fermented meat products. In this context, Ingham et al. (2004) screened 15 different fermented meats packed in vacuum or air for their ability to inactivate postfermentation inocula of *L. monocytogenes* during storage at 21 or 4°C for 11 weeks. The pH of the products ranged from 4.7 to 5.6 and a_w from 0.91 to 0.98; products with 50–60% water phase salt (a_w 0.27 and pH above 6.0) were also evaluated (Table 42.3). The results of the study showed that inactivation of *L. monocytogenes* was higher on products with lower pH and a_w as well as at ambient temperatures. The authors recommended 1-week postpackaging storage at room temperature prior to shipment to ensure elimination of the pathogen (Ingham et al. 2004) and challenge testing to validate various manufacturing processes for the lack of growth of *L. monocytogenes*. However, a significant body of scientific evidence related to the survival of *L. monocytogenes* in various dry-fermented sausages has been produced in the last decade (Table 42.3). The findings of these studies are reviewed in the next paragraphs.

Glass and Doyle (1989) reported a greater than 4 log units reduction in *L. monocytogenes* during fermentation and drying of pepperoni fermented with starter cultures. Farber et al. (1993) inoculated *L. monocytogenes* (approximately 10^4 CFU/g) in three types of sausages: German-style smoked sausage fermented with *Pediococcus pentocaceus* starter (pH 4.8) and dried to a_w 0.80, American-style smoked sausage fermented with *Pediococcus acidilactici* starter (pH 4.6–4.7) and dried to a_w 0.87–0.89, and naturally fermented Italian-style sausage (pH 4.9–5.5) dried to a_w 0.88–0.89. After drying, which lasted 42 days for all three types of products, all sausages were stored at 4°C for 70 days. In sausages fermented with starters (i.e., American and German-style), the population of *L. monocytogenes* reduced by 2–3 logarithms by the end of fermentation; smoking, drying, and storage caused another 1–2-log reduction, which dropped the pathogen close to the detection limit. In contrast, in the Italian-style sausage, *L. monocytogenes* showed slight growth during fermentation, remained constant during drying, and reduced slightly during a 4-week holding period. Campanini et al. (1993) investigated the behavior of *L. monocytogenes* (10^3–10^4 CFU/g) during maturation (49 days at 14–18°C) of Italian salami (pH after maturation 5.5–5.8) without starters, with a bacteriocin-producing *Lactobacillus plantarum*, or with a nonbacteriocin-producing variant of *L. plantarum*. In sausages without starters, *L. monocytogenes* showed considerable growth throughout maturation period. In sausages with bacteriocin-negative starter, survival without reduction of *L. monocytogenes* was observed, whereas significant decrease in numbers of the pathogen were noticed in sausages with the bacteriocin-positive starter. Similar findings in relation to the effect of bacteriocin-producing starters on prefermentation inoculation of *L. monocytogenes* in dry sausage (pH 4.5, a_w 0.75–0.87 after 49 days of fermentation-drying) have been reported by Lahti et al. (2001), too. However, the inadequacy of Italian-style salami to reduce *L. monocytogenes* was also confirmed by the challenge studies by Nightingale et al. (2006). The application of bacteriocins as dipping solutions has also been evaluated as an alternative preservation and safety intervention following the ripening and slicing of fermented sausages. Mattila et al. (2003) reported that immersion of sliced cooked sausages, bearing more than 2 logs of *L. monocytogenes*, into a pediocin preparation 0.9% reduced *L. monocytogenes* to undetectable levels

Table 42.3. Data related to prevalence, survival, and growth of *Listeria* spp. (1–6) and *Listeria monocytogenes* (7–52) in traditional meat products.

			Process Conditions							Final Product				
a/a	Type of Sausages	Type of Study§	Pre-valence	Concen-tration Data	Starter Cultures (Y/N)	Sugars (Y/N)	NO_3^-/ NO_2^- (Y/N)	Fermen-tation T°C	Drying T°C	pH	a_w/ RH %	Conditions of Storage T°C/	Packaging	Pathogen Response
1	Soudjouck	P	9%	Presence in 25g	nd¥	nd	nd	nd	nd	4.9–6.6	nd	nd	nd	nd
2	Salami	P	14.8%	Presence in 25g	N	nd	nd	nd	nd	4.8–5.2	0.85–0.90	nd	nd	nd
3	Chorizo picante	S-NP	nd	3.36 Log_{10} CFU/g	N	nd	nd	nd	nd	5.19	nd	nd	nd	Present at 2.90 Log_{10} CFU/g
4	Chorizo dulce	S-NP	nd	3.57 Log_{10} CFU/g	N	nd	nd	nd	nd	5.10	nd	nd	nd	Present at 2.98 Log_{10} CFU/g
5	Chorizo dulce	S-NP	nd	1.17 Log_{10} CFU/g	N	nd	nd	nd	nd	4.80	nd	nd	nd	Present at 1.54 Log_{10} CFU/g
6	Salami	C[a]	nd	6 Log_{10} CFU/g	Y	Y	Y	18–23	15	4.5–4.6	24.7 ± 1.36	nd	nd	3.9–1.7 Log_{10} CFU/g reduction depending on irradiation
7	Fermented sausage	P	10%	<3 CFU/g	Y	Y	N	22–24	12	4.7–5.4	0.78–0.90	nd	nd	nd
8	Italian salami	P	13.3%	Presence in 25g	nd	nd	nd	nd	nd	4.8–5.2	0.85–0.90	nd	nd	nd

9	Soudjouck	P	7%	Presence in 25g	nd	nd	nd	nd	nd	4.9–6.7	nd	nd	nd	nd
10	Fermented sausages	P	3.25%	Presence in 25g	nd	nd	nd	nd	nd	nd	nd	nd	nd	nd
11	Fermented sausages	P	20%	Presence in 25g	nd	nd	nd	nd	nd	nd	nd	nd	nd	nd
12	Fermented sausages	P	19.05%	Presence in 25g	nd	nd	nd	nd	nd	nd	nd	nd	nd	nd
13	Fermented sausages	P	44%	Presence in 25g	nd	nd	nd	nd	nd	nd	nd	nd	nd	nd
14	Fermented sausages	P	20%	Presence in 25g	nd	nd	nd	nd	nd	nd	nd	nd	nd	nd
15	Sausages	P	25%	Presence in 25g	nd	nd	nd	nd	nd	nd	nd	nd	nd	nd
16	Spreadable fermented sausage	P	11.30%	Presence in 25g	nd	nd	nd	nd	nd	nd	nd	nd	nd	nd
17	Sliceable fermented sausage	P	4.8%	Presence in 25g	nd	nd	nd	nd	nd	nd	nd	nd	nd	nd
18	Salami	P	16.67%	<100 CFU/g	nd	nd	nd	nd	nd	nd	nd	nd	nd	nd
19	Raw sausage, sliced	P	1.59%	Presence in 25g	nd	nd	nd	nd	nd	nd	nd	nd	nd	nd
20	Salsiccia	P	11.54%	Presence in 25g	nd	nd	nd	nd	nd	nd	nd	nd	nd	nd
21	Fermented sausage	P	20%	Presence in 25g	nd	nd	nd	nd	nd	nd	nd	nd	nd	nd
22	Salami	P	10%	Presence in 25g	nd	nd	nd	nd	nd	nd	nd	nd	nd	nd
23	Salami	P	16%	Presence in 25g	nd	nd	nd	nd	nd	nd	nd	nd	nd	nd

(*Continues*)

Table 42.3. (*Continued*)

					Process Conditions					Final Product				
a/a	Type of Sausages	Type of Study[§]	Pre-valence	Concen-tration Data	Starter Cultures (Y/N)	Sugars (Y/N)	NO_3^-/ NO_2^- (Y/N)	Fermen-tation T°C	Drying T°C	pH	a_w/ RH %	Conditions of Storage T°C/	Packaging	Pathogen Response
24	Salami	P	5%	20 CFU/g	nd	nd	nd	nd	nd	nd	nd	nd	nd	nd
25	Dry cured	P	10%	Presence in 25g	nd	nd	nd	nd	nd	nd	nd	nd	nd	nd
26	Fermented sausages	P	10%	Presence in 25g	nd	nd	nd	nd	nd	nd	nd	nd	nd	nd
27	Salami	P	40%	Presence in 25g	nd	nd	nd	nd	nd	nd	nd	nd	nd	nd
28	Spanish-style sausage	P	3,70%	Presence in 25g	nd	nd	nd	nd	nd	nd	nd	nd	nd	nd
29	Smoked sausage	P	0,00%	Presence in 25g	nd	nd	nd	nd	nd	nd	nd	nd	nd	nd
30	Italian-style salami	C[a]	nd	7.4 Log_{10} CFU/g	Y	Y	Y	30	12	4.3–4.8	0.87–0.95	nd	nd	<2.0 Log_{10} CFU/g reduction
31	Naples-type fermented sausage	C[a]	nd	10^3–10^4 CFU/g	N	Y	Y	17–22	17–22	5.2–5.5	0.78–0.80	nd	nd	Not detected
32	Spanish-type fermented sausage	C[a]	nd	10^3–10^5 CFU/g	Y	Y	Y	20–22	15	5.1–5.2	nd	nd	nd	HI: detected after 8d; LI: detected after 8d
33	Summer sausage	C[b]	nd	3.6 Log_{10} CFU/ sample	Y	Y	Y	nd	nd	4.7	0.96	5	V	2.7 Log_{10} CFU/g reduction

34	Summer sausage	C^b	nd	3.9–4.2 Log_{10} CFU/g	Y	Y	Y	nd	nd	4.9	0.95	5,21	V	~3.0–3.3 Log_{10} CFU/g reduction
35	Summer sausage	C^b	nd	3.4 Log_{10} CFU/ sample	Y	Y	Y	nd	nd	4.8	0.96	5,21	V	2.5 Log_{10} CFU/g reduction
36	Buffalo summer sausage	C^b	nd	3.1–3.7 Log_{10} CFU/ sample	N	Y	Y	nd	nd	5.2	0.95	5,21	V	2.2 Log_{10} CFU/g reduction
37	Elk summer sausage	C^b	nd	3.8–4.0Y Log_{10} CFU/ sample	Y	Y	nd	nd	5.2	0.95	5,21	V		1.4–3.1 Log_{10} CFU/g reduction
38	Lebanon bologna luncheon slices	C^b	nd	6.7 Log_{10} CFU/g	Y	nd	nd	37.7	nd	4.4–4.6	nd	3.6,13	V	$D_{3.6}$ = 5.02d, D_{13} = 2 .97d
39	Fermented sausage American-style	$C^{a/b}$	nd	10^4 CFU/g	Y	Y	Y	24	16–18	4.6–5.1	0.87–0.88	4	A	Detected after 28d
40	Fermented sausage German-style	$C^{a/b}$	nd	10^3–10^5 CFU/g	Y	Y	Y	24	16–18	4.6–4.8	0.80 –0.81	4	A	Detected after 28d
41	Fermented sausages Italian-style	$C^{a/b}$	nd	10^5 CFU/g	N	Y	Y	16–24	14	4.9–5.5	0.88–0.89	4	A	Detected (10^2–10^4 CFU/g) after 28d
42	Lebanon bologna (model sausages)	$C^{a/b}$	nd	6.3 log CFU/g	Y	nd	nd	37.7	nd	4.4–4.6	nd	3.6	V	$D_{3.6}$ = 5.02

(Continues)

Table 42.3. (*Continued*)

			Process Conditions							Final Product			
a/a	Type of Sausages	Type of Study§	Pre-valence	Concen-tration Data	Starter Cultures (Y/N)	Sugars (Y/N)	NO_3^-/ NO_2^- (Y/N)	Fermen-tation T°C	Drying T°C	pH	a_w/ RH %	Conditions of Storage T°C/ Packaging	Pathogen Response
43	Chorizo (experimental)	C[a/b]	nd	7.51 ± 1.38 Log_{10} CFU/g	N	nd	nd	nd	nd	4.9–5.1	0.85	7,25,30 A	2.77–5.33 Log_{10} CFU/g reduction depending on T°C
44	Chorizo (experimental)	C[a/b]	nd	7.51 ± 1.38 Log_{10} CFU/g	N	nd	nd	nd	nd	4.9–5.1	0.90	7,25,30 A	3.0–6.13 Log_{10} CFU/g reduction depending on T°C
45	Chorizo (experimental)	C[a/b]	nd	7.51 ± 1.38 Log_{10} CFU/g	N	nd	nd	nd	nd	4.9–5.1	0.93	7,25,30 A	2.38–4.62 Log_{10} CFU/g reduction depending on T°C
46	Chorizo (experimental)	C[a/b]	nd	7.51 ± 1.38 Log_{10} CFU/g	N	nd	nd	nd	nd	4.9–5.1	0.95	7,25,30 A	1.31–3.14 Log reduction depending on T°C
47	Chorizo (experimental)	C[a/b]	nd	7.51 ± 1.38 Log_{10} CFU/g	N	nd	nd	nd	nd	4.9–5.1	0.97	7,25,30 A	0.68–2.16 Log_{10} CFU/g reduction depending on T°C

48	Dry-fermented sausage	$C^{a/b}$	nd	2.92–3.35, 5.1–5.4 Log_{10} CFU/g	Y	Y	Y	20–24	17	4.4–4.6	0.76–0.88	15–17	A	HI: detected after 35–49d; LI: not detected after processing
49	Dry sausage	$C^{a/b}$	nd	10^3–10^7 CFU/g	Y	Y	Y	27	14	4.9	0.93	4,20	V	Detected after 46d
50	Dry sausage	$C^{a/b}$	nd	4.0–4.6 Log_{10} CFU/g	Y	Y	Y	20–24	13–14	5.8–5.9	0.76–0.81	10	A	Not detected after 60d
51	Greek dry salami	S-NP	nd	Presence in 25g	N	Y	Y	20–24	15–16	5–5.2	27.7–30.3%	nd	nd	Not detected
52	Fermented sausage	S-NP	nd	nd	N	Y	Y	nd	14	5.62–5.7	nd	nd	nd	Not detected

[§]P: prevalence study; C^a: preprocessing inoculation (challenge) without storage; C^b: postprocessing inoculation (challenge); $C^{a/b}$: preprocessing inoculation (challenge) and storage; S-NP: survival study on naturally contaminated products.

[¥]nd: no data, HI: high inoculum; LI: low inoculum.

A: aerobically packaged; V: vacuum packaged.

1–6: Based on Siriken et al. (2006); Borges et al. (1999); Encinas et al (1999); Samelis et al. (2005).

7–52: Based on Thevenot et al. (2005 a,b); Borges et al. (1999); Siriken et al. (2006); Levine et al. (2001); FDA/CFSAN (2003); Mena et al. (2004); Villani et al. (1997); Hugas et al. (2002); Ingham et al. (2004); Chikthimmah and Knabel (2001); Farber et al. (1993); Hew et al. (2005); Lahti et al. (2001); Nissen and Holck (1998); Samelis et al. (1998); Rantsiou et al. (2005).

within 21 days of storage of slices in vacuum packs. In contrast, *L. monocytogenes* remained stable on the surface of nondipped samples. Nissen and Holck (1998) monitored the survival of two inoculation levels of *L. monocytogenes* (10^3–10^4 and 10^5–10^7 CFU/g) during the fermentation, drying, and storage of Norwegian fermented dry sausages (pH 4.8 and a_w 0.89 after drying) at 4 and 20°C for 5.5 months. Inoculation was carried out in the center of stuffed sausages. In accordance to the findings of Farber et al. (1993), the fermentation caused a limited reduction of 2–3-logs in *L. monocytogenes* numbers, whereas it was the storage at the ambient temperature that totally eliminated the pathogen as opposed to cold storage. In the study of Encinas et al. (1999), 21 lots of five varieties of naturally contaminated Spanish chorizo with (pH 4.27–4.57) or without (pH 4.8–5.20) starter cultures were surveyed for presence and enumeration of *L. monocytogenes* during maturation. The authors concluded that in the naturally fermented chorizo varieties, the initial contamination of *L. monocytogenes* survived at the same levels throughout maturation, in contrast to the chorizos fermented with starter cultures, where *L. monocytogenes* reduced to undetectable levels. In another study, Hew et al. (2005) performed artificial contamination of laboratory-made chorizos of different a_w (0.85–0.97) with 10^6 CFU/g *L. monocytogenes* and stored them at ambient temperatures (21–30°C) and cold temperatures (6–8°C). The pH of sausages was adjusted with vinegar and the a_w by addition of NaCl. It was found that low water activity and ambient temperatures caused the highest reductions of the pathogen. Chikthimmah and Knabel (2001) demonstrated that prefermentation *L. monocytogenes* contamination of Lebanon bologna batter survives longer than postfermentation contamination. This was also evident for *E. coli* O157:H7 (see above) and *Salmonella* Typhimurium. However, the survival potential of *L. monocytogenes* during fermentation and drying has also been recognized as strain-dependent (Thévenot et al. 2005b). The outcome of the aforementioned studies can be summarized from a safety standpoint as follows: the use of starters in combination with good manufacturing practices, drying to low water activity, and storage of sausages at room temperature minimize the risk of listeriosis by consumption of these products.

Salmonella spp.

In contrast to *L. monocytogenes*, *Salmonella* species have been involved in outbreaks linked to the consumption of fermented meats and specifically of dry salami, Lebanon bologna sausages, and fresh sausages (Sauer et al. 1997; Nissen and Holck 1998; Moore 2001; Giovannini et al. 2004).

Proposed lethality performance standards for RTE products by the USDA require a 6.5-log CFU/g reduction of viable *Salmonella* populations for finished products (USDA 2001). However, concern has been expressed for the survival of postprocessing *Salmonella* contamination, especially on sliced fermented meats, such as Lebanon bologna, as well as in naturally fermented meat products. Thus, validation of manufacturing processes and challenge testing on finished products for survival of *Salmonella* is essential. For example, Escartin et al. (1999) demonstrated the reduction of native salmonellas, present at levels up to 900 CFU/g in chorizos, to undetectable levels during storage at ambient and cold temperatures. Detailed description of available challenge studies is provided in Table 42.4.

Ellajosyula et al. (1998) demonstrated a 5-$\log_{10}$ CFU reduction of *S.* Typhimurium in cooked Lebanon bologna sausages. In the same product, Chikthimmah et al. (2001) showed that *S.* Typhimurium exhibited the most rapid reduction compared to *E. coli* O157:H7 and *L. monocytogenes* inoculated both pre- and postfermentation. Likewise, in a comparative evaluation of the survival of *L. monocytogenes*, *E. coli* O157:H7 and *S. kentucky* during manufacture and storage of Norwegian dry salami, *S. kentucky* was the most sensitive pathogen, which reduced to undetectable levels after 46 days of storage both at ambient and cold temperatures. The same observation accounted for two inoculation levels, one low (10^3 CFU/g) and one high (10^6 CFU/g). The higher sensitivity of *Salmonella* to the manufacturing process of dry-fermented sausages, in comparison to other pathogens of concern, was also recently observed in Italian-style salami of pH 4.4–4.8 and a_w of 0.869–0.935 (Nightinghale et al. 2006). However, in the latter study, lower reductions in *Salmonella* serotype Enteritidis population were obtained in fermented nondried sausages than in fermented dried products, whereas drying to a 1.4:1 moisture protein ratio (MPR) ensures higher reductions than drying to a 1.9:1 MPR. In contrast to the above studies, *S.* Typhimurium and *S. dublin* have been reported to survive the fermentation and drying process of pepperoni at 5°C with starters (pH 4.5, a_w 0.798 in the finished product) or without starters (pH 5.0, a_w 0.805 in the finished product) (Smith et al. 1975). A total of a 3-log reduction was achieved in the presence of starters (*Lactobacillus* and *Pediococcus*) as opposed to only a 2-log reduction in naturally

Table 42.4. Data related to prevalence, survival, and growth of *Salmonella* spp. (1–10), *Salmonella kentucky* (11) *S. tymphimurium* (12–14), *S. dublin* (15), and *S. Enteritidis* (16) in traditional meat products.

			Process Conditions							Final Product				
a/a	Type of Sausages	Type of Study§	Pre-valence	Concen-tration Data	Starter Cultures (Y/N)	Sugars (Y/N)	NO_3^-/ NO_2^- (Y/N)	Fermen-tation T°C	Drying T°C	pH	a_w/ RH %	Conditions of Storage T°C/ Packaging		Pathogen Response
1	Dry/semidry sausages	P	0.08%	Presence in 25g	nd	nd	nd	nd	nd	4.0–6.0	0.85–0.94	nd	nd	nd
2	Fermented	S-NP	nd[¥]	Presence in 25g	N	Y	Y	nd	nd	5.62–5.70	nd	nd	nd	Not detected
3	Chorizo	P	61.4%	nd	nd	nd	nd	nd	nd	nd	nd	nd	nd	nd
4	Chorizo	S-NP	nd	0.04–900 CFU/g	nd	nd nd	nd	nd	nd	nd	nd	5, 18–22	nd	HI; detected after 42d– LI: not detected after 35d
5	Soudjouck	P	7%	nd	nd	nd	nd	nd	nd	4.9–6.5	nd	nd	nd	nd
6	Fermented	P	1.43%	nd	nd	nd	nd	nd	nd	nd	nd	nd	nd	nd
7	Greek dry salami	S-NP	nd	nd	N	Y	Y	20–24	15–16	5.0–5.2	27.7–30.3	nd	nd	Absence in 25g
8	Spreadable fermented sausage	P	0%	nd	nd	nd	nd	nd	nd	nd	nd	nd	nd	Absence in 25g
9	Fermented minced meat, fermented meat cubes	C[a]	nd	10^5 CFU/g	Y	Y	Y	30, 37	nd	4.0–4.2	nd	nd	nd	Not detected

(*Continues*)

Table 42.4. *(Continued)*

				Process Conditions						Final Product			
a/a	Type of Sausages	Type of Study§	Pre-valence	Concen-tration Data	Starter Cultures (Y/N)	Sugars (Y/N)	NO_3^-/ NO_2^- (Y/N)	Fermen-tation T°C	Drying T°C	pH	a_w/ RH %	Conditions of Storage T°C/ Packaging	Pathogen Response
10	Dry sausage	S-NP	nd	Presence in 25g	N	nd	nd	12	12	nd	nd	nd nd	Detected during production
11	Dry sausage	C[a/b]	nd	10^3–10^4 CFU/g	Y	nd	nd	27	14	nd	nd	4,20 V	Not detected
12	Lebanon bologna luncheon slices	C[b]	nd	8 Log_{10} CFU/g	nd	nd	nd	nd	nd	4.4–4.6	nd	3.6–13 V	$D_{3.6}$ = 4.99d and D_{13} = 4.89d
13	Lebanon bologna (model sausages)	C[a/b]	nd	7.5 Log_{10} CFU/g	Y	nd	nd	nd	nd	4.4–4.6	nd	3.6 V	$D_{3.6}$ = 4.99d
14	Beef-pork pepperoni	C[a]	nd	10^4 CFU/g	Y/N	Y	Y	35	12	4.6–5.0	0.798–0.805	nd A	With starter culture: not detected after 42d; With natural flora: detected after 42d

15	Beef-pork pepperoni	C^a	nd	10^4 CFU/g	Y/N	Y	Y	35	12	4.6–5.0	0.798–0.805	nd	A	~2–3 Log_{10} CFU/-reduction depending on starter culture; Detected after 42d
16	Italian-style salami	C^a	nd	7.4 Log_{10} CFU/g	Y	Y	Y	30	10~13	4.3–4.9	0.87–0.96	nd	nd	5–6.9 Log_{10} CFU/g reduction

[§]P: prevalence study; C^a: preprocessing inoculation (challenge) without storage; C^b: postprocessing inoculation (challenge); $C^{a/b}$: preprocessing inoculation (challenge) and storage; S-NP: survival study on naturally contaminated products.

[¥]nd: no data, HI: high inoculum; LI: low inoculum.

A: aerobically packaged; V: vacuum packaged.

1–10: Based on Little et al. (1998); Rantsiou et al. (2005); Escartin et al. (1999); Siriken et al. (2006); Levine et al. (2001); Samelis et al. (1998); Sakhare and Rao (2003); Chevallier et al. (2006).

11: Based on Nissen and Holck (1998).

12–15: Based on Chikthimmah and Knabel(2001); Smith et al. (1975).

16: Based on Nightingale et al. (2006).

Table 42.5. Data related to prevalence, survival, and growth of *Staphylococcus aureus* in traditional meat products.

				Process Conditions						Final Product				
a/a	Type of Sausages	Type of Study[§]	Prevalence	Concentration Data	Starter Cultures (Y/N)	Sugars (Y/N)	NO_3^-/ NO_2^- (Y/N)	Fermentation T°C	Drying T°C	pH	a_w/ RH %	Conditions of Storage T°C/ Packaging		Pathogen Response
1	Chorizo	C[a]	nd[¥]	10^6 CFU/g	Y/N	Y	Y	20,30	15	4.6–4.8	0.82–0.89	nd	nd nd	2–3 Log_{10} CFU/g reduction depending on formulation
2	Salchichon	C[a]	nd	10^5 CFU/g	Y/N	Y	Y	20,30	15	4.6–4.8	0.82–0.89	nd	nd	1.5–1.9 Log_{10} CFU/g reduction depending on formulation
3	Fermented sausage	S-NP	nd	3.18–3.48 Log_{10} CFU/g	nd	nd	nd	nd	nd	6.2–6.5	76.5–86.0	nd	nd	Present at >0.30–2.51 Log_{10} CFU/g
4	Dry/semi-drysausages	P	1%	≥10^2 CFU/g	nd	nd	nd	nd	nd	4.0–6.0	0.85–0.94	nd	nd	nd
5	Spreadable fermented sausage	P	100%	<10^1 CFU/g	nd	nd	nd	nd	nd	4.0–6.0	0.85–0.94	nd	nd	nd
6	Beef sausage	C[a]	nd	10^7 CFU/g	Y/N	Y	Y	35	nd	nd	nd	nd	nd	Not detected
7	Fermented sausage (salsiccia)	S-NP	nd	<2.0–3.40 Log_{10} CFU/g	nd	nd	nd	nd	nd	5.6–6.2	0.86–0.95	nd	nd	<2 or 4 Log_{10} CFU/g depending on the producer

8	Fermented sausage (soppres-sata lucana)	S-NP	nd	<2.0–3.40 Log_{10} CFU/g	nd	nd	nd	nd	nd	5.2–6.4	0.89–0.95	nd	nd	<2 or 4.16 Log_{10} CFU/g depending on the producer
9	Greek dry salami	S-NP	nd	<2.0–3.0 Log_{10} CFU/g	N	Y	Y	20–24	15–16	5–5.2	27.7–30.3%	nd	nd	<2 or 2.6 Log_{10} CFU/g
10	Fermented sausage	C[a]	nd	4.1–4.2 Log_{10} CFU/g	Y	Y	Y	20,35	nd	4.4–5.5	90–95	nd	nd	At °C: 0.3–4.6 Log_{10} CFU/g increase depending on pH, at 20°C: 0.3 Log_{10} CFU/g reduction to 2.6 Log_{10} CFU/g increase, depending on pH
11	Fermented minced meat, Fermented meat cubes	C[a]	nd	10^5 CFU/g	Y	Y	nd	30,37	nd	4.0–4.2	nd	nd	nd	Not detected
12	Fermented sausage	S-NP	nd	<100 CFU/g	N	Y	Y	nd	0–14	5.62–5.70	nd	nd	nd	<100 CFU/g

[§]P: prevalence study; C[a]: preprocessing inoculation (challenge) without storage; C[b]: postprocessing inoculation (challenge); C [a/b]: preprocessing inoculation (challenge) and storage; S-NP: survival study on naturally contaminated products.

[¥]nd: no data.

Based on Gonzalez-Fandos et al. (1999); Little et al. (1998); Smith et al. (1978); Blaiotta et al. (2004); Samelis et al. (1998); Sameshima et al. (1998); Sakhare and Rao (2003); Chevallier et al. (2006); Rantsiou et al. (2005).

fermented products. However, a cooking step at 60°C could eliminate the pathogen from the product.

Staphylococcus aureus

Different types of fermented sausages have been implicated in staphylococcal food poisoning outbreaks (Lücke 1985). *Staph. aureus* may be commonly found in industrial equipment surfaces and is salt- and nitrite-tolerant, thus being capable of growing and producing enterotoxin during the initial stage of fermentation (Gonzalez-Fandos 1999). Based on available reports, growth and enterotoxin production of *Staph. aureus* is highly dependent on the starter cultures used and the fermentation conditions, such as duration, temperature, and aeration (Smith and Palumbo 1978; Adams and Nicolaides 1997; Sameshima et al. 1998; Gonzalez-Fandos 1999). According to Gonzalez-Fandos (1999), growth and toxin production was not prevented at 20 and 30°C in chorizos and salchichon, two traditional Spanish sausages (pH 5.5–5.8 and a_w 0.82–0.88), regardless of the presence of starter cultures. The combination of lower fermentation temperature and activity of starter cultures simply restricted growth and toxin production. However, *Staph. aureus* significantly reduced throughout ripening. Conversely, Sameshima et al. (1998) found that three starters, added at an initial level of 10^7 CFU/g (some of them being potential probiotics) were capable of inhibiting both growth of 10^3 CFU/g *Staph. aureus* and production of staphylococcal toxin in salami fermented at 20 or 30°C. Likewise, Smith and Palumbo (1978) addressed the importance of starters and the addition of glucose in meat fermentations in order to enhance the injury of *Staph. aureus* by the metabolic activity of starter cultures. Overall, a common feature of previous reports is that challenge tests with *St. aureus* do not lead to standard conclusions regarding the inhibition of growth and toxin production (Adams and Nicolaides 1997), as is the case with the other three pathogens. Thus, prevalence studies and thorough validation of separate processes are required to ensure minimization of risk by staphylococcal food poisoning due to consumption of fermented sausages (Table 42.5).

CONTROL MEASURES

The ability of the above-mentioned pathogens to survive in many low acid as well as low water activity meat products makes it unlikely that complete suppression can be achieved by the application of control measures at a single source. Thus, effective control strategies must consider the multiple points at which pathogens can gain access to the human food chain. The persistence and the ability of very small numbers of these organisms to establish life-threatening infections with serious long-term clinical consequences, particularly among at-risk sections of the human population, mean that many elements of our food safety strategies have to be improved.

Measures to control pathogens during food production, processing, and distribution, at retail level and during commercial/domestic preparation should be considered in detail.

Therefore, the best approach to control pathogens is to implement HACCP principles into the food safety management systems at all stages of food production and distribution. However, due to their unusual tolerance to low pH, of some pathogens, i.e., *Listeria*, *E. coli*, and *Salmonella* should be considered at higher risk than other pathogens (e.g., *Staph. aureus*) in meat products (sausages), which have a low pH and are minimally processed or are not cooked before consumption.

The basic control measures should include the following:

- A quantitative microbiological hazard analysis or quantitative microbial risk assessment on current practices should be carried out for products in these categories.
- Training of personnel working in food preparation and food service industries coupled with consumer education on hygienic handling and adequate cooking of food can play a role in reducing the incidence of pathogenic infection.

REFERENCES

M Adams, R Mitchell. 2002. Fermentation and pathogen control: A risk assessment approach. Int J Food Microbiol 79:75–83.

MR Adams, L Nicolaides. 1997. Review of the sensitivity of different foodborne pathogens to fermentation. Food Contr 8:227–239.

S Ananou, M Garriga, M Hugas, M Maqueda, M Martinez-Bueno, A Galvez, E Valdivia. 2005. Control of *Listeria monocytogenes* in model sausages by enterocin AS-48 Int J Food Microbiol 103:179–190.

G Blaiotta, C Pennacchia, F Villan, A Ricciardi, R Tofal, E Parente. 2004. Diversity and dynamics of communities of coagulase-negative staphylococci in traditional fermented sausages. J Appl Microbiol 97:271–284.

MF Borges, RS de Siqueira, AM Bittencourt, MCD Vanetti, LAM Gomide. 1999. Occurrence of *Listeria monocytogenes* in salami. Revista de microbiologia 30:362–364.

RL Buchanan, SG Edelson. 1996. Culturing enterohemorrhagic *Escherichia coli* in the presence and absence of glucose as a simple means of evaluating the acid tolerance of stationary-phase cells. Appl Environ Microbiol 62:4009–4013.

M Calicioglu, NG Faith, DR Buege, JB Luchansky. 1997. Viability of *Escherichia coli* O157:H7 in fermented semidry low-temperarure-cooked beef summer sausage. J Food Prot 60:1158–1162.

———. 2001. Validation of a manufacturing process for fermented, semidry Turkish soudjouk to control *Escherichia coli* 0157:H7. J Food Prot 64: 1156–1161.

———. 2002. Viability of *Escherichia coli* O157:H7 during manufacturing and storage of a fermented, semidry Soudjouk-style sausage. J Food Prot 65: 1541–1544.

M Campanini, I. Pedrazzoni, S. Barbuti, P. Baldini. 1993. Behaviour of *Listeria monocytogenes* during the maturation of naturally and artificially contaminated salami: Effect of lactic-acid bacteria starter cultures. Int J Food Microbiol 20:169–175.

Centers for Disease Control and Prevention (CDC). 1994. *Escherichia coli* 0157:H7 outbreak linked to commercially distributed dry-cured salami—Washington and California. Morbid Mortal Weekly Rep 44:157–160.

———. 1995. Community outbreak of hemolytic uremic syndrome attributable to *Escherichia coli* 0111:NM South Australia. Morbid Mortal Weekly Rep 44:550–551,557.

I Chevallier, S Ammor, A Laguet, S Labayle, V Castanet, E Dufour, R Talon. 2006. Microbial ecology of a small-scale producing traditional dry sausages. Food Contr 17:446–453.

N Chikthimmah, SJ Knabel. 2001. Survival of *Escherichia coli* 0157:H7, *Salmonella typhimurium*, and *Listeria monocytogenes* in and on vacuum packaged Lebanon bologna stored at 3.6 and 13.0°C. J ood Protect 64:958–963.

MRS Clavero, LR Beuchat. 1996. Survival of *Escherichia coli* O157:H7 in broth and processed salami as influenced by pH, water activity and temperature and suitability of media for its recovery. Appl Environ Microbiol 62:2735–2740.

S Cosansu, K Ayhan. 2000. Survival of enterohaemorrhagic *Escherichia coli* O157:H7 strain in Turkish soudjouck during fermentation, drying and storage periods. Meat Sci 54:407–411.

G Duffy, DCR Riordan, JJ Sheridan, BS Eblen, RC Whiting, IS Blair, DA McDowell. 1999. Differences in thermotolerance of various *Escherichia coli* 0157:H7 strains in a salami matrix. Food Microbiol 16:83–91.

KR Ellajosyula, S Doores, EW Mills, RA Wilson, RC Anantheswaran, SJ Knabel. 1998. Destruction of *Escherichia coli* O157:H7 and *Salmonella typhimurium* in Lebanon bologna by interaction of fermentation pH, heating temperature and time. J Food Prot 61:152–157.

JP Encinas, JJ Sanz, ML Garcia-Lopez, A Otero. 1999. Behaviour of *Listeria* spp. in naturally contaminated chorizo (Spanish fermented sausage). Int J Food Microbiol 46:167–171.

S Erkkila, M Venalainen, S Hielm, E Petaja, E Puolanne T. Mattila-Sandholm. 2000. Survival of *Escherichia coli* 0157:H7 in dry sausage fermented by probiotic lactic acid bacteria. J Sci Food Agric 80:2101–2104 (available online: 2000).

EF Escartin, A Castillo, A Hinojosa-Puga and J Saldana-Lozano. 1999. Prevalence of Salmonella in chorizo and its survival under different storage temperatures. Food Microbiol 16:479–486.

European Union (EU). 2005. Commission Regulation 2073/2005/EC of 15 November 2005 on the microbiological criteria for foodstuffs. Official Journal of the European Communities, 22.12.2005, L338/1–25.

NG Faith, N Parniere, T Larson, TD Lorang, CW Kaspar, JB Luchansky. 1998a. Viability of *Escherichia coli* O157:H7 in salami following conditioning of batter, fermentation and drying of sticks and storage of slices. J Food Prot 61:377–382.

NG Faith, RK Wierzba, AM Ihnot, AM Roering, TD Lorang, CW Kaspar, JB Luchansky. 1998b. Survival of *Escherichia coli* O157:H7 in full- and reduced-fat pepperoni after manufacture of sticks, storage of slices at 4°C or 21°C under air and vacuum, and baking slices on frozen pizza at 135, 191 and 246°C. J Food Prot 61:383–389.

FAO/WHO. 2004. Risk Assessment; Appendix 4—Prevalence and incidence of L. monocytogenes in fermented meat products (QMRA).

JM Farber, E Daley, R Holley, WR Usborne. 1993. Survival of *Listeria monocytogenes* during the production of uncooked German, American and Italian-style fermented sausages. Food Microbiol 10:123–132.

JM Farber, PI Peterkin. 1991. *Listeria monocytogenes*, a food-borne pathogen. Microbiol Rev 55:476–511.

Food and Drug Administration, Center for Food Safety and Applied Nutrition (FDA/CFSAN). 2003. Quantitative assessment of the relative risk to public health from foodborne *Listeria monocytogenes* among selected food categories of ready-to-eat foods. Available at: http://www.fsis.usda.gov/OA/news/2003/rtedata.htm.

V Freeman. 1996. The E. coli equation. Meat Poultry 44.

CGM Gahan, B O'Driscoll, C Hill. 1996. Acid adaptation of *Listeria monocytogenes* can enhance survival in acidic foods and during milk fermentation. Appl Environ Microbiol 62:3128–3132.

A Giovannini, V Prencipe, A Conte, L Marino, A Petrini, F Pomilio, V Rizzi, G Migliorati. 2004. Quantitative risk assessment of *Salmonella* spp. infection for the consumer of pork products in an Italian region. Food Contr 15:139–144.

KA Glass, MP Doyle. 1989. Fate and thermal inactivation of *Listeria monocytogenes* in beaker sausage and pepperoni. J Food Prot 52:226–231.

KA Glass, JM Loeffelholz, JP Ford, MP Doyle. 1992. Fate of *Escherichia coli* 0157:H7 as affected by pH or Sodium chloride and in fermented, dry sausage. Appl Environ Microbiol 58(8):2513–2516

ME Gonzalez-Fandos, M Sierra, ML Garcia-Lopez, MC Garcia-Fernandez, A. Otero. 1999. The influence of manufacturing and drying conditions on the survival and toxogenesis of *Staphylococcus aureus* in two Spanish dry sausages (chorizo and salchichon). Meat Sci 52:411–419.

GH Grau. 1986. Microbial ecology of meat and poultry. In: Advances in Meat Research. Meat and Poultry Microbiology. Westport, Connecticut: AVI Publishing Company, Inc., pp. 1–48.

CM Hew, MN Hajmeer, TB Farver, JM Glover DO Cliver. 2005. Survival of *Listeria monocytogenes* in experimental chorizos. J Food Prot 68:324–330.

JC Hinkens, NG Faith, TD Lorang, PH Bailey, D Buege, CW Kaspar, JB Luchansky. 1996. Validation of pepperoni processes for control of *Escherichia coli* O157:H7. J Food Prot 59:1260–1266.

M Hugas, M Garriga, M Pascual, MT Aymerichand, JM Monfort 2002. Enhancement of sakacin K activity against *Listeria monocytogenes* in fermented sausages with pepper or manganese as ingredients. Food Microbio 19:519–528.

K Incze. 1998. Dry fermented sausages. Meat Sci 49:S169–S177.

SC Ingham, DR Buese, BK Dropp, JA Losinski. 2004. Survival of *Listeria monocytogenes* during storage of ready-to-eat meat products processed by drying, fermentation, and/or smoking. J Food Prot 67: 2698–2702.

E Lahti, T Johansson, T Honkanen-Buzalski, P Hill, E Nurmi. 2001. Survival and detection of *Escherichia coli* 0157:H7 and *Listeria monocytogenes* during the manufacture of dry sausage using two different starter cultures. Food Microbiol 18:75–85.

P Levine, B Rose, S Green, G Ransom W Hilli. 2001. Pathogen testing of ready-to-eat meat and poultry products collected at federally inspected establishments in the United States, 1990–1999. J Food Protect 64:1188–1193.

GJ Leyer, L-L Wang, EA Johnson. 1995. Acid adaptation of *Escherichia coli* O157:H7 increases survival in acidic foods. Appl Environ Microbiol 61:3752–3755.

CL Little, HA Monsey, GL Nichols, J de Louvois. 1998. The microbiological quality of ready-to-eat dried and fermented meat products. Int J Environ Health Res 8:277–284.

FK Lücke. 1985. Fermented sausages. In: Microbiology of Fermented Foods. BJB Woods, ed. London: Elsevier Publishers, pp. 41–83.

BM Mackey, CM Derrick. 1979. Contamination of the deep tissues of carcasses by bacteria present on the slaughter instruments or in the gut. J Appl Bacteriol 46:355–366.

K Mattila, P Saris S Tyopponen. 2003. Survival of *Listeria monocytogenes* on sliced cooked sausage after treatment with pediocin AcH. Int J Food Microbiol 89:281–286.

C Mena, G Almeida, L Carneiro, P Teixeira, T Hogg, and PA Gibbs. 2004. Incidence of *Listeria monocytogenes* in different food products commercialized in Portugal. Food Microbiol 21:213–216.

J Metaxopoulos, C Genigeorgis, MJ Fanelli, C Franti, E Cosma. 1981a. Production of Italian dry salami I. Initiation of staphylococcal growth in salami under commercial manufacturing conditions. J Food Prot 44:347–352.

———. 1981b. Production of Italian salami: Effect of starter culture and chemical acidulation on staphylococcal growth in Salami under commercial manufacturing conditions. Appl Envir Microb 42:863–871.

TE Minor, EH Marth. 1972a. *Staphylococcus aureus* and staphylococcal food intoxication. A review II. Enterotoxin and epidemiology. J Milk Food Technol 35:21–29.

———. 1972b. *Staphylococcus aureus* and staphylococcal food intoxication. A review: IV Staphylococci in meat bakery products and other foods J Milk Food Technol 35:228–241.

JE Moore. 2004. Gastrointestinal outbreaks associated with fermented meats. Meat Sci 67:565–568.

F Naim, D. Leblanc, S. Messier, L. Saucier, G. Piette, A Houde. 2006. Shiga toxin production by sausage-borne *Escherichia coli* 0157:H7 in response to a post-processing in vitro digestion challenge. Food Microbiol 23:231–240.

F Naim, S Messier, L Saucier, G Piette. 2003. A model study of *Escherichia coli* O157:H7 survival in fermented dry sausages—Influence of inoculum preparation, inoculation procedure, and selected process parameters. J Food Prot 66:2267–2275.

———. 2004. Postprocessing in vitro challenge to evaluate survival of *Escherichia coli* O157:H7 in fermented dry sausages. Appl Environ Microbiol 70:6637–6642.

KK Nightingale, H Thippareddi, RK Phebus, JL Marsden, AL Nutsch. 2006. Validation of a traditional Italian-style salami manufacturing process for control of Salmonella and *Listeria monocytogenes*. J Food Protect 69(4):794–800.

H Nissen, A. Holck. 1998. Survival of *Escherichia coli* O157:H7, *Listeria monocytogenes* and Salmonella kentucky in Norwegian fermented, dry sausage. Food Microbiol 15:273–279.

G-JE Nychas, JS Arkoudelos. 1990. Staphylococci: Their role in fermented sausages. In: Staphylococci. D Jones, RG Board, C Collins, eds. Symposium Series No. 19, Society for Applied Bacteriology. London: Blackwell Scientific Publishers.

G-JE Nychas, D Marshall, J Sofos. 2007. Chapter 6: Meat poultry and seafood in food microbiology fundamentals and frontiers. Doyle, Beuchat, and Montville, eds. Herndon, Virginia: ASM Press.

TJ Pond, DS Wood, IM Mumin, S Barbut, MW Griffiths. 2001. Modelling the survival of *Escherichia coli* 0157:H7 in uncooked, semidry, fermented sausage. J Food Prot 64:759–766.

MM Pullen, CA Genigeorgis. 1977. A study of coagulase-positive staphylococci in salami before fermentation. J Food Protect 40:704–708.

K. Rantziou, R Urso, L Iacumin, C Cantoni, PCG Comi, L Cocolin. 2005. Culture-dependent and -independent methods to investigate the microbial ecology of Italian fermented sausages. Appl Environ Microbiol 71:1977–1986.

CA Reed. 1995. Challenge study-*Escherichia coli* O157:H7 in fermented sausage. USDA Food Safety and Inspection Service, Washington, DC, 21 August, letter to Plant Managers.

DCR Riordan, C Duffy, JJ Sheridan, RC Whiting, IS Blair, DA McDowell. 2000. Effects of acid adaptation, product pH, and heating on survival of *Escherichia coli* O157:H7 in pepperoni. Appl Environ Microbiol 66:1726–1729.

M. Rocelle, S Clavero, LR Beuchat 1996. Survival of *Escherichia coli* O157:H7 in broth and processed salami as influenced by pH, water activity, and temperature and suitability of media for its recovery. Appl Environ Microbiol 62:2735–2740.

PZ Sakhare, DN Rao. 2003. Microbial profiles during lactic fermentation of meat by combined starter cultures at high temperatures. Food Control 14:1–5.

J Samelis, A Kakouri, IN Savvaidis, K Riganakos, MG Kontominas. 2005. Use of ionizing radiation doses of 2 and 4kGy to control *Listeria* spp. and *Escherichia coli* 0157:H7 on frozen meat trimmings used for dry fermented sausages. Meat Sci 70:189–195.

J Samelis, J Metaxopoulos 1999. Incidence and principal sources of *Listeria* spp. and *Listeria monocytogenes* contamination in processed meats and a meat processing plant. Food Microbiol 16: 465–477.

J Samelis, J Metaxopoulos, M Vlassi, A Pappa. 1998. Stability and safety of traditional Greek salami—A microbiological ecology study. Int J Food Microbiol 44:69–82.

T Sameshima, C Magome, K Takeshita, K. Arihara, M Itoh, Y Kondo. 1998. Effect of intestinal Lactobacillus starter cultures on the behaviour of *Staphylococcus aureus* in fermented sausage. Int J Food Microbiol 41:1–7.

C Sauer, J Majkowski, S Green, R Eckel. 1997. Foodborne illness outbreak associated with a semi-dry fermented sausage product. J Food Prot 60:1612–1617.

C Shadbolt, T Ross, TA McMeekin, 2001. Differentiation of the effects of lethal pH and water activity: Food safety implications. Lett Appl Microbiol 32:99–102.

B Siriken, S Pamuk, C Ozakin, S Gedikoglu, M Eyigot. 2006. A note on the incidences of *Salmonella* spp., *Listeria* spp. and *Escherichia coli* O157:H7 serotypes in Turkish sausage (Soudjouck). Meat Sci 72:177–181.

JL Smith, CN Huhtanen, JC Kissinger, SA Palumbo. 1975. Survival of Salmonellae during pepperoni manufacture. Appl Microbiol 30:759–763.

JL Smith, SA Palumbo. 1978. Injury to *Staphylococcus aureus* during sausage fermentation. Appl Environ Microbiol 36:857–860.

D Thévenot ML Delignette-Muller, S Christieans, C Vernozy-Rozand. 2005a. Prevalence of *Listeria monocytogenes* in 13 dried sausage processing plants and their products Int J Food Microbiol 102: 85–94.

———. 2005b. Fate of *Listeria monocytogenes* in experimentally contaminated French sausages. Int J Food Microbiol 101:189–200.

J Tilden, Jr., W Young, A-M McNamara, C Custer, B Boesel, MA Lambert-Fair, J Majkowski, D Vugia, SB Werner, J Hollingsworth, JG Morris, Jr. 1996. A new route of transmission for *Escherichia coli*: Infection from dry fermented salami. Am J Pub Health 86:1142–I 145.

RB Tompkin. 2002. Control of *Listeria monocytogenes* in the food processing environment. J Food Prot 65:709–725.

RB Tompkin, LN Christiansen, AB Shaparis, H Bolin. 1974. Appl Microbiol 28:262–266

US Department of Agriculture, Food Safety and Inspection Service. 2001. Performance standards for the production of processed meat and poultry products; proposed rule. Fed Regist 66:12590–12636.

———. 2003. Control of *Listeria monocytogenes* in ready-to-eat meat and poultry products; final rule. Fed Regist 68:34207–34254.

———. 2004. Compliance guidelines to control *Listeria monocytogenes* in post-lethality exposed ready-to-eat meat and poultry products. Available at: http://www.fsis.usda.gov/OPPDE/rdad/FRPubs/97013F/CompGuidelines.pdf. Accessed April 19, 2004.

M Uyttendaele, S Vankeirsbilck, J Debevere. 2001. Recovery of heat-stressed *E. coli* 0157:H7 from ground beef and survival of *E. coli* 0157:H7 in refrigerated and frozen ground beef and in fermented sausage kept at 7°C and 22°C. Food Microbiol 18: 511–519.

F Villani, L Sannino, G Moschetti, G Mauriello, O Pepe, R Amodio-Cocchieri, S Coppola. 1997. Partial characterization of an antagonistic substance produced by *Staphylococcus xylosus* 1E and determination of the effectiveness of the producer strain to inhibit *Listeria monocytogenes in* Italian sausages. Food Microbiol 14:555–566.

T Zhao, MP Doyle, RE Besser. 1993. Fate of enterohemorragic *Escherichia coli* O157:H7 in apple cider with and without preservatives. Appl Environ Microbiol 59:2526–2530.

43
Biogenic Amines: Risks and Control

M. Carmen Vidal-Carou, M. Teresa Veciana-Nogués, M. Luz Latorre-Moratala, and Sara Bover-Cid

INTRODUCTION: BIOGENIC AMINE CLASSIFICATION AND RELEVANCE

Biogenic amines, also known as *biologically active amines*, are nitrogenous low-molecular weight substances of biological origin found in a wide range of concentrations in nearly all types of foods. In food products, these compounds can be classified in three groups according to their origin and chemical structure: aromatic biogenic amines, aliphatic diamines, and natural polyamines (Figure 43.1). Biogenic amines are produced mainly by decarboxylation of precursor amino acids by specific microbial enzymes. This group includes aromatic amines (tyramine, phenylethylamine, histamine, and tryptamine) and aliphatic amines (putrescine, cadaverine, and agmatine). The physiologically natural polyamines, spermine and spermidine, are not associated with microbial activity and their biosynthesis follows a more complex process. Low levels of putrescine can also be considered of physiological or natural origin and are a precursor of spermidine (Bardócz 1995; Izquierdo-Pulido et al. 1999).

Interest in biogenic amines arose in the 1960s after reports of the so-called "cheese reaction," a severe hypertensive crisis after the consumption of tyramine-rich cheese. This reaction is caused by the interaction of these compounds with monoamine-oxidase inhibitor (MAOI) drugs and, in some cases, may have lethal consequences (Blackwell 1963). Indeed, studies on biogenic amines in food have traditionally focused on their potentially toxicological effects caused by vasoactive and/or psychoactive properties. Although these effects are well documented, information about the mechanisms and toxicological doses as well as individual factors that determine the severity of reactions is still inconsistent. Therefore, it is difficult to establish the maximum tolerable limits of amines at which a potentially toxic food products should be rejected. However, the severity of the above mentioned reaction is now reduced, mainly because of three factors: (1) most MAOI drugs currently used are selective inhibitors of one MAO isoenzyme, thereby leaving the other available for biogenic amine metabolism; (2) patients treated with MAOI drugs generally follow a tyramine-restricted diet; and (3) biogenic amine content in food is now lower than in the past thanks to improved manufacturing practices.

Furthermore, biogenic amines are relevant from the hygienic and technological points of view (Brink et al. 1990; Mariné-Font et al. 1995; Izquierdo-Pulido et al. 1999). Several biogenic amines (mainly diamines and histamine) may result from the activity of contaminant bacteria, for which these compounds are potentially useful chemical indicators of defective hygienic conditions of raw materials and/or manufacturing processes. As microbial metabolites, biogenic amines (mainly tyramine) have been closely associated with milk and meat fermentation, and it was even believed that the occurrence of tyramine was inherent to the manufacturing process of fermented products. However, fermented foods can also show low levels of this amine in particular and biogenic amines in general. Therefore, it is of

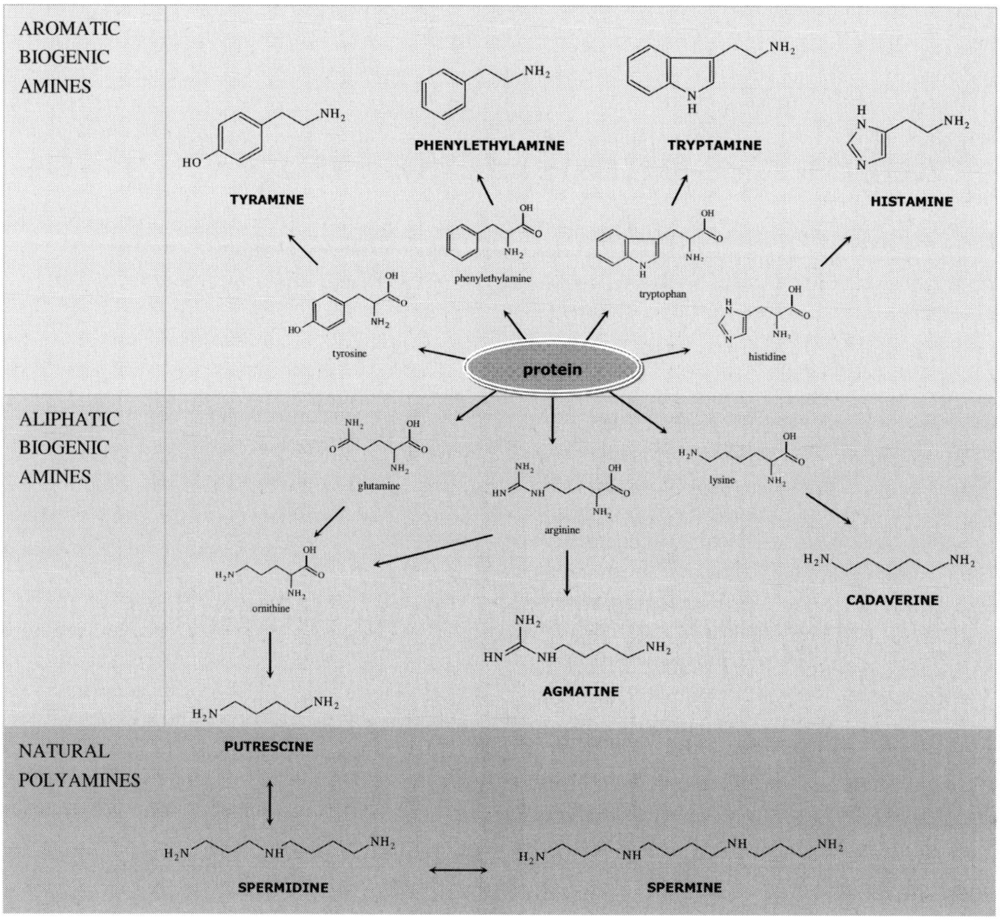

Figure 43.1. Chemical structure and formation pathway of biologically active amines.

interest to determine the levels of biogenic amines (as a sum of amines or of a particular amine) that could be regarded as normal and/or unavoidable, and thereby to establish a threshold value that would indicate improper hygienic practices.

In this chapter, biogenic amines in fermented sausages are reviewed in relation to safety, hygienic, and technological issues. The toxicological risks for consumers as well as the prospects for manufacturers to reduce aminogenesis during fermentation of this product are discussed.

HEALTH RISKS OF BIOGENIC AMINES IN FERMENTED SAUSAGES

Tyramine, histamine, and to a lesser extent phenylethylamine are the main dietary biogenic amines associated with several acute adverse reactions. The mechanisms of action of these compounds involve vasoactive (vasoconstriction for tyramine and phenylethylamine, and vasodilatation for histamine) and psychoactive reactions. In addition to interaction with MAOI drugs, the ingestion of biogenic amines may cause other health complaints, such as histaminic intoxication, histamine intolerance due to enteral histaminosis, and food-induced migraine (Taylor 1986; Bardócz 1995; Mariné-Font et al. 1995; Shalaby 1996; Amon et al. 1999). The diamines putrescine and cadaverine are not toxic by themselves, but they may enhance the absorption of vasoactive amines because of saturation of intestinal barriers. Competition for the mucine attachment sites and detoxification enzymes has been proposed to explain the potentiating effect of these diamines (Chu and Bjeldanes 1981; Hui and Taylor 1985; Taylor 1986).

Despite the recognized wide inter- and intra-individual variability of susceptibility to biogenic amines, the enzymatic intestinal and hepatic barriers to these compounds in nonmedicated healthy individuals are efficient, and dietary sources of these compounds are not a health risk under normal circumstances. However, certain genetic factors, some intestinal diseases, the occurrence of dietary potentiators (alcohol, other biogenic amines, etc.), and several drugs that inhibit MAO and diamino oxidase (DAO) enzymes, responsible for biogenic amine metabolism, may notably increase susceptibility to biogenic amines; consequently the risk of adverse reactions may be significant (Mariné-Font et al. 1995; Shalaby 1996; Lehane and Olley 2000).

Unfortunately, little information on the incidence of intoxication by dietary amines is available. Moreover, the wide variability of biogenic amine content in food products hinders the estimation of the toxic dose on the basis of epidemiological data.

Although the precise amount of tyramine required to produce adverse effects is not known, according to the literature, 6 mg of tyramine could cause mild crisis and 10–25 mg may lead to a severe headache with intracranial hemorrhage in patients treated with classical MAOI drugs (McCabe 1986). In contrast, from 50–150 mg of tyramine would be well tolerated by patients receiving a new generation of MAOI agents (Korn et al. 1988; Dingemanse et al. 1998; Patat et al. 1995). Neither is there consensus regarding the dose of histamine that provokes toxic reactions. In the risk assessment review of histamine fish poisoning, Lehane and Olley (2000) proposed 30 mg of histamine as a safe dose, calculated from the 100 mg/kg histamine level established to protect public health and safety (e.g., Council Directive 91/493/EEC of the EC 1991) and considering 300 g as a fish serving. The same authors, however, consider this calculation of limited value.

Biogenic amines in food products are precursors of nitroso compounds with potential carcinogenic activity, which constitutes an indirect additional risk associated with dietary amines. Nitrosamines result from the action of nitrite on secondary amines. The primary amines, such as the aliphatic diamines and polyamines, may cyclize to secondary amines under certain conditions (Warthesen et al. 1975; Brink et al. 1990; Shalaby 1996). Cured and heated/smoked meat products (mainly cooked and fried bacon) are frequently described as nitrosamine sources. The manufacturing process of these products involves nitrosating agents (such as nitrites and their precursor nitrates), mild acidic pH, and high temperatures, all conditions that favor nitrosamine formation. On few occasions, fermented sausages have also been reported to contain small amounts of nitrosamines. For instance, examination of the content of four volatile nitrosamines in dry sausages made in Finland showed very low or even undetectable amounts (Eerola et al. 1997, 1998a). *N*-nitrosopyrrolidine was the main and relatively most frequently detected nitrosamine in correlation with the amount of putrescine (its precursor). Some aromatic amines, such as tyramine, have been also studied as possible precursors of the mutagenic diazotyramine (Ochiai et al. 1984). Nevertheless, the limiting factor in nitrosamine formation is not the availability of precursors but the amount of nitrosating agent (McIntyre and Scanlan 1993). The consequences of dietary exposure to nitroso compounds have not been accurately established. These compounds show carcinogenic properties in laboratory animals and probably induce cancer in humans, although convincing data and epidemiological evidence are still lacking. Furthermore, nitrosamines may also be formed endogenously from precursors in the gastrointestinal tract (Vermeer et al. 1998). According to several risk assessments (i.e., Gangolli et al. 1994; Fernlöf and Darnerud 1996), dietary sources of nitrite and N-nitroso compounds make a small contribution to the body pool of nitroso compounds, and the main source of these compounds would be attributed to the intestinal bacterial metabolism of ingested nitrates.

AMINOGENESIS IN FERMENTED SAUSAGES AND MEASURES FOR ITS CONTROL

The fermentation of food in general, and meat sausages in particular, offers optimal conditions for biogenic amine accumulation: growth of microorganisms (some with recognized aminogenic capacity), and proteolytic phenomena that increase the amounts of precursor amino acids. The production of weak organic acids is an additional factor favoring the aminogenic activity of microorganisms, which produce biogenic amines as a defense mechanism against an unfavorable acidic environment. Although a wide number of bacteria have aminogenic potential, amino acid decarboxylases are not homogeneously distributed in all microorganisms, but are strain-dependent (Bover-Cid and Holzapfel 1999). Indeed, sausages with comparable microbial profiles may differ greatly in their amine content, indicating that the accumulation of these compounds is governed by a complex interaction of microbial, physicochemical and technological factors (Suzzi and Gardini 2003).

Biogenic Amine Content in Fermented Sausages of the Retail Market

Fermented sausages and cheese are the two food products that show the highest biogenic amine contents. Table 43.1 summarizes the reports describing the occurrence of biogenic amines in fermented meat products retailed in several countries. According to the literature, levels of biogenic amines in fermented sausages show a large variation among types of products, manufacturers, and also samples from distinct batches of the same kind of product made by the same producer. However, tyramine is usually the most frequent and most abundant biogenic amine found in fermented sausages. The occurrence of diamines, putrescine and cadaverine, is also quite common but more variable than tyramine. Most samples show relatively low amounts of diamines and only a few accumulate higher amounts than tyramine. By contrast, histamine is rare in fermented sausages, but in some particular samples it may reach quite high levels, usually accompanied by high amounts of other biogenic amines. Similarly, the contents of phenylethylamine and tryptamine are relatively low. These two amines could be considered minor amines in fermented sausages and their accumulation appears to depend on the occurrence of high contents of tyramine.

Fermented sausages are significant sources of physiological polyamines. However, these types of polyamines have received less attention than other amines (Pavel and Krausová 2005). The natural origin of these polyamines in raw meat implies that their levels are less variable than those of biogenic amines derived from microbial metabolism. Also, spermine levels are higher than those of spermidine in food products of animal origin.

Biogenic Amine Production by Meat-Related Microorganisms

Microbial flora, either present in raw meat or also added as starter culture, has a strong effect on biogenic amine formation during the fermentation and ripening of meat products. The production of these compounds in meat and meat products has often been related to a variety of microorganisms, including enterobacteria, pseudomonads, lactobacilli, enterococci, carnobacteria, and staphylococci (Table 43.2). Indeed, many meat enterobacteria and pseudomonads strains produce putrescine and cadaverine. Particular strains, especially among enterobacteria, have been also described as histamine producers (Roig-Sagués et al. 1996), although less often and to a lower extent than the enterobacteria isolated from fish products. *Enterobacteriaceae* and *Pseudomonas* are contaminant gram-negative bacteria. They are usually present in low numbers in the final sausage product or even totally inhibited in the first days of fermentation because of unfavorable environmental conditions for growth during production and storage. Consequently, these two strains of bacteria contribute to biogenic amine accumulation mainly during the early steps of sausage production, particularly when raw meat is subjected to lengthy storage or high temperatures or when hygienic practices are not followed (Paulsen and Bauer 1997; Bover-Cid et al. 2000a; Bover-Cid et al. 2003). It has been also suggested that a considerable contribution to biogenic amine accumulation could derive from the activities of viable but nongrowing cells, or even from the decarboxylase enzymes released at the beginning of fermentation by these contaminating flora (Halász et al. 1994; Suzzi and Gardini 2003). Nevertheless, no specific studies confirm these hypotheses. In any case, the occurrence of excessive levels of biogenic amines, such as diamines and histamine, could be indicative of contamination of raw materials or hygienic failure during processing, regardless of the presence of aminogenic microorganisms in the final product.

Among gram-positive bacteria, lactobacilli strains belonging to *L. curvatus*, *L. buchneri*, and *L. brevis* are the most potent tyramine producers. Some of these species may also produce other aromatic amines (phenylethylamine or tryptamine) simultaneously to tyramine, or even diamines (putrescine and/or cadaverine) (Straub et al. 1995; Bover-Cid et al. 2001a; Aymerich et al. 2006). Histidine-decarboxylase capacity is delimited to some strains of a reduced number of species (for instance *L. hilgardii,*), which are not commonly found but arise by specific contaminations (Maijala and Eerola 1993; Paulsen and Bauer 1997). By contrast, *L. sakei*, *L. bavaricus* and *L. plantarum* show a lower proportion of aminogenic strains (Bover-Cid et al. 2001a; Aymerich et al. 2006). A very high number of enterococci strains produce considerable amounts of tyramine, and the strongest tyrosine-decarboxylase-positives also have the capacity to decarboxylate phenylethylamine (Bover-Cid et al. 2001a; Gardini et al. 2001). Much less information is available about the capacity to produce biogenic amines of the gram-positive catalase-positive cocci, namely staphylococci, micrococci, and kocuria. Although some strains produce putrescine, tyramine, phenylethylamine, or histamine (Straub et al. 1994; Montel et al. 1999; Martín et al. 2006), they are usually described as weak or negative decarboxylase microorganisms (Masson et al. 1996; Martuscelli et al. 2000; Bover-Cid et al. 2001a; Martín et al. 2006).

Table 43.1. Biogenic amine contents (mg/kg fresh matter) in fermented sausages of the retail market from several countries.

	Reference	Product	n	Tyramine	Histamine	Phenylethyl-amine	Tryptamine	Cadaverine	Putrescine	Spermidine	Spermine
Spanish Sausage	Vidal-Carou et al. (1990)	Chorizo	11	131 Md[a] 176 ± 149[b] (2–509)[c]	46 Md 76 ± 80 (2–249)	–[d]	–	–	–	–	–
		Salchichón	19	135 Md 133 ± 62 (35–270)	5 Md 18 ± 27 (1–103)	–	–	–	–	–	–
		Salami	5	5 Md 6 ± 3 (3–12)	73 Md 66 ± 39 (2–102)	–	–	–	–	–	–
		Sobrasada	3	8 Md 8 ± 6 (3–14)	75 Md 55 ± 36 (14–78)	–	–	–	–	–	–
	Hernandez-Jover et al. (1997a)	Chorizo	20	282 ± 129 (30–627)	18 ± 27 (0–314)	1 ± 3 (0–52)	16 ± 20 (0–88)	20 ± 16 (0–658)	60 ± 141 (3–416)	4 ± 3 (2–10)	26 ± 8 (14–44)
		Salchichon	22	281 ± 109 (53–513)	7 ± 14 (0–151)	7 ± 6 (0–35)	9 ± 11 (0–65)	12 ± 23 (0–342)	103 ± 76 (6–400)	5 ± 3 (1–14)	15 ± 8 (7–43)
		Fuet	11	191 ± 73 (32–743)	2 ± 40 (0–358)	2 ± 4 (0–34)	9 ± 8 (0–68)	19 ± 18 (5–51)	72 ± 41 (2–222)	5 ± 3 (1–11)	17 ± 7 (9–30)
		Sobrasada	7	332 ± 131 (58–501)	9 ± 17 (3 –143)	2 ± 6 (0–39)	12 ± 23 (0–65)	13 ± 14 (3–42)	65 ± 50 (2–501)	3 ± 2 (2–7)	14 ± 7 (10–18)
	Bover-Cid et al. (1999a)	Thin fuet	15	68 Md 92 ± 72 (1–218)	0.6 Md 1 ± 2 (0–5)	1 Md 4 ± 8 (0–29)	0 Md 5 ± 11 (0–39)	11 Md 43 ± 48 (1–115)	15 Md 80 ± 152 (1–513)	5 Md 7 ± 5 (1–21)	30 Md 25 ± 14 (2–44)
		Fuet	23	103 Md 119 ± 64 (22–272)	2 Md 12 ± 34 (0–158)	3 Md 8 ± 13 (0–47)	4 Md 8 ± 11 (0–36)	8 Md 28 ± 42 (2–156)	34 Md 49 ± 43 (1–169)	6 Md 8 ± 9 (2–45)	23 Md 24 ± 16 (2–84)

(*Continues*)

Table 43.1. *(Continued)*

	Reference	Product	n	Tyramine	Histamine	Phenylethyl-amine	Tryptamine	Cadaverine	Putrescine	Spermidine	Spermine
		Salchichón	19	98 Md 141 ± 124 (3–490)	5 Md 14 ± 20 (0–59)	5 Md 12 ± 28 (0–126)	4 Md 15 ± 33 (0–142)	8 Md 18 ± 30 (0–127)	86 Md 99 ± 96 (0–325)	3 Md 4 ± 3 (1–13)	20 Md 25 ± 13 (7–52)
	Ruiz-Capillas and Jiminéz-Comenero (2004)	Chorizo	3	129 ± 100 (19–214)	6 ± 9 (1–16)	nd	nd	103 ± 113 (9–229)	92 ± 92 (0.8–185)	8 ± 1 (7–8)	46 ± 11 (39–59)
	Miguélez-Arrizado et al. (2006)	Dry sausage (industrial)	33	128 Md 143 ± 120 (6–675)	2 Md 28 ± 50 (0–186)	4 Md 10 ± 32 (0–186)	5 Md 14 ± 34 (0–194)	11 Md 47 ± 90 (2–466)	41 Md 91 ± 110 (0–410)	4 Md 6 ±5 (1–20)	22 Md 22 ± 9 (8–44)
		Dry sausage (traditional)	67	118 Md 139 ± 93 (1–433)	1 Md 14 ± 37 (0–205)	3 Md 8 ± 14 (0–61)	3 Md 7 ± 12 (0–52)	9 Md 26 ± 38 (0–156)	33 Md 63 ± 89 (0–537)	5 Md 6 ± 5 (1–29)	21 Md 24 ± 15 (2–78)
French Sausages	Montel et al. (1999)	Saucisson (industrial)	5	220 (172–268)	71 (16–151)	4 (0–8)	4 (0–9)	103 (31–192)	279 (195–410)	5 (4–6)	91 (59–119)
		Saucisson (traditional)	3	164 (84–217)	15 (15–16)	1 (0–4)	nd	71 (39–110)	223 (61–317)	4 (2–6)	84 (82–86)
Italian Sausages	Parente et al. (2001)	Soppressata	9	178 (0–557)	22 (0–101)	3 (0–20)	–	61 (0–271)	99 (0–416)	40 (0–91)	36 (0–98)
		Salsiccia	10	77 (0–339)	nd	nd	–	7 (0–39)	20 (0–78)	19 (0–57)	3 (0–28)
	Coisson et al. (2004)	Salamini Italiani	10	194 205 ± 105 (60–372)	19 Md 46 ± 54 (8–165)	0 Md 14 ± 20 (nd–53)	9 Md 20 ± 25 (nd–69)	–	–	–	–
Dutch Sausages	Brink et al. (1990)		14	110 (40–310)	11 (1–63)	14 (5–45)	–	63 (1–150)	52 (1–190)	–	–

Finnish Sausages	Eerola et al. (1998b)	Finnish sausage	11	88 (4–200)	54 (0–180)	13 (2–248)	14 (0–43)	50 (0–270)	79 (0–230)	4 (2–7)	31 (19–46)
		Russian sausage	4	110 (6–240)	89 (0–200)	11 (1–33)	22 (0–43)	10 (3–18)	93 (3–310)	5 (2–8)	33 (23–40)
		Danish sausage	8	54 (5–110)	9 (1–56)	2 (0–4)	27 (0–91)	180 (0–790)	130 (0–450)	7 (3–9)	37 (23–47)
		Meatwurst	12	72 (5–320)	21 (0–170)	3 (0–5)	18 (0–54)	6 (0–16)	77 (2–580)	6 (3–11)	29 (22–38)
		Lubeck	9	73 (9–150)	6 (0–40)	4 (0–7)	10 (0–20)	3 (0–8)	49 (0–220)	5 (3–7)	33 (20–40)
		Salami	13	93 (3–200)	3 (0–9)	5 (0–8)	20 (0–51)	14 (0–71)	54 (0–210)	5 (1–8)	30 (19–45)
		Pepperoni	11	94 (5–190)	21 (0–200)	6 (0–48)	18 (0–42)	82 (0–390)	61 (0–230)	6 (3–11)	33 (21–48)
Egyptian Sausages	Shalaby (1996)	Egyptian sausages	50	14 (10–53)	5 (7–41)	10 (2–81)	13 (3–34)	19 (6–39)	39 (12–100)	2 (5–12)	2 (2–5)
Turkish Sausages	Ekici et al. (2004)	Turkish dry sausage	46	–	32 ± 17 (20–87)	–	–	–	–	–	–
	Erkmen and Bozkurt (2004)	Sucuk—factory-made	19	62 ± 69 (1–189)	69 ± 83 (4–255)	9 ± 20 (0–87)	11 ± 14 (0–47)	–	75 ± 123 (0–383)	–	13 ± 14 (0–50)
		Sucuk—butcher-made	31	77 ± 92 (2–316)	94 ± 151 (2–478)	6 ± 9 (0–32)	25 ± 31 (0–7)	1 ± 2 (0–7)	121 ± 239 (0–919)	–	9 ± 9 (0–42)
Thai Sausages	Riebroy et al. (2004)	Som-fug	7	87 ± 72 (19–228)	120 ± 82 (55–291)	–	49 ± 25 (19–86)	161 ± 111 (20–328)	127 ± 90 (17–275)	–	–

[a] Md = median value shown when available; [b] mean ± standard deviation when available; [c] range (minimum-maximum); [d] –, not reported.

Table 43.2. Microorganisms, with a recognized capacity to produce one or more biogenic amines associated with meat and meat products.

Species	Tyramine	Phenylethylamine	Tryptamine	Histamine	Putrescine	Cadaverine	Reference
Carnobacterium divergens	+	+					9,13
Carnobacterium piscicola	+	+					9,13
Enterococcus faecalis	+	+					3,11
Enterococcus faecium	+	+					11
Lactobacillus bavaricus	+			+	+		3,13
Lactobacillus brevis	+		+	+	+	+	3,13,14
Lactobacillus buchneri	+			+	+		13,14
Lactobacillus curvatus	+	+	+	+	+	+	2,3,9,10, 11,13,14
Lactobacillus hilgardii				+			6,14
Lactobacillus plantarum	+	+					9,11
Pediococcus pentosaceus	+						9
Staphylococcus carnosus	+	+					1,5,7
Staphylococcus epidermidis		+			+	+	7
Staphylococcus xylosus	+						8
Staphylococcus warneri	+						7
Kocuria varians	+	+			+		1,9
Citrobacter freundii					+	+	4
Enterobacter aerogenes				+	+	+	4,10
Enterobacter cloacae				+	+	+	3,10,12
Escherichia coli				+			14
Hafnia alvei				+	+	+	12
Klebsiella oxytoca				+		+	3,6,10,14
Morganella morganii				+			6,14
Serratia liquefaciens				+	+	+	3,4, 11,12,14
Pseudomonas spp.					+		12

Sources: (1) Ansorena et al. (2002); (2) Aymerich et al. (2006); (3) Bover-Cid et al. (2001a); (4) Durlu-Özkaya et al. (2001); (5) Hammes et al. (1995); (6) Kranner et al. (1991); (7) Martín et al. (2006); (8) Marstuscelli et al. (2000); (9) Masson et al. (1996); (10) Roig-Sagués et al. (1996); (11) Roig-Sagués et al. (1997); (12) Slerm (1981); (13) Straub et al. (1995); (14) Tschabrun et al. (1990).

It should be highlighted that the results reported in the literature about aminogenic bacterial activity obtained in laboratory media do not imply the same behavior in true fermentation processes, because environmental conditions and microbial interactions are not reproducible. For instance, the production of putrescine by enterobacteria in meat is conditioned to a previous deamination of arginine by lactic acid bacteria, which yields the precursor ornithine, a substance that is not naturally present in food protein structures (Dainty et al. 1986; Brink et al. 1990). It is feasible to associate diamines with the activity of undesired contaminant gram-negative bacteria, although fermenting flora, such as some lactobacilli and staphylococci, may occasionally be involved. Histamine formation would be shared by specific strains of species belonging mainly to enterobacteria but also to some lactic acid bacteria, which occur rarely in fermented sausages. In contrast, tyramine accumulation would be related mainly to lactic acid bacteria, enterococci, and specific strains of lactobacilli species (especially *L. curvatus*) showing a remarkable contribution.

Factors Affecting Biogenic Amine Accumulation in Fermented Sausages

Meat constitutes a protein-rich product with high water activity and a variety of nutrients that support the growth of a wide number of microorganisms. During the fermentation and ripening of sausages, the initially dominant gram-negative flora (such as *Enterobacteriaceae* and *Pseudomonas*) is replaced by the more salt- and nitrite-tolerant and competitive gram-positive species (such as lactic acid bacteria and staphylococci and micrococci). Among the variables affecting aminogenesis in fermented sausages, those related to microbial contamination of raw materials as well as the conditions governing the selection of the microbial communities and the activity of amino acid decarboxylase bacteria are of crucial relevance.

Therefore, in fermented sausage production, particular attention should be paid to the hygienic quality of meat and ingredients and also to the use and type of starter culture (Maijala et al. 1995a; Paulsen and Bauer 1997; Bover-Cid et al. 2000a,b; Ansorena et al. 2002). Indeed, high amounts of cadaverine and putrescine, as well as histamine, have been related to the poor hygienic quality of raw materials. In particular, enterobacteria contributes to cadaverine accumulation during the ripening of fermented sausages made from meat stored for long periods (Paulsen and Bauer 1997; Bover-Cid et al. 2000a) and also from intentionally contaminated raw meat (Bover-Cid et al. 2003).

In addition, the ingredients (sugars, additives such as curing agents, species, etc.) and diameter of the sausage, as well as the temperature and humidity during ripening affect biogenic amine accumulation (Maijala et al. 1995b; Bover-Cid et al. 1999a, 2001b; Parente et al. 2001; González-Fernández et al. 2003; Suzzi and Gardini 2003; Bozkurt and Erkmen 2004; Komprda et al. 2004; Garriga et al. 2005; Latorre-Moratalla et al. 2007). However, most of these variables can show paradoxical effects on the accumulation of these compounds because they exert influences through several phenomena associated with aminogenesis, including growth and interaction among microbial communities, acidification, proteolysis, decarboxylase enzyme production and activity, etc. For instance, the pH regulates the activity of amino acid decarboxylases as a physiological system of bacteria to neutralize an unfavorable acidic environment. Indeed, the maximum rate of tyramine accumulation during sausage fermentation usually occurs when pH falls. A relationship between acidification and biogenic amine content has been reported (Santos-Buelga et al. 1986; Buncic et al. 1993; Teodorovic and Buncic 1999; Bover-Cid et al. 1999b; Miguélez-Arrizado et al. 2006). However, the effect of pH on overall aminogenesis in fermented sausages is controversial, because it is also accepted that rapid and sharp acidification inhibits contaminant bacteria and the consequent formation of biogenic amines (Maijala et al. 1993, 1995a; Bover-Cid et al. 2000b, 2001c; González-Fernández et al. 2003). In summary, the type and the amount of biogenic amines in a product depend on multiple and complex variables, all of which interact, making difficult to characterize the effects of each technological factor on aminogenesis during sausage fermentation and ripening. Therefore, it is important to study each particular case in order to elucidate when and why the aminogenesis occurs and then specifically apply the best measures of control.

Control Measures to Reduce Biogenic Amine Accumulation in Fermented Sausages

Literature reported biogenic amine levels for fermented sausages ranging from very small amounts to nondetectable levels. These observations indicate that it is possible to produce fermented meat products free or nearly free of biogenic amines. With this aim, several alternatives have been put forward.

Several authors have proposed the removal of amines after their formation. Some microorganisms of the sausage environment, such as a number of micrococci and staphylococci strains, break down tyramine and/or histamine in vitro by means of amine oxidase enzymes (Leuschner and Hammes 1998; Martuscelli et al. 2000). However, under real conditions of sausage fermentation, amine-oxidizing microorganisms show a limited effect on tyramine and histamine levels (Leuschner and Hammes 1998; Gardini et al. 2002). This ineffectiveness may be explained by low oxygen availability in the sausage and also amine-oxidizing bacteria below the minimum 10^7 cfu/g required for amine degradation (Leuschner and Hammes 1998). Degradation of biogenic amines has also been achieved by means of gamma irradiation. Doses from 2.5 kGy to 25 kGy can destroy from 5% to 100% of biogenic amines dissolved in distilled water (Kim et al. 2004). In sliced pepperoni sausages, radiolysis causes a slight reduction of putrescine, tyramine, spermidine, and spermine, but no effect on phenylethylamine and cadaverine (Kim et al. 2005). Nevertheless, given that biogenic amines may indicate poor hygienic conditions, their removal a posteriori should be considered inappropriate or nonethical because it would imply lack of control over the quality of raw materials and processing conditions.

Therefore, to reduce or prevent biogenic amine accumulation in fermented sausages "prevention is better than cure." It would be more suitable to apply measures focused on the minimization of potentially aminogenic microorganisms, their growth, and their amino acid decarboxylase activity during production and merchandising. Given the importance of the hygienic quality of raw meat material, the selection of these materials, the control of storage time, and temperature and freezing-thawing process are crucial to prevent excessive accumulation of biogenic amines during sausage manufacture (Maijala et al. 1995a; Bover-Cid et al. 2000a, 2001d). In contrast to other fermented products, such as cheese, the raw materials for fermented sausages can not be pasteurized nor sterilized because these processes lead to detrimental changes from both sensorial and technological points of view. However, attempts to optimize the hygienic quality of raw materials by reducing the loads of potentially aminogenic microorganisms in raw materials have been made by means of freezing meat and lard (Bover-Cid et al. 2006) and also by applying high hydrostatic pressure to meat batter before fermentation (Latorre-Moratalla et al. 2007). These two strategies inhibit enterobacteria development and, as a result, contribute to a notable reduction of cadaverine accumulation.

However, the microorganisms responsible for fermentation, such as lactic acid bacteria and gram-positive catalase-positive cocci (staphylococci and micrococci), can also produce biogenic amines. Enterococci may also be present, and even multiply, during sausage fermentation because these strains are tolerant to acid, sodium chloride, nitrite, and other environmental stress factors, and they have also been associated with the production of biogenic amines, mainly tyramine (Bover-Cid et al. 2001a; Ansorena et al. 2002). Therefore, to control these gram-positive aminogenic microorganisms, which are inevitably in raw materials, specific technological measures are required. In this field, the use of selected amino acid decarboxylase-negative starter cultures has been repeatedly recommended as one of the most reliable tools by which to control fermentation and, in turn, biogenic amine accumulation. Several studies have examined the usefulness of starter cultures, including commercial strains and single and mixed cultures of experimental strains, to reduce aminogenesis during sausage fermentation. A number of these studies failed to demonstrate a consistent efficiency of *Lactobacillus plantarum*, *Pediococcus acidilatici*, *P. pentosaceus*, or *Staphylococcus carnosus* to reduce aminogenesis (Rice and Koehler 1976; Buncic et al. 1993; Paulsen and Bauer 1997; Bozkurt and Erkemn 2002). Starter cultures consisting of *L. pentosus*, *S. carnosus*, *S. xylosus*, or *Kocuria varians* reduce the accumulation of some amines, but fail to inhibit the production of others (Maijala et al. 1995a; Hernández-Jover et al. 1997b; Bover-Cid et al. 1999b; Ayhan et al. 1999). The species *L. sakei* and *L. curvatus* are well adapted to the meat fermentation environment, which make them good candidates for starter cultures because they are highly competitive to outgrow spontaneous fermenting flora and can inhibit gram-negative contaminants (Hugas and Monfort 1997). Indeed, the literature confirms that starter cultures, including decarboxylase-negative strains of *L. sakei* are the most protective because they reduce the overall amine accumulation by up to 95% in comparison with other species commercially used as starters, such as *L. plantarum*, *Pediococcus* spp., and *S. carnosus* (González-Fernández et al. 2003; Bover-Cid et al. 2000b, 2000c; Latorre-Moratalla et al. 2007). Nevertheless, the effectiveness of the starter culture is strongly conditioned by raw materials of appropriate hygienic standards (Maijala et al. 1995a; Eerola et al. 1996; Paulsen and Bauer 1997; Bover-Cid et al. 2001d). Moreover, optimization of the technological conditions favors proper implantation and development of the starter. Therefore, the extent of aminogenesis reduction also depends on sausage formulation, the type and amount of sugar added (González-Fernández et al. 2003), the addition of additives (Bozkurt and Erkemn 2002), and the temperature and relative humidity of the ripening conditions (Maijala et al. 1995b; Bover-Cid et al. 2001e).

CONCLUDING REMARKS

A multidisciplinary approach combining the chemical, microbiological, and technological aspects of biogenic amine accumulation in foods has been a cornerstone of the research in this field. This approach has provided reliable answers about the origin and significance of amines as well as suitable technological measures to minimize their occurrence. Biogenic amines have no favorable effect on foodstuffs; on the contrary, in addition to the potential risks for sensitive consumers, they can reduce the overall hygienic quality of the product. Consequently, the food industry faces the challenge of producing foods with low levels of amines, thereby minimizing health risks and assuring the quality of the final product.

REFERENCES

Amon U, Bangha E, Kuster T, Menne A, Vollrath IB, Gibbs BF. 1999. Enteral histaminosis: Clinical implications. Inflamm Res 48(6):291–295.

Ansorena D, Montel MC, Rokka M, Talon R, Eerola S, Rizzo A, Raemaekers M, Demeyer D. 2002. Analysis of biogenic amines in northern and southern European sausages and role of flora in amine production. Meat Sci 61(2):141–147.

Ayhan K, Kolsarici N, Ozkan GA. 1999. The effects of a starter culture on the formation of biogenic amines in Turkish soudjoucks. Meat Sci 53(3): 183–188.

Aymerich T, Martín B, Garriga M, Vidal-Carou MC, Bover-Cid S, Hugas M. 2006. Safety properties and molecular strain typing of lactic acid bacteria from slightly fermented sausages. J Appl Microbiol 100: 40–49.

Bardócz S. 1995. Polyamines in food and their consequences for food quality and human health. Trends Food Sci Tech 6:341–346.

Blackwell B. 1963. Hypertensive crisis due to monoamine-oxidase inhibitors. Lancet 2(731):849.

Bover-Cid S, Holzapfel W. 1999. Improved screening procedure for biogenic amine production by lactic acid bacteria. Int J Food Microbiol 53:33–41.

Bover-Cid S, Schoppen S, Izquierdo-Pulido M, Vidal-Carou MC. 1999a. Relationship between biogenic amine contents and the size of dry fermented sausages. Meat Sci 51:305–311.

Bover-Cid S, Izquierdo-Pulido M, Vidal-Carou MC. 1999b. Effect of proteolytic starter cultures of *Staphylococcus* spp. on biogenic amine formation during the ripening of dry fermented sausages. Int J Food Microbiol 46:95–104.

Bover-Cid S, Izquierdo-Pulido M, Vidal-Carou MC. 2000a. Influence of hygienic quality of raw materials on biogenic amine production during ripening and storage of dry fermented sausages. J Food Protect 63(11):1544–1550.

Bover-Cid S, Hugas M, Izquierdo-Pulido M, Vidal-Carou MC. 2000b. Reduction of biogenic amine formation using a negative amino acid-decarboxylase starter culture for fermentation of fuet sausages. J Food Protect 63(2):237–243.

Bover-Cid S, Izquierdo-Pulido M, Vidal-Carou MC. 2000c. Mixed starter cultures to control biogenic amine production in dry fermented sausages. J Food Protect 63(11):1556–1562.

Bover-Cid S, Hugas M, Izquierdo-Pulido M, Vidal-Carou MC. 2001a. Amino acid-decarboxylase activity of bacteria isolated from fermented pork sausages. Int J Food Microbiol 66:185–189.

Bover-Cid S, Miguélez-Arrizado MJ, Vidal-Carou MC. 2001b. Biogenic amine accumulation in ripened sausages influenced by the addition of sodium sulfite. Meat Sci 59(4):391–396.

Bover-Cid S, Izquierdo-Pulido M, Vidal-Carou MC. 2001c. Changes in biogenic amine and polyamine contents in slightly fermented sausages manufactured with and without sugar. Meat Sci 57:215–221.

Bover-Cid S, Izquierdo-Pulido M, Vidal-Carou MC. 2001d. Effectiveness of a *Lactobacillus sakei* starter culture in the reduction of biogenic amines accumulation as a function of the raw material quality. J Food Protect 64:367–373.

Bover-Cid S, Izquierdo-Pulido M, Vidal-Carou MC. 2001e. Effect of the interaction between a low tyramine-producing *Lactobacillus* and proteolytic staphylococci on biogenic amine production during ripening and storage of dry sausages. Int J Food Microbiol 65(1–2):113–123.

Bover-Cid S, Hernández-Jover T, Miguélez-Arrizado MJ, Vidal-Carou M. 2003. Contribution of contaminant enterobacteria and lactic acid bacteria to biogenic amine accumulation in spontaneous fermentation of pork sausages. Eur Food Res Technol 216:477–482.

Bover-Cid S, Miguélez-Arrizado MJ, Latorre-Moratalla ML, Vidal-Carou MC. 2006. Freezing of meat raw materials affects tyramine and diamine accumulation in spontaneously fermented sausages. Meat Sci 72:62–68.

Bozkurt H, Erkmen O. 2002. Effects of starter cultures and additives on the quality of Turkish style sausage (sucuk). Meat Sci 61(2):149–156.

———. 2004. Effects of temperature, humidity and additives on the formation of biogenic amines in Sucuk during ripening and storage periods. Food Sci Technol Int 10(1):21–28.

Brink B, Damink C, Joosten HM, Huis in 't Veld JH. 1990. Occurrence and formation of biologically active amines in foods. Int J Food Microbiol 11(1):73–84.

Buncic S, Paunovic L, Radisic D, Vojinovic G, Smiljanic D, Baltic M. 1993. Effects of gluconodeltalactone and *Lactobacillus plantarum* on the production of histamine and tyramine in fermented sausages. Int J Food Microbiol 17:303–309.

Chu C, Bjeldanes F. 1981. Effect of diamines, polyamines and tuna fish extracts on the binding of histamine to mucin in vitro. J Food Sci 47:79–80,88.

Coïsson JD, Cerutti C, Travaglia F, Arlorio M. 2004. Production of biogenic amines in "Salamini italiani alla cacciatora PDO." Meat Sci 67:343–349.

Dainty R, Edwards R, Hibbard C, Ramantanis S. 1986. Bacterial sources of putrescine and cadaverine in chill stored vacuum-packaged beef. J Appl Bacteriol 61:117–123.

Dingemanse J, Wood N, Guentert T, Oie S, Ouwerkerk M, Amrein R. 1998. Clinical pharmacology of moclobemide during chronic administration of high doses to healthy subjects. Psychopharmacology 140(2):64–172.

Durlu-Özkaya F, Ayhan K, Vural N. 2001. Biogenic amine produced by *Enterobacteriaceae* isolated from meat products. Meat Sci 58:163–166.

EC (European Commission). 1991. Council Directive 91/493/EEC of 22 July 1991 laying down the health conditions for the production and the placing on the market of fishery products. Official Journal of the European Communities L 268:15–34.

Eerola S, Maijala R, Sagues AXR, Salminen M, Hirvi T. 1996. Biogenic amines in dry sausages as affected by starter culture and contaminant amine-positive *Lactobacillus*. J Food Sci 61(6):1243–1246.

Eerola S, Roig-Sagués A, Lilleberg L, Aalto H. 1997. Biogenic amines in dry sausages during shelf-life storage. *Z* Lebensm Unters Forsch A 205:351–355.

Eerola S, Otegui I, Saari L, Rizzo A. 1998a. Application of liquid chromatography atmospheric pressure chemical ionization mass spectrometry and tandem mass spectrometry to the determination of volatile nitrosamines in dry sausages. Food Addit Contam 15(3):270–279.

Eerola HS, Roig-Sagués AX, Hirvi TK. 1998b. Biogenic amines in Finnish dry sausages. J Food Safety 18:127–138.

Ekici K, Sekeroglu R, Sancak YC, Noyan T. 2004. Note on histamine levels in Turkish style fermented sausages. Meat Sci 68(1):123–125.

Erkmen O, Bozkurt H. 2004. Quality Characteristics of retailed sucuk (Turkish dry-fermented sausage). Food Technol Biotechnol 42:63–69.

Fernlöf G, Darnerud PO. 1996. *N*-nitroso compounds and precursors in food—Level, intake and health effect data and evaluation of risk. *Livsmedelsverkets rapport* 15/96. Uppsala, Sweden.

Gangolli SD, Brandt PA, Feron VJ, Janzowsky C, Koeman JH, Speijers GJA, Spiegelhandler B, Walter R, Wishnow JS. 1994. Assessment: Nitrate, nitrite and *N-nitroso compounds*. Eur J Pharmac Environ Toxic Pharmac Section 292:1–38.

Gardini F, Martuscelli M, Caruso MC, Galgano F, Crudele MA, Favati F, Guerzoni ME, Suzzi G. 2001. Effects of pH, temperature and NaCl concentration on the growth kinetic, proteolytic activity and biogenic amine production of *Enterococcus faecalis*. Int J Food Microbiol 64:105–117.

Gardini F, Martuscelli M, Crudele MA, Paparela A, Suzzi G. 2002. Use of *Staphylococcus xylosus* as a starter culture in dried sausage: Effect on the biogenic amine content. Meat Sci 61:275–283.

Garriga M, Marcos B, Martín B, Veciana-Nogués MT, Bover-Cid S, Hugas S, Aymerich T. 2005. Starter cultures and high pressure processing to improve the hygiene and safety of slightly fermented sausages. J Food Protect 68(11):2341–2348.

González-Fernández C, Santos E, Jaime I, Rovira J. 2003. Influence of starter cultures and sugar concentrations of biogenic amine contents in chorizo dry sausage. Food Microbiol 20:275–284.

Halász A, Baráth A, Simon-Sarkadi L, Holzapfel W. 1994. Biogenic amines and their production by microorganisms in food. Trends Food Sci Tech 5:42–49.

Hammes W, Bosch I, Wolf G. 1995. Contribution of *Staphylococcus carnosus* and *Staphylococcus piscifermentans* to the fermentation of protein foods. J Appl Bact Symposium Suppl 79:76S–83S.

Hernández-Jover T, Izquierdo-Pulido M, Veciana-Nogués MT, Mariné-Font A, Vidal Carou MC. 1997a. Biogenic amine and polyamine contents in meta and meat products. J Agric Food Chem 45:2098–2102.

———. 1997b. Effect of starter cultures on biogenic amine formation during fermented sausage production. J Food Protect 60(7):825–830.

Hugas M, Monfort J. 1997. Bacterial starter cultures for meat fermentation. Food Chem 59(4):547–554.

Hui JY, Taylor SL. 1985. Inhibition of in vivo histamine-metabolism in rats by foodborne and pharmacologic inhibitors of diamine oxidase, histamine n-methyltransferase, and monoamine-oxidase. Toxicol Appl Pharm 81(2):241–249.

Izquierdo-Pulido M, Veciana-Nogués MT, Mariné-Font A, Vidal-Carou MC. 1999. Polyamine and biogenic amine evolution during food processing. In: Bardócz S, White A, eds. Polyamines in Health and Nutrition. London, UK: Kluwer Academic Publishers, pp. 139–159.

Kalac P, Krausová P. 2005. A review of dietary polyamines: Formation, implications for growth and health and occurrence in foods. Food Chem 90(1–2): 219–230.

Kim JH, Ahn HJ, Jo C, Park HJ, Chung YJ, Byun MW. 2004. Radiolysis of biogenic amines in model system by gamma irradiation. Food Control 15(5):405–408.

Kim JH, Ahn HJ, Lee JW, Park HJ, Ryu GH, Kang IJ, Byun MW. 2005. Effects of gamma irradiation on the biogenic amines in pepperoni with different packaging conditions. Food Chem 89(2):199–205.

Komprda T, Smela D, Pechova P, Kalhotka L, Stencl J, Klejdus B. 2004. Effect of starter culture, spice mix and storage time and temperature on biogenic amine content of dry fermented sausages. Meat Sci 67(4): 607–616.

Korn A, Da Prada M, Raffesberg W, Allen S, Gasic S. 1988. Tyramine pressor effect in man: Studies with moclobemide, a novel, reversible monoamine oxidase inhibitor. J Neural Transm Suppl 26:57–71.

Kranner P, Bauer F, Hellwig E. 1991. Investigations on the formation of histamine in raw sausages. Proc 37th Int Congr Meat Sci Tech (ICoMST) 2:889–891.

Latorre-Moratalla ML, Bover-Cid S, Aymerich T, Marcos B, Vidal-Carou MC, Garriga M. 2006. Aminogenesis control in fermented sausages manufactured with pressurized meat batter and starter culture. Meat Sci 75:460–469.

Lehane L, Olley J. 2000. Histamine fish poisoning revisited. Int J Food Microbiol 58(1–2):1–37.

Leuschner RGK, Hammes WP. 1998. Tyramine degradation by micrococci during ripening of fermented sausage. Meat Sci 49(3):189–196.

Maijala R, Eerola S. 1993. Contaminant lactic acid bacteria of dry sausages produce histamine and tyramine. Meat Sci 35:287–395.

Maijala R, Eerola S, Aho M, Him J. 1993. The effect of GDL-induced pH decrease on the formation of biogenic amines in meat. J Food Prot 56:125–129.

Maijala R, Eerola S, Lievonen S, Hill P, Hirvi T. 1995a. Formation of biogenic amines during ripening of dry sausages as affected by starter culture and thawing time of raw materials. J Food Sci 60(6):1187–1190.

Maijala R, Nurmi E, Fischer A. 1995b. Influence of processing temperature on the formation of biogenic amines in dry sausages. Meat Sci 39:9–22.

Mariné-Font A, Vidal-Carou MC, Izquierdo-Pulido M, Veciana-Nogués MT, Hernández-Jover T. 1995. Les amines biògenes dans les aliments: Leur signification, leur analyse. Ann Fals Exp Chim 88:119–140.

Martín B, Garriga M, Hugas M, Bover-Cid S, Veciana-Nogués MT, Aymerich T. 2006. Molecular, technological and safety characterization of Gram-positive catalase-positive cocci from slightly fermented sausages. Int J Food Microbiol 107:148–158.

Martuscelli M, Crudele M, Gardini F, Suzzi G. 2000. Biogenic amine formation and oxidation by *Staphylococcus xylosus* from artisanal fermented sausages. J Appl Microbiol 31:228–232.

Masson F, Talon R, Montel M. 1996. Histamine and tyramine production by bacteria from meat products. Int J Food Microbiol 32:199–207.

McCabe BJ. 1986. Dietary tyramine and other pressor amines in MAOI regimens: A review. J Am Diet Assoc 86(8):1059–1064.

McIntyre T, Scanlan R. 1993. Nitrosamines produced in selected foods under extreme nitrosation conditions. J Agric Food Chem 41:101–102.

Miguélez-Arrizado MJ, Bover-Cid S, Vidal-Carou MC. 2006. Biogenic amine contents in Spanish fermented sausages of different acidification degree as a result of artisanal or industrial manufacture. J Sci Food Agric 86:549–557.

Montel M, Masson F, Talon R. 1999. Comparison of biogenic amine content in traditional and industrial French dry sausages. Sci Aliment 19: 247–254.

Ochiai M, Wakabayashi K, Nagao M, Sugimura T. 1984. Tyramine is a major mutagen precursor in soy sauce, being convertible to a mutagen by nitrite. Gann 75(1):1–3.

Parente E, Martuscelli M, Gardini F, Grieco S, Crudele M, Suzzi G. 2001. Evolution of microbial populations and biogenic amine production in dry sausages produced in Southern Italy. J Appl Microbiol 90:882–891.

Patat A, Berlin I, Durrieu G, Armand P, Fitoussi S, Molinier P, Caille P. 1995. Pressor effect of oral tyramine during treatment with befloxatone, a new reversible monoamine oxidase-A inhibitor, in healthy subjects. J Clin Pharmacol 35(6):633–643.

Paulsen P, Bauer F. 1997. Biogenic amines in fermented sausages. 2. Factors influencing the formation of biogenic amines in fermented sausages. Fleischwirtsch Int 4:32–34.

Rice SL, Koehler PE. 1976. Tyrosine and histidine decarboxylases activities of *Pediococcus cerevisiae* and *Lactobacillus* species and the production of tyramine in fermented sausages. J Milk Food Tech 39(3):166–169.

Riebroy S, Benjakul S, Visessanguan W, Kijrongrojana K, Tanaka M. 2004. Some characteristics of commercial Som-fug produced in Thailand. Food Chem 88:527–535.

Roig-Sagués A, Hernández-Herrero M, López-Sabater E, Rodríguez-Jerez J, Mora-Ventura M. 1996. Histidine decarboxylase activity of bacteria isolated from raw and ripened salchichón, a Spanish cured sausage. J Food Protect 59(5):516–520.

Roig-Sagués AX, Hernández-Herrero M, Rodríguez-Jerez JJ, López-Sabater EI, Mora-Ventura MT. 1997. Occurrence of tyramine producing microorganisms in "salchichón" and tyramine production in sausages inoculated with a tyramine producing strain of *Lactobacillus brevis*. J Food Safety 17:13–22

Ruiz-Capillas C, Jiménez-Colmenero F. 2004. Biogenic amine content in Spanish retail market meta products treated with protective atmosphere and high pressure. Eur Food Res Technol 218:237–241.

Santos-Buelga C, Peña-Egido MJ, Rivas-Gonzalo JC. 1986. Changes in tyramine during Chorizo-sausages ripening. J Food Sci 51:518–527.

Shalaby AR. 1996. Significance of biogenic amines to food safety and human health. Food Res Int 29: 675–690.

Slerm J. 1981. Biogene Amine als potentieller chimischer qualitätsindikator für Fleish. Fleischwirtsch 61: 921–926.

Straub BW, Kicherer M, Schilcher SM, Hammes WP. 1995. The formation of biogenic amines by fermentation organisms. Z Lebensm Unters Forsh A 201: 79–82.

Straub BW, Tichaczek PS, Kicherer M, Hammes WP. 1994. Formation of tyramine by *Lactobacillus curvatus* LTH-972. Z Lebensm Unters Forsh A 199: 9–12.

Suzzi G, Gardini F. 2003. Biogenic amines in dry fermented sausages: A review. Int J Food Microbiol 88:41–54.

Taylor SL. 1986. Histamine food poisoning—Toxicology and clinical aspects. Crit Rev Toxicol 17(2):91–128.

Teodorovic V, Buncic S. 1999. The effect of pH on tyramine production in fermented sausages. Fleischwirtsch 79(5):85–88.

Tschabrun R, Sick K, Bauer F, Kranner P. 1990. Bildung von Histamin in schnittfesten Rohwürsten. Fleischwirtsch 70:448–452.

Vermeer IT, Pachen DM, Dallinga JW, Kleinjans JC, van Maanen JM. 1998. Volatile N-nitrosamine formation after intake of nitrate at the ADI level in combination with an amine-rich diet. Environ Health Perspect 106(8):459–463.

Vidal-Carou MC, Izquierdo ML, Martín MC, Mariné A. 1990. Histamina y tiramina en derivados cárnicos. Rev Agroquim Tecnol Aliment 30:102–108.

Warthesen JJ, Scanlan RA, Bills DD, Libbey LM. 1975. Formation of heterocyclic N-nitrosamines from the reaction of nitrite and selected primary diamines and amino acids. J Agric Food Chem 23(5):898–902.

44
Chemical Origin Toxic Compounds

Fidel Toldrá and Milagro Reig

INTRODUCTION

The development of modern analytical technologies linked to epidemiologic studies and investigations on safety aspects of food components have revealed in the latest decades that some toxic compounds may be present or generated in certain types of meat processing. This is the case of N-nitrosamines when using nitrite as a preservative under certain conditions, the polycyclic aromatic hydrocarbons generated in certain smoking processes, oxidation of lipids and proteins, and other products that may be present in the raw materials used for processing (veterinary drug residues, environmental contaminants, etc.). Even though meat fermentation is an old technology used for generations in many countries, corrective measures (i.e., reduction in the addition of nitrite, control of raw materials, etc.) have been taken to minimize this problem. All these hazardous compounds are briefly described in this chapter. Biogenic amines, which are generated by microbial decarboxylation of certain amino acids are dealt with in Chapter 43, "Biogenic Amines and Nitrosamines."

N-NITROSAMINES

Nitrite is the main additive used as a preservative in fermented meats because of its powerful inhibition of the outgrowth of spores of putrefactive and pathogenic bacteria like *Clostridium botulinum*. Nitrite also has some other technological roles, such as its contribution to the color through the formation of nitrosylmyoglobin, its contribution to the oxidative stability of lipids, and its contribution to the development of typical and distinctive cured meat flavor.

Nitrite can be converted to nitric oxide, which is a nitrosating agent. This agent can react with secondary amines and produce potent carcinogenic nitrosamines. Compounds such as N-nitrosodimethylamine have been shown to be carcinogenic in a wide range of animal species. The presence of volatile N-nitrosamines have been largely studied in cured meat products. Some of the most important are N-nitrosodimethylamine, N-nitrosopirrolidine, N-nitrosopiperidine, N-nitrosodiethylamine, N-nitrosodi-n-propylamine, N-nitrosomorpholine and N-nitrosoethylmethylamine. In addition, a large number of nonvolatile nitroso compounds, higher in molecular weight and more polar, have also been reported. Some of the most important are N-nitrosoaminoacids such as N-nitrososarcosine and N-nitrosothiazolidine-4-carboxylic acid, hydroxylated N-nitrosamines, N-nitroso sugar amino acids, and N-nitrosamides such as N-nitrosoureas, N-nitrosoguanidines, and N-nitrosopeptides (Pegg and Shahidi 2000). The amount of N-nitrosamines in meat products depends on many variables, such as the amount of added and residual nitrite, processing conditions, amount of lean meat in the product, heating if any, and presence of catalysts or inhibitors (Hotchkiss and Vecchio 1985; Walker 1990). Table 44.1 lists some carcinogenic nitrosamines.

The presence of N-nitrosamines in certain cured meat products was the reason for a serious debate

Table 44.1. List of N-nitrosamines with carcinogenic properties.

N-nitrosamines
N-nitrosodiethylamine
N-nitrosodiethanolamine
N-nitrosopirrolidine
N-nitrosopiperidine
N-nitrosodi-n-propylamine
N-nitrosodi-n-butylamine
N-nitrosomethylbenzylamine

held in the 1970s on residual nitrite in cured meats. As a consequence, the levels of nitrites were lowered, and the use of ascorbate or erythorbate were recommended to inhibit nitrosamines formation (Cassens 1997). Ascorbate reacts with nitrite 240 times faster than ascorbic acid does, and this is the reason why it is preferred (Pegg and Shahidi 2000). Erythorbic and ascorbic acids compete with amines for nitrite.

It must be mentioned that nitrite and nitric oxide are normal human metabolites. Nitric oxide can be generated in the human organism by enzymatic (nitric oxide synthase) reaction of arginine. This nitric oxide can then be converted to nitrate and nitrite, which then can be excreted (Leaf et al. 1989). In addition, the endogenous generation of N-nitrosamines from precursors has also been reported in the gastrointestinal tract (Vermeer et al. 1998).

A key factor in the N-nitrosamines generation is the amount of nitrite added to the meat product. The fate of nitrite is dependent on many variables, such as the pH of the product, the processing temperatures used, and the addition of reducing substances such as ascorbate or isoascorbate. In general, the levels of residual nitrite decrease quite rapidly during processing, remaining at low percentages in the final product in relation to the initial added nitrite (Hill et al. 1973). According to different surveys, made in the late 1990s and early 2000s in European countries, the residual nitrite content in fermented sausages was found to be below 20 mg/kg in most of the products (EFSA 2003). Nitrosodimethylamine and nitrosopiperidine were the main nitrosamines found at levels above 1 μg/kg. In other surveys made on European fermented sausages, the levels of nitrosamines were rather poor or even negligible (Demeyer et al. 2000). Some N-nitrosamines have also been reported in hams packaged in elastic rubber nettings. These nitrosamines were generated because the reaction of nitrite with amine additives in the rubber nettings (Sen et al. 1987).

Even though nitrate has no preservative effect, it acts as a reservoir of nitrite in traditional dry meat products, typical of the Mediterranean countries, like long-ripened dry-fermented sausages and dry-cured ham. Long-ripening processes, with a slow and mild pH drop, are typical of traditional dry-fermented sausages. In these cases, the added nitrate is progressively reduced to nitrite and contributes to safety as well as to particular sensory properties.

The detection of nitrosamines is complex. Volatile N-nitrosamines can be extracted from the meat matrix by aqueous distillation and then analyzed by gas chromatography coupled to a thermal energy analyzer. Other extraction methods can be solid phase extraction with specific solvents (Raoul et al. 1997) or supercritical fluid extraction (Fiddler and Pensabene 1996). When specific identification and confirmation of N-nitrosamines are necessary, mass spectrometry detectors are coupled to the gas chromatographs or, in other cases, liquid chromatography atmospheric pressure chemical ionization mass spectrometry and tandem mass spectrometry (Eerola et al. 1998).

POLYCYCLIC AROMATIC HYDROCARBONS (PAH)

Traditionally, smoking has consisted of the exposition of meat products to the smoke generated by controlled combustion of certain natural hardwoods, sometimes accompanied by aromatic herbs and spices. The pyrolysis of wood generates smoke through different oxidation routes. Moist wood chips can also be used for direct generation of smoke. The smoke is condensed and adsorbed on the surface of the meat product. It can penetrate to a certain depth into the product depending on the conditions of the process. The smoke contains flavoring substances that exert typical smoke flavor but also contain some health-hazardous compounds like polycyclic aromatic hydrocarbons (PAH), phenols and formaldehyde (Bem 1995). Formaldehyde has been identified as a cause of cancerous tumours. Some polycyclic aromatic hydrocarbons, especially benzo-a-pyrene, are known to possess cancer-inducing and carcinogenic properties. The most important polycyclic aromatic hydrocarbons are listed in Table 44.2.

Furthermore, some smoke phenols could react to form highly toxic nitrosophenols that could further react to form toxic reaction products like nitrophenols,

Table 44.2. List of polycyclic aromatic compounds (PAH) potentially present in primary products used for the production of smoke flavorings.

Polycyclic Aromatic Compounds (PAH)
Anthanthrene
Benz[a]anthracene
Benzo[b]fluoranthene
Benzo[j]fluoranthene
Benzo[k]fluoranthene
Benzo[ghi]perylene
Benzo[a]pyrene
Chrysene
Cyclopenta[cd]pyrene
Dibenz[a,h]anthracene
Dibenzo[a,e]pyrene
Dibenzo[a,h]pyrene
Dibenzo[a,i]pyrene
Dibenzo[a,l]pyrene
Indeno[1,2,3-cd]pyrene
5-Methylchrysene
Perylene
Phenanthrene
Pyrene

polymeric nitrosic compounds, and other toxic compounds or even catalyzed the formation of nitrosamines (Bem 1995). The worse situations are found in heavily smoked meat products with old or inadequate smokehouses. In any case, the content in PAH is highly variable because it depends on many variables like the use of direct or indirect smoking, the type of generator used, the type and composition of wood and herbs, the accessibility to oxygen, and the temperature and time of the process.

Some alternative processes have been designed to reduce the contamination of the smoked meat products with hazardous compounds. Some of these strategies, which can reduce significantly the PAH content in smoked meat products, consist of the filtration of particles, use of cooling traps, use of lower temperatures, and/or reduction of the duration of the process. Alternatively, liquid smoke can be obtained through distillation and subsequent condensation of volatile compounds and then applied to the surface of the meat product. Another strategy with extended use consists of the use of smoke flavorings incorporated at concentrations within the range of 0.1–1.0%. These flavorings are produced from primary products obtained from different woods after specific pyrolisis conditions and extraction protocols. In addition to forming part of the smoke flavorings, these primary products can be used as such in foods. Smoke flavorings have a wide variability of compounds, including polycyclic aromatic hydrocarbons (Jennings 1990; Maga 1987). The toxicological effects of smoke flavorings can vary significantly among preparations because these effects depend on many factors, such as the production process of the primary products, the qualitative and quantitative composition, the concentration used in the flavoring, and the final use levels (SCF 1995). Recently, the application of smoke flavoring primary products is controlled in the European Union through the Council Regulation 2065/2003 of 10 November 2003 on smoke flavorings used or intended for use in or on foods (EC 2003). Under this regulation, the use of a primary product in and on foods shall be authorized only if it is sufficiently demonstrated that it does not present risks to human health. The European Food safety Authority (EFSA) will issue a list of Primary Products allowed for use as such in/or on food and/or for the production of derived smoke flavorings. Studies on subchronic toxicity and genotoxicity must be performed to evaluate the potential toxicological effects of the primary products used for the smoke flavoring.

Detection of PAHs can be performed with gas chromatography coupled to a flame ionization detector or high-performance liquid chromatography coupled to ultraviolet or fluorescence detectors. In other cases, mass spectrometry detectors are coupled to both types of chromatographies for the identification and confirmation of PAHs.

OXIDATION

Lipid-derived Compounds

Triacylglycerols, phospholipids, lipoproteins, and cholesterol are the main lipid compounds. Phospholipids are very susceptible to oxidation due to its high content in polyunsaturated fatty acids. Lipid oxidation follows a free radical mechanism consisting in three steps: initiation, propagation, and termination. Hydroperoxides are the primary products of oxidation but they are flavorless, whereas the secondary products of oxidation can contribute to off-flavors, color deterioration, and potential generation of toxic compounds (Kanner 1994). Lipid oxidation may also be induced by hydrogen peroxide generated by peroxide-forming bacteria grown during meat fermentation. Some products of lipid oxidation may be chronic toxicants and have been reported to contribute to aging, cancer, and cardiovascular diseases (Hotchkiss and Parker 1990). In any case, the low levels are far from any acute toxicity.

Cholesterol oxidation may occur through an autooxidative process or in conjunction to fatty acid oxidation (Hotchkiss and Parker 1990). Cholesterol oxides are considered as prejudicial for health due to its role in arteriosclerotic plaque and also as a mutagenic, carcinogenic, and cytotoxic (Guardiola et al. 1996). No cholesterol oxides have been detected after heating of pork sausages (Baggio and Bragagnolo 2006). Studies made on European sausages revealed the generation of cholesterol oxides up to 1.5 μg/g even though the percentage of cholesterol oxidation was below 0.17. The major cholesterol oxide found in Italian sausage was reported to be 7-ketocholesterol whereas 5,6α-5,6-epoxycholesterol was the major end product in the other analyzed sausages (Demeyer et al. 2000). In any case, the values were quite below those considered as toxic based on in vivo tests with laboratory animals (Bösinger et al. 1993).

Protein-derived Compounds

The generation of hydrogen peroxide by certain bacteria during meat fermentation may induce oxidation of proteins. The main modifications of amino acids by oxidation consist of the formation of carbonyl groups, thiol oxidation, and aromatic hydroxylation (Morzel et al. 2006). Sulfur amino acids of proteins are those more susceptible to oxidation by peroxide reagents, like hydrogen peroxide. So, cystine is oxidized only partly to cysteic acid, whereas methionine is oxidized to methionine sulfoxide and methionine sulfone in a minor amount (Slump and Schreuder 1973). Sulfinic and cysteic acids can also be produced by direct oxidation of cysteine (Finley et al. 1981). The oxidation of homocystine can generate homolanthionine sulfoxide as the main product (Lipton et al. 1977). Peptides like the reduced glutathione can also be oxidized by hydrogen peroxide. The oxidation rates increase with the pH, and most of the cysteine in the glutathione is oxidized to the monoxide or dioxide forms.

VETERINARY DRUG RESIDUES

Veterinary pharmaceutical drugs have been used in animals for different purposes (Dixon 2001) such as the following:

- *Therapeutic agents*: medicines used to control infectious diseases
- *Prophylactic agents*: drugs given to prevent outbreaks of diseases and control parasitic infections
- *Growth-promoting agents—Anabolic agents*: substances added to improve the feed conversion efficiency by increasing the lean:fat ratio
- *Growth-promoting agents—Antimicrobial agents*: substances added to make more nutrients available to the animal and not to the gut bacteria

The abuse of antibiotics has increased the concern on the development of bacteria-resistant to antibiotics as detected in recent years (Butaye et al. 2001).

The administration of most veterinary drugs was banned in the European Union because of fears about health effects from residues. Some of them may exert genotoxic, immunotoxic, carcinogenic, or endocrine effects on consumers. These substances can be administered to animals only for therapeutic purposes under strict control of a responsible veterinarian (Van Peteghem and Daeselaire 2004). Main veterinary drugs and substances with anabolic effect are listed in Table 44.3. If these substances are illegally added to farm animals to improve feed conversion rate and promote animal growth, they could remain in the treated-animal–derived foods and could constitute a health concern. The meat processing industry also faces some problems when using treated meat as raw materials for the manufacture of fermented meat products. Some of these problems include lower quality of the resulting products, problems in fermentation if there are antibiotic residues, and less fat with the subsequent loss in juiciness and poorer flavor development (Brockman and Laarveld 1986). For all these reasons, but very especially the health concern, the presence of these substances in farm animals and foods of animal origin must be monitored (Croubels et al. 2004). The development of new methodologies for the analysis of antimicrobials and hormone residues was recently supported by the EC Quality of Life Programme (Boenke 2002).

In the European Union, the presence of these substances in foods and the number of samples to be tested each year are regulated by the EC Directive 96/23/EC on measures to monitor certain substances and residues in live animals and animal products. The analytical methodology for the monitoring of compliance was given in Decisions 93/256/EEC and 93/257/EEC. The Council Directive 96/23/EC was recently implemented by the Commission Decision

Table 44.3. List of veterinary drugs and substances with anabolic effect according to classification in Council Directive 96/23/EC.

Group A: Substances Having Anabolic Effect	Some Main Substances within Each Subgroup
1. Stilbenes	Diethylstilbestrol
2. Antithyroid agents	Thiouracils, mercaptobenzimidazoles
3. Steroids	
Androgens	Trenbolone acetate
Gestagens	Melengestrol acetate
Estrogens	17-β-estradiol
4. Resorcycilic acid lactones	Zeranol
5. β-agonists	Clenbuterol, mabuterol, salbutamol
6. Other substances	Nitrofurans
Group B: Veterinary Drugs	
1. Antibacterial substances	Sulfonamides, tetracyclines, β-lactam, macrolides (tylosin), quinolones, aminoglycosides, carbadox, olaquindox
2. Other veterinary drugs	
Antihelmintics	Benzimidazoles, probenzimidazoles, piperazines, imidazothiazoles, avermectins, tetrahydropyrimidines, anilides
Anticoccidials	Nitroimidazoles, carbanilides, 4-hydroxyquinolones, pyridinols, ionophores
Carbamates and pyrethroids	Esters of carbamyc acid, type 1 and 2 pyrethroids
Sedatives	Butyrophenones, promazines, β-blocker carazolol
Nonsteroideal antiinflammatory drugs	Salicylates, pyrazolones, nicotinic acids, phenamates, arylpropionic acids, pyrrolizines
Other pharmacologically active substances	Dexamethasone
Group B: Contaminants	
3. Environmental contaminants	
Organochlorine compounds	PCBs, compounds derived from aromatic, ciclodiene, or terpenic hydrocarbons
Organophosphorous compounds	Malathion, phorate
Chemical elements	Heavy metals
Mycotoxins	Aflatoxins, deoxynivalenol, zearalenone
Dyes	
Others	

2002/657/EC, which has been in force since September 1, 2002. This decision provided rules for the analytical methods to be used in testing of official samples and specific common criteria for the interpretation of analytical results of official control laboratories for such samples. In the United States, the Food Safety and Inspection Services (FSIS) establishes the surveillance programs, including the National Residue Program, the exploratory residue testing programs, and inspector generated in-plant residue test samples (Croubles et al. 2004). The FDA Center for Veterinary Medicine issues the analytical criteria. The USDA designed an Additional Testing Program to ensure the control of residues of these substances in the meats exported to the European Union (Croubles et al. 2004).

The controls for the detection of these substances are usually based on screening tests, most of the times ELISA test kits or antibody-based automatic techniques. In the case of antibiotics, microbiological tests such as the European Four Plate Test are used. Screening tests are relatively rapid but give only qualitative or semiquantitative data that require further confirmation. Thus, those samples suspected to be noncompliant are confirmed through gas or high-performance liquid chromatography coupled to mass spectrometry or other sophisticated methodologies for accurate characterization and confirmation (Toldrá and Reig 2006). Immunochromatography has been used recently for specific veterinary drugs for further cleaning of samples in order to increase sensitivity.

ENVIRONMENTAL CONTAMINANTS

There are numerous environmental contaminants that may be present in the meats used as raw materials for the manufacture of fermented meat products. These contaminants include dioxins, organophosphorous and organochlorine compounds including PCBs, mycotoxins, and heavy metals among others. In the case of PCBs (polychlorinated biphenyls), its use was phased out but they remain in the environment for many years, especially in stream and lake sediments (Moats 1994).

The contaminants described above can be present in feeds given to the animals as a direct cause of contamination of the meat. The contamination of feeds may have many reasons, such as inappropriate formulation, lack of control, use of contaminated ingredients, inadequate processes, etc. (Croubels et al. 2004). For instance, some molds like species of *Fusarium, Aspergillus*, and *Penicillium* can grow in feed grains and meals and contaminate them with mycotoxins. Some of these mycotoxins are toxic, for example, aflatoxin B_1, which is a genotoxic and carcinogenic substance.

The assessment of environmental contaminants differs from the previous residues because the latest are subject of safety evaluation (analytical controls), whereas some contaminants are unavoidable even though they can exert great potential toxicity (Heggum 2004).

REFERENCES

SR Baggio, N Bragagnolo. 2006. The effect of heat treatment on the cholesterol oxides, cholesterol, total lipid and fatty acid contents of processed meat products. Food Chem 95:611–619.

Z Bem. 1995. Desirable and undesirable effects of smoking meat products. Die Fleischerei 3:3–8.

A Boenke. 2002. Contribution of European research to anti-microbials and hormones Analytica Chimica Acta 473:83–87.

S Bösinger, W Luf, E Brandl. 1993. Oxysterols: Their occurrence and biological effects. Inter Dairy J 3:1–33.

RP Brockman, R Laarveld. 1986. Hormonal regulation of metabolism in ruminants. Rev. Livestock Prod Sci 14:313–317.

P Butaye, LA Devriese, F Haesebrouck. 2001. Differences in antibiotic resistance patterns of Enterococcus faecalis and Enterococcus faecium strains isolated from farm and pet animals. Antimicrob Agents Chemother 45:1374–1378.

RG Cassens. 1997. Composition and safety of cured meats in the USA. Food Chem 59:561–566.

S Croubels, E Daeselaire, S De Baere, P De Backer, D Courtheyn. 2004. Feed and drug residues. In: Encyclopedia of Meat Sciences. W Jensen, C Devine, M Dikemann, eds. London, UK: Elsevier Applied Science, pp. 1172–1187.

DI Demeyer, M Raemakers, A Rizzo, A Holck, A De Smedt, B Ten Brink, B Hagen, C Montel, E Zanardi, E Murbrek, F Leroy, F Vanderdriessche, K Lorentsen, K Venema, L Sunesen, L Stahnke, L De Vuyst, R Talon, R Chizzolini, S Eerola. 2000. Control of bioflavor and safety in fermented sausages: First results of a European project. Food Res Inter 33:171–180.

SN Dixon. 2001. Veterinary drug residues. In: Food Chemical Safety. Volume 1: Contaminants. DH Watson, ed. Cambridge. England: Woodhead Publishing, Ltd., pp. 109–147.

EC. 2003. European Parliament and Council Regulation 2065/2003 of 10 November 2003 on smoke flavourings used or intended for use in or on foods.

S Eerola, I Otegui, L Saari, A Rizzo. 1998. Application of liquid chromatography atmospheric pressure chemical ionization mass spectrometry and tandem mass spectrometry to the determination of volatile nitrosamines in dry sausages. Food Addit Contam 15:270–279.

EFSA. 2003. The effects of nitrites/nitrates on the microbiological safety of meat products. The EFSA J 14:16.

W Fiddler, JW Pensabene. 1996. Supercritical fluid extraction of volatile N-nitrosamines in fried bacon and its drippings: Method comparison. J AOAC Inter 79:895–901.

JW Finley, EL Wheeler, SC Witt. 1981. Oxidation of glutathione by hydrogen peroxide and other oxidizing agents. J Agric Food Chem 29:404–407.

F Guardiola, R Codony, PB Addis, M Rafecas, P Boatella. 1996. Biological effects of oxysterols: Current status. Food Chem Toxicol 34:193–198.

C Heggum. 2004. Risk analysis and quantitative risk management. In: Encyclopedia of Meat Sciences. W Jensen, C Devine, M Dikemann, eds. London, UK: Elsevier Applied Science, pp. 1192–1201.

LH Hill, NB Webb, LD Mongol, AT Adams. 1973. Changes in residual nitrite in sausages and luncheon meat products during storage. J Milk Food Technol 36:515–519.

JH Hotchkiss, RS Parker. 1990. Meat and Health. AM Pearson, TR Dutson, eds. London, UK: Elsevier Applied Science, pp. 105–134.

JH Hotchkiss, AL Vecchio. 1985. Nitrosamines in fired-out bacon fat and its use as a cooking oil. Food Technol 39:67–73.

WG Jennings. 1990. Analysis of liquid smoke and smoked meat volatiles by headspace gas chromatography. Food Chem 37:135–144.

J Kanner. 1994. Oxidative processes in meat and meat products: Quality implications. Meat Sci 36: 169–189.

CD Leaf, JS Wishnok, SR Tannenbaum. 1989. L-arginine is a precursor for nitrate biosynthesis in humans. Biochimica Biophysica Res Commun 163:1032–1037.

SH Lipton, CE Bodwell, AH Coleman, Jr. 1977. Amino acid analyzer studies of the products of peroxide oxidation of cystine, lanthionine and homocystine. J Agric Food Chem 25:624–628.

JA Maga. 1987. The flavour chemistry of wood smoke. Food Rev Inter 3:139–183.

WA Moats. 1994. Chemical residues in muscle foods. In: Muscle Foods. Meat, Poultry and Seafood Technology. DM Kinsman, AW Kotula, BC Breidenstein, eds. New York: Chapman and Hall, pp. 288–295.

M Morzel, P Gatellier, T Sayd, M Renerre, E Laville. 2006. Chemical oxidation decreases proteolytic susceptibility of skeletal muscle myofibrillar proteins. Meat Sci 73:536–543.

RB Pegg, F Shahidi. 2000. Nitrite curing of meat. Trumbull, Connecticut: Food & Nutrition Press, pp. 175–208.

S Raoul, E Gremaud, H Biaudet, RJ Turesky. 1997. Rapid solid-phase extraction method for the detection of volatile nitrosamines in food. J Agric Food Chem 45:4706–4713.

SCF. 1995. Smoke flavorings. Report of the Scientific Committee for Food. Opinion adopted on 23 June 1993. 34 series Food Science Techniques. European Commission.

NP Sen, PA Baddoo, SW Seaman. 1987. Volatile nitrosamines in cured meats packaged in elastic rubber nettings. J Agric Food Chem 35:346–350.

P Slump, HAW Schreuder. 1973. Oxidation of methionine and cystine in foods treated with hydrogen peroxide. J Sci Food Agric 24:657–661.

F Toldrá, M Reig. 2006. Methods for rapid detection of chemical and veterinary drug residues in animal foods. Trends Food Sci Technol 17:482–489.

C Van Peteguem, E Daeselaire. 2004. Residues of Growth Promoters. In: Handbook of Food Analysis, 2nd ed. LML Nollet, ed. New York: Marcel Dekker, Inc., pp. 1037–1063.

IT Vermeer, DM Pachen, JW Dallinga, JC Kleinjans, JM van Maanen. 1998. Volatile N-nitrosamine formation after intake of nitrate at the ADI level in combination with an amine-rich diet. Environ Health Perspect 106:459–463.

R Walker. 1990. Nitrates, nitrites and nitrosocompounds: A review of the occurrence in food and diet and the toxicological implications. Food Addit Contam 7:717–768.

45
Disease Outbreaks

Colin Pierre

INTRODUCTION

Foodborne disease is caused by consuming foods contaminated by microbes (bacteria, viruses, or parasites) or poisonous chemicals, including toxins produced by some bacteria such as *Clostridium botulinum* or *Staphylococcus aureus*.

A foodborne outbreak occurs when a group of people consume the same contaminated food and two or more of them come down with the same illness. But it could be mentioned that the majority of reported cases of foodborne illness are not part of recognized outbreaks and occur as individual or "sporadic" cases.

Some fermented products are sometimes associated with foodborne outbreaks. The main example should be, in the past, the relationship between "home-made" fermented pork meat and the presence of *Clostridium botulinum* producing a powerful paralytic toxin in the food during the shelf life (Hauschild and Gauvreau 1985; Anon. 2003). More recently, in Alaska in 2001, 3 persons out of 14 consuming fermented beaver tail and paw had symptoms suggestive of botulism, and type E toxin was detected in serum specimen and in stools. The type E toxin was also detected in three beaver paws implicated in this meal (Anon. 2001). The "fermented" tail and paws had been wrapped in a paper rice sack and stored for up to 3 months. It seems that changing the food preparation, and particularly using plastic or glass containers for fermentation of food in this country, increase the risk of botulism (Shaffer et al. 1990; Chiou et al. 2002). It is hypothesized that use of these containers promotes an anaerobic environment in which *C. botulinum* thrives.

STAPHYLOCOCCUS

Staphylococcal food poisoning is an intoxication caused by the consumption of foods containing enterotoxins produced by certain strains of *Staphylococcus aureus* (Wieneke et al. 1993) and is of major concern for public health. It seems that the type of foods involved differ widely among countries (Le Loir et al. 2003). Nevertheless, ham, sausages, and other fermented meat could be associated with outbreaks (Anon. 1977, 1979; Bergdoll 1989; Wieneke et al. 1993).

SALMONELLA, CAMPYLOBACTER

Salmonellosis and campylobacteriosis remain the two most common human foodborne outbreaks in industrialized countries.

In 2004, 5,067 outbreaks (out of 6,860) caused by *Salmonella* spp. were reported in the European Union, affecting a total of 30,638 persons, with 10.8% hospitalized and 12 deaths (Anon. 2005). There is a very strong relationship between outbreaks due to the presence of *Salmonella* Enteritidis and the consumption of eggs and egg products. The second serovar involved is *Salmonella* Typhimurium, mainly associated with the consumption of meat and meat products. Unfortunately, no information on the type of meat (fresh, cured, fermented) was done. Nevertheless, during the last decades, some outbreaks involving *Salmonella* spp. and fermented meat occurred in Europe:

- In 1986, Van Netten et al. described a small outbreak in August, 1985 (17 patients, without deaths or complications) due to the presence of *Salmonella* Typhimurium (10^6 per gram) in Bologna sausage.
- In 1987–88, an outbreak of *Salmonella* Typhimurium DT 124 affected 101 patients in England. The epidemiological and microbiological investigations identified a salami stick as the vehicle of infection (Cowden et al. 1989).
- In Italy, Pontello et al. (1998) described a community-based outbreak of *Salmonella* Typhimurium PT193 associated with the consumption of salami.
- More recently, an outbreak due to *Salmonella* Typhimurium associated with a traditional salted, smoked, and dried ham, after a family party, was described (Mertens et al. 1999). *Salmonella* Typhimurium PT20 was found in Coburger ham and in feces of patients. Only the consumption of Coburger ham and "bone ham," originated from the same batch of meat and prepared in the same manner in the same salt bath, was significantly related to a positive patient's stools culture. It was concluded that "traditional salting, drying, and smoking of raw pork meat was not antimicrobiological-effective against *Salmonella* Typhimurium.
- In 2001, another serovar, *Salmonella* Goldcoast, was identified as the causative agent of an outbreak in Germany, linked to the consumption of fermented sausage (Bremer et al. 2001). Twenty-four cases were identified with a correlation with the consumption of a raw fermented sausage manufactured by a local company; a part of this product was sold after only 4 days of fermentation.

In the European Union, *Campylobacter jejuni* and *Campylobacter coli* are the second agents associated with outbreaks: 1,243 (out of 6,860), affecting 3,749 persons, have been reported in 2004 (Anon. 2005). It is well known that these bacteria are present in the pork and poultry chain productions but, to date, no outbreaks associated with the consumption of fermented meat have been reported.

VEROTOXIGENIC STRAINS OF *ESCHERICHIA COLI*

Verotoxigenic strains of *Escherichia coli* (VTEC), and in particular *E. coli* O157:H7, have been identified in some raw fermented finished products associated with foodborne outbreaks:

- For Tilden et al. (1996), studying an outbreak in Washington and California involving *E. coli* O157:H7, dry-fermented salami can serve as a vehicle of transmission of these bacteria which can survive "currently accepted processing methods."
- In 1995, a large outbreak of hemolytic-uremic syndrome (HUS) occurred in Australia (Paton et al. 1996). Twenty-one cases with one fatality, from 4 months to 12 years old, were identified. Different strains of *E. coli* were isolated, proving that outbreaks are not necessarily caused by contamination with a single "outbreak" strain. Nevertheless, a large number of these strains, and others, were isolated from samples from a semidry-fermented sausage (mettwurst) suspected and collected from the homes of patients and from retail outlets. This product was made from a mixture of raw pork, beef, and lamb.
- In Canada, in 1998, 39 cases of *E. coli* O157:H7 infection were identified, and 2 children were diagnosed with HUS. Consumption of Genoa salami was significantly associated with the illness. Phage typing and Pulse Field Gel Electrophoresis (PFGE) patterns of *E. coli* strains isolated from patients' stools and from samples of salami produced by the most commonly identified manufacturer were identical (Williams et al. 2000).
- Finally, in British Columbia (Canada), in 1999, a new outbreak of *E. coli* O157:H7 was reported, involving 143 cases, including 6 HUS. Again, the case-control study found an association with a salami production. The mixture of raw beef, raw beef suet, raw pork, and ingredients was fermented and dried without a cooking step during the production (MacDonald et al. 2004). Strains of *E. coli* O157:H7 isolated from the stools of patients and from salami samples has the same PFGE patterns.

YERSINIA, LISTERIA

Interestingly, a case-control study found an epidemiologic association between human cases of yersiniosis and consumption of salami (Anon. 2000), confirming a Norwegian study concerning different types of sausages (Ostroff et al. 1994).

The risk of human listeriosis from fermented ready-to-eat meat is very low (Anon. 2004). However, Schwartz et al. (1989) reported an outbreak due to the consumption of salami in Philadelphia (U.S.A.).

PARASITES

Finally, the consumption of meat and meat products may present a risk of parasitic infection (Pozio 2003), mainly by *Trichinella spiralis*, *Taenia solium*, and *Toxoplasma gondii* (Gamble 1997). Nevertheless, these parasites are inactivated by cooking, freezing, or curing and dry-cured and fermented meat present little danger (Van Sprang 1984).

CONCLUSION

Following this review, it can be concluded that outbreak cases due to the consumption of fermented meat represent the minority of all laboratory-confirmed cases of gastrointestinal infections (Moore 2004). In general, on a worldwide basis, *Salmonella* spp., verotoxigenic strains of *Escherichia coli* (VTEC) and *Staphylococcus aureus* seem to be the main concerns (Adams and Mitchell 2002). In particular, raw fermented sausage (salami), contaminated with verotoxigenic strains of *Escherichia coli*, has recently been implicated in several outbreaks, including hemolytic-uremic syndrome.

REFERENCES

M Adams, R Mitchell. 2002. Fermentation and pathogen control: A risk assessment approach. Int J Food Microbiol 79:75–83.

Anonymous. 1977. The Staphylococcal Enterotoxin Problem in Fermented Sausage. U.S. Department of Agriculture. Task Force Report.

Anonymous. 1979. Staphylococcal food poisoning associated with Genoa and hard salami. U.S. Department of Health, Education and Welfare. Morbid Mortal Weekly Report 28:179–180.

Anonymous. 2000. Case-control study assessing the association between yersiniosis and exposure to salami. Canada Communicable Disease Report 26(19).

Anonymous. 2001. Botulism outbreak associated with eating fermented food—Alaska 2001. CDC, MMWR Weekly, 50(32):680–682.

Anonymous. 2003. The effects of nitrites/nitrates on the microbiological safety of meat products. EFSA J 14:1–31.

Anonymous. 2004. FAO/WHO Risk assessment of *Listeria monocytogenes* in ready to eat foods—Interpretative summary. Microbiological Risk Assessment series 4. http://www.fao.org.

Anonymous. 2005. Trends and sources of zoonoses, zoonotic agents and antimicrobial resistance in the European Union in 2004. EFSA J—310.

MS Bergdoll. 1989. *Staphylococcus aureus*. In: MP Doyle, ed. Foodborne Bacterial Pathogens. New York: Marcel Dekker, Inc., pp. 463–523.

V Bremer, K Leitmeyer, F Jensen, U Metzel, H Meczulat, E Weise, D Werber, H Tschaepe, L Kreienbrock, S Glaser, A Ammon. 2001. Outbreak of *Salmonella* Goldcoast infections linked to consumption of fermented sausage, Germany 2001. Epidemiol Infect 132(5):881–887.

LA Chiou, TW Hennessy, A Horn, G Carter, JC Butler. 2002. Botulism among Alaska natives in the Bristol Bay area of southwest Alaska: A survey of knowledge, attitudes, and practices related to fermented foods known to cause botulism. Int J Circumpolar Health 61(1):50–60.

JM Cowden, M O'Mahony, CL Bartlett, B Rana, B Smyth, D Lynch, H Tillet, L Ward, D Roberts, RJ Gilbert. 1989. A national outbreak of *Salmonella* Typhimurium DT 124 caused by contaminated salami sticks. Epidemiol Infect 103(2):219–225.

HR Gamble. 1997. Parasites associated with pork and pork products. Rev Sci Tech 16(2):496–506.

AHW Hauschild, L Gauvreau. 1985. Food-borne botulism in Canada, 1971–84. Can Med Assoc J 133(December 1):1141–1146.

Y Le Loir, F Baron, M Gautier. 2003. *Staphylococcus aureus* and food poisoning. Genet Mol Res 2(1):63–76.

DM MacDonald, M Fyfe, A Paccagnella, A Trinidad, K Louie, D Patrick. 2004. *Escherichia coli* O157:H7 outbreak linked to salami, British Columbia, 1999. Epidemiol Infect 132:283–289.

PL Mertens, JF Thissen, AW Houben, F Sturmans. 1999. An epidemic of *Salmonella* Typhimurium associated with traditional salted, smoked, and dried ham. Ned Tijdschr Geneeskd 143(20): 1046–1049.

JE Moore. 2004. Gastrointestinal outbreaks associated with fermented meats. Meat Sci 67:565–568.

SM Ostroff, G Kapperud, LC Hutwagner. 1994. Sources of sporadic *Yersinia enterocolitica* infections in Norway: A prospective case-control study. Epidemiol Infect 112:133–141.

AW Paton, RM Ratcliff, RM Doyle, J Seymour-Murray, D Davos, JA Lanser, JC Paton. 1996. Molecular microbiological investigation of an outbreak of Hemolytic-Uremic Syndrome caused by dry fermented sausage contaminated with Shiga-like toxin-producing *Escherichia coli*. J Clin Microbiol 34(7): 1622–1627.

M Pontello, L Sodano, A Nastasi, C Mammina, M Astuti, M Domenichini, G Belluzi, E Soccini, MG Silvestri, M Gatti, E Gerosa, A Montagna. 1998. A community-based outbreak of *Salmonella enterica* serotype Typhimurium associated with salami consumption in Northern Italy. Epidemiol Infect 120(3): 209–214.

E Pozio. 2003. Foodborne and waterborne parasites. Acta Microbiol Pol 52:83–96.

B Schwartz, D Hexter, CV Broome, AW Hightower, RB Hirschhorn, JD Porter, PS Hayes, WF Bibb, B Lorber, DG Faris. 1989. Investigation of an outbreak of listeriosis: New hypotheses for the etiology of epidemic *Listeria monocytogenes* infections. J Infect Dis 159:680–685.

N Shaffer, RB Wainwright, JP Middaugh. 1990. Botulism among Alaska Natives: The role of changing food preparation and consumption practices. West J Med 153:390–393.

J Tilden, Jr, W Young, AM McNamara, C Custer, B Boesel, MA Lambert-Fair, J Majkowski, D Vugia, SB Werner, J Hollingsworth, JG Morris. 1996. A new route of transmission for *Escherichia coli*: Infection from dry fermented salami. Am J of Public Health 86(8):1142–1145.

P Van Netten, J Leenaerts, GM Heikant, DA Mossel. 1986. A small outbreak of salmonellosis caused by Bologna sausage. Tijdschr Diergeneeskd 111 (24): 1271–1275.

AP Van Sprang. 1984. Possibilities of survival of various parasites in meat and meat products. Tijdschr Diergeneeskd 109(1):344–348.

AA Wieneke, D Roberts, RJ Gilbert. 1993. Staphylococcal food poisoning in the United Kingdom, 1969–1990. Epidemiol Infect 110:519–531.

RC Williams, S Isaacs, ML Decou, EA Richardson, MC Buffett, RW Slinger, MH Brodsky, BW Ciebin, A Ellis, J Hockin, and the *E. coli* O157:H7 Working group. 2000. Illness outbreak associated with *Escherichia coli* O157:H7 in Genoa salami. CMAJ 162(10):1409–1413.

Part XI

Processing Sanitation and Quality Assurance

46
Basic Sanitation

Stefania Quintavalla and Silvana Barbuti

INTRODUCTION

Basic sanitation within food processing facilities, often referred to as *prerequisite programs* or *standard operating procedures (SOPs)*, are the foundation of a successful food safety management system (Wallace and Williams 2001). They consist of hygienic practices designed to maintain a clean and healthy environment for the food product. The application of proper sanitation technique is important in maintaining food safety. The significance of hygienic design of plants and equipment has been recently recognized in European (EC 2004) and United States legislations (USDA 2000; FDA 2003, 2004). The WHO puts emphasis on the role of prerequisite programs defined as "practices and conditions needed prior to and during the implementation of HACCP and which are essential for food safety" (WHO 1993).

Sanitation is a dynamic and ongoing function and cannot be sporadic; first, it is necessary that the full meaning of sanitation and its wide economic scope be accepted by everyone concerned in the food system, including management.

Sanitation means preventing product contamination by microorganisms, allergens, and foreign materials.

Poor sanitation practices can contribute to outbreaks of foodborne illnesses. Many of the microbiological food safety problems result from basic sanitation failures that occur in food production, processing, storage, transportation, retailing, and handling in the home (CDC 2002). This may indicate that there is the need to improve on general food safety practices.

Meat companies have to write and follow basic sanitation programs in order to ensure effective cleaning and sanitizing of the manufacturing equipment and environment.

Examples of concerns addressed by the programs mentioned above include the following: sanitary design and construction of food facilities, cross-contamination control, personal hygiene and sanitary food handling, waste product disposal, and pest control.

SANITARY DESIGN AND CONSTRUCTION OF FOOD FACILITIES

Sanitary design and layout permit appropriate maintenance, cleaning, and disinfection and minimize airborne contamination (Codex Alimentarius Commission 1999).

Establishment buildings, including their structures, rooms, and compartments, must be kept in good repair and be of sufficient size to allow processing, handling, and storage of product in a manner that does not result in product adulteration (Schmidt 1997; AMI 2003). Rooms or compartments in which edible product is processed, handled, or stored must be separate and distinct from rooms or compartments in which inedible product is processed, handled, or stored to prevent product adulteration by unsanitary conditions.

Because there is the possibility that the raw material may carry a low number of pathogens, e.g., *Salmonella* spp. and *Listeria monocytogenes*, primary building construction should be designed to provide separate areas for raw materials and a finished product with a flow pattern being in one direction and following a logical sequence from raw material handling to finished product storage. In the layout of the plant, an area to store raw material on reception must be arranged (Cramer 2003).

There should be a physical separation between processing areas by the installation of walls and doorways and by an air handling system to provide positive pressure in the finished product rooms. Additional steps—for example, separation of processing times—should be taken to minimize the risk of cross-contamination if complete separation is not possible. Color-coding is recommended, with different colors identifying different areas. This can be applied to clothing, cleaning supplies, containers, and any other equipment. Separation can also be accomplished by the installation of sanitizer systems (foot baths, spray systems) inside entrance doors to critical areas.

The establishment buildings should have floor and wall surfaces constructed of materials that are durable, smooth, easily cleaned, and, where necessary, disinfected. Drains should be operative and clean without standing water; pipes and insulation should be dry and in a good repair; overheads and conveyors should be accessible, cleanable, and free of condensate (Lelieveld 2000).

Equipment for food processing and handling operations should be cleanable and maintained in such a manner as to prevent contamination.

Chapter II of European Regulation 852/2004 (EC 2004) lists the specific requirements for nonfood contact and food contact surfaces in rooms where foodstuffs are prepared.

The basic principles of sanitary design are also explained in the U.S. document *FDA Food Code* (2005): All surfaces in contact with food must be inert, smooth, and not porous, visible for the inspection, readily accessible for normal cleaning; the nonfood contact surfaces should be arranged to prevent harboring of soils, bacteria, or pests; equipment must be designed to protect the contents from external contamination.

Adequate space must be provided within and around equipment, and equipment must be accessible for cleaning, sanitizing, maintenance, and inspection.

All food contact and nonfood contact surfaces, including utensils, should be cleaned as frequently as necessary to avoid contamination of food. It is important to establish a regular schedule for cleaning and sanitizing based on microbiological monitoring: daily or more often when required for food contact surfaces, and daily or weekly for nonfood contact surfaces.

An adequate supply of hot and/or cold potable water should be available throughout the establishment; if nonpotable water is used, for example, for fire control, no cross-connections between potable and nonpotable water supplies must be present.

Adequate natural or artificial lighting of sufficient intensity to ensure that sanitary conditions are maintained must be provided in areas where food is processed, handled, or stored; where equipment and utensils are cleaned; and in hand-washing areas, dressing and locker rooms, and toilets. Lighting apparatus should be protected to prevent food from contamination by breakages.

Positive air pressure is required in microbiological sensitive areas—for example, in slicing and packaging areas.

Dressing rooms and toilet rooms must be sufficient in number, ample in size, conveniently located, and maintained in a sanitary condition and in good repair to ensure cleanliness of all persons handling product. They must be separate from, and not directly entered from, food processing and handling areas.

Lavatories with running hot and cold water, soap, and towels must be placed in or near toilets and in food processing areas.

CROSS-CONTAMINATION CONTROL

Cross-contamination occurs when biological, chemical, or physical contaminants are transferred to food products.

The type of cross-contamination most frequently implicated in foodborne illnesses occurs when pathogenic bacteria are transferred to ready-to-eat (RTE) foods. Contaminants, especially bacteria, are transferred by people with dirty hands, clothing, or shoes; by dirty utensils, such as knives; and by dirty processing equipment—for example, mincing machine, mixing machine, stuffing machine, tables, and conveyer belts.

Many pathogens, like *Escherichia coli*, *Salmonella*, and *L. monocytogenes*, enter the food processing environment via contaminated raw materials. Raw material contamination can affect any industry, but it is more common in industries that use animal-derived products or products at risk of cross-contamination by animal feces (Franco et al. 1995).

During processing of fermented meat products, fresh meat comes in contact with many solid surfaces: examples include conveyors, peelers, collators, belts, slicers, and tables (Mafu et al. 1990). Both employees and equipment that touch raw meat can transmit pathogenic bacteria to the finished product; raw and RTE products should be physically separated in coolers or other storage areas.

Bacteria from fresh meat can affix to wet equipment surfaces, form biofilm and/or niche environments, and supply a source of cross-contamination. Biofilms occur when bacteria form a slime layer upon a surface and provide an environment for pathogens to proliferate. The adhesion of pathogenic bacteria to a biofilm is a food safety hazard because the biofilm can detach and become a significant source of food contamination. Cleaning to remove biofilms prior to sanitation is often sufficient to prevent this problem. However, studies have shown that attached bacteria may survive conventional cleaning methods (Stopforth et al. 2002). Adequate cleaning prior to sanitizing is therefore paramount to controlling this problem (Chmielewski and Frank 2003).

Niche environments are sites within the manufacturing environment where bacteria can get established, multiply, and contaminate the processed food. These sites may be impossible to reach and to clean with normal cleaning and sanitizing procedures: examples include hollow rollers on conveyors, cracked tubular support rods, the space between close-fitting metal-to-metal or metal-to-plastic parts, worn or cracked rubber seals around doors, and on-off valves and switches. Manufacturers must identify and eliminate niches in order to minimize the risk of postprocessing contamination from niche environments (Tompkin 2002; AMI 2003).

Conveyors, peelers, collators, belts, gloves, slicers, and tables can be sources of direct product contamination. Slicers must be cleaned at the end of the shift, and if the slicer is used for different dry-cured meat products, it must be cleaned and sanitized before reuse.

Some equipment may need to be disassembled to be cleaned. Where possible, equipment should be running during cleaning and sanitizing for complete exposure to chemicals. The cleaning process should be verified primarily by visual inspection, and the effectiveness of sanitation procedures should be confirmed by microbiological monitoring, enzyme-linked immunosorbent assay (ELISA) tests, and ATP testing (Griffiths 1996; Griffiths et al. 1997; Shumaker and Feirtag 1997).

Do not limit sampling to flat surfaces. Microbiological sampling of the environment and equipment can detect a niche.

A product can become cross-contaminated with allergens on the production line. To minimize the risk of cross-contamination, equipment must be cleaned and sanitized to remove all traces of allergens when the next run includes product that should not contain allergens. Wash-down techniques may need adjustment to ensure that they remove allergens as well as pathogens (Higgins 2000). Rinsing with water only or cleaning only at the end of the day could be inadequate. Manufacturers may choose to physically separate lines for allergen- and nonallergen-containing products (Morris 2002).

Visitors, suppliers, laboratory personnel, truck drivers, inspectors, and management should be made aware of operating procedures; tours should be done in a traffic direction counter to production, starting with finished product rooms and ending in the raw material handling areas.

PERSONAL HYGIENE

Good personnel hygiene practices are the foundation for successful food safety and quality assurance in all food manufacturing facilities. Bacteria causing diseases or spoilage may be carried and transmitted to surfaces and food by plant personnel handling the food products.

The transfer of contaminants can occur through a direct route, such as bacteria transferred from the body skin, mouth, hands, or hair to the product, or indirectly via personal equipment, such as clothing, footwear, utensils, and other tools used in the daily tasks.

Two important areas of focus when setting personal hygiene policies to prevent contamination are protective outer clothing (working clothes, footwear, and hair covering) and hand hygiene.

Protective Outer Clothing

The clothing of workers must be clean. The purpose is not to protect the worker against contamination but to protect the meat products against contamination. Working clothes must be used exclusively in the working area and nowhere else. If possible, it is advisable to avoid admittance from the unclean area to the clean area without changing clothes.

Working clothes should be comfortable and easy to wash. Their design should encourage good hygiene habits: light-colored working clothes show the need for cleaning sooner than dark-colored ones.

Factory clothing should be hygienically designed to prevent foreign bodies from being lost directly (buttons) or indirectly (objects falling out of pockets).

During work, jewelry, wristwatches, etc., are prohibited, because these objects may be sources for contamination and make hand-washing difficult.

Laundering has to be controlled by the company in order to ensure the sanitary condition of the protective clothing.

Footwear can be a vehicle for the transfer of pathogens from production areas considered at high risk to low-risk areas. Footwear should be made of nonporous material that is cleanable. Foamers are the most commonly used footwear decontamination method in the industry. Foot dips/baths and boot-washers are also common; in this case it is essential that they are monitored to make sure that a "culture broth" for bacteria has not been created.

Human hair and beards are normally heavily contaminated with bacteria; to prevent contamination of food, a hair or beard covering in the process area is a necessary part of work clothes. Many different types of hair coverings are seen in the food industry. Disposable or washable hair and beard coverings are recommended.

Hand Hygiene

Hand-washing policies should require employees to wash after any type of activity that could contaminate the hands with pathogens, including using the restroom; blowing the nose; touching body parts, waste, or nonfood contact surfaces, such as light switches or pipes; smoking; and eating. Bacteria may be transmitted to the hands by these acts and thereafter transmitted to meat products that are handled by hand.

Employees also should wash before entering food-handling areas, changing clothing, and putting on gloves.

Careful and frequent hand-washing will do much to reduce contamination. Therefore, hand-washing facilities must be sufficient if the water supply is adequate. Basically there should be two sites where the staff can wash their hands, the restroom and the working area, where sufficient hand-washing facilities must be placed close to the working places. If hand-washing stations are not placed in convenient locations, employee might skip washing.

Hand-washing

Use of an effective soap, vigorous rubbing of hands during the washing process, and thorough hand-drying are key to good sanitary hygiene.

Both literature and experimental data were used to develop a quantitative risk assessment model to assess the risk associated with different hand-washing techniques (Monville et al. 2002). Soap with an antimicrobial agent was observed to be more effective than regular soap, whereas there was little difference in the efficacy of alcohol and alcohol-free sanitizers. Hot air-drying had the capacity to increase the amount of bacterial contamination on hands, whereas paper towel drying caused a slight decrease in contamination.

The frictional aspects of hand-drying can help remove microorganisms missed during hand-washing on precisely the parts of the hands that are missed (Blackmore 1989).

Hand-drying by contact with dispenser mechanisms should be avoided because it can cause cross-contamination.

For some operations, the use of a special bacteriostatic soap or dipping of the hands after washing in a germicidal rinse could be required. Use of a nail brush is recommended because bacteria often hide along and under the nails.

Gloves

The use of gloves does not eliminate the need for sanitary hand-washing and drying. Gloves can provide a false sense of security.

If the use of gloves is indicated, they must be kept in the same good hygienic conditions as hands; otherwise, it is better to avoid their use. Gloves should be changed at regular intervals (at least once per hour), after touching contaminated surfaces and utensils, or if punctured.

Gloves may be of rubber or plastic and they are used to protect the meat and meat products against contamination. They may also be used to protect the hands against knife cuts and will then be made of steel. Great care should be taken to keep a certain hygienic standard of these gloves.

A study was conducted to determine whether the level of selected microorganisms differed on foods handled by gloved and bare hands at fast food restaurants (Lynch et al. 2005). There were no significant differences between samples handled by gloved or bare hand. The observed tendency of food workers to wear the same pair of gloves for extended periods might account for the apparent failure of glove use to reduce or prevent bacterial contamination.

Personnel Health

Good health is important for workers in the meat industry. Ill persons will often be carriers of more microorganisms (pathogenic microorganisms) than usual and these microorganisms may then be transmitted to the meat products with the risk of causing disease to the consumers. Illness must always be reported to the manager and/or the meat inspector of the plant who will decide whether the worker can stay or has to leave.

Likewise, persons afflicted with infected wounds, skin infections, sores, etc., must also be restricted from food handling areas. Any persons with open cuts or wounds should not handle food unless the injury is completely protected by a secure, waterproof covering.

WASTE PRODUCT DISPOSAL

Handling of liquid and solid waste influences both hygiene in processing and of the environment, the latter depending on the precautions taken to avoid contamination with liquid and solid waste.

Whenever food is processed, wastewater is generated. Sewage must be disposed of in a sanitary manner; it must not become a source of contamination to the plant environment. Of particular importance is the prevention of microbiological cross-contamination and pests. Sewage lines must not connect to drainage lines within the plant.

Garbage disposal and the handling of waste from within the plant during operations and the waste collection facilities outside the plant must be handled in a manner to reduce the risk of bacterial contamination and the pest infestation and harboring.

Within the plant there must be a sufficient number of garbage containers so that they are accessible to all employees. Covered containers are required. These containers must be clearly marked and identified, leakproof, emptied regularly, and cleaned and sanitized prior to use.

Waste disposal facilities that are located outside of the plant must not attract pests. They must have covers and be kept closed and in good conditions.

Environmental Protection—IPPC Directive

In 1996, Council Directive 96/61/EC on integrated pollution prevention and control was published (EC 1996). The purpose of the Directive is to achieve integrated prevention and control of pollution arising from European industrial activities, such as the activities of the FDM sector (Food, Drink and Milk industries), leading to a high level of protection of the environment as a whole. Central to this approach is the general principle given in Article 3 that operators should take all appropriate preventative measures against pollution through the application of best available techniques (BAT). On January 2006, a Reference Document on Best Available Techniques (BREF) in the Food, Drink and Milk industries was published (EC 2006). Both general BAT for the whole FDM sector and additional BAT for some processes and unit operations in the individual FDM sectors are listed. Their scope is to reduce the consumption of water, energy, and packaging.

For example, for meat and poultry processing installations, BAT is to thaw meat in the air to reduce water consumption and wastewater production; to dose spices and other solid ingredients from a bulk container rather than from plastic bags to minimize the use of plastic packaging material; and to stop the water supply automatically when sausage fillers and similar equipment are not used at breaks or at production stops to reduce waste of materials, products, and water.

PEST CONTROL

One of the simplest, yet essential, steps in food safety is pest prevention. Pests, including cockroaches, flies, stored products insects, rodents, and birds, harbor and spread disease-carrying organisms that can easily contaminate food products. However, pest control is often ignored until pests and their damage are discovered. For example, if rodents or insects are found in a food storage room, temporary measures are taken to eliminate them. The real trouble, however, is not corrected. The method of treating a single outbreak is a poor concept of sanitation.

Control of pests and use of pesticides are particularly critical in places where food is processed and/or packaged. Top management is ultimately responsible for identifying a competent person to develop a pest prevention and control program. Persons who apply pesticides in industrial facilities have a responsibility to use the needed pesticide, to apply it correctly (according to label instructions), and to be certain there is no hazard to man or the environment.

Integrated Pest Management (IPM)

The many drawbacks and problems associated with chemical pesticides have led pest control experts to reevaluate control procedures. As a result, pest control programs based on predicted ecological and economic consequences have been developed

(Thomas 1999). *Integrated pest management (IPM)* is a dynamic combination of varied control practices designed and implemented to meet the need of maintaining control of pests using a variety of techniques without total reliance upon pesticides.

Economic, social, and environmental advantages may be attained through IPM. The apparent benefits are related through lower costs, increased pest control, and reduced pesticide usage by up to 60% (Paschall et al. 1992).

A few steps are necessary for implementing an ongoing IPM strategy in a food processing facility (Pest Management Regulatory Agency 1998). When designing and building a new facility or renovating an existing one, it is essential that the IPM strategy be incorporated at the earliest stages.

The first step of the IPM strategy is an assessment of the actual or potential pest problem. Identification of potential pest problems can be based on past experience, knowledge of similar facilities, and knowledge of particular pests in the region in which the plant is located.

Inspections are used to identify the type and variety of pests that are or may be present. Inspections should include raw materials, prepared products, interior and exterior of the facility, and equipment. Monitoring, by trapping for example, could be used to determine the intensity and the distribution of a pest problem in and around a facility. Observations can be used to confirm the results of inspection and monitoring.

The next stage is to develop a pest management plan for implementation, based on the assessment findings. At this stage, information is needed on the biology of the pest organisms and the management of that particular pest.

Elements that should be included in the pest management plan for effective IPM are:

- *Building and materials design and retrofitting*: The facility, its equipment and its exterior surroundings must not promote pest populations.
- *Exclusion practices*: Infestations in incoming food and ingredients should be reduced or eliminated through strict purchase specifications, audits of suppliers, and inspection of incoming material.
- *Good sanitation practices*: They include thorough and regular cleaning, prevention of dust generation and accumulation, and removal of food sources and harbors for pests.
- *Building maintenance*: Holes and cracks in floors, walls, ceilings, roofs, doors, and windows that allow access for pests and allow dust to collect should be eliminated.
- *Inspections and monitoring*: These activities are necessary to guide schedules and locations of treatments, and to monitor the effectiveness of the overall management strategy.
- *Pest identification*: Pest identification is necessary to select the most appropriate control methods.
- *Physical and chemical controls*: Examples are rodent traps, glue boards, electric fly traps, and temperature manipulation.

These methods are noncontaminating and can fill some of the gaps left by a restricted pesticide use.

As for HACCP, implementation of the pest management plan starts with the commitment of the facility's management and decision makers. Employees must be trained with an awareness of the IPM program and its elements, particularly sanitation. They should learn to recognize pests, pest habitats, and pest conditions, and how to deal with them.

Written procedures should be available for all aspects of an IPM program, including cleaning, monitoring, identifying, and correcting of problems, and treatments.

The overall effectiveness of the pest management plan must be reviewed based on monitoring of pests and inspection reports.

Changes to the pest management plan can be made to achieve the desired level of control; to adjust to new situations, such as new equipment, new products, and new pests; or to incorporate new pest management techniques.

PERSONNEL TRAINING

Regardless of the type of processing or food handling operation, people play a key rule in food sanitation. That is why the most important aspect of a sanitation program is ongoing personnel training. It is essential that the full meaning of sanitation and its wide economic scope be accepted by everyone concerned in the food system including management. Personnel training should include appropriate sanitation principles and food handling practices, manufacturing controls, and personal hygiene practices.

Personnel training should encourage an understanding of the processing steps and technologies for the product manufactured and where potential problems exist. Production personnel should be trained in the critical elements of the operations for which they are responsible, in the importance of these operations, monitoring these operations, and in action to be taken when these operations are not controlled.

REFERENCES

AMI (American Meat Institute). 2003. Sanitary equipment design. AMI Fact Sheet. March. EDIS website at http://www.sanitarydesign.org/pdf/Sanitary.

MA Blackmore. 1989. A comparison of hand drying methods. Catering & Health 1:189–198.

CDC. 2002. Preliminary Foodnet data on the incidence of foodborne illnesses—Selected sites, United States, 2001. Morbid Mortal Weekly Rep 51:325–329.

RAN Chmielewski, JF Frank. 2003. Biofilm formation and control in food processing facilities. Comprehens Rev Food Sci/Food Saf 2:22–32.

Codex Alimentarius Commission. 1999. Recommended international code of practice general principles of food hygiene. CAC/RCP 1-1969, Rev. 3 (1997), Amended 1999.

MM Cramer. 2003. Building the self-cleaning food plant: Six steps to effective sanitary design for the food plant. Food Saf Mag February/March. EDIS website at http://www.foodsafetymagazine.com/master/webex.htm.

EC (European Commission). 1996. Council Directive 96/61/EC of 24 September 1996 concerning integrated pollution prevention and control. OJ L257, 10/10/1996, 0026-0040.

———. 2004. Regulation (EC) No 852/2004 of the European Parliament and of the Council of 29 April 2004 on the hygiene of foodstuff. Official Journal of the European Union L139/1 30.4.2004, 1–54.

———. 2006. Integrated pollution prevention and control. Reference Document on Best Available Techniques in the Food, Drink and Milk Industries. EDIS website at http://eippcb.jrc.es.

FDA. 2003. 21 CFR Part 110—Current good manufacturing practice in manufacturing, packing or holding human food. Code of Federal Regulations, Title 21, Volume 2, revised as of April 1, 2003.

———. 2004. Center for Food Safety and Applied Nutrition—Good Manufacturing Practices (GMPs) for the 21st Century—Food Processing—Final Report.

———. 2005. U.S. Department of Health and Human Service. Food Code—Equipment, utensils, and linens, Chapter 4, pp. 101–143.

CM Franco, EJ Quinto, C Fente, JL Rodriguez-Otero, L Dominguez, A Cepeda. 1995. Determination of the principal sources of *Listeria* spp. Contamination in poultry meat and poultry processing plant. J Food Prot 58:1320–1325.

MW Griffiths. 1996. The role of ATP bioluminescence in the food industry: New light on old problems. Food Technol 50:63–72.

MW Griffiths, CA Devidson, AC Peters, LM Fielding. 1997. Towards a strategic cleaning assessment programme: Hygiene monitoring and ATP luminometry, an options appraisal. Food Sci Technol Today 11:15–24.

KT Higgins. 2000. A practical approach to allergen control. Food Engineering, July. EDIS website at http://www.foodengineering.org/CDA/ArticleInformation.

HLM Lelieveld. 2000. Hygienic Design of Factories and Equipment. In: Lund, Baird-Parker, Gould, eds. The Microbiological Safety and Quality of Food. Gaithersburg, Maryland: Aspen Publishers, pp. 1656–1690.

RA Lynch, ML Phillips, BL Elledge, S Hanumanthaiah, DT Boatright. 2005. A preliminary evaluation of the effect of glove use by food handlers in fast food restaurants. J Food Prot 68:187–190.

AA Mafu, D Roy, J Goulet, P Magny. 1990. Attachment of *Listeria monocytogenes* to stainless steel, glass, polypropylene, and rubber surfaces after short contact times. J Food Prot 53:742–746.

R Monville, Y Chen, DW Schaffner. 2002. Risk assessment of hand washing efficacy using literature and experimental data. Int J Food Microbiol 73:305–313.

CE Morris. 2002. Best practices for allergen control. Food Engineering. March. EDIS website at http://www.foodengineering.org/CDA/ArticleInformation.

MJ Paschall, C Hollingsworth, WM Coli, NL Cohen. 1992. Washington bugs out. Integrated pest management saves crops and the environment. J Am Diet Assoc 92:931.

Pest Management Regulatory Agency. 1998. Integrated Pest Management in food processing: Working without methyl bromide. EDIS website at http://www.hcsc.gc.ca/pmra-arla/.

RH Schmidt. 1997. Basic elements of a Sanitation Program For Food Processing and Handling, Document FS15, Florida Cooperative Extension Service, Institute of Food and Agriculture Sciences, University of Florida. EDIS website at http://edis. ifas.ufl.edu.

SN Shumaker, JM Feirtag. 1997. Environmental analysis methods utilized to determine the contamination sources in a sausage processing plant. Dairy Food Environ Sanitation 17:274–280.

JD Stopforth, J Samelis, JN Sofos, PA Kendall, GC Smith. 2002. Biofilm formation by acid-adapted and nonadapted *Listeria monocytogenes* in fresh beef decontamination washings and its subsequent inactivation with sanitizers. J Food Prot 65: 1717–1727.

MB Thomas. 1999. Ecological approaches and the development of "truly integrated" pest management. Proc Natl Acad Sci USA. May 25 96:5944–5951.

RB Tompkin. 2002. Control of *Listeria monocytogenes* in the food processing environment. J Food Prot 65:709–725.

USDA, Food Safety and Inspection Service. 2000. FSIS Directive 11,000.1, Sanitation Performance Standards. Jan 25.

C Wallace, T Williams. 2001. Pre-requisites: A help or a hindrance to HACCP. Food Contr 12:235–240.

WHO. 1993. Training consideration for the application of the Hazard Analysis Critical Control Points System to food processing and manufacturing. WHO document, WHO/FNU/FOS/93.3, Division of Food and nutrition. Geneva.

47
Processing Plant Sanitation

Jordi Rovira and Dorota Puszczewicz

INTRODUCTION

Sanitation is a word that comes from the Latin *sanitas,* which means *health*. In that sense, in the food industry, plant sanitation refers to all the measures that can be taken to create or maintain hygienic and healthful conditions in the production plant, to avoid the entrance of contaminants to foodstuffs that can give rise to potential risks for consumers. There are three main types of food contaminants according to their nature: physical, chemical, and microbiological. These are of major importance in processing plant sanitation. These contaminants can enter into the food chain by different routes: with the raw material, ingredients, or other auxiliary material needed for food elaboration; through the environment surrounding the processing line, such as contact surfaces, air, or water, by pests; or by personnel food manipulation (Table 47.1). During different processing steps, new contaminants can appear as well if they are not done properly or are done without taking some preventive measures. Smoking, for example, can produce some hazardous substances known as Polycyclic Aromatic Compounds (PAH), such as benzo(α)pyrene, which has been shown to be carcinogenic in experimental animals (Anonymous 2002).

As a consequence of the presence of these contaminants, the shelf life of the foodstuffs could be reduced and, even worse, consumer health could be compromised. All measures should be taken to avoid food contamination and decrease the risk of foodborne illnesses.

Plant sanitation programs are an important part of the prerequisite programs, which are the base for the implementation of a HACCP plan and the food quality management systems, all of them acting as different barriers to avoid the penetration of contaminants in the foodstuffs. In that sense, the more barriers that can be added to the processing line, the more difficult it will be for contaminants to enter it and the safer will be the food product manufactured. So, the primary aim in processing plant sanitation should be to set up effective barriers to putative contaminants. Table 47.2 shows different barriers that can be considered for this purpose.

FERMENTED MEAT PRODUCTS AND POULTRY

As has been stated in other chapters of this book, fermented meat products are a group of meat products whose main characteristic is that their sensory characteristics (flavor, texture, and shelf life) have been subjected to the action of microorganisms. They are ready-to-eat products, although most of them do not receive any heat treatment during their processing, at least in European countries. Some other fermented meats, such as pepperoni, can suffer a slight heat treatment at 43–53°C (110–128°F) in the core of the sausage during 1 (or several) hour(s), according to the casing diameter, in the last stages of processing (Hinkens et al. 1996). The lack of a heat treatment does not mean that these products are not safe enough for consumption.

Table 47.1. Sources of contaminants in fermented meat products.

Source	Examples
Raw Material	
Meat and fat	*Enterobacteria, E. coli, Listeria* spp., *Salmonella* Small pieces of bones
Spices	Microorganisms
Environment	
Air	Airborne microorganisms "Light" foreign bodies Chemical taints
Water supply	Microorganisms Chemical pollution
Unwanted waters	Microorganisms and insects
Surfaces	Biofilms development Cleaning detergent and sanitizer residues Condensations in walls and ceilings
Personnel Food Manipulation	Microorganisms associated with human activity Small personal belongings, hair, fingernails
Processing	Machinery lubricants Chemical compounds as N-nitrosamines, smoke components as benzo-α-pyrenes Biogenic amines Mycotoxins production by surface molds

Table 47.2. Several barriers to avoid penetration of contaminants in food processing plants.

Barrier	Action
Suppliers control	Raw material inspection and specifications Ingredient specifications Air control quality Water supply quality
Factory design	Plant location Building design Plant distribution
Sanitation program	Cleaning and sanitation
Equipment maintenance plan	Careful revision of equipment Lubricant control
Personnel GMPs	Training
Waste management	Updated plan
Pest control	Updated plan
HACCP plan	Safety processing control
Quality management system	Implementation of a quality system

Dry and semidry-fermented sausages are generally considered shelf-stable, safe meat products, and they have rarely been involved in food contamination outbreaks (Barbuti and Parolari 2002). These products are a clear example of application of the hurdle technology. Among these hurdles, several are related to the growth of lactic acid bacteria (LAB), such as pH reduction, lactic acid, and other organic acids production, and nutrient competitiveness; others are due to the drying process, such as moisture loss and a_w decrease, as well as the use of nitrates and/or nitrites as preservatives. Moreover, some of these products are smoked, and although they are shelf-stable at room temperature, most of them are sold vacuum- or MAP-packaged and in refrigerated condition. However, although fermented meat products are traditionally considered as very safe ready-to-eat meat products, some recent outbreaks, mainly due to *E. coli* O157:H7 have been detected, in the U.S. (1994), Australia (1995), and Canada (1998, 1999). In all these cases, products were elaborated totally or partially from beef meat. It has been proved that some strains of *E. coli* can survive the acidic conditions required for manufacturing of raw fermented sausages (Anonymous 2000). *Salmonella* has been detected as well as the main source of two outbreaks in Italy due to these products (Pontello et al. 1998). At the same time, studies on the behavior of *L. monocytogenes* have shown that it may be reduced but not necessarily eliminated from meat during the fermentation process, particularly when present in high numbers in the raw meat (Glass and Doyle 1989). Nevertheless, according to a risk assessment of *L. monocytogenes* in ready-to-eat foods carried out by the Food and Agricultural Organization (FAO), the risk of listeriosis from fermented meats is very low (2.1×10^{-12} cases/serving) and there have been no reported outbreaks of listeriosis from such products to this point.

In summary, although there are limited epidemiological data related to foodborne disease caused by dry-fermented products, sausage makers must ensure that their products are not contaminated by pathogens. Although processing techniques normally used to manufacture fermented sausages appear to be effective in pathogen control, there is evidence that raw materials are still a major source of bacterial contamination, and current cleaning and sanitation procedures may fail to prevent such pathogens entering the production line (Barbuti and Parolari 2002).

Although some authors consider to a broad extent that dry-cured ham is a whole fermented meat product, in this chapter only dry and semidry-fermented sausages are considered and discussed.

FERMENTED SAUSAGE PROCESSING PLANT SANITATION

As stated previously, the aim of the processing plant sanitation programs is to put as many barriers as possible to avoid the entrance of contaminants into the foodstuffs along the processing line. This section describes general fermented sausage processing and the four main sources that can contaminate the product during the process: raw material, processing environment, personnel manipulation, and some processing steps.

Fermented Sausage Processing

A flow chart of a typical dry-fermented sausage processing is shown in Figure 47.1. For the manufacturing of dry-fermented sausages, different trimmings of pork meat are used, mainly shoulder, belly, and backfat, although a percentage of beef can also be used. Today, and for dietetic reasons, some poultry and turkey dry-fermented sausages are made. Trimmings are kept chilled or frozen prior to use in fabrication. Most of the time these trimmings are used in blocks that are cut in small pieces by guillotines before mincing.

Depending on the type of sausage, meat is ground to different sizes, and with different machines. Normally for dry-fermented sausages with a particle size higher than 3 mm, a traditional plate grinder is used—for instance, in chorizo making, and for sausages with lower particle size, such as some types of salami, it is possible to mince the meat with a bowl chopper. Minced meat and fat are mixed together with the rest of the ingredients and additives in vacuum mixers, which promote air leaving the meat mixture, improve the absorption of additives by the meat mixture, and reduce the redox potential to create favorable conditions for lactic acid bacteria development. After mixing, some traditional factories keep the meat mixture resting in refrigeration for a couple of days before stuffing; for this purpose, vacuum plate stuffers are used and sausages can be stuffed in natural (pork or beef casings) or in synthetic collagen or fibrous casings. Natural casings should be profusely cleaned before use, and synthetic casings should be dipped in a salty solution prior to use.

Once dry-fermented sausages are made, some of them can be dipped in a solution with mold spores or yeast that later will give sausages a whitish surface. If the idea is to produce sausages without molds on its surface, it is usual to dip them into a solution of 20% sodium sorbate, or with a solution

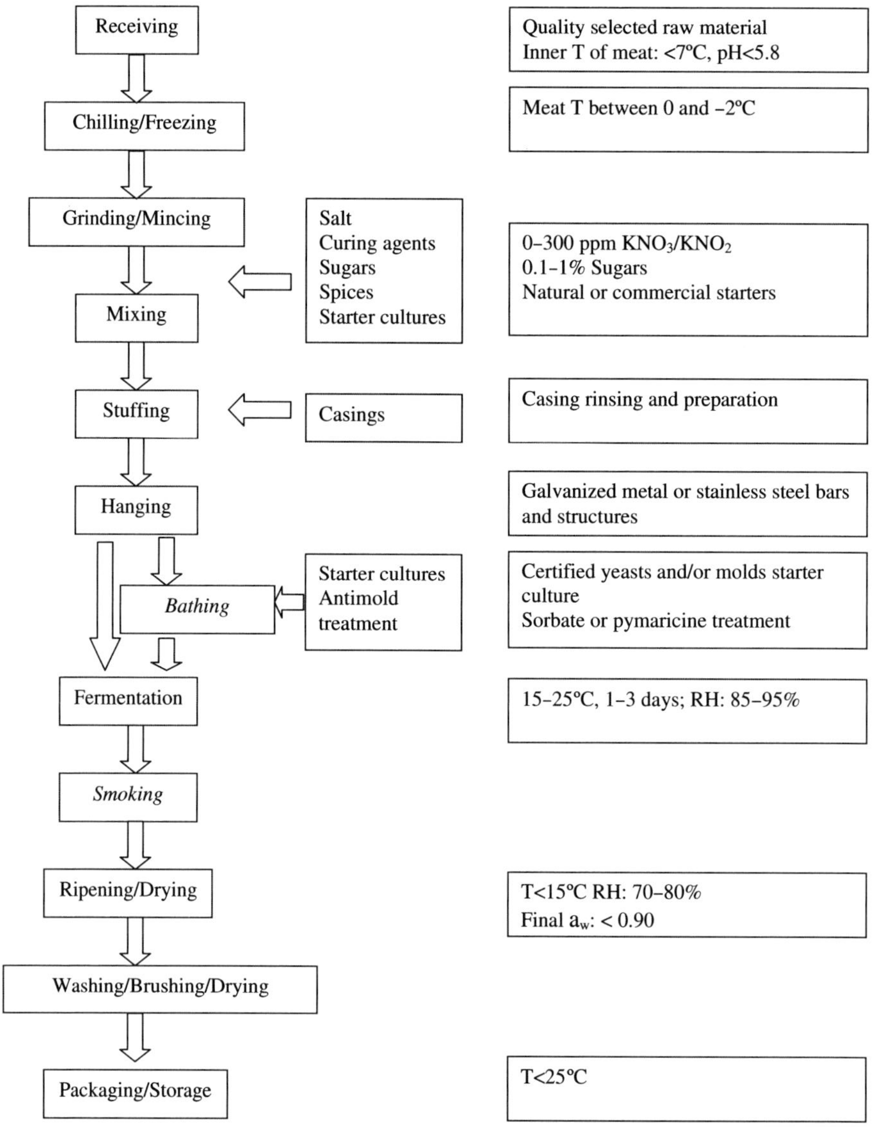

Figure 47.1. Dry-fermented meat processing flow chart (adapted from Martín 2005).

of pymaricine. After bathing, they are hung and placed in bars and transported to the ripening room (green room). Ripening rooms provide the sausages with the proper conditions for fermenting, drying, and ripening throughout a certain period of time, depending on the conditions and mainly on the sausage diameter. Fermentation takes place with more or less intensity depending on the temperature conditions in the first steps of the ripening process; normally the fermentation step occurs during the first 48–72 hours at 20–25°C, and after that decrease temperature to 13–15°C. In most craft industries, fermentation occurs at lower temperatures (below 12–15°C) and for a longer period of time, and the result used to be a milder fermented sausage. Ripening rooms can be natural, when no forced air is used for the drying process, or artificial, when forced air is used. For more details about fermented sausage production, see previous chapters.

CONTAMINATION SOURCES

Raw Material

The quality state of the raw material is of great importance to assure a good quality and safety of the final product. In that sense, it is necessary to reduce to a great extent the contamination level of the raw material and other ingredients related with product formulation. It must be taken into account that the passage of meat trimmings and other ingredients over the processing surfaces leaves residual food debris, which encourages the growth of microorganisms. With time, these microorganisms can multiply to relative high numbers and act as a contamination focus that compromises the safety quality of the meat product manufactured. It is also well known that contaminated raw materials are prone to give more fabrication defects (abnormal fermentations) and to shorten the shelf life of the final product. In that sense, it is very important to control the raw material, proving that meat has been transported at the right temperature: less than 7°C in chilled condition or below −18°C in the case of frozen meat. In both cases, meat trimmings should be properly packaged with plastic films inside the boxes, to avoid the contact of the meat with any dirty surface. Control plant inspectors should check the overall appearance, off-odors, slime production, and strange colorations in meat lots before unloading them from the trucks. Some samples of meat could be taken periodically for microbiological analysis. It is also very important to have a supplier that is certified to follow HACCP plans during slaughter and cutting of animals. The main contaminants of raw meat are *E. coli* O157:H7 (in beef) and *Salmonella*, *Listeria* and *Staphylococcus aureus*.

Processing Environment

The term *processing environment* refers to all elements that surround, or that can enter in contact with, the meat along the processing line in the plant. It includes proper plant design, the quality of the air and water, and the contact surfaces (mainly production machinery).

Hygienic Plant Design

The primary aim of hygienic plant design should be to set up effective barriers to microbial and other contaminants, providing an appropriate environment for processing operations, while ensuring compliance with all applicable building safety and environment regulations (Wierenga and Holah 2003). In that sense, three main barriers can be established, from the point of view of factory design, with the aim of lowering the risk of contamination: the siting of the factory, the factory building design, and the internal plant barriers used to separate the manufacturing processes into different sections.

The design, construction, and maintenance of the site surrounding the factory provide an opportunity to establish the first outer barrier to protect production operations from contamination (Wierenga and Holah 2003). In that sense, an excellent location will be an area with good air quality and no pollution problems (far from other polluting industries). Traditionally, fermented sausage plants were located in dry, cold places to obtain a good air quality and favor the drying process. Normally they used a low-temperature fermenting process with long ripening periods, and sometimes, when the temperature was too low, they used a slight smoking stage or heater fans to raise the temperature. Now, these conditions have changed due to the use of artificial ripening rooms that are able to reproduce practically any natural condition for the ripening stage. It is also necessary to take care of the surroundings of the factory, avoiding the accumulation of waste material and stagnant water that can attract birds, animals, and insects. In that sense, the entire factory should be provided with a good pest control plan in order to avoid the entrance of animals into the factory.

The building structure is the second major barrier, and the factory should be constructed with the aim of providing protection for raw materials, processing facilities, and manufactured product from contamination or deterioration. To review with much more detail all elements regarding the factory building, see Wierenga and Holah (2003).

The final sets of barriers to contamination are those within the factory itself, where two levels of barriers are required: the first level separates processing from nonprocessing areas, and the second one separates high-risk from low-risk processing areas. To avoid cross-contamination, it is very important that the processing flow moves in one straight line of production operations from raw materials to finished products, with the aim of minimizing the possibility of contamination between processed and raw materials. This concept is also more efficient in terms of handling, and it is easier to separate clean and dirty process operations and restrict movement of personnel from dirty to clean areas. High-risk areas are those that process food products that have suffered a decontamination or preservation process and where there is a risk of product contamination. Low-risk areas are those where food components have not yet suffered a decontamination or preservation process. In that sense, in dry-fermented sausage production, most of the process is developed in low-risk areas, with the exception of the packaging area after the ripening process—although in that stage there is also a low risk of cross-contamination by bad manipulation by personnel. Nevertheless, there should be strict separation between the production room, from meat grinding to sausage hanging, and the ripening rooms, where wild microflora can grow on the surface of the products.

In dry-fermented sausages industries, the concept in building design and distribution of the different stages of the processes changed some decades ago, mainly in the fermentation and drying steps, with the introduction of the artificial ripening rooms. Table 47.3 shows the most relevant differences between traditional and modern fermented sausages facilities, pointing out some cleaning and sanitation issues.

Air

Air plays a very important role in the processing of dry-fermented sausages, because of the drying process that takes place during the fermentation and ripening stages of production. It also serves as transfer media of contamination to food products, and unless the air is treated, microorganisms will be present in it and can be transported from different reservoirs outside or inside the plant—for instance, from nonproduct contact surfaces such as floors, ceilings or walls, to the processing line. Through the air "light" foreign bodies such as dust and insects can move around the plant and contaminate foodstuffs. Finally, chemical taints can enter the production area through airborne transmission (Lelieveld 2003).

Water

In the food industry, water has several important applications. It is used as an ingredient, as a production aid, and as a medium for cleaning. In all cases, water can serve as a microbial and chemical contamination dissemination media, which is why it is very important to check the water quality every day. Unwanted water, such as steam or water vapor, condensation, leaking pipes or drains, or rainwater can also be a vector for contamination. Particularly hazardous is stagnant water, because microorganisms can multiply quickly under favorable conditions and serve as a contamination focus (Dawson 2000).

Cleaning and Sanitation

These are two operations that have a great importance for avoiding the entrance of contaminants to the process line, and they are discussed broadly in this chapter.

There is a good review of this matter from Holah (2003). According to this author, cleaning and disinfection are applied with the intention of removing microorganisms or material that supports microbial growth; removing materials that could lead to foreign body contamination or could provide food or shelter for pests; removing food debris, which can deteriorate product quality, from the production line; extending the life and preventing damage to surfaces and equipment; providing a safe and clean working

Table 47.3. Differences between traditional and modern dry-fermented sausage facilities.

	Traditional	Modern
Plant site	Dry and cold places	Everywhere
Building design	Several floors in height	One or two floors
Ripening rooms	Big spaces with many windows in opposite walls to facilitate air flow	Small spaces without windows
	Sometimes several productions or products mixed	Batches or lots more homogenous
	Fixed structures for hanging sausages	Mobile structures for hanging sausages, such as pallets or mobile frames by ceiling rails
	Wooden or metal hanging supports and bars for hanging sausages	Hanging structures of galvanized metal or stainless steel
	Difficult to clean properly	Easy to clean

place for employees; and showing a good impression to visitors. For all these reasons, sanitation programs are of great importance and form an important pillar to support the HACCP plan.

Generally, in sanitation programs different factors should be taken into account, such as timing of cleaning and disinfection, the sequence in which equipment and environmental surfaces must be cleaned and disinfected within the processing area (see Table 47.4), types of cleaners and disinfectants, applying method, frequency, type of inspection and confirmation.

To make an effective cleaning and sanitation procedure, it is necessary to remember that sanitation programs employ a combination of four major factors: mechanical or kinetic energy, chemical energy, temperature or thermal energy, and time. Mechanical energy is of great importance for removing biofilms (Gibson et al. 1999). Chemical energy removes soils from surfaces and suspends them in the cleaning solution to aid rinsing, following a chemical sanitizer attack against microorganisms remaining on the surfaces after cleaning to reduce their viability. Temperature increases the effectiveness of cleaning and disinfection. Finally, the longer the application times of the chemicals, the more effective the process.

Because dry-fermented processing is considered mostly a low-risk–producing area, cleaning and sanitation programs are of great importance, together with raw material and personnel manipulation control, to avoid the presence of contaminants throughout the production line. This production line could be divided into four different areas: (1) chilling rooms and stores, (2) the main production room (from meat grinding to hanging of the sausages), (3) ripening rooms (aging), and (4) the packaging area.

Chilling rooms and stores are the places where raw materials and other ingredients are stocked before they are used in production. These facilities are very difficult to clean and disinfect because they are always full of materials or ingredients that will be used in the next productions. So, emptying these facilities for a full cleaning and sanitation is difficult for a company. In the case of chilling rooms, the low temperatures themselves act as a barrier for microbial contaminants. Nevertheless, cleaning these facilities should be done in another way. It helps to have products well organized, and all of them well packaged or covered to avoid cross-contamination. Accidental spillages should be cleaned and disinfected as soon as possible by the production cleaning team. In chilling rooms, special cleaners adapted to lower temperatures should be used.

The major part of the processing operation for manufacturing dry-fermented sausages is done in one room, where the meat is ground/minced, mixed, and stuffed, and where sausages can receive a cover treatment to inhibit or facilitate the growth of molds and yeast. These operations are still done in a very traditional way, which means that they are still operating in a batch-discontinuous system. This type of production determines the cleaning system; because the use of CIP (cleaning in place) systems is impossible, a COP (cleaning on place) system is used instead.

Table 47.4. Sanitation sequence useful to control the proliferation of undesirable microorganisms.

1. Remove all food and tidy up the production areas.
2. Remove gross soil from production equipment and dismantle if required.
3. Remove gross soil from environmental surfaces.
4. Rinse down environmental surfaces (usually to a minimum of 3 m in height for walls).
5. Rinse down equipment and flush to drain.
6. Clean environmental surfaces, usually in this order: drains, walls, floors.
7. Rinse environmental surfaces.
8. Clean equipment.
9. Rinse equipment.
10. Disinfect equipment and rinse if required.
11. Dry equipment surfaces to avoid condensation.
12. Fog if required (optional, or do it periodically, for instance, once a week).
13. Clean the cleaning equipment.

Adapted from Holah (2003).

To effectively clean the main production room, it is very important to follow a cleaning and sanitation program sequence as is described in Table 47.4. The following sequence should be done daily after production activities:

Step 1. The first step consists of removing and keeping the remaining raw materials and the other ingredients and auxiliary materials (casings, clips, etc.) not used in production. It is very important to switch off all the equipment and cover properly all the electrical parts in order to protect them and the employees.

Step 2. It is important to remove all gross soil and debris from the equipment and conveyors and deposit it in wastebaskets. This operation is done manually. Most of the equipment used in dry-fermented sausage processing can be dismantled to a certain extent; this operation facilitates the following cleaning and disinfection step. All parts of the equipment removed should be kept in trays for later cleaning.

Step 3. Gross soil and debris should be removed as well from the environmental surfaces such as walls and floors.

Steps 4 and 5. Before applying the cleaners, it is important to rinse down all surfaces (equipment and environmental). Normally walls are rinsed up to 3 m in height. For rinsing, temperate water at 40–60°C is used and low-pressure machines (max 25 bar). This temperature range is enough for melting most fat debris and aids in its removal. Hotter water could be used, although it is not advised to do so, because it is not economical, cooks the protein debris and produces microdrops that expand contamination, and produces some problems of condensation. The rinsing operation should be done following an appropriate order, for instance, from up to down and from one side of the machine to the other.

Steps 6–8. Normally for cleaning the different surfaces of the production area, which are rich in fat and protein soils, an alkaline detergent to which active chlorine is added is used to aid the removal of proteinaceous deposits. These detergents are applied at low pressure at concentrations ranging from 1–5% and as foams, from top to bottom, covering all surfaces. Foams have the advantage of not producing aerosol, and they allow the use of more concentrated detergents, can remain on vertical surfaces for longer periods (10–15 min), can cope with high levels of soils, and make it is easy to identify "missed cleaning" areas. Using foams, however, requires more water for rinsing. Cleaning agents should be left for 10 to 30 minutes on the surfaces to clean. A sporadic acid cleaning (once a week) is recommended to remove inorganic stains that can accumulate on the equipment surfaces.

Step 9. Foam is rinsed with low-pressure water at 40–50°C from top to bottom. After this step, all organic soils should have been eliminated from all surfaces. The removal of organic material is extremely important for avoiding the formation of biofilms.

Step 10. Normally in the meat industry, disinfectants based on hypochlorites or quaternary ammonium compounds are used. Disinfectant solutions are sprayed over the surfaces, from top to bottom, and left for 10–30 minutes. This step should eliminate, or at least reduce to safety levels, the microbial population. Some disinfectants are left to air-dry, but normally they are rinsed off, again with low-pressure water. This rinsing step is better at lower temperatures (40°C), in order to avoid condensations.

Step 11. All equipment surfaces must be dried to avoid condensation. Floors are cleaned and dried with the aid of mechanical equipment.

Step 12. Environmental decontamination is recommended at least once a week, and it is done producing a fog (smoke) of disinfectants and/or an insecticide that covers all the production area. Cleaning and disinfection of walls higher than 3 m and ceilings should be done periodically, depending on the splashes and dirt accumulation, normally every 3–6 months.

Step 13. Finally, it is very important to dismantle, clean, and disinfect all cleaning tools after every cleaning to prevent cross-contamination.

Ripening rooms are a very particular part of the dry-fermented sausage process, where a soft meat mixture converts to a dry-fermented sausage. The time sausages spend in the ripening rooms depends on the type of product, and mainly on the casing diameter. In modern plants, it is much easier to clean the ripening room. It is important to empty the ripening room of products and hanging structures, so the cleaning occurs between two batch productions. First the floor and walls are cleaned and disinfected, following a sanitation program sequence very similar to the one described for the main production room. Environmental decontamination is very important when cleaning the ripening rooms. Disinfection and disinfestations are done by applying the sanitizer and insecticides, fogging them in closed ripening rooms, for a period of 2–4 hours. After that period of time, ripening rooms are kept closed for at least 24 hours more, and nobody should enter during that time. Fog is produced by means of a self-producing smoke can or by using some mechanical devices that produce a thermonebulization of the sanitizer products (Aerobrumer; José Collado, S.A. Barcelona; Spain). The hanging structures, which could be pallets or metal frames that move by ceiling rails, are cleaned manually using high-pressure water jets or with special washing machines. In some traditional industries, where natural ripening rooms have fixed hanging structures, fogging is the only way to sanitize the facilities.

When dry-fermented sausages end their ripening period, most of them are covered in different grades by molds that could be inoculated or wild. In both cases, it is normal that, before packaging, this mold cover will be reduced or skipped off. This operation is made in water brushing machines, which remove this cover almost completely. After this operation, and before the sausages go to packaging, they are air-dried for a couple of days in a ripening room. The waste produced by these washing machines should be canalized and sent to the drains.

In packaging areas, it is normal to find different packaging materials, such as cardboard boxes, plastic foils, etc. It is not possible to use wet cleaning here to a great extent, so it is necessary to use a dry cleaning system, which is more manual. Periodically, every week, an environmental disinfection using fog methods, as described above, is recommended.

EVALUATING THE SANITATION STATE OF THE PLANT

Because microorganisms are not visible directly to human eyes, it is very important to assess the effectiveness of the sanitation programs to ensure the safety and the quality of the final products. The cleaning and disinfection programs should be monitored by different methods. It is possible to distinguish between two different types of methods, which are complementary.

The first group of methods is related to visual inspection of the cleaning and sanitizing processes, and it includes auditing the application of the sanitation program sequences described previously. This auditing should be done at least once a month. In this method, a visual inspection of the cleaning process should be done every day after sanitizing to check the effectiveness of the procedure; it involves inspecting surfaces under good lighting, smelling for taints and off-odors, and looking for greasy and encrusted surfaces.

The second monitoring group uses microbiological analysis to evaluate sanitation programs. These methods can be used to monitor surfaces or the environmental state of the production areas. For surface analysis, classical microbiological methods are normally used. Total Viable Count (TVC) is typically used to check the number of microorganisms remaining after cleaning and disinfection procedures. Samples are taken from the surface with a sterile cotton swab and/or sponges. Attached microorganisms are resuspended in a recovery medium, to which a disinfectant neutralizer can be added to stop the disinfectant effect. In the lab, these samples are processed as usual, making serial dilutions, plating in PCA (Plate Count Agar) or TSA (Tryptose Soy Agar) or other selective media, depending on the microorganism search. Another way to monitor the remaining microbial population of the surface is by using self-prepared commercial plates, such as contact plates (Rodac plates) or contact slides. In all cases, results are given as cfu/cm^2.

Traditional hygiene monitoring techniques do not provide a satisfactory solution because they are very time consuming. Increasing attention is being paid to development of rapid methods able to detect microbial and/or organic contamination of surfaces. This is possible using the technique of ATP bioluminescence. ATP is present in all living organisms, including bacteria, and in large quantities in foodstuffs. Usually, there is more nonmicrobial ATP in swab samples coming from processing surfaces than microbial ATP (Hawronskj and Holah 1997). A bioluminescent reaction is produced when ATP entrapped in a swab enters in contact with a solution that contains luciferin and luciferase, which reacts to produce the fluorescent compound oxyluciferin (Deluca and McElroy 1978). Light output is measured in a luminometer, and results are given as relative light units (RLUs). This method gives, in a few minutes, a precise indication of the cleaning and/or disinfection state of a processing surface. A new development of this technique, using an amplification system of ATP, is capable of detecting very low levels of ATP down to 26 pM (Hawronskj and Holah 1997).

Other rapid techniques to detect protein soil in surfaces have been developed based on the Biuret reaction. In that case, a swab rubbed on a testing surface changes its color, indicating the presence of dirt on it. All these techniques could be applied every day and on every surface.

Finally, an indication of the contamination level of the processing environment can be checked by air samplers, which force the passage of air through an agar plate, which is incubated later. Results are given as cfu/m^3.

Table 47.5. Personnel as source of food contamination.

Route	Source	Example of Contaminants
Direct (host)	Gastrointestinal tract Bad hygienic practices	Fecal microorganisms *Enterobacteria, Enterococci* Intestinal parasites
	Skin Touching foods or desquamation	Skin host bacteria *Staphylococcus*
	Hair Hair loss	*Staphylococcus aureus* *E. coli* *Streptococci*
	Mouth and nose Nasal cavity	*Staphylococcus aureus*
	Ears and eyes Eye infections	*Staphylococcus aureus* *Streptococci*
Indirect (vector)	Clothing and footwear Knives Personal small belongings	Soil and microorganisms Microorganisms Pieces of metal

PERSONNEL FOOD MANIPULATION

According to the processing plant sanitation, there is a need to take into account people who are working in the factories. Personnel, and in general people that can move around the production line (mechanics, visitors, managers), are a major source of contamination because they are reservoirs and vectors of microorganisms and other contaminants (physical). Personnel can act as a direct or indirect source of contamination. Direct contamination implies the transfer of microorganisms from people to the food product by direct physical contact; the indirect route involves the transportation (vector) of contaminants by the human actions from one area or surface to another (Holah and Taylor 2003). Table 47.5 shows some examples of direct and indirect human contamination of foods.

To avoid personnel contamination, it is important that factories implement a system of Good Manufacturing Practices (GMP), and all personnel are trained in their right application. Moreover, strict hygienic and cleaning regulations, appropriate clothing, cleaning and sanitizing methods, and devices should be provided to all personnel before they enter a production area.

Production personnel who are in direct contact with food must pass a medical screening to pinpoint any transmissible illness before they start working in a food plant. Personnel with an acute infectious disease or with wounds in their skin should be kept out of the production line.

FINAL CONSIDERATIONS

The aim of all prerequisite programs, which are the basis for a good HACCP plan, is to reduce at a maximum the risks of food contamination. This chapter has discussed briefly some programs, but there are more that should be taken into account, such as plant pest control or maintenance of production machinery, which also contribute to reducing food contaminants.

REFERENCES

Anonymous. 2000. Interim guidelines for the control of verotoxinogenic Escherichia coli including *E. coli* 157:H7 in ready to eat fermented sausages containing beef or a beef product as an ingredient. Health Canada Report.

Anonymous. 2002. Opinion of Scientific Committee of Food on the risks to human health of Polycyclic Aromatic Hydrocarbons in food. European Community.

S Barbuti, G Parolari. 2002. Validation of manufacturing process to control pathogenic bacteria in typical dry fermented products. Meat Sci 62:323–329.

D Dawson. 2000. Water quality for the food industry: Management and microbiological issues. Campden & Chorleywood Food Research Association.

M Deluca, WD McElroy. 1978. Purification and properties of firefly luciferase. Methods Enzymol 57:3–15.

FAO/WHO. 2004. Risk assessment of Listeria *monocytogenes* in ready-to-eat foods. Technical Report.

H Gibson, JH Taylor, KE Hall, JT Holah. 1999. Effectiveness of cleaning techniques used in the food industry in terms of the removal of bacterial biofilms. J Appl Microbiol 87:41–48.

KA Glass, MP Doyle. 1989. Fate and thermal inactivation of Listeria *monocytogenes* in beaker sausage and pepperoni. J Food Protec 52:231–235.

JM Hawronskj, J Holah. 1997. ATP: A universal hygiene monitor. Trends Food Sci Technol 8:79–84.

JC Hinkens, NC Faith, TD Lorang, P Bailey, D Buege, C Kaspar, JB Luchansky. 1996. Validation of pepperoni processes for control of E. coli O157:H7. J Food Protec 59:1260–1266.

JT Holah. 2003. Cleaning and disinfection. In: HLM Lelieveld, MA Mostert, J Holah, B White, eds. Hygiene in Food Processing. Cambridge: Woodhead Publishing Limited.

JT Holah, JH Taylor. 2003. Personal hygiene. In: HLM Lelieveld, MA Mostert, J Holah, B White, eds. Hygiene in Food Processing. Cambridge: Woodhead Publishing Limited.

HLM Lelieveld. 2003. Sources of contamination. In: HLM Lelieveld, MA Mostert, J Holah, B White, eds. Hygiene in Food Processing. Cambridge: Woodhead Publishing Limited.

B Martín. 2005. Estudio de las comunidades microbianas de embutidos fermentados ligeramente acidificados mediante técnicas moleculares, estandarización, seguridad y mejora tecnológica. PhD dissertation. Universitat de Girona, Girona, Cataluña. Spain.

M Pontello, L Sodano, A Nastasi, C Mammina, M Astuti, M Domenichini, E Gerosa, A Montagna. 1998. A community-based outbreak of *Salmonella* enterica serotype typhimurium associated with salami consumption in Northern Italy. Epidemiol Infect 120:209–214.

G Wierenga, JT Holah. 2003. Hygienic plant design. In: HLM Lelieveld, MA Mostert, J Holah, B White, eds. Hygiene in Food Processing. Cambridge: Woodhead Publishing Limited.

48
Quality Control

Fidel Toldrá, M-Concepción Aristoy, Mónica Flores, and Miguel A. Sentandreu

INTRODUCTION

Quality is an important and decisive factor for the commercial transactions throughout the world market and within each particular country and region. Meat processing industries are thus required to implement control systems to guarantee their products and comply with the exigent market demands and legislative requirements (Toldrá et al. 2004).

The final quality of a fermented meat product depends on many factors, such as the quality of the initial raw materials, ingredients, and additives; the processing conditions, such as fermentation conditions, extent of drying and ripening and packaging; and, finally the storage conditions at the factory and during the commercial distribution (Toldrá 2006a). It is necessary to control the quality, not only in the final product but also at different stages of the process. This allows the possibility to trace back to find out and check whether the conditions used for processing are correct and satisfy the demands of the consumer in terms of safety, nutritional value, and sensory characteristics (Toldrá 2006b).

Quality control is essential for the standardization of fermented meat products, which is a traditional consumer demand for acceptability. It is important to remember that these products are bought in supermarket displays or at retail shops, and consumers basically evaluate their general appearance (size, shape, distribution of lean and fat, color, and general presentation). This chapter deals with the quality controls that are generally performed on fermented and ripened meat products.

QUALITY CONTROLS AT EACH STAGE

The quality control of fermented meat products can be done at different stages of the processing. The quality control can be applied to 1) the raw materials and ingredients used for processing, 2) the process, and 3) the final product (Curt et al. 2004). Controls for each one are shown in Table 48.1 and are briefly described below.

Controls of Raw Materials

Several types of control measurements, including instrumental and sensory measurements, are performed. Periodic microbiological controls to check the hygienic conditions of the raw materials are also applied. Main controls (see Table 48.1) are as follows:

- *Lean meat*: The controls mainly consist of the measurement of pH, temperature, weight, hygiene inspection, and visual examination of lean color and appearance. These controls are useful to verify the technological quality of the lean meat.
- *Fat*: These controls consist mainly of instrumental measurement of temperature and visual examination of color and aroma.
- *Casings*: These controls involve checking the thickness, pore size, and absence of fat as indicated by suppliers. This can affect the ripening stage by affecting the drying process.

Table 48.1. Quality controls to be performed at each stage.

Stage	Products	Controls
Raw materials	Lean meat	Hygiene pH, temperature Weight, color, and appearance
	Fat	Hygiene Temperature Aroma, color, and appearance
	Casings	Hygiene Thickness, pore size, and lack of fat
	Other ingredients and additives	Compliance with specifications
Process: Fermentation	Curing chamber Sausages	Temperature, air speed, and relative humidity pH, temperature, and a_w Microbial growth
Process: Drying	Curing chamber Sausages	Temperature, air speed, and relative humidity General appearance pH, temperature, and a_w Microbial growth, presence of molds, metabolites Weight losses and moisture content
Final product	Composition analysis Microbial quality Sensory quality	Protein, fat, carbohydrates, moisture, and ashes Control of microbial flora General appearance Color: L, a, and b parameters Texture: texture profile analysis Flavor: aroma and taste Sensory analyses: evaluation of sensory properties

- *Starter cultures and external molds*: These controls verify growth ability and compliance with the specifications indicated by suppliers.
- *Spices and seasonings, sugars, and additives*: These controls verify the specifications indicated by suppliers.

Controls during Processing

The fermentation stage is usually considered the critical control point concerning the safety and sensory quality of the final product. Both can be either improved or worsened during drying/ripening (Toldrá 2004).

The processing conditions to be checked during fermentation are temperature, time of fermentation and hygrometry of the chamber. Generally, the control parameters are temperature and hygrometry conditions in the chamber as well as the rate of microbial growth and evolution of water activity and pH in the product (see Table 48.1). The processing conditions to be checked during drying and ripening are temperature, air speed, extent of drying, and hygrometry. Generally, the control parameters are temperature and hygrometry conditions in the chamber as well as the weight losses, temperature, and evolution of water activity and the pH of the product. Some products from proteolysis and lipolysis may be analyzed. The growth of molds on the external surface must be controlled, too (Toldrá 2006b).

Controls in the Final Product

Once the product is ready for distribution and sale, there are several important controls that must be performed to verify its final quality. The most important are given in Table 48.1 and briefly described below.

Composition Analysis

The analysis of moisture, fat, protein, and collagen is usually performed according to AOAC methods. The moisture/protein ratios are required by USDA for adequate classification of U.S. dry or semidry products (Sebranek 2004). The measurement of water activity is used in Europe for product standardization (Toldrá 2006a). Insufficient dryness may be due to incorrect drying conditions, as will be discussed in the section "Control of Drying," below. The measurement of pH is convenient in order to control the extent of acidification during fermentation and its final value by the end of the process.

Microbial Quality

Microbial quality is essential from the safety point of view. A number of successive hurdles protect the fermented meat product against undesirable growths (Leistner 1992). The generation of off-odors may be indicating some kind of microbial spoilage. This phenomena may be accompanied by green or gray colorations resulting from pigment degradation as well as holes due to gas generation. The growth of undesirable microorganisms is generally linked to poor hygienic conditions of the raw materials and/or processing conditions, incorrect refrigeration during processing, or insufficient pH drop (Toldrá et al. 2004; Toldrá 2006a). The generation of amines must be checked and the growth of undesirable molds must be prevented.

Sensory Quality

The final appearance depends on the manufacturing conditions. Therefore, defects such as fissures due to a rapid drying at the initial stages, dark stains due to molds, or yellow stains due to the use of rancid fat affect the appearance. Sensory quality is important for consumer attraction toward the product and is further discussed in the later section "Control of Sensory Quality."

CONTROL OF DRYING

The conditions of the curing chamber (temperature, relative humidity, and air rate) are very important for an adequate distribution of the water (balance between external and internal amounts) in the sausage (Baldini et al. 2000). The equilibrium between diffusion of the water through the sausage and evaporation rate from the outer sausage is difficult to reach in sausages of larger diameter. An excessive drying rate may give color fading in the center of the sausage and a dark color on the outside surface, which is associated with hardening. On the other hand, poor drying conditions are detected when the product contains excessive moisture (Demeyer and Toldrá 2004).

The moisture content of sausages and hams constitutes a good control of processing performance. Therefore, dry-fermented sausages may reach water losses above 20%, whereas dry-cured ham may reach nearly 30–34% of weight loss (Toldrá 2006c).

CONTROL OF SENSORY QUALITY

The sensory quality of fermented meat products is essential for attracting the consumer. Therefore, it is essential to study the factors that affect the appearance, color, aroma, taste, and texture of the product and its relationship to the sensory characteristics perceived by the consumers. Chapters in Part IV of this book, "Sensory Attributes," discuss these sensory characteristics. Some controls are given below.

Color

The assessment of color is a useful tool for processing control. The development of the typical cured color is due to the production of nitrosomyoglobin by the reaction of nitric oxide, resulting from the enzymatic reduction of nitrite, with the muscular protein myoglobin (Sanz et al. 1997). This reaction depends on the activity of starter cultures, myoglobin content, and muscle pH. The content in myoglobin, which is the major pigment in meat, varies according to the physiological role of the muscle but substantially increases with the age of the animal (Miller 1994; Aristoy and Toldrá 1998). This is one of the reasons why older pigs give better red color than meats from younger animals and are thus preferred for the production of cured meats.

The use of nonappropriate meats like pale, soft, and exudative (PSE) pork meat or meat from very young pigs may give insufficient color development. Sometimes, and when nitrate is used, the lack of correct color development may be due to an insufficient nitrosomyoglobin formation due to the inhibition of nitrate-reductase by rapid pH drop (Toldrá et al. 2001). The development of brown color can be due to an incorrect nitrification. The hollow center defect can be due to an overrapid drying, poor binding ability of the batter due to the lack of acidification, and poor protein solubilization (use of exudative meats, insufficient salt content, etc). Some color deterioration and/or the development of green areas may be caused by an excessive acid production or by certain strains of *Lactobacilli* or *Streptococci* (Roca and Incze 1990; Toldrá et al. 2006b).

TEXTURE

Consistency is related with the pH and water loss (Roca and Incze 1990). Some typical defects are related to defective drying processes. When the sausages are of small diameter, it is usual to find a hard texture due to an excessive drying as a result of the use of high temperature, low relative humidity, and/or high air rate. These processing parameters must be controlled as consumers tend to reject tough sausages. When the sausages have larger diameters, a hard dry area on the outer sausage surface may be formed under similar circumstances and act as a shield retaining most of the moisture and leaving a soft area inside the sausage. This is known as *case hardening*.

Texture measurements in the form of texture profile analysis may be performed, obtaining parameters such as hardness, springiness, cohesiveness, adhesiveness, and chewiness (Honikel 1997). Another type of test may consist of the assessment of resistance to penetration as an index of hardness during mastication (Tabilo et al. 1999).

An excessive softness inside the sausage may be due to different causes (Toldrá 2002):

- The application of nonefficient drying (i.e., high relative humidity), which results in poor water loss and an excess of moisture inside the sausage
- Fat smearing, which retains water
- Insufficient pH drop
- An excess of fat in the mix
- A very fast drying (high temperature, low relative humidity, and high air velocity), which produces hardening of the outer sausage surface but retains most of the moisture inside the sausage.

Fermentation (correct sugar addition, temperature, and starter cultures function) and drying, especially relative humidity and air rate within the chamber, must be carefully controlled. The microbial activity produces a decrease in pH value approaching the isoelectrical point of meat proteins. The water holding capacity is reduced, producing an increase in consistency that also will be accelerated during the drying process. This facilitates the sliceability typical of the product. The salt added in the formula gives a suitable cohesion and texture during drying by solubilizing proteins, which act as a bridge between the constituting meat fragments. However, several defects can appear in the final product. This is the case of the sticky defect or softness (due to either an excessive amount of fat or high moisture in the final product), dry-edge defect (due to an overrapid drying), and inconsistency (stuffing carried out at low pressure producing holes and air inclusions, which also can produce oxidation, rancidity, and mold development).

Table 48.2. Possible origin and odor descriptions of compounds detected in dry-fermented sausages.

	Dry Fermented Sausage	
	Compound	Odor
Amino acid degradation	Methional	Cooked potato
	3-Methyl-butanal	Sour cheese
	2-Methyl-butanal	Nail polish
	2-acetyl-1-pyrroline	Roasty, popcorn
	3-Methyl-butanoic acid	Sweaty socks
Lipid Oxidation	Hexanal	Green leaves
	Heptanal	Potatoes
	Decanal	Cucumber, dry grass
	1-Octen-3-ol	Mushroom
Carbohydrate fermentation	2,3-Butandione	Butter
Staphylococci Esterification	Propyl acetate	Sourish apple, candy
	Ethyl isobutanoate	Pineapple, fruit
	Ethyl butanoate	Fruit candy
	Ethyl 2-methylbutanoate	Sweet pineapple
	Ethyl 3-methylbutanoate	Chutney, spicy
Spices	Myrcene	Lemon, fruity
	Limonene	Menthol
	Terpinolene	Fruity, eucalyptus

From Meynier et al. (1999); Stahnke (1994); Blank et al. (2001).

Flavor

The characteristic flavor of fermented sausages is developed by the action of microbial and endogenous meat enzymes on carbohydrates, lipids, and proteins. In addition, the salt and spices added to the meat batter should be taken into account because of their important contribution to flavor (Demeyer and Toldrá 2004).There also are other nonenzymatic pathways that contribute to the generation of flavor compounds, such as autooxidation.

The carbohydrate fermentation is responsible for the typical sausage tangy or sour taste (Lücke 1985). The interactions between carbohydrates and protein metabolites during meat fermentation determine the rate of pH decline and flavor development (Demeyer 1992). There are internal and external parameters that influence flavor (Verplaetse 1994). The internal parameters are chemical (added sugars or spices) or microbiological (starter cultures), whereas external parameters are physical, such as the temperature and humidity during the process.

In the case of dry-fermented aroma, because no compound or group of compounds with the unique odor of meat has been isolated, it seems likely that meaty aromas depend upon a subtle quantitative balance of various components (Mottram and Edwards 1983). The relative contribution of the chemical compounds to the meat aroma depends on its odor threshold, which gives an idea of the odor impact. Therefore, if the concentration of an odorant in a sausage is higher than its threshold, probably the odorant will contribute to the overall flavor. In fermented sausages, different odors have been detected, and their compounds and possible origin were identified (see Table 48.2). The development of electronic noses has found important applications for the aroma control of cured meats (Eklov et al. 1998). The main objective on aroma analysis is the determination of those volatile compounds that have a high impact on the food flavor. However, these compounds can contribute not only to the general dry-fermented aroma but also can contribute to specific aroma that in same cases can be detrimental (Demeyer et al. 2000). This is the case of an excessive lipid oxidation, where the inadequate processing conditions produce the generation of a high

Table 48.3. Sensory evaluation techniques used in dry-fermented sausages.

Sensory Techniques	Panel	Scale	References
Free choice profiling (selection)	Trained sensory panel	6-point intensity scale	Berry et al. (1979)
	Untrained panel	8-point hedonic scale	Berry et al. (1979)
Fixed choice profile	Trained panel	Unstructured scale with end points defined by consensus	Beilken et al. (1990)
	Semitrained assessors	9-point intensity scales	Roncales et al. (1991)
Quantitative descriptive method	Trained panel	Unstructured line scale of 15 cm	Stahnke (1995)
Flavor profile	Trained panel	Nonstructured scales	Rousset-Akrim et al. (1997)
Acceptance test	Trained panel	Nonstructured 10 cm hedonic scale	Bruna et al. (2000, 2001) Selgas et al. (2003) Herranz et al. (2005, 2006)
Acceptance test	Trained panel	Nonstructured 10 cm hedonic scale	de la Hoz et al. (2004)
Quantitative descriptive analysis	Trained panel		Benito et al. (2004)
Acceptance test	Semitrained panel	6-point intensity scale	Moretti et al. (2004)
Descriptive sensory panel	Trained panel	5-point intensity scale	Ruiz Perez-Cacho et al. (2005)
Quantitative descriptive analysis	Trained panel	6-point intensity scale	Valencia et al. (2006)

Table 48.4. Sensory descriptive terms used for fermented sausages.

Sensory Terms	Beilken et al. (1990)	Roncales et al. (1991)	Stahnke (1995)	de la Hoz et al. (2004)	Benito et al. (2004)	Moretti et al. (2004)	Ruiz Perez-Cacho et al. (2005)
Aroma	Acidic/sour Off-aroma Smoke Acceptability of aroma	External smell intensity External smell quality Smell intensity Smell quality	Odor intensity Salami Cheese Sourdough Sourish Fatty Rancid Nauseous Burned Solvent	Odor	Aroma intensity Cured aroma Rancid		Black pepper Lactic acid Mold Spice odor
Flavor	Acidic/sour Hot/spice Off-flavor Smoke Acceptability of flavor	Flavor intensity Flavor quality				Rancidity	Black pepper Mold Spice aroma
Taste				Taste Sweet Bitter Sour Salty Meat impression	Saltiness Sweetness Bitterness Acidness	Saltiness Acidity	Acid Salty
Texture	Initial bite/adhesion Chewiness Greasiness	Toughness Juiciness Overall mouth perception		Texture Juiciness	Hardness Softness Fibrousness Juiciness	Hardness Elasticity Cohesiveness	Hardness Juiciness

	Acceptability of texture					
Other Terms	Appearance: Color Particle size Fat content Acceptability of appearance Overall acceptability	Appearance: Visual external Mold cover Resistance to pressure Overall external perception Visual cut appearance Total fat Ease of skin peeling Color Overall cut perception Overall acceptability	Color	Aftertaste	Color intensity Overall acceptability	Appearance: Exudate Fat/lean connection Luminance Presence of crust

concentration of aldehydes, compounds responsible for rancid aromas. Other important flavor defects that can appear in fermented products are excessive acidity (due to large amounts of lactic acid), rancidity (high oxidation due to the use of previous rancid fat as a consequence of undesirable storage conditions such as high temperatures and long times), putrid aromas (developed by undesirable bacteria due to an inappropriate fermentation or by the use of highly contaminated raw materials), and mold aromas (due to contamination by molds). Unpleasant urinelike off-flavor is due to the use of raw meats from adult pigs containing androstenone, the male hormone 5-α-androsten-16-en-3, at amounts above the detection threshold level.

Sensory Analyses

The appropriate evaluation of sensory properties involves the use of attribute difference tests, such as paired comparison, scoring, or ratio-scaling tests (Meilgaard et al. 1991). The descriptive sensory analysis gives details on sensory changes and is useful in investigating treatment differences, process changes, quality control, and defining sensory properties of a product (Bett 1993).

The idea of descriptive analysis is based on the division of total flavor into flavor terms, notes, or descriptors. Therefore, the flavor of a product is characterized using descriptive terms and it is also measured giving the descriptor intensity. The panel consists on 5 to 20 members, and they are selected based on their abilities to smell. Common descriptive flavor methods are the flavor profile (Caul 1957), a nonnumerical intensity scale (simple universal scale), with which the panel discusses each sample and forms a consensus score; the quantitative descriptive analysis (QDA), which is a continuous line scale converted to numbers for statistical analysis (Stone et al. 1974); and the spectrum method (Meilgaard et al. 1991), which is a universal intensity scale with defined reference points of commercially available foods.

In dry-fermented sausages many different techniques have been used for sensory evaluation (see Table 48.3). Frequently, quantitative descriptive techniques have been applied, although many authors prefer the use of acceptance tests with hedonic scales.

On the other hand, many different sensory descriptive terms have been used in fermented sausages (see Table 48.4). In general, the aroma descriptive terms are related to the typical sausage aroma, the presence of defects such as rancidity, the use of different spices, and the presence of molds. The taste terms are the basic tastes, and the texture terms used are related to the hardness, softness, and cohesiveness of the sausage. In the evaluation of the appearance, the external appearance has been generally evaluated.

REFERENCES

MC Aristoy, F Toldrá. 1998. Concentration of free amino acids and dipeptides in porcine skeletal muscles with different oxidative patterns. Meat Sci 50:327–332.

P Baldini, E Cantoni, F Colla, C. Diaferia, L Gabba, E Spotti, R Marchelli, A Dossena, R Virgili, S Sforza, P Tenca, A Mangia, R Jordano, MC Lopez, L Medina, S Coudurier, S Oddou, G Solignat. 2000. Dry sausages ripening: Influence of thermohygrometric conditions on microbiological, chemical and physico-chemical characteristics. Food Res Inter 33:161–170.

SL Beilken, LM Eadie, PN Jones, PV Harris, 1990. Sensory and other methods for assessing salami quality. CSIRO Food Res Quarterly 50:54–66.

MJ Benito, M Rodríguez, A Martin, E Aranda, JJ Córdoba. 2004. Effect of the fungal protease EPg222 on the sensory characteristics of dry fermented sausage "salchichon" ripened with commercial starter cultures. Meat Sci 67:497–505.

BW Berry, HR Cross, AL Joseph, SB Wagner, JA Maga. 1979. Sensory and physical measurements of dry fermented salami prepared with mechanically processed beef product and structured soy protein fiber. J Food Sci 44:465–468.

KL Bett. 1993. Measuring sensory properties of meat in the laboratory. Food Technol 11:121–126,134.

I Blank, S Devaud, LB Fay, C Cerny, M Steiner, B Zurbriggen. 2001. Odor-active compounds of dry-cured meat: Italian type salami and Parma ham. In: Aroma Active Compounds in Foods. GR Takeoka, M Günter, KH Engel, eds. pp. 9–20, Washington D.C.: American Chemical Society.

———. 2000. Combined used of Pronase E and a fungal extract (*Penicillium aurantiogriseum*) to potentiate the sensory characteristics of dry fermented sausages. Meat Sci 54:135–145.

JM Bruna, M Fernández, EM Hierro, JA Ordóñez, L de la Hoz. 2001. The contribution of *Penicillium aurantiogriseum* to the volatile composition and sensory quality of dry fermented sausages. Meat Sci 59:97–107.

JF Caul. 1957. The profile method of flavor analysis. Adv Food Res 7:1–40.

C Curt, G Trystram, H Nogueria-Terrones, J Hossenlop. 2004. A method for the analysis and control of sensory properties during processing—Application to the dry sausage process. Food Contr 15:341–349.

L de la Hoz, M D'Arrigo, I Cambero, JA Ordoñez. 2004. Development of an n-3 fatty acid and alpha-tocopherol enriched dry fermented sausage. Meat Sci 67:485–495.

D Demeyer. 1992. Meat fermentation as an integrated process. In: New Technologies for Meat and Meat Products. FJM Smulders, F Toldrá, J Flores, M Prieto, eds. Nijmegen, The Netherlands: Audet, pp. 21–36.

DI Demeyer, M Raemakers, A Rizzo, A Holck, A De Smedt, B Ten Brink, B Hagen, C Montel, E Zanardi, E Murbrek, F Leroy, F Vanderdriessche, K Lorentsen, K Venema, L Sunesen, L Stahnke, L De Vuyst, R Talon, R Chizzolini, S Eerola. 2000. Control of bioflavor and safety in fermented sausages: First results of a European project. Food Res Inter 33: 171–180.

DI Demeyer, F Toldrá. 2004. Fermentation. In: Encyclopedia of Meat Sciences. W Jensen, C Devine, M Dikemann, eds. London, UK: Elsevier Science Ltd., pp. 467–474.

T Eklov, G Johansson, F Windquist, I Lundström. 1998. Monitoring sausage fermentation using an electronic nose. J Sci Food Agric 76:525–532.

B Herranz, L de la Hoz, E Hierro, M Fernandez, JA Ordonez. 2005. Improvement of the sensory properties of dry-fermented sausages by the addition of free amino acids. Food Chem 91:673–682.

B Herranz, M Fernandez, L de la Hoz, JA Ordonez. 2006. Use of bacterial extracts to enhance amino acid breakdown in dry fermented sausages. Meat Sci 72:318–325.

KO Honikel. 1997. Reference methods supported by OECD and their use in Mediterranean meat products. Food Chem 59:573–582.

F Leistner. 1992. The essentials of producing stable and safe raw fermented sausages. In: New Technologies for Meat and Meat Products. FJM Smulders, F Toldrá, J Flores, M Prieto, eds. Nijmegen, The Netherlands: Audet, pp. 1–19.

FK Lücke. 1985. Fermented sausages. In: Microbiology of Fermented Foods. BJB Wood, ed. London, UK: Elsevier Applied Science Pub, pp. 41–81.

M Meilgaard, GV Civille, BT Carr. 1991. Sensory Evaluation Techniques. Boca Raton, Florida: CRC Press, Inc.

A Meynier, E Novelli, R Chizzolini, E Zanardi, G Gandemer. 1999. Volatile compounds of commercial Milano salami. Meat Sci 51:175–183.

RK Miller. 1994. Quality characteristics. In: Muscle Foods. DM Kinsman, AW Kotula, BC Breideman, eds. New York: Chapman and Hall, pp. 296–332.

VM Moretti, G Madonia, C Diaferia, T Mentasti, MA Paleari, S Panseri, G Pirone, G Gandini. 2004. Chemical and microbiological parameters and sensory attributes of a typical Sicilian salami ripened in different conditions. Meat Sci 66:845–854.

DS Mottram, RA Edwards. 1983. The role of triglycerides and phospholipids in the aroma of cooked beef. J Sci Food Agric 34:517–522.

M Roca, K Incze. 1990. Fermented sausages. Food Rev Inter 6:91–118.

P Roncalés, M Aguilera, JA Beltrán, I Jaime, JM Peiró. 1991. The effect of natural or artificial casings on the ripening and sensory quality of a mould-covered dry sausage. Inter J Food Sci Technol 26:83–89.

S Rousset-Akrim, JF Martin, MC Bayle, JL Berdagué. 1997. Comparison between an odour profile and a flavour profile of dry fermented sausages. Inter J Food Sci Technol 32:539–546.

MPR Ruiz Perez-Cacho, H Galan-Soldevilla, F Leon-Crespo, G Molina-Recio. 2005. Determination of the sensory attributes of a Spanish dry-cured sausage. Meat Sci 71:620–633.

Y Sanz, R Vila, F Toldrá, P Nieto, J Flores. 1997. Effect of nitrate and nitrite curing salts on microbial changes and sensory quality of rapid ripened sausages. Inter J Food Microbiol 37:225–229.

JG Sebranek. 2004. Semidry fermented aausages. In: YH Hui, L Meunier-Goddik, ÅS Hansen, J Josephsen, W-K Nip, PS Stanfield, F Toldrá, eds. Handbook of Food and Beverage Fermentation Technology. New York: Marcel Dekker, Inc., pp. 385–396.

MD Selgas, J Ros, ML Garcia. 2003. Effect of selected yeast strains on the sensory properties of dry fermented sausages. Eur Food Res Technol 217:475–480.

LH Stahnke. 1994. Aroma components from dried sausages fermented with staphylococcus xylosus. Meat Sci 38:39–53.

———. 1995. Dried sausages fermented with Staphylococcus-Xylosus at different temperatures and with different ingredient levels. 2. Volatile components. Meat Sci 41:193–209.

H Stone, J Sidel, S Oliver, A Woolsey, RC Singleton. 1974. Sensory evaluation by quantitative descriptive analysis. Food Technol 28:24–34.

G Tabilo, M Flores, SM Fiszman, F Toldrá. 1999. Postmortem meat quality and sex affect textural properties and protein breakdown of dry-cured ham. Meat Sci 51:255–260.

F Toldrá. 2002. Dry cured meat products. Trumbull, Connecticut: Food & Nutrition Press, pp. 27–220.

———. 2004. Fermented meats. In: Food Processing: Principles and Applications. YH Hui, JS Smith, eds. Ames, Iowa: Blackwell Publishing, pp. 399–415.

———. 2006a. Biochemistry of fermented meat. In: Food Biochemistry and Food Processing. YH Hui, WK Nip, ML Nollet, G Paliyath, BK Simpson, eds. Ames, Iowa: Blackwell Publishing, pp. 641–658.

———. 2006b. Meat fermentation. In: Handbook of Food Science, Technology and Engineering, Volume 4. YH Hui, E Castell-Perez, LM Cunha, I Guerrero-Legarreta, HH Liang, YM Lo, DL Marshall, WK Nip, F Shahidi, F Sherkat, RJ Winger, KL Yam, eds. Boca Raton, Florida: CRC Press, Inc., pp. 181-1–181-12.

———. 2006c. The role of muscle enzymes in dry-cured meat products with different drying conditions. Trends Food Sci Technol 17:164–168.

F Toldrá, M Flores, Y Sanz. 2001. Meat fermentation technology. In: Meat Science and Applications. YH Hui, WK Nip, RW Rogers, OA Young, eds. New York: Marcel Dekker, Inc., pp. 537–561.

F Toldrá, R Gavara, JM Lagarón. 2004. Packaging and quality control. In: Handbook of Food and Beverage Fermentation Technology. YH Hui, LM Goddik, J Josephsen, PS Stanfield, AS Hansen, WK Nip, F Toldrá, eds. New York: Marcel Dekker, Inc., pp. 445–458.

I Valencia, D Ansorena, I Astiasarán. 2006. Nutritional and sensory properties of dry fermented sausages enriched with n-3 PUFAs. Meat Sci 72:727–733.

A Verplaetse. 1994. Influence of raw meat properties and processing technology on aroma quality of raw fermented meat products. Proc 40th Int Congr Meat Sci and Technol, The Hague, The Netherlands, p. 45.

49
HACCP

Maria Joao Fraqueza, Antonio S. Barreto, and Antonio M. Ribeiro

THE HAZARD ANALYSIS CRITICAL CONTROL POINT CONCEPT: WHY USE IT

The major goal of a *food operator* is to produce and put on the market tasty, convenient, and safe products in order to establish and maintain a trust relationship with consumers and be rewarded for his effort. The concepts of safety and genuineness imply that food will not cause harm to the consumer, from when it is prepared to when it is eaten in accordance with its intended use, and that its nature and origin may be retrieved along the food chain. These are the primary concerns of consumers, and therefore constitute the utmost priorities of EU policy *Agenda 2000* reported on the White Paper on Food Safety (COM 2000) and supported by EU food safety authorities as well as those of the United States through *Healthy People 2010* (USDHHS 2000) supported by the Food and Drug Administration (FDA), Center for Disease Control and Prevention (CDC), Department of Health and Human Services (DHHS) and the Food Inspection Systems of the U.S. Department of Agriculture (USDA) (FDA 2005). In a global market, food safety is a top priority for food trade, and given even greater emphasis by the aging of the population, the increased number of immunocompromised individuals and changes in consumer eating habits and production practices.

If the big food plants in the developed world claim to have elevated the production of some fermented foods to a large industrial and technologically sophisticated level, in Europe there are a great number of regions where products are manufactured in a traditional cottage-style or farmhouse manner (Caplice and Fitzgerald 1999; Tradisausage 2006). It is also true that consumers prefer traditional fermented meat products produced in small and farmhouse units due to their characteristics of taste and aroma. However, for large food corporations or small farm or cottage manufactures, the safety of the products must be assured and has to be achieved by different preventive proactive measures assisted by mandatory legal requirements.

The proactive methodology such as HACCP (*Hazard Analysis and Critical Control Point*) implies a systematic application of preventive measures to be managed in all the fermented meat production steps according to an established plan. It is a document where identification and evaluation of food safety hazards are reported along with preventive measures for their control (Mortimore and Wallace 1997; FDA 2005).

Originally, this system was developed and has been applied to food production since the early 1960s by the U.S. Pillsbury Company with the cooperation and participation of the National Aeronautic and Space Administration (NASA), Natick Laboratories of the U.S. Army, and the U.S. Air Force Space Laboratory Project Group. It was considered the only way to ensure 100% food safety for the astronauts of the Apollo program (Mortimore and Wallace 1997).

The terms of reference for food safety include biological, chemical, or physical agents that may be present in a food and which are hazards likely to cause illness or injury to consumers if not put under control.

The HACCP system has been recognized worldwide as an effective system to warrant food safety. The system must integrate every step in the food chain from producer to consumer to assure "*food safety from farm to fork.*" Before being enforced by law and praised by global free trade, HACCP was recognized by food operators as the most reliable and least expensive system to assure food safety.

Foundations of System Development and Implementation

The development of a HACCP program depends on the will of the upper administration and free acceptance by the work force. Work starts by organizing a team of workers previously selected and trained in hazard food analysis and HACCP principles. One single person cannot carry out a HACCP plan; it has to be worked by a multidisciplinary team with competences in different areas such as quality, commodity purchasing, production, engineering, distribution, microbiology, toxicology, and auditing. When all such expertise is not available, as in the case of small units producing fermented meat products, expert advice should be obtained from outer sources such as trade and industry associations, independent experts, regulatory authorities, and assisted HACCP training with the work force being guide by sector-specific operative manuals.

The team leader should have organizational and communication skills, familiarity with plant processes, and absolute authority to pursue this program in the company. The development of HACCP program will include:

- *Product description and distribution*, including formulation and screening of potential abuse through distribution or by consumers
- *Identification of projected uses and consumers of the product*, raw or cooked, and whether segments of the population are at increased risk, such as infants, elderly, immunocompromised, etc.
- *Construction of a flow diagram*, providing a brief description of all operative steps involved, from the reception of raw materials to finished products distribution
- *A hazard analysis based on stated terms of reference* is the foundation of HACCP construction, coupled with the critical control points (CCPs) to put them under control

Construction and implementation of this system is a big expensive task, the success and financial return of which implies both the support and full involvement of upper management. HACCP development and implementation supported only by middle management or technical personnel is an impossible mission.

The Seven HACCP Principles

Principle 1 is the identification of biological, chemical, and physical hazards in all materials and process steps (Goodfellow 1995; Mortimore and Wallace 1997; FAO 1997).

After assessing a consistent risk and preventive measures designed to control hazards, *Principle 2* implies the identification of adequate critical control points (CCPs) in materials and process steps to control the hazards.

Principle 3 is the definition of critical limits for each CCP.

Principle 4 is monitoring to warrant fulfillment of procedures at each stated CCP.

When a deviation from the outlined limits occurs at a CCP, *Principle 5* states that corrective action is immediately applied to restore control.

Principle 6 is verification to test compliance of the plan with the HACCP system.

Principle 7 is a record of operations including procedures for monitoring, corrective actions, and verification providing an effective demonstration of the system at work.

The final element for HACCP development is to *update the plan* when changes in the process or new legal or trade requirements are introduced.

The HACCP plan is specific to a certain product, process, and enterprise. Nevertheless, in many cases, several product lines are so similar that they can be grouped together in generic models. The use of a generic HACCP model to build specific HACCP plans always needs creative adaptation and tuning. In this case, system validation is always needed for each specific plan.

To accomplish with success the implementation of HACCP plans, some prerequirements need to be fulfilled. They are mandatory for food safety. In small food workshops, they can per se be the foundation of so called "light" HACCP. It may be assumed in extremis that a well-trained person with access to guidance is able to implement in-house HACCP methods.

HACCP MODEL FOR FERMENTED MEAT SAUSAGES: TOWARD A GENERIC MODEL FOR HACCP IMPLEMENTATION IN TRADITIONAL WORKSHOPS AND SMALL FERMENTED MEAT SAUSAGE PLANTS

PREREQUIREMENTS

Standard requirements of hygiene for the production, handling, packaging, storing, and distribution of fermented meat sausages are mandatory to assure a healthy and wholesome supply of such products and the successful construction and implementation of a HACCP program.

Layout and operative conditions of premises and equipment coupled with Good Hygiene Practices (GHP) and Good Manufacturing Practices (GMP) programs are critical to small meat and poultry processing plants, and are required to fulfill a HACCP plan for fermented meat sausages. The traceability of meat and any other substances used in traditional fermented sausages is also required and is to be assessed at all stages of production and distribution.

General GHP and GMP codes for different food process categories have already been performed by regulatory food standard organizations (FAO/WHO 2005b). However, enterprises must define and elaborate their own GHP and GMP codes as operative guides.

An evaluation of prerequirements in small units and farmhouses producing traditional fermented meat products was performed by the *Tradisausage project* (2006) in southern European countries, including France, Greece, Italy, Spain, Portugal, and Slovakia. The data from a checklist designed to evaluate prerequirements, GHP, and GMP (Table 49.1) led to the conclusion that approximately 80% of the small units in southern Europe have the main prerequirements essential to implement the HACCP system. However, they have to overcome the problem related to documented evidence and validation of GHP. The project revealed that the integration of existing quality control facilities into an operative HACCP plan by a small team is a difficult task.

It is important to note that Codex Alimentarius Commission (FAO/WHO 2005) and European Commission Health and Consumer Protection Directorate–General (EC 2005) permit, when applying HACCP, flexibility when appropriate, depending on the context of the application and taking into account the nature and the size of the operation. So for small units, the application of a validated Code of Hygiene practices will be effective to reduce the risk of hazards and is fundamental for subsequent implementation of a HACCP-based approach.

PRODUCT DEFINITION AND INTENDED USES

A great diversity of products which are fermented dry or semidry sausages with singular organoleptic characteristics are found in different countries, but they have in common a sequence of process steps to assure their stability such as a decrease in pH and a_w, promoted by environmental or technological dedicated flora (starters), and drying coupled sometimes with particular ripening agents, preservatives and smoke (Leistner and Gould 2003).

Thus, fermented sausages are the end product of comminuted pork meat stuffed into casings (salami, the Italian fermented sausages also contain beef) mixed with fat, salt, sugar, preservatives (nitrite/nitrate), and seasonings (wine, garlic, pepper, paprika, and other spices), more-or-less acidified by lactic fermentation (Caplice and Fitzgerald 1999; Leistner and Gould 2003). The particular seasoning, structure (meat particles size), type of casing, acidity, and a_w gives rise to stability and specific organoleptic characteristics of the sausages coming from a traditional knowhow of different countries and regions from central and southern Europe.

These sausages are marketed without the assurance of a lethal heat treatment. It is important to a HACCP team to assume what kind of product actually is concerned (Table 49.2) and how and by whom it will be consumed (Mortimore and Wallace 1997).

PROCESS DESCRIPTION AND FLUX DIAGRAM

A technological flow diagram of operation depicting all pertinent manufacturing steps from the reception of raw materials to final product, is outlined by the HACCP team (Figure 49.1). This flow diagram is used for hazard analysis after being validated on the premises.

The technology of process description must comply with all specified good hygiene and manufacturing practices for the elaboration of a specific fermented product, as defined in the prerequirements.

TERMS OF REFERENCE FOR HAZARDS: RISK EVALUATION

A hazard is any factor that may be present in the product causing harm to the consumer either through injury or illness. The identification of hazards in the production of fermented sausages must be related to

Table 49.1. A checklist for prerequirements evaluation in small units and farmhouse facilities producing traditional fermented meat sausages in southern European countries.

Questions	Answer (% of Processing Unit) n = 55		
	Yes[a] Bad[b]	No Sufficient	Good
Premises: Buildings and Equipment			
Competent authority approved the plant?[a]	96.4	3.6	
There is waste treatment?[a]	70.9	29.1	
Selection and separation of solid trash?[a]	60.0	40.0	
Water supply? With treatment or not?[a]	96.4	3.6	
Evaluation of the layout: Are there crossed lines?[a]	32.7	67.3	
Facilities: Materials used in the construction (walls, floor, ceilings, doors, windows) are appropriate for easy cleaning?[a]	94.5	5.5	
Walls and floor material conservation?[b]	11.1	38.9	50.0
Walls and floor hygiene level?[b]	16.4	32.7	50.9
Do open windows have fine mesh to keep out insects?[a]	80.0	20.0	
Are there electric flytraps?[a]	70.9	29.1	
Are all lights in preparation or processing areas equipped with shields or proper covers?[a]	87.3	12.7	
Is there adequate or sufficient ventilation or equipment to minimize odors and vapors and prevent water condensation?[a]	76.4	23.6	
Are there foot-washing devices at the entrance of your workshop?[a]	25.5	74.5	
Sanitary Facilities			
Is the establishment producing in compliance with some Code of Hygiene Good Practices?[a]	70.9	29.1	
In the process area are the hand-washing facilities adequate and sufficient?[a]	96.4	3.6	
Hand-washing facilities before entrance?[a]	83.6	16.4	
Effective hand-cleaning and sanitizing preparations?[a]	87.3	12.7	
Presence of hand-drying devices?[a]	76.4	23.6	
Are there toilets, urinals?[a]	94.5	5.5	
Are there toilet rest areas?[a]	58.2	41.8	
Are the toilets, urinals, and rest areas kept clean?[b]	10.9	38.2	50.9
Are there bird, insect, and rat control plans provided by a specialized pest control enterprise?[a]	40.0	60.0	
Equipment			
Equipment and utensils are designed and constructed with material and workmanship as to be adequately cleanable?[a]	98.2	1.8	
Equipment and utensils are made of innocuous material that will not cause food contamination?[a]	94.5	5.5	
Hygiene and Sanitation			
Is an individual health certificate issued?[a]	93.6	36.4	
Do personnel have routine medical examinations?[a]	78.2	21.8	
Do all workers use suitable protective clothing (hair nets/caps, boots)?[a]	81.8	18.2	

Table 49.1. (*Continued*)

Questions	Answer (% of Processing Unit) n = 55		
	Yes[a] Bad[b]	No Sufficient	Good
Do all workers keep protective clothing in good hygienic condition before beginning to work?[a]	87.3	12.7	
Are jewels, watches, and other adornments removed before entering working premises?[a]	83.3	16.7	
How frequently is the protective clothing changed?[b]	9.1	58.2	32.7
Are workers with illnesses, open lesions, or any other abnormal source of microbial contamination not in direct contact with food?[a]	83.6	16.4	
Are specific and adequate education and training provided to workers (hygiene and process)?[a]	80.0	20.0	
Cleaning-Disinfecting Methods			
Facilities cleaning schedule—floor. Daily, twice a week, others?[b]	92.6 Daily	3.7 Twice a week	3.7 Others
Facilities cleaning schedule—walls. Daily, twice a week, others?[b]	29.1 Daily	21.8 Twice a week	49.1 Others
Ceilings cleaning schedule. Twice a week, every week, others?[b]	5.5 Twice a week	9.1 Every week	85.5 Others
Equipment cleaning/disinfection frequency. At the end of each utilization, daily, others?[b]	67.3 At the end of each utilization	30.9 Daily	1.8 Others
Is removal of solid residues done prior to cleaning operations?[a]	100		
Is there specific equipment used for sterilization of knives and other utensils?[a]	45.5	54.5	
Are chemicals (pesticides, herbicides, cleaning agents, lubricants, boiler compounds) stored in concealed designated areas?[a]	89.1	10.9	
Production and Process Controls			
Are there specific demands for raw materials characteristics?[a]	61.8	38.2	
Are raw materials and other ingredients checked on their preemption or "best before" date?[a]	72.7	27.3	
Are records of raw materials and other ingredients received kept on file?[a]	45.5	54.5	
Are food additives stored in concealed areas, more reserved?[a]	76.4	14.5	
Is there manipulation of other meat species apart from pork or beef?[a]	41.8	58.2	
Is there an operative process control system?[a]	54.5	45.5	
Is there a sampling analysis control plan?[a]	52.7	47.3	
Is there some sort of regular monitoring of the process's time and temperature?[a]	60.0	40.0	
Is there a plan for metrology calibration control with a specialized enterprise?[a]	25.5	74.5	
Is a record kept on file with this information?[a]	34.5	65.5	
Are relevant documents kept in a specific record?[a]	47.3	52.7	
Is there some kind of documentation control?[a]	38.2	61.8	

[a] Answer in two classes.

[b] Answer in three classes.

Table 49.2. Product definition of some fermented dried sausages.

Product Names	Saucisson, Chouriço, Llonganissa, Loukanika xoriatika, Salami
Product composition (depends on the recipes of different countries and regions)	Pork (70–80% lean meat and 20–30% fat), beef, salt, sugar, condiments (wine, pepper, paprika, garlic, other spices), additives (ac. ascorbic, nitrite) casings
Product characteristics	Size and weight: 40 cm, 3 cmø, 270 g–2 kg pH: 4.8–6.0 a_w: 0.85–0.90
How it is to be used	Raw ready-to-eat by healthy uncompromised people
Packaging	No packaging or in modified atmosphere or vacuum packages
Shelf life	Best before 4 months
Places/premises of commercialization or marketing	Preferentially in open markets and butcher's shops, but also in restaurants, retail dealers, and directly in the workshop
Conditions of storage	Dry and cool room without incidence of light
Labeling instructions	According to label legal requirements
Commercialization or marketing conditions	Keep in dry and cool space or in a refrigerated display

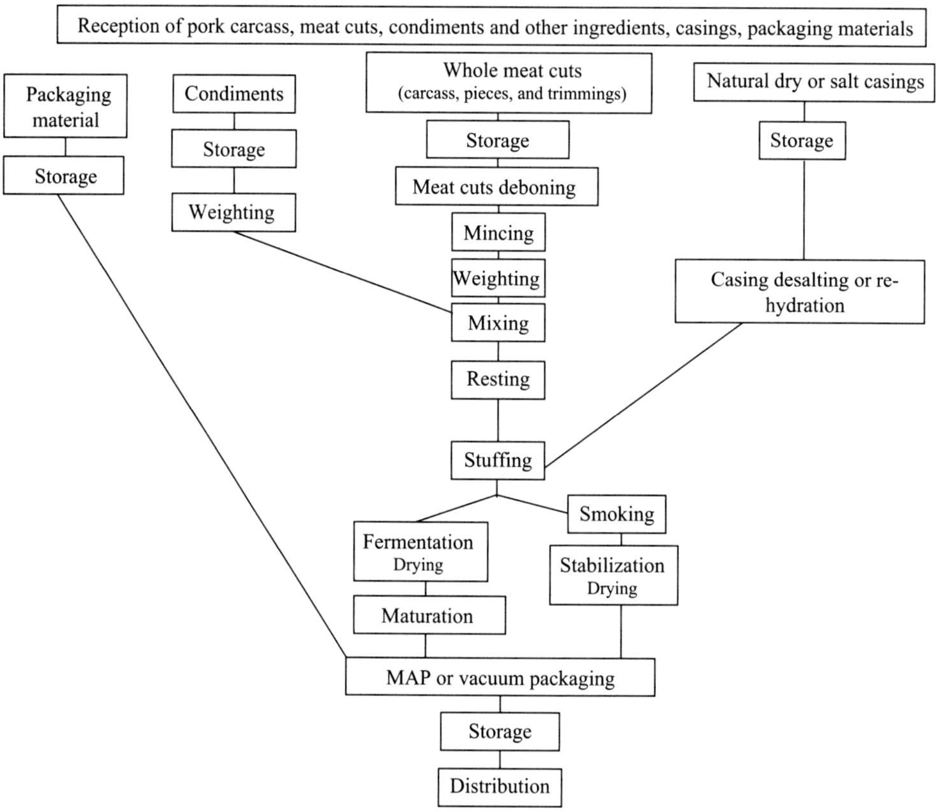

Figure 49.1. General flow diagram of dry/smoked-fermented sausages.

all ingredients, incoming materials and steps of process.

In Table 49.3 potential biological, chemical, and physical hazards are shown. They are usually based on literature references and experimental and epidemiological historic data. After considering the probability of occurrence and severity of a hazard, its risk must be assessed, which then determines which hazards are to be addressed in the HACCP plan.

The microbial hazards in processing units of fermented sausages depend on the prevalence of pathogens in meats and premises, considering their specific microbial ecology (Tompkin 2002). Inadequate hygiene practices may result in a loss of microbial control and become a hazard (Metaxopoulos et al. 2003). According to Tompkin (2002) and Eisel et al. (1997), the complete elimination of pathogens from raw material and the food processing environment is rather difficult if not impossible; however, monitoring of their control is badly needed within the preventive practices prerequired by HACCP methodology.

The Tradisausage project assessed the probability of occurrence of potential microbial hazards (*E. coli*, *Salmonella* spp., *S. aureus*, *Listeria monocytogenes*) associated with the production of traditional fermented sausages. The occurrence of pathogens in 314 environmental samples (machines, cutting tables, knives, and other utensils) from 54 different processing units covering central and southern European countries was evaluated: *Salmonella* spp. was absent in all but 3 located in Greece; *L. monocytogenes* were found in 2.2% of the environmental samples; and *S. aureus* were encountered in 6.1%.

There are no epidemiological data on chemical residues of products used in sanitation and maintenance programs in fermented sausages or raw meat (Alaña et al. 1996). Albeit these chemicals may be in the final product and reach the consumer, they are not considered an actual risk because they are prevented by the implemented actions of GHP and GMP.

The frequency of physical hazards occurring on fermented sausages is unknown; there is no data related to this kind of hazards, but bone and metal fragments must be addressed. In spite of its low frequency, there is a high social repercussion with economic losses to the producer.

Critical Control Point Identification

A critical control point (CCP) is defined as a point, step, or procedure where control can be applied and a hazard can be prevented, eliminated, or reduced to acceptable levels (Mortimore and Wallace 1997). CCPs are essential for product safety because they are the points where control of hazards is performed. An action taken at the CCP designed to control the hazard is called a *preventive measure*. These include physical, chemical, or biological factors or other hurdles required to control a hazard likely to occur at particular stages of the production of fermented sausages (Table 49.4). More than one preventive measure may be required to control a hazard. On the other hand, more than one hazard might be effectively controlled by one single preventive measure. Control of some preventive measures designed to eliminate or reduce potential hazards may be assured by validated GHP/GMP. Critical Control points are those requiring strict and expensive monitoring, so they are to be maintained at low numbers—no more than three or four in each plan.

Effective monitoring operations of CCPs are fundamental to the safety of the product. CCPs can be found by using team knowledge about fermented meat products, only for real and likely occurring hazards and where preventive measures are available for their control.

To assist in finding CCPs, a Decision Tree can be used (Figure 49.2). Several versions published by FAO/WHO (2005a) provide a tool for structured thinking and ensure a consistent approach (or embarrassing difficulty) to CCP finding in process steps for the real control of each hazard identified. The same reasoning is performed for raw materials (Figure 49.3). Table 49.5 summarizes CCPs identification in fermented dry sausage process steps using the CCP Decision Tree.

For the purpose of Decision Tree reasoning, steps where safety risk is assured by GHP/GMP are called *Control Point-GHP/GMP*.

Fermentation and/or smoking and maturation/stabilization are important sequential steps in the process of fermented traditional sausages contributing all together, as a clustered CCP, to the reduction of likely occurring biological (growth of pathogens) and chemical (biogenic amines) hazards.

Establishment of Critical Limits

After identification of all CCPs, the next step is to decide how to control them and establish the criteria that indicate the difference between a safe and an unsafe product. The absolute tolerance at a CCP is known as a *critical limit*.

It is important to note that the critical limit must be associated with a measurable factor that can be routinely monitored according to a fixed schedule.

These critical limits are established based on published data (scientific literature, in-house and

Table 49.3. List of potential biological, chemical, and physical hazards to be addressed in fermented sausages.

	Potential Hazards Identification		
	Biological (Bacteria, Parasites, Virus)	Chemical	Physical
Incoming Materials			
Raw meat: pork and beef	Nonsporulating bacteria: *E. coli*, *Salmonella* spp., *S. aureus*, *L. monocytogenes*, *Campylobacter* spp., *Yersinea enterocolitica.* Sporulating bacteria: *C. botulinum*, *C. perfringens* Parasites: *Trichinella*; *Cysticercus cellulosae*, *C. bovis*, *Toxoplasma gondii*	Antibiotic/drug residues Hormones, pesticides Biogenic amines of microbial origin (tyramine and histamine)	Bone, plastic, wood, metal particles
Spices	*B. cereus*, *C. botulinum*, *Salmonella* spp., *L. monocytogenes*	Nonfood chemical Pesticides, mycotoxines	Plastic, sand and wood particles, stones
Salt	*E. coli*, *Salmonella* spp., *Vibrio* spp., S. aureus; *Clostridia*	Nonfood chemical	Metal, sand and soil particles, stones
Casings natural	*E. coli*, *Salmonella* spp., *L. monocytogenes*, *Clostridia*		
Water	Not potable, *E. coli*, *Salmonella* spp., *Vibrio* spp., *Clostridia, Cryptosporidium*, virus	Nonfood chemical	
Starter cultures	Contamination by pathogens		
Package		Nonfood grade packaging material or improper printed labels	
Process Steps			
Reception	Reception of noncompliant material with legal or predefined requirements Growth of pathogens on fresh meat due to time/temperature abuse	Nonfood grade packaging material or improper printed labels	Reception of noncompliant material with legal or predefined requirements
Storage	Growth of pathogens on meat due to time/temperature abuse Bacterial contamination of spices during storage, growth of fungi producers of mycotoxines due to humidity abuse.		

Table 49.3. (*Continued*)

	Potential Hazards Identification		
	Biological (Bacteria, Parasites, Virus)	Chemical	Physical
Pork cuts and trimmings and beef cuts	Contamination and growth of pathogens due to time/temperature abuse		Bone and metal particles falling into meat during cutting/trimming
Mincing	Contamination and growth of pathogens due to time/temperature abuse	Detergent, disinfectant, and lubricants residues	Bones and metal particles falling into meat during mincing
Casings desalting or rehydration	Growth of pathogens due to time/temperature abuse		
Weighting ingredients and additives		Excess of nitrites/nitrates or other additives due to weighing errors	
Mixing	Growth of pathogens due to time/temperature abuse	Detergent, disinfectant, and lubricants residues	
Stuffing	Contamination with pathogens by poor hygiene practices	Detergent, disinfectant, and lubricants residues	Casing bursts, bones particles, metal clips, cotton string falling into meat during stuffing
Resting	Growth of pathogens due to time/temperature abuse		
Fermentation/drying	Growth of pathogens due to time/temperature/humidity abuse, improper fermentation, insufficient reduction of pH and a_w	Biogenic amines of microbial origin (tyramine and histamine)	
Smoking	Growth of pathogens due to time/temperature/humidity abuse, improper fermentation, insufficient reduction of pH and a_w	Biogenic amines of microbial origin (tyramine and histamine), 3,4-benzopyrene, dioxins	
Maturation	Growth of pathogens due improper time/temperature/humidity, insufficient reduction of pH and a_w	Biogenic amines of microbial origin (tyramine and histamine)	
Packaging vacuum and MAP and labeling	Pathogen bacterial contamination by environment or handling during packaging Growth of pathogens by improper package sealing or gases concentration	Allergies to an ingredient due to wrong labeling of a product or unsatisfied lack of advice on allergenic substances	

(*Continues*)

Table 49.3. *(Continued)*

	Potential Hazards Identification		
	Biological (Bacteria, Parasites, Virus)	Chemical	Physical
Storage end product	Growth of molds due to improper time/ temperature/ humidity	Mycotoxines	
Distribution	Pathogenic bacterial contamination through damaged packages Growth of pathogens due to improper time/ temperature/ humidity	Nonfood chemical residues cross contamination through damaged packages	Plastic, wood, metal particles into damaged packages

Sources: Feng (1995); FESP (1997); Foster (1997); Paukatong and Kunawasen (2001); Hoornstra et al. (2001); Oiye and Muroki (2002); Hugas et al. (2003); Roy et al. (2003); FDA/CFSAN (2005); CDC (2006).

Table 49.4. Preventive measures to control potential hazards identified in the production of fermented dry sausages.

Process Steps	Hazard Identification	Preventive Measure
Reception: raw meat	Biological Chemical Physical	Suppliers selected and homologated and/or certified Correct temperature of delivery (<7°C) Visual control
Other ingredients		Suppliers selected homologated and/or certified Visual control of shelf life dates according to specified requirements
Water		According to specified and legal requirements of potability
Package materials		Suppliers selected homologated and/or certified According to specified legal requirements
Storage	Biological	GHP and GMP Corrected temperature and relative humidity
Pork cuts and trimming and beef cuts	Physical Biological	GHP and GMP Visual control Corrected temperature and relative humidity
Casing desalting or rehydration	Biological	GHP, GMP Corrected time/temperature
Mincing	Biological Chemical Physical	GHP, GMP Preventive equipment maintenance Use of nontoxic food-compatible cleaning compounds
Weighting of ingredients and additives	Chemical	Adequate weighting control, safe operating practices according to additive instructions and legal requirements, calibration of scales
Mixing	Biological Chemical	GHP, correct time/temperature, preventive equipment maintenance, use of nontoxic food-compatible cleaning compounds

Table 49.4. (*Continued*)

Process Steps	Hazard Identification	Preventive Measure
Stuffing	Biological Chemical Physical	GHP, GMP, correct time/temperature, use of nontoxic food-compatible cleaning compounds, preventive equipment maintenance, correct stuffing pressure machine Visual check, metal detector
Resting	Biological	GHP Correct time/temperature, batch meat pH
Fermentation/drying	Biological Chemical	GHP, GMP Correct time/temperature/relative humidity
Smoking	Biological Chemical	GHP, GMP Corrected time/temperature/relative humidity
Maturation/stabilization	Biological Chemical	GHP, GMP Corrected time/temperature/relative humidity, control of pH and a_w of end product
Packaging (vacuum and MAP) and package labeling	Biological Chemical	GHP, GMP, preventive equipment maintenance, control of sealing vacuum and CO_2/O_2 concentration into packages Label monitoring by photoelectric cell or visual check
Storage end product	Biological	GHP, GMP Correct time/temperature/humidity
Distribution	Biological Chemical Physical	GHP, GMP Correct time/temperature/humidity

supplier records, regulatory guidelines), experimental data, expert advice, and mathematical modelling (Mortimore and Wallace 1997) and designated to be monitored at CCPs. Verification procedures assure they have been respected.

So-called target limits are often established at CCPs to act before a deviation occurs. Recently, in Europe, new legal standards were defined concerning food microbial criteria (Commission Regulation [EC] 2073/2005) that must be used in verification actions at CCPs. These specifications (end-product criteria) are intended to prevent adverse health effects. It is assumed, as a starting point, that values below the fixed criteria do not result in significant health effects and those above that limit lead to an increased probability of an adverse health effect.

Monitoring Measurements and Other Tests

Monitoring measurements are immediate measurements or sensorial or analytical quotable determinations at a CCP assuring that the process is operating within the critical limits (Mortimore and Wallace 1997). The monitoring procedures must be able to detect immediately loss of control at a CCP. The frequency of monitoring will depend on the nature of the CCP and the type of monitoring procedure.

The procedure could be a continuous on-line measurement where critical data is continuously recorded, or a discontinuous off-line monitoring system. This last system could be disadvantageous because the sample taken to analyze may not be fully representative of the whole batch and periodicity of sampling could create gaps without information about the process.

Preferentially, methods used for monitoring must produce rapid answers, permitting the previously stated corrective action to be started when necessary.

According to the CCPs established for fermented sausage (Table 49.6), the measurement of temperature, relative humidity, time, pH, and a_w are objective methods that provide a rapid answer to control the process. Also scheduled sensorial inspections can be used for monitoring some preventive measures at CCPs; however, there is some restriction because the objectivity and accuracy of human sensorial

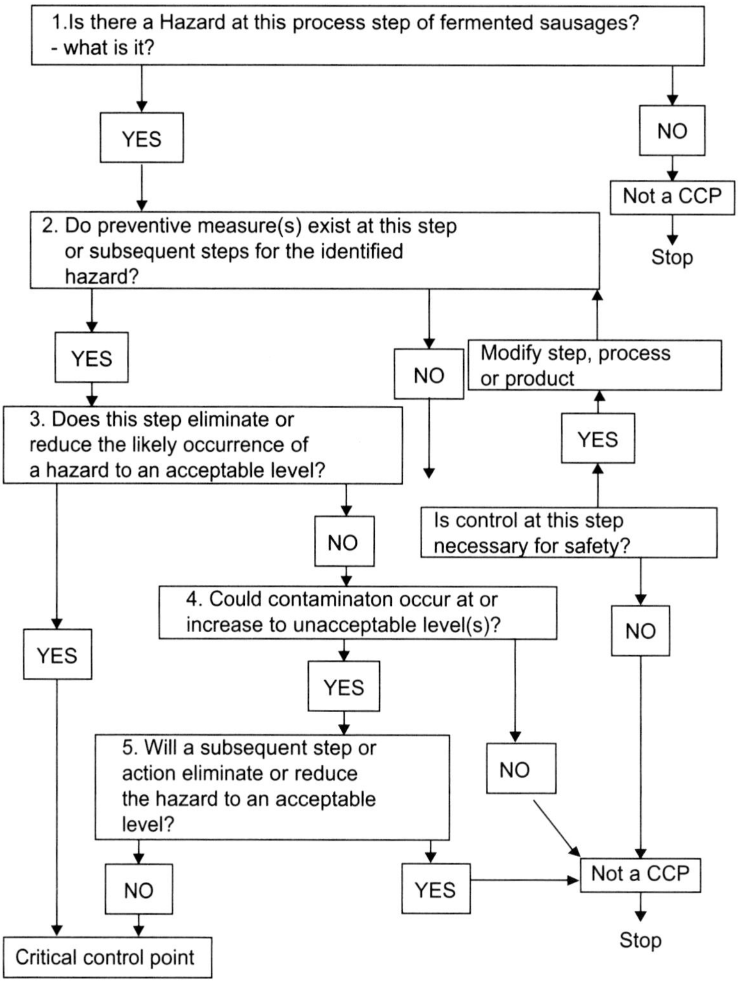

Figure 49.2. CCP Decision Tree for establishment of critical control points, adapted from FAO/WHO (2005).

evaluation are influenced by several factors, some of them avoidable by training. The frequency of monitoring will depend on the nature of the CCP, the type of monitoring procedure, and the amount of production. Records of monitoring data are necessary to prove that the process is under control and provides a pool of data that will contribute, after trend analysis, to the establishment of new criteria and new safety objectives, improving the implemented system.

Corrective Actions for Deviations at CCPs

A corrective action must be taken when monitoring data shows a deviation from critical limits at a CCP; it is mandatory to act quickly.

The actions to be taken when a deviation occurs include those related to the adjustment of the process to bring it back under control and those related to the amount of product that might not be complying with hygiene and safety requirements. They include the segregation of any suspect product and holding of it for the period of time needed to seek advice and more information from the HACCP team or outside experts and to perform analysis to assess safety. All this information will lead to different decisions: rejection and destruction of product, product rework, and product release. All these actions must be kept on record.

Table 49.6 gives some examples related to corrective actions in a fermented sausage process.

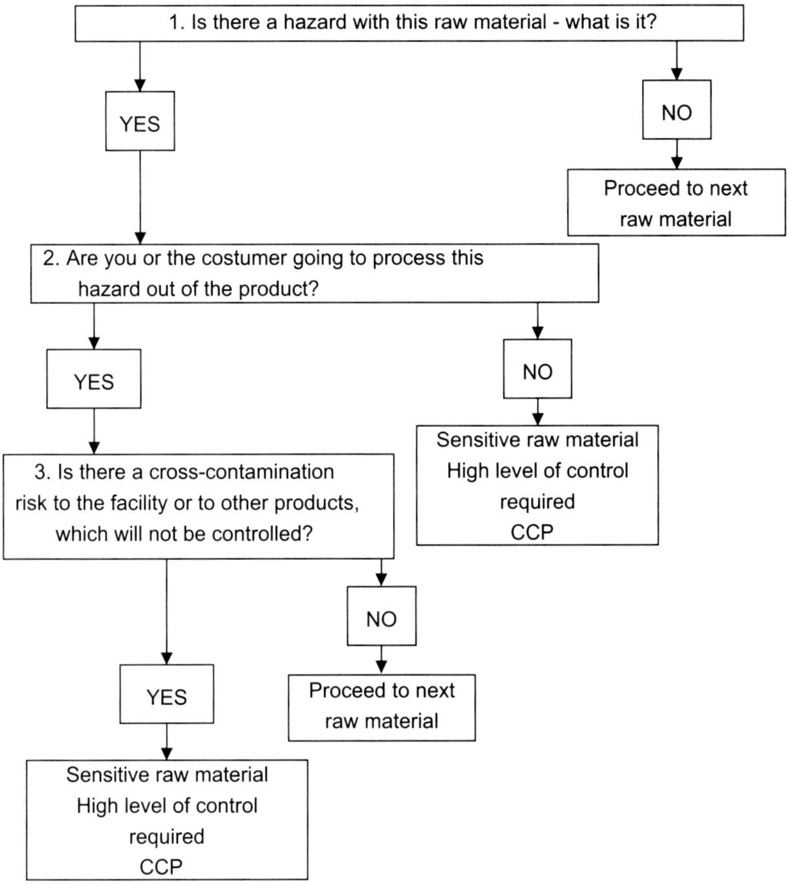

Figure 49.3. CCP Decision Tree for raw material (Mortimore and Wallace 1997).

HACCP Control Chart Foundation of Written HACCP Plan

Fulfillment of the seven principles of HACCP leads to a finished control chart that has been drawn up, step by step, during the system development of the practical application of each principle (Table 49.6). It is the foundation, in synthesis, of every specific HACCP plan.

Verification Procedures

The HACCP system must testify that it ensures that appropriate measures and related monitoring procedures at critical control points established to prevent all known potential hazards are working effectively and according to the plan.

Verification includes procedures (others than those used in monitoring) that ensure that the HACCP study has been carried out correctly and that the HACCP plan remains effective.

Verification can be carried out by different actions. Table 49.6 summarizes some of the possible verification actions related to a HACCP plan for fermented sausages.

When CCPs are required for suppliers or distribution operators, regular audits to evaluate their quality systems, GHP/GMP, and HACCP should be performed.

Verification also includes inspective analysis of operations to assure that CCPs are under control, scrutiny of CCPs monitoring records, examination of metrology certificates of all equipment used for measurements, and examination of records relating to corrective actions taken when deviations have occurred. To underscore product uses, random samplings for physicochemical, sensorial, or

Table 49.5. Control points and critical control points identification in the production of fermented dry sausages.

Process Steps	Is there a hazard at this process step of fermented sausages? What is it?	Do preventive measure(s) exist at this step or subsequent steps for the identified hazard?	Does this step eliminate or reduce the likely occurrence of a hazard to an acceptable level?	Could contamination occur at or increase to unacceptable level(s)?	Will a subsequent step or action eliminate or reduce the hazard to an acceptable level?	Final Answer
Reception	Yes—Bi.	Yes	No	Yes	Yes	Not a CCP
	Yes—Ch.	Yes	No	No		Not a CCP
	Yes—Ph.	Yes	No	No		Not a CCP
Storage	Yes—Bi.	Yes	Yes	Yes	Yes	Not a CCP
Deboning and trimming of pork meat cut	Yes—Ph.	Yes	No	Yes	Yes	Not a CCP
	Yes—Bi.	Yes	No	Yes	Yes	Not a CCP
Casing desalting or rehydration	Yes—Bi.	Yes	No	Yes	Yes	Not a CCP
Mincing	Yes—Bi.	Yes	No	Yes	Yes	Not a CCP
	Yes—Ch.	Yes	No	Yes	Yes	Not a CCP
	Yes—Ph.	Yes	No	Yes	Yes	Not a CCP
Weighting of ingredients and additives	Yes—Ch.	Yes	No	No		Not a CCP
Mixing	Yes—Bi.	Yes	No	Yes	Yes	Not a CCP
	Yes—Ch.	Yes	No	No		Not a CCP
Stuffing	Yes—Ph.	Yes	No	Yes	No	Yes CCP
	Yes—Bi.	Yes	No	No		Not a CCP
	Yes—Ch.	Yes	No	No		Not a CCP
Resting	Yes—Bi.	Yes	Yes			Yes CCP
Fermentation/ drying	Yes—Bi.	Yes	Yes			Yes CCP*
	Yes—Ch.	Yes	Yes			Yes CCP*

Smoking	Yes—Bi.	Yes	Yes		Yes CCP*
	Yes—Ch.	Yes	Yes		Yes CCP*
Maturation/ stabilization	Yes—Bi.	Yes	Yes		Yes CCP
	Yes—Ch.	Yes	Yes		Yes CCP
Packaging vacuum and MAP	Yes—Bi.	Yes	No	No	No	Not a CCP
Labeling	Yes—Ch.	Yes	Yes		Yes	CCP
Storage end product	Yes—Bi.	Yes	No	No		Not a CCP
Distribution	Yes—Bi.	Yes	No	No		Not a CCP
	Yes—Ch.	Yes	No	No		Not a CCP
	Yes—Ph.	Yes	No	No		Not a CCP

* All these steps could be monitored by a single CCP at stabilization.
Bi. = biological; Ch. = chemical; Ph. = physical.

Table 49.6. HACCP control chart for fermented/dry sausages.

Process Steps	Preventive Action	CCP/CP	Critical Limits	Monitoring Procedure	Monitoring Frequency	Corrective Actions	Verification
Reception of raw materials (carcasses/meat cuts, casings, ingredients/ additives)	Suppliers selected and homologated and/or certified: agreed specification (maximum acceptable levels) Visual check ("best before" dates) Temperature of meat (4 ± 2°C)	CP–GHP/GMP	Presence of certificates or stamp of homologation, legal limits Data "best before" Temperature of meat (4 ± 2°C)	20% Visual check of certificates (or stamp of homologation) and data 20% "best before" check ingredients Measurement of meat temperature	100%	Reject batch; Contact supplier Change supplier	Raw materials microbial and chemical analysis according to planned Suppliers audits Calibration of thermometers Check temperature against calibrated thermometer Data record analysis
Storage of raw materials	Temperature and relative humidity of room refrigeration Effective stock rotation, time of storage (first-in/ first-out)	CP–GHP/GMP	Room temp. 2±2°C; RH = 85 ± 5%; time of meat storage = 5 ± 2 days Data "best before"	Measurement of room temperature and relative humidity Visual check of labels to ensure stock rotation	On line 20%	Correction of temperature and relative humidity Segregation of meat and evaluation of its hygiene and safety Reject product on end shelf life	Calibration of thermometer and hygrometer
Casing desalting or rehydration	Time of rehydration or desalting GMP	CP–GHP/GMP	Time of rehydration or desalting <2 h	Measurement of time	100%	Reject batch Training operators	Data record monitoring analysis
Weighting of ingredients and additives	Safe and accurate weight practice of each ingredient in each batch GHP/GMP	CP–GHP/GMP	Weight according to legal standards Salt concentration ≤ 2%	Weight Visual check control of each ingredient and record	100%	Correction of weight, segregation of batch Recalibration of scales	Chemical additives analysis Calibration of scales

						Correction of sausage formula	Dates record monitoring analysis
Resting period	Temperature of refrigeration room, time of resting, relative humidity	CCP	Room temp. 4 ± 2°C; RH = 85 ± 5%; time = 3 ± 1 days	Measurement of room temperature, relative humidity, and resting time	On line	Correction of temperature and relative humidity Segregation of suspected product	Calibration of thermometer and hygrometer
Stuffing	Preventive maintenance of equipment, effective metal detection, visual inspection	CCP	0% metal particles; 0% bones	Metal detection, check, visual check	On line	Reject product Notify maintenance, operators training	Calibration of metal detector, maintenance equipment plan
Fermentation/drying	Temperature, time relative humidity GHP/GMP	CP–GHP/GMP	Room temp. = 12–16°C, time = 25 days, relative humidity ≤ 85 ± 2% according to particular/traditional practices	Measurement of room temperature, relative humidity, and fermentation time	On line or periodically scheduled	Correction of temperature and relative humidity	Calibration of thermometer and hygrometer
Smoking	Temperature, time, relative humidity control GHP/GMP	CP–GHP/GMP	Room temp. = 20 ± 5°C; RH = 75 ± 5%; time = 3 ± 2 days according to particular/traditional practices	Measurement of room temperature, relative humidity and smoking time	On line or periodically scheduled	Correction of temperature, relative humidity, and smoking time Segregation of suspected product	Calibration of thermometer and hygrometer

(*Continues*)

Table 49.6. *(Continued)*

Process Steps	Preventive Action	CCP/CP	Critical Limits	Monitoring Procedure	Monitoring Frequency	Corrective Actions	Verification
Maturation, stabilization, drying, air-conditioned	Temperature, time, relative humidity, pH and a_w control	CCP	Room temp. 12 ± 2°C; RH = 75 ± 5%; time of maturation according to desirable loss of weight related to product pH = 4.8–6 according to particular traditional fermented sausage a_w 0.85–0.90	Measurement of room temperature, relative humidity, maturation time, and weight of products till desirable loss of weight 10% of end product batch measurement of pH and a_w end sausage sampling	On line or periodically scheduled	Correction of temperature, relative humidity or period of stabilization Segregation of suspected product Reject product	Calibration of scales, thermometer and hygrometer Microbiological and chemical end product analysis according to plan
Packaging vacuum and MAP	Control of selling and vacuum and CO_2/O_2 concentration in packages	CP–GHP/GMP	0% fail of seal package, 0% fail of vacuum, O_2 residual head space package <2%;	Measure of residual CO_2/O_2 concentration in packages	On line	Stop line, replace package and notify maintenance	Calibration of residual CO_2/O_2 concentration equipment analysis
	GHP/GMP		legal limits for MAP packaging	Visual check of vacuum package Selling test			Maintenance equipment plan, data record monitoring analysis
Labeling	Control of label with cell photoelectric or visual	CCP	0% absence of label	Visual or photoelectric cell labeling check	On line	Label product or replace label	Maintenance equipment plan, data record monitoring analysis

Storage end product	Time/ temperature control GHP/GMP	CP–GHP/GMP	Room temp. = 18±2°C "Best before" date	Measurement of room temperature and storage time	On line or periodically scheduled	Correction of temperature Segregation of end product and evaluation of its hygiene and safety Reject product on end shelf life	Calibration of thermometer Check temperature
Distribution	Temperature control, GHP, GMP	CP–GHP/GMP	Temp. 18±2°C or other specified for MAP packed sausages	Measurement of car temperature	On line	Correct temperature Training drivers	Calibration of thermometer Check temperature Audit records of car temperature

CP–GHP/GMP = control point of GHP and GMP.

microbiological analyses are further examples of verification procedures.

Microbial analysis of fermented sausages, namely those related to the quantification of certain microbial groups that are indicators of spoilage and hygiene level and also those for detection/quantification of pathogens, offer information to evaluate whether the system is reaching the targets. However, an efficient system must establish the real significance of each parameter analyzed in order to avoid wasting time and money.

All planned verification actions must be stated on the HACCP plan; however, others can be taken without any previous communication. These actions must take in account any new safety information about the product or when the product has been implicated as a disease vehicle.

Verification reports must be produced containing all the information from the above-mentioned actions, modifications introduced into the plan, and the training of the persons responsible for CCP monitoring and surveillance.

Record-keeping of HACCP Plan Data and Extraordinary Events

As stated above, all data produced by the HACCP operating procedures has to be kept on file for easy consultation by the team, auditors, and government food inspection authority agents. It is of great value to prove existing workable HACCP and provides due diligence in court trials.

VALIDATION OF THE OPERATIVE HACCP PLAN

All the evidence that preventive measures are effective in eliminating or reducing hazards to an acceptable level in the HACCP plan at CCPs, GHP/GMP, or other prerequirements are used for validation. If the selected preventive measures are not able to achieve the control of identified hazards, they must be modified and reevaluated.

According to Hoornstra et al. (2001) quantitative risk assessment is a powerful tool for food companies for the validation of a HACCP system, particularly for validation of critical limits at critical control points. The authors give examples for risk assessment referring to the fact that for raw fermented products the risk manager is the government. Quantitative risk assessment needs hazard identification, hazard characterization, exposure assessment, and risk characterization. Hazards characterization is the qualitative or quantitative evaluation of the nature of the adverse health effects associated with biological, chemical, and physical agents that may be present in the food. Factors such as virulence, microbial variability, and consumer sensitivity influence the nature of the adverse health effect. Exposure assessment is the qualitative or quantitative evaluation of the likely intake of hazardous biological, chemical, and physical agents via foods. Risk factors in exposure assessment must be focused on the quality of the raw materials, the process steps, and the process environment as well as on the composition, packaging, and storage conditions of the product.

Gonzalez-Miret et al. (2001) applied statistical univariate and multivariate methods to identify the most useful and discriminative microbial parameters (total counts, *Pseudomonas*, *Enterobacteriaceae,* and *Staphylococcus aureus*) used in verification related to poultry meat production, such as refrigeration, cutup, and packaging operations. The authors concluded that *Enterobacteriaceae* and total counts were the most discriminating to be used for validating parameters used on HACCP system at CCPs/GHPs/GMPs. Similarly, microbiological parameters can be used for verification of goodness of monitoring pH, a_w, and time/temperature measurements at CCPs to assure safety control after the Maturation/Drying/Stabilization step.

Quantitative microbial risk assessment with *L. monocytogenes* counts performed with data obtained in southern European fermented sausages (Tradisausage 2006) revealed that if initial counts are higher than 10^2 CFU/g, no significant reduction is to be expected, mainly because of the insufficient lowering of pH whose mean value was higher than pH 5.0 in final products. However, *L. monocytogenes* is not able to grow to more than two logs in any product due to the low a_w of traditional fermented sausages, which is the most crucial hurdle for safety of these products. According to these findings, a_w and pH coupled with time/temperature are the most important parameters to assure safety and stability and have to be used for monitoring CCP at the maturation/stabilization step.

REVISION OF THE HACCP PLAN

HACCP plan needs must be revised when any modification in production or equipment occurs. The HACCP team is responsible for the revision based on internal and external audit reports, records of corrective actions, and client complaints. Revision will contribute to a continual improvement of the plan.

CERTIFICATION OF FOOD SAFETY MANAGEMENT SYSTEMS

Audits of all data recorded and direct measurements produced in a HACCP working system have to be performed with technical competence and compliance with international standards (ISO/FDIS 22000: 2005) or other normative references for food safety management systems. These audits conducted by an independent and qualified institution certify food safety management systems, giving recognition to clients and consumers that the producer is able to control food safety hazards. Unfortunately, these external audits are too expensive for small companies to afford. However, the above-mentioned normative reference can be used by small developed industries to evaluate their own safety management system.

REFERENCES

GH Alaña, JC Crespo, AH Marteache, M Aguado, JG Gonzalez. 1996. Implantacion del sistema HACCP en la industria cárnica. Ed. Servicio Central de Publicationes del Gobierno Vasco. Vitoria-Gasteiz. España, 222 pp.

E Caplice, GF Fitzgerald. 1999. Food fermentations: Role of microorganisms in food production and preservation. Inter J Food Microbiol 50:131–149.

CDC. 2006. Preliminary foodNet data on the incidence of infection with pathogens transmitted commonly through food—10 states, United States, 2005. Department of health and human services. Centers for disease control and prevention. Morbid Mortal Weekly Rep 55:14,392–395.

COM. 2000. White paper on food safety. Commission of the European Communities. COM (1999) 719 final. pp. 1–52. http://ec.europa/comm/dgs/health_consumer/library/pub/pub06_en.pdf.

Commission Regulation (EC) 2073/2005 of 15 November 2005 on microbial criteria for foodstuffs. Official Journal of the European Union L 338. 26 pp.

WG Eisel, RH Linton, PM Muriana. 1997. A survey of microbial levels for incoming raw beef, environmental sources and ground beef in a red meat processing plant. Food Microbiol 14:273–282.

European Commission (EC). 2005. Guidance document on the implementation of procedures based on the HACCP principles, and on the facilitation of the implementation of the HACCP principles in certain food business. European Commission, Health & Consumer Protection Directorate General. pp. 1–29.

FAO. 1997. Hazard analysis and critical control point principles and application guidelines. Food and Agriculture Organization of the United States, U.S. Department of Agriculture, National Advisory Committee on Microbiological Criteria for Foods. http://www.cfsan.fda.gov/~comm/nacmcfp.html.

FAO/WHO. 2003. Recommended International Code of Practice General Principles of Food Hygiene. Codex Alimentarius Commission. Food and Agriculture Organization of the United States, World Health Organization, CAC/RCP1-1969, Rev.4-2003:1–31.

———. 2005a. Codex Alimentarius, Food Hygiene. Basic texts. Third Edition. Codex Alimentarius Commission. Food and Agriculture Organization of the United States, World Health Organization.

———. 2005b. Code of hygiene practice for meat. Codex Alimentarius Commission. Food and agriculture Organization of the United States, World Health Organization, CAC/RCP 58-2005, 1–52.

FDA. 2005. Food Code 2005. Recommendations of the United States Public Health Services, Food and Drug Administration, National Technical Information Service Publication PB-2005-102200. Springfield. http://www.cfsan.fda.gov/~dms/fc05-toc.html.

FDA/CFSAN. 2005. Foodborne Pathogenic microorganisms and Natural Toxins handbook. Center for Food Safety & Applied Nutrition. U.S. Food & Drug Administration. http://www.cfsan.fda.gov/~mow/.

P Feng. 1995. *Escherichia coli* serotype O157:H7: Novel vehicles of infection and emergence of phenotypic variants. Emerg Infect Dis 1:2,47–52.

EM Foster. 1997. Historical overview of key issues in food safety. Emerging Infectious diseases, 3: 4,481–482.

FSEP. 1997. HACCP Generic Model: Fermented smoked sausages. Food Safety Enhancement Program. Canadian Food Inspection Agency. http://www.inspection.gc.ca/english/fssa/polstrat/haccp/smsaufum/smsaufumie.shtml.

ML Gonzalez-Miret, MT Coello, FJ Heredia. 2001. Validation of parameters in HACCP verification using univariate and multivariate statistics. Application to the final phases of poultry meat production. Food Contr 12:261–268.

SJ Goodfellow. 1995. Implementation of HACCP program by meat and poultry slaughterers. In: HACCP in Meat, Poultry and Fish Processing. Advances in Meat Research—Volume 10. AM Pearson, TR Dutson, eds. London: Blackie Academic & Professional, pp. 58–71.

E Hoornstra, MD Northolt, S Notermans, AW Barendsz. 2001. The use of quantitative risk assessment in HACCP. Food Contr 12:229–234.

M Hugas, M Garriga, MT Aymerich. 2003. Functionality of enterococci in meat products. Inter J Food Microbiol 88:223–233.

ISO/FDIS 22000:2005. Food safety management systems—Requirements for any organization in the food chain. International standard organization. Geneva. 31 pp.

J Mexatopoulos, D Kritikos, EH Drosinos. 2003. Examination of microbial parameters relevant to the implementation of GMP and HACCP system in Greek meat industry in the production of cooked sausages and cooked cured meat products. Food Contr 14:323–332.

S Mortimore, C Wallace. 1997. HACCP. A practical approach. London: Chapman & Hall, 296 pp.

National Advisory Committee on Microbiological Criteria for Foods, 1992. Hazard analysis and critical control point system. Inter J Food Microbiol 16:1–23.

SO Oiye, NM Muroki. 2002. Use of spices in foods. J Food Technol Africa 7:33–44.

KV Paukatong, S Kunawasen. 2001. The hazard analysis and critical control points (HACCP) generic model for the production of Thai fermented pork sausage (Nham). Berliner und Munchener Tierarztliche Wochenschrift 114:9–10, 327–330.

SL Roy, AS Lopez, PM Schantz. 2003. Trichinellosis surveillance—United States 1997–2001. Morbid Mortal Weekly Rep 52:SS–6,1–12.

RB Tompkin. 2002. Control of *Listeria monocytogenes* in the food processing environment. J Food Protect 65:709–725.

Tradisausage. 2006. Final report. Assessment and improvement of safety dry sausages from producers to consumers. Tradisausage, QLK1 CT-2002-02240. 189 pp.

U.S. Department of Health and Human Services. 2000. Healthy people 2010: Understanding and improving health, 2nd ed. Washington, DC: U.S. Government Printing Office, 76 pp.

50
Quality Assurance Plan

Friedrich-Karl Lücke

INTRODUCTION

The objective of food quality assurance is to make sure that the consumer receives foods that meet his/her needs in terms of safety, nutritional value, convenience (including shelf life), and eating quality. An ever-increasing number of consumers also demand "authentic" foods, i.e., foods of defined origin (regional foods) and/or foods manufactured and distributed by use or nonuse of certain production and processing methods (e.g., according to regulations on organic and fair-trade foods, to requirements for kosher foods; without application of ionizing irradiation or genetic technology, etc.). These requirements (*process quality*) are also quality criteria according to ISO 9000:2000 standards, even though these properties of a food cannot always be readily detected by sensory testing or in the laboratory. Official regulations require that food must be safe, and that the consumer must not be misled by inadequate labeling about weight, composition (constituents and additives), shelf life, nutritive value, origin (traceability), and production method. Official regulations on food hygiene (such as EU regulation no. 852/2004) require that food businesses implement and maintain a system to ensure safety, and specify that such a system must be based on HACCP. For information on the implementation and maintenance of HACCP in meat processing, the reader is referred to official documents (e.g., the Generic HACCP model for not heat-treated, shelf-stable meat and poultry products in the USA [USDA-FSIS 1999]) or the literature (Brown 2000a,b; Sheridan 2000; see also Chapter 49, "HACCP"). Even though HACCP refers to the protection of consumer's health only, a similar approach should also be used to assure overall product quality. Hence, food businesses should establish and maintain an integrated, process-oriented system that covers all criteria of product quality. An effective system of quality assurance also reduces the producer's risk (of economic losses due to process failure) and cuts production costs by minimizing unnecessary work and inputs.

This chapter provides the key elements of quality assurance plans specific for the manufacture of fermented sausages and raw hams, respectively. The information on processes is based on the publications of Stiebing (1996, 1997), Lücke (2000), Toldrá (2002), and Lautenschläger (2006). An example for a quality assurance plan for the design and the production of a Mediterranean-type salami is provided by Thorn (2006). Reference is also made to other contributions to this book. The reader will note some bias on products common in Germany, but the system presented may be easily adapted to fermented meat products common in other countries. However, it is essential to note that not all standards and specifications given are transferable to all types of fermented meats. For example, it is possible to avoid the use of curing agents if proper raw material and salting/ripening conditions are selected.

Finally, this chapter gives some information on the structure of the documents on the processes (plans and records).

GENERAL REMARKS ON THE PURCHASE AND SELECTION OF THE RAW MATERIAL

Specifications should include the following:

- Hygienic aspects, in particular temperature (including temperature history) of the meat and fat, integrity of containers and transport equipment
- Traceability
- pH of the meat: generally, 5.8 or below, but definitely not higher than 6.0

The suppliers of meat and fat should be carefully selected on the base of their ability to reliably meet the specifications established. This is a complex process, which should involve responsible staff from the processing, purchase, and quality assurance department. A Standard Operation Procedure (SOP) should be in place to organize this process.

On reception, the documents indicating the origin of the raw material should be inspected, and the temperature and cleanliness within the transport equipment, the integrity of the containers, and the overall appearance of the meat and fat should be examined and recorded. Measuring and recording the internal temperature and pH value of the cuts is also advisable even though experienced butchers are able to select cuts of appropriate pH by visual inspection only. The same is true for the firmness and whiteness of fatty tissue, which correlates well with the content of saturated fatty acids in the tissue. To avoid unnecessary paperwork, it is advisable to stamp and fill in a "miniature checklist" on the delivery note. The supplier should also be able to provide evidence that the process hygiene criteria specified in the EU Regulation 2073/2005 on microbiological criteria for foodstuffs has been fulfilled, and it is advisable to verify this by regularly testing the microbiological quality of the meat supplied. In the United States, suppliers must provide certificates indicating that the meat has been sampled and tested negatively for salmonellae and *E. coli* O157:H7 if the manufacturing process of fermented sausages does not guarantee a 5 log reduction of these organisms (USDA-FSIS 1999).

QUALITY ASSURANCE PLANS AND RECORDS FOR FERMENTED SAUSAGES

Figure 50.1 summarizes the processing steps in the manufacture of fermented sausages. Control at these steps is important to assure quality of the final product even though not all of these steps are "critical control points" in terms of the HACCP concept.

Purchase and Selection of Meat and Fat (Table 50.1)

Table 50.1 lists quality criteria for the raw material, indicates why these factors are important for the quality of the final product, and suggests how the

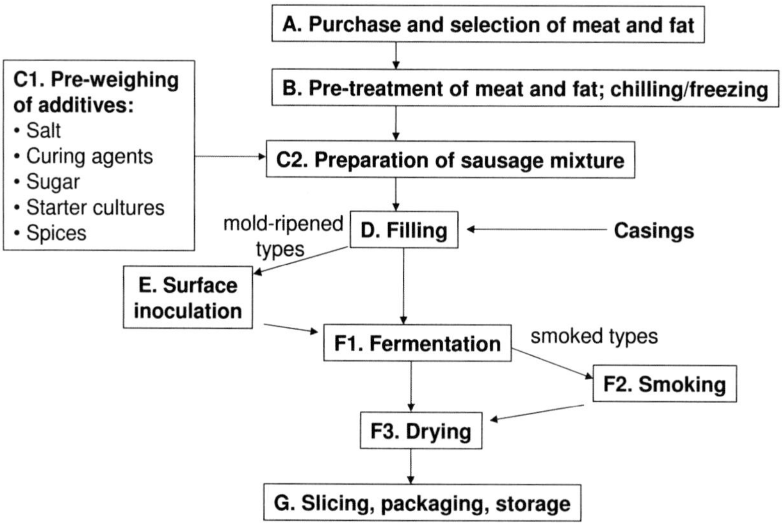

Figure 50.1. General processing scheme for fermented sausages.

Table 50.1. Purchase and selection of raw material for fermented sausages (process step A).

Specification	Objective	Monitoring and Documentation
Origin of the material	Traceability in case of defects caused by process failure in primary production, slaughtering, and/or butchering	Check of documents and stamps on cuts/containers
Type of cut	pH affects product safety; fat, collagen, and myoglobin content affects sensory properties of the final product	Visual inspection
pH of lean meat ≤5.8	Slower growth of undesirable or hazardous microorganisms Reduction of water binding capacity	Measurement of pH or visual inspection
Microbiological quality	Reduction of microbiological risk	Visual inspection Temperature measurement Check of hygiene-related documentation provided by supplier
Maximum mol% of polyunsaturated fatty acids	Less fat oxidation, slower rancidity	Visual inspection Check of documents on feeding regime and age of slaughter animals
Maximum content of fat and connective tissue	Sensory and nutritive value	Sorting of cuts on the basis of visual appearance (e.g., using the German GEHA scheme based on the Leitsätze), and/or analysis of preblends (e.g., by NIR)
Age of slaughter animals	Age affects myoglobin content and thereby appearance of the final product	Visual inspection Reflectometry

quality of the raw material should be assessed and recorded.

Pretreatment of Meat and Fat: Chilling/Freezing

Trimmings with the appropriate content of lean meat, fat, and connective tissue are sorted; at this stage, it is important to clearly record the identity, origin, and classification of the material (e.g., according to the German Code of Practice [Leitsätze]). The material is subsequently chilled and partly frozen; freezing the meat may also somewhat reduce water binding capacity and improve salt diffusion. The flow of raw material should be carefully controlled and documented, in order to ensure traceability, to avoid mislabeling, and to minimize storage time according to the "first in, first out" principle. Extended chill storage results in poor microbiological quality, and extended frozen storage results in fat deterioration.

Preparation of Sausage Mixture (Table 50.2)

The final temperature of the mix should be low, preferably between –5 and 0°C. This is important to avoid fat deterioration and "smearing" and the resulting sensory defects and problems associated with drying the sausages. For some sausage varieties with a coarse, crumbly structure, a meat grinder is used instead of a cutter. In this case, higher temperatures of the meat (up to 5°C) are still acceptable.

The appropriate choice of raw material and nonmeat ingredients, such as salt, nitrite, ascorbate, sugars, starter cultures, and spices, is critical for the quality and safety of the final product. Hence,

Table 50.2. Preweighing of ingredients, preparation of sausage mixture (process step C).

Subprocess; Examples for Specifications/Critical Limits	Objective	Monitoring and Documentation
Addition of meat and fat	Fat, total protein, and collagen within specified limits	Weighing protocol
Addition of ⩾0.3% of a rapidly fermentable sugar	Sufficient rate and extent of acid formation	Weighing protocol
Addition of lactic acid bacteria (10^5– 10^6/g) capable of growing and metabolizing in sausage mixture		Weighing protocol; specifications issued by manufacturer of starter culture
Addition of catalase-positive cocci (10^5– 10^6/g) capable of growing and metabolizing in sausage mixture	Nitrate reduction, aroma formation, protection from deleterious effects of O_2	Weighing protocol; specifications issued by manufacturer of starter culture; lot identification
Addition of 2.4– 3.0% NaCl	Reduction of microbiological risk texture formation	Weighing protocol
Addition of nitrite (100–150 mg $NaNO_2$/kg, as nitrite curing salt, in combination with 300– 500 mg sodium ascorbate/kg)	Formation of NO myoglobin (curing color) Inhibition of oxidative changes in unsaturated lipids Inhibition of salmonellae at early stages of fermentation	Specification of the nitrite content in curing salt; weighing protocol
Addition of nitrate (⩽300 mg KNO_3/kg)	Improved color and aroma of some long-ripened products	Weighing protocol
Addition of defined spice mixture	Aroma and taste	Weighing protocol; specifications issued by manufacturer; lot identification
Comminution to specified particle size	Appearance of the final product; prevention of fat deterioration and drying problems	Appearance of the mixture; final temperature of the mixture; lot identification

appropriate records (weighing protocols) are necessary. Moreover, the quality of additives (e.g., nitrite content in curing salt, activity of starter cultures under the fermentation conditions selected) should be defined in specifications issued by the manufacturers.

Selection of Casings and Filling

The choice of casings has a strong effect on product quality and safety. Specifications should include mechanical stability and—especially for dry sausages—moisture permeability and ability to shrink during drying. Use of poorly purified and pretreated natural casings may lead to bacterial contamination of the surface and to insufficient drying. For convenience, the casing must be easy to peel off. The meat processor should set up specifications and select a supplier able to comply with them.

During filling, inclusion of oxygen should be minimized in order to prevent hole formation with subsequent oxidative deterioration and microbial spoilage. This can be achieved by using a vacuum filler or manually.

Surface Inoculation (Table 50.3)

Most French- and Mediterranean-type sausages are covered with surface molds. Some manufacturers rely on the "house flora"; others dip the sausages (after temperature equilibration) into a suspension of starter culture consisting of molds and sometimes yeasts. The starter organisms should be able to rapidly colonize the surface after completion of lactic fermentation, and this and other relevant properties (sensory and safety aspects) should be defined in specifications issued by the manufacturers.

Table 50.3. Sausage fermentation, smoking, drying, storage of unpackaged product (process steps E, F).

Subprocess; Examples for Specifications/Critical Limits	Objective	Monitoring and Documentation
Equilibration at 20–25°C and 60% RH for about 6 hours	Prevention of moisture condensation on sausage surface	Time-temperature protocol
Surface inoculation with molds/yeasts	Standardization of sensory properties	Specifications issued by manufacturer of starter culture; preparation of suspension
Fermentation: • 20–25°C • RH ca. 90% • air velocity ca. 1 m/s • 2–3 days until pH ≤5.3	Acid formation to • Inhibit undesired bacteria • Transform nitrite to nitrous acid and NO • Reduce water binding capacity and facilitation of drying	Adjustment of ripening chamber; monitoring of temperature, humidity, and air velocity; monitoring of pH decrease
Smoking	Prevention of mold growth Inhibition of oxidative processes at the sausage surface Aroma formation	Specifications for wood; adjustment and monitoring of charring and chamber temperature and smoking time
Drying/aging: • 10–15°C • RH 70–80% • air velocity ca. 0.2 m/s until a_w ≤0.93	Prevention of • Undesired mold growth • Uneven drying	Adjustment and monitoring of climate (temperature, humidity, and air velocity) in ripening chamber
Storage: • 10–15°C • RH 70–80% • Air velocity ca. 0.05 m/s	Prevention of • Undesired mold growth • Uneven drying	Adjustment and monitoring of climate (temperature, humidity, and air velocity) in storage room

Fermentation, Smoking, and Drying (Table 50.3)

The climate in the fermentation and ripening chambers is critical for product safety and uniform drying of the sausages. Fermentation and drying/aging conditions (temperature, relative humidity, air velocity) should be clearly specified and monitored. The proper function of measuring devices should be checked regularly and the results documented. For the manufacture of smoked sausages, the smoking process (usually performed after fermentation) should be well controlled by specifying and recording the type of the wood (or liquid smoke preparations), the charring temperature, and the time and temperature during smoking. Poor control may result in elevated amounts of polycyclic hydrocarbons and other undesired compounds, and in poor sensory quality.

QUALITY ASSURANCE PLANS AND RECORDS FOR RAW DRY HAMS

Figure 50.2 summarizes the processing steps in the manufacture of raw dry hams.

Purchase and Selection of Meat and Fat and Pretreatment of Cuts (Table 50.4)

As outlined in the section on fermented sausages, the origin of the meat and its quality must be specified and recorded. For raw hams, it is even more important to carefully select cuts and inspect them for integrity, to reduce the risk of internal contamination and of poor appearance of the final product. To achieve this, it is very important to establish and maintain fair, long-term cooperation with reliable

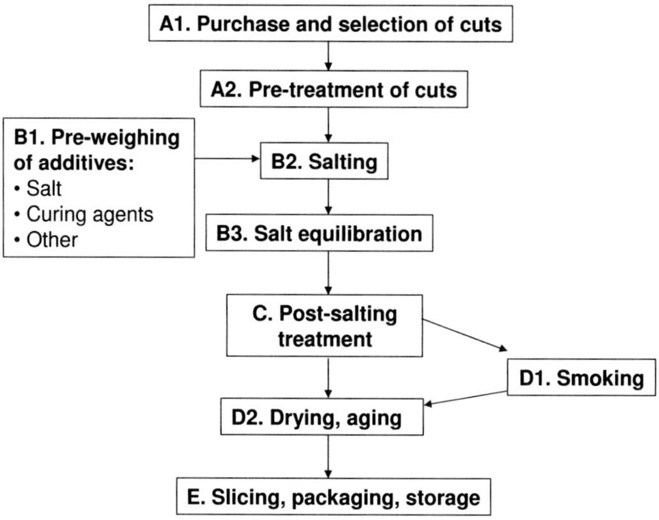

Figure 50.2. General processing scheme for dry raw hams.

Table 50.4. Purchase and selection of raw material for raw dry hams (process step A).

Specification	Objective	Monitoring and Documentation
Origin of the material	Traceability in case of defects caused by process failure in primary production, slaughtering, and/or butchering	Check of documents and stamps on cuts/containers
Age and weight of slaughter animals (pigs: > 7 months)	Age affects myoglobin content and thereby appearance of the final product	Visual inspection Reflectometry
Type and integrity of cuts	pH affects product safety Fat, collagen and myoglobin content affects sensory properties of the final product	Visual check inspection
pH ≤5.8 (pork: 24 hours, beef: 36 hours postmortem)	Reduction of water binding capacity	Visual inspection; pH measurement
Fat quality: maximum mol% of polyunsaturated fatty acids	Less fat oxidation, slower rancidity	Visual inspection Check of documents on feeding regime and age of slaughter animals
Microbiological quality	Reduction of microbiological risk	Visual inspection Temperature measurement (core temperature below 4°C) Check of hygiene-related documentation provided by supplier

suppliers. A low pH of the meat is essential for product safety; in general, it should not exceed 5.8, and only in certain pork muscles, pH values up to 6.0 may be tolerated. Cuts should be trimmed to obtain a smooth, even surface. Freezing and thawing the cuts may increase the salt diffusion rate somewhat, but this process must be carefully controlled to avoid spoilage and fat deterioration.

Salting and Salt Equilibration (Table 50.5)

Hams are salted by rubbing or tumbling them with salt (dry salting), by immersing them in a brine, by injection of brine, or a combination of these methods. Salt diffusion through the meat is slow, resulting in the risk of growth of pathogens and spoilage bacteria in the core of the cut if it has been contaminated during slaughtering and butchering. Hence, salting and salt equilibration should be performed at 5°C or below until the target water activity of 0.96 is reached in all parts of the ham.

A work instruction should specify the composition of the salt (in particular, the levels of nitrite and/or nitrate in the salt), the salt content of the brine (if applicable), the amount of salt and/or brine to be added per kg of meat, the size of the cut (maximum distance from the surface to the geometric center), the temperature (≤5°C). and the minimal time for salting and salt equilibration. Records should include weighing and time-temperature protocols. Also, the results of visual inspections and/or measurements of the final salt content or a_w value in the core should be recorded because the salt diffusion rate may vary between individual cuts.

Postsalting Treatment

Some hams are washed and subsequently dried in order to remove excessive salt from the surface layers and to obtain a dry surface before smoking. For quality and safety of the product, extensive washing should be avoided, and the surface should be dried at low temperatures; otherwise, there is a risk of growth of *Staphylococcus aureus* and molds or yeasts during drying. Temperatures and times of the washing and drying process should be specified and monitored.

Smoking and Drying/Aging

Before smoking and/or aging, it has to be ascertained that the water activity has been reduced to below 0.96 in all parts of the cut. Most German-type hams are smoked. An appropriate time-temperature combination is critical for the quality and safety of the products. At 18°C and below, products may be smoked for longer periods, whereas at higher temperatures, shorter smoking times should be applied to avoid microbiological problems, and a smoking temperature of 25°C should not be exceeded. The

Table 50.5. Manufacture of raw ham: salting and salt equilibration (process step B).

Subprocess; Examples for Specifications/Critical Limits		Objective	Monitoring and Documentation
Purchase and preparation of salt		Correct levels of curing agents in salt and/or brine to obtain curing color without exceeding legal limits	Weighing protocol
Preparation of brine: 8–20% salt, depending on salting method and size of the cut			Weighing protocol
Application of salt	Dry cure: e.g., ca. 60 g/kg meat Brine cure: e.g., ca. 1 liter brine per 2 kg meat Injection cure: depending on salt content in brine	Maximum rate of salt diffusion, minimal "oversalting" of surface layers; even distribution of the salt and curing agents	Amount of salt or brine per kg of meat; time and temperature of tumbling
Reduction of water activity to < 0.96 in all parts by	Salting (at < 5° C) Salt equilibration (at < 5° C and < 85% RH)	Inhibition of microorganisms possibly present in the interior of the meat	Time and temperature in salting room Time, temperature, and relative humidity in salt equilibration room

precautions against undesirable aroma and excessive levels of polycyclic hydrocarbons have been outlined in the section on smoking of fermented sausages.

Mediterranean-type raw hams are usually subjected to extensive aging. By drying, this process improves shelf life at ambient temperatures but, even more important, leaves enough time for enzymatic processes, which makes the hams tender and tasty. Aging normally takes place at temperatures above 15°C. The hams are stable at ambient temperatures if the water activity is below 0.90. Aging time, temperature, and relative humidity, as well as the intended and measured weight loss should be specified and recorded.

SLICING, PACKAGING, AND STORAGE OF FERMENTED SAUSAGES AND RAW DRY HAMS (TABLE 50.6)

The climate in the slicing and packaging room should be adjusted so as to avoid moisture condensation and undesired surface growth of microorganisms. As a rule, the relative humidity in the slicing and packaging room should be below 60%. The temperature depends on the type of product and the time the product is held in the packaging room. A cleaning and disinfection plan for the slicing device should be in place to avoid contamination of the product. To prevent mold growth, the residual oxygen level in a package of sliced fermented sausage should be kept below 0.5%. This is achieved by packaging under vacuum or—increasingly common—under a modified atmosphere containing about 70% N_2 and 30% CO_2. Moreover, the oxygen permeability of the packaging material should be below 25 ml m^{-2} d^{-1} or even lower if a shelf life of more than 1 month is desired (Stiebing 1992). On the other hand, mold-ripened sausages should be packaged in material sufficiently permeable to oxygen. At any rate, the properties of the packaging material must be clearly specified.

The necessary storage temperature depends on the type of products. Most fermented sausages and hams are stored at 10–15°C; undried sausages may require a storage temperature at 7°C or below, whereas products dried to lower water activities (e.g., sausages with a_w below 0.90) may be stored at ambient temperatures (≤25°C). Illumination in display cabinets may be detrimental to fat quality, and the light intensity should be adjusted to below 600 lux.

END PRODUCT TESTING

Specifications for the final product may include pH, water activity (or weight loss), selected sensory properties (such as firmness), and the compliance with official or internal microbiological standards. Testing whether the final products meet their specifications is useful to verify that the production process was under control, but it cannot replace control measures and process monitoring. The parameters used should be fit for the purpose, i.e., provide a maximum of information with minimum input. For example, it is pointless to carry out total microbial counts or coliform counts of fermented sausages, and it is much more effective to test for pH than to count lactic acid bacteria. In contrast, *Enterobacteriaceae* counts could be an indicator of process hygiene, in particular for undried, fresh sausages. Standards referring to the absence of pathogens (salmonellae, *Listeria monocytogenes*) from the final product are defined in the EU Regulation 2073/2005 on microbiological criteria for foodstuffs.

Table 50.6. Slicing and packaging (in this case, under modified atmosphere as an example) of fermented sausages and raw hams (process steps G and E, respectively).

Subprocess; Examples for Specifications/Critical Limits	Objective	Monitoring and Documentation
Removal of Casings		
Slicing	Prevention of moisture condensation on sausage surface	Adjustment and monitoring of climate in packaging room
Packaging	Protection against recontamination, oxygen, and weight loss	Specifications for packaging material (permeability, etc.); weighing protocol; gas composition; lot identification
Bulk Packaging		Lot identification

GENERAL REMARKS ABOUT THE STRUCTURE AND EXTENT OF DOCUMENTATION

The purpose of process plans is to ascertain that processes work reliably even though the responsible persons may change. Moreover, they are necessary to provide evidence that the management has delegated responsibility in an appropriate manner. Basically, the better trained and experienced the workers are, the less paperwork is necessary. The purposes of records on the processes are to provide evidence of appropriate process control, e.g., compliance with "due diligence" as defined in product liability legislation. Records such as the Shewhart control charts are also a tool to identify problems and start corrective actions before critical limits are exceeded and it becomes necessary to reject the batch.

On the other hand, overdocumentation and problems related to information flow and distribution of documents are a major cause of frustration for staff and of noncompliance with the system. In particular, overdocumentation

- Prevents effective use of documents (increased time for searching)
- Increases the workload necessary for updating
- Reduces flexibility

The key question must be "what could go wrong if we do not have this document." Hence, each document should be checked using the following questionnaire:

- Is the document necessary to reach its purpose?
- Can the purpose of the document be reached with less words/pages?
- Can the document replace other documents?
- Is the style and layout of the document suitable for use by its target group?
- Can the purpose of the document also be reached by training and education?

To avoid confusion, it is also essential that documents are paginated (page x of y) and that every page of a document has a header or footer indicating

- What it covers
- Who has written it
- Who has approved it
- The date and number of the version

It is also important to make sure that everyone has exactly those documents available that he/she needs, and that any previous versions of the document are removed.

REFERENCES

Literature Cited

MH Brown. 2000a. Implementing HACCP in a meat plant. In: MH Brown, ed. HACCP in the Meat Industry. Cambridge, UK: Woodhead Publishing, Ltd., pp. 177–2001.

———. 2000b. Validation and verification of HACCP plans. In: MH Brown, ed. HACCP in the Meat Industry. Cambridge (UK): Woodhead Publishing, Ltd., pp. 231–272.

R Lautenschläger. 2006. Rohpökelware. In: W Branscheid, ed. Qualität von Fleisch und Fleischwaren, 2nd ed. Frankfurt/M.: Deutscher Fachverlag.

FK Lücke. 2000. Fermented meats. In: BM Lund, AC Baird-Parker, GW Gould, eds. The Microbiological Safety and Quality of Food. Gaithersburg, Maryland: Aspen Publishers pp. 420–444.

JJ Sheridan. 2000. Monitoring CCPs in HACCP systems. In: MH Brown, ed. HACCP in the Meat Industry. Cambridge, UK: Woodhead Publishing, Ltd., pp. 203–230.

A Stiebing. 1992. Verpackung—Anforderung bei Fleisch und Fleischerzeugnissen. Fleischwirtschaft 72:564–575.

———. 1996. Herstellung von Rohpökelware. In: F Wirth, J Barciaga, UM Krell, ed. Handbuch Fleisch und Fleischwaren. Hamburg: Behr's Verlag, Chapter A 6.6.

———. 1997. Herstellung von Rohwurst. In: F Wirth, J Barciaga, UM Krell, ed. Handbuch Fleisch und Fleischwaren. Hamburg: Behr's Verlag, Chapter A 6.5.

F Toldrá, ed. 2002. Dry-cured meat products. Trumbull, Connecticut: Food & Nutrition Press.

V Thorn. 2006. Qualitätsmanagement in der Lebensmittelindustrie. In: W Frede, ed. Taschenbuch für Lebensmittelchemiker. Berlin: Springer, pp. 183–213.

Official Documents

Commission Regulation (EC) no. 2073/2005 of 15 November 2005 on microbiological criteria for foodstuffs. 2005. Official Journal of the EU L338:1–26.

Regulation (EC) no. 852/2004 of the European Parliament and Council of 29 April 2004 on the hygiene of foodstuffs. 2004. Official Journal of the EU L226:1–19.

USDA-FSIS. 1999. Generic HACCP model for not heat treated, shelf stable meat and poultry products. http://www.fsis.usda.gov/OPPDE/nis/outreach/models/HACCP-15.pdf.

Index